THEOR

THE THEORY OF MACHINES

A Text-book for Engineering Students

By

THOMAS BEVAN

M.Sc. Tech. (Manchester), A.M.I. Mech. E.

Senior Lecturer in Mechanical Engineering in the University and in the College of Technology, Manchester.

WITH DIAGRAMS

CBS Publishers & Distributors Pvt. Ltd.

New Delhi • Bangalore • Pune • Cochin • Chennai (India)

ISBN : 81-239-0874-1

First Indian Edition : 1996
Reprint : 1998, 1999, 2000, 2002, 2003, 2004, 2005

This edition has been published in India by arrangement with Longman Group Limited, London

Published by Satish Kumar Jain and produced by V.K. Jain for
CBS Publishers & Distributors Pvt. Ltd.,
CBS Plaza, 4819/XI Prahlad Street, 24 Ansari Road, Daryaganj,
New Delhi - 110002, India. • Website: www.cbspd.com
e-mail: delhi@cbspd.com, cbspubs@airtelmail.in
Ph.: 23289259, 23266861, 23266867 • Fax: 011-23243014

Branches:

- ***Bengaluru:*** Seema House, 2975, 17th Cross, K.R. Road, Bansankari 2nd Stage, Bengaluru - 560070
  • Ph.: +91-80-26771678/79 • Fax: +91-80-26771680
  • E-mail: cbsbng@gmail.com, bangalore@cbspd.com
- ***Pune:*** Bhuruk Prestige, Sr. No. 52/12/2+1+3/2, Narhe, Haveli (Near Katraj-Dehu Road By-pass), Pune - 411041
  • Ph.: +91-20-64704058/59, 32342277 • E-mail: pune@cbspd.com
- ***Kochi:*** 36/14, Kalluvilakam, Lissie Hospital Road, Kochi - 682018, Kerala • Ph.: +91-484-4059061-65
  • Fax: +91-484-4059065 • E-mail: cochin@cbspd.com
- ***Chennai:*** 20, West Park Road, Shenoy Nagar, Chennai - 600030
  Ph.: +91-44-26260666, 26208620 • Fax: +91-44-42032115
  • E-mail: chennai@cbspd.com

Printed at :
J.S. Offset Printers, Delhi

PREFACE TO THE THIRD EDITION

In this edition the general arrangement of the book remains unchanged, but the opportunity has been taken of revising and amplifying certain sections and of redrawing some of the diagrams. Most of the additional matter will be found in the sections on gyroscopic motion, velocity and acceleration diagrams, toothed gearing, epicyclic trains, inertia forces in mechanisms and vibrations.

The author is grateful to all those who have pointed out errors or misprints in the earlier editions, or who have offered suggestions for the improvement of the book. He will welcome any similar intimations in connection with the new edition.

PREFACE TO THE FIRST EDITION

In writing this book I have chiefly had in mind the needs of the student who is preparing for a University degree in engineering or who intends to sit for the membership examination of one of the Engineering Institutions, but I hope that many of the sections will appeal to the draughtsman and designer. In order to allow space for a detailed treatment of the various sections of the Theory of Machines, I have cut down that devoted to elementary mechanics to a minimum, although certain fundamental principles have been emphasised and extended where it appeared desirable.

The book is largely based on lectures given at the Manchester College of Technology. The lectures cover a period of one hour a week for three sessions. It seemed better, however, not to adhere to the order followed in the lectures, but to group both the elementary and the more advanced treatments of each section in the same chapter.

As so many of the problems which arise may be solved more quickly and easily by graphical methods, particular care has been taken to draw the diagrams correctly to scale. Sometimes, however, the scale is unavoidably small, and the student is advised in his own interests to redraw these diagrams to a larger scale. I have carried out the whole of the work involved in drawing the diagrams and in working out the numerical answers to the problems. I do not expect that there will be no mistakes, but I hope that they will be few and not serious ones.

My thanks are due to the Senates of the Universities of London and Manchester for permission to include questions taken from the final papers set for their degree examinations; to the Controller of H.M. Stationery Office for permission to include questions from the papers set for the Whitworth Scholarship and Senior Scholarship Examinations; to the British Standards Institution for permission to include abstracts from their Standard Specifications and to Messrs. Heenan and Froude for permission to include the illustration of their well-known hydraulic dynamometer, Fig. 179. Other acknowledgments are made in the text, and I sincerely hope that none have been overlooked.

Finally, I have great pleasure in expressing my indebtedness to Professor Dempster Smith for his help and encouragement; to my colleagues, Dr. R. O. Boswall and Mr. J. C. Brierley, for criticisms and suggestions on parts of the manuscript, and to my former student Mr. Johnson Ball for his very painstaking work in reading through the proofs.

T. B.

CONTENTS

CHAPTER I

DEFINITIONS. SIMPLE MECHANISMS

1. This book is about machines and it is therefore necessary to understand at the outset what is meant by a machine. A detailed definition will have to be deferred until later in the chapter, but for present purposes it will suffice to define a machine as a contrivance which receives energy in some available form and uses it to do some particular kind of work. Let us take one or two simple illustrations. A crowbar together with its fulcrum forms a machine, which enables the muscular energy of a man to be employed in raising a heavy weight. Again, a petrol engine is a machine which may use the heat energy derived from the combustion of the fuel to propel a vehicle along the road. Or again, a lathe is a machine which receives mechanical energy from the lineshaft through the belt and uses that energy to remove metal from a bar or other piece of work.

The *theory of machines* comprises the study of the relative motion between the parts of a machine and the study of the forces which act on those parts. The study of the relative motion between the parts is known as *kinematics*, while the study of the forces which act on the parts is known as *dynamics*. Dynamics may be subdivided into *statics*, which deals with the forces which act on the various parts when those parts are assumed to be without mass, and *kinetics*, which deals with the inertia forces arising from the combined effect of the mass and the motion of the parts.

These divisions of the subject may be illustrated by reference to a reciprocating engine. In this engine the piston is made to reciprocate in the cylinder by the pressure which the steam or gas exerts, and the reciprocating motion of the piston is converted into rotary motion of the crankshaft by means of the connecting rod and the crank. So far as the relative motion between the parts is concerned it is only necessary to represent each part by its centre line and to know the length and the speed of rotation of the crank, the length of the connecting rod and the position of the line of stroke relative to the crankshaft centre. A kinematical problem, such as the determination of the velocity or acceleration of the piston, is largely a problem in geometry, although an additional factor, time, also enters in. But for every position of the parts of the engine the thrust exerted by the steam or gas on the piston produces a thrust along the connecting rod and is balanced by a tangential force at the crankpin. To determine the relative magnitudes of these forces is a problem in statics, and requires that the machine shall be treated as a structure or framework with pin-joints. All the possible relative positions of the parts must be examined in order to find the maximum values of the forces which each part is required to transmit. The approximate cross-sections of the parts may then be fixed so as to enable these forces to be transmitted with safety.

It is afterwards necessary to examine the inertia forces which act on each part, since the parts are subject to acceleration and retardation during the working of the engine. The net force on each part due to the combined effect of the static and kinetic forces may then require some modification to the cross-section which was provisionally fixed from a consideration of the static force only.

2. Link or Element. Each part of a machine which has motion relative to some other part is termed an *element* or *link*. It is important to notice that each link or element may consist of several parts which are manufactured as separate units. Thus, for instance, the piston rod and crosshead of a steam engine are manufactured separately, but when assembled in the engine they are rigidly fastened together and therefore constitute one element or link. Similarly, the connecting rod, complete with big and little-end brasses, caps and bolts, constitutes a second element or link; the crankpin, crankshaft and flywheel a third element or link; and the cylinder, bedplate and main bearings a fourth element or link.

A link need not necessarily be a rigid body, but it must be a resistant body, i.e. it must be capable of transmitting the required force with negligible deformation. Examples of links which are resistant, but not rigid, are to be found in: (a) liquids, which are resistant to compressive forces and are used as links in hydraulic presses, hydraulic brakes and hydraulic jacks; and (b) chains, belts and ropes, which are resistant to tensile forces and are used for transmitting motion and force.

3. Kinematic Pair. Two elements or links which are connected together in such a way that their relative motion is completely constrained form a *kinematic pair*. This definition requires to be modified in order to include those pairs in which the form of the connection between the elements is not in itself such as to give complete constraint, but in which the constraint is completed by some other means. The constraint in such pairs is said to be successful rather than complete.

Several different pairs of elements are shown diagrammatically in Fig. 1. These will serve to illustrate the differences between complete, incomplete and successful constraint. For instance, (a) is an example of incomplete constraint. The round bar or shaft A passes through the cylindrical hole in the other element B. If relative motion is possible, then A may either slide through B, it may rotate about the axis of B, or it may have a motion relative to B which consists partly of sliding and partly of rotation. There is nothing in the form of the connection between A and B to determine which of the three types of relative motion will take place. But if the form of the connection is modified as at (b), (c) and (d), the element A can only slide relative to B and the

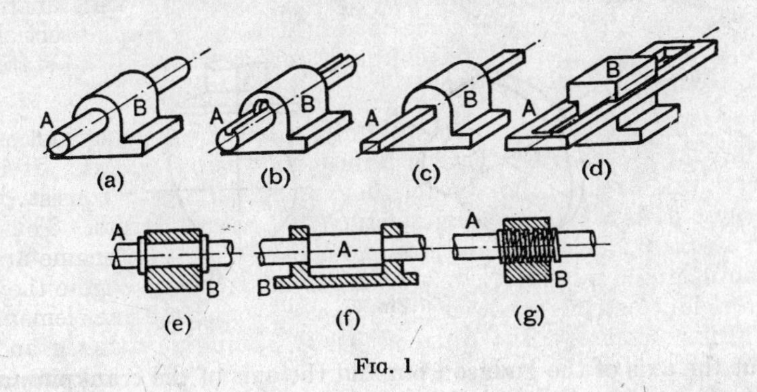

FIG. 1

constraint is therefore complete. At (b) a feather key is fitted to the element B and this key is a sliding fit in the corresponding groove in the element A; at (c) the cross-section of the element A and the shape of the hole in the element B are both rectangular; while at (d) the element A takes the form of a slotted bar and the element B that of a crosshead or die-block. Despite the considerable differences in the appearance of the pairs (b), (c) and (d), they are kinematically identical and form what are known as *sliding* pairs. Similarly, if the form of the connection between A and B is modified as shown at (e) and (f) the relative motion is limited to one of rotation and the constraint is complete. These pairs are known as *turning* pairs. Finally, if the connection between A and B is modified as shown at (g), so that the contact surfaces are screw threads, the constraint is again complete and we have what is termed a *screw* pair. It should be noted that, although both rotation and sliding of A with respect to B take place, a given amount of rotation of A relative to B is accompanied by a strictly proportionate amount of axial displacement of A relative to B.

It is clear that the arrangement (e) could be used without the right-hand collar, if it is known that an axial force which will prevent A from sliding axially towards the left will always act towards the right. In vertical turbines the connection between the shaft and the thrust bearing does not as a rule prevent axial movement of the shaft in the upward direction, but the weight of the turbine is far in excess of any upward force which is likely to arise during the operation of the turbine. The pair formed by the shaft and the thrust bearing is force-closed and is an example of successful constraint. In a similar way the connection between the piston and the cylinder of a petrol engine corresponds to Fig. 1 (a) and the relative motion is not completely constrained.

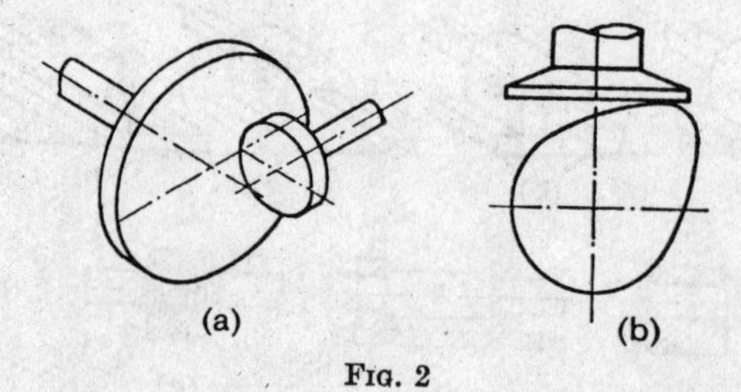

Fig. 2

But the axis of the gudgeon pin and the axis of the crankpin are maintained parallel to each other by the connecting rod, so that the piston and the cylinder form a successfully constrained pair.

There is one feature common to all three of the above types of kinematic pairs, namely, the two elements have surface contact and, when relative motion takes place, the surface of one element slides over the surface of the other element. Such pairs are called *lower* pairs. All other types of pairs are known as *higher* pairs. The elements of higher pairs generally have line or point contact and the pair must be force-closed in order to provide completely constrained motion. Examples of higher pairs are shown in Fig. 2, where (a) represents a pair of friction discs and (b) a cam and follower. Other examples are provided by toothed gearing, belt and rope drives and ball and roller bearings.

4. Kinematic Chain. A kinematic chain is a combination of kinematic pairs in which each element or link forms part of two pairs and in which the relative motion is completely constrained.

Let us consider the possible combinations of turning pairs. Fig. 3 shows three arrangements in which each element or link forms part of two turning pairs, so that the first part of the definition of a kinematic chain is satisfied by each arrangement. In the arrangement (a) the three pin-jointed links clearly form a rigid frame, so that no relative motion between the links is possible. In the arrangement (b) the relative motion is completely constrained, since if link AD is fixed and a definite displacement is given to AB, the resulting displacements of the remaining two links, BC and CD, are also perfectly definite. But if five bars are connected together by pin-joints as shown at (c), the relative motion is not completely constrained, since, with AE fixed and AB displaced to AB_1, the resulting displacements of the remaining links BC, CD and DE cannot be predicted. Hence only the arrangement (b) constitutes a kinematic chain.

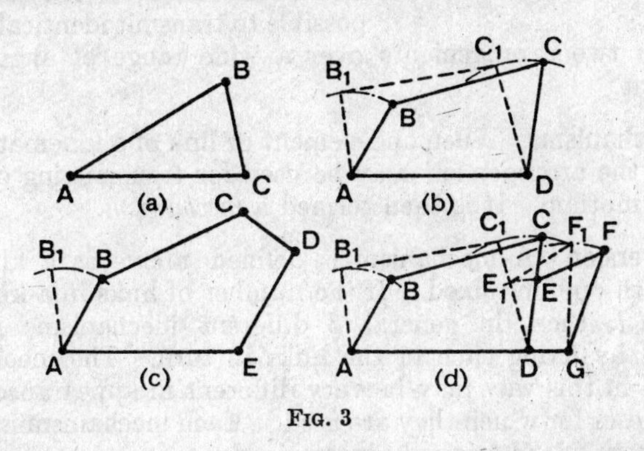

FIG. 3

In order to build up a more complicated kinematic chain from the simple four-bar chain shown in Fig. 3 (b), two more elements or links may be added, provided that these two elements form pairs with two of the existing links as well as forming in themselves a pair—see Fig. 3 (d). Proceeding in this way, pairs may be added indefinitely until the chain becomes very complex. The relation between the number of pairs and the number of links may be expressed in the form of an equation. If each link is reckoned as forming a part of two pairs, and two pairs only, then for the four-link chain the number of pairs is also four and for each additional pair two links have to be added, so that, if l is the number of links and p the number of pairs, the general equation may be written:

$$l = 2p-4 \quad \quad (1.1)$$

Alternatively, the relation between the number of links and the number of joints may be expressed in the form of an equation. Thus, if j is the number of joints, then:

$$l = \tfrac{2}{3}(j+2) \quad . \quad . \quad . \quad . \quad . \quad (1.2)$$

The above equations only apply to kinematic chains in which lower pairs are used. If they are applied to chains which contain higher pairs, each higher pair must be taken as equivalent to two lower pairs and an additional element or link. This may be seen from Fig. 4, in which the arrangement of three links, OA, AB and BQ, shown in dotted lines, is kinematically equivalent to the friction discs, shown in full lines, over a short range of movement. It would, of course, be impossible to transmit identical motion with the two arrangements over a wide range of angular displacement.

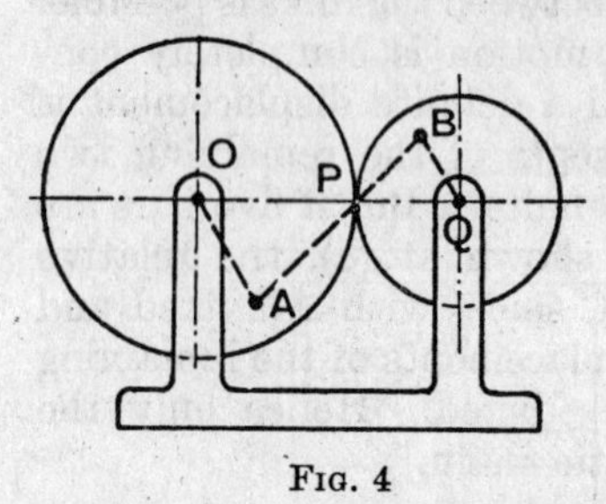

FIG. 4

5. Mechanism. When one element or link of a kinematic chain is fixed, the arrangement may be used for transmitting or transforming motion. It is then termed a *mechanism*.

6. Inversion A mechanism is defined above as a kinematic chain with one link fixed. If the number of links in a kinematic chain is l, then, in general, l different mechanisms may be obtained by fixing each of the links in turn. The mechanisms obtained in this way may be very different in appearance and in the purposes for which they are used. Each mechanism is termed an *inversion* of the original kinematic chain.

7. Machine. When a mechanism is required to transmit power or to do some particular kind of work, the various elements or links have to be designed so as to carry with safety the forces, both static and kinetic, to which they are subjected. The arrangement then becomes a *machine*. A mechanism may therefore be regarded as a machine, in which each part is reduced to the simplest form required in order to transmit the desired motion. In the preceding pages the building up of a machine has been traced from the individual elements or links, through kinematic pairs, a kinematic chain and a mechanism to the final machine. The reader should now be in a position to understand the following definition of a machine:

A machine is a combination of resistant bodies, with successfully constrained relative motions, which is used for transmitting or transforming available energy so as to do some particular kind of work.

8. Kinematic Chains with Three Lower Pairs. Although it is impossible to have a kinematic chain which consists of only three turning pairs, it is quite possible to have one which consists of three sliding pairs and one which consists of a turning, a sliding and a screw pair. These two chains are shown in Fig. 5. At (a) each

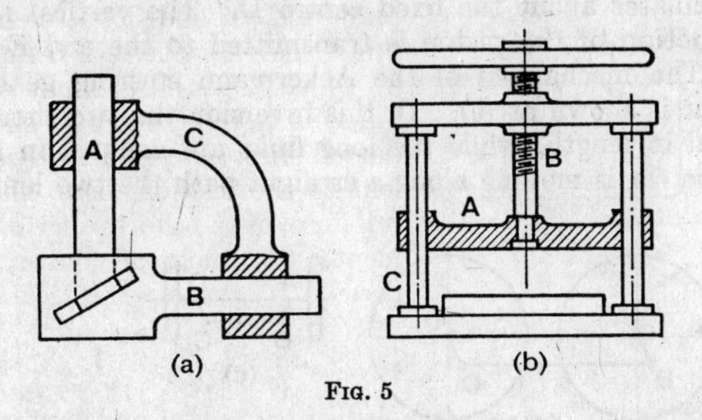

FIG. 5

of the two elements A and B forms a sliding pair with the frame C, and, in addition, the elements A and B themselves form a sliding pair. If the frame C is fixed and the element A is given a definite displacement, it will cause the element B to receive a definite displacement, so that the relative motions of A, B and C are completely constrained. At (b) the element A forms a sliding pair with C and a turning pair with B, and the element B forms a screw pair with C. Hence each element forms part of two pairs, and if the frame C is fixed and B is rotated on its axis, the displacement of A relative to C is proportional to the rotation of B, so that the relative motions of A, B and C are completely constrained. When the link C is fixed the mechanism of the hand or fly-press is obtained.

9. Kinematic Chains with Four Lower Pairs. The most important kinematic chains from the practical point of view are those which consist of four lower pairs, each pair being either a sliding pair or a turning pair. It will be found that many complicated machines are based on combinations of the different inversions of these simple chains.

10. The Four-bar Chain. This kinematic chain is shown in Fig. 3 (b), and each of the four pairs is a turning pair. The four elements or links may be of different lengths and the use to which the various inversions of the mechanism are put will depend solely on the relative lengths of the links. There are many practical mechanisms which are based on the four-bar chain. Some of these are shown in Fig. 6 and others will be found in Chapter IV. The mechanism of the coupling rod of a locomotive is shown at (a). In this inversion the opposite links are equal in length. Part of the mechanism of a beam engine is shown at (b). The engine crank AB turns about the fixed centre A, while the beam CDE oscillates about the fixed centre D. The vertical reciprocating motion of the piston is transmitted to the end E of the beam. The mechanism of the Ackermann steering gear for a motor-car is shown at (c). In this inversion the two short links are equal in length, while the long links are unequal in length. When the car is moving along a straight path the two long links

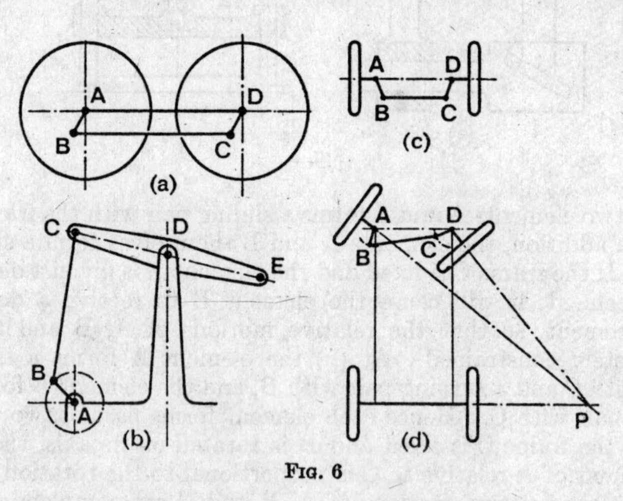

Fig. 6

AD and BC are parallel. When the car moves along a curved path the mechanism takes up the position shown at (d) and the proportions of the links are so fixed that the axes of all four wheels intersect at the same point P. This ensures that the relative motion between the tyres and the road surface shall be one of pure rolling (see Article 60).

11. The Slider-crank Chain. This chain consists of three turning pairs and one sliding pair. The most usual form in which the slider-crank chain appears is that of the reciprocating-engine mechanism. This is shown at (a) in Fig. 7. In the same figure are illustrated all the possible inversions of the slider-crank chain. Corresponding pairs are indicated by the same letter in each inversion, so that there should be no difficulty in recognising that each mechanism is based on the same kinematic chain. The oscillating-cylinder engine mechanism shown at (b) is derived from the slider-crank chain by fixing the link CP, which in the reciprocating engine mechanism forms the connecting rod. As the crank OC revolves about an axis through C, the slotted link OQ slides over the block which is pivoted to the fixed link at P. The actual form of the mechanism as used in the oscillating-cylinder engine is shown at the right. The cylinder is carried on

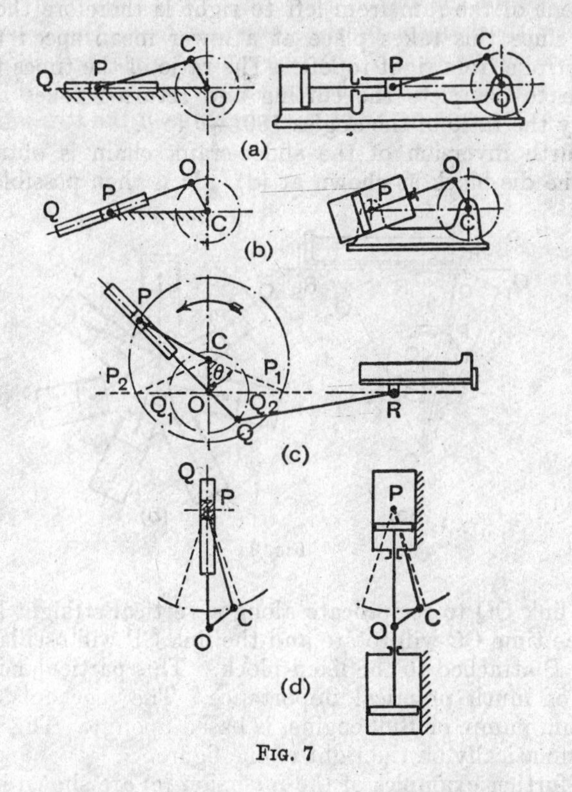

FIG. 7

trunnions at P, and as the piston slides inside the cylinder the crank revolves and the cylinder oscillates about the axis of the trunnions.

In a similar way the mechanism shown at (c) is obtained by fixing the link OC, which in both the inversions (a) and (b) forms the driving crank. This inversion is known as the Whitworth quick-return motion and is used on slotting and shaping machines. CP is the driving crank and rotates at uniform speed, the die-block attached to the crankpin P slides along the slotted link OQ and causes this link to revolve about O with a variable angular velocity. From the pin Q on the slotted link a connecting rod passes to a pin R on the ram which carries the toolbox, and R reciprocates along a line of stroke which passes through O and is normal to OC.

Obviously the two extreme positions of the ram will correspond to the two positions OQ_1 and OQ_2 of the slotted link, and if CP rotates counter-clockwise, the time taken to turn from CP_1 to CP_2 will be greater than the time taken to turn from CP_2 to CP_1. Movement of the ram from left to right is therefore the cutting stroke, since this takes place at a lower mean speed than the return stroke from right to left. The ratio of the times taken by the ram to complete the cutting and return strokes is clearly given by the ratio of the angles $180° - \theta$ to θ.

A fourth inversion of the slider-crank chain is obtained by fixing the die-block as shown at (d). It is then possible for the

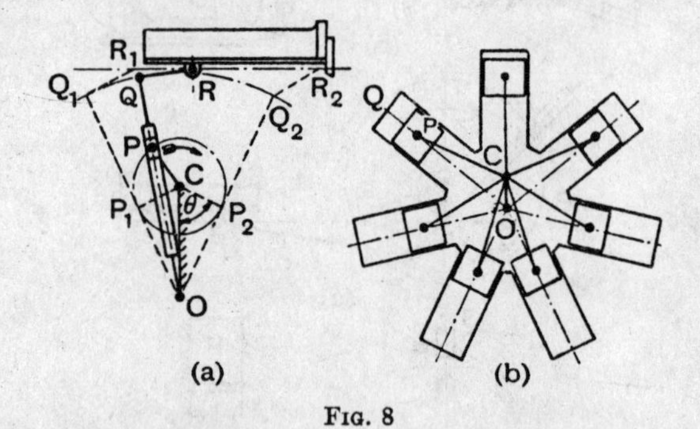

FIG. 8

slotted link OQ to reciprocate along a vertical straight line. At the same time OC will rotate and the link CP will oscillate about the pin P attached to the fixed block. This particular inversion is not of much practical importance. The mechanism of the pendulum pump, or Bull engine, is based upon it. This is shown diagrammatically on the right of the figure.

Two further examples of the inversion (c) are shown in Fig. 8. The mechanism shown at (a) is the crank and slotted lever quick-return mechanism. The only difference between this quick-return motion and the one described above lies in the different proportions adopted for the driving crank CP and the fixed link OC. The fact that CP is shorter in length than OC results in the slotted link OQ oscillating between the two extreme positions OQ_1 and OQ_2, while the crank CP revolves about the centre C. The pin Q is connected to a pin R on the ram of a slotting or shaping machine and causes R to reciprocate along a path normal to OC. Evidently, as in the Whitworth quick-return motion, the ratio of the times taken by the ram on the cutting and return strokes is given by the ratio $180° - \theta$ to θ.

The same inversion of the slider-crank chain is also used in the rotary internal-combustion engine which contributed so largely to the development of mechanical flight in the early years of aviation. In this engine, which is shown diagrammatically in Fig. 8 (b), the crank OC is fixed. The complete assembly of cylinders and crankcase rotates about the centre O and the pistons reciprocate along their respective cylinders. It will be seen that each piston, connecting rod and cylinder form with the fixed crank OC an inversion of the slider-crank chain.

12. The Double Slider-crank Chain. This kinematic chain consists of two turning and two sliding pairs. Referring to Fig. 9, two die-blocks slide along slots in a frame and the pins P and Q on the die-blocks are connected by the link PQ. Each of the die-blocks forms a sliding pair with the frame and a turning pair with

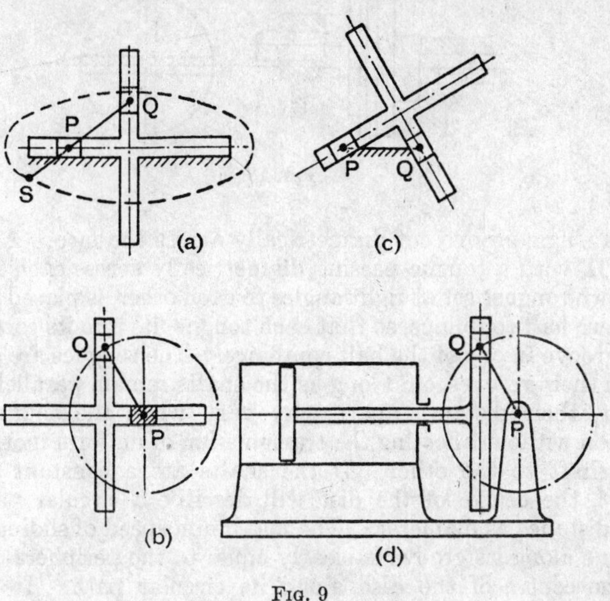

FIG. 9

the link PQ. Three inversions are possible. These are shown at (a), (b) and (c) in the figure. At (a) the slotted frame is fixed. Any point, such as S, on the link PQ will trace out an ellipse as the blocks P and Q slide along their respective slots. Clearly QS and PS are respectively the semi-major and semi-minor axes of the ellipse. This inversion is known as the ellipse trammels. In the second inversion one of the two blocks P or Q is fixed. As shown at (b), the block P is fixed, so that PQ can rotate about P as centre and thus cause the frame to reciprocate. The fixed block P guides the frame. The Scotch yoke, which is illustrated diagrammatically at (d) is the same inversion of the double slider-crank chain. It is used for converting rotary into reciprocating motion.

The third inversion (c) is obtained by fixing the link PQ. Each of the two die-blocks may then turn about the pins P and Q. If one block is turned through a definite angle, the frame and the other block must turn through the same angle, and, as rotation takes place, the frame will slide relative to each of the two blocks. The Oldham shaft coupling is an example of this inversion. Referring to Fig. 10, each half-coupling is identical in form and

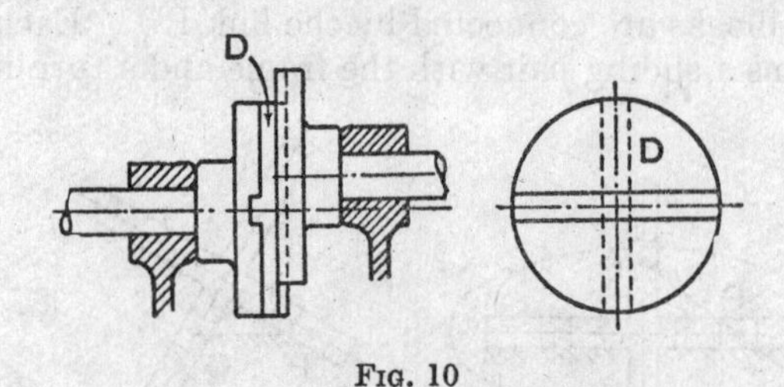

FIG. 10

has a single groove cut diametrically across the face. A circular disc D, with a tongue passing diametrically across each face and the two tongues set at right angles to each other, is placed between the two half couplings, so that each tongue fits into its corresponding groove in one of the half couplings. The tongues are a sliding fit in their grooves. So long as the shafts remain parallel to each other, their distance apart may vary while the shafts are in motion without affecting the transmission of uniform motion from one shaft to the other. If the shafts are a constant distance apart, the centre of the disc will describe a circular path with this distance as diameter. The maximum speed of sliding of each tongue along its groove is clearly equal to the peripheral velocity of the centre of the disc along its circular path. It may be expressed in terms of the distance apart of the two shafts and the angular velocity of rotation of the shafts.

13. Compound Kinematic Chains. A kinematic chain in which there are more than four pairs is known as a compound chain. Compound chains may be built up from any of the simple chains which have been described above by adding further kinematic pairs. Any addition to the number of pairs must, of course, be such as to retain completely constrained motion. The number of pairs and links must therefore satisfy the equations given in Article 4. In Chapters III, IV and V many examples of compound kinematic chains will be found.

EXAMPLES I

1. Define the following terms, illustrating with sketches where possible, element or link, higher pair, lower pair, kinematic chain.

2. Distinguish between complete, incomplete and successful constraint of the relative motion between two elements or links.

3. Distinguish between a kinematic chain, a mechanism and a machine. What are the most commonly used kinematic chains consisting of lower pairs?

4. What is meant by: (a) a resistant body, (b) a lower kinematic pair, (c) a higher kinematic pair, (d) a kinematic chain?

5. Different mechanisms may be obtained by *inversion* of the same kinematic chain. Explain, with sketches, the meaning of this statement.

6. What is the difference between the slider-crank chain and the double slider-crank chain? Give diagrammatic sketches of three mechanisms which are inversions of each of the above chains and state the purpose for which each mechanism is used.

7. What is the relation between the number of links and the number of pairs in a kinematic chain, when the pairs are all lower pairs? Show that the Stephenson valve gear, Fig. 120, the Hackworth valve gear, Fig. 126, and the Joy valve gear, Fig. 127, satisfy the relation.

8. In a crank and slotted lever quick-return motion, the distance between the fixed centres O and C is 6 in. and the driving crank CP is $2\frac{3}{4}$ in. long. Find the ratio of the times taken on the cutting and the return strokes. Sketch the complete mechanism, showing the ram and the direction of rotation of the crank.

9. Describe the construction of the Oldham shaft coupling and state for what purpose it is used. In a coupling of this type the distance between the shaft axes is 1 in. and the speed of rotation is 300 r.p.m. What is the maximum speed of sliding of each tongue in its slot?

10. Write down a definition of a machine and explain the meaning of the terms which enter into the definition.

11. Describe with neat sketches a quick-return motion suitable for a small slotting or shaping machine. Show how the ratio of the times taken on the two strokes is determined.

12. Give diagrammatic sketches of the following mechanisms and state on which kinematic chain each one is based: (a) ellipse trammels, (b) Whitworth quick-return motion, (c) oscillating cylinder engine, (d) Oldham shaft coupling.

CHAPTER II

MOTION. INERTIA

14. The theory of machines, as already pointed out, is concerned with the motion of the parts of machines and with the forces which act on those parts. It is therefore a branch of the wider science of mechanics, which deals with the interaction of force and motion. It will be assumed that the reader already possesses some knowledge of the mechanics of particles and rigid bodies, and in this chapter only certain fundamental ideas and relations will be emphasised and in some cases extended.

15. Displacement. Velocity. Acceleration. Every particle of every link of a machine is constrained to move along a definite path, which may be either curved or straight. Generally there are certain links the particles of which all move along straight lines, and other links the particles of which all move along circular arcs, while for the remaining links the particles move along curved paths which are not circular arcs. For example, in the reciprocating engine all particles of the piston, piston rod and crosshead move along parallel straight lines and all particles of the crank and crankpin move along circular arcs, the centres of which lie on the axis of the crankshaft. A particle on the connecting rod, however, moves along an oval path, the radius of curvature of which changes from instant to instant. Whatever the path followed, it is possible to determine successive displacements of the particles for equal intervals of time.

Since displacement involves direction as well as magnitude, it is a vector quantity, and can therefore be represented by a straight line. The length of the line indicates the amount of the displacement to some convenient scale, its inclination shows the direction in which the displacement has taken place and the arrowhead shows the sense of the displacement.

The velocity of a particle is defined as the rate of change of its displacement with respect to time. It follows, therefore, that velocity also is a vector quantity. If the displacement is constant in direction, the velocity will obviously be in the same direction. If the displacement is not constant in direction, then the velocity at a given instant will be in the same direction as the displacement at that instant, and will therefore be tangential to the path of the particle.

Thus, if the displacement takes place along the circular path ABC, Fig. 11, the velocities at the instants when the particle occupies the positions A, B and C will be in the directions given by the tangents AD, BE and CF respectively.

In certain problems we are concerned merely with the rate at which the magnitude of the displacement is changing, and the change of direction is immaterial, as, for instance, in the case of a train which is travelling between two stations. The rate of change of magnitude of the displacement with respect to time is termed the speed.

The acceleration of a particle is defined as the rate of change of the velocity with respect to time. This also is a vector quantity, but it is important to bear in mind that its direction is not necessarily the same as that of the velocity and the displacement vectors. This will be made clear in the following Article, in which the general case of the acceleration of a particle which moves along a circular path is considered.

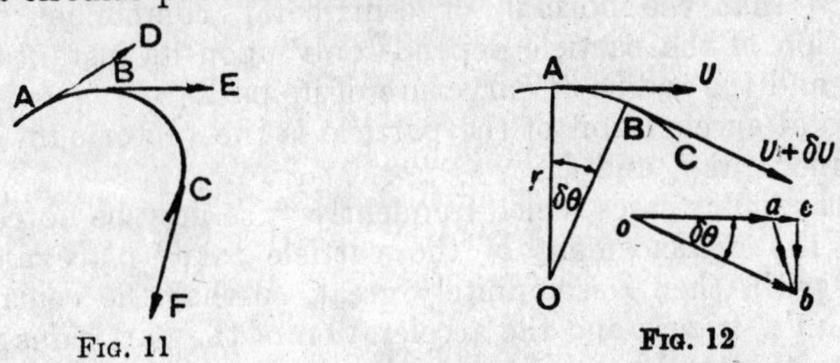

FIG. 11 FIG. 12

16. Acceleration of a Particle displaced along a Circular Path. Let A and B, Fig. 12, be two positions of a particle which is displaced along the circular path ABC. Let δt be the time required by the particle to move from A to B. Let r be the radius of curvature of the path, and let v and $v+\delta v$ represent the velocities of the particle when at A and B respectively.

The change of velocity as the particle moves from A to B may be found by drawing the vector triangle oab, in which oa represents the velocity v of the particle when at A and ob represents the velocity $v+\delta v$ of the particle when at B, so that ab represents the change of velocity in the time δt. Resolve ab into two components ac and cb respectively parallel and perpendicular to oa.

Then
$$\text{ac} = \text{ob} \cos \delta\theta - \text{oa}$$
$$= (v+\delta v) \cos \delta\theta - v$$

and
$$\text{cb} = \text{ob} \sin \delta\theta$$
$$= (v+\delta v) \sin \delta\theta$$

Therefore the component acceleration of the particle in the direction oa, i.e. in the tangential direction, is given by:

$$f_t = \frac{ac}{\delta t} = \frac{(v+\delta v)\cos\delta\theta - v}{\delta t}$$

and, in the limit, as δt approaches zero, this reduces to:

$$f_t = dv/dt \quad . \quad . \quad . \quad . \quad . \quad . \quad (2.1)$$

Similarly, the component acceleration of the particle in a direction normal to oa, i.e. the centripetal component, is given by:

$$f_c = cb/\delta t = (v+\delta v)(\sin\delta\theta/\delta t)$$

and, in the limit, this reduces to:

$$f_c = v.d\theta/dt = v\omega = v^2/r = \omega^2 r \quad . \quad . \quad (2.2)$$

It is clear from equation (2.1) that the tangential component of the acceleration of the particle is equal to the rate of change of the magnitude of the velocity of the particle, and from equation (2.2) that the normal, or centripetal, component of the acceleration of the particle depends only upon its instantaneous velocity and the radius of curvature of its path.

The total acceleration of the particle is the vector sum of the two components f_t and f_c.

Two particular cases which frequently arise may be noted:

(a) If the displacement of the particle takes place along a straight path, then r is infinitely great, so that the centripetal component f_c is zero and the acceleration of the particle is in the same direction as its velocity and its displacement. It is given by $f_t = dv/dt$.

(b) If the displacement of the particle takes place with constant speed along a circular path, the tangential component is zero and the acceleration of the particle is normal to its velocity and its displacement. It is given by:

$$f_c = v^2/r = v\omega = \omega^2 r$$

17. Angular Displacement, Velocity and Acceleration. Let a line OP, Fig. 13, rotate about the centre O and let its inclination to the fixed line OX be ϕ radians. If at the end of a short interval of time the line has moved to the position OQ, then the angle $\delta\phi$ is the angular displacement of the line. Angular displacement is a vector quantity, since it has not only magnitude but also direction. In order completely to specify an angular displacement by a vector, the vector must fix: (a) the direction of the axis of rotation in space; (b) the sense of the angular displacement, i.e.

whether clockwise or counter-clockwise; and (c) the magnitude of the angular displacement. To fix (a) and (c), the vector may be drawn at right angles to the plane in which the angular displacement takes place, say along the axis of rotation, and its length may be made to represent the magnitude of the angular displacement to some convenient scale. The conventional way of representing (b) is to use the right-handed screw rule; the arrowhead points along the vector in the same direction as a right-handed screw would move, relative to a fixed nut, if given an angular displacement of the same sense. According to the above convention, the angular displacement $\delta\phi$ would be represented by a vector perpendicular to the plane of the paper. The length of the vector would represent the magnitude of $\delta\phi$ to some convenient scale and the arrowhead would point upwards from the paper, since the sense of the displacement is counter-clockwise.

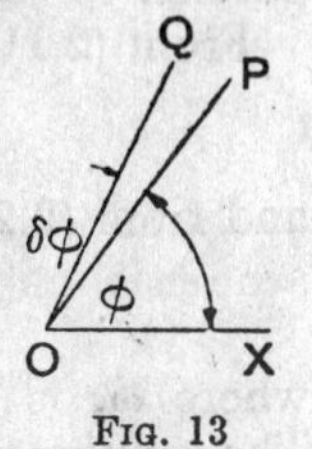

FIG. 13

Angular velocity is defined as the rate of change of angular displacement with respect to time. It has direction as well as magnitude, and it may be represented by a vector if the same convention is followed as that just described for angular displacement. If the direction of the angular displacement vector is constant, i.e. if the plane of the angular displacement does not change its direction, then we are concerned merely with the rate of change of the magnitude of the angular displacement with respect to time. This ought strictly to be called the angular speed of the line OP.

Similarly, angular acceleration is defined as the rate of change of angular velocity with respect to time. This also is a vector quantity, but it is important to bear in mind that its direction is not necessarily the same as that of the angular displacement and the angular velocity vectors.

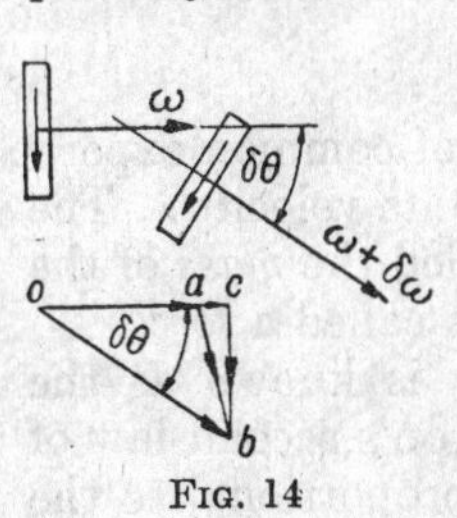

FIG. 14

For instance, let us suppose that, at a given instant, a disc is spinning with angular velocity ω in a plane at right angles to the paper, Fig. 14, and that, after a short interval of time δt, it is spinning with angular velocity $\omega+\delta\omega$ and the axis of spin has changed direction by the amount $\delta\theta$. Then, applying the right-handed screw rule, the angular velocities at the two instants are represented by the vectors oa and ob and the change of angular velocity in time δt is represented by the vector ab. This may be resolved into two components ac and cb, which are respectively parallel and perpendicular to oa. Clearly the conditions are analogous to those for linear motion, which were examined in the preceding Article. Hence it follows that the angular acceleration of the disc has two components, one parallel to oa and the other perpendicular to oa.

From (2.1) the component parallel to oa is given by:

$$\alpha_t = d\omega/dt \quad . \quad . \quad . \quad . \quad . \quad (2.3)$$

and from (2.2) the component perpendicular to oa is given by:

$$\alpha_c = \omega . d\theta/dt = \omega\omega_p \quad . \quad . \quad . \quad . \quad (2.4)$$

where ω_p is the rate of change of direction of the vector oa. The component α_t is the rate of change of *magnitude* of the angular velocity ω of the disc, while the component α_c depends only upon the instantaneous value of ω and the rate at which the direction of ω, and therefore of the plane of rotation of the disc, is changing.

The total angular acceleration of the disc is the vector sum of α_t and α_c.

Two particular cases should be noted:

(a) If the plane of rotation of the disc is constant in direction, then ω_p is zero and the component acceleration α_c is zero. The angular acceleration of the disc is then given by (2.3).

(b) If the angular velocity of the disc is constant in magnitude but the plane of rotation changes direction at the rate ω_p radians per second, then the angular acceleration of the disc is given by (2.4). The direction of this acceleration vector is at right angles to the angular velocity vector and it lies in the plane of motion of the velocity vector.

The change in direction of the plane of rotation of the disc is known as *precessional* motion and ω_p is known as the angular velocity of precession. The angular acceleration α_c is termed the *gyroscopic* acceleration of the disc.

18. Mass. Force. Weight. It is a matter of common experience that a body resists any attempt to change its velocity. The property which determines the resistance is called the *mass* of the body and the cause of the change of velocity is called a *force*.

The product of the mass and the velocity is known as the *momentum* of the body, and according to Newton's second law of motion the magnitude of the applied force is proportional to the rate of change of momentum which it produces, or

$$F \propto (d/dt)(mv)$$

Generally the mass m is constant, so that the rate of change of momentum is $m . dv/dt$ or mf.

The units of force and mass are so chosen that the equation may be written:

$$F = m.f \quad . \quad . \quad . \quad . \quad . \quad . \quad (2.5)$$

In the British system of units the unit of length is the foot and the unit of acceleration is therefore one foot per second per second (1 ft/s^2).

The unit of mass is the mass of a piece of platinum which is kept in the Standards Department of the Board of Trade in London. It is called the *pound mass.* The force which gives to this mass an acceleration of 1 ft/s^2 is defined as the unit of force. It is known as the *poundal.*

The poundal is a small unit of force, and it is much more convenient in practice to have a larger unit. This larger unit is the *pound weight.* It is the force of gravity on the standard lump of platinum, the *pound mass.* But the pull of gravity gives to a body which is allowed to fall freely an acceleration g ft/s^2, so that the weight of the pound mass is a force of g poundals. Since the value of g varies by approximately one part in 200 for points on the earth's surface, the weight of the one-pound mass will vary in the same way. In order to ensure that the pound weight is an invariable unit of force, it is therefore necessary to specify the particular conditions under which the force is measured.

It has now been agreed that the *pound weight* shall be defined as the pull of gravity on the *pound mass* (in vacuo) at a place where the acceleration of gravity has the internationally accepted standard value of 32·1741 ft/s^2 (980·665 cm/s^2). With this definition the *pound weight* is a unit of force exactly equal to 32·1741 *poundals.*

At a place where the local acceleration of gravity is g ft/s^2, the pull of gravity on a body of mass m lb (i.e the weight W of the body) will be given by:

$$W = mg/32{\cdot}1741 \text{ pounds weight} \quad . \quad . \quad (2.6)$$

Bearing in mind the small variation in the value of g, so long as bodies close to the earth's surface are being considered, it follows that the weight of a body in pound weight is for practical purposes *numerically* equal to its mass in pounds.

From equation (2.5) we have:

$$F \text{ (poundals)} = m \text{ (lb)} \times f \text{ (ft/s}^2\text{)} \quad . \quad . \quad . \quad (2.7)$$

or, expressing the force F in terms of the larger unit, the pound weight (lb wt),

$$F \text{ (lb wt)} = \frac{m \text{ (lb)} \times f \text{ (ft/s}^2\text{)}}{32{\cdot}1741} \quad . \quad . \quad . \quad (2.8)$$

which may be written:

$$F \text{ (lb wt)} = m \text{ (slugs)} \times f \text{ (ft/s}^2\text{)} \quad . \quad . \quad (2.9)$$

where one slug = a mass of 32·1741 lb.

But from (2.6):

$$W/g = m/32{\cdot}1741$$

and, substituting in (2.8),

$$F \text{ (lb wt)} = \frac{W \text{ (lb wt)}}{g \text{ (ft/s}^2\text{)}} \times f \text{ (ft/s}^2\text{)} \quad . \quad . \quad . \quad (2.10)$$

In (2.10) g is the local acceleration of gravity and therefore has the units of acceleration. It is not a pure number like the denominator of (2.8). Although it varies in magnitude it may be assumed for practical purposes to have the standard value of 32·1741 ft/s². In all equations the figure 32·1741 must appear in order to give the force in terms of the standard pound weight, and the approximate value 32·2 is generally adopted.

It is a matter of opinion which of the various forms of the fundamental equation should be used. One important branch of engineering has decided on (2.9) with the slug as the unit of mass, but this has not found general acceptance. Many engineers prefer (2.8) while others prefer (2.10). Whatever may be said against the form (2.10), the author believes that engineering students find it the least confusing, partly because in practical problems it is the weight of a body rather than its mass which is more often specified.

19. Centripetal Force. We have seen in Article 16 that velocity is a vector quantity and that, in general, the velocity of a particle may change both in magnitude and in direction. If the velocity is constant in magnitude but changes in direction, the acceleration is perpendicular to the velocity and its magnitude is given by $f_c = \omega v = v^2/r$, where r is the radius of curvature of the path along which the particle is moving and ω is the angular velocity of the velocity vector. Hence a force must act radially inwards in order to constrain the particle to follow the curved path. This force is called a *centripetal* force and its magnitude is given by:

$$F_c = mf_c = m.v^2/r = m\omega^2 r = mv\omega \quad . \quad (2.11)$$

It will be seen that the magnitude of the centripetal force is given by the product of the linear momentum of the particle and the angular velocity of the momentum vector.

Referring to Fig. 15 (a), let a particle of mass m be moving along a circular path of radius r with velocity v, and let the necessary constraint be applied by an inextensible string which joins the particle to the centre O of the circular path. Then the string must exert a radially inward centripetal force F_c on the particle. Since, by Newton's third law of motion, action and reaction are equal and opposite, the particle must exert a radially outward force on the string of equal magnitude. The radially outward reaction is generally referred to as the *centrifugal* force.

It must be emphasised that the centrifugal force is not a force applied to the particle. It is the reaction of the particle on the string and arises from the inertia of the particle or its resistance to

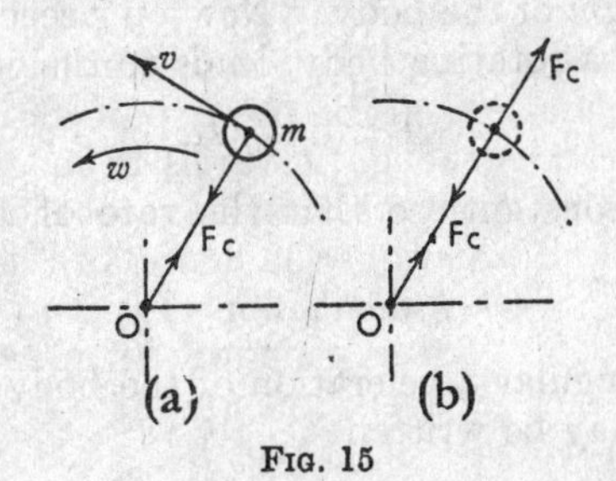

Fig. 15

the change of motion represented by the centripetal acceleration. So far as the tension in the string is concerned, conditions are the same as if the particle were at rest or, alternatively, moving in the specified way but without mass, and were acted upon by a radially outward force equal in magnitude to the centripetal force, as shown in Fig. 15 (b). This is perhaps the most common application of a general principle, known as d'Alembert's principle, by means of which problems in dynamics are reduced to equivalent problems in statics. (See Article 28.)

20. Mass Moment of Inertia. Couple. So far as linear motion is concerned, the mass of a rigid body may be concentrated at a point. This point is known as the centre of mass or, more usually, the centre of gravity. But although the centre of gravity may be fixed so that there is no motion of translation of the body as a whole, it is still possible for the body to rotate about an axis which passes through the fixed centre of gravity. Again, it is a matter of experience that the body resists any attempt to change its angular velocity ω about the axis. It is the angular inertia or the *mass moment of inertia* of the body which determines the resistance to a given rate of change of the angular velocity. In order to cause the angular acceleration a *couple* must be applied.

A body of mass m when rotating behaves as if all its mass were concentrated in a ring at a distance k from the axis of rotation. The radius k is known as the radius of gyration of the body, and the product mk^2 is known as the *mass moment of inertia* of the body. The symbol I will be used to denote the mass moment of inertia mk^2.

A *couple* consists of two equal, opposite and parallel forces with different lines of action. The magnitude of the couple is given by the product of one of the forces and the perpendicular distance between the lines of action of the two forces.

The product of the mass moment of inertia and the angular velocity of a body is known as the *moment of momentum* or the *angular momentum* of the body. Newton's second law of motion, when applied to a rotating body, leads to the equation:

$$T \propto (\mathrm{d}/\mathrm{d}t)(I\omega)$$

Generally I is constant, so that the rate of change of angular momentum is:

$$I \, . \, \mathrm{d}\omega/\mathrm{d}t \quad \text{or} \quad I\alpha$$

where α is the angular acceleration of the body and the equation for the couple may be written:

$$T = I\alpha \quad . \quad . \quad . \quad . \quad . \quad . \quad (2.12)$$

21. Work. Power. Energy. If a force F acts on a body so as to produce a displacement x of the body in the direction of the force, the *work* done is defined as the product of the force and the displacement, i.e. work $= F.x$. Where the magnitude of the force changes continuously during the displacement, the work done during any small displacement δx is given by $F.\delta x$, and the total work done during a finite displacement is given by $\Sigma F.\delta x$, or, using the infinitesimal notation, the total work done is equal to $\int F.\mathrm{d}x$.

Similarly, if a couple T acts on a body so as to produce an angular displacement θ about an axis perpendicular to the plane of the couple, the work done is given by the product $T.\theta$. If the magnitude of the couple changes continuously during the displacement, the work done is given by $\int T.\mathrm{d}\theta$.

In engineering the unit of work is the foot-pound, i.e. the work done by a force of one pound acting through a distance of one foot, or the work done by a couple or torque of one pound-foot acting through an angular displacement of one radian.

Power is the rate of doing work, or the work done in unit time. The unit of power is 1 ft lb per second, but a larger and more convenient unit for many purposes is the horse-power, which is equal to 550 ft lb per second.

Energy is usually defined as the capacity for doing work. There are many different forms of energy, such as heat, light, electric, potential, kinetic, etc. The mechanical forms of energy are *potential*, *strain* and *kinetic energy*.

If a body of weight W is raised through a vertical distance h above some datum level, it is said to possess *potential energy* of amount $W.h$, since it is able to do an amount of work $W.h$ in falling to the datum level.

Similarly, if a spring of stiffness s lb per unit extension or compression is extended or compressed by the amount x, it is said to possess *strain energy* of amount $\frac{1}{2}s.x^2$, since it is able to do an amount of work $\frac{1}{2}s.x^2$ in returning to the unstrained condition.

N.B.—The mean force exerted by the spring is $\frac{1}{2}s.x$ and the distance through which it acts is x, so that the work done is $\frac{1}{2}s.x.x$ or $\frac{1}{2}sx^2$. A torsional spring of stiffness q lb ft per unit of angular displacement, when twisted through an angle of θ radians, possesses strain energy of amount $\frac{1}{2}q.\theta^2$, since it is able to do an amount of work $\frac{1}{2}q.\theta^2$ in returning to the unstrained condition.

Again, if a body of mass m is moving with a velocity v, it is said to possess *kinetic energy* of amount $\frac{1}{2}mv^2$, since it is able to do an amount of work $\frac{1}{2}mv^2$ in being brought to rest.

This may be proved as follows. Since force = mass × acceleration and work done = force × displacement, we have, during a small displacement δx, which takes place in time δt with change of velocity δv:

$$\delta W = F.\delta x = m(\delta v/\delta t)\delta x = mv\delta v$$

In the limit, $$dW = mv dv$$

and, integrating, $$W = \int_0^v mv dv = \frac{1}{2}mv^2$$

Similarly, a body of mass moment of inertia I about a given axis, when rotating about that axis with angular velocity ω, possesses kinetic energy of amount $\frac{1}{2}I\omega^2$, since it is able to do an amount of work $\frac{1}{2}I\omega^2$ in being brought to rest.

22. Conservation of Energy and Conservation of Momentum. Energy exists in many different forms, but it is subject to a conservation law, i.e. although it can be converted from one form to another all the available evidence goes to show that it cannot be created or destroyed. Thus we may say that the total energy possessed by a system of moving bodies is at every instant constant, provided that no energy is rejected to or received from a source external to the system. Many problems in mechanics may be very readily solved by an application of the principle of the conservation of energy. The chief danger lies in the possibility of overlooking the fact that the conditions of the problem may lead to a change of form of some of the energy possessed by the system, such as, for example, a change from mechanical to heat energy.

The principle of the conservation of momentum is a direct consequence of Newton's laws of motion. It is, in fact, only another way of stating the third law, that action and reaction are equal and opposite. For instance, imagine a system consisting of two bodies completely isolated from their surroundings, so that the only forces acting on the bodies are their mutual interactions. Then, clearly, since force is equal to rate of change of momentum and action and reaction are equal and opposite, the rate of change of momentum of one of the two bodies must at any instant be equal in magnitude but opposite in direction to the rate of change of momentum of the other body. Hence, during a finite interval of time the momenta of the two bodies will change by equal and opposite amounts, from which it follows that the total momentum of the system will remain unchanged. Generalising from the above, we may say that, for a system of moving bodies which is not acted upon by any external forces, the sum of the momenta remains constant. Or, stated in another way, the velocity of the centre of gravity of a system of moving bodies, which is not acted upon by any external forces, remains invariable. The application of the principles of the conservation of energy and the conservation of momentum to problems on impact will be considered in the following article.

23. Impulse. Impact. Force is equal to rate of change of momentum, i.e $F = (\mathrm{d}/\mathrm{d}t)(mv)$, from which $F\mathrm{d}t = \mathrm{d}(mv)$, and integrating both sides:

$$\int F\mathrm{d}t = \int \mathrm{d}(mv)$$

The left-hand side of this equation is the time integral of the force, while the right-hand side is the change of momentum produced. A given finite change of momentum may be produced by a small force acting for an appreciable interval of time or by a very large force acting for a very short interval of time. Where the interval of time is so short as to border on the infinitesimal, the force is termed an *impulsive* force or blow and the time-integral of the force is known as the *impulse*. Impulsive forces occur in collisions, in explosions, in the striking of a nail by a hammer or of a pile by a tup or monkey.

Let us consider the impact between two bodies which move with different velocities along the same straight line. It will be assumed that the point of impact lies on the line joining the centres of gravity of the two bodies, as it would do if the bodies were spheres. The behaviour of the colliding bodies during the complete period of impact will depend on the properties of the materials of which they are made. For the purposes of analysis the materials may be assumed to be (a) perfectly elastic, or (b) perfectly inelastic.

In either case the first effect of impact will be approximately the same. The parts of each body adjacent to the point of impact will be deformed and the deformation will continue until the centres of gravity of the two bodies are moving with the same velocity. What the velocity then is may be found by applying the principle of the conservation of momentum. Assuming that there are no external forces acting on the system, the total momentum must remain constant.

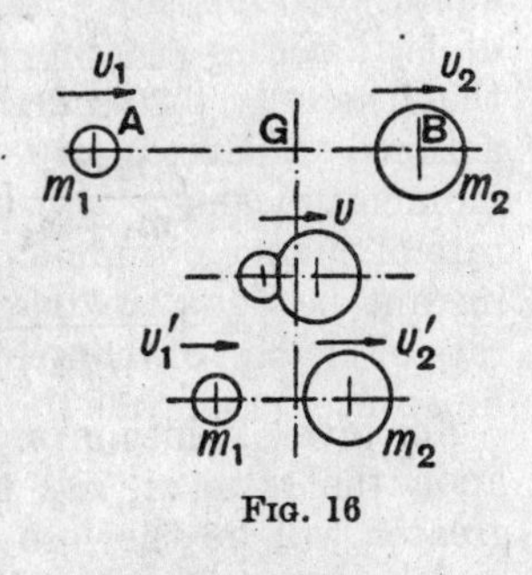

FIG. 16

Referring to Fig. 16, let m_1, v_1 be the mass and velocity of one body, m_2, v_2 the mass and velocity of the other body before impact, where $v_1 > v_2$; also, let v be the common velocity of the centres of gravity at the instant of impact. Then:

$$(m_1+m_2)v = m_1v_1+m_2v_2$$

or

$$v = \frac{m_1v_1+m_2v_2}{m_1+m_2} \quad . \quad . \quad . \quad (2.13)$$

Note that this is equal to the velocity of the c.g. of the combined masses before impact. This follows because the c.g. of the combined masses divides the distance between the c.g. of the individual masses inversely as the magnitudes of the two masses, i.e.

$$\text{AG} = \frac{m_2}{m_1+m_2}\,\text{AB}$$

also the velocity of G relative to A is clearly equal to AG/AB times the velocity of B relative to A, i.e. $(\text{AG}/\text{AB})(v_2-v_1)$. Hence, the velocity of G

$$= v_1 + \frac{\text{AG}}{\text{AB}}(v_2-v_1) = v_1 + \frac{m_2}{m_1+m_2}(v_2-v_1) = \frac{m_1v_1+m_2v_2}{m_1+m_2}$$

Before impact the total kinetic energy of the system was $\frac{1}{2}m_1v_1^2+\frac{1}{2}m_2v_2^2$. At the instant the two c.g. are moving with the same velocity, the kinetic energy is $\frac{1}{2}(m_1+m_2)v^2$. Substituting for v from (2.13) and subtracting, the loss of kinetic energy of the system is given by·

$$\text{loss of K.E.} = \tfrac{1}{2}m_1v_1^2+\tfrac{1}{2}m_2v_2^2-\tfrac{1}{2}(m_1+m_2)\left(\frac{m_1v_1+m_2v_2}{m_1+m_2}\right)^2$$

$$= \tfrac{1}{2}\frac{1}{m_1+m_2}\{(m_1+m_2)(m_1v_1^2+m_2v_2^2)-(m_1v_1+m_2v_2)^2\}$$

$$= \tfrac{1}{2}\frac{1}{m_1+m_2}(m_1m_2v_1^2+m_1m_2v_2^2-2m_1m_2v_1v_2)$$

$$= \tfrac{1}{2}\frac{m_1m_2}{m_1+m_2}(v_1-v_2)^2 \quad . \quad . \quad . \quad . \quad . \quad . \quad . \quad (2.14)$$

$$= \tfrac{1}{2}\frac{m_1}{m_1/m_2+1}(v_1-v_2)^2 \quad . \quad . \quad . \quad . \quad . \quad . \quad (2.15)$$

For given values of m_1, v_1 and v_2, the loss of K.E. will depend upon the ratio m_1/m_2; the smaller the value of this ratio, the greater will be the loss of K.E. If $m_1 = m_2$, the loss of K.E. $= \tfrac{1}{4}.m_1(v_1-v_2)^2$. But if m_2 is very large in comparison with m_1, the loss of K.E. approaches the limit $\tfrac{1}{2}m_1(v_1-v_2)^2$. In other words, when a body of mass m_1 impinges on a second body of equal mass, the loss of K.E. is only one-half as great as when it impinges on a second body of infinitely large mass.

It follows from the prineiple of the conservation of energy that this loss of kinetic energy is equal to the work done in deforming the two bodies. If the two bodies are perfectly inelastic, the work of deformation will be absorbed in overcoming internal friction of the material. There will be no strain energy stored up in the material due to elastic deformation and therefore there will be no tendency for either body to regain its original shape. Hence the two bodies will adhere together and will move on with reduced kinetic energy after impact. The reduction of kinetic energy will appear as heat energy because of the work done in overcoming the internal friction during deformation. The impact between two lead spheres or two clay spheres approximates to inelastic impact.

If the colliding bodies are perfectly elastic, the whole of the work done in deforming the bodies will be stored up as strain energy. No energy will be absorbed in overcoming internal friction and there will be no conversion of kinetic energy into heat energy. Immediately after the instant at which the two centres of gravity are moving with the same velocity, the bodies will begin to regain their original shape, the strain energy being reconverted into kinetic energy and the two bodies ultimately separating. In this case the impulse on each of the colliding bodies will have exactly the same magnitude during the second phase of impact, i.e. while the centres of gravity are separating. as it had during the first stage, i.e. while the centres of gravity were approaching. Hence the change of momentum of each body during the second phase will be exactly equal to the change of momentum during the first phase.

Let v_1', v_2' be the velocities of the respective c.g. at the instant when contact between the colliding bodies ceases. Then the change of momentum of one body during the second phase of impact $= m_1(v_1'-v)$ and the corresponding change of momentum of the same body during the first stage of impact $= m_1(v-v_1)$. These two are equal, so that:

$$v_1'-v = v-v_1 \quad \text{or} \quad v_1' = 2v-v_1 \qquad (2.16)$$

Similarly, for the second body, the change of momentum during the second phase of impact $= m_2(v_2'-v)$ and the corresponding change of momentum during the first stage of impact $= m_2(v-v_2)$, so that:

$$v_2'-v = v-v_2 \quad \text{or} \quad v_2' = 2v-v_2 \quad . \quad (2.17)$$

Subtracting (2.17) from (2.16), we get:

$$v_1'-v_2' = v_2-v_1 = -(v_1-v_2) \quad . \quad . \quad (2.18)$$

Hence, the relative velocity of the colliding bodies after impact is equal and opposite to the relative velocity of the two bodies before impact. The impact between two glass or steel spheres approximates to elastic impact.

In all practical problems on impact, we have to deal with materials that are neither perfectly inelastic nor perfectly elastic. The more nearly perfect the elasticity of the material, the smaller will be the amount of energy converted into heat energy at impact and the more closely will the relative velocity of the two bodies after impact approach equality with the relative velocity of the two bodies before impact. Actually the former is always less, than the latter and the ratio of the two, viz, $(v_1'-v_2')/(v_2-v_1)$ is termed the *coefficient of restitution* for the particular material, and is denoted by e. Because some energy is absorbed in overcoming internal friction of the materials during the period of impact, the total kinetic energy after impact is always less than the total kinetic energy before impact. The momentum, however, is always the same after impact as before impact. Where momentum has apparently been destroyed, what has really happened is that the momentum has been imparted to a body of such large mass that it makes a negligible difference to the momentum already possessed by that body, as, for instance, when a moving body strikes the earth.

Where the coefficient of restitution is e, we have:

$$v_1'-v_2' = -e(v_1-v_2) \quad . \quad . \quad . \quad (2.19)$$

The final velocities of the colliding bodies after impact and the loss of kinetic energy during impact may be most easily found as follows: During the second phase of impact the change of velocity of each body is now only e times the change of velocity during the first phase of impact,

$$\therefore \; v_1'-v = e(v-v_1) \quad \text{or} \quad v_1' = (1+e)v-ev_1 \quad . \quad (2.20)$$

and $$v_2'-v = e(v-v_2) \quad \text{or} \quad v_2' = (1+e)v-ev_2 \quad . \quad (2.21)$$

These two equations reduce to (2.16) and (2.17), when $e = 1$. It follows that the energy returned to the system as kinetic energy during the second phase of impact is only e^2 times the energy absorbed during the first phase of impact. Hence the net loss of kinetic energy during impact is $1-e^2$ times the energy absorbed during the first phase. Or, from (2.14):

$$\text{net loss of K.E.} = \frac{1-e^2}{2} \cdot \frac{m_1 m_2}{m_1+m_2}(v_1-v_2)^2 \qquad (2.22)$$

Example 1. A sphere of mass 100 lb moving at 10 ft/s overtakes and collides with another sphere of mass 50 lb moving at 5 ft/s in the same direction. Find the velocities of the two masses after impact and the loss of kinetic energy during impact when (a) the impact is inelastic, (b) when it is elastic, (c) when e is 0·6.

(a) *Inelastic Impact*. The two masses adhere after impact and move with a common velocity v.

From (2.13), $$v = \frac{100.10+50.5}{100+50} = \frac{1250}{150} = 8{\cdot}333 \text{ ft/s}$$

The total kinetic energy before impact

$$= \frac{1}{2.32{\cdot}2}(100.10^2+50.5^2) = 174{\cdot}6 \text{ ft lb}$$

The total kinetic energy after impact

$$= \frac{150}{2.32{\cdot}2} \cdot \left(\frac{25}{3}\right)^2 = 161{\cdot}7 \text{ ft lb}$$

Loss of kinetic energy during impact

$$= 174{\cdot}6-161{\cdot}7 = 12{\cdot}9 \text{ ft lb}$$

or, from (2.14), loss of kinetic energy

$$= \frac{100.50}{2.32{\cdot}2.150}(10-5)^2 = 12{\cdot}93 \text{ ft lb}$$

(b) *Elastic Impact.* Just as for inelastic impact when the two bodies have a common velocity, that velocity is given by $v = 8{\cdot}333$ ft/s. Immediately after impact ends, the velocity of the 100-lb sphere is given by (2.16):

$$v_1' = 2v - v_1 = 2.8{\cdot}333 - 10 = 6{\cdot}667 \text{ ft/s}$$

Similarly, the velocity of the 50-lb sphere is given by (2.17):

$$v_2' = 2v - v_2 = 2.8{\cdot}333 - 5 = 11{\cdot}667 \text{ ft/s}$$

In this case there is no loss of kinetic energy during impact.

(c) *Coefficient of Restitution* $e = 0{\cdot}6$.

From (2.20), $v_1' = (1+e)v - ev_1 = 1{\cdot}6.8{\cdot}333 - 0{\cdot}6.10$
$= 13{\cdot}333 - 6 = 7{\cdot}333$ ft/s

and from (2.21),

$$v_2' = (1+e)v - ev_2 = 13{\cdot}333 - 3{\cdot}0 = 10{\cdot}333 \text{ ft/s}$$

The total kinetic energy after impact

$$= \frac{1}{2.32{\cdot}2}(100.7{\cdot}333^2 + 50.10{\cdot}333^2) = \frac{10\,720}{64{\cdot}4} = 166{\cdot}3 \text{ ft lb}$$

$$\therefore \text{ loss of kinetic energy during impact} = 174{\cdot}6 - 166{\cdot}3 = 8{\cdot}3 \text{ ft lb}$$

Or, from (2.22),

$$\text{loss of kinetic energy} = (1 - 0{\cdot}6^2)12{\cdot}93 = 8{\cdot}28 \text{ ft lb}$$

Example 2. A loaded railway truck weighs 15 tons and moves along a level track at 12 m.p.h. It overtakes and collides with an empty truck which weighs 5 tons and which is moving along the same track at 8 m.p.h. If the four buffer springs affected each have a stiffness of 2 tons/in., find the maximum deflection of each spring during impact and the speeds of the trucks immediately after impact ends.

If the coefficient of restitution for the buffer springs were only 0·5, how would the final speeds be affected and what amount of energy would be dissipated during impact?

This example serves to show how the shock between two colliding bodies may be softened, or cushioned, by means of buffers. The purpose of the buffers is to increase the duration of impact, by allowing considerable local deformation of the colliding bodies, and thus to reduce the magnitude of the force which acts between the bodies during impact. Energy is absorbed by the buffer springs during the interval of time required for the speeds of the colliding bodies to be equalised, and is returned, either wholly or in part, during the remainder of the period of impact.

At the instant when the two trucks are moving at the same speed during impact, their linear momentum will be equal to the sum of the linear momenta of the trucks before impact. Hence the speed at this instant is given by:

$$(15+5)v = 15.12+5.8 \quad \text{or} \quad v = 220/20 = 11 \text{ m.p.h.}$$

The difference between the kinetic energy before impact and the kinetic energy at the instant the two trucks are moving at the same speed may be calculated from (2.14). Neglecting all losses, it must of course be equal to the strain energy stored up in the buffer springs.

$\therefore$ strain energy stored in the springs

$$= \frac{1}{2}\cdot\frac{15.5}{32\cdot2.20}\left(\frac{88}{60}\right)^2 (12-8)^2 = 2\cdot00 \text{ ft tons}$$

Let x inches be the maximum deflection of each buffer spring during impact. Then the strain energy stored up in each spring equals the work done in compressing the spring.

$$= \tfrac{1}{2}.2.x.x = x^2 \text{ in. tons}$$

Since there are four buffer springs to absorb the energy,

$$4x^2 = 24\cdot0 \quad \text{or} \quad x = 2\cdot45 \text{ in.}$$

Hence the maximum force which acts between each pair of buffers during impact = stiffness of spring × deflection = $2.2\cdot45 = 4\cdot90$ tons.

Neglecting all losses and assuming the buffer springs to be perfectly elastic, the speeds of the trucks immediately after impact has ended may be calculated from equations (2.16) and (2.17).

For the loaded truck,

$$v_1' = 2.11-12 = 10 \text{ m.p.h.}$$

and for the unloaded truck,

$$v_2' = 2.11-8 = 14 \text{ m.p.h.}$$

If the coefficient of restitution for the buffer springs is 0·5, the final speeds of the trucks may be obtained from equations (2.20) and (2.21). For the loaded truck,

$$v_1' = (1+0\cdot5)11-0\cdot5.12 = 10\cdot5 \text{ m.p.h.}$$

and for the unloaded truck,

$$v_2' = (1+0\cdot5)11-0\cdot5.8 = 12\cdot5 \text{ m.p.h.}$$

The net loss of kinetic energy during impact may be calculated from (2.22). It amounts to

$$(1-e^2)2\cdot00 = (1-0\cdot25)2\cdot00 = 1\cdot50 \text{ ft tons}$$

Example 3. Fig. 17 shows a flywheel A connected through a torsionally flexible spring S to one element C of a dog clutch. The other element D of the clutch is free to slide along, but must revolve with the shaft to which the flywheel B is keyed. The

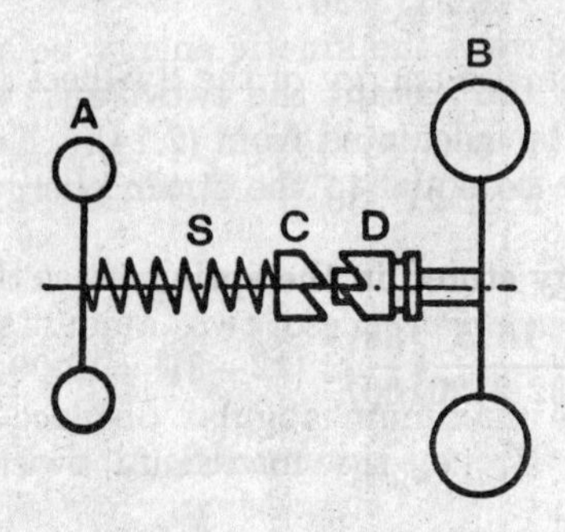

FIG. 17

moments of inertia of A and B are 500 lb ft² and 1500 lb ft² and the torsional stiffness of the spring S is 150 lb ft per radian. When the flywheel A is revolving at 150 r.p.m. and the flywheel B is at rest, the dog clutch is suddenly engaged. Neglecting all losses, find:

(a) the maximum twist of the spring S,

(b) the speeds of the flywheels at the instant the twist in S is a maximum,

(c) the speed of each flywheel when the spring regains its initial unstrained condition.

Immediately after the clutch is engaged, the element C is brought momentarily to rest. But the spinning flywheel A starts to wind up the spring S, thus causing equal and opposite torques to act on A and B. The magnitude of the torque increases continuously until the speeds of A and B are equalised, energy being stored in the spring during this interval. Beyond this point the spring begins to unwind and the strain energy stored in the spring is reconverted into kinetic energy of rotation of the flywheels.

(a) Since there is no external torque acting on the system the angular momentum remains constant throughout.

Let N be the angular velocity of both flywheels at the instant their speeds are equalised. Then

$$(500+1500)N = 500 . 150$$
$$N = 37{\cdot}5 \text{ r.p.m.}$$

The kinetic energy of the system at this instant

$$= \frac{1}{2} \cdot \frac{500+1500}{32 \cdot 2} \left(\frac{\pi N}{30}\right)^2$$

$$= \frac{1}{2} \cdot \frac{2000}{32 \cdot 2} \left(\frac{\pi . 37 \cdot 5}{30}\right)^2 = 480 \text{ ft lb}$$

But the initial kinetic energy of the flywheel A

$$= \frac{1}{2} \cdot \frac{500}{32 \cdot 2} \left(\frac{\pi . 150}{30}\right)^2 = 1915 \text{ ft lb}$$

The strain energy stored in the spring must therefore be equal to the difference between these two amounts of energy, i.e. 1435 ft lb.

(b) Let θ be the maximum angular displacement of wheel A relative to wheel B, i.e. the maximum twist of the spring, measured in radians.

Then the mean torque exerted by the spring during this displacement is $\frac{1}{2}.150.\theta$ lb ft and the work done on the spring, i.e. the strain energy stored in the spring, is $\frac{1}{2}.150.\theta^2$ ft lb.

$$\therefore \ \frac{1}{2}.150.\theta^2 = 1435$$

$$\therefore \ \theta = \sqrt{(1435/75)} = 4 \cdot 38 \text{ radians} = 250°$$

(c) Neglecting all losses, the change of momentum of each flywheel while the spring is unwinding must be exactly the same as the change of momentum while the spring is being wound up, since the impulse on each flywheel must be the same in each period. The speeds of the two flywheels when the spring regains its initial unstrained condition may therefore be calculated from equations of the same form as (2.16) and (2.17).

Thus, if N_a, N_a' are the initial and final speeds of the flywheel A and N_b, N_b' those for flywheel B, the equation corresponding to (2.16) is

$$N_a' = 2N - N_a$$
$$= 2.37 \cdot 5 - 150 = -75 \text{ r.p.m.}$$

and that corresponding to (2.17) is

$$N_b' = 2N - N_b$$
$$= 2.37 \cdot 5 - 0 = 75 \text{ r.p.m.}$$

Hence at the instant the spring regains its initial unstrained condition the flywheel A will be revolving at 75 r.p.m. in the opposite sense to its initial motion, while the flywheel B will be revolving at 75 r.p.m. in the same sense as the initial motion of A.

Note that it is theoretically possible for the whole of the initial kinetic energy of the flywheel A to be transmitted to the flywheel B, the former being brought to rest. But this can only occur if the moments of inertia of the two flywheels are identical.

This example illustrates the principle of the inertia starter which is sometimes used for starting internal-combustion engines. The flywheel A is set in motion either by hand or by electric motor. The clutch is then engaged thus coupling the starter to the engine flywheel and crankshaft, represented by the rotor B. In this way the kinetic energy of the starter flywheel is used to spin the engine crankshaft for the first few revolutions until firing begins. In practice the starter flywheel would require to be much too heavy if it were directly connected to the spring as shown in Fig. 17. A smaller flywheel is therefore used and is geared to the starter dog so as to revolve at a very much higher speed. If G is the speed reduction from the flywheel to the starter dog, then the equivalent moment of inertia at the starter dog is G^2 times the actual moment of inertia (see Article 30). But if the whole of the initial kinetic energy of the flywheel A is to be transmitted to the flywheel B, the moments of inertia of the two flywheels should be identical. Hence it follows that to obtain the highest efficiency from a starter with a geared flywheel the moment of inertia of the starter flywheel ought to be $1/G^2$ times the moment of inertia of the equivalent engine flywheel. (See Question, 6 p. 596.)

24. Simple Harmonic Motion. If a body oscillates about an equilibrium position in such a way that its acceleration towards the equilibrium position is directly proportional to its displacement from the equilibrium position, it is said to have simple

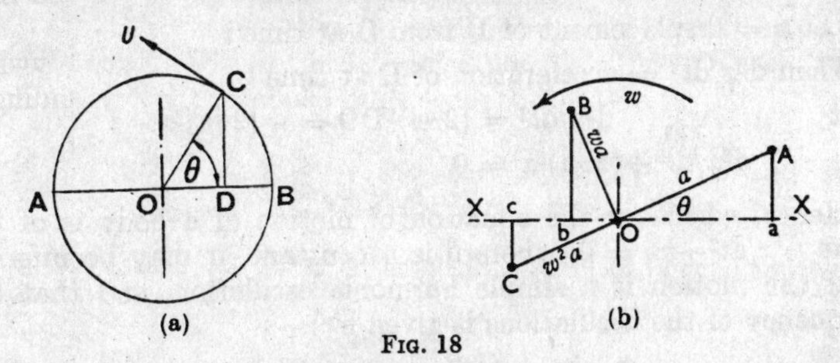

FIG. 18

harmonic motion. Oscillatory motion of the above kind is of frequent occurrence in practice and its characteristics may be deduced from Fig. 18 (a). It is easily shown that, if the point C moves with uniform speed round the circumference of a circle, the projection D of the point C on a diameter AB has simple harmonic motion. The amplitude of the oscillations of D is equal to the radius of the circle along which C moves, and the frequency of the oscillations of D is equal to the peripheral speed of C divided by the circumference of the circle along which C moves.

Let v = the peripheral speed of C,
a = the radius of the circle along which C moves
and n = the frequency of the oscillations of D.

Then
$$v = 2\pi na$$
and the centripetal acceleration of C is
$$f_c = v^2/a = (2\pi n)^2 . a$$

When CO makes an angle θ with AB, the velocity and acceleration of D are respectively equal to the components of the velocity and acceleration of C parallel to AB.

$$\therefore\ v_d = v \sin\theta = 2\pi na \sin\theta = 2\pi n . \mathrm{CD}$$
and
$$f_d = f_c \cos\theta = (2\pi n)^2 a \cos\theta = (2\pi n)^2 \mathrm{DO}$$

From this last equation
$$f_d/\mathrm{DO} = (2\pi n)^2$$
or
$$\frac{\text{Acceleration of D}}{\text{Displacement of D}} = (2\pi n)^2 = \text{constant} \quad . \quad (2.23)$$
$$\therefore\ n = \frac{1}{2\pi}\sqrt{\frac{\text{Acceleration of D}}{\text{Displacement of D}}} \quad . \quad . \quad (2.24)$$

The equation of motion of a simple harmonic motion may be obtained as follows:

Let x = displacement of D from O at time t

Then $\mathrm{d}^2x/\mathrm{d}t^2$ = acceleration of D at time t

But
$$\mathrm{d}^2x/\mathrm{d}t^2 = (2\pi n)^2\mathrm{DO} = -(2\pi n)^2 x$$
or
$$\mathrm{d}^2x/\mathrm{d}t^2 + (2\pi n)^2 x = 0$$

Hence, whenever the equation of motion of a body is of the form $\mathrm{d}^2x/\mathrm{d}t^2 + bx = 0$, where b is a constant, it may be inferred that the motion is a simple harmonic oscillation, and that the frequency of the oscillations is given by:

$$(2\pi n)^2 = b \quad \text{or} \quad n = (1/2\pi)\sqrt{b}$$

It is possible, and sometimes convenient, to represent the changes in displacement, velocity and acceleration of a body with simple harmonic motion by the changes in projected length of three vectors. The vectors are of constant length, occupy fixed relative positions and rotate at uniform angular speed, ω. Thus in Fig. 18 (b), the displacement vector OA is of length a equal to the amplitude of the vibration, the velocity vector OB is of length ωa $(2\pi na)$ equal to the maximum velocity of the vibrating body, and it leads the vector OA by 90°, the acceleration vector OC is of length $\omega^2 a$ $\{(2\pi n)^2 a\}$ equal to the maximum acceleration of the vibrating body and it leads the vector OA by 180°.

If time is measured from the instant at which the body has maximum displacement towards the right, then at time t the vectors will have turned through the angle $\theta = \omega t$ from their initial positions.

The projected lengths along the horizontal displacement line XX are:

$$\text{Oa} = \text{OQ} \cos \theta = a \cos \theta = \text{displacement}$$
$$\text{Ob} = \text{OB} \cos (90^\circ + \theta) = -\omega a \sin \theta = \text{velocity}$$
$$\text{Oc} = \text{OC} \cos (180^\circ + \theta) = -\omega^2 a \cos \theta = \text{acceleration.}$$

The representation in this way of a simple harmonic motion by means of vectors forms the basis of a method of solving vibration problems. (See Article 199.)

25. The Simple Pendulum. A heavy bob of negligible dimensions when suspended vertically by means of an inextensible weightless cord forms a simple pendulum, Fig. 19.

L θ W

FIG. 19

Let W be the weight of the bob, m the mass and L the length of the cord from the point of suspension to the centre of the bob. Let the cord be displaced through a small angle θ, Fig. 19, and the bob allowed to swing.

Then the couple tending to restore the bob to the equilibrium position $= T = WL \sin \theta$; and, since the angle θ is small, $\sin \theta \simeq \theta$ and $T \simeq WL\theta$.

The mass moment of inertia of the bob about an axis through the point of suspension $= I = mL^2$.

$$\therefore \text{ angular acceleration of the cord} = \alpha = \frac{T}{I} \simeq \frac{WL\theta}{mL^2} \simeq \frac{g}{L}\theta$$

$$\therefore \frac{\text{Angular acceleration}}{\text{Angular displacement}} = \frac{\alpha}{\theta} \simeq \frac{g}{L} \simeq \text{constant}$$

The motion of the pendulum is therefore approximately simple harmonic, and from equation (2.24) the frequency of oscillation

$$n = \frac{1}{2\pi}\sqrt{\frac{g}{L}} \quad . \quad . \quad . \quad . \quad . \quad (2.25)$$

and the periodic time

$$t = \frac{1}{n} = 2\pi\sqrt{\frac{L}{g}} \quad . \quad . \quad . \quad . \quad (2.26)$$

26. The Compound Pendulum. The Torsion Pendulum. A rigid body suspended vertically so as to oscillate with small amplitude under the action of gravity is termed a compound pendulum.

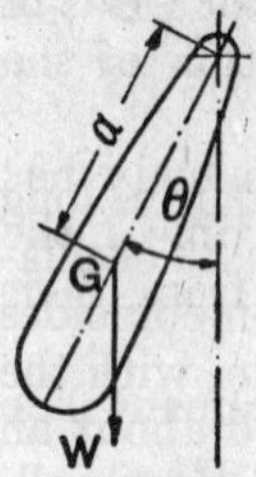

Let W be the weight of the body, m the mass, k the radius of gyration about an axis through the centre of gravity perpendicular to the plane of motion and a the distance of the point of suspension from the centre of gravity, Fig. 20.

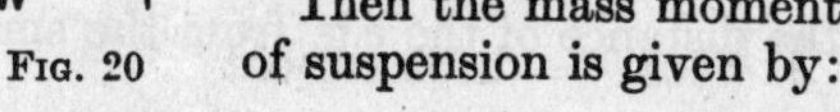

FIG. 20

Then the mass moment of inertia about the axis of suspension is given by:

$$I = m(k^2 + a^2)$$

The restoring couple

$$T = Wa \sin\theta \simeq Wa\theta$$

The angular acceleration of the pendulum

$$= \alpha = \frac{T}{I} \simeq \frac{Wa\theta}{m(k^2+a^2)} \simeq \frac{ga}{k^2+a^2}.\theta \simeq \text{constant}.\theta$$

The motion of the pendulum is therefore approximately simple harmonic, and from (2.24):

$$n = \frac{1}{2\pi}\sqrt{\frac{g.a}{k^2+a^2}} \quad . \quad . \quad . \quad . \quad (2.27)$$

The length of a simple pendulum which would have the same frequency is given by:

$$L = \frac{k^2+a^2}{a} = \frac{k^2}{a} + a \quad . \quad . \quad . \quad (2.28)$$

Example 4. The connecting rod of an engine weighs 150 lb and is 3 ft long between centres. Its c.g. is 25 in. from the centre of the small end, about which the connecting rod oscillates in a vertical plane. The rod is found by experiment to complete 50 oscillations in 92·5 sec. Find the moment of inertia of the rod about an axis through the c.g. What is the length of the equivalent simple pendulum?

The number of oscillations per second $= n = 50/92{\cdot}5 = 0{\cdot}541$.

From equation (2.25) the length of the simple pendulum which has the same frequency is given by:

$$L = \frac{g}{(2\pi n)^2} = \frac{32{\cdot}2}{(2\pi . 0{\cdot}541)^2} = 2{\cdot}79 \text{ ft}$$

The distance a of the c.g. from the point of suspension $= 25/12 = 2{\cdot}08$ ft.

But from (2.28):

$$k^2/a+a = L, \quad \text{so that} \quad k^2 = 2{\cdot}08(2{\cdot}79-2{\cdot}08) = 1{\cdot}48 \text{ ft}^2$$

The moment of inertia of the rod $= 150 \,.\, 1{\cdot}48 = 222$ lb ft^2.

Example 5. A connecting rod of weight 90 lb and 30 in. long between centres is suspended vertically. The time for 50 oscillations is found to be 84·4 sec when the axis of oscillation coincides with the small-end centre, and 80·3 sec when it coincides with the big-end centre. Find the moment of inertia of the rod about an axis through the c.g. and the distance of the c.g. from the small-end centre.

Let L_1, L_2 = length of equivalent simple pendulum when the axis of oscillation coincides with the small-end and the big-end centres respectively.

Let n_1, n_2 = corresponding frequencies of oscillation per second, so that $n_1 = 50/84{\cdot}4$ and $n_2 = 50/80{\cdot}3$.

Let a_1, a_2 = distances of c.g. from small-end and big-end centres respectively.

Then from (2.25):

$$L_1 = g/(2\pi n)^2 = 32{\cdot}2 \,.\, 12(84{\cdot}4/100\pi)^2 = 27{\cdot}9 \text{ in.}$$

Similarly, $\quad L_2 = 32{\cdot}2 \,.\, 12(80{\cdot}3/100\pi)^2 = 25{\cdot}3$ in.

But, from (2·28): $\quad k^2 = a_1(L_1-a_1)$

and $\quad k^2 = a_2(L_2-a_2)$

$$\therefore\ a_1(L_1-a_1) = a_2(L_2-a_2)$$

Also a_1+a_2 = distance between centres = 30 in.

Substituting for a_2 in terms of a_1 and also for L_1 and L_2, we get:

$$a_1(27{\cdot}9-a_1) = (30-a_1)(25{\cdot}3-30+a_1)$$
$$\therefore\ 27{\cdot}9a_1-a_1^2 = (30-a_1)(a_1-4{\cdot}7)$$
$$= 30a_1+4{\cdot}7a_1-141-a_1^2$$
$$\therefore\ 6{\cdot}8a_1 = 141$$
$$\therefore\ a_1 = 20{\cdot}7 \text{ in.}$$

and from (2.28):

$$k^2 = 20{\cdot}7(27{\cdot}9-20{\cdot}7) = 149 \text{ in}^2$$

$$\therefore \text{ Moment of inertia of rod} = 90 \,.\, 149/144 = 93{\cdot}3 \text{ lb ft}^2$$

Torsional Pendulum. A disc or flywheel is suspended with its axis vertical from a ceiling by three long flexible parallel wires of equal length l. The wires are attached at the corners of an equilateral triangle which are each distant a from the axis of the disc. If the disc is twisted about its axis through a small angle θ and then released it will oscillate with frequency n, which may be expressed in terms of the dimensions of the system.

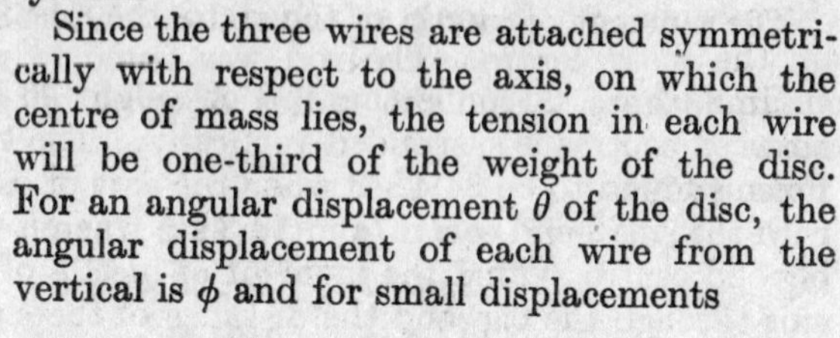

FIG. 21

Since the three wires are attached symmetrically with respect to the axis, on which the centre of mass lies, the tension in each wire will be one-third of the weight of the disc. For an angular displacement θ of the disc, the angular displacement of each wire from the vertical is ϕ and for small displacements

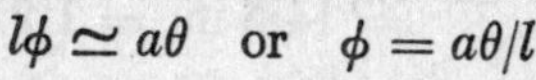

$$l\phi \simeq a\theta \quad \text{or} \quad \phi = a\theta/l$$

The horizontal component of the tension in each wire

$$= F = (W/3) \tan \phi$$

and the moment of this component about the axis of the disc

$$= Fa \cos (\theta/2) = (W/3)a \tan \phi . \cos (\theta/2) \simeq (W/3)a\phi$$

since θ and ϕ are small, so that $\cos (\theta/2) \simeq 1$ and $\tan \phi \simeq \phi$.

The total moment applied to the disc

$$= Wa\phi = Wa^2\theta/l$$

This is a pure couple, since the three horizontal forces have zero resultant, and it tends to restore the disc to its initial equilibrium position.

The angular acceleration towards the equilibrium position:

$$\alpha = \frac{\text{Applied torque}}{I} = \frac{Wa^2}{l} . \theta . \frac{g}{Wk^2}$$

$$= \frac{g}{l} . \left(\frac{a}{k}\right)^2 \theta$$

The acceleration towards the equilibrium position is therefore directly proportional to the displacement from the equilibrium position and the oscillation is simple harmonic.

From (2.24):

$$n = \frac{1}{2\pi}\sqrt{\frac{\alpha}{\theta}} = \frac{1}{2\pi} \frac{a}{k} \sqrt{\frac{g}{l}} \quad . \quad . \quad . \quad (2.29)$$

The radius of gyration of a flywheel, airscrew or similar body may be experimentally determined in this way. For a connecting rod two wires would be used and the same equation would apply. The wires would, of course, be attached to the rod at equal distances from the mass centre, so that the tension is the same in each wire and a pure couple is applied to the rod.

Example 6. The connecting rod of Example 4 was suspended on two wires 8·5 ft long attached to the rod at points 14 in. from the mass centre. The rod was found to make 25 oscillations in 83·2 sec. Find the radius of gyration about the mass centre.

From (2.29):

$$k = \frac{a}{2\pi n}\sqrt{\frac{g}{l}} = \frac{14.83{\cdot}2}{2\pi.25}\sqrt{\frac{32{\cdot}2}{8{\cdot}5}}$$

$$= 14{\cdot}5 \text{ in.} = 1{\cdot}21 \text{ ft}$$

$\therefore$ The moment of inertia of the rod $= 150.1{\cdot}21^2$
$= 219$ lb ft^2

27. Equivalent Dynamical System. In many problems on the dynamics of a rigid body it is convenient to replace the body by two masses assumed to be concentrated at points and connected rigidly together.

In order that the two-mass system shall be dynamically equivalent to the rigid body it must react to a given system of forces in exactly the same way as the rigid body reacts.

Obviously the conditions which must be satisfied by the two-mass system are:

(a) The total mass must be equal to that of the rigid body.
(b) The c.g. must coincide with that of the rigid body.
(c) The total moment of inertia about an axis through the c.g. must be equal to that of the rigid body.

For the rigid body shown in Fig. 22, let m be the mass and k the radius of gyration about an axis through G; also let m_a, m_b be

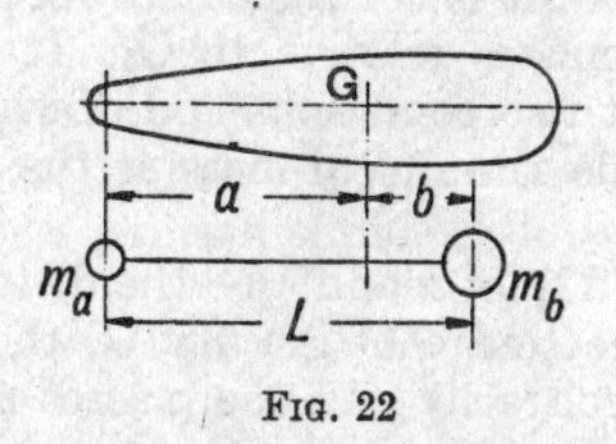

FIG. 22

two masses which form an equivalent dynamical system and a, b be the distances of m_a and m_b respectively from G.

Then the three conditions stated above lead to the following equations:

$$m_a+m_b = m \qquad (1)$$

$$m_a a = m_b b \qquad (2)$$

$$m_a a^2+m_b b^2 = mk^2 \qquad (3)$$

Substituting for $m_b b$ from (2) in (3), we get:

$$m_a a^2+m_a ab = mk^2$$

$$\therefore\ m_a = \frac{mk^2}{a(a+b)} \qquad (4)$$

But from (1) and (2):

$$m_a = \frac{mb}{a+b} \qquad (5)$$

Equating (4) and (5):

$$\frac{mk^2}{a(a+b)} = \frac{mb}{a+b}$$

$$\therefore\ k^2 = ab \quad . \quad . \quad . \quad . \quad (2.30)$$

From (2.30) it is clear that either a or b, but not both, may be chosen arbitrarily. If a is fixed, then the distance b will follow from the above equation.

A simple and convenient way of finding the position of the second mass, when the radius of gyration k is unknown, is to suspend the body vertically, so as to be free to swing about an axis through A. The length L of a simple pendulum which has the same period of oscillation as the body will then be equal to $a+b$. This means that the second mass is situated at the centre of percussion of the body. The proof of this statement is as follows:

From (2.28) $L = k^2/a+a$ and from (2.30) $k^2/a = b$,

$$\therefore\ L = a+b$$

For some problems, as, for instance, when considering the effect of the inertia of the connecting rod on the crankshaft turning moment (Article 154), it is convenient to fix arbitrarily the positions of both the masses relative to G. It is then possible to satisfy only the first two conditions laid down above, i.e. the two masses may have the same total mass as the body and the same centre of gravity.

Referring to Fig. 23, let a and c be the distances of two masses m_a and m_c from the centre of gravity G, the distances a and c being fixed quite arbitrarily and the masses m_a and m_c having a total mass m and centre of gravity G.

Then $$m_a = \frac{c}{a+c}m \quad \text{and} \quad m_c = \frac{a}{a+c}m$$

The mass moment of inertia about an axis through G of the system of two masses m_a and m_c is given by $I_1 = m_a a^2 + m_c c^2$, and, substituting for m_a and m_c, we have:

$$I_1 = \frac{m}{a+c}(a^2c+c^2a) = mac$$

Let k_1 be the radius of gyration of the two-mass system. Then $I_1 = mk_1{}^2$, so that:

$$k_1{}^2 = ac \quad . \quad . \quad . \quad . \quad . \quad (2.31)$$

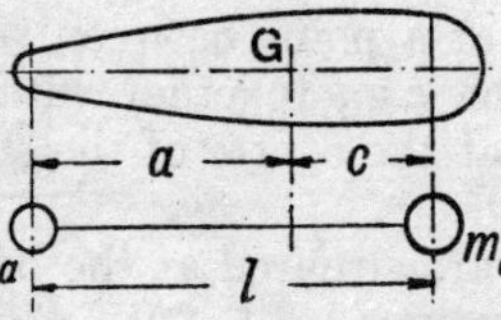

FIG. 23

The mass moment of inertia of the system in which the two distances a and c are fixed arbitrarily differs from the mass moment of inertia of the body by the amount:

$$I_1 - I = m(k_1{}^2 - k^2)$$

If α is the angular acceleration of the body, then the difference between the torque required to accelerate the two-mass system and that required to accelerate the body is given by:

$$T' = m(k_1{}^2 - k^2)\alpha \quad . \quad . \quad . \quad . \quad (2.32)$$

The two-mass system must therefore have a correction couple T' applied to it before it can be considered as dynamically equivalent to the body.

It may be pointed out that, if l is the distance between the two masses m_a and m_c and L is the distance between the two masses m_a and m_b which form a true dynamically equivalent system, then $c - b = (a+c)-(a+b) = l - L$.

But from (2.30) $k^2 = ab$ and from (2.31) $k_1{}^2 = ac$, so that

$$I_1 - I = mac - mab = ma(c-b) = ma(l-L)$$

and

$$T' = ma(l-L)\alpha \quad . \quad . \quad . \quad . \quad (2.33)$$

Incidentally L is the length of the equivalent simple pendulum when the body is suspended from an axis which passes through the position of the mass m_a.

Example 7. A connecting rod weighs 2·5 lb and the length between centres is 9 in. The distance of the c.g. from the small-end centre is 6 in. and the radius of gyration about an axis through the c.g. is 3·8 in. Find the equivalent dynamical system if one of the masses coincides in position with the small-end centre.

If the rod is replaced by two masses, one at the small-end centre and the other at the big-end centre, and the angular acceleration of the rod is 22,500 rad/s^2 clockwise, what correction couple must be applied to this system in order that its effect may be identical with that of the rod ?

From (2.30) the distance b of the second mass of the equivalent system from the c.g. of the rod $= k^2/a = 3{\cdot}8^2/6 = 2{\cdot}41$ in.

The equivalent system therefore consists of a mass of $2{\cdot}5 . 6/8{\cdot}41 = 1{\cdot}783$ lb situated at a point between the c.g. and the big end and 2·41 in. from the c.g., together with a mass of $2{\cdot}5-1{\cdot}783 = 0{\cdot}717$ lb situated at the centre of the small end, i.e. 6 in. from the c.g.

If the two masses are situated at the bearing centres, then the mass at the big-end centre$=2{\cdot}5 . 6/9 = 1{\cdot}667$ lb and the mass at the small-end centre $= 2{\cdot}5-1{\cdot}667 = 0{\cdot}833$ lb.

The radius of gyration k_1 of this system about an axis through the c.g. is given by (2.31):

$$k_1{}^2 = ac = 6.3 = 18 \text{ in.}^2$$

The correction couple which must be applied in order that the two-mass system may be dynamically equivalent to the rod is given by (2.32):

$$\begin{aligned} T' &= m(k_1{}^2-k^2)\alpha \\ &= \frac{2{\cdot}5}{32{\cdot}2} . \frac{18-3{\cdot}8^2}{12.12} . 22\,500 = 43{\cdot}2 \text{ lb ft} \end{aligned}$$

This correction couple has the same sense as the angular acceleration and therefore acts clockwise.

28. The Resultant Effect of a System of Forces which acts on a Rigid Body. D'Alembert's Principle. If a rigid body is acted upon by a system of forces, that system may be reduced to a single resultant, whose magnitude, direction and line of action may be found by the methods of graphic statics. In general the line of action of the resultant F, Fig. 24 (a), will not pass through the c.g. but will be at some distance x from it. Two equal and opposite forces of magnitude F may be applied through G parallel to the resultant without influencing the effect of the resultant, since the two forces acting through G are in equilibrium. It follows that the given resultant is equivalent to an equal and parallel force through G together with a couple consisting of the original force and the equal and opposite force through G. The moment of this couple is the product $F . x$.

The force F through G causes linear acceleration of the c.g. and the couple of moment $F.x$ causes angular acceleration of the body about an axis through G perpendicular to the plane in which the couple acts.

Then $$F = m.f \quad \text{and} \quad F.x = mk^2\alpha$$

where m = mass of body,
k = radius of gyration about an axis through G,
f = linear acceleration of G
α = angular acceleration.

From these two equations, given F, x, m and k, the accelerations f and α may be calculated.

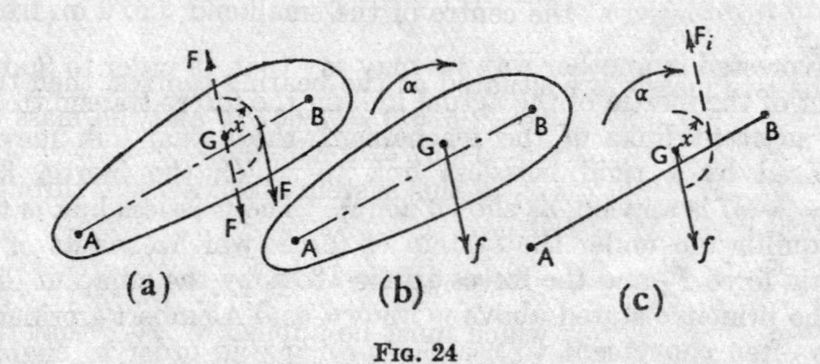

FIG. 24

In dynamics of machines, the problem usually presents itself in the reverse way. Thus each part of a machine has a definite motion imposed upon it through its connection with the adjacent moving parts, and the linear acceleration of the c.g. and the angular acceleration may be determined by methods given in the next chapter. It will then be possible, given m, k, f and α, to find the magnitude, direction and line of action of the resultant applied force F. The magnitude of $F = mf$ and the distance of its line of action from the c.g. is $x = mk^2\alpha/F$. The direction of F and the position of its line of action in relation to G must, of course, correspond to the known senses of f and α.

D'Alembert's Principle. As we have just seen, if F is the resultant of a system of forces applied to a rigid body of mass m, the magnitude of F is given by the product of m and the linear acceleration f of the centre of mass of the body, or

$$F = mf$$

This may be written

$$F - mf = 0 \quad . \quad . \quad . \quad . \quad . \quad (2.34)$$

From this equation it may be inferred that if we regard $-mf$ as a force, equal, opposite and with the same line of action as the resultant F, and include this force with the system of forces of which F is the resultant, the complete system of forces thus obtained will be in equilibrium. The ordinary rules of statics for a system of forces in equilibrium will then apply. In effect the inclusion of the force $-mf$ with the system of applied forces converts the problem in dynamics to one in statics. The force F which causes the acceleration, both linear and angular, of the rigid body is termed the *effective force* acting on the body. The equal and opposite force $-mf$ is variously termed the *reversed effective force*, the *kinetic reaction* or the *inertia force*. If we denote this inertia force by F_i, then substituting $F_i = -mf$ in (2.34), we get

$$F+F_i = 0 \quad . \quad . \quad . \quad . \quad . \quad . \quad (2.35)$$

Expressed in another way we may say that, in order to find the effect of the inertia of the actual link on the forces transmitted to the adjacent links of the mechanism, the actual link may be replaced by a rigid massless link to which the inertia force $F_i = -mf$ is applied, as shown at (c). The massless link is then in equilibrium under the system of forces which consists of the inertia force F_i and the forces applied to it by the adjacent links.

The principle stated above is known as d'Alembert's principle. It is often convenient to use this principle in order to enable a problem in dynamics to be reduced to an equivalent problem in statics. As already pointed out in Article 19 the usual treatment of problems involving centripetal acceleration is an application of this principle and there are many other examples in the later chapters of this book.

Example 8. The link AB, Fig. 25, weighs 20 lb. It is 18 in. long between the bearing centres, the c.g. is $7\frac{1}{2}$ in. from B and the radius of gyration about an axis through the c.g. is 7 in. The linear acceleration of G is 200 ft/s^2 and the angular acceleration is 120 rad/s^2 clockwise.

Find the forces which must be applied through the pins at A and B in order to accelerate the link, if the line of action of the force through A is given.

The effective force applied to the link:

$$F = (20/32{\cdot}2)200 = 124{\cdot}2 \text{ lb}$$

This force acts parallel to the acceleration f_g.

The couple required in order to provide the angular acceleration:

$$T = I\alpha = (20/32{\cdot}2)(7/12)^2.120 = 25{\cdot}4 \text{ lb ft}$$

The line of action of F is therefore at a distance from G given by:

$$x = T/F = 25{\cdot}4/124{\cdot}2 = 0{\cdot}204 \text{ ft}$$

and in order to correspond to the specified directions of f and α, the line of action of F must be as shown at (a).

Since there are only two forces F_a and F_b applied to the link, and F is their resultant, it follows that F_a, F_b and F must all intersect at the same point. The direction of F_a is given and the magnitude and line of action of F has just been found, so that the line of action of F_b must pass through the point of intersection O of F_a and F. The force F may then be resolved along the appropriate lines of action to give the magnitudes of F_a and F_b. Scaled from the diagram, $F_a = 65$ lb and $F_b = 91$ lb.

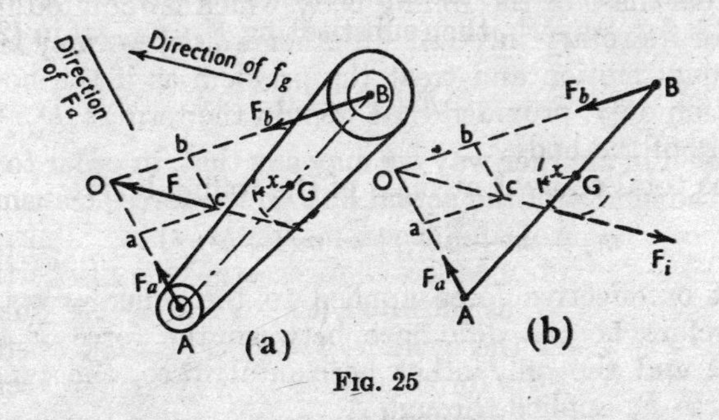

FIG. 25

Alternatively, the actual link AB is equivalent to a massless link, which has exactly the same motion and to which the inertia force F_i is applied Fig. 25 (b). The massless link is then in equilibrium under the three forces F_i, F_a and F_b, which therefore intersect at a point. The triangle of forces may then be drawn to find the magnitudes of F_a and F_b.

29. Force required to accelerate a Body which rolls without slipping on a Horizontal Plane (Fig. 26). Let m be the mass of the body and k its radius of gyration about an axis through the c.g. Let r be the radius of the cylindrical surface in contact with the plane and f the linear acceleration of the c.g.

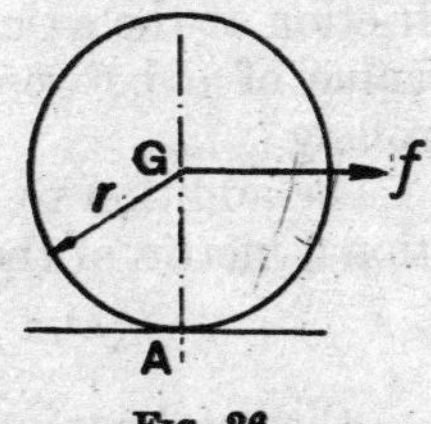

FIG. 26

Since the body rolls without slipping, the point of contact A is at rest. This point is, in fact, the instantaneous centre for the relative motion of the rolling body and the plane (see Article 41).

The mass moment of inertia of the body about an axis through A perpendicular to the plane of rotation

$$I_a = m(k^2+\mathrm{AG}^2) = m(k^2+r^2)$$

The angular acceleration $\alpha = f/r$, so that the couple which must be applied to the rolling body $= m(k^2+r^2)\alpha = m(k^2+r^2)(f/r)$.

Let F be the force applied through G.

Then, taking moments about A,

$$Fr = m(k^2+r^2)(f/r)$$

$$F = m(k^2+r^2)(f/r^2) = m(k^2/r^2+1)f \quad . \quad (2.36)$$

The expression $mk^2/r^2 = I/r^2$ may be regarded as an addition to the actual mass of the rolling body which is required in order to allow for its rotary inertia. In other words, we may suppress the rotational motion and treat the problem as if the body had linear motion only, provided that we add the term $m.k^2/r^2$ to the actual mass of the body.

Then the total equivalent mass of the rolling body:

$$m_e = m+mk^2/r^2 = m(1+k^2/r^2) \quad . \quad . \quad (2.37)$$

The net or effective force applied to the roller $= mf$. This must therefore be the difference between the force F applied through G and the only other horizontal force, the tangential friction force F_a applied through A.

$$\therefore \ F_a = F-mf = mf.k^2/r^2 \quad . \quad . \quad (2.38)$$

If there is to be pure rolling, with no slip at A, $F_a<\mu W$, where μ is the coefficient of friction between the surfaces and W is the weight of the roller.

$$\therefore \ mf.k^2/r^2<\mu W \quad \text{or} \quad \mu>(f/g)k^2/r^2$$

Example 9. A four-wheel truck has a total weight of 10 tons and each pair of wheels with the axle weighs 1000 lb and has a radius of gyration of 15 in. The diameter of the wheel treads is 3 ft 6 in. What force must be exerted on the truck in order to give to it an acceleration of 3 ft/s² along a horizontal track, if friction at the axle journals is neglected, and what is the smallest value of μ between wheel tyres and rails consistent with pure rolling?

The addition to the actual mass in order to allow for the rotational inertia of the wheels and axles

$$mk^2/r^2 = 2(1000/2240)(15/21)^2 = 0{\cdot}455 \text{ ton}$$

$\therefore$ the total equivalent mass to be accelerated

$$= 10{\cdot}46 \text{ tons}$$

and the accelerating force required

$$= 3.10{\cdot}46/32{\cdot}2 = 0{\cdot}975 \text{ ton} = 2180 \text{ lb}$$

The total tangential force required in order to provide the angular acceleration of the wheels and axles is, from (2.38),

$$F_a = \frac{2.1000}{2240}\cdot\frac{3}{32\cdot 2}\cdot\left(\frac{15}{21}\right)^2 = 0\cdot 042 \text{ ton}$$

But the limiting friction force $= \mu W$ where W is the total weight of the truck, so that, if there is to be pure rolling with no slip,

$$\mu . 10 > 0\cdot 042$$

or

$$\mu > 0\cdot 0042$$

30. The Acceleration of a Geared System. In Fig. 27 two shafts A and B are geared together, so that B rotates at G times the speed of A, i.e $G = N_b/N_a$. The total mass moment of inertia of the masses attached to A is I_a and of those attached to B is I_b.

If the angular acceleration of shaft A is α, what torque must be applied to the shaft A?

Since the shaft B turns at G times the speed of shaft A, the rate of change of the speed of shaft B with respect to time must necessarily be G times the rate of change of the speed of shaft A with respect to time, or $\alpha_b = G\alpha_a$.

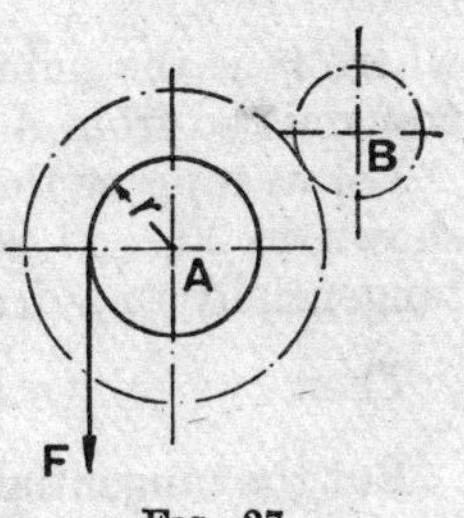

FIG. 27

It follows that the torque required for the angular acceleration of B is given by $T_b = I_b\alpha_b = GI_b\alpha_a$.

But to provide a torque T_b on shaft B, the torque applied to shaft A must be $G.T_b$, i.e. a torque $G^2.I_b\alpha_a$ must be applied to shaft A in order to accelerate shaft B.

In addition the torque required to accelerate shaft A by itself is equal to $I_a\alpha_a$.

Hence the total torque which must be applied to shaft A in order to accelerate the geared system is given by:

$$T = I_a\alpha_a + G^2 I_b\alpha_a = \alpha_a(I_a + G^2.I_b) = \alpha_a.I$$

where,

$$I = I_a + G^2.I_b \quad . \quad . \quad . \quad . \quad (2.39)$$

I may be regarded as the equivalent mass moment of inertia of the system referred to shaft A.

If the efficiency of the gearing between the two shafts A and B is η, then the torque which must be applied to A in order to accelerate B

$$= GT_b/\eta = G^2 I_b\alpha_a/\eta,$$

the total torque applied to shaft A in order to accelerate the geared system

$$= T = I_a\alpha_a + G^2 I_b\alpha_a/\eta,$$

and the equivalent mass moment of inertia of the geared system referred to shaft A

$$= I = I_a + G^2 I_b / \eta \quad . \quad . \quad . \quad . \quad (2.40)$$

For a system in which a number of shafts are geared together in series, the equivalent inertia referred to shaft A is evidently given by:

$$I = I_a + \Sigma(G_x{}^2 I_x / \eta_x) \quad . \quad . \quad . \quad (2.41)$$

where I_x is the mass moment of inertia of shaft X, G_x is the ratio of the speed of shaft X to the speed of shaft A and η_x is the overall efficiency of the gearing from shaft A to shaft X.

If each pair of gear wheels is assumed to have the same efficiency η, the overall efficiency from shaft A to shaft X is given by:

$$\eta_x = \eta^m$$

where m is the number of gear pairs through which the power is transmitted from A to X.

Let us suppose that the torque required to accelerate the system shown in Fig. 27 is applied by means of a force F which acts tangentially to a drum or pulley of radius r.

Then $$F.r = \alpha_a I = \alpha_a(I_a + G^2.I_b)$$

But the tangential acceleration of the drum $f = \alpha_a.r$ or $\alpha_a = f/r$ so that

$$F.r = (f/r)(I_a + G^2.I_b)$$

$$\therefore\ F = f.(1/r^2)(I_a + G^2.I_b) = f.m_e$$

where $$m_e = (1/r^2)(I_a + G^2.I_b) \quad . \quad . \quad . \quad . \quad (2.42)$$

This may be regarded as the equivalent mass of the system referred to the line of action of the accelerating force F.

Example 10. The moment of inertia of A, Fig. 27, is 200 lb ft^2, and the moment of inertia of B is 15 lb ft^2. The shaft B runs at five times the speed of shaft A. A mass of 150 lb is hung from a rope wrapped round a drum of effective radius 8 in., which is keyed to the shaft A. If the mass is allowed to fall freely, find its acceleration.

What would be the acceleration, if the efficiency of the gearing were 90%?

The equivalent mass of the geared system referred to the circumference of the drum may be found from (2.42):

$$m_e = (1/r^2)(I_a + G^2.I_b) = (12/8)^2(200 + 5^2.15) = 1293 \text{ lb}$$

The total equivalent mass to be accelerated $= 1293 + 150$

$= 1443$ lb. But the accelerating force is provided by the pull of gravity on the mass suspended from the rope.

$$\therefore \text{ acceleration } = f = (150/1443)32{\cdot}2 = 3{\cdot}35 \text{ ft/s}^2$$

If the efficiency of the gearing is 90%, the equivalent mass of the geared system referred to the circumference of the drum

$$= (1/r^2)(I_a + G^2 I_b/\eta) = (3/2)^2(200 + 5^2 . 15/0{\cdot}9) = 1388 \text{ lb}$$

and the total equivalent mass to be accelerated

$$= 1388 + 150 = 1538 \text{ lb}$$

$$\therefore \text{ acceleration } = f = (150/1538)32{\cdot}2 = 3{\cdot}14 \text{ ft/s}^2$$

Example 11. A motor-car weighs $13\frac{1}{2}$ cwt, and the moment of inertia of the wheels and back axle is 150 lb ft² and of the engine parts is 5 lb ft². The gear ratios provided are 5·4, 7·3, 12·5 and 22·5 to 1, and the effective diameter of the road wheels is 25·2 in. If the engine torque is 31·0 lb ft, find the maximum acceleration of the car on each gear.

The overall efficiency of the transmission is 90% in top gear and 82% in each of the other gears.

This problem will be solved in a different way from that given above, by applying the principle of the conservation of energy.

Let S = displacement of car from rest with uniform acceleration f, the engine torque T being assumed to remain constant,
v = final speed of car,
G = gear ratio,
r = effective radius of road wheels,
η = efficiency of transmission from engine to road wheels,
M = mass of car,
and I_a, I_b = moments of inertia of road wheels and engine

Since the speed of the car increases from 0 to v with uniform acceleration f in distance S,

$$\therefore v^2 = 2fS$$

At the end of the displacement, the angular velocity of the wheels $= \omega_a = v/r$, and the angular velocity of the engine crankshaft $= \omega_b = G . \omega_a = G . v/r$.

The total angle through which the wheels have turned $= S/r$, and the total angle through which the engine crankshaft has turned $= \theta_b = G . S/r$.

The total work done by the engine torque $= T . \theta_b = T . G . S/r$.

The increase of kinetic energy of translation of the car $= \frac{1}{2}Mv^2$.

The increase of K.E. of rotation of the wheels $= \frac{1}{2}I_a\omega_a^2 = \frac{1}{2}I_a.v^2/r^2$, and the increase of K.E. of rotation of the engine parts $= \frac{1}{2}I_b\omega_b^2 = \frac{1}{2}I_bG^2.v^2/r^2$.

The energy available for transmission from the engine to the wheels is the difference between the total work done by the torque and the increase of K.E. of the engine parts, i.e.

$$T.G.S/r - \tfrac{1}{2}I_bG^2.v^2/r^2$$

Because of the losses in the drive the energy available at the rear axle

$$= \eta(T.G.S/r - \tfrac{1}{2}I_bG^2.v^2/r^2)$$

This must be equal to the sum of the increase of K.E. of translation of the car and the increase of K.E. of rotation of the wheels.

$$\therefore\ \eta(T.G.S/r - \tfrac{1}{2}I_bG^2.v^2/r^2) = \tfrac{1}{2}Mv^2 + \tfrac{1}{2}I_a.v^2/r^2$$

$$\therefore\ \eta(T.G.S/r) = \tfrac{1}{2}v^2(M + I_a/r^2 + \eta G^2.I_b/r^2)$$

Substituting $v^2 = 2fS$, we get:

$$\eta(G.T/r) = f(M + I_a/r^2 + \eta G^2.I_b/r^2)$$

In this equation, the left-hand side represents the total tractive force F at the rim of the road wheels, and the expression in the parentheses on the right-hand side represents the total equivalent mass M_e of the car. From the equation the acceleration for each gear ratio may be calculated.

Then $\quad F = \eta G(31{\cdot}0.12/12{\cdot}6) = 29{\cdot}5\eta G$ lb wt

and $\quad M_e = 13{\cdot}5.112 + 150(12/12{\cdot}6)^2 + \eta G^2 5(12/12{\cdot}6)^2$

$= 1512 + 136 + 4{\cdot}54\eta G^2$

$= 1648 + 4{\cdot}54\eta G^2$ lb

$\therefore\ f = 32{\cdot}2.F/M_e$ ft/s²

The results of the calculations are set down in the following table:

Gear	Top	Third	Second	First
G	5·4	7·3	12·5	22·5
η	0·90	0·82	0·82	0·82
F	143·5	176·8	303	545
$4{\cdot}54\eta G^2$	119·2	198·5	582	1882
M_e	1767	1847	2230	3530
f	2·62	3·08	4·37	4·98

This example shows how important is the effect of the rotary inertia of the engine parts on the acceleration of the car, particularly when first or second gear is engaged and the ratio of engine speed to road wheel speed is high.

31. Gyroscopic Couple. In Article 17 we saw that angular velocity is a vector quantity and that angular acceleration is involved if either the magnitude or the direction of the angular velocity changes. But, in order to produce angular acceleration of a rotating body, a couple must be applied. So far we have only considered the particular case in which the couple produces a change in the magnitude of the angular velocity, but no change in its direction. We have now to consider the other particular case in which the couple produces a change in the direction of the angular velocity, but no change in its magnitude. Although the magnitude of the angular acceleration has already been found in Article 17, it is worth while to start from first principles.

(a) *Plane Disc.* Suppose that a disc, Fig. 28, is spinning in a vertical plane parallel to plane YOZ with an angular velocity ω and that the axis of spin is at the same time rotating in a horizontal plane XOZ with an angular velocity ω_p. Then, applying the right-handed screw rule, the angular momentum of the disc, when in the position shown by full lines, may be represented by the vector Oa and, when in the position shown by dotted lines, by the vector Ob. The change of momentum in the interval of time δt, during which the disc moves from one position to the other, is therefore represented by the vector ab. But change of angular momentum can only be produced by the application of a couple to the disc. Since the applied couple is equal to the rate of change of angular momentum, we have

$$T = \delta(I\omega)/\delta t$$

But $\delta(I\omega) = \text{ab} \simeq \text{Oa}.\delta\theta$, where $\delta\theta$ is the angle through which the axis of spin rotates in the time δt.

$$\therefore\ T \simeq \text{Oa}.\delta\theta/\delta t \simeq I\omega.\delta\theta/\delta t$$

and in the limit, when $\delta\theta$ is very small:

$$T = I\omega.\mathrm{d}\theta/\mathrm{d}t = I\omega\omega_p \quad . \quad . \quad . \quad (2.43)$$

It will be seen that the couple T is given by the product of the angular momentum $I\omega$ and the angular velocity ω_p of the angular momentum vector. Equation (2.43) should be compared with equation (2.11), which gives the centripetal force required in order to cause the linear velocity of a body to change in direction without changing in magnitude.

Referring to Fig. 28, the vector ab lies in the plane XOZ and in the limit, when $\delta\theta$ is very small, its direction is perpendicular to Oa and therefore to the plane XOY. The applied couple T must therefore act in the plane XOY and, to conform to the right-handed screw rule, its sense must be clockwise when viewed in the direction ab, i.e. when viewed in the direction OZ.

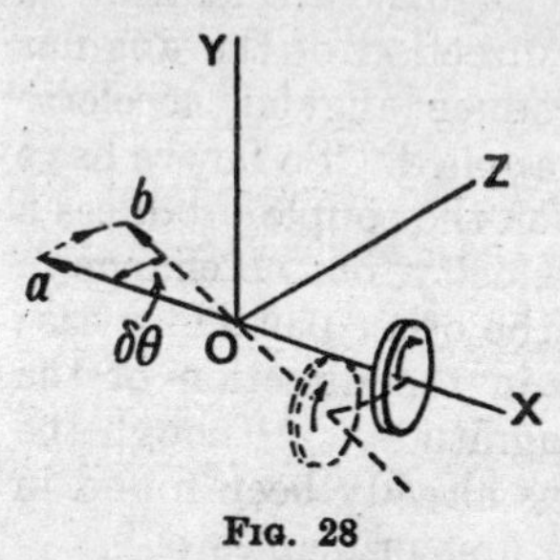

FIG. 28

As already pointed out in Article 17, the angular acceleration which is involved, when the angular velocity changes in direction but remains constant in magnitude, is known as gyroscopic acceleration and the couple which gives rise to it is known as a *gyroscopic couple.* The rotation of the axis of spin is called *precessional* motion.

It will be seen that the plane of rotation or spin is parallel to plane YOZ, the plane of precession is the plane XOZ and the plane of the gyroscopic couple is the plane XOY. These three planes are mutually perpendicular.

It may also be pointed out that the spin vector, or the angular momentum vector, Oa and the couple vector Oc, both lie in the plane of precession XOZ, and that the sense of the precession is such as to tend to bring the spin vector into the position occupied by the couple vector by the shortest possible route.

Whenever the axis of rotation or spin of a body changes its direction, a gyroscopic couple must be applied to it. The couple is usually applied through the bearings which support the shaft. The reaction of the shaft on each bearing is of course equal and opposite to the action of the bearing on the shaft. Hence the precession of the axis of rotation causes a gyroscopic reaction couple to act on the frame to which the bearings are fixed. The conditions are analogous to those which exist when the linear momentum of a body changes in direction. Then it is the radially inward force which acts on the body, but the reaction of the body is radially outward.

The couple which must be applied to a spinning body in order to maintain the precessional motion of the axis of rotation may be derived in a different way. Suppose the disc, Fig. 29 (a), is spinning counter-clockwise with angular velocity ω about an axis through O and the axis of rotation is precessing counter-clockwise with angular velocity ω_p, as shown in plan. Consider a particle P in the upper half of the disc at a distance y from the horizontal diameter. Then the velocity of P in the plane of the disc is ω.OP at right angles to OP, and the component velocity of P parallel to XX is $v = \omega . y$. As seen in plan the particle P will appear to oscillate along AB with simple harmonic motion, while AB rotates counter-clockwise with angular velocity ω_p. Because of the rotation of AB, the velocity v is changing its direction and P has a centripetal acceleration $v\omega_p$ at right angles to AB acting downwards. But P also has a velocity $\omega_p x$ in the upward direction as seen in plan, which decreases as x decreases, so that there is an additional acceleration of P in the downward direction amounting to $(d/dt)(\omega_p x)$, or $\omega_p v$, since ω_p is constant and $dx/dt = v$.

The total acceleration of P perpendicular to the plane of the disc is therefore $2\omega_p v$. This is known as the Coriolis acceleration of P. (See also Article 53, p. 102.)

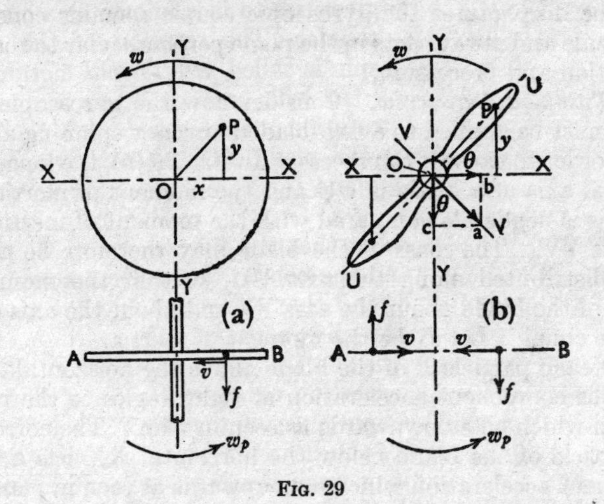

FIG. 29

The Coriolis acceleration of P depends only upon the magnitudes and senses of ω_p and v. So long as P is in the upper half of the disc and v is therefore in the direction from right to left, the acceleration is downwards in plan independent of the position of P. Similarly, for all positions of P in the lower half of the disc v is in the direction from left to right and the acceleration is upwards in plan.

To provide this acceleration a force must be applied to P perpendicular to the plane of the disc. If δm is the mass of the particle, the applied force

$$\delta F = \delta m . 2\omega_p v = \delta m . 2\omega_p \omega y$$

The total accelerating force applied to the upper half of the disc will clearly be equal and opposite to the total accelerating force applied to the lower half of the disc, and the two forces will constitute a couple.

The moment of the force applied to P about the horizontal diameter XX

$$= \delta T = \delta F . y = \delta m . 2\omega_p \omega . y^2 = 2\omega_p \omega . \delta m . y^2$$

where $\delta m . y^2$ is the moment of inertia of the particle P about XX.

For the complete disc, the applied couple

$$= T = 2\omega_p \omega . I_x$$

But for a disc $2I_x = I$ the polar moment of inertia, so that, as before,

$$T = I\omega\omega_p \quad . \quad . \quad . \quad . \quad . \quad (2.44)$$

As the disc rotates the gyroscopic couple remains constant in magnitude and always acts in the plane perpendicular to the planes of rotation and precession.

(b) *Two-bladed airscrew.* Consider now the gyroscopic couple which must be applied to a two-bladed airscrew spinning about its axis in order to cause it to precess. In Fig. 29 (b), the longitudinal principal axis of a blade is UU and the moment of inertia about this axis is negligible compared with the moment of inertia about the axis VV. The mass of the blade may therefore be assumed to be distributed along the axis UU, so that the moments of inertia of the blade about the axis VV and about the axis of rotation are equal. Let I_1 be the moment of inertia.

Then each particle P of the blade above the horizontal XX has a Coriolis component acceleration at right angles to the plane of rotation which acts downwards as seen in plan. The corresponding particle of the blade below the horizontal XX has a Coriolis component acceleration which acts upwards as seen in plan. The forces required to produce these accelerations will be equal and opposite and give rise to a couple which acts in the plane containing the axis of rotation and the longitudinal axis UU.

$$\text{Force applied to P} = \delta m . 2\omega\omega_p y$$

and moment of this force about VV

$$= \delta m . 2\omega\omega_p . yr$$
$$= \delta m . 2\omega\omega_p r^2 \sin\theta$$

and the total moment for one blade

$$= 2I_1\omega\omega_p \sin\theta$$

since $\Sigma\delta m . r^2 = I_1$.

For the two blades, the applied couple

$$= T = 2I\omega\omega \ \sin\theta \quad . \quad . \quad . \quad (2.45)$$

where $I = 2I_1$ = polar moment of inertia of the airscrew about the axis of rotation.

The plane of the couple rotates with the airscrew and the magnitude of the couple varies from nil, when $\theta = 0°$, to a maximum of $2I\omega\omega_p$, when $\theta = 90°$.

The couple vector is shown by Oa in Fig. 29 (b). It will be clear from considerations of symmetry that when the positions of the two blades are interchanged, i.e. for $\theta + 180°$, the couple vector Oa will have exactly the same magnitude and position. In other words, the couple vector Oa points into the first or fourth quadrants for all angular positions of the airscrew.

The total applied couple may be resolved into component couples which act in vertical and horizontal planes perpendicular to the plane of rotation. These couples are represented by the vectors Ob and Oc.

The component couple applied in the vertical plane, i.e. with axis XX

$$\begin{aligned} &= T_x = \text{Ob} = \text{Oa} \sin\theta \\ &= 2I\omega\omega_p \sin^2\theta \\ &= I\omega\omega_p(1 - \cos 2\theta) \quad . \quad . \quad . \quad . \quad (2.46) \end{aligned}$$

It varies in magnitude from 0, when $\theta = 0°$, to $2I\omega\omega_p$, when $\theta = 90°$, and back to 0, when $\theta = 180°$. It has a mean value $I\omega\omega_p$ when $\theta = 45°$ and $135°$.

The component couple applied in the horizontal plane, i.e. with axis YY

$$\begin{aligned} &= T_y = \text{Oc} = \text{Oa} \cos\theta \\ &= 2I\omega\omega_p \sin\theta\cos\theta \\ &= I\omega\omega_p \sin 2\theta \quad . \quad . \quad . \quad . \quad (2.47) \end{aligned}$$

This couple is nil, when $\theta = 0°, 90°, 180°$. It has a maximum value $I\omega\omega_p$, clockwise as seen in plan, when $\theta = 45°, 225°$, and counter-clockwise as seen in plan, when $\theta = 135°, 315°$.

N.B.—From equation (2.46), the mean value of T_x is equal to the couple which would have to be applied to a disc with the same moment of inertia about the axis of spin as the airscrew, but the maximum value of T_x is twice as great.

The vectors Oa, Ob, Oc give the resultant and the two component couples applied to the airscrew shaft through the reactions of the bearings on the shaft. The vectors which represent the resultant and the two component couples applied to the bearings by the airscrew will be equal in magnitude but opposite in sense.

(c) *Multi-bladed Airscrew.* Let n be the number of blades spaced at angle $\alpha = 2\pi/n$. Then assuming as before that the mass of each blade is concentrated along the longitudinal axis UU, the moments of inertia about the transverse axis VV and the axis of rotation are each equal to I_1.

The total moment of inertia of the airscrew about the axis of rotation

$$= I = nI_1$$

Let one blade be inclined to the horizontal at the angle θ.

Then with ω and ω_p as shown in Fig. 29 (b), the total moment about XX required to accelerate this blade

$$= T_{x_1} = 2\omega\omega_p \Sigma \delta m y^2 = 2\omega\omega_p I_1 \sin^2\theta$$
$$= I_1\omega\omega_p(1 - \cos 2\theta)$$

Similar expressions apply for each of the blades, if the appropriate angle $\theta+\alpha$, $\theta+2\alpha$, etc., is substituted.

Therefore, total moment about XX for n blades

$$= T_x = I_1\omega\omega_p[n - \{\cos 2\theta + \cos 2(\theta+\alpha) + \ldots + 2(\theta + \overline{n-1}\,.\,\alpha)\}]$$

The expression in brackets { } is the sum of a cosine series of n terms in which the angles increase in arithmetical progression. The sum of such a series is

$$S = \cos 2\left(\theta + \frac{\overline{n-1}}{2}\,.\,\alpha\right)\frac{\sin n\alpha}{\sin \alpha}$$

Since $n\alpha = 2\pi$, $\sin n\alpha = 0$ and $\alpha = 2\pi/n$, so that $\sin\alpha \neq 0$ except when $n = 2$.

$$\therefore\ S = 0, \quad \text{for all values of } n > 2$$

and $$T_x = n\,.\,I_1\omega\omega_p$$

But $nI_1 = I$, the total moment of inertia of the multi-bladed airscrew about the axis of rotation.

$$\therefore\ T_x = I\omega\omega_p \quad . \quad . \quad . \quad . \quad (2.48)$$

The moment about the vertical axis YY for a blade which makes angle θ with XX from (2.47) is

$$T_{y_1} = I_1\omega\omega_p \sin 2\theta$$

Therefore, total moment about YY for n blades

$$= T_y = I_1\omega\omega_p\{\sin 2\theta + \sin 2(\theta+\alpha) + \ldots + \sin 2(\theta + \overline{n-1}\,.\,\alpha)\}$$

The sum of the sine series in the brackets is zero for all values of $n > 2$, so that the resultant couple required to accelerate the multi-bladed airscrew is given by equation (2.48). This couple is constant in magnitude and it acts in the vertical plane perpendicular to the planes of rotation and precession. It is therefore identical with the couple required for a plane disc with the same moment of inertia about the axis of rotation.

Example 12. The moment of inertia of the disc in Fig. 28 is 40 lb ft^2 and it is spinning at 500 r.p.m. If the shaft precesses through one revolution in 5 sec, what couple must be applied to the shaft?

The angular velocity of spin of the disc

$$= \omega = \pi . 500/30 = 50\pi/3 \text{ rad/s}$$

The angular velocity of precession

$$= \omega_p = 2\pi/5 \text{ rad/s}$$

Substituting in (2.40), the gyroscopic couple T

$$= I\omega\omega_p = 40/32{\cdot}2 . 50\pi/3 . 2\pi/5 = 81{\cdot}7 \text{ lb ft}$$

With the directions of spin and precession which are shown on Fig. 28, we have seen that the applied couple must act in the clockwise sense. The reaction couple exerted by the shaft on the bearings and transmitted to the frame is of equal magnitude but of opposite sense, i.e. it acts in the counter-clockwise sense.

Example 13. The moment of inertia of the airscrew of an aeroplane is 250 lb ft^2, and the direction of rotation is clockwise when looking at the front of the machine. The speed of rotation of the airscrew is 1600 r.p.m. when the speed of flight is 150 m.p.h. If the aeroplane makes a right-handed turn on a path of 500 ft radius, find the gyroscopic reaction of the airscrew on the aeroplane, when:

(a) the airscrew has three blades,
(b) the airscrew has two blades.

The angular velocity of rotation

$$= \omega = \pi . 160/3 \text{ rad/s}$$

The angular velocity of precession

$$= \omega_p = \frac{150 . 88}{60 . 500} = 0{\cdot}44 \text{ rad/s}$$

(a) With a three-bladed airscrew, the gyroscopic couple remains constant as the airscrew rotates and always acts in a plane perpendicular to the plane of rotation and to the plane of precession. Its magnitude from (2.43) $= T = I\omega\omega_p$

$$= \frac{250}{32{\cdot}2} . \frac{\pi . 160}{3} . 0{\cdot}44 = 572 \text{ lb ft}$$

Referring to the plan view shown in Fig 30, the angular momentum is represented by the vector Oa.

Since the aeroplane is making a right-hand turn, the angular momentum vector will change in direction in a short interval of time from Oa to Ob, so that the change of momentum, and therefore the applied couple, is represented by the vector ab. The plane of the applied couple is perpendicular to ab and therefore vertical and its sense is counter-clockwise when looking on the left-hand side of the machine as shown in the elevation. The reaction of the rotating parts on the machine is opposite to the applied couple and thus tends to raise the nose and depress the tail of the machine.

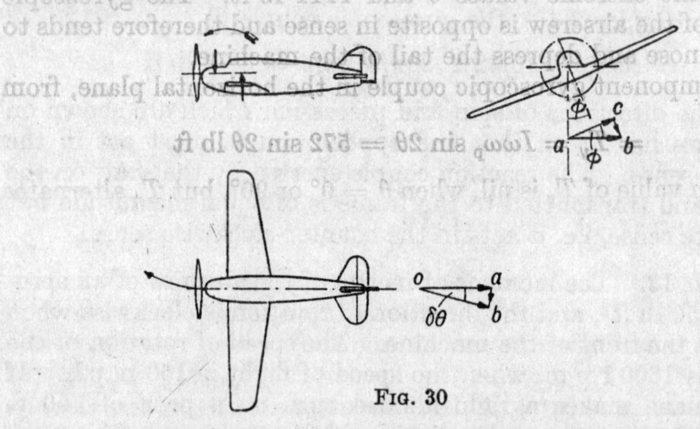

Fig. 30

It is easily seen that a left-hand turn would have the opposite effect. It should be noted, however, that the aeroplane will actually be banked over on the turn as shown in the end elevation, so that the reaction couple should be resolved into two components. One component will act in the longitudinal plane normal to the wings and will have the effect already explained, while the other component couple will act in the plane of the wings. The effect of this latter component will be to tend to turn the machine towards the outside of the curved path. The resultant effect of the gyroscopic reaction couple, when turning to the right, is therefore to tend to raise the nose of the machine and also to make it turn less sharply.

(b) With a two-bladed airscrew the applied gyroscopic couple varies in magnitude, but the plane in which it acts is the plane containing the axis of rotation and the longitudinal axes of the blades. It therefore rotates with the airscrew.

Its magnitude is given by equation (2.45),

$$T = 2I\omega\omega_{\mathrm{p}} \sin\theta$$

where θ is the inclination of the longitudinal axis of the blades to the plane of precession.

The component gyroscopic couple in the vertical plane, i.e. the plane of elevation, Fig. 30, from (2.46)

$$= T_x = I\omega\omega_p(1-\cos 2\theta)$$
$$= 572(1-\cos 2\theta) \text{ lb ft}$$

T_x has a mean value of 572 lb ft when the axis of the blades is at 45° or 135° to the plane of precession, but it is nil when $\theta = 0°$ and a maximum 1144 lb ft when $\theta = 90°$.

It has the sense shown by the arrow in the elevation, but varies between the extreme values 0 and 1144 lb ft. The gyroscopic reaction of the airscrew is opposite in sense and therefore tends to raise the nose and depress the tail of the machine.

The component gyroscopic couple in the horizontal plane, from (2.47),

$$= T_y = I\omega\omega_p \sin 2\theta = 572 \sin 2\theta \text{ lb ft}$$

The mean value of T_y is nil, when $\theta = 0°$ or 90°, but T_y alternates between maximum values of 572 lb. ft in opposite senses twice per revolution of the airscrew. The maximum values occur when $\theta = 45°$ or 135°.

When θ lies between 0° and 90°, T_y is clockwise as seen in plan; and when θ lies between 90° and 180°, it is counterclockwise. The gyroscopic reaction in the former case tends to make the aircraft turn less sharply and in the latter case more sharply.

The gyroscopic reaction of a two-bladed airscrew therefore tends to set up vibrations in the aircraft structure. For this reason three-bladed airscrews which give a gyroscopic reaction of constant magnitude are much to be preferred and are generally used.

32. Gyroscopic Stabilisation. At various times proposals have been put forward for using one or more gyroscopes for the stabilisation of ships or single-track vehicles. Some success has been achieved with their application to ships, but none of the applications to single-track vehicles has survived the experimental stage. A ship of course is intrinsically stable, and the purpose of the gyroscope is to reduce the amplitude of the oscillations of the ship in a heavy sea. In this connection it should be noted that fore-and-aft pitching about a transverse axis is much smaller in amplitude than rolling about a longitudinal axis, and the gyroscope has been applied in order to reduce the amplitude of rolling.

So far as the stabilisation of single-track vehicles is concerned, the requirements are much more exacting. The vehicle is inherently unstable and the gyroscope has not merely to provide the initial stability but it must preserve the stability under changing conditions, such as the action of the wind on the vehicle, displacement of the dead load carried by the vehicle, the inertia effects when rounding a curve, etc.

It is only possible here to outline briefly the main principles on which gyroscopic stabilisation depends. The fundamental requirement is that the gyroscope shall be made to precess in such a way that the reaction couple exerted by the rotor shall oppose any disturbing couple which may act on the frame. If at every instant the reaction couple of the gyroscope and the applied, or disturbing, couple are equal, then complete stabilisation will be obtained.

For instance, consider the application of a gyroscope to a ship in order to limit the amplitude of rolling. The couple which tends to cause rolling arises from the effect of the difference in buoyancy on the two sides of the centre line of the ship when on the wave slope, Fig. 31. This couple is a periodic couple which has its maximum value when the ship is on either side of the wave at the point of maximum slope. It has zero value when the ship is

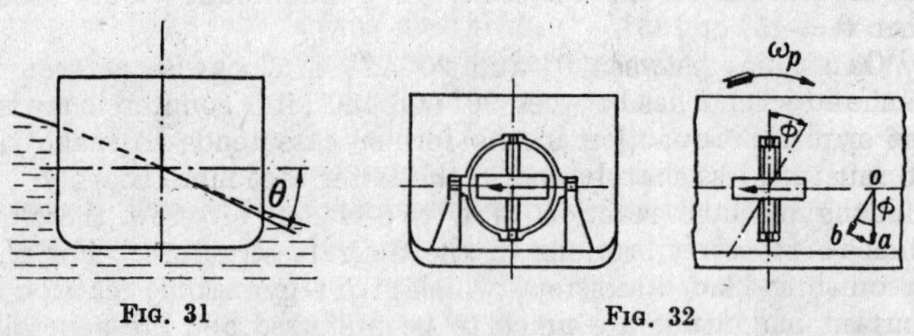

FIG. 31 FIG. 32

at the peak or in the trough of the wave. The way in which the magnitude of the couple varies will of course depend on the shape of the wave profile. If we assume a sinusoidal profile for the wave, then the couple will also vary sinusoidally. In order to maintain the ship on an even keel the gyroscope must at every instant provide a reaction couple on the ship which is equal in magnitude and of opposite sense to that exerted on the ship by the wave.

In Article 31 it was shown that the plane of spin, the plane of precession and the plane of the applied couple are mutually perpendicular. In the case of the rolling motion the plane of the applied couple is a transverse plane. Hence spin and precession must take place the one in a horizontal plane and the other in a vertical longitudinal plane. The choice as to whether the spin shall be in the horizontal plane and the precession in the vertical longitudinal plane, or vice versa, is governed by practical considerations. Fig. 32 shows a gyroscope mounted with its plane of rotation horizontal, and Fig. 33 a gyroscope with its plane of rotation vertical and longitudinal. It is easily seen that with the former arrangement a change of course of the ship either to port or to starboard has no tendency to change the direction of the axis of rotation and therefore introduces no gyroscopic effects. But if the ship pitches, the direction of the axis of rotation will be changed and a gyroscopic effect will be introduced. With the latter arrangement, on the other hand, a change of course will, and pitching will not, introduce a gyroscopic effect. In practice the arrangement with the plane of rotation horizontal is preferred.

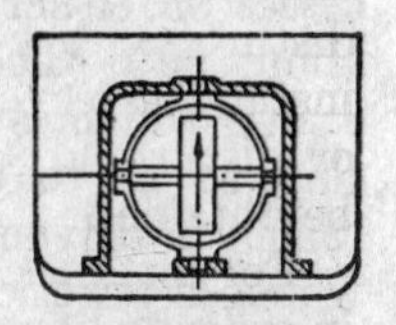

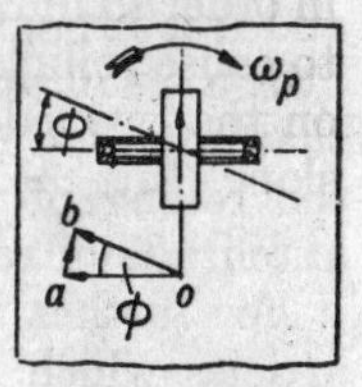

FIG. 33

The direction in which precession must take place may be found as follows: let the couple applied by the wave be clockwise, when looking towards the bow of the ship and the direction of the spin be clockwise when viewed from above, as in Fig. 32. Then the angular momentum vector is vertically downwards as shown by Oa and the couple vector is horizontal and to the left in the side elevation (looking on the left or port side of the ship). The change of angular momentum ab must also be represented by a horizontal vector to the left, so that the angular momentum vector turns clockwise from Oa to Ob. Hence the precession of the gyroscope must be clockwise, as seen in the side elevation, in order to provide a gyroscopic reaction couple on the ship which will oppose the applied couple.

But it has to be remembered that the gyroscopic reaction couple is always perpendicular to the plane of rotation, so that, as the plane of rotation of the gyroscope precesses from the horizontal, the plane of the reaction couple will precess from the vertical. Only the component of this couple in the vertical plane will actually be available for balancing the couple applied by the wave.

Thus, if $T_w \sin\theta$ is the couple applied by the wave, where θ is the slope of the wave, T is the gyroscopic reaction couple and ϕ is the inclination of the plane of rotation to the horizontal, then

$$T \cos\phi = T_w \sin\theta$$

But
$$T = I\omega\omega_p$$

so that
$$I\omega\omega_p \cos\phi = T_w \sin\theta \quad . \quad . \quad . \quad (2.49)$$

The maximum reaction couple will clearly be required when the ship is at the point of maximum slope, so that the plane of rotation of the gyroscope should then be horizontal ($\phi = 0$).

The couple required when the ship is at the peak or the trough of the wave is zero, and theoretically this condition could be satisfied if the gyroscope precessed at such a rate that the angle ϕ became equal to 90° at the end of a quarter period of the wave. It is, however, impracticable to do this.

In practice I and ω are constant and the axis of rotation precesses on either side of the vertical through an angle limited to about 60°. The vertical shaft of the gyroscope is carried in a casing which is supported in bearings fixed to the frame of the ship so as to allow of the desired precessional motion. The casing is driven by an electric motor at an angular velocity ω_p which is practically constant over most of the arc of precession, and it is brought to rest and accelerated in the opposite direction over a small arc at the end of each outward swing. This means that equation (2.49) cannot be satisfied at every instant, so that the rolling of the ship is not entirely prevented, although the amplitude of roll is reduced.

For further information the reader should refer to a series of articles which appeared in the *Engineer*, 1930, and which have been reprinted in book form: *The Automatic Stabilisation of Ships*, by T. W. Chalmers.

There is a very full description of the gyroscopic stabilising equipment of the Italian liner *Conte di Savoia* in the *Engineer* Jan. 1932, and an account of the results obtained with the equipment in the *Engineer*, Sept. 1936.

EXAMPLES II

1. Show that, when a particle is displaced along a circular path, its acceleration has a component perpendicular to the path as well as a component tangential to the path. Deduce the magnitudes of the two components in terms of the instantaneous velocity of the particle and the radius of curvature of the path.

2. Explain the conventional method of defining an angular displacement by means of a vector. Then show that the angular acceleration of a spinning body may arise from a change in the magnitude of the angular velocity, a change in the direction of the axis of rotation or a change in both the magnitude and the direction.

3. The direction of motion of a body changes through an angle of 90° in an interval of 8 sec, the initial and final speeds being 20 ft/s. Find the force required to produce this change: (a) when the force is constant both in magnitude and direction; (b) when the body moves along a circular arc with constant speed during the change. In the former case sketch the path of the body and show how it may be drawn to scale. The mass of the body is such that a force of 8 lb produces an acceleration of 10 ft/s^2. W.S.

4. A body weighing 3 lb is known to change its velocity in 2 sec from 20 ft/s. due east to 10 ft/s 30° north of east Find the change in velocity, the acceleration and the uniform force capable of causing the acceleration. W.S.

5. A body weighs 20 lb and has a moment of inertia of 15 lb ft^2. At a given instant it is acted upon by a force of 5 lb whose line of action is 10 in. from the c.g. of the body. What effect will the given force have on the motion of the body?

6. A gas engine has two flywheels each of which weighs 1600 lb and has a radius of gyration of 27 in. At the full speed of 275 r.p.m. the engine develops 25 b.h.p. Assuming that the useful work done per revolution is independent of the speed, find the time required to increase the speed of the engine, when running light, from 50 to 275 r.p.m. What accelerating torque will act on the crankshaft during this period?

7. A petrol engine connecting rod weighs 2·2 lb and is suspended in a vertical plane from a horizontal knife-edge which passes through the small end and coincides with the small-end centre. The distance of the c.g. from the point of suspension is 6·6 in. The rod is found to make 50 oscillations in 47·8 sec. What is its moment of inertia about an axis through the c.g.?

8. In order to find the moment of inertia of a small flywheel, it is suspended in a vertical plane as a compound pendulum. The distance of the c.g. from the knife-edge support is 10 in. and the flywheel makes 100 oscillations in 134·4 sec. Find the moment of inertia about an axis through the c.g. if the weight of the flywheel is 160 lb.

9. The connecting rod of an oil engine weighs 116·5 lb, the distance between the bearing centres is $33\frac{3}{4}$ in., the diameter of the big-end bearing is $4\frac{3}{4}$ in. and of the small-end bearing is 3 in. When suspended vertically with a knife-edge through the small end it makes 100 oscillations in 181 sec, and with the knife-edge through the big end it makes 100 oscillations in 166 sec. Find the moment of inertia of the rod and the distance of the c.g. from the small-end centre.

10. For the connecting rod of Question 7, find: (a) the equivalent dynamical system when one of the masses is placed at the small-end centre; (b) the correction couple required when the two masses are placed one at the small-end centre and the other at the big-end centre and the angular acceleration of the rod is 18 000 rad/s^2. The length between centres is 9·5 in.

11. In order to find the moment of inertia of the armature of a small dynamo, a weight of 5 lb attached to a cord wound round the 3-in. dia. shaft was found to be just sufficient to overcome the friction of the bearings. An additional weight of 6 lb, making altogether 11 lb, was attached to the cord and allowed to fall freely from rest. At the end of 10·2 sec it had fallen through a distance of 5 ft. If the friction of the bearings is assumed to remain constant, find the moment of inertia of the armature and shaft.

12. A cage weighs 1 ton and is raised by a rope which is wound round a drum 42 in. dia. The drum has a moment of inertia of 2500 lb ft^2. What torque must be applied to the drum in order to give to the cage an acceleration of 8 ft/s^2? Friction may be neglected.

If the rotating parts of the hoisting motor, including an allowance for the intermediate gearing, have a moment of inertia of 20 lb ft^2 and the gear reduction between motor and drum is 20 to 1, what torque must be exerted by the motor in order to give to the cage the same acceleration?

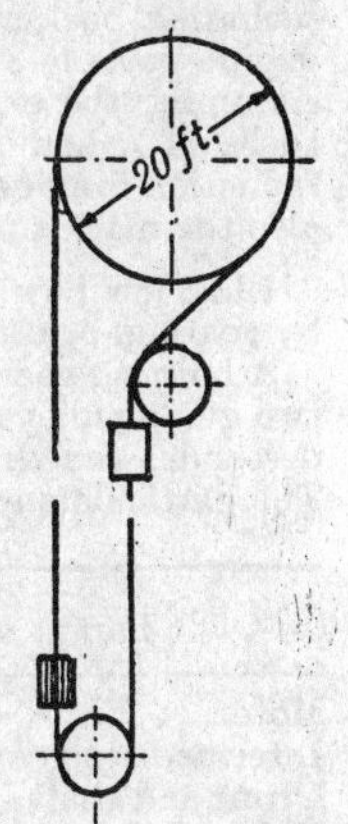

FIG. 34

13. Material is lifted from a deep mine by the balanced hoist shown in Fig. 34. The weight of the unloaded cage is 15 000 lb and of the loaded cage 25 000 lb. The weight of the rope is 42 000 lb. The head pulley is 20 ft in diameter and its moment of inertia is 50 000 lb ft^2. During hoisting operations the pulley is first uniformly accelerated acquiring a speed of 5 rad/s in 20 sec, then revolves at constant speed and finally is uniformly retarded and brought to rest in 20 sec.

Calculate the torque required during acceleration and during retardation. Neglect the effects of friction, of the inertia of the smaller pulleys and of the flexing of the rope. W.S.

14. The rotor of a motor is observed to drop from a speed of 420 to 380 r.p.m. in 60 sec when the current has been switched off; the same drop of speed occupies 70 sec when a ring of mass moment of inertia I has been attached concentrically to the rotor. Find the mass moment of inertia of the rotor and the resistance which it encounters in the neighbourhood of 400 r.p.m. on the assumption that this resistance is not appreciably affected by the ring. W.S.S.

15. Two shafts are geared together as in Fig. 27. The moments of inertia are 750 lb ft^2 for shaft A and 30 lb ft^2 for shaft B. The gear ratio is 120 to 19. If a weight of 300 lb is hung from a rope wound round a drum of 20 in. dia. which is keyed to shaft A, find the acceleration of the weight when it is allowed to fall freely. Frictional resistances may be assumed to be equivalent to a torque of 40 lb ft on shaft A.

16. A motor-cycle engine gives a torque of 15 lb ft at 1800 r.p.m. The speed reduction from engine to rear wheel is 9 to 1 on second gear and the efficiency of the transmission is 88%. The weight of machine and rider is 400 lb and rolling and windage resistance at the corresponding road speed amounts to 20 lb. The moment of inertia of the road wheels is 60 lb ft^2 and of the engine parts is 2 lb ft^2. The effective diameter of the driving wheel is 25·6 in. Find the road speed and the acceleration of the motor-cycle under the above conditions.

17. A motor A exerts a constant torque and is geared to a shaft B, the speed of the motor being G times the speed of the shaft B. Show that the angular acceleration of the shaft B is a maximum when $G = \sqrt{(I_b/I_a)}$, where I_a, I_b are the total mass moments of inertia of the revolving parts attached to the respective shafts.

If the torque exerted by the motor is 20 lb ft, the moment of inertia of the parts attached to the motor shaft is 15 lb ft^2 and that of the parts attached to the other shaft is 240 lb ft^2, find the gear ratio which gives maximum acceleration and the corresponding angular acceleration of each shaft.

18. A motor-car weighs 1 ton, the moment of inertia of the road wheels, including back-axle, differential and propeller shaft, is 200 lb ft^2, that of the engine parts is 8 lb ft^2 and the effective diameter of the road wheels is 26·4 in. Assuming the engine torque to be 62·5 lb ft, the efficiency of the transmission to be 90% and the resistance to motion of the car to be 80 lb, find the gear reduction from engine to driving wheels that will give maximum acceleration and also the magnitude of the acceleration.

19. Show how the inertia of a system of masses positively connected together by gearing, or otherwise, may be referred to any one selected line or shaft.

A hoisting gear with a 5-ft dia. drum operates two cages by ropes passing over two guide pulleys, 3 ft 6 in. dia. One cage, loaded, rises, while the other, empty, descends; the drum is driven by a motor through double-reduction gearing. The particulars of the various parts are as follows:

Part	Max. speed	Weight, lb	k, in.	Friction
Motor	900	400	3·5	—
Intermediate gear . .	275	750	9	10·0 lb ft
Drum and shaft . .	50	4500	24	75·0 lb ft
Guide pulley (each) .	—	400	17	10·0 lb ft
Rope out	—	500	—	—
Ascending cage . .	—	1800	—	100 lb
Descending cage . .	—	800	—	70 lb

Determine the motor torque required to produce a cage acceleration of 3 ft/s^2. L.U.A.

20. The cage of a goods hoist weighs 9 cwt and carries a maximum load of 15 cwt. It is raised by a rope passing over a 4-ft dia. drum of weight 800 lb and radius of gyration 18 in. The other end of the rope is connected to a balance weight, the cage being overbalanced to the extent of 40% of the full load. If the drive, when raising the maximum load, is to be capable of a performance equivalent to an acceleration of 4 ft/s² at a speed of 10 ft/s, calculate the drum torque and the power necessary for the masses given. L.U

21. A truck of mass 8 tons moving at 6 ft/s collides with a truck of mass 10 tons at rest. Find: (a) the common velocity of the trucks when for an instant they move together; (b) the velocities of the trucks just after the action of the buffers is completed. What is the relation between the relative velocities of the trucks before and after collision?

If each buffer is compressed 1 in. by a force of 1·5 tons, find the maximum reaction between the trucks. Neglect friction. W.S.

22. An engine whose effective mass is n times that of a truck is attached to it by a loose coupling which allows a free movement of l in. The engine moves l in. with an acceleration of a ft/s², starting from rest in contact with the truck. Find the velocity with which the truck moves after the impact, assuming that the coefficient of restitution is e.

If the engine maintains its acceleration after the impact and the truck is subject to a retardation of b ft/s², show that the truck cannot overtake the engine and find the interval before the next impact. W.S.S.

23. Two masses, m, m', are moving in the same straight line with velocities u, u'. Find the velocity of each mass relative to their mass centre. Hence, or otherwise, verify that their total kinetic energy exceeds the energy, R, of their motion relative to their mass centre by the energy which they would have if collected at that centre.

Show also that, if impact occurs with coefficient of restitution e, the loss of kinetic energy during the impact is $(1-e^2)R$. W.S.S.

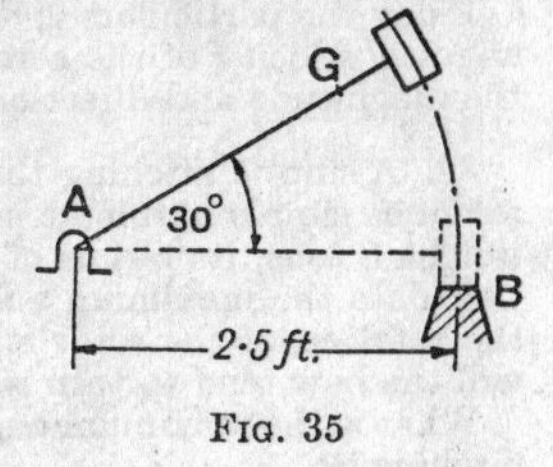

FIG. 35

24. Fig. 35 represents a tilt hammer hinged at A and raised 30° ready to strike an object B. If the total mass of the hammer is 10 lb, the distance of its centre of gravity G from A is 2·0 ft and its radius of gyration about the axis of the hinge is 2·2 ft, calculate the force of the blow on B which may be assumed to take place in 0·004 sec. Also calculate the reaction at the hinge. L.U.A.

25. During shunting operations a loaded truck weighing 12 tons and moving at 4 m.p.h. overtakes a truck weighing 10 tons moving at 1 m.p.h. The four buffer springs which are affected are made of steel wire of circular cross-section and have a safe shear stress value of 50 000 lb/in². If the mean diameter of the coils is 5 in., calculate the length and diameter of wire in each buffer spring which will take the shock safely with a deflection of 2 in. Neglect the obliquity of the coils and take $C = 12.10^6$ lb/in². L.U.A.

26. A motor-car weighs 1800 lb; each of the four road wheels has an effective diameter of 25 in. and a moment of inertia of 40 lb ft². The rotating parts of the engine have a moment of inertia of 8 lb ft². The car is coasting at 15 m.p.h. with the bottom gear of 20 to 1 engaged and the clutch pedal depressed. If the clutch pedal is suddenly released, find the final speed of the car when the engine is (a) initially at rest, (b) idling at a speed of 700 r.p.m.

27. Explain why a gyroscope precesses and obtain the relation between the rate of precession, the speed of rotation, the moment of inertia of the flywheel and the applied torque.

A flywheel weighs 15 lb and has a radius of gyration of 8 in. It is given a spin of 900 r.p.m. about its axis, which is horizontal and is suspended at a point distant 6 in. from the plane of rotation of the flywheel. Investigate the subsequent motion of the wheel.

28. Explain the meaning of the terms "gyroscopic torque" and "precessional motion". Deduce an equation for gyroscopic torque in terms of the moment of inertia of the spinning body, the angular velocity of spin and the angular velocity of precession.

The wheels of a motor-cycle have a moment of inertia of 50 lb ft^2 and the engine parts a moment of inertia of 2·5 lb ft^2. The axis of rotation of the engine crankshaft is parallel to that of the road wheels. If the gear ratio is 5 to 1, the diameter of the road wheels is 25·5 in. and the motor cycle rounds a curve of 100 ft radius at 35 m.p.h., find the magnitude and direction of the gyroscopic couple.

29. Deduce an expression for the couple that is called into play in the case of a wheel rotating with uniform angular velocity in order to maintain a given rate of precession.

The rotary engine of an aeroplane weighs 750 lb and has a radius of gyration of 1 ft. When viewed from in front the engine rotates in a clockwise sense at 1800 r.p.m. When flying at 90 m.p.h. the aeroplane loops the loop in a circle of 100 ft dia. Find the magnitude and direction of the gyroscopic couple acting on the aeroplane. L.U.A.

30. The rotor of the turbine of a yacht makes 1200 r.p.m. Its weight is 1500 lb and its radius of gyration is 10 in. If in a seaway the yacht pitches with a maximum angular velocity of 1 rad/s, what gyroscopic couple will be transmitted to the hull? The turbine rotates clockwise when viewed from the stern.

31. Investigate the effect of the gyroscopic couple due to the rotating parts of the engine and airscrew when an aeroplane loops the loop.

Using the particulars given in Example 13, p. 57, and assuming the maximum angular velocity of the aeroplane when describing the loop to be 1·5 rad/s, find the magnitude and direction of the gyroscopic couple.

32. A ship is pitching through a total angle of 15°. The oscillations may be taken as simple harmonic and the complete period as 32 sec. The turbine rotor weighs 5 tons, its radius of gyration is 17·6 in. and it is rotating at 2000 r.p.m. Calculate the maximum value of the gyroscopic couple set up by the rotor. If the rotation of the rotor is clockwise when looking from aft in which direction will the bow tend to turn when falling?

What is the maximum angular acceleration to which the ship is subjected while pitching? L.U.A.

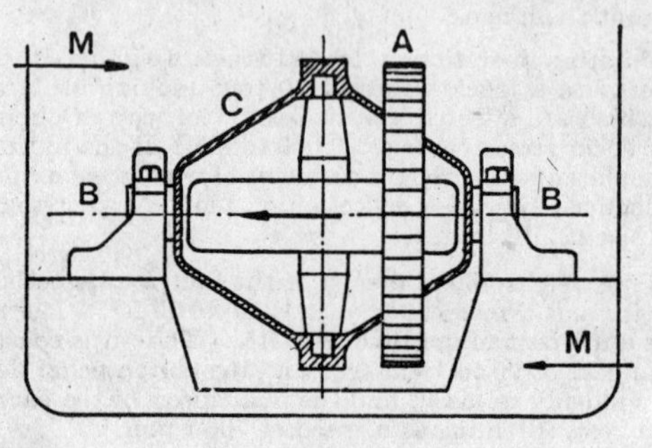

Fig. 36

33. The arrangement of a stabilising gyroscope in a ship is shown in Fig. 36. A toothed wheel A, surrounding the casing C and lying in the fore and aft plane of the ship, oscillates backwards and forwards through a given angle; rotation of A causes precession about the athwartship axis BB and enables a varying couple to be opposed to the rolling moment due, at any instant, to a train of waves.

In a particular case the heeling moment at any instant is given by $M = 3140 \cos t$ ft ton. The flywheel weighs 100 tons and spins at 850 r.p.m.; the radius of gyration about the spinning axis is 4·6 ft and about the axis BB is 3·4 ft. Assuming that A moves so that the hull is maintained upright:

(a) Calculate the angular velocity of A at the instant when $t = 0$, if, at this instant, the flywheel is horizontal.

(b) Calculate the couple required on A to make the wheel precess with angular acceleration 0·67 rad/s^2 when $t = \pi/2$, if, at this instant, the flywheel is at the end of an oscillation.

Assuming A locked with the spinning axle in the vertical plane.

(c) Calculate the couple tending to cause pitching if the ship rolls through the upright position with angular velocity 0·1 rad/s. L.U.A.

CHAPTER III

VELOCITY AND ACCELERATION

33. In Article 16 we saw that when a particle or point moves along a curved path its velocity is tangential to its path, while its acceleration usually has two components. One component is tangential to the path and its magnitude is equal to the rate of change of magnitude of the velocity; the other is normal to the path and its magnitude is given by the product of the instantaneous value of the velocity and the rate of change of direction of the velocity. The former is referred to as the tangential component of the acceleration and the latter as the centripetal component of the acceleration.

Before considering the methods which may be applied to determine the velocity and acceleration of a point which moves along a curved path, we shall first of all give methods which can only be applied either (a) when the point has straight-line motion or (b) when it is not desired to take into account the effect on the velocity and acceleration of a change in the direction of the displacement of the point. It must be clearly understood that the following methods can only be applied to problems in which we are concerned with nothing but the time rate of change of the magnitude of the displacement, i.e. the speed of the point, and the time rate of change of the speed, i.e. the tangential component of the acceleration of the point.

34. Displacement, Speed and Acceleration-time Curves. The curve (a), Fig. 37, shows the displacements of a point from a given initial position plotted as ordinates with the corresponding time intervals plotted as abscissæ. Since speed is defined as the rate of change of the displacement with respect to time, it follows that the speed at a given instant is represented to scale by the slope of the tangent to the displacement curve at the same instant. If, therefore, a second curve is drawn as at (b), the ordinates of which at every instant are proportional to the corresponding slopes of the displacement-time curve, this second curve will show the variation of the speed of the point with time. In a similar way, since tangential acceleration is rate of change of speed with respect to time, the variation of acceleration with time is given by the curve (c), the ordinates of which at every instant are proportional to the corresponding slopes of the speed-time curve.

The above method of deriving the speed and tangential acceleration curves depends for its accuracy upon the precision with which a tangent may be drawn to a given curve at a particular point. In view of the difficulty of this operation and of the fact that any error in deriving the speed curve will be magnified in deriving the tangential acceleration curve, it is usual to adopt the method given in the next paragraph.

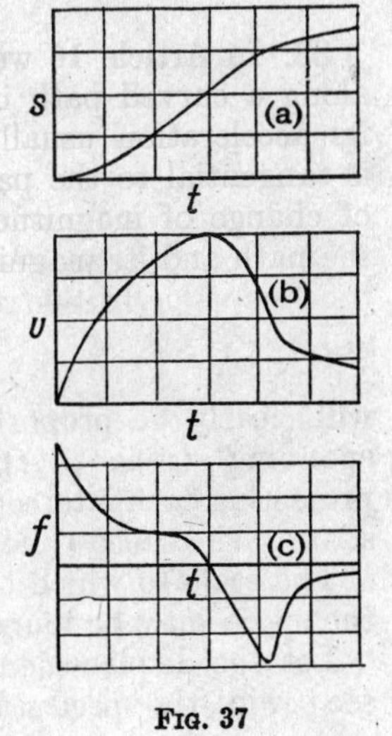

FIG. 37

If two ordinates AB, CD are drawn on the displacement-time curve, Fig. 38, the slope of the chord BD will be approximately equal to the slope of the tangent to the curve at the instant represented by E, mid-way between A and C, and will therefore be proportional to the speed at E. The error in this approximation will be smaller the shorter the distance between the two ordinates AB and CD. On the other hand, the shorter this distance the greater will be the percentage error in determining the difference of length of the two ordinates AB and CD. In any actual example the spacing of the ordinates must therefore be a matter for compromise.

A better method of drawing the speed-time and acceleration-time curves is as follows:

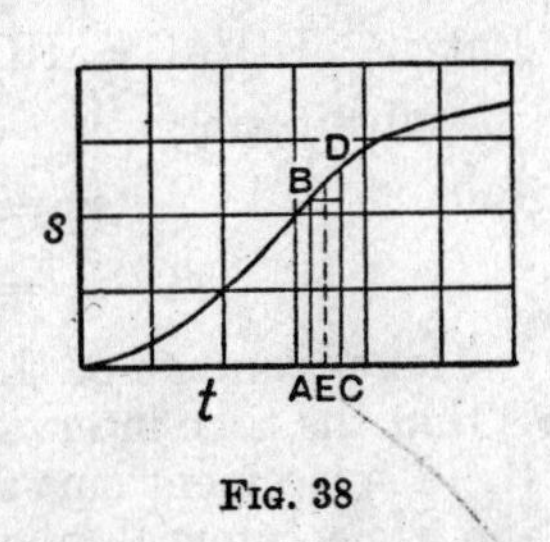

FIG. 38

Referring to the displacement-time curve shown in Fig. 39, draw a line ab parallel to the displacement axis. Choose a convenient polar distance oa, preferably to represent an even interval on the time scale and also to give an open speed scale, i.e. a line drawn through o parallel to the tangent to the displacement curve at the point of maximum slope should give a distance along ab approximately equal to the maximum displacement.

Then, for any small interval of time δt, the increase of displacement is equal to ED, the difference between the ordinates CD and AB, and the mean speed is equal to ED/BE = ED/AC.

If, through the pole o, a line is drawn parallel to BD to cut ab at c, then triangles oca, BDE are similar and the mean speed during the interval of time δt is given by

$$v_m = \delta s/\delta t = ED/BE = ac/ao$$

$$\therefore ac = ao . v_m$$

If the diagram is divided into a number of vertical strips, not necessarily all of the same width, lines radiating from o may be drawn in the same way for each strip. The intercepts along ab

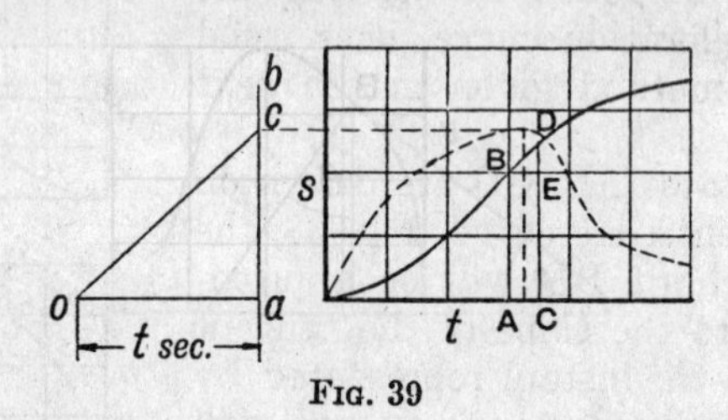

FIG. 39

will clearly be proportional to the mean speeds for the corresponding strips. A speed-time curve may be quickly drawn by projecting each intercept on the mid-ordinate of the corresponding strip. The speed-time curve is shown dotted in Fig. 39.

The scale to which the ordinates of the dotted curve represent the speed may be found as follows:

Let the displacement scale be k_s ft per in., the time scale k_t sec per in., the speed scale k_v ft/s per in. and the polar distance t sec.

Then $$v_m = \delta s/\delta t = (k_s . ED)/(k_t . AC)$$

But $$ED/AC = ED/BE = ac/ao \quad \text{and} \quad oa = t/k_t \text{ in.}$$

$\therefore$ substituting, $$v_m = (k_s/k_t)(k_t/t)ac = k_s/t . ac$$

But $$v_m = k_v . ac$$

$$\therefore k_v = k_s/t$$

The advantage of this method over those previously described is that the time intervals may be varied from point to point along the displacement curve. Where the radius of curvature is small, the time interval may be shortened in order to increase the accuracy. On the other hand, where the radius of curvature is large, a longer time interval may be taken without sacrificing appreciably the accuracy of the results obtained.

35. Speed-displacement and Acceleration-displacement Curves. Sometimes a curve showing the variation of speed with displacement is given and it is required to draw a curve showing the variation of acceleration with displacement.

Since $\delta v = f_m \delta t$ and $\delta s = v_m \delta t$, we may eliminate δt from these two equations and write $f_m = v_m . \delta v/\delta s$. In the limit $f = v . \mathrm{d}v/\mathrm{d}s$, or the acceleration is equal to the product of the speed and the rate of change of speed with respect to displacement.

Referring to Fig. 40, let BN be drawn normal to the curve at point B. Then AN = AB tan θ.

But AB represents to scale the speed of the point and tan θ is proportional to $\mathrm{d}v/\mathrm{d}s$, so that AN is proportional to $v . \mathrm{d}v/\mathrm{d}s$ and therefore to f.

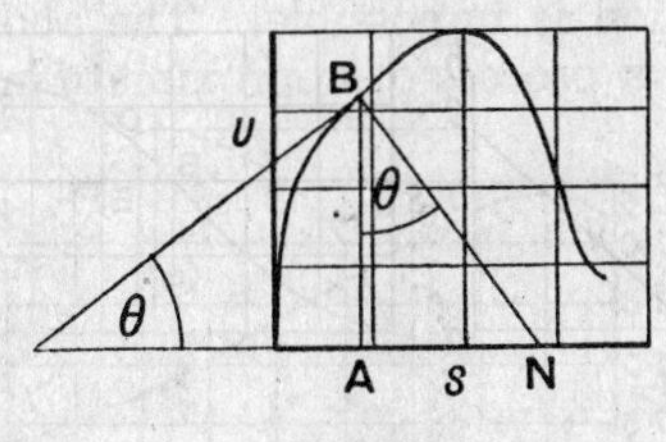

FIG. 40

The scale to which AN represents the acceleration of the point may be found as follows:

Let k_s = displacement scale, i.e. 1 in. = k_s ft,

k_v = speed scale, i.e. 1 in. = k_v ft/s

and k_f = acceleration scale, i.e. 1 in. = k_f ft/s².

Then the acceleration of the point at A $= f = k_f . \mathrm{AN}$. But

$$\mathrm{AN} = \mathrm{AB} \tan \theta = \frac{v}{k_v} . \frac{\mathrm{d}v}{k_v} . \frac{k_s}{\mathrm{d}s} = \frac{k_s}{k_v^2} . v \frac{\mathrm{d}v}{\mathrm{d}s} = \frac{k_s}{k_v^2} . f$$

$$\therefore f = k_v^2/k_s . \mathrm{AN}$$

and

$$k_f = k_v^2/k_s$$

Hence, a curve drawn with ordinates which are everywhere proportional to the corresponding subnormals of the speed-displacement curve will give the acceleration-displacement curve for the point.

36. Speed-time and Displacement-time Curves from the Acceleration-time Curve. Let us suppose that a body of known mass has plane motion and is acted upon by a force the magnitude of which varies from instant to instant. Since the acceleration is directly proportional to the applied force, the acceleration-time curve may be drawn if the force-time curve is given. But $f = \mathrm{d}v/\mathrm{d}t$ or $\mathrm{d}v = f . \mathrm{d}t$, so that the change of speed during an interval of time t

may be found by integrating the above equation. Let v_0 be the initial speed and v the final speed, then we have:

$$v - v_0 = \int_0^t f dt$$

If the law of variation of f with t is known, the change of speed for different time intervals may be found by substituting for f in terms of t and integrating. In most practical cases, however, the variation of f with t cannot be expressed by a simple equation and direct integration is impossible. The above equation then has to be solved by a process of approximate integration.

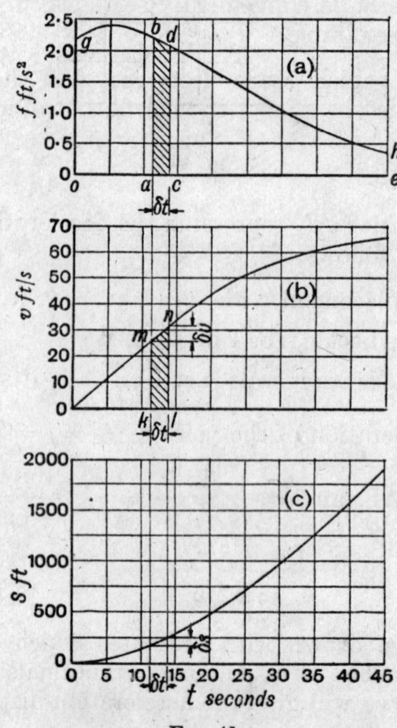

Fig. 41

For a small finite interval of time δt, the increase of speed δv is given by $\delta v = f_m \delta t$, where f_m is the mean acceleration during the interval of time δt. If, then, the acceleration-time diagram is divided into a number of vertical strips, the increase of speed during the time interval corresponding to the width of any one strip is proportional to the area of that strip. For example, referring to Fig. 41 (a), the increase of speed during the time interval ac is directly proportional to the shaded area abdc. Hence, for the time interval oe, the total increase of speed is proportional to the whole area oghe. The speed-time curve, therefore, takes the form shown at (b), in which each ordinate is proportional to the area under the acceleration-time curve up to the time corresponding to that ordinate.

In a similar way the increase of displacement during the interval of time δt is proportional to the area of the strip kmnl of the speed-time curve. The displacement-time curve (c), is obtained from the speed-time curve in exactly the same way as the speed-time curve is obtained from the acceleration-time curve. The data and calculations may be conveniently set down in tabular form, as in the following example.

Example 1. A vehicle starts from rest and the acceleration varies with the time as shown by the first two columns of the table. Find the variation of speed with time and of displacement with time and plot curves of acceleration, speed and displacement on a time base.

f, ft/s²	t, sec	f_m, ft/s²	δt, sec	$\delta v = f_m \delta t$, ft/s	$v = \Sigma \delta v$, ft/s	v_m, ft/s	$\delta s = v_m \delta t$, ft	$s = \Sigma \delta s$, ft
2·20	0				0			0
		2·30	5	11·5		5·75	28·8	
2·40	5				11·5			28·8
		2·33	5	11·6		17·3	86·5	
2·25	10				23·1			115·3
		2·13	5	10·6		28·4	142·0	
2·00	15				33·7			257·3
		1·85	5	9·3		38·4	192·0	
1·70	20				43·0			449·3
		1·53	5	7·63		46·8	234·0	
1·35	25				50·6			683·3
		1·18	5	5·88		53·6	268·0	
1·00	30				56·5			951·3
		0·87	5	4·33		58·7	293·5	
0·73	35				60·8			1245
		0·62	5	3·08		62·4	312·0	
0·50	40				63·9			1557
		0·43	5	2·13		65·0	325·0	
0·35	45				66·0			1882

The values of f, v and s are plotted to scale against the corresponding values of the time in Fig. 41.

37. Speed-time and Displacement-time Curves from the Acceleration-speed Curve. One further example of problems of the above class may be considered. In self-propelled vehicles the information available generally consists of a curve showing the variation of tractive force with speed. Since the acceleration of the vehicle is directly proportional to the tractive force, a curve of acceleration against speed may at once be drawn. The problem then is to determine the time and distance required in order to reach a given speed.

Since $f = dv/dt$, we have for a small finite change of speed $f_m = \delta v/\delta t$, so that if, during the small increase of speed δv, the mean value of the acceleration is f_m, then the time δt required for the increase of speed δv is given by $\delta t = \delta v/f_m$.

Similarly, $v_m = \delta s/\delta t$ and the distance δs travelled in time δt will be given by the product $v_m . \delta t$.

The various steps in the calculations may be conveniently set down in tabular form, as shown in the example below.

Curves may then be drawn showing the variation of f, t and s with v, or, if desired, curves of f, v and s may be plotted against t.

Example 2. A motor-car has a total weight, including an allowance for rotational inertia, of 4500 lb. In top gear the tractive force available for acceleration varies with the speed of the car, as shown in the following table. Draw the displacement, velocity and acceleration-time curves for the increase of speed from 10 to 60 m.p.h

Since 60 m.p.h. = 88 ft/s, the speed in ft/s = 88/60 times the speed in m.p.h.

Also $F = mf$, so that $f = 32{\cdot}2/4500 . F$.

v, m.p.h.	v, ft/s	v_m, ft/s	δv, ft/s	F, lb	f, ft/s²	f_m, ft/s²	$\delta t = \delta v/f_m$	t, $=\Sigma\delta t$	δs, ft $=v_m\delta t$	s, ft $=\Sigma\delta s$
10	14·7			310	2·22			0		0
		18·3	7·33			2·255	3·25		59·6	
15	22·0			320	2·29			3·25		59·6
		25·7	7·33			2·305	3·18		81·7	
20	29·3			325	2·32			6·43		141·3
		33·0	7·33			2·285	3·21		105·9	
25	36·7			315	2·25			9·64		247·2
		40·4	7·33			2·18	3·37		136·0	
30	44·0			295	2·11			13·01		383·2
		47·7	7·33			2·015	3·64		173·6	
35	51·3			268	1·92			16·65		556·8
		55·0	7·33			1·765	4·15		228	
40	58·7			225	1·61			20·80		784·8
		62·3	7·33			1·43	5·13		320	
45	66·0			175	1·25			25·92		1105
		69·7	7·33			1·055	6·95		484	
50	73·3			120	0·86			32·87		1589
		77·0	7·33			0·65	11·3		870	
55	80·7			62	0·44			44·2		2459
		84·3	7·33			0·22	33·3		2810	
60	88·0			0	0			77·5		5269

The curves of f, v and s are plotted to scale against t in Fig. 42.

38. The Velocities of Points in Mechanisms. As mentioned in Chapter I, the relative motions of the links of a mechanism are completely constrained. The displacement of one link of the mechanism brings about corresponding displacements of all the other links. Every point of every link of a mechanism is therefore compelled to follow a definite path. Hence it is possible to plot a curve to show how the magnitude of the displacement of a given point on one of the links varies with time. From this curve the speed-time and the tangential acceleration-time curves may be obtained by the methods of Article 34. But we saw in Article 16 that, if the displacement takes place along a circular path the change of direction as well as the change of magnitude of the displacement has to be taken into account. The velocity of the point is tangential to the curved path and the acceleration has both a tangential component and a centripetal component. For the present we shall confine ourselves to a consideration of the methods which may be used to find the velocity of one point in a

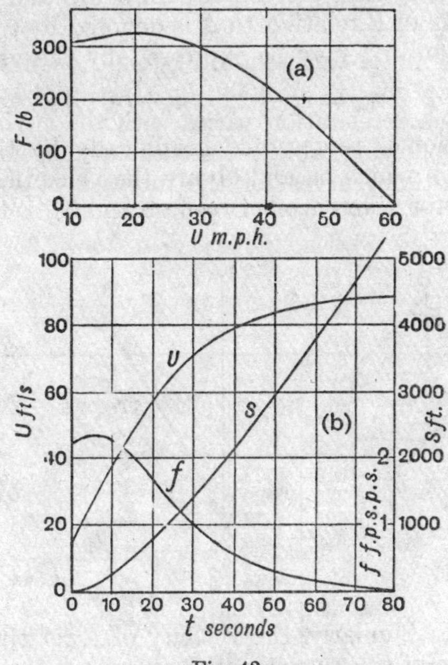

Fig. 42

mechanism given the velocity of another point in the mechanism at the same instant. Methods of finding the acceleration will be dealt with later. Although it is possible to obtain the velocity by direct calculation, it is not practicable to do so except for certain simple mechanisms. Generally one of two graphical methods is used. These two methods are known as (a) the *relative velocity* method and (b) the *instantaneous centre* method.

39. The Relative Velocity Method. The principles which underlie the application of the relative velocity method will be made clear if we first consider the general question of relative motion.

Referring to Fig. 43, let A and B be two independent particles which are moving with velocities v_a and v_b respectively. Then the velocity of B relative to A is the velocity with which B appears to be moving to an observer situated at, and moving with, A. Let velocities equal and opposite to v_a be applied to both the particles. Then the particle A will be brought to rest and the particle B will have a resultant velocity given by the vector sum of v_b and $-v_a$. This resultant will clearly be the velocity with which, to the observer on A, B appears to be moving. In the figure v_b is represented by the vector BC and $-v_a$ by the vector BD or the vector CE, so that the velocity of B relative to A is represented by BE, the vector sum of BC and CE.

If the velocity of B relative to A is denoted by v_{ba}, the relation between v_a, v_b and v_{ba} may be expressed by the vector equation

$$v_{ba} = v_b \rightarrow v_a = v_b +\!\!\!\!\rightarrow (-v_a) \quad . \quad . \quad . \quad (3.1)$$

and this equation is represented graphically by the triangle of velocities oab, in which oa and ob are the velocities of A and B, and ab, the vector difference between ob and oa, is the velocity of

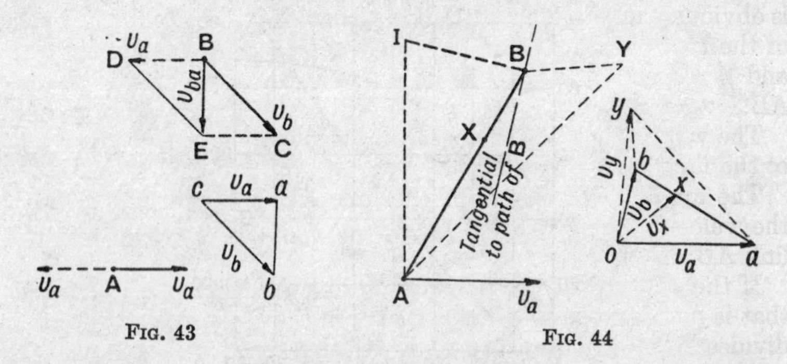

Fig. 43

Fig. 44

B relative to A. Similarly, the velocity of A relative to B is the vector difference of the velocity of A and the velocity of B, or

$$v_{ab} = v_a \rightarrow v_b = v_a +\!\!\!\!\rightarrow (-v_b) \quad . \quad . \quad . \quad (3.2)$$

This equation is represented by the same triangle oab and ba, the vector difference between oa and ob, is the velocity of A relative to B. Hence the side ab of the triangle oab taken in the sense ab represents the velocity of B relative to A and taken in the sense ba represents the velocity of A relative to B.

Suppose now that A and B are points on the same link at a fixed distance apart, Fig. 44. Then it will not be possible to assign arbitrary velocities to both A and B. If the velocity of A is fixed in magnitude and direction, then the velocity of B can be fixed only in direction. This will be clear when it is remembered that A and B are now at a fixed distance apart, so that B can only move, relative to A, along the circular arc which has A as centre and AB as radius. Hence the direction of the velocity of B relative to A must be perpendicular to the link AB. The problem of finding the velocity of B resolves itself into drawing a triangle one side of which is known in magnitude and direction and the other two in direction only.

Let oa represent v_a to scale. Through o, draw ob tangential to the path of B. Through point a draw a line perpendicular to AB, corresponding in direction to the velocity of B relative to A. The intersection of these two lines will give point b. Then v_b will be represented to scale by ob and v_{ba} will be represented to scale by ab.

The following alternative explanation of the above construction may be given. Since AB is a rigid link, then, whatever may be the velocities of the points A and B, the components of these velocities parallel to the line AB must be equal. Otherwise an increase or decrease of the length of AB would take place, which is obviously impossible. It follows, therefore, that the extremities of the two vectors oa and ob, which represent the velocities of A and B respectively, must lie on a line ab that is perpendicular to AB.

The vector ab which represents v_{ba} is known as the *velocity image* of the link AB.

The angular velocity ω of the link AB is found by dividing the velocity of B relative to A, i.e. v_{ba}, by the length of the link AB.

If the velocity of the point X on link AB is required, then all that is necessary is to divide ab at x in the same proportion as X divides AB. The velocity of X relative to A will obviously be equal to $(AX/AB)v_{ba}$ and is represented to scale by ax. Since the velocity of X is the vector sum of v_a and v_{ba} it is represented to scale by the line ox.

The velocity of the point Y on the link AB may be found by drawing a triangle aby similar to triangle ABY. This is most easily done by drawing a line through point a perpendicular to AY, and a line through b perpendicular to BY. The point of intersection of these two lines will fix y. Then oy will represent the velocity of Y. The proof is as follows. The velocity of Y is the vector sum of the velocity of A and the velocity of Y relative to A. Bus the velocity of Y relative to $A = v_{ya} = (YA/BA)v_{ba}$ and it is perpendicular to AY. By construction $YA/BA = ya/ba$ and ya i, perpendicular to AY, so that ay represents v_{ya} to scale and oyt which is the vector sum of oa and ay, therefore represents v_y to scale.

Suppose that I is a point fixed to AB in such a position that triangles IAB, oab are similar, then clearly the velocity of I is zero, since the velocity of I relative to A is represented by ao and this is equal and opposite to the velocity of A, which is represented by oa. Hence I is the point on the link AB which has zero velocity, i.e. it is the centre about which the link AB is turning at the instant it occupies the given position. The point I is termed the *instantaneous centre* of the link AB. It will be seen that IA is perpendicular to oa and therefore to v_a, while IB is perpendicular to ob and therefore to v_b. The position of I may be found directly by drawing lines normal to the velocities of points A and B, when I will lie at the point of intersection of these two lines. An alternative proof of the construction for finding I is given later.

Where a complex mechanism has to be dealt with, the complete velocity diagram may be obtained by drawing the velocity triangles for each link in turn. Starting with the link on which the point of known velocity is situated, the velocity triangle may be drawn and the velocity of the point which is common to the first and to a second link may be found. The velocity triangle for the second link may then be drawn and from it the velocity of the point which is common to the second and to a third link may be found. This process may be continued indefinitely until all the links in the mechanism have been dealt with.

40. Applications of the Relative Velocity Method. (a) *The Four-bar Chain*: Fig. 45. AD is the fixed link and AB is the driving link, which turns about centre A. The peripheral velocity of B is known and is represented by the vector ob.

The pin C moves along a circular path which has D as centre and CD as radius, so that the velocity of C is perpendicular to CD.

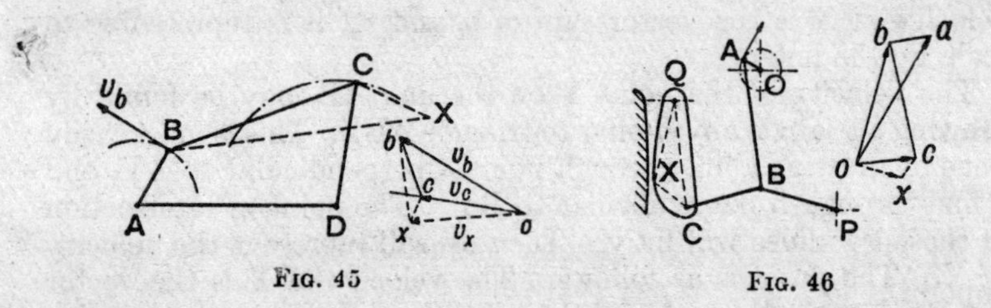

FIG. 45 FIG. 46

Draw through o a line perpendicular to CD. The velocity of C relative to B is at right angles to link BC. Draw through b a line perpendicular to BC. Let these two lines meet at c. Then bc will be the velocity of C relative to B and oc will be the velocity of C. The vector bc is the velocity image of link BC. The velocity of any point X on the link BC is found by reproducing on the velocity image bc a triangle bcx, which is similar to the triangle

BCX in the mechanism. This is most easily done by drawing bx perpendicular to BX and cx perpendicular to CX. Then in the velocity diagram bx will represent the velocity of X relative to B, and ox, which is the vector sum of ob and bx, will represent the velocity of X.

(b) *The Stone-crusher Mechanism*: Fig. 46. Let it be required to find the velocity of point X on link CQ, given the r.p.m. of the crank OA and the lengths of the various links, and given also that O, P and Q are fixed centres.

Set off the vector oa to represent the peripheral velocity of pin A. Then, since the velocity of B relative to A is perpendicular to AB and the velocity of B is tangential to the path of B, i.e. perpendicular to BP, the velocity triangle may be completed by drawing ab perpendicular to AB and ob perpendicular to BP. The vector ob will then represent to scale the velocity of B. Having found the velocity of B, we proceed to find the velocity of C in a similar way. The velocity of C is obviously at right angles to CQ and the velocity of C relative to B is at right angles to BC, so that we must draw oc normal to CQ and bc normal to BC. Then oc represents to scale the velocity of C. The velocity of the point X is found as follows:

Since the link QC turns about pin Q, any point on the link moves with a velocity which is proportional to its distance from Q and in a direction which is normal to the line joining that point to Q. Thus the velocity of point X is at right angles to XQ and its magnitude v_x is equal to $(XQ/CQ)v_c$.

The vector ox which represents the velocity of X is therefore found by drawing ox perpendicular to QX and cx perpendicular to CX.

With this construction the triangles QXC, oxc are similar, so that ox/oc = QX/QC.

But $$QX/QC = v_x/v_c \quad \therefore \ ox/oc = v_x/v_c$$

and the vector ox represents to scale the velocity of X.

41. The Instantaneous Centre Method. This method is based upon the fact that at any particular instant the motion of a rigid body is equivalent to a rotation of the body as a whole about a fixed point in space.

Suppose the link AB, Fig. 47, moves to the position A_1B_1. Then this displacement may be effected by turning the whole link about a centre I, the position of which is given by the point of intersection of the two lines which bisect AA_1 and BB_1 at right angles. But if the displacement were actually brought about in this way, the paths followed by the ends A and B of the link would have to be circular arcs with I as centre. In other words, the centre I can only remain fixed in space for a finite displacement of the link AB if the ends A and B—and, incidentally, all other points on the link—move along concentric circular arcs. In general, if AB is a link of a mechanism, the ends A and B will not move along concentric circular arcs during a finite interval of time.

Suppose, however, that the interval of time between the two positions of AB is reduced, then in the limit, when the displacements AA_1 and BB_1 are infinitesimal, the lines which bisect AA_1 and BB_1 at right angles become the normals to the paths of A and B respectively. Hence at the instant the link occupies the position AB, Fig. 48, the centre about which the link is turning is given by I, the point of intersection of the lines AI and BI drawn respectively normal to the velocities of A and B. The point I is termed the *instantaneous centre* of the link AB.

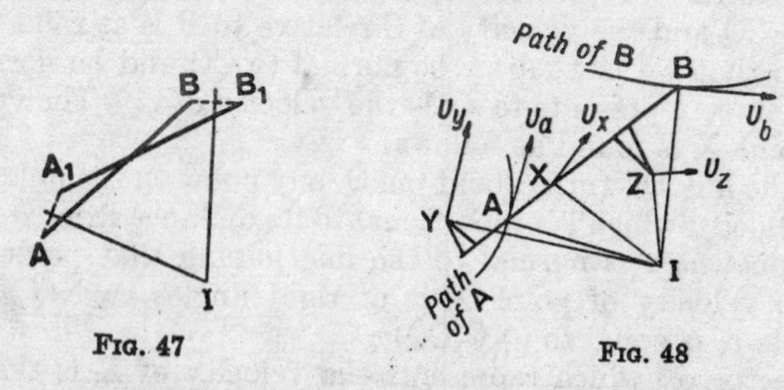

FIG. 47

FIG. 48

The usefulness of the instantaneous centre of a body lies in the fact that the velocity of any point on the body must be proportional to its distance from the instantaneous centre and, further, the direction of the velocity must be perpendicular to the line which joins that point to the instantaneous centre. Thus, in Fig. 48, the velocities of the points X, Y and Z, which are rigidly attached to AB, are at right angles to XI, YI and ZI and their magnitudes are such as to satisfy the relationship

$$v_x/XI = v_y/YI = v_z/ZI = v_a/AI = v_b/BI = \text{angular velocity of AB}$$

Since the paths of A and B will not usually be concentric circles, the position of I will change for every position of the link AB.

42. Body-centrode and Space-centrode. The locus of I is called a centrode. If the locus of I on the body is drawn, it is termed the body-centrode. If the locus of I in space is drawn, this is termed the space-centrode. The two curves are generally of different shapes, but at every instant they have one point in common, the instantaneous centre of the body at that instant. In Fig. 49 the paths of the points A and B on a rigid body are represented by the curves $A_1A_2A_3A_4$ and $B_1B_2B_3B_4$. Four different positions of the body are represented by A_1B_1, A_2B_2, A_3B_3 and A_4B_4, and the corresponding instantaneous centres are indicated by I_1, I_2, I_3 and I_4. The curve $I_1I_2I_3I_4$ is the space-centrode of the rigid body, i.e. it is the curve traced out in space by the instantaneous centre of the rigid body AB. But when the body is in the position A_4B_4, the point on the body which coincided with I_1 for the first position A_1B_1 of the body will now be at O_1. Similarly, the point on the body which coincided with I_2 for the second position A_2B_2 of the body will now be at O_2 and so on. The curve $O_1O_2O_3I_4$, therefore, represents the curve traced on the body itself by the instantaneous centre. This curve is the body-

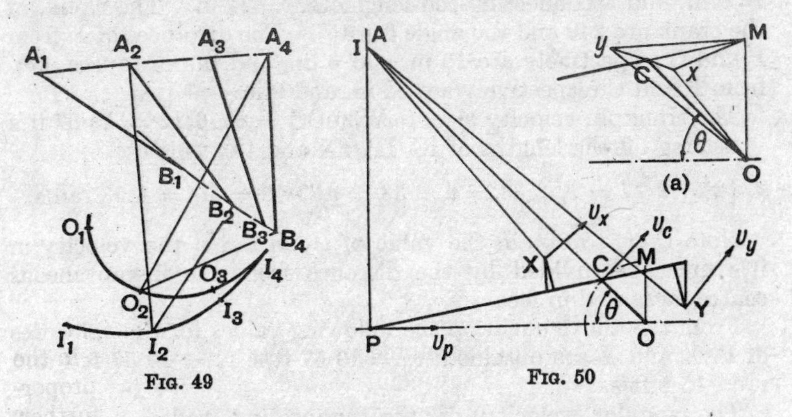

FIG. 49

FIG. 50

centrode. At the instant when the body is in the position A_1B_1, the body-centrode and the space-centrode are in contact at I_1 and O_1 and I_1 coincide. As the body moves from A_1B_1 to A_4B_4, the point of contact of the two centrodes will move from I_1 to I_4 and the continuous motion of the body AB corresponds to a rolling of the body-centrode on the space-centrode.

43. Applications of the Instantaneous Centre Method. (a) *The Reciprocating Engine Mechanism.* The instantaneous centre method will be used to find the velocity of the pin P in the reciprocating engine mechanism shown in Fig. 50. The velocity of the crankpin C is at right angles to the crank OC, so that the instantaneous centre of the connecting rod lies on OC produced. The velocity of the gudgeon pin P is along the line PO, so that the instantaneous centre of the connecting rod PC also lies on the line PA drawn through P at right angles to OP. The position of the instantaneous centre is thus given by I, the point of intersection of PA and OC produced, and for the given position of the mechanism the connecting rod PC is turning about the centre I. Hence every point on the rod is moving with a velocity which is proportional to its distance from I, and, further, it is moving in a direction perpendicular to the line joining that point to I.

If ω_1 = angular velocity of CP, then

$$\omega_1 = v_c/CI = v_p/PI = v_x/XI = v_y/YI$$

where X and Y are points fixed to PC, as shown in Fig. 50.

The directions of the velocities of X and Y are at right angles to IX and IY, as shown in the figure.

Example 3. Fig. 50 is drawn to scale for a crank length OC = 6 in. and a connecting rod length CP = 24 in. The r.p.m. of the crank are 240 and the angle θ is 45°. The distances of X from P and C respectively are 19 in. and 6 in., and the distances of Y from P and C respectively are 32 in. and 9 in.

The crankpin velocity $v_c = (\pi N/30)OC = 8\pi.6/12 = 12{\cdot}57$ ft/s

Scaling off the lengths of IC, IP, IX and IY, we have:

$$12{\cdot}57/2{\cdot}77 = v_p/2{\cdot}33 = v_x/2{\cdot}33 = v_y/3{\cdot}48 = \omega_1 = 4{\cdot}53 \text{ rad/s}$$

Note that, to obtain the value of ω_1 in rad/s, the velocity in ft/s must be divided by the distance from the instantaneous centre measured in feet.

From the above equation the following values for the velocities of P, X and Y are obtained: $v_p = 10{\cdot}57$ ft/s, $v_x = 10{\cdot}57$ ft/s and $v_y = 15{\cdot}8$ ft/s.

The angular velocity of the connecting rod $= \omega_1 = 4{\cdot}53$ rad/s.

It may happen that the instantaneous centre I is inaccessible. If so, a triangle similar to triangle ICP must be drawn. Let M be the point of intersection of PC produced with the line through O perpendicular to the line of stroke of P, Fig. 50 Then the triangles OCM, ICP are similar, so that:

$$v_p/v_c = IP/IC = OM/OC \quad \text{and} \quad v_{pc}/v_c = CP/IC = CM/OC$$

But if ω is the angular velocity of the crank OC, then $v_c = \omega.OC$ and therefore $v_p = \omega.OM$ and $v_{pc} = \omega.CM$.

The triangle OCM is drawn to a larger scale in Fig. 50 (a). The side CM is the velocity image of the connecting rod CP and the velocities of the points X and Y may be found by constructing triangles CMx and CMy respectively similar to triangles CPX and CPY. To find x draw Cx and Mx respectively parallel to CX and PX, and to find y draw Cy and My respectively parallel to CY and PY. Then,

$$\text{velocity of X} = v_x = \omega.\text{Ox}$$

and

$$\text{velocity of Y} = v_y = \omega.\text{Oy}$$

The directions of v_x and v_y are of course perpendicular to Ox and Oy.

(b) *Wrapping Machine Mechanism*: Fig. 51. This mechanism is used in wrapping machines. The pins O, P and Q are fixed. OA is the driving crank, which turns at uniform speed and oscillates the link BP through the link AB. The link CD transmits motion from the pin C on link AB to the bell-crank lever DQE. It will be found that the motion of E is intermittent and this motion is used to feed the paper into the wrapping machine.

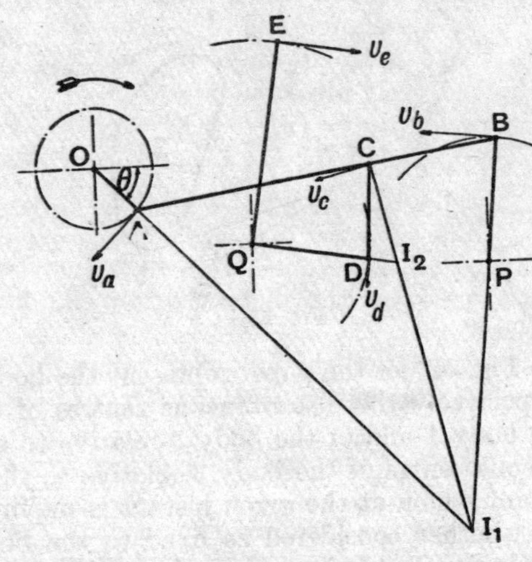

FIG. 51

It is required to find the velocity of pin E in terms of the velocity of the crankpin A for the given configuration of the mechanism.

Since the velocity of A is perpendicular to OA and the velocity of B is perpendicular to BP, the instantaneous centre of the link AB is given by I_1, the point of intersection of OA produced and BP produced.

$\therefore$ the velocity of C is perpendicular to I_1C and its magnitude is given by

$$v_c = (I_1C/I_1A)v_a$$

But the velocity of D is perpendicular to DQ, so that the

instantaneous centre of link CD is given by I_2, the point of intersection of QD produced with CI_1.

$$\text{velocity of D} = v_d = (I_2D/I_2C)v_c$$

Since DQE turns about the fixed centre Q, the velocity of E is given by $v_e = (QE/QD)v_d$.

For clockwise rotation of the crank, the directions of the velocities of B, C, D and E are as shown by the arrows on the figure.

44. The Three-centres-in-line Theorem. If two rigid bodies 2 and 3 are moving in the same plane relative to a third rigid body 1, then the instantaneous centres of 3 relative to 1, of 2 relative to 1 and of 3 relative to 2 all lie in a straight line. The theorem may be proved as follows:

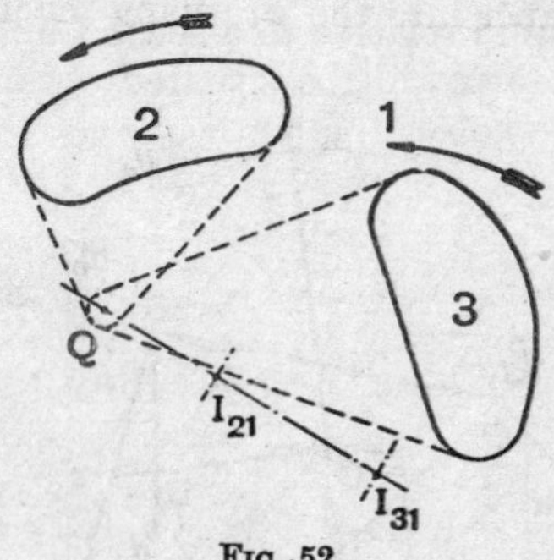

Fig. 52

Referring to Fig. 52, let the paper represent the body 1 and let I_{21}, I_{31} be respectively the instantaneous centres of the body 2 relative to the body 1 and of the body 3 relative to the body 1. The instantaneous centre of the body 3 relative to the body 2 is clearly that point which at the given instant is moving with the same velocity whether considered as fixed to the body 3 or as fixed to the body 2. But it is obvious that only those points on the bodies 2 and 3, which lie on the line $I_{21}I_{31}$, produced if necessary, can be moving in the same direction at the given instant. Let Q be a point on the line $I_{21}I_{31}$ produced. Then v_q is at right angles to QI_{21}, when Q is considered as a point fixed to the body 2. Also v_q is at right angles to QI_{31}, when Q is considered as a point fixed to the body 3.

If v_q is the same for the point Q on body 2 as for the point Q on body 3, then

$$v_q = \omega_2 . QI_{21} = \omega_3 . QI_{31} \quad \text{and} \quad QI_{21}/QI_{31} = \omega_3/\omega_2$$

where ω_2 and ω_3 are the angular velocities of the bodies 2 and 3 relative to the body 1.

If this condition is satisfied, then the point Q coincides with the instantaneous centre I_{32} of the body 3 relative to the body 2.

It should be noted that if the body 1 is itself moving the above relationship remains unaffected.

45. Applications of the Three-centres-in-line Theorem. For many problems in kinematics the application of the above theorem will give a simple solution.

For a kinematic chain with l links, the total number of instantaneous centres will be equal to the number of different combinations of the links in pairs, that is $l(l-1)/2$.

Obviously, where two links are connected together by a pin-joint, that pin-joint will be the instantaneous centre for the motion of the one link relative to the other, and where one link has straight-line motion relative to a second link, the corresponding instantaneous centre will be situated at an infinite distance along the normal to the relative path.

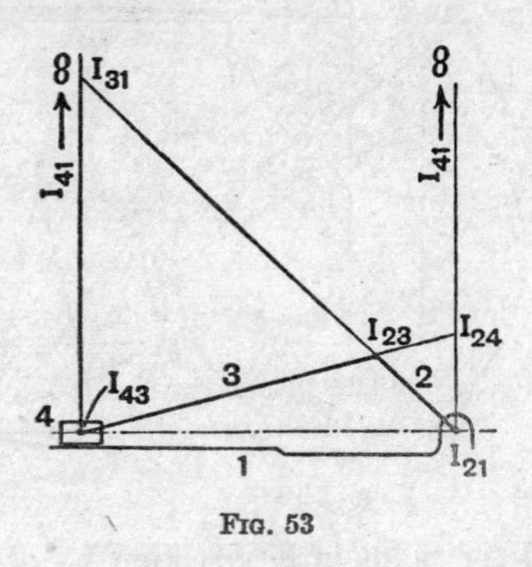

Fig. 53

(a) *The Reciprocating Engine Mechanism*: Fig. 53. Here there are four links, and therefore $4(4-1)/2 = 6$ instantaneous centres Four of these are already known. Thus, with the links numbered as shown in the figure, the crankshaft centre is I_{21}, the crankpin centre is I_{23}, the gudgeon-pin centre is I_{34} and I_{41} is along the line normal to the path of 4 and at an infinite distance away.

To find the remaining two centres, we have shown that the three instantaneous centres for any three links which have plane motion lie on the same straight line.

For the links 1, 2 and 3, I_{21} and I_{23} are known, and therefore I_{31} must lie somewhere along the line $I_{21}I_{23}$. But I_{31} must also lie along the line $I_{43}I_{41}$. It must, therefore, be situated at the point of intersection of these two lines, as shown in the figure. Similarly,

I_{24} must lie at the point of intersection of the line $I_{21}I_{41}$ and the line $I_{43}I_{23}$.

Then, since I_{24} is the instantaneous centre for the relative motion of links 2 and 4, the velocity of the point I_{24} must be the same whether regarded as a point on link 2 or as a point on link 4. But, regarded as a point on link 2, its velocity is easily seen to be $\omega . I_{21}I_{24}$, where ω is the angular velocity of link 2. Since link 4 has straight-line motion, all points fixed to link 4 must be moving with the same velocity. From this it follows that the velocity of link 4 is given by $\omega . I_{21}I_{24}$.

This indicates a very simple method of finding the velocity of the piston in the reciprocating-engine mechanism. All that is

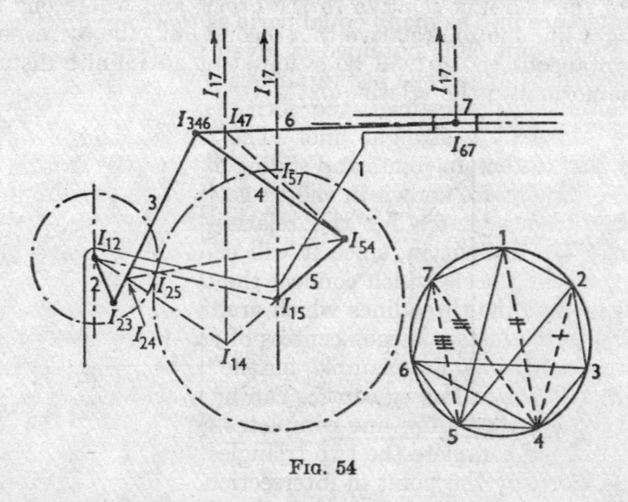

Fig. 54

required is to find the point of intersection I_{24} of the connecting rod, produced if necessary, with the line through the crankshaft centre I_{21} normal to the line of stroke. The velocity of the piston is then given by $\omega . I_{21}I_{24}$. Note that the distance $I_{21}I_{24}$ is the same as the distance OM, Fig. 50.

In measuring the length of the intercept $I_{21}I_{24}$, the scale to which the diagram is drawn must, of course, be taken into account.

It is not always necessary to find the position of all the instantaneous centres for the mechanism. In the example just given, for instance, no use is made of the instantaneous centre I_{31} and therefore its position need not be determined. In the examples which follow, only those instantaneous centres will be found which are actually required in solving the problems.

(b) *The Andreau Differential-stroke Engine*: Fig. 54. The purpose of this mechanism is to obtain four strokes of different lengths for the piston (link 7) for one revolution of the crank (link 5) or two revolutions of the crank (link 2). The crank 2 is only one-half the length of the crank 5, but the two shafts are geared together, so that the shorter crank rotates with twice the speed of the longer crank but in the opposite sense. The two cranks are coupled by two equal links 3 and 4 to the connecting rod (link 6). It is required to find the velocity of the piston for the given position of the mechanism and for given angular velocities of the cranks 2 and 5. This involves finding the position of either I_{57} or I_{27}. There are seven links in this mechanism and therefore $7(7-1)/2 = 21$ instantaneous centres.

The following method[1] will be found useful for indicating and checking the instantaneous centres. First, draw a circle and divide its circumference into as many equal parts as there are links in the chain. Each point thus obtained is given the same number as one of the links. When the instantaneous centre for a given pair of links has been determined, the corresponding points on the circle are joined by a straight line. The instantaneous centres corresponding to the pin-joints and sliding connections can be put in at once. These are shown in the circle, Fig. 54, by full lines. The instantaneous centre for the relative motion of the two cranks may also be put in, since it will coincide with the pitch point of the gear wheels which connect the two shafts.

It is clear that the three lines which are drawn in the circle to correspond to the instantaneous centres of any three of the links, will form a triangle, as, for example, for the three links 1, 2 and 5. Further, one of the unknown centres can be found only if, in this diagram, the corresponding line completes two triangles. Thus, the line 24 would complete the two triangles 234 and 254, so that I_{24} will be given by the point of intersection of $I_{32}I_{34}$ and $I_{25}I_{45}$. The points 2 and 4 may then be joined as shown by the dotted line. In a similar way the line 14 would complete the two triangles 154 and 124, so that the centre I_{14} may be found. Next we may find the centre I_{47} and finally either I_{27} or I_{57}. In this case it is better to find I_{57}. The dotted lines of Fig 54 (a) are numbered in the order in which the corresponding centres are found. All the instantaneous centres for the mechanism could be determined in the same way, but no further centres are required for the solution of the problem.

Since I_{57} is the instantaneous centre for the relative motion of links 5 and 7, it follows that the points on links 5 and 7 which coincide with I_{57} are both moving with the same velocity. But link 7 has straight line motion, so that the velocity of the

[1] Reprinted by permission from *Kinematics of Machines*, by Guillet, published by John Wiley & Sons, Inc.

piston 7 is given by $\omega_5 . I_{15}I_{57}$, where ω_5 is the angular velocity of crank 5.

(c) *The Crank and Slotted-lever Quick-return Mechanism.* As a further example, consider the crank and slotted-lever quick-return mechanism shown in Fig. 55. This mechanism has already been described in Article 11. There are six links, numbered as shown, so that there are $6(6-1)/2 = 15$ instantaneous centres. It is required to find the velocity of the ram (link 6), given the angular velocity of the driving crank (link 2).

As in the last example, draw a circle and divide the circumference into six equal parts. Number each point thus obtained from 1 to 6. Where the instantaneous centre for two links is known, join the corresponding points on the circle by a straight

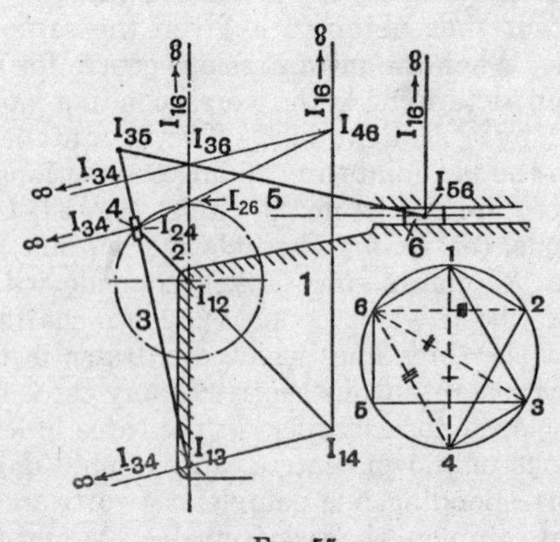

FIG. 55

line. All pin-joints are instantaneous centres. In addition, the instantaneous centres I_{34} and I_{16} are known, since link 4 has straight-line motion relative to link 3 and link 6 has straight-line motion relative to the frame 1.

In order to determine the velocity of the ram (link 6), we require to find the centre I_{26}.

The dotted line 14, Fig. 55, completes the triangles 124 and 134, so that the centre I_{14} is given by the point of intersection of the lines $I_{12}I_{24}$ and $I_{13}I_{34}$. Similarly line 36 completes the triangles 316 and 356 and the centre I_{36} lies at the point of intersection of the lines $I_{13}I_{16}$ and $I_{35}I_{56}$.

It is then possible to find first I_{46}, since line 46 completes the triangles 436 and 416, and afterwards I_{26}, since line 26 completes the triangles 216 and 246.

Then the velocity of the points, rigidly fixed to links 2 and 6, which coincide with I_{26} is given by $\omega_2 . I_{12}I_{26}$, where ω_2 is the angular velocity of link 2. Since link 6 has straight-line motion, the velocity of the ram is given by $\omega_2 . I_{12}I_{26}$.

46. The Acceleration Diagram for a Link. It was pointed out in Article 16 that, in general, the velocity of any particle of a link in a mechanism changes both in direction and magnitude, so that the acceleration of the particle at any instant has two components, viz. (a) a centripetal component, which is at right angles to the velocity at the given instant and (b) a tangential component which is parallel to the velocity of the particle at the given instant.

In Fig. 56 (a), A and B are two points on the same link of a mechanism and ω is the angular velocity and α the angular acceleration of the link.

Let f_{ba} = total acceleration of B relative to A,

f^c_{ba} = centripetal component of the acceleration of B relative to A

and f^t_{ba} = tangential component of the acceleration of B relative to A.

Then
$$f_{ba} = f^c_{ba} \nrightarrow f^t_{ba}$$

But $f^c_{ba} = \omega^2 . BA$ and $f^t_{ba} = \alpha . BA$, so that
$$f_{ba} = \omega^2 . BA \nrightarrow \alpha . BA = (\omega^2 \nrightarrow \alpha)BA$$

Set off ap, Fig. 56 (b), to represent to some convenient scale the centripetal component and pb to represent to the same scale the

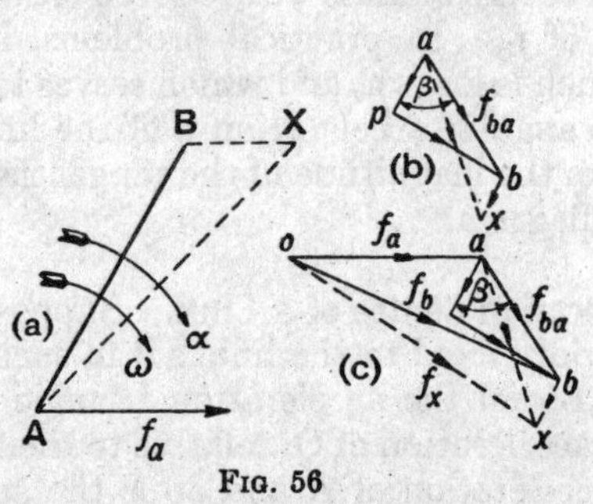

Fig. 56

tangential component, where ap is parallel to BA and pb is perpendicular to BA. Then ab the vector sum of ap and pb must represent to scale the total acceleration of B relative to A.

The vector ab is known as the *acceleration image* of the link AB. It is inclined to AB at the angle β, where $\tan \beta = pb/ap = \alpha/\omega^2$.

If X is a point rigidly fixed to AB, then the acceleration of X relative to A has a centripetal component $\omega^2 . AX$ and a tangential component $\alpha . AX$, so that f_{xa} is inclined to XA at the same angle β. Once f_{ba} is known, f_{xa} is most easily determined by finding the point x which occupies the same position relative to the ends a and b of the acceleration image as the point X occupies relative to the ends A and B of the link. This means that the triangles abx, ABX must be similar. The vector ax will then represent the acceleration of X relative to A, and, similarly, the vector bx will represent the acceleration of X relative to B.

If the point A itself has an acceleration f_a relative to the fixed link of the mechanism, or to some fixed plane of reference, then the acceleration f_b of B relative to the fixed link is given by the vector sum of f_a and f_{ba}. In Fig. 56 (c), let oa represent f_a and ab represent f_{ba}, then ob which is the vector sum of oa and ab, will represent f_b. Similarly ox, the vector sum of oa and ax, will represent the acceleration f_x of X relative to the fixed link.

When constructing the acceleration diagram for an actual mechanism two things should be noted. First, the angular velocity of any link of a mechanism may be determined by one or other of the methods given in Articles 39 and 41, so that the centripetal component of the acceleration of one point on the link relative to another point on the same link can be calculated. Secondly, although the tangential component cannot be directly calculated, since the magnitude and sense of α are both unknown, its direction must be at right angles to the centripetal component. Referring to Fig. 56 (c) it is clear that if α is unknown the position of point b cannot be fixed unless we are given either the magnitude or the direction of f_b. In practical problems, it is usually the direction of f_b which is known, and which serves to fix the position of point b. The angular acceleration α of the link AB may then be calculated from the magnitude of the tangential component pb, scaled from the diagram.

47. The Acceleration Centre of a Link. Suppose that the point O_a, Fig. 57, is rigidly fixed to the link AB in such a position that the triangle O_aAB and the acceleration triangle oab are similar. Then clearly the acceleration of O_a relative to the fixed link is zero, since oa is the acceleration of A and ao is the acceleration of O_a relative to A. The point O_a is known as the *instantaneous centre of acceleration* of the link AB, or simply the *acceleration centre.* The position of O_a relative to the ends A and B of the link will, of course, change from instant to instant but, at a given instant, the accelerations of all points on the link AB will be directly proportional to their distances from the acceleration centre. The directions of the accelerations will be inclined at the angle β to the lines which join the points to the acceleration centre.

It will be shown later (in Article 153) that, when the position of the acceleration centre is known, the line of action of the accelerating force on the link may easily be found.

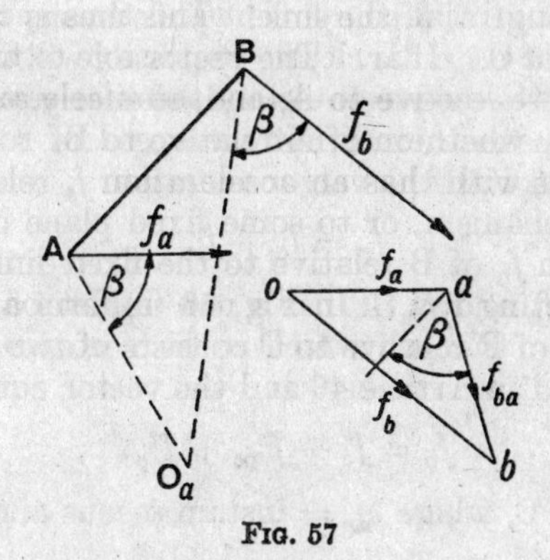

FIG. 57

48. The Acceleration Diagram for the Reciprocating-engine Mechanism.

Referring to Fig. 58, OC is the crank and CP the connecting rod of a reciprocating-engine mechanism. It is required to find the acceleration of the gudgeon pin P given the requisite dimensions and data.

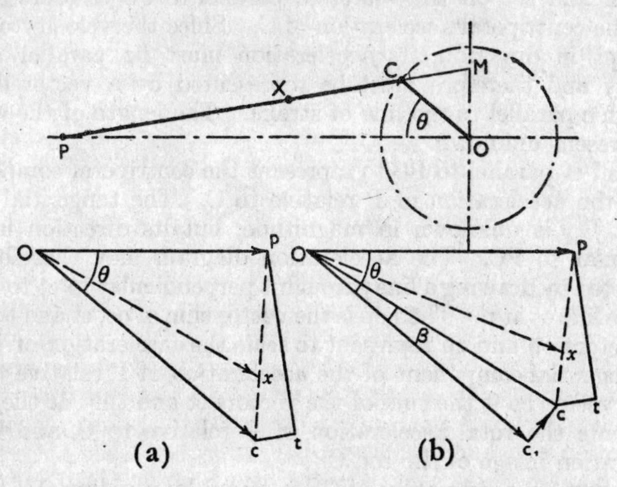

FIG. 58

The dimensions required are the lengths of OC and CP and the angle θ. In addition the velocity and acceleration of C must both be known. Generally, the crank OC is specified to rotate at a given uniform speed, in which case the acceleration of C is centripetal along CO. But, if the speed of OC is not uniform, it is necessary to know the rate at which it is changing in order to determine the tangential component and thus specify completely the acceleration of C. It is clearly impossible to find the acceleration of P unless we know completely the acceleration of C.

In either case, whether or not the speed of rotation of OC is uniform, we start with the vector equation:

$$f_p = f_c \nrightarrow f_{pc}$$

Since the connecting rod CP has a non-uniform angular velocity, the acceleration of P relative to C consists of two components, as already explained in Article 46 and the vector equation becomes:

$$f_p = f_c \nrightarrow f^c_{pc} \nrightarrow f^t_{pc}$$

But $f^c_{pc} = \omega_r^2 . PC$, where ω_r = instantaneous angular velocity of PC.

We have already shown in Example 3, Article 43, that the velocity of the piston is given by $\omega . OM$ and the velocity of P relative to C by $\omega . CM$, so that the angular velocity of the connecting rod is given by $\omega_r = \omega . CM/PC$.

The vector polygon of accelerations may now be drawn Fig. 58 (a). Assume the angular velocity of the crank to be uniform and set off the vector oc parallel to CO to represent to scale the centripetal acceleration of C. Since the velocity of P is constant in direction, its acceleration must be parallel to its velocity and therefore must be represented by a vector drawn through o parallel to the line of stroke. The length of the vector is at present unknown.

Set off ct parallel to PC to represent the centripetal component f^c_{pc} of the acceleration of P relative to C. The tangential component f^t_{pc} is unknown in magnitude, but its direction is perpendicular to PC. The acceleration diagram may therefore be completed by drawing a line through t perpendicular to ct to intersect the line os at p. Then op is the vector sum of oc, ct and tp, and the vectors op and tp represent to scale the acceleration of P and the tangential component of the acceleration of P relative to C.

The vector cp is the sum of the vectors ct and tp. It therefore represents the total acceleration of P relative to C, and is the acceleration image of the rod CP.

The angular acceleration of CP is found by dividing f^t_{pc}, represented to scale by tp, by the length of CP.

If the angular velocity of the crank is not uniform, the total acceleration of C consists of a centripetal component f^c_c and a tangential component f^t_c, so that the vector equation is

$$f_p = f^c_c \mathrel{+\!\!\!>} f^t_c \mathrel{+\!\!\!>} f^c_{pc} \mathrel{+\!\!\!>} f^t_{pc}$$

The acceleration diagram, Fig. 58 (b), is set out in a similar way to that for uniform speed of rotation of the crank, except that oc is now inclined to the crank at an angle β ($= \tan^{-1} \alpha/\omega^2$). The total acceleration of C is represented by oc, where oc is the vector sum of oc_1, which represents $f^c_c (= \omega^2 . CO)$ and c_1c which represents f^t_c ($= \alpha . CO$). The rest of the construction is unchanged. Fig. 58 is drawn to scale for the following example.

Example 4. The crank of an engine is 9 in. long and the connecting rod is 36 in. long. Find the acceleration of the piston, the acceleration of a point X on the rod, 12 in. from C and the angular acceleration of the rod, when θ is 40° and:

(a) the crank turns at a uniform speed of 240 r.p.m.,
(b) the instantaneous speed of rotation is 240 r.p.m. clockwise and it is increasing at the rate of 100 rad/s².

The centripetal acceleration of C

$$= f^c_c = (\pi . 8)^2 . \tfrac{3}{4} = 474 \text{ ft/s}^2$$

The angular velocity of CP

$$= \omega_r = \omega . CM/CP = 8\pi . 6{\cdot}98/36 = 4{\cdot}87 \text{ rad/s}$$

and the centripetal component of the acceleration of P relative to C

$$= f^c_{pc} = 4{\cdot}87^2 . 3 = 71 \text{ ft/s}^2$$

(a) From Fig. 58 (a):

$$f^t_{pc} = tp = 296 \text{ ft/s}^2, \quad f_{pc} = cp = 306 \text{ ft/s}^2$$
$$\alpha_r = f^t_{pc}/PC = 296/3 = 98{\cdot}7 \text{ rad/s}^2 \text{ clockwise}$$

To find the acceleration of X, divide the acceleration image cp at x in the same proportion as X divides CP. Then

$$f_x = ox = 422 \text{ ft/s}^2$$

(b) From Fig. 58 (b):

$$oc_1 = f^c_c = 474 \text{ ft/s}^2, \quad c_1c = f^t_c = \alpha . OC = 100 . \tfrac{3}{4} = 75 \text{ ft/s}^2$$

and $$f_c = oc = 480 \text{ ft/s}^2$$

$$f^t_{pc} = tp = 238 \text{ ft/s}^2, \quad f_{pc} = cp = 246 \text{ ft/s}^2$$
$$\alpha_r = f^t_{pc}/PC = 238/3 = 79{\cdot}3 \text{ rad/s}^2 \text{ clockwise}$$
$$f_x = ox = 452 \text{ ft/s}^2$$

49. Klein's Construction for determining the Acceleration of the Piston: Fig. 59. A simple graphical construction for determining a quadrilateral similar in shape to the acceleration diagram octp of the last Article is that given by Professor Klein.[1] The construction is carried out as follows. Produce, if necessary, the line PC to cut a line through O perpendicular to the line of stroke at M. Draw a circle on CP as diameter; with centre C and radius CM draw a second circle. Then KL, the chord common to these two circles, will cut CP at Q and OP at N, such that the quadrilateral OCQN is similar to the acceleration diagram octp.

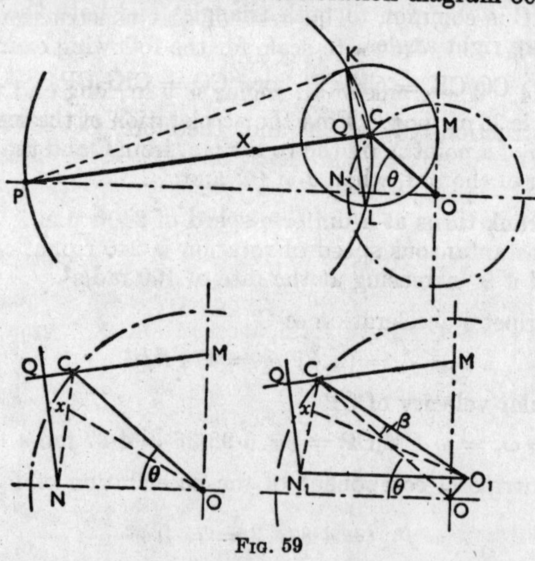

FIG. 59

Proof. It is easily seen that corresponding angles in the two quadrilaterals are equal, so that in addition to this it is only necessary to show that two sides of OCQN bear the same ratio to each other as do the corresponding sides of the acceleration diagram octp. Since, in the construction of the acceleration diagram, the lengths of the sides oc and ct represent to scale the acceleration of the crankpin and the centripetal component of the acceleration of P relative to C, we must show that

$$CQ/CO = ct/co$$

We have already seen, Article 43, that the triangle OCM is similar to the velocity triangle for the mechanism, so that, if ω is the angular velocity of the crank OC, it follows that $v_c = \omega . OC$, $v_p = \omega . OM$ and $v_{pc} = \omega . CM$.

[1] For alternative graphical methods of finding the points Q and N the reader is referred to Example 24, p. 118.

The acceleration of the crankpin $= \omega^2.OC$ and the centripetal component of the acceleration of P relative to C $= v_{pc}{}^2/PC$. On substituting for v_{pc} this becomes $\omega^2.CM^2/PC$. If, therefore, a point on the connecting rod CP is found, such that its distance from C is equal to CM^2/CP, this distance will represent the centripetal component of the acceleration of P relative to C to the same scale $1/\omega^2$ as that to which OC represents the acceleration of C. That the point Q, as determined by Klein's construction, is such that $CQ = CM^2/PC$ may be shown by joining CK and KP, when it will be seen that the triangles CQK and CKP are similar, since the angle at C is common to both triangles and both the angles CQK, CKP are right angles.

$$\therefore \; CQ/CK = CK/CP \quad \text{or} \quad CQ = CK^2/CP$$

But by construction, $CK = CM$ and, therefore, $CQ = CM^2/CP$. Hence $CQ/CO = ct/co$ and the quadrilateral OCQN obtained by Klein's construction is similar to the acceleration diagram octp.

Therefore, the acceleration of the piston

$$f_p = \omega^2.NO \quad . \quad . \quad . \quad . \quad . \quad (3.3)$$

and the tangential component of the acceleration of P relative to $C = \omega^2.NQ$, so that the total acceleration of P relative to $C = \omega^2.NC$.

The acceleration of the point X on the rod may be found by drawing XX_1 parallel to the line of stroke to cut CN at X_1, then $f_x = \omega^2.X_1O$.

The linear scale to which the diagram is drawn must, of course be taken into account when measuring NO etc.

It should be noted that the acceleration of P is in the direction NO. If the crank position is such that N lies to the right of O instead of to the left as in Fig. 59 the acceleration of P is negative, that is to say, P is undergoing retardation.

The tangential component of the acceleration of P relative to C is given by $\omega^2.NQ$ and its sense is from N to Q, so that the angular acceleration of the connecting rod is given by:

$$\alpha_r = \omega^2.NQ/CP$$

and its sense is such as to tend to reduce the inclination of the connecting rod to the line of stroke.

Klein's construction may equally well be applied when the crank has an angular velocity ω and an angular acceleration α. The point Q on CP is found in exactly the same way as already described. Then, referring to Fig. 59 (b), draw a line from C, at an angle $\beta = \tan^{-1} \alpha/\omega^2$ to CO, to intersect at O_1 a line through O perpendicular to CO. Finally draw through O_1 a line parallel to the line of stroke of P to intersect at N_1 the common chord KL. With this construction,

$$f^c_c = \omega^2.CO, \quad f^t_c = \omega^2.OO_1, \quad f_c = \omega^2.CO_1$$
$$f^c_{pc} = \omega^2.QC, \quad f^t_{pc} = \omega^2.N_1Q \quad \text{and} \quad f_p = \omega^2.N_1O_1$$

Fig. 59 is drawn to scale for the particulars given in Example 4, Article 48.

Klein's construction may be applied for all crank positions. For the inner and outer dead-centre positions it is clear that M coincides with O and the radius of the circle drawn with centre C

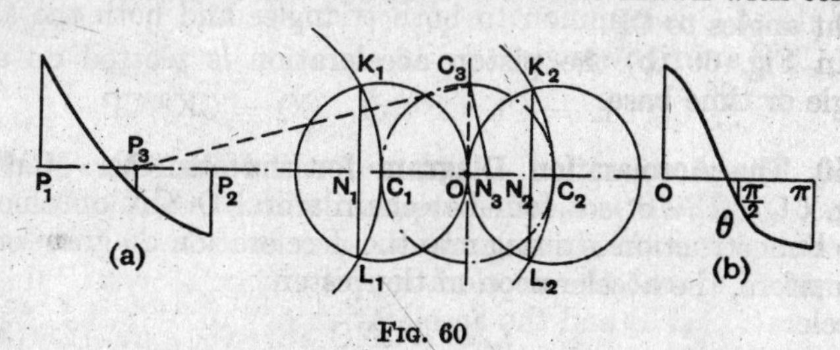

FIG. 60

is equal to the crank length CO. The corresponding positions of the point N are denoted respectively by N_1 and N_2 in Fig. 60. C_1N_1 is obviously equal to C_2N_2 and each is equal to CO^2/CP, so that we get:

At the beginning of the stroke

$$\begin{aligned} f_p = \omega^2.N_1O &= \omega^2(N_1C_1+C_1O) \\ &= \omega^2.CO(CO/CP+1) \\ &= \omega^2.CO(1+1/n) \qquad \qquad (3.4) \end{aligned}$$

where $n = CP/CO$.

Similarly, at the end of the stroke

$$\begin{aligned} f_p &= \omega^2.N_2O \\ &= -\omega^2.CO(1-1/n) \qquad \qquad (3.5) \end{aligned}$$

When the crank is at right angles to the line of stroke, the points M and C coincide at C_3 and the common chord becomes tangential to the circle drawn on C_3P_3 as diameter. N_3 is therefore found by drawing C_3N_3 at right angles to C_3P_3 and the corresponding acceleration of the piston is given by $\omega^2.N_3O$.

If several different crank positions are taken and through each position of the gudgeon pin P ordinates are set up with lengths equal to the corresponding distances of N from O, a curve is obtained which shows the variation of piston acceleration throughout the stroke. Such a curve is shown in Fig. 60 (a).

The acceleration of the piston is zero, and its velocity a maximum, when N coincides with O. There is no simple graphical method of finding the corresponding crank position, but it can be shown that for N and O to coincide the angle between the crank and the connecting rod must be slightly less than 90°. Thus, if OC is at right angles to CP, then the common chord KL will be parallel to OC and will cut the line of stroke at a point very close to, but to the left of, O. The piston, therefore, has a small positive acceleration. However, for most practical purposes it may be assumed that the acceleration of P is zero when OC is at right angles to CP.

In Fig. 60 (b) the piston acceleration is plotted on a crank angle or time base.

50. The Acceleration Diagram for the Four-bar Mechanism: Fig. 61. ABCD is a four-bar-chain with the link AD fixed, and the link AB as the driving crank. For the pins B and C we have the relation that the acceleration of C is the vector sum of the acceleration of B and the acceleration of C relative to B. Or,

$$f_c = f_b \dotplus f_{cb}$$

In general, each of the three links AB, BC and CD, has at any given instant both an angular velocity and an angular acceleration.

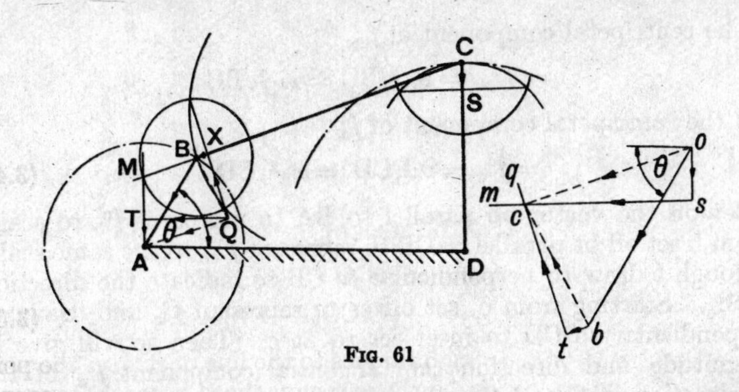

FIG. 61

It follows that the three accelerations f_c, f_b and f_{cb} each have a centripetal and a tangential component and the vector equation may be written:

$$f^c{}_c \dotplus f^t{}_c = f^c{}_b \dotplus f^t{}_b \dotplus f^c{}_{cb} \dotplus f^t{}_{cb}$$

The polygon of accelerations is therefore a six-sided figure and the directions of all six sides are known. In addition the magnitudes of the three centripetal components can be calculated, since they depend on the instantaneous angular velocities of the links, which may be found from the velocity diagram. This still leaves the three tangential components and, before the acceleration polygon can be drawn, one of these must be known in magnitude. Usually, the driving crank AB is specified to rotate at uniform speed, in which case f^t_b is zero and the polygon reduces to a five-sided figure. If the speed of AB is not uniform its angular acceleration must of course, be given. In what follows it will be assumed that AB turns at uniform speed, so that the vector equation reduces to

$$f^c_c \nrightarrow f^t_c = f^c_b \nrightarrow f^c_{cb} \nrightarrow f^t_{cb}$$

Velocity Diagram. This may be drawn in the usual way, but it is simpler to make use of the similar triangle ABM, where M is found by producing CB to intersect a line drawn through A parallel to DC. The three sides of this triangle are perpendicular to the sides of the velocity triangle, so that:

$$v_b = \omega.AB, \quad v_c = \omega.AM \quad \text{and} \quad v_{cb} = \omega.BM$$

The angular velocity of BC $= \omega_1 = v_{cb}/BC$ and the angular velocity of CD $= \omega_2 = v_c/CD$

The acceleration of B is wholly centripetal

$$= f^c_b = v_b^2/AB = \omega^2.AB$$

The centripetal component of f_{cb}

$$= f^c_{cb} = v_{cb}^2/BC = \omega_1^2.BC$$

and the centripetal component of f_c

$$= f^c_c = v_c^2/CD = \omega_2^2.CD$$

Set off the vector ob parallel to BA to represent f^c_b to scale. From b set off bt parallel to CB to represent f^c_{cb} to the same scale. Through t draw tq perpendicular to CB to indicate the direction of f^t_{cb}. Starting from o, set off os to represent f^c_c and draw sm perpendicular to CD to intersect tq at c. Then sc will give in magnitude and direction the tangential component f^t_c of the acceleration of C and tc will give the tangential component f^t_{cb} of the acceleration of C relative to B.

The dotted line bc $= f_{cb}$ and is the acceleration image of BC. Similarly, oc $= f_c$ and is the acceleration image of CD.

Fig. 61 is drawn to scale for the following example:

Example 5. The lengths of the links AB, BC, CD and DA are respectively 2·5 in., 7 in., 4·5 in. and 8 in. The r.p.m. of AB are 100 and the angle BAD is 60°.

The angular velocity of AB $= \omega = \pi N/30 = 10{\cdot}47$ rad/s.
From the triangle ABM:

$$v_b = \omega.\text{AB} = 2{\cdot}18 \text{ ft/s}$$
$$v_c = \omega.\text{AM} = 1{\cdot}47 \text{ ft/s}$$
$$v_{cb} = \omega.\text{BM} = 1{\cdot}25 \text{ ft/s}$$
$$\text{ob} = f_b = \omega^2.\text{AB} = 22{\cdot}8 \text{ ft/s}^2$$
$$\text{bt} = v_{cb}{}^2/\text{BC} = 2{\cdot}68 \text{ ft/s}^2$$

and
$$\text{os} = v_c{}^2/\text{CD} = 5{\cdot}76 \text{ ft/s}^2$$

Then, by measurement from the acceleration diagram, sc $= 19{\cdot}1$ ft/s^2, oc $= f_c = 20$ ft/s^2, tq $= 14{\cdot}4$ ft/s^2 and bc $= f_{cb} = 14{\cdot}6$ ft/s^2.

Therefore angular acceleration of CD (counter-clockwise)

$$= 19{\cdot}1/\text{CD} = 19{\cdot}1.12/4{\cdot}5 = 51{\cdot}0 \text{ rad/s}^2$$

and the angular acceleration of BC (counter-clockwise)

$$= 14{\cdot}4/\text{BC} = 14{\cdot}4.12/7 = 24{\cdot}7 \text{ rad/s}^2$$

A complete graphical construction for the acceleration diagram is shown on the figure. This is an extension of Klein's construction as already given for the reciprocating-engine mechanism. A circle is drawn with centre B and radius BM; a second circle is drawn on the link BC as diameter; the chord common to these two circles intersects BC at X and is produced to Q. Two further circles are drawn, one with centre C and radius AM, the other on CD as diameter. The chord common to these two circles cuts CD at S. Along AM mark off AT equal to CS Through T draw TQ at right angles to AM.

With this construction

$$f_c = \omega^2.\text{QA} \quad \text{and} \quad f_{cb} = \omega^2.\text{QB}$$

Proof. It follows from the proof of Klein's construction given on p. 94 that BX $=$ BM^2/BC and AT $=$ CS $=$ AM^2/CD.

But the centripetal component of $f_{cb} = v_{cb}{}^2/\text{BC}$. On substituting $\omega.\text{BM}$ for v_{cb} this becomes $\omega^2.\text{BM}^2/\text{BC} = \omega^2.\text{BX}$. Similarly, the centripetal component of $f_c = v_c{}^2/\text{CD} = \omega^2.\text{AM}^2/\text{CD}$ $= \omega^2.\text{CS} = \omega^2.\text{TA}$.

We have then that BA, XB and TA represent to the scale $1/\omega^2$ the acceleration of B, the centripetal component of f_{cb} and the centripetal component of f_c respectively. It follows that QX and QT will represent to the same scale the tangential components of f_{cb} and f_c respectively, so that the total acceleration of C relative to B is represented by QB and the total acceleration of C by QA. The arrows on the figure indicate the directions of the various accelerations.

51. Analytical Determination of the Velocity and Acceleration of the Piston of a Reciprocating Engine. There are many problems which arise in connection with the dynamics of the reciprocating engine that make it desirable to be able to calculate the acceleration of the piston. Examples of such problems are given in later chapters where the effect of the inertia of the reciprocating parts on the turning moment diagram and on the balance of the engine is considered.

Let r = length of the crank,
ω = angular velocity of the crank,
$l = n.r.$ = length of the connecting rod,
θ = inclination of the crank to the i.d.c.,
x = displacement of the piston from the beginning of its stroke,
v_p = velocity of the piston,
f_p = acceleration of the piston,
ω_r = angular velocity of the connecting rod
and α_r = angular acceleration of the connecting rod.

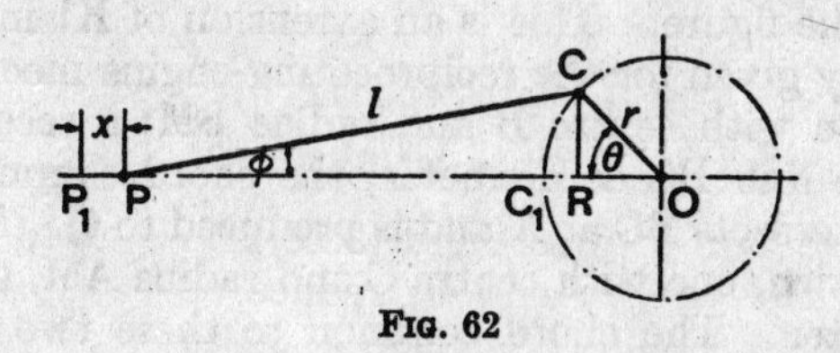

FIG. 62

Then, referring to Fig. 62, in which CR is perpendicular to the line of stroke, the displacement

$$x = P_1P = P_1O - PO = l + r - (l \cos\phi + r\cos\theta)$$

$$= r(1-\cos\theta) + l(1-\cos\phi) \quad \ldots \quad (3.6)$$

But $$CR = r\sin\theta = l\sin\phi$$

$$\therefore \sin\phi = (r/l)\sin\theta = (\sin\theta)/n \quad \ldots \quad (3.7)$$

and $$\cos\phi = \sqrt{(1-\sin^2\phi)} = (1/n)\sqrt{(n^2-\sin^2\theta)} \quad \ldots \quad (3.8)$$

On substituting in (3.6), we get

$$x = r(1-\cos\theta) + nr\{1-(1/n)\sqrt{(n^2-\sin^2\theta)}\}$$

$$= r\{1+n-\cos\theta-\sqrt{(n^2-\sin^2\theta)}\}$$

Differentiating with respect to time:

$$v_p = \frac{dx}{dt} = \frac{d\theta}{dt}\cdot\frac{dx}{d\theta} = \omega r\left\{\sin\theta + \frac{\sin 2\theta}{2\sqrt{(n^2-\sin^2\theta)}}\right\} \quad (3.9)$$

and $$f_p = \frac{dv_p}{dt} = \frac{d\theta}{dt} \cdot \frac{dv_p}{d\theta} = \omega^2 r \left\{ \cos\theta + \frac{n^2 \cos 2\theta + \sin^4\theta}{(n^2 - \sin^2\theta)^{3/2}} \right\} \quad (3.10)$$

Since $\sin^2\theta$ is small in comparison with n^2, we may write:

$$v_p \simeq \omega r\{\sin\theta + (\sin 2\theta)/2n\} \quad . \quad . \quad . \quad (3.11)$$

and $$f_p \simeq \omega^2 r\{\cos\theta + (\cos 2\theta)/n\} \quad . \quad . \quad . \quad (3.12)$$

The angular velocity and the angular acceleration of the connecting rod may be obtained from (3.7). Differentiating both sides with respect to time:

$$\cos\phi . \frac{d\phi}{dt} = \frac{\cos\theta}{n} \cdot \frac{d\theta}{dt}$$

$$\therefore \omega_r = \frac{d\phi}{dt} = \frac{\omega}{n} \frac{\cos\theta}{\cos\phi} = \omega \frac{\cos\theta}{\sqrt{(n^2 - \sin^2\theta)}} \quad . \quad (3.13)$$

and $$\alpha_r = \frac{d\omega_r}{dt} = \frac{d\theta}{dt} \cdot \frac{d\omega_r}{d\theta} = \frac{-\omega^2(n^2-1)\sin\theta}{(n^2 - \sin^2\theta)^{3/2}} \quad (3.14)$$

Since n^2 is large in comparison with both unity and $\sin^2\theta$, we may write:

$$\omega_r \simeq (\omega/n)\cos\theta \quad . \quad . \quad . \quad . \quad . \quad (3.15)$$

and $$\alpha_r \simeq (-\omega^2/n)\sin\theta \quad . \quad . \quad . \quad (3.16)$$

The negative sign in (3.14) and (3.16) shows that the sense of α_r is always such as to tend to reduce the inclination of the connecting rod to the line of stroke.

52. Fourier Series for the Velocity and Acceleration of the Piston of a Reciprocating Engine. Although the approximate expressions (3.11) and (3.12) are sufficiently accurate for most practical purposes it is necessary to use the exact expressions when considering the extent to which the reciprocating parts of a high-speed engine are balanced. For this purpose it is much more convenient to have these expressions in the form of Fourier series.

Substituting for $\cos\phi$ from (3.8) in (3.6):

$$x = r(1-\cos\theta) + l\{1 - (1/n)\sqrt{(n^2 - \sin^2\theta)}\}$$
$$= r(1-\cos\theta) + l\{1 - \sqrt{(1 - (\sin^2\theta)/n^2)}\}$$

If $\sqrt{\{1-(\sin^2\theta)/n^2\}}$ is expanded by the binomial theorem,

$$x = r(1-\cos\theta) + l\left(\frac{\sin^2\theta}{2n^2} + \frac{\sin^4\theta}{8n^4} + \frac{\sin^6\theta}{16n^6} + \cdots\right)$$

Since $l = nr$

$$\therefore x = r\left(1 - \cos\theta + \frac{\sin^2\theta}{2n} + \frac{\sin^4\theta}{8n^3} + \frac{\sin^6\theta}{16n^5} + \cdots\right)$$

The velocity of the piston

$$= v_p = \omega . dx/d\theta$$

$$= \omega r\left\{\sin\theta + \frac{1}{2n}\frac{d}{d\theta}(\sin^2\theta) + \frac{1}{8n^3}\frac{d}{d\theta}(\sin^4\theta) + \ldots\right\} \quad . \quad (3.17)$$

But $(d/d\theta)(\sin^2\theta) = 2\sin\theta\cos\theta = \sin 2\theta$

$$(d/d\theta)(\sin^4\theta) = (d/d\theta)(\sin^2\theta)^2 = 2\sin^2\theta\sin 2\theta$$
$$= (1-\cos 2\theta)\sin 2\theta = \sin 2\theta - (\sin 4\theta)/2$$
$$(d/d\theta)(\sin^6\theta) = (d/d\theta)(\sin^2\theta)^3 = 3(\sin^2\theta)^2(d/d\theta)\sin^2\theta$$
$$= 3\sin^2\theta . \sin^2\theta . \sin 2\theta$$
$$= (3/4)(1-\cos 2\theta)\{\sin 2\theta - (\sin 4\theta)/2\}$$
$$= (3/16)(5\sin 2\theta - 4\sin 4\theta + \sin 6\theta)$$

Substituting in (3.17):

$$v_p = \omega r(\sin\theta + A\sin 2\theta + B\sin 4\theta + C\sin 6\theta + \ldots) \quad . \quad (3.18)$$

where $A = \frac{1}{2n} + \frac{1}{8n^3} + \frac{15}{256n^5} + \frac{7.5}{8.128n^7} + \ldots$

$$B = -\frac{1}{16n^3} - \frac{3}{4}.\frac{1}{16n^5} - \ldots$$

$$C = \frac{3}{16}.\frac{1}{16n^5} + \frac{3}{8}.\frac{5}{128n^7} + \ldots$$

$$D = -\frac{1}{16}.\frac{5}{128n^7} - \ldots$$

The acceleration of the piston

$$= f_p = \omega . dv_p/d\theta$$

$$= \omega^2 r(\cos\theta + A_1\cos 2\theta + B_1\cos 4\theta + C_1\cos 6\theta + \ldots) \quad (3.19)$$

Values of the coefficients for different values of n are given in the table below.

n	3·0	3·5	4·0	4·5	5·0
A	0·171 6	0·145 9	0·127 0	0·112 5	0·101 0
B	0·002 53	0·001 55	0·001 03	0·000 70	0·000 53
C	0·000 048	0·000 022	0·000 011	0·000 006 3	0·000 003 7
A_1	0·343 1	0·291 8	0·254 0	0·225 0	0·202 0
B_1	0·010 1	0·006 2	0·004 1	0·002 8	0·002 1
C_1	0·000 29	0·000 134	0·000 068	0·000 038	0·000 022

N.B.—The coefficients B and B_1 are negative.

53. The Coriolis Component Acceleration. In Article 46 the relative acceleration of two points which are at a fixed distance apart on a moving link was considered. It frequently happens that it is necessary to know the acceleration, relative to a fixed point on the link, of a second point which slides along the link so that its distance from the fixed point varies. The conditions of the problem are then represented in their simplest form in Fig. 63. The link OPR turns about the fixed centre O, while at the same time the point P slides along OR. It is required to determine the velocity and the acceleration of P relative to the fixed point O. If Q is the point, fixed to OR, with which P coincides at a given instant, then the velocity of P relative to O will be the vector sum of the velocity of P relative to Q and the velocity of Q relative to O. Similarly the acceleration of P relative to O will be the vector sum of the acceleration of P relative to Q and the acceleration of Q relative to O.

The velocity v_q of the coincident point Q is perpendicular to OR and its magnitude is $\omega . OQ = \omega r$, where ω is the angular velocity of OR and r is the distance of P, and therefore of Q, from O. The velocity v_{pq} of P relative to Q is parallel to OR and is equal to the velocity of sliding of P along OR. In the general case both ω and v_{pq} will vary, but we shall first of all consider the simpler case in which they are both constant.

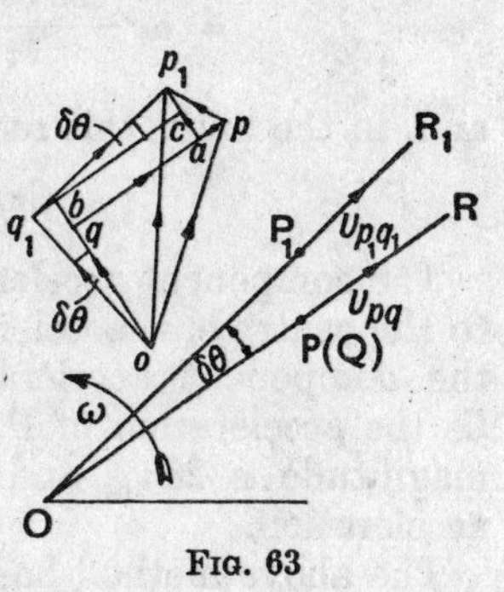

Fig. 63

Referring to Fig. 63, let OR rotate counter-clockwise and let P slide radially outwards along OR. When in the position P, the velocity of the sliding point is represented by op, the vector sum of oq and qp, where oq represents v_q and qp represents v_{pq}. Similarly when in the position P_1, the velocity of the sliding point is represented by op_1, the vector sum of oq_1 and q_1p_1, where oq_1 represents v_{q_1} and q_1p_1 represents $v_{p_1q_1}$.

The change of velocity of P in the time δt during which it moves from P to P_1 is therefore given by the vector pp_1. This may be resolved into two components, one parallel and the other perpendicular to OR.

Draw q_1c parallel to OR and produce oq to b.

Then the component of pp_1 parallel to $RO = pa = q_1b + qp - q_1c$

$$= oq_1 \sin \delta\theta + v_{pq} - v_{p_1q_1} \cos \delta\theta$$

But $oq_1 = \omega . OP_1 = \omega(r + \delta r)$ and $v_{p_1q_1} = v_{pq}$, since the velocity of sliding of P along OR is constant.

$$\therefore \ ap = \omega(r + \delta r) \sin \delta\theta + v_{pq}(1 - \cos \delta\theta)$$

The component acceleration of P parallel to RO

$$= \omega(r+\delta r)(\sin \delta\theta)/\delta t + v_{pq}.(1-\cos \delta\theta)/\delta t$$

and, in the limit, this reduces to $\omega r.d\theta/dt$ or $\omega^2 r$.

Also the component of pp_1 perpendicular to OR

$$= ap_1 = ac+cp_1 = ob-oq+cp_1 = oq_1 \cos \delta\theta - oq + q_1p_1 \sin \delta\theta$$

But $\quad oq_1 = \omega(r+\delta r), \quad oq = \omega r \quad$ and $\quad q_1p_1 = v_{p_1q_1} = v_{pq}$

$$\therefore\ ap_1 = \omega(r+\delta r)\cos \delta\theta - \omega r + v_{pq} \sin \delta\theta$$
$$= \omega r(\cos \delta\theta - 1) + \omega\delta r \cos \delta\theta + v_{pq} \sin \delta\theta$$

The component acceleration of P perpendicular to OR

$$= \omega r\frac{(\cos \delta\theta-1)}{\delta t} + \omega\frac{\delta r}{\delta t}\cos \delta\theta + v_{pq}\frac{\sin \delta\theta}{\delta t}$$

and, in the limit, this reduces to

$$\omega.dr/dt + v_{pq}d\theta/dt = 2\omega v_{pq}$$

The component acceleration of P parallel to RO is clearly equal to the centripetal acceleration of the coincident point Q. Hence the component acceleration of P perpendicular to OR must be the acceleration of P relative to the coincident point Q. Its magnitude is $2\omega v_{pq}$ and it is known as the Coriolis component acceleration.

The above method has been used to find the acceleration of P, when ω and v_{pq} are constant, with the object of showing more clearly how the Coriolis component acceleration arises.

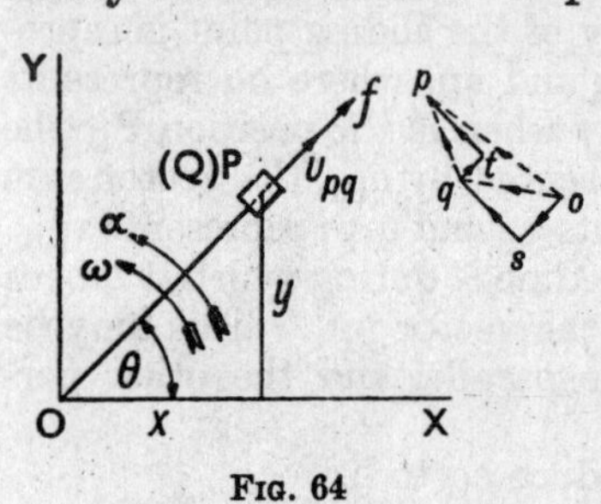

FIG. 64

If the sense of v_{pq} is reversed, i.e. if P slides radially inwards along RO, it is easily seen that the sense of the Coriolis component acceleration will be reversed. The sense of the Coriolis component is, indeed, such as to rotate v_{pq} about its origin in the same sense as ω.

The acceleration of P in the general case, when ω and v_{pq} both vary, could be found in the same way, but it is much more convenient to use an analytical method.

Through O, Fig. 64, draw any pair of co-ordinate axes OX and OY. Let θ be the inclination of OP to OX, x, y the rectangular co-ordinates of P at the given instant, and r the distance of the coincident point Q from O.

The ordinate $y = \text{OP} \sin \theta = \text{OQ} \sin \theta = r \sin \theta$.

The component of the velocity of P parallel to the y-axis is given by:

$$\mathrm{d}y/\mathrm{d}t = (\mathrm{d}/\mathrm{d}t)(r \sin\theta) = (\mathrm{d}r/\mathrm{d}t)\sin\theta + r(\mathrm{d}\theta/\mathrm{d}t)\cos\theta$$

But $\mathrm{d}r/\mathrm{d}t$ is obviously the velocity of sliding of P along OR, i.e. the velocity of P relative to the coincident point Q. This will be denoted by v_{pq}. Also $\mathrm{d}\theta/\mathrm{d}t$ is the angular velocity ω of OR.

$$\therefore\ \mathrm{d}y/\mathrm{d}t = v_{pq}\sin\theta + \omega r\cos\theta \quad . \quad . \quad (3.20)$$

The component of the acceleration of P parallel to the y-axis is given by:

$$\frac{\mathrm{d}^2y}{\mathrm{d}t^2} = \frac{\mathrm{d}}{\mathrm{d}t}(v_{pq}\sin\theta + \omega r\cos\theta)$$

$$= \sin\theta.\frac{\mathrm{d}v_{pq}}{\mathrm{d}t} + v_{pq}\cos\theta.\frac{\mathrm{d}\theta}{\mathrm{d}t} + r\cos\theta.\frac{\mathrm{d}\omega}{\mathrm{d}t} + \omega\cos\theta.\frac{\mathrm{d}r}{\mathrm{d}t} - \omega r\sin\theta.\frac{\mathrm{d}\theta}{\mathrm{d}t}$$

But $\frac{\mathrm{d}v_{pq}}{\mathrm{d}t}$ is the radial acceleration of P along OR, which will be denoted by f, and $\frac{\mathrm{d}\omega}{\mathrm{d}t}$ is the angular acceleration of OR, which will be denoted by α.

$$\therefore\ \mathrm{d}^2y/\mathrm{d}t^2 = f\sin\theta + 2\omega v_{pq}\cos\theta + \alpha r\cos\theta - \omega^2 r\sin\theta$$

$$= (f - \omega^2 r)\sin\theta + (2\omega v_{pq} + \alpha r)\cos\theta \quad . \quad (3.21)$$

Equations (3.20) and (3.21) apply, whatever the directions chosen for the co-ordinate axes, so that if we substitute $\theta = \pi/2$ in (3.20), we get:

$$\left.\begin{array}{l}\text{the component of the velocity}\\ \quad\text{of P parallel to OP}\end{array}\right\} = v_{pq} \quad . \quad . \quad . \quad (3.22)$$

and, substituting in (3.19), we get:

$$\left.\begin{array}{l}\text{the component of the accelera-}\\ \quad\text{tion of P parallel to OP}\end{array}\right\} = f - \omega^2 r \quad . \quad . \quad (3.23)$$

Similarly, substituting $\theta = 0$ in (3.18), we get:

$$\left.\begin{array}{l}\text{the component of the velocity}\\ \quad\text{of P at right angles to OP}\end{array}\right\} = \omega r = v_q \quad . \quad (3.24)$$

where v_q is the velocity of the coincident point Q.

Also, substituting $\theta = 0$ in (3.21), we get:

$$\left.\begin{array}{l}\text{the component of the accelera-}\\ \quad\text{tion of P at right angles to OP}\end{array}\right\} = 2\omega v_{pq} + \alpha r \quad . \quad (3.25)$$

But the acceleration of Q relative to O consists of two components, namely:

(a) the centripetal component $f^c_q = \omega^2 r$

and (b) the tangential component $f^t_q = \alpha r$

Hence it follows that the acceleration of P relative to the coincident point Q also consists of two components, namely:

(c) the component parallel to OP, f,

and (d) the component perpendicular to OP, $2\omega v_{pq}$

In connection with these four components it should be emphasised that:

(i) Components (a) and (d) depend only upon the instantaneous values of the velocities, so that they may be completely determined from information provided by the velocity diagram.

(ii) The counter-clockwise sense has been taken as positive for ω and α and the radially outward direction as positive for v_{pq} and f.

(iii) The sense of the Coriolis component acceleration (d) will be changed if the sign of either ω or v_{pq} is changed, but its sense will be unchanged, if the signs of both ω and v_{pq} are changed. It is convenient to note that this conforms to the following simple rule: the direction of the component acceleration (d) is such as to rotate the vector v_{pq} about its origin in the same sense as that of the angular velocity ω.

In Fig. 64 the complete acceleration diagram is drawn. The four components (a), (b), (c) and (d) are respectively represented by the four vectors os, sq, qt and tp, so that oq, the vector sum of os and sq, represents the acceleration of Q, and qp, the vector sum of qt and tp, represents the acceleration of P relative to Q. Hence op, the vector sum of oq and qp, represents the acceleration of P.

The original diagram was drawn to scale for the following example:

Example 6. The distance OP is 2 ft; ω is 2 rad/s counter-clockwise; α is 5 rad/s^2 counter-clockwise; v_{pq} is 3 ft/s radially outward; and f is 4 ft/s^2 radially outward. Find the acceleration of P relative to the fixed point O.

Component (a) $= \text{os} = \omega^2 r = 2^2.2 = 8$ ft/s^2
Component (b) $= \text{sq} = \alpha r = 5.2 = 10$ ft/s^2
Component (c) $= \text{qt} = f = 4$ ft/s^2
Component (d) $= \text{tp} = 2\omega v_{pq} = 2.2.3 = 12$ ft/s^2

The acceleration of Q is the vector sum of components (a) and (b). It is represented by oq and is equal to 12·8 ft/s^2.

The acceleration of P relative to Q is the vector sum of components (c) and (d). It is represented by qp and is equal to 12·65 ft/s^2.

The acceleration of P is the vector sum of the acceleration of P relative to Q and of Q relative to O. It is represented by op and is equal to 22·4 ft/s^2.

54. Examples on the Construction of the Acceleration Diagrams for Mechanisms that are Inversions of the Slider-crank Chain. In order to draw the acceleration diagrams for mechanisms such as the crank and slotted-lever quick-return mechanism, the Whitworth quick-return mechanism, the rotary-engine mechanism and the oscillating-cylinder-engine mechanism, the principles outlined in the preceding article must be used. The method will be made clear if one or two typical examples are considered

(a) *The Crank and Slotted-lever Quick-return Mechanism*: Fig. 65. In this mechanism the driving crank CP revolves with uniform angular velocity about the fixed centre C. A die-block attached to the crankpin P slides along the slotted link ON, and thus causes ON to oscillate about the fixed centre O. A short link NR transmits the motion from ON to the ram which carries the tool-box and which reciprocates along the line R_1R_2. The line of stroke R_1R_2 of the ram is perpendicular to the line of centres OC. The problem is to find the acceleration of the ram.

To find the acceleration of R it is first necessary to find the acceleration of the pin N on the slotted link. The acceleration of R may then be determined by applying the method already given in Article 46.

Let ω_1 = angular velocity of CP (constant); let ω = angular velocity of ON (variable).

Then the peripheral velocity of the crankpin P = $v_p = \omega_1 . CP$.

Velocity Diagram Fig. 65 (a). The velocity v_p of the crankpin is perpendicular to CP and is represented to scale by op. This may be resolved into components oq and qp perpendicular and parallel to the slotted link PO. The component oq is the velocity of the coincident point Q and qp is the velocity of sliding of the block P along ON.

The velocity of N is perpendicular to ON and is represented by on where on = ON/OP.oq.

The velocity of R is horizontal and the velocity of R relative to N is perpendicular to NR, so that the velocity diagram is completed by drawing a line through n perpendicular to NR to intersect the horizontal line through o at r. Then or represents to scale the velocity of the ram R.

Acceleration Diagram Fig. 65 (b). The first step is to find the acceleration of the coincident point Q.

Since $f_p = f_q \nrightarrow f_{pq}$ and f_q and f_{pq} both consist of two components as explained in Article 53, we have

$$f_p = f^c{}_q \nrightarrow f^t{}_q \nrightarrow f^s{}_{pq} \nrightarrow f^{Cor}{}_{pq}$$

Of these five vectors, three are known in magnitude and direction and two are known in direction only.

Thus:

$f_p = \omega_1{}^2.PC$, $f^c{}_q = \omega^2.QO = v_q{}^2/QO$ and $f^{Cor}{}_{pq} = 2\omega v_{pq}$

f_p is parallel to PC, $f^c{}_q$ is parallel to QO and $f^{Cor}{}_{pq}$ is perpendicular to QO. The Coriolis component acts towards the right so as to rotate v_{pq} about its origin in the same sense as ω.

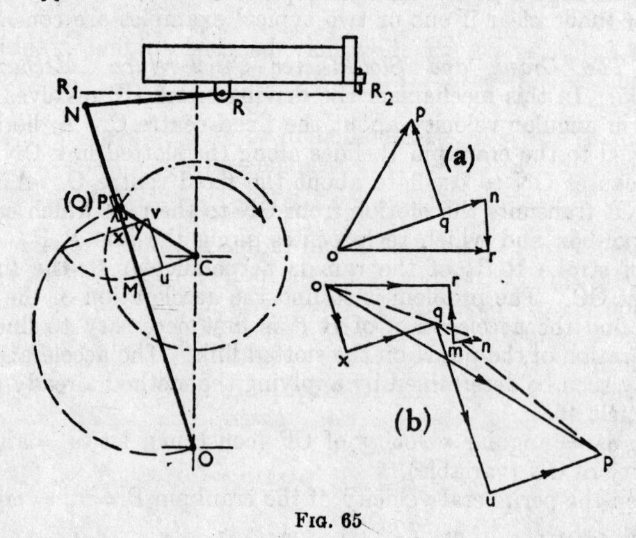

Fig. 65

Set off op parallel to PC to represent to scale the acceleration $f_p = \omega_1{}^2.PC$.

From o set off ox parallel to PO to represent $f^c{}_q = \omega^2.QO = v^2{}_q/QO$.

The Coriolis component of $f_{pq} = 2\omega v_{pq}$. It acts perpendicular to QO towards the right and must finish at p. It is represented to scale on the acceleration diagram by up.

The other two vectors $f^t{}_q$ and $f^s{}_{pq}$ are known in direction, the former perpendicular to QO and the latter parallel to QO, so that the vector diagram may be completed by drawing xq perpendicular to QO and qu parallel to QO.

The total acceleration f_q of the coincident point Q is then given to scale by oq, the vector sum of ox and xq, and the total acceleration f_{pq} of P relative to Q by qp, the vector sum of qu and up.

Since N and Q are fixed points on the link ON, the acceleration of N will be parallel to f_q and of magnitude $f_n = (ON/OQ)f_q$. It is therefore given to scale on the acceleration diagram by on, where on/oq = ON/OQ.

The acceleration of the ram R is the vector sum of the acceleration of N and the acceleration of R relative to N. Since R has horizontal straight-line motion, its acceleration is also horizontal, while the acceleration of R relative to N is partly centripetal, parallel to RN, and partly tangential, perpendicular to RN. The centripetal component $= v_{rn}^2/RN$, and is represented by nm. A line drawn through m perpendicular to NR intersects the horizontal through o at r. Then or represents to scale the acceleration of R.

The following construction may be used to find graphically without any calculation a polygon PXVUC similar to the acceleration diagram oxqup.

From C draw CM perpendicular to ON. Then the triangle PCM is similar to the velocity triangle opq Fig. 65 (a).

$$\therefore\ v_p = \omega_1 . PC, \quad v_q = \omega_1 . PM, \quad v_{pq} = \omega_1 . MC$$

Draw a circle with centre P and radius PM and a second circle on PO as diameter. Let X be the point of intersection of the common chord of these two circles with PO, and Y its point of intersection with PC. Along MC set off UC = 2XY, and draw UV parallel to OP to intersect XY, produced if necessary, at V.

Then the polygons PXVUC, oxqup are similar.

Proof. By the construction given $PX = PM^2/PO$. But $v_q = \omega_1 . PM$, so that

$$f^c_q = v_q^2/PO = \omega_1^2 . PM^2/PO = \omega_1^2 . PX$$

Also the angular velocity of OP

$$= \omega = v_q/OP = \omega_1 . PM/OP$$

so that the Coriolis component of the acceleration of P relative to Q

$$= 2\omega v_{pq} = 2\omega_1(PM/PO) . \omega_1 MC = 2\omega_1^2 . (PM . MC)/PO$$

From the similar triangles PXY, PMC,

$$XY = PX . MC/PM$$

and, substituting $PX = PM^2/PO$,

$$XY = (PM^2/PO)MC/PM = (PM . MC)/PO$$

$\therefore$ the Coriolis acceleration $= 2\omega_1^2 . XY = \omega_1^2 . 2XY = \omega_1^2 . UC$, since by construction UC = 2XY.

Hence in the polygon PXVUC, we have

$$f_p = \omega_1^2 PC, \quad f^c_q = \omega_1^2 PX, \quad f^{Cor}_{pq} = \omega_1^2 UC$$

and the other two component accelerations f^t_q and f^s_{pq} must be given to the same scale by XV and VU.

Note that with this construction the Coriolis component is always in the sense from X to Y.

Fig. 65 is drawn to scale for the following example.

Example 7. The distance between the fixed centres O and C of the quick-return mechanism, Fig. 65, is 8 in., the length of the driving crank CP is 4 in. and it makes 60 r.p.m. The length of the slotted link ON is 15 in. and that of NR is 6 in. The angle OCP is 120°. Draw the velocity and acceleration diagrams and find the velocity and acceleration of the ram R.

The values of the velocities and accelerations involved in the construction of Fig. 65 (a) and (b) are given below.

Velocity Diagram Fig. 65 (a)

	Vector	ft/s	
v_p	op	2·09	$\omega_1 = 2\pi$ rad/s
v_q	oq	1·56	
v_{pq}	qp	1·37	OP = OQ and scales 10·6 in.
v_n	on	2·21	$\therefore\ \omega = v_q/OQ = 1{\cdot}77$ rad/s
v_{rn}	nr	0·74	$\omega_{rn} = v_{rn}/RN = 1{\cdot}48$ rad/s
v_r	or	2·14	

Acceleration Diagram. Fig. 65 (b)

		Vector	ft/s²
f_p	$\omega_1^2.PC$	op	13·2
f^c_q	$\omega^2.PO = v_q^2/PO$	ox	2·77
f^t_q	$\alpha.PO$	xq	3·74
f_q		oq	4·65
f^s_{pq}		qu	7·0
f^{cor}_{pq}	$2\omega v_{pq}$	up	4·87
f_{pq}		qp	8·53
f_n		on	6·59
f^c_{rn}	v_{nr}^2/RN	nm	1·09
f^t_{rn}	$\alpha_1.RN$	mr	2·03
f_r		or	5·10

$$\alpha = \text{angular acceleration of ON} = f^t_q/PO = (3{\cdot}74.12)/10{\cdot}6 = 4{\cdot}25 \text{ rad/s}^2$$

$$\alpha_1 = \text{angular acceleration of RN} = f^t_{rn}/RN = (2{\cdot}03.12)/6 = 4{\cdot}06 \text{ rad/s}^2$$

(b) *The Rotary Engine Mechanism*: Fig. 66. In this mechanism the crank OC is fixed. The cylinder axis ON turns about the fixed centre O with uniform angular velocity ω, while the connecting rod CP turns about the fixed centre C with variable angular velocity ω_1. If, as in the earlier examples, Q is the point on ON that is coincident with P, then the velocity of Q $= v_q = \omega . OQ = \omega . OP$. The velocity of P is the vector sum of the velocity of Q and the velocity of P relative to Q.

But the velocity of P is perpendicular to CP and the velocity of P relative to Q is along OP, hence the triangle Pab is the triangle of velocities. If OM is drawn perpendicular to PO to meet PC produced at M, the triangles Pab, POM are similar, and therefore

$$v_q = \omega . PO, \quad v_p = \omega . PM \quad \text{and} \quad v_{pq} = \omega . OM$$

The acceleration of P is the vector sum of the acceleration of Q and the acceleration of P relative to Q.

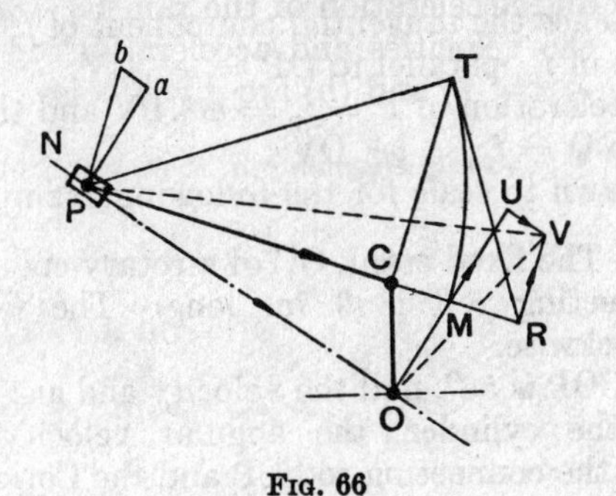

FIG. 66

Since the angular velocity of ON is constant, the acceleration of the coincident point Q is wholly centripetal, so that $f_q = \omega^2 . QO = \omega^2 . PO$. This is the component (a) and there is no component (b).

The acceleration of P is, however, partly centripetal and partly tangential, since the angular velocity of CP varies from instant to instant.

The centripetal component of the acceleration of P $= \omega_1{}^2 . PC = v_p{}^2/PC = \omega^2 . PM^2/PC = \omega^2 . PR$, where R is found by drawing CT perpendicular to PC, marking off PT = PM and drawing TR perpendicular to PT. Then, from the similar triangles PCT, PTR:

$$PR/PT = PT/PC \quad \text{and} \quad PR = PT^2/PC = PM^2/PC$$

The tangential component of the acceleration of P is at present unknown in magnitude, but its direction must, of course, be at right angles to PC. Draw through R a line perpendicular to PR to indicate the direction of this tangential component.

The acceleration of P relative to Q consists of the two components (c) and (d) of Article 53.

The component (d) of the acceleration of P relative to Q is the Coriolis component and its magnitude is given by:

$$2\omega v_{pq} = 2\omega^2 . OM = \omega^2 . OU$$

where OU = 2OM.

The direction of component (d) is such as to rotate v_{pq} about its origin in the same sense as ω and is therefore in the direction OU.

The component (c) of the acceleration of P relative to Q is parallel to OP, but is unknown in magnitude. But if UV is drawn parallel to PO to meet at V the line previously drawn through R at right angles to PC, then the acceleration diagram will be completed. Hence the tangential component of $f_p = \omega^2 . RV$ and the component of f_{pq} parallel to OP $= \omega^2 . UV$.

Then the acceleration of P $= f_p = \omega^2 . PV$ and the acceleration of P relative to Q $= f_{pq} = \omega^2 . OV$.

Fig. 66 is drawn to scale for the following example.

Example 8. The fixed crank OC of a rotary engine is 3 in. long and the connecting rod is 9 in. long. The cylinders make 1200 r.p.m. clockwise.

If the angle COP is 55°, find the velocity and acceleration of the piston along the cylinder, the angular velocity and angular acceleration of the connecting rod CP and the Coriolis component of the acceleration of P.

The angular velocity of the cylinder centre line OP

$$= \omega = (\pi . 1200)/30 = 40\pi = 125{\cdot}7 \text{ rad/s}$$

The velocity of Q

$$= v_q = \omega . OQ = \omega . OP = 40\pi . 10{\cdot}35/12 = 108{\cdot}2 \text{ ft/s}$$

The velocity of P

$$= v_p = \omega . PM = 40\pi . 10{\cdot}74/12 = 112{\cdot}4 \text{ ft/s}$$

The angular velocity of CP

$$= \omega_1 = v_p/CP = 112{\cdot}4 . 12/9 = 150 \text{ rad/s}$$

The velocity of sliding of the piston along the cylinder

$$= v_{pq} = \omega . OM = 40\pi . 2{\cdot}95/12 = 30{\cdot}9 \text{ ft/s}$$

The centripetal acceleration of Q

$$= f_q = \omega^2 . OQ = \omega^2 . OP = (40\pi)^2 . 10{\cdot}35/12 = 13\,600 \text{ ft/s}^2$$

The centripetal component of the acceleration of P

$$= \omega_1^2.\text{PC} = \omega^2.\text{PR} = (40\pi)^2.12{\cdot}84/12 = 16\,900 \text{ ft/s}^2$$

The tangential component of the acceleration of P

$$= \alpha.\text{PC} = \omega^2.\text{RV} = (40\pi)^2.2{\cdot}49/12 = 3280 \text{ ft/s}^2$$

so that the angular acceleration of the connecting rod CP

$$= \alpha = 3280/\text{CP} = 3280.12/9 = 4370 \text{ rad/s}^2$$

The component (c) of the acceleration of P relative to Q

$$= f = \omega^2.\text{UV} = (40\pi)^2.1{\cdot}29/12 = 1700 \text{ ft/s}^2$$

This component gives the acceleration of the piston along the cylinder.

The component (d) of the acceleration of P relative to Q

$$= 2\omega v_{pq} = \omega^2.\text{OU} = (40\pi)^2.5{\cdot}90/12 = 7760 \text{ ft/s}^2$$

The resultant acceleration of P

$$= \omega^2.\text{PV} = (40\pi)^2.13{\cdot}05/12 = 17\,170 \text{ ft/s}^2$$

and the acceleration of P relative to Q

$$= \omega^2.\text{OV} = (40\pi)^2.6{\cdot}06/12 = 7970 \text{ ft/s}^2$$

The directions of the various accelerations are indicated by the arrows on Fig. 66.

It may be pointed out that the existence of the Coriolis component (d) of the acceleration of P relative to Q will result in a considerable side thrust between the piston and the cylinder wall due to the inertia of the piston. In addition there will be a side thrust due to the inertia of the connecting rod and, of course, a side thrust due to the gas pressure and the inclination of the connecting rod to the line of stroke.

EXAMPLES III

1. The following table gives the displacement of the valve in a Joy valve gear, Fig. 127, for different crank angles:

θ . .	22·5°	45°	67·5°	90°	112·5°	135°	157·5°	180°	202·5°	225°	247·5°	270°	292·5°	315°	337·5°	360°
x, in. .	5·42	6·16	6·60	6·46	5·64	4·42	2·90	1·60	0·78	0·50	0·72	1·22	1·92	2·76	3·66	4·54

Plot the displacement curve on a base of crank angle. If the crank turns at a uniform speed of 150 r.p.m., draw the corresponding speed and acceleration curves, also on a base of crank angle.

2. The variation of the lift of a petrol-engine valve with the angle of rotation of the camshaft is shown in the following table:

θ . . .	0°	5°	10°	15°	20°	25°	30°	35°	40°	45°	50°	55°
x, in. . .	0	0·017	0·068	0·149	0·229	0·299	0·360	0·410	0·449	0·477	0·494	0·500

Plot the displacement curve on a base of θ. If the speed of rotation of the camshaft is 900 r.p.m., draw the speed and acceleration diagrams on a base of θ.

3. The speed (v), ft/s, of the ram of a small shaping machine on both the cutting and return strokes is given in the following table for points at different distances (x) from the beginning of each stroke:

x, in. . .	0·5	1·0	1·5	2·0	2·5	3·0	3·5	4·0	4·5	5·0	5·5	6·0
v (cutting stroke) .	0·500	0·642	0·723	0·775	0·810	0·830	0·830	0·810	0·766	0·682	0·532	0
v (return stroke) .	0·854	1·285	1·581	1·780	1·900	1·935	1·868	1·720	1·505	1·215	0·812	0

Plot the curves of v against x and find the acceleration of the ram when at a distance of 2 in. from the beginning and the end of each stroke.

4. A vehicle starts from rest and its speed varies with the time as shown in the following table:

t, min.	0·5	1·0	1·5	2·0	2·5	3·0	3·5	4·0	4·5	5·0
v, m.p.h. . . .	12·0	22·5	31·0	38·0	42·5	45·0	44·0	41·0	39·5	40·5
t, min.	5·5	6·0	6·5	7·0	7·5	8·0	8·5	9·0	9·5	10·0
v, m.p.h. . . .	45·0	52·0	57·5	60·0	58·5	54·0	45·0	33·0	17·5	0

Plot curves to show the variation of acceleration with time and of displacement with time.

5. Particulars of the full-throttle power curve of a petrol engine are given in the following table:

N	700	1000	1500	2000	2500	3000	3500	4000	4500
B.h.p.. . .	6·9	11·0	17·7	24·2	29·8	35·0	39·2	41·8	40·0

The engine is fitted to a car which, in the top gear of 5·13 to 1, has a deadweight, allowing for rotary inertia, of 22 cwt. The effective diameter of the road wheels is 26·5 in., the efficiency of the transmission is 90% and the resistance to motion is given by R (lb) $= 33+0{\cdot}035v^2$, where v is the speed in m.p.h.

Plot a curve to show the variation of acceleration with road speed in top gear and find the minimum time required in order to increase the speed from 20 to 60 m.p.h. and the distance through which the car moves in this time.

6. In a certain trial, a ship of 1150 tons diplacement was towed at a uniform speed of 20ft/s. The towing rope was slipped and the retardation obtained at various speeds as the ship came to rest. Corresponding retardation and speed figures are given in the table:

Speed, ft/s	20	19	18	17	16	15	14	13	12	11	10	9	8	7
Retardation, ft/s²	0·127	0·125	0·120	0·112	0·098	0·082	0·068	0·056	0·046	0·037	0·030	0·024	0·020	0·017

Construct the speed-time and the displacement-time curves for the motion and determine the time that elapsed and the distance covered during the decrease of speed from 20 ft/s to 10 ft/s. W.S.S.

7. A motor-car weighs 25 cwt and the engine develops 40 b.h.p. at 3200 r.p.m. The efficiency of the transmission is 90% in the top gear of 5 to 1 and 80% in the second gear of 9 to 1. When the engine speed is 3200 r.p.m., (a) the car reaches its maximum speed of 60 m.p.h. in top gear on a level road and (b) the car is just capable of climbing a gradient of 1 in 11 in second gear. If the resistance to motion in lb is given by $R = a + bv^2$, where v is the speed in m.p.h., find the values of the constants a and b. M.U.

8. In a four-bar chain ABCD, AB is the driving link, CD the driven link and AD the fixed link. Show that the angular velocity of CD is to that of AB as QA is to QD, where Q is the point of intersection of BC and AD, produced if necessary.

When the links AB, BC, CD and DA are respectively 2·5, 7, 4·5 and 8 in. long, the angle BAD is 60°, AB and DC are on opposite sides of AD and the velocity of B is 10 ft/s., find the velocity of C and the angular velocity of BC.

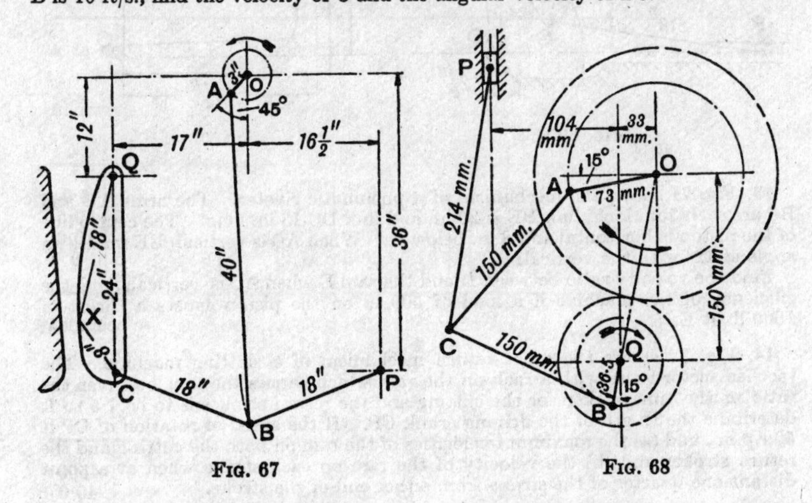

FIG. 67 FIG. 68

9. The dimensions of the mechanism of a stone-crusher, Fig. 67, are as follows. The horizontal distances of P and Q from O are respectively 16·5 in. and 17 in.; the vertical distances of P and Q from O are respectively 36 in. and 12 in. The lengths of the links are OA 3 in., AB 40 in., BP 18 in., BC 18 in. and CQ 24 in. The point X on the jaw is 8 in. from C and 18 in. from Q. If OA turns at a uniform speed of 60 r.p.m. and is inclined at 45° to the horizontal as shown, find: (a) the velocity of point X, (b) the torque required on OA to overcome a horizontal force of 3 tons at X.

10. The dimensions of the Andreau differential-stroke-engine mechanism, Fig. 68, are: OA = 73 mm, QB = 36·5 mm, AC = BC = 150 mm, CP = 214 mm. OA and QB are geared together so that QB turns at twice the speed of OA and in the opposite sense to OA. Find the velocity of the piston P for the given configuration, when OA makes 700 r.p.m.

11. The dimensions of the Atkinson-cycle-engine mechanism, Fig. 69, are: OA = 6 in., QB = 8 in., AB = 15 in., AC = 16 in., BC = 2·5 in., CP = 18 in. If the crank OA makes 150 r.p.m., find for the given configuration the velocity of the piston P and the angular velocities of the links ABC and CP.

12. In Fig. 70 OA is a uniformly rotating crank, ABC is a continuous link, DE is a slotted lever and D, O and Q are fixed centres. The die-block attached to the pin C slides along the slot in the link DE. If the speed of rotation of OA is 120 r.p.m., find for the given position the angular velocity of DE and the velocity of sliding of the die-block along the slot.

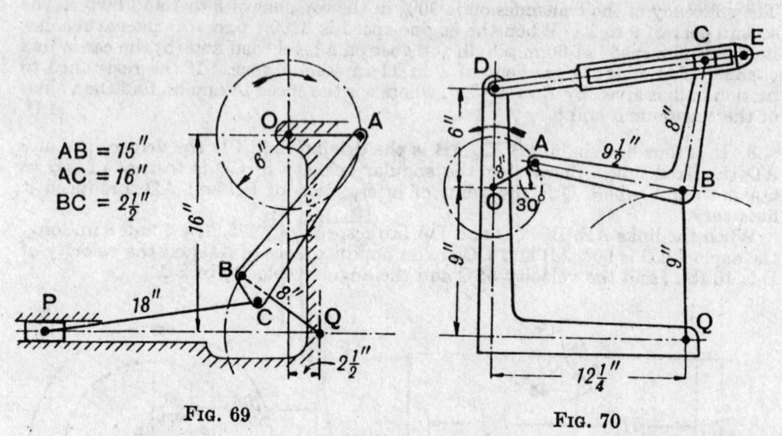

FIG. 69 FIG. 70

13. Fig. 71 shows the mechanism of a pneumatic riveter. The arms AB and BC are each 7 in. long, link BE is 20 in. and link DC 13 in. long. The centre line of the piston is horizontal and 8 in. below A. When AC is vertical, BE makes an angle of 12° with the vertical.

Find the velocity ratio between D and the ram E when AC is vertical, and the efficiency of the machine if a load of 500 lb on the piston causes a thrust of 1000 lb at E.

L.U.A.

14. Fig. 72 shows the quick-return mechanism of a slotting machine. The toothed sector gears with a rack on the ram which carries the tool box. If the ratio of the times taken for the cutting and the return strokes is to be 1·5 to 1, determine the length of the driving crank CP. If the speed of rotation of CP is 50 r.p.m., find (a) the maximum velocities of the ram on both the cutting and the return strokes and (b) the velocity of the ram on each stroke when at a point distant one-quarter of the stroke from either end of the stroke.

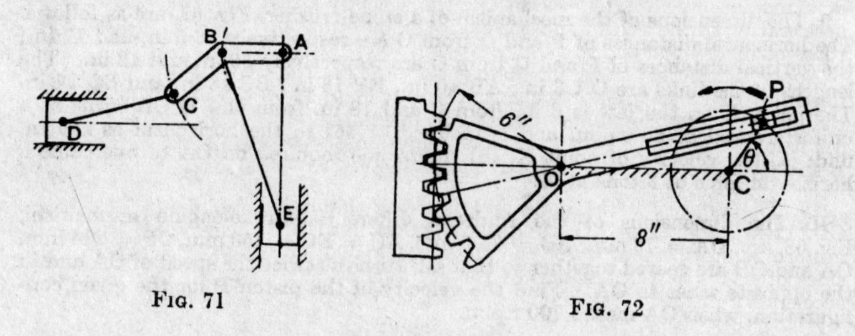

FIG. 71 FIG. 72

15. In a crank and slotted-lever quick-return motion the distance between the fixed centres O and C is 8 in. The driving crank CP is 3 in. long and makes 90 r.p.m. The pin Q on the slotted lever, 14½ in. from the fulcrum O, is connected by a link QR, 4 in. long, to the pin R on the ram. The line of stroke of R

is perpendicular to OC and intersects OC produced at a point 6 in. from C. Find:

(a) the ratio of the times taken on the two strokes of R;
(b) the maximum velocities of R on the cutting and the return strokes;
(c) the velocities of R when 3 in. from the beginning and the end of each stroke.

16. Mitchell's luffing gear is shown in Fig. 73. The lower end, B, of the jib

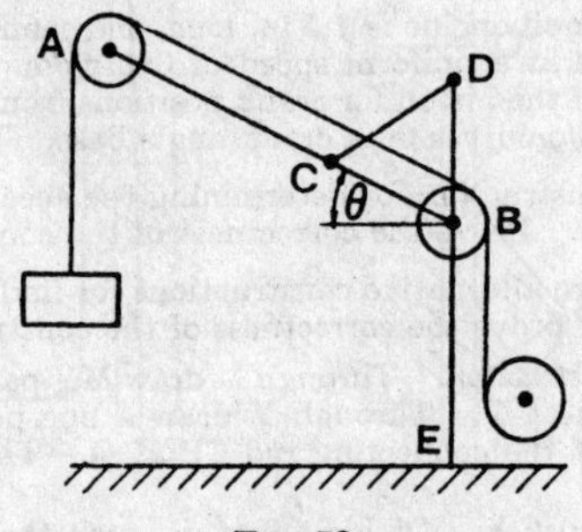

Fig. 73

AB is pivoted to a carriage, which is moved up and down the vertical mast DE to luff the load horizontally. Find the relation which must exist between the length of the jib and that of the link CD and find the ratio between the lengths AB and CB.

Determine also the ratio between the speed of the carriage and that of the load, when luffing, for a given angle of inclination of the jib.

Show how the mechanism may be balanced, approximately, so as to reduce the bending stresses in the mast. W.S.S

17. AB, DC, AC and BD are four separate links pin-jointed as shown in Fig. 74. AB and DC are 2 ft long, BD and AC are 2·5 ft long. If BD remains fixed and A is made to approach D, find the velocity ratio of A and C at the instant when AB becomes perpendicular to DC. W.S.

18. When a link is transmitting motion from one part of a machine to another, show how the velocity and acceleration of the driven end can be found in terms of those of the driving end.

Apply the method to a four-bar chain. AD is the fixed link, 3·25 ft long; AB the driver, 1 ft long; CD the driven, 2 ft long; and BC the connecting link, 3·5 ft long. Find the velocity and acceleration of the point X midway between B and C when the angle BAD is 135° and the velocity of B is 6 ft/s.

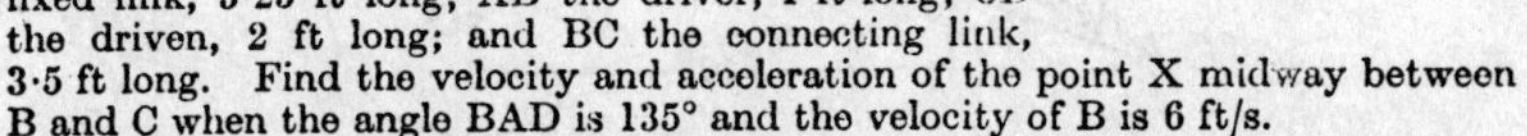

Fig. 74

19. The lengths of the links of a four-bar chain are: AB, 6 in.; BC, 18 in.; CD, 12 in.; and DA, 21 in. The link AD is fixed and the link AB turns with uniform angular velocity. When angle BAD is 90° and B and C are on opposite sides of AD, find the position of the point E on BC which, at that instant, is accelerated along BC.

20. ABCD is a four-bar chain with the link AD fixed. The lengths of the links are: AB, 2·5 in.; BC, 7 in.; CD, 4·5 in.; and DA 8 in. The crank AB makes 180 r.p.m. Find the acceleration of C and the angular accelerations of BC and CD when (a) angle BAD is 15° and B and C lie on opposite sides of AD, (b) angle BAD is 60° and B and C lie on the same side of AD.

21. Explain what is meant by (a) the instantaneous centre of a link, (b) the acceleration centre of a link. How may the positions of these two centres be found?

AB is a link of a mechanism 4 ft long. The acceleration of A is 10 ft/s² along AC, where angle BAC = 60°. The acceleration of B is 30 ft/s² along BD, where angle ABD is 70° and the velocity of B is 10 ft/s along BE, where angle ABE is 40°. C and D lie on the same side of AB and E on the opposite side. Find the velocity of A, the angular velocity and the angular acceleration of AB, and the positions of the instantaneous centre and the acceleration centre of AB.

22. The crank of an oil engine is 7·5 in. long, the connecting rod is 33 in. long and the crank rotates at a uniform speed of 310 r.pm. Calculate the velocity and the acceleration of the piston for crank positions from 0° to 180°, at intervals of 30°, and plot the two curves on a crank angle base.

23. Give Klein's construction for determining the acceleration of the piston of a reciprocating engine. Prove the correctness of the construction.

24. The following are alternative constructions for finding the points Q and N, Fig. 59. In each case prove the correctness of the construction.

(a) *Ritterhaus's Construction.* Through M draw MY parallel to OP to meet OC, produced if necessary, at Y. Through Y draw a line perpendicular to the line of stroke of P to meet the connecting rod CP at Q. Through Q draw QN perpendicular to CP.

(b) *Bennett's Construction.* Divide CP at R, such that $CR = CO^2/CP$. This may be done graphically by taking OC at right angles to the line of stroke of P and dropping the perpendicular OR on to the corresponding position of CP. For any other crank position, draw RS perpendicular to CP to meet the line of stroke at S, draw SQ perpendicular to the line of stroke to meet CP at Q and draw QN perpendicular to CP.

25. The crank of a reciprocating engine is 9 in. long, the connecting rod is 36 in. long and the r.p.m. are 150. Find the velocity and acceleration of the piston and the angular velocity and angular acceleration of the connecting rod when the angle which the crank makes with the i.d.c. is (a) 30°, (b) 120°.

26. A petrol engine has a stroke of 5 in. a connecting rod 10 in. long and runs at 2000 r.p.m. The crankshaft is offset ¾ in. from the cylinder centre line. Determine the velocity and acceleration of the piston when at one-quarter of the stroke from the crank end on both strokes. M.U.

27. In the mechanism shown in Fig. 75, the crank OC is 3 in. long and makes 240 r.p.m. The link CP is 10 in. long, Q lies on an extension of CP, 2½ in. from

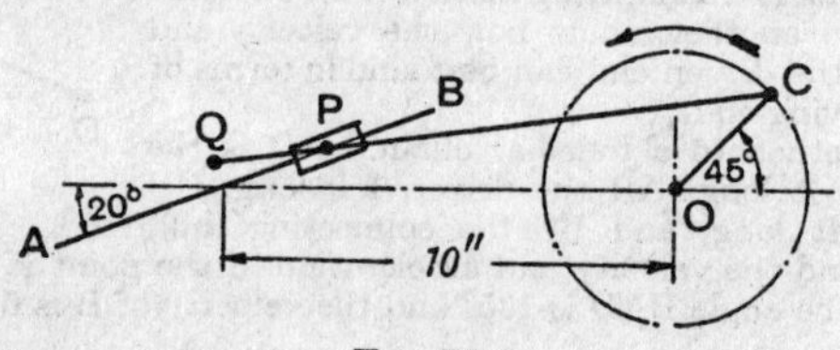

FIG. 75

P, and the pin P is attached to a block which slides along AB. Find the velocity and acceleration of Q and the angular velocity and angular acceleration of CQ for the position shown and also when the crank has turned through an angle of 225° from the horizontal.

28. A petrol engine has a crank 1·75 in. long and a connecting rod 7 in. long. At the instant when the crank makes an angle of 60° with the i.d.c., its angular velocity is 40 rad/s and its angular acceleration is 400 rad/s². Find the acceleration of the piston and the angular acceleration of the connecting rod.

29. A gas engine has a stroke of 17 in. and a connecting rod 40 in. long. The crankshaft carries two flywheels each of which weighs 1200 lb and has a radius of gyration of 27 in. When the crank makes an angle of 30° with the i.d.c. on the firing stroke, there is an unbalanced turning moment on the crankshaft of 5400 lb ft in the direction of motion of the crank. If at this instant the angular velocity of the crank is 8·38 rad/s (80 r.p.m.), find (a) the acceleration of the piston and (b) the angular acceleration of the connecting rod.

30. In the mechanism shown in Fig. 76, O and Q are fixed centres. If the crank OC revolves at a uniform speed of 120 r.p.m., find the angular accelerations of the links CP, PA and AQ.

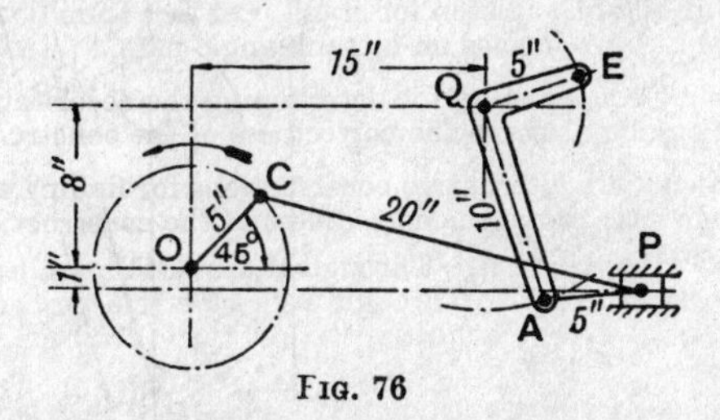

Fig. 76

31. For the Atkinson-cycle-engine mechanism, Question 11, Fig. 69, find the angular acceleration of the link ABC and the linear acceleration of the piston P.

32. An oscillating-cylinder engine has a stroke of 2 ft, the distance between the trunnion and crankshaft centre lines is 4 ft and the crank turns at a uniform speed of 90 r.p.m. Find the acceleration of the piston along the cylinder and the angular acceleration of the cylinder for the two positions of the mechanism in which the crank is inclined at (a) 45° and (b) 135° to the inner dead centre. M.U.

33. Referring to the quick-return motion of Question 14, Fig. 72, find the acceleration of the ram on each stroke when at a point distant one-quarter of the stroke from one end of the stroke.

34. The mechanism of a Whitworth quick-return motion is shown in Fig. 77. The distance between the fixed centres O and C is 1·5 in.; the driving crank CP is 5 in. long, the slotted link OQ is 4 in. long and the connecting link QR is 15 in.

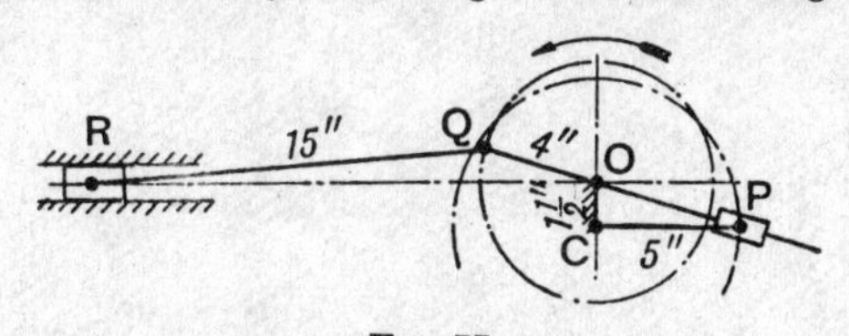

Fig. 77

long. The pin R is attached to the ram which carries the tool box and reciprocates along a line which passes through O and is perpendicular to OC. If CP makes 60 r.p.m., find for the given position the acceleration of R. What is the acceleration of R when it occupies the same position but is on the return stroke?

35. The driving crank AB of the quick-return mechanism shown in Fig. 78 revolves at a uniform speed of 200 r.p.m. Find the velocity and acceleration of the tool box R in the position shown, when the crank makes an angle of 60° with the vertical line of centres PA. What is the acceleration of sliding of the block at B along the slotted lever PQ? L.U.

36. The cylinders of an aeroplane engine are arranged radially round the crankshaft and they rotate at uniform speed round the fixed crank. If the stroke of each piston is 5 in., the length of each connecting rod is 9 in. and the speed is 1000 r.p.m., find the components of the acceleration of the piston, parallel to and at right angles to the cylinder centre line and the angular acceleration of the connecting rod for a cylinder which has turned through 45° from the i.d.c. position.

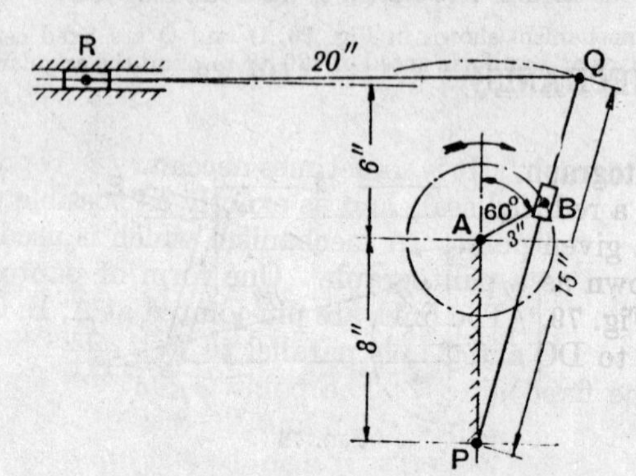

FIG. 78

37. In a rotary aero engine in which the cylinders rotate about a fixed crank, there is a side thrust, due to inertia, between each piston and its cylinder.

If the stroke of such an engine is 6 in., the effective weight of each piston is 1·5 lb and the length of each connecting rod is 12 in., determine the magnitude of this side thrust at 1200 r.p.m. in a cylinder which has turned through (a) 0° and (b) 90° from the i.d.c. L.U.A.

38. Two slotted links APC, BPD are pivoted on centres A, B, 4 ft apart. The links cross and carry a common block P. Determine the velocity and the acceleration of the block at the instant when it is 30 in. from A and 24 in. from B, the angular velocity of AC being 2 rad/s and that of BD being 1·5 rad/s, both links moving outward from AB. L.U.A.

CHAPTER IV

MECHANISMS WITH LOWER PAIRS

55. The Pantograph. It is sometimes necessary to reproduce to an enlarged or a reduced scale and as exactly as possible the path described by a given point. A mechanism which is used for this purpose is known as a pantograph. One form of pantograph is illustrated in Fig. 79. The links are pin-jointed at A, B, C and D. AB is parallel to DC and AD is parallel to BC. The link BA is extended to the fixed pin O. The point Q on the link AD and the point P on the extension of the link BC both lie on a straight line which passes through O. It can be shown that in these circumstances the point P will reproduce the motion of the point Q to an enlarged scale, or, alternatively, the point Q will reproduce the motion of the point P to a reduced scale.

When the mechanism is in the position shown by full lines the triangles OAQ, OBP are similar, so that $OQ/OP = OA/OB$.

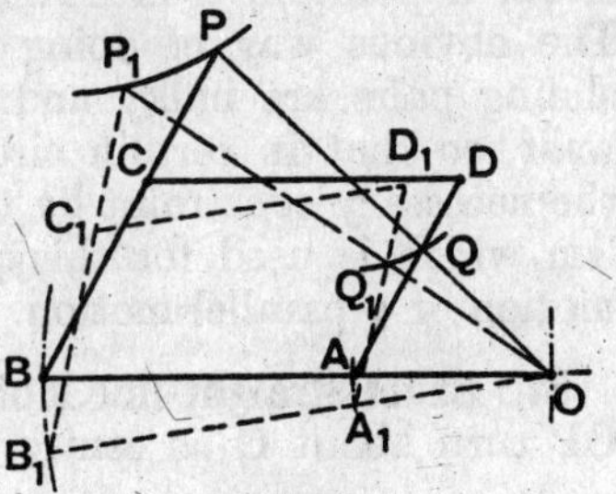

Let P move to P_1 along the path shown, so that the mechanism occupies the position shown by dotted lines. Then the triangles OA_1Q_1, OB_1P_1 are similar and therefore $OQ_1/OP_1 = OA_1/OB_1$. But $OB_1 = OB$ and $OA_1 = OA$, so that $OQ_1/OP_1 = OQ/OP = OA/OB$.

Fig. 79

Hence the triangles OPP_1, OQQ_1 are similar and PP_1 and QQ_1 are parallel.

The displacement of Q is therefore parallel to the displacement of P and is smaller than that of P in the proportion OA: OB. This is true however small, or however large, the displacement of P may be within the limits allowed by the mechanism. Hence P and Q must trace out similar paths.

The pantograph is sometimes used as an indicator rig in order to reproduce to a small scale the displacement of the crosshead, and therefore of the piston, of a reciprocating engine. When applied for this purpose, say, to a vertical engine, the pin O is fixed to the frame of the engine and the pin P moves with the crosshead. The cord which oscillates the indicator drum is fixed to the link AD at the point Q. Since the pin Q is required to reproduce the motion of P to a much reduced scale, the ratio OB : OA must be large.

The pantograph was used by Watt in his beam engine, Fig. 80, in order to enable the motion of the end P of the piston rod to be

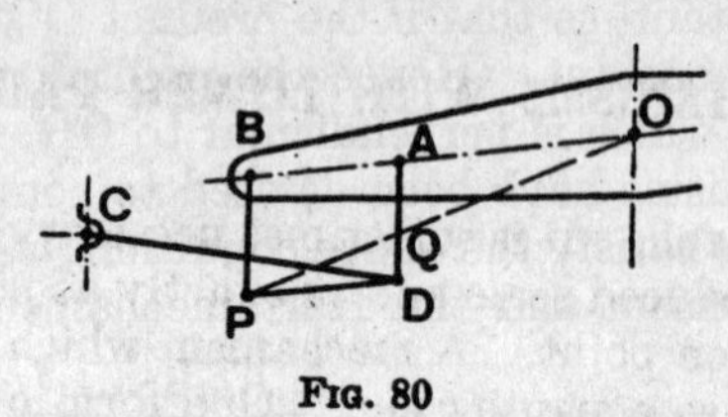

FIG. 80

an enlarged copy of that of a point Q, which is constrained to move along an approximately straight path. In the figure O and C are fixed centres, the links OA, AD and DC determine the path of Q, while the links OB, AD, DP and PB form the pantograph. The path of P is a reproduction of the path of Q to an enlarged scale, the ratio of the two displacements being OP : OQ.

56. Straight-line Motions. It is frequently necessary to constrain a point in a mechanism to move along a straight path. The obvious way of doing this is to use a sliding pair. But sliding pairs are bulky and are subject to comparatively rapid wear, so that in certain circumstances it is desirable to obtain the necessary constraint by the use of turning pairs. A mechanism which is used for this purpose is known as a straight-line motion or a parallel motion.

57. Exact Straight-line Motions. Referring to Fig. 81, let a line OP turn about O as centre and let the position of the point P

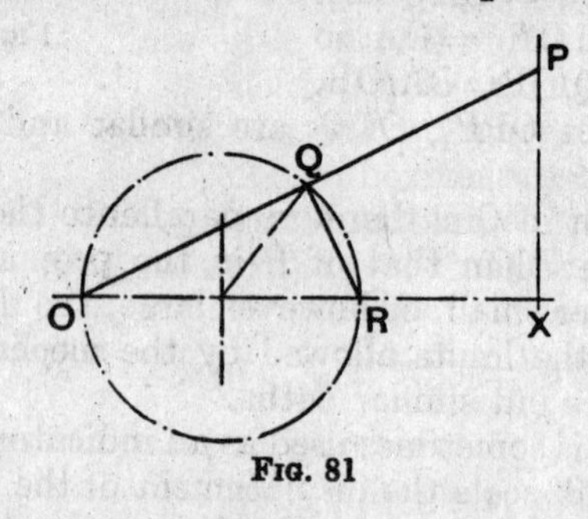

FIG. 81

be such that the product OQ . OP is constant. Then the path of P will be a straight line perpendicular to the diameter OR of the circle along the circumference of which Q moves. This may be proved as follows:

Since the triangles OQR, OXP are similar (angle QOR is common and angles OQR, OXP are each right-angles), therefore $OQ/OR = OX/OP$ and $OX = (OQ.OP)/OR$.

But OR is constant, so that if the product OQ.OP is constant, OX will also be constant. Hence the point P moves along the straight path XP which is perpendicular to OR.

Several mechanisms have been devised to connect O, P and Q in such a way as to satisfy the above condition. Two of them, the Peaucellier mechanism and the Hart mechanism, are shown in Figs. 82 and 83.

(a) *The Peaucellier Mechanism*: Fig. 82. The pin Q is constrained to move along the circumference of a circle with a fixed diameter OR by means of the link QA. The link QA and the

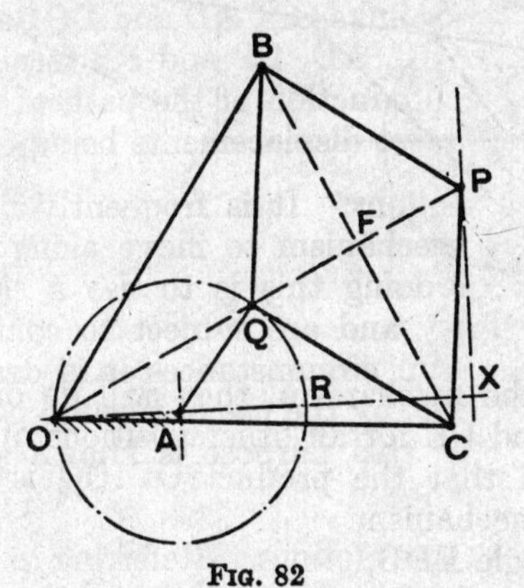

FIG. 82

fixed link OA are equal in length. The pins P and Q are at opposite corners of a four-bar chain which has all four links QB, BP, PC and CQ of equal length. The pins B and C are connected by links of equal length to the fixed pin O. That the product OQ.OP remains constant as the link QA rotates may be proved as follows:

Join BC to bisect PQ at F; then, from the right-angled triangles OBF, BPF, we have

$$OB^2 = OF^2+FB^2 \quad \text{and} \quad BP^2 = BF^2+FP^2$$

Subtracting,

$$OB^2-BP^2 = OF^2-FP^2 = (OF-FP)(OF+FP)$$
$$= OQ.OP$$

But OB and BP are of constant length, so that the product OQ.OP is constant and therefore the point P traces out a straight path normal to OR.

(b) *The Hart Mechanism*: Fig. 83. The four links BC, CD, DE and EB form a crossed parallelogram. The links BC and DE are equal in length, as are also the links CD and EB. It is clear that with this arrangement the lines joining B to D and C to E are parallel for all possible positions of the links. Also, if the three points O, Q and P lie on a straight line parallel to BD and CE for

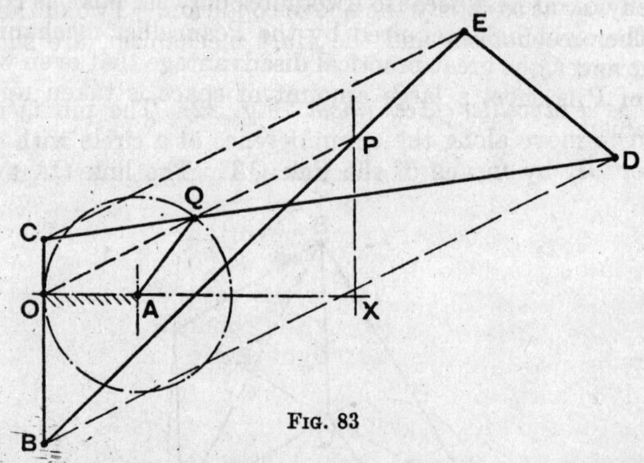

FIG. 83

one position of the mechanism, they will lie on a straight line parallel to BD and CE for all other positions of the mechanism. It will be shown that the product OP.OQ is constant for all positions of the mechanism.

From the triangle EDB,

$$BE^2 = BD^2 + DE^2 - 2BD.DE \cos EDB$$

But, from symmetry,

$$\cos EDB = (BD - CE)/2DE$$

$$\therefore \; BE^2 = BD^2 + DE^2 - BD(BD - CE) = DE^2 + BD.CE$$

so that

$$BD.CE = BE^2 - DE^2 = \text{constant} \quad . \quad . \quad (4.1)$$

From the similar triangles CBE, OBP,

$$CE/CB = OP/OB \quad \text{or} \quad CE = CB.OP/OB$$

and from the similar triangles BCD, OCQ,

$$BD/BC = OQ/OC \quad \text{or} \quad BD = BC.OQ/OC$$

Substituting for CE and BD in (4.1):

$$\therefore \; OQ.OP.BC^2/(OB.OC) = \text{constant}$$

But BC, OB and OC are all constant, so that the product OP.OQ must be constant.

It follows that if O is a fixed centre and Q is constrained to move along a circular arc which passes through O, then the path of P will be a straight line normal to the diameter of the circular path of Q which passes through O. These conditions will be satisfied if Q is attached to a link which turns about the fixed centre A, provided that the distance between the fixed centres A and O is equal to the length of the link AQ. The Hart mechanism is chiefly of interest because it requires only six links as compared with the eight links required by the Peaucellier mechanism. It has, however, the great practical disadvantage that even when the path of P is short a large amount of space is taken up by the mechanism.

(c) *The Scott-Russell Mechanism.* This straight-line motion is shown in Fig. 84. It differs from the two mechanisms which have just been described in one important respect—the straight line is not generated but is merely copied. The mechanism is essentially the same as that of the reciprocating engine. The crank OC is equal in length to the connecting rod CP and P is constrained to move along a straight path by a crosshead and guide

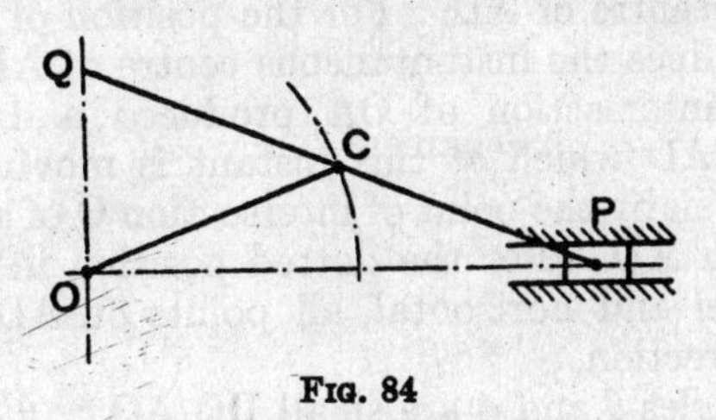

FIG. 84

bars. The connecting rod is extended to Q, such that CQ = CP, and it is easily seen that Q then moves along a straight path normal to OP. The exactness with which Q follows a straight path obviously depends upon the exactness with which P is guided along a straight path. This straight-line motion is not of much practical value, since it makes use of a sliding pair, the friction and wear of which are always higher than those of a turning pair.

58. Approximate Straight-line Motions. A large number of mechanisms have been devised in order to give a path which is approximately straight. Most of these mechanisms are derived from the four-bar chain.

(a) *The Watt Mechanism.* The type of motion used by Watt to guide the piston rod of many of his early steam engines is shown in Fig. 80 and is drawn to an enlarged scale in Fig. 85. The links OA and DC oscillate about the fixed centres O and C and the point Q on the connecting link AD describes the figure-of-eight path which is shown in dotted lines. When the position of Q along AD is suitably chosen, part of the path of Q is approximately straight. The length of the link AD is such that, for the position of the mechanism in which OA and DC are parallel, it is approximately perpendicular to OA and DC.

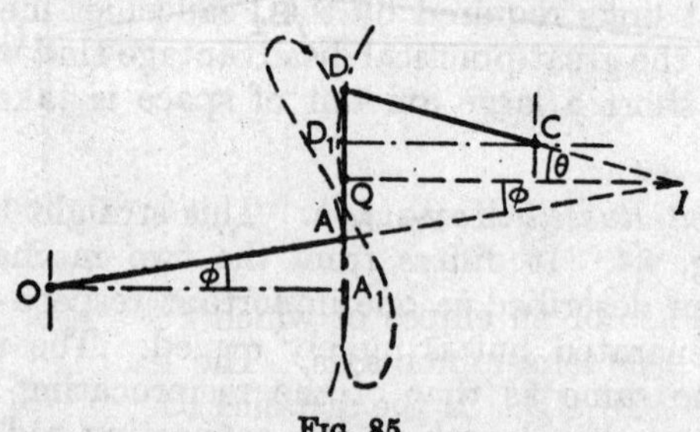

Fig. 85

The best position of Q may be found by making use of the instantaneous centre of AD. For the position of the mechanism shown by full lines the instantaneous centre of AD is given by I, the point of intersection of OA produced and DC produced. The point on AD which at this instant is moving in a vertical direction is given by the point of intersection Q of a horizontal line through I with AD. For the dotted position in which OA and DC are parallel and horizontal, all points on AD are moving in the vertical direction.

Since the angles θ and ϕ are small $DQ/AQ \simeq \theta/\phi$.

But $\theta = DD_1/DC$ and $\phi = AA_1/OA \simeq DD_1/OA$

$$\therefore\ DQ/AQ \simeq OA/DC$$

The point Q should therefore divide AD such that

$$AQ:QD::DC:AO$$

(b) *The Grasshopper Mechanism.* The grasshopper straight-line motion is shown in Fig. 86. The centres O and Q are fixed. The pin A moves along a curved path with Q as centre and QA as radius and the pin B moves along a curved path with O as centre and OB as radius. The point P on an extension of the link AB describes an approximately straight path for small angular displacements of OB on each side of the horizontal. The degree of approximation involved in this mechanism may be seen by comparing it with the ellipse trammels, Fig. 9 (a). In order that P may describe an exact straight line, the point A ought to move along a straight path which passes through O, and B ought to

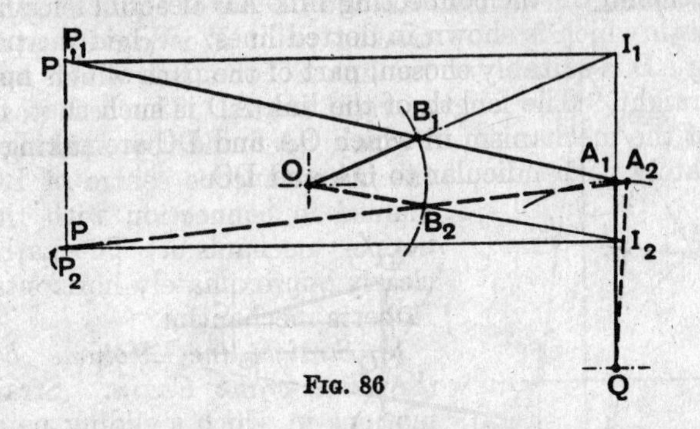

FIG. 86

move along the arc of an ellipse of which PB is the semi-major axis and BA is the semi-minor axis. The length of PB would then be equal to AB^2/OB. If the distance PB in the grasshopper mechanism is made equal to this length, the path of P will be approximately straight and perpendicular to AO. A better way of fixing the point P is to find the instantaneous centre of PA in several positions of the mechanism. For each position find the point at which a horizontal line drawn through the instantaneous centre intersects the link AB produced. The points thus obtained will clearly be moving in a vertical direction at the given instants, but their distances from B will be different and the mean distance will give the best position for P. In the figure this construction has been applied in two positions of the mechanism. I_1 and I_2 are the respective instantaneous centres and P_1 and P_2 the respective points which are moving in the vertical direction. The distance PB is intermediate between P_1B_1 and P_2B_2.

(c) *The Tchebicheff Straight-line Motion*: Fig. 87. This is a four-bar chain in which the crossed links AB and CD are equal in length. The tracing point P is situated at the mid-point of the link BC. The proportions of the links are usually such that P is directly above A or D in the extreme positions of the mechanism, i.e. when CB lies along AB or when CB lies along CD. It can easily be shown that in these circumstances the tracing point P will lie on

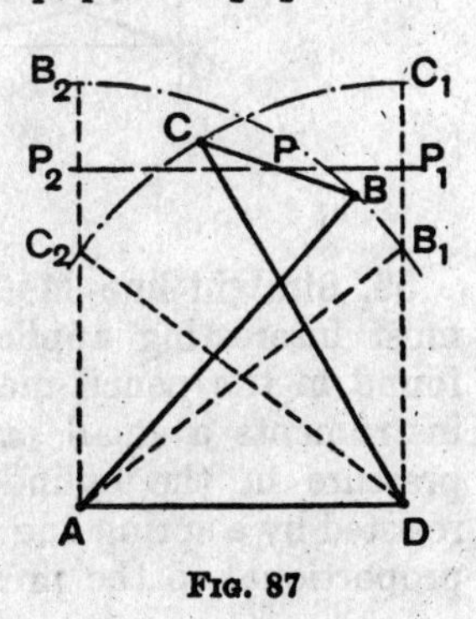

FIG. 87

a straight line parallel to AD in the two extreme positions and in the mid-position if BC:AD:AB::1:2:2·5.

(d) *The Roberts Straight-line Motion*: Fig. 88. This also is a four-bar chain. The links AB and CD are of equal length and the tracing point P is rigidly attached to the link BC on a line which bisects BC at right angles. The best position for P may be found by making use of the instantaneous centre of BC as explained in connection with the grasshopper mechanism. The path of P is clearly approximately horizontal in the Roberts mechanism.

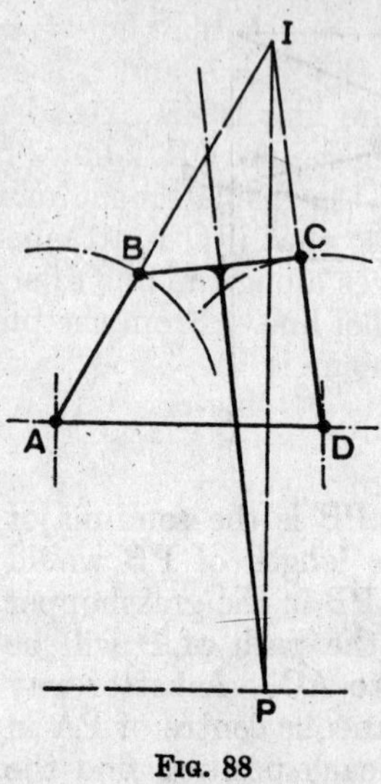

FIG. 88

(e) *Straight-line Motions based on the Slider-crank Chain.* Straight-line motions in which a sliding pair is used are not very often employed, although inversions of the slider-crank chain are possible as shown in Fig. 89. The type of motion shown at (a) is similar to the grasshopper mechanism, except that A has a straight-line motion instead of describing a circular arc. The type of motion shown at (b) is essentially the same inversion of the slider-crank chain as the oscillating-cylinder engine. As the crank OC revolves, the link OP slides in the pivoted block Q. The best position for the tracing point P may be found in each of these mechanisms by using the instantaneous centre as already explained.

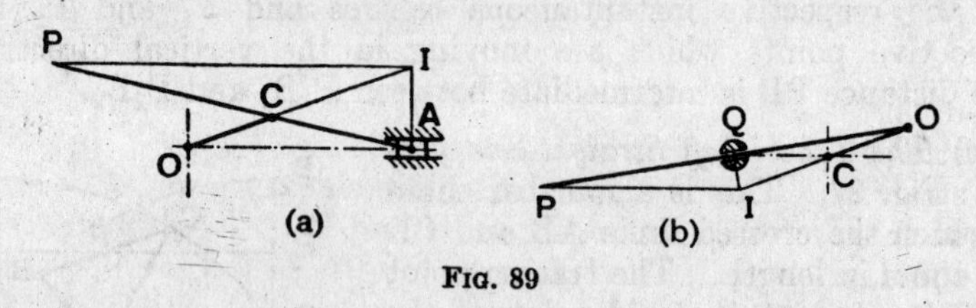

FIG. 89

59. Straight-line Motions for Engine Indicators. One of the most interesting applications of straight-line motions is to be found in the pencil mechanisms of engine indicators. In these instruments a small piston is acted upon by the steam or gas pressure in the cylinder. The displacement of the piston is resisted by a spring and the amount of the displacement is directly proportional to the pressure exerted by the steam or gas. The pencil which records on a diagram the variation of pressure in the cylinder is required to reproduce to an enlarged scale the displacement of the indicator piston. The most direct way of doing this is to use a pantograph, as shown in Fig. 90. Here O is a fulcrum fixed to the body of the indicator and A, B, C and Q are pin-joints. The distance between the pins A and B is equal to that between the pins Q and C and the distance between the pins A and Q is equal to that between the pins B and C. The link BC is extended to P such that O, Q and P lie in one straight line. It follows from Article 55 that with this arrangement the displacement of Q is reproduced to an enlarged scale by P, so that if Q is attached to the piston rod of the indicator and moves along the line of stroke of the piston, P will move along a parallel line. From the purely theoretical point of view this mechanism is ideal. It satisfies the two essential conditions, viz. (a) that the line of stroke of the pencil shall be parallel to that of the piston, and (b) that the displacement of the pencil shall be directly proportional to the displacement of the piston. From the practical point of view, however, there are serious objections to the mechanism. In the first place the accuracy of the straight-line motion of P depends upon the accuracy of the motion of Q. Any deviation of Q from a straight path involves a proportionate deviation of P from a straight path. In the second place slackness due to wear in any of the five pin-joints A, B, C, O and Q destroys the accuracy of the motion of P, and in view of the relatively short distance between pins A and O the effect of wear in the former is particularly harmful.

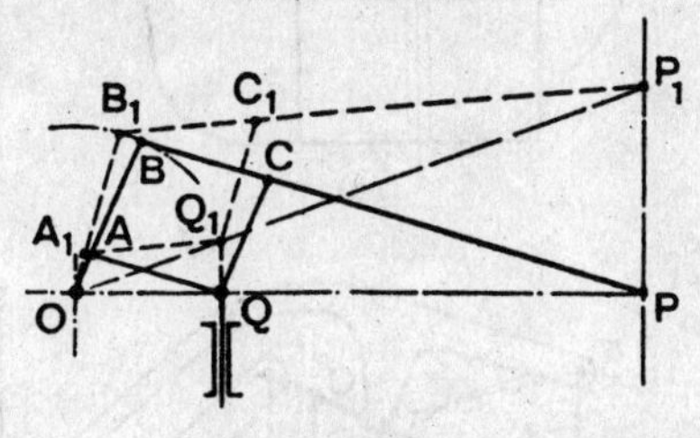

FIG. 90

It is more satisfactory in practice to have the mechanism so arranged that the pencil moves along a straight or an approximately straight path quite independently of the path followed by Q. This reduces the number of pin-joints on whose tightness the accuracy of the path followed by P depends. The mechanisms of two indicators in which the path of P is independent of that of

Q are drawn diagrammatically in Fig. 91. That shown at (a) is the mechanism of the Thompson indicator, while (b) shows the mechanism of the Dobbie McInnes indicator. In each case it will be seen that the links OA, AB, BR and RO constitute a straight-line motion of the grasshopper type. The best position for the tracing point P may be found by making use of the instantaneous centre of the link AB as already explained. There are only four pin-joints in the mechanism which determines the path of P and none of the links is very short, so that the effect of wear at the joints is minimised.

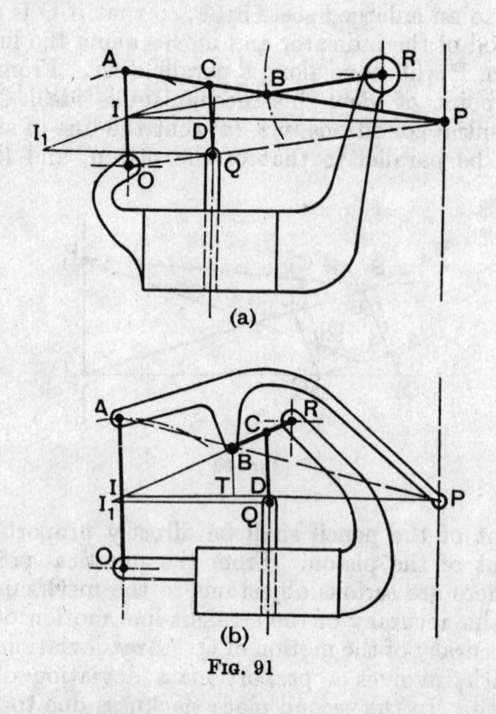

FIG. 91

The only real difference between the two mechanisms is that in one the motion of the indicator piston is transmitted from Q to the pin C on the link AB, while in the other it is transmitted from Q to the pin C on the link BR. The links QC and OA should be approximately parallel.

The approximate value of the ratio of the displacement of P to the displacement of Q may be expressed in terms of the lengths of the links. Since the ratio of the displacements is to be constant, it follows that the ratio of the velocities of P and Q at any given instant must be constant.

For the given position of each mechanism the link AB is turning about the centre I, which lies at the point of intersection of RB produced with AO, produced if necessary, and IP is approximately perpendicular to the line of stroke of the piston.

For the Thompson mechanism the pin C is moving in a direction normal to IC, while the pin Q is moving along the line of stroke, so that I_1 is the centre about which the link CQ is turning, where QI_1 is perpendicular to the line of stroke of Q. Let CQ intersect IP at D. Then:

$$v_c/v_q = I_1C/I_1Q = IC/ID$$

$$v_p/v_c = IP/IC$$

$$\therefore\ v_p/v_q = v_p/v_c . v_c/v_q = IP/ID$$

If the links QC and OA are parallel, the triangles PCD, PAI are similar, so that IP/ID = AP/AC and $v_p/v_q = AP/AC$.

The links QC and OA cannot be exactly parallel, nor can the line IP be exactly perpendicular to the line of stroke of the piston for all positions of the mechanism. Hence the ratio of the displacement of P to the displacement of Q cannot be quite constant. The variations from the value AP : AC are, however, for all practical purposes negligible.

Similarly, for the Dobbie McInnes mechanism I and I_1 are respectively the instantaneous centres of AB and CQ, while R is the centre about which BR turns.

$$\therefore\ v_c/v_q = I_1C/I_1Q = IC/ID$$

Also $\quad v_b/v_c = RB/RC \quad \text{and} \quad v_p/v_b = IP/IB$

$$\therefore\ v_p/v_q = v_p/v_b . v_b/v_c . v_c/v_q = IP/IB . RB/RC . IC/ID$$

But if BT is drawn parallel to CQ,

$$IT/IB = ID/IC \quad \text{or} \quad IT = (IB . ID)/IC$$

Substituting, $\quad v_p/v_q = IP/IT . RB/RC$

If QC and AO are parallel, the triangles PBT, PAI are similar,

$$\therefore\ IP/IT = AP/AB, \quad \text{so that} \quad v_p/v_q = AP/AB . RB/RC$$

This expression gives approximately the ratio of the displacement of P to the displacement of Q.

60. The Motor-car Steering Gear. The relative motion between the wheels of a self-propelled vehicle and the road surface should be one of pure rolling. In order to satisfy this condition when the vehicle is moving along a curved path, the steering gear must be so designed that the paths of the points of contact of each wheel with the ground are concentric circular arcs. Steering is usually effected by turning the axes of rotation of the two front wheels relative to the chassis or body of the vehicle, and, to satisfy the above condition, the axis of the wheel on the inside of the curve must be turned through a larger angle than the axis of the wheel on the outside of the curve. The front wheels are mounted on

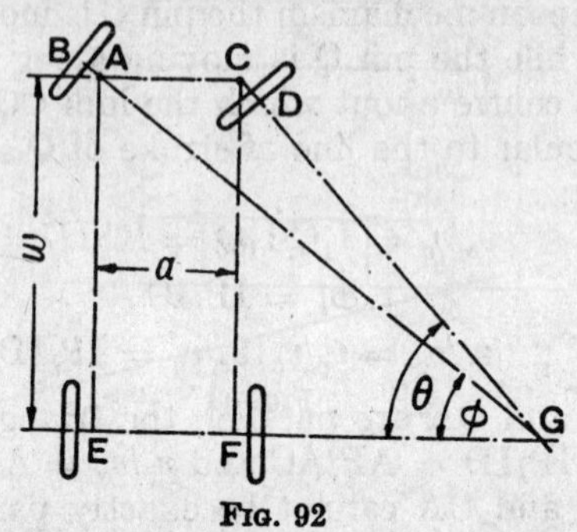

FIG. 92

short separate axles which are pivoted to the chassis of the car. In Fig. 92 a plan view is shown in which AB and CD are the two axles with pivots at A and C. When turning to the right the axes AB and CD intersect the common axis EF of the rear wheels at the point G, so that the path of contact of each wheel with the ground is a circular arc with centre G.

From the figure

$$AC = EF = EG - FG = AE \cot \phi - CF \cot \theta$$

$$\therefore \cot \phi - \cot \theta = AC/AE = a/w \quad . \quad . \quad (4.2)$$

The two front axles must therefore be operated by the steering gear in such a way that this equation is satisfied whatever the radius of curvature of the path followed by the car. Two different steering mechanisms will be described: (a) The Davis steering gear, and (b) the Ackermann steering gear.

(a) *The Davis Steering Gear*: Fig. 93. In this mechanism the arms AK and CL are fixed to the axles so as to form bell-crank levers and the angles BAK, DCL are equal. The arms are slotted and slide relative to two die-blocks which are pivoted to the link MN. The link MN is supported in guides so as to be able to move parallel to the link AC. Steering is effected by sliding MN either to the right or to the left. When the gear is in the mid-position and the car is moving along a straight path the steering arms AK and CL are each inclined at the angle α to the centre line of the car. If, now, the link MN is moved through a distance x to the right relative to the chassis, the bell-crank levers BAK and DCL are moved to the dotted positions and BA and CD when produced intersect at G.

Let ϕ and θ be the angles through which the arms AK and CL are turned by the displacement of MN.

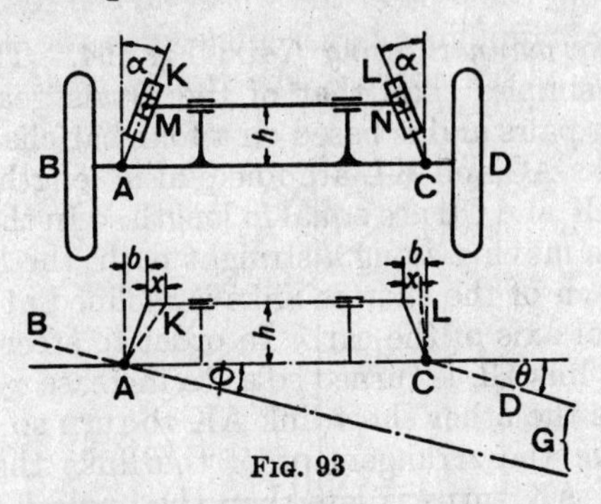

FIG. 93

Let h be the distance of MN from AC and $2b$ the difference between AC and MN.

Then

$\tan \alpha = b/h$, $\tan(\alpha+\phi) = (b+x)/h$ and $\tan(\alpha-\theta) = (b-x)/h$

But
$$\tan(\alpha+\phi) = \frac{\tan\alpha+\tan\phi}{1-\tan\alpha\tan\phi} = \frac{b+x}{h}$$

Substituting for $\tan\alpha$ and simplifying, we get:

$$\tan\phi = \frac{xh}{h^2+b^2+bx} \quad . \quad . \quad . \quad . \quad (4.3)$$

Also
$$\tan(\alpha-\theta) = \frac{\tan\alpha-\tan\theta}{1+\tan\alpha\tan\theta} = \frac{b-x}{h}$$

Substituting for $\tan\alpha$ and simplifying, we get:

$$\tan\theta = \frac{xh}{h^2+b^2-bx} \quad . \quad . \quad . \quad . \quad (4.4)$$

From (4.3) and (4.4):

$$\cot\phi-\cot\theta = \frac{h^2+b^2+bx}{xh} - \frac{h^2+b^2-bx}{xh}$$
$$= 2b/h = 2\tan\alpha \quad . \quad . \quad . \quad . \quad (4.5)$$

But, for the steering to be correct, we have from (4.2) that:

$$\cot\phi-\cot\theta = a/w$$
$$\therefore\ 2\tan\alpha = a/w$$

or
$$\tan\alpha = a/2w \quad . \quad . \quad . \quad . \quad (4.6)$$

The disadvantages of the Davis gear are that, owing to the number of sliding pairs, friction is high and the wear which takes place at the contact surfaces rapidly impairs the accuracy of the mechanism.

(b) *The Ackermann Steering Gear*: Fig. 94. This mechanism is very much simpler than that of the Davis gear. It consists only of turning pairs and is based on a four-bar chain in which the two longer links AC and KL are unequal in length while the two shorter links AK and CL are equal in length. In the mid-position, when the car is moving along a straight path, the longer links are parallel and each of the shorter links is inclined at the angle α to the longitudinal axis of the car. In order to steer the car to the right the short link CL is turned so as to increase α, while the long link LK causes the other short link AK to turn so as to reduce α. It is clear from the arrangement of the links that the angle ϕ through which AK turns is less than the angle θ through which

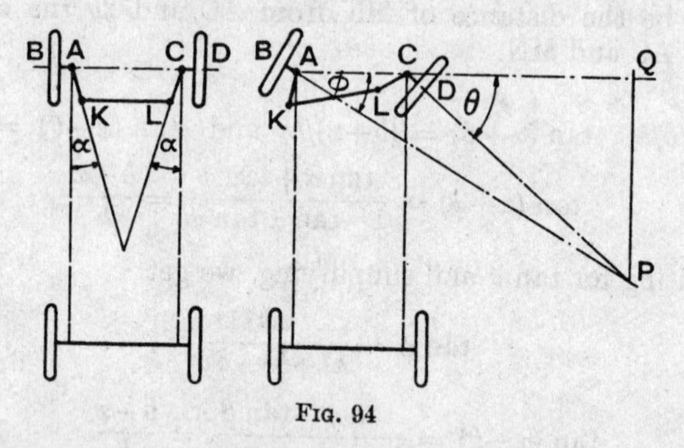

FIG. 94

CL turns, and therefore the left front axle turns through a smaller angle than the right front axle. It is also clear that the value of ϕ obtained for a given value of θ will depend upon the ratio AK/AC and the angle α. For given values of AK/AC and α, corresponding values of θ and ϕ may be obtained either graphically or by calculation. The difference between cot ϕ and cot θ for each pair of corresponding values of θ and ϕ will be found to increase, at first slowly and then more rapidly, as θ increases. As an example, corresponding values of θ, ϕ and $\cot\phi - \cot\theta$ have been calculated for a mechanism in which AK/AC $= 1/8{\cdot}5$ and $\alpha = 18°$, and are shown in the table below.

θ	10°	20°	30°	40°	50°
ϕ	9° 25′	17° 43′	24° 49′	30° 34′	34° 43′
$\cot\phi - \cot\theta$.	0·356	0·383	0·431	0·501	0·604
ϕ_0	9° 21′	17° 38′	25° 8′	32° 8′	38° 54′

But $\cot\phi - \cot\theta = AC/QP$, where P is the point of intersection of BA and CD produced.

Since for correct steering P ought to lie on the common axis of the rear wheels, it follows that the Ackermann mechanism can only give correct steering for one value of θ apart from the value $\theta = 0$. There will, of course, be a corresponding value of θ when the car is turning to the left. If the distance AC is 0·4 times the wheelbase of the car, which is approximately correct for a private car, a mechanism of the above proportions would give correct steering for $\theta \simeq 24°$. For smaller values of θ the angle ϕ given by the mechanism would be too high for correct steering and for larger values of θ it would be too low. The values of ϕ required for correct steering are shown in the last line of the table. It will be seen that the errors in the values of ϕ given by the mechanism are negligible except when θ is large. But when θ is large the speed of the car will of necessity be low and a small error in the angle ϕ will not seriously affect the wear of the tyres.

For a given value AK/AC a reduction in the value of α causes a reduction in the value of AC/QP. Similarly, for a given value of α a reduction in the value of AK/AC causes a reduction in the value of AC/QP, although the effect of a change in the ratio AK/AC is relatively small. When the gear is in the mid-position and the car is moving along a straight path, it will be found that the distance of the point of intersection of the arms AK and CL from the line AC should be about 0·7 of the wheelbase in order to give the best results. Increasing this distance causes an increase in the value of θ at which correct steering is obtained.

61. Hooke's Joint: Fig. 95. This joint is used to connect two non-parallel, intersecting shafts. The end of each shaft is forked and each fork provides two bearings for the arms of a cross. The arms of the cross are at right angles and the cross serves to transmit motion from the driving to the driven shaft. The inclination of the driven shaft to the driving shaft may be constant, but usually it varies while the motion is being transmitted. Examples of the application of Hooke's joint, or the universal joint, as it is more frequently called, are to be found in the transmission from the gear-box to the back axle of automobiles and in the transmission of the drive to the spindles of multiple drilling machines.

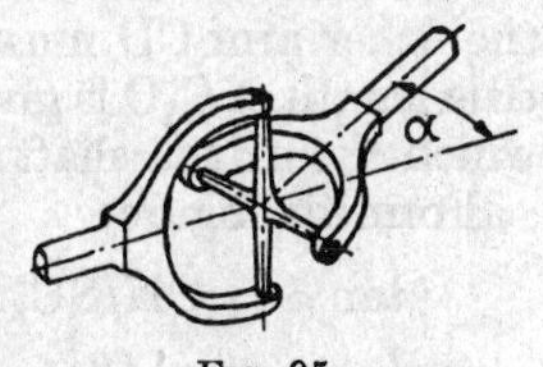

FIG. 95

In Fig. 96, an end elevation looking along the axis of the driving shaft and a plan are shown. The planes of rotation of the two arms of the cross are represented by the traces PP and QQ in plan. The plane of rotation of the arm attached to the driving shaft is represented by the plane of the paper in elevation.

Let the initial position of the cross be such that both arms lie in the plane of the paper in elevation, while the arm AB attached to the driving shaft lies in the plane containing the axes of the two

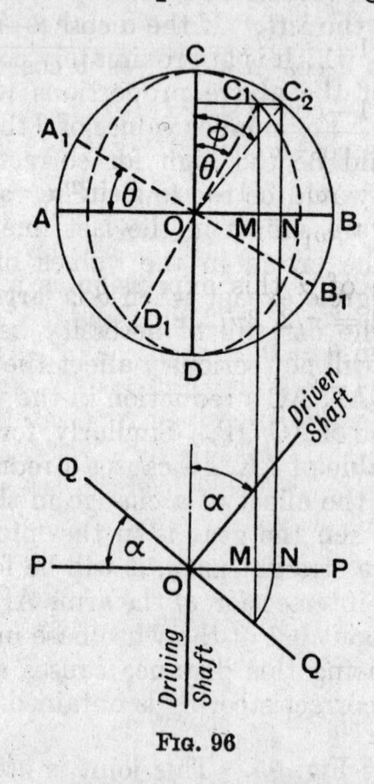

FIG. 96

shafts. Let the driving shaft turn through the angle θ, so that the arm AB is displaced to A_1B_1. Then the projection C_1D_1 of the other arm CD must also turn through the angle θ. But the true length of C_1O is given by C_2O and therefore the angle through which the driven shaft has turned is given by ϕ.

From the figure,

$$\tan \phi = ON/NC_2 \quad \text{and} \quad \tan \theta = OM/MC_1 = OM/NC_2$$

$$\therefore \tan \phi/\tan \theta = ON/NC_2 . NC_2/OM = ON/OM$$

If α is the inclination of the driven shaft to the driving shaft, then clearly $OM/ON = \cos \alpha$, so that $\tan \phi/\tan \theta = 1/\cos \alpha$,

or

$$\tan \theta = \cos \alpha \tan \phi \quad . \quad . \quad . \quad . \quad (4.7)$$

Let ω = angular velocity of the driving shaft $= d\theta/dt$

and ω_1 = angular velocity of the driven shaft $= d\phi/dt$.

Differentiating both sides of (4.7):

$$\sec^2\theta . \mathrm{d}\theta/\mathrm{d}t = \cos\alpha \sec^2\phi . \mathrm{d}\phi/\mathrm{d}t$$

$$\therefore \quad \omega/\omega_1 = \cos\alpha \sec^2\phi/\sec^2\theta = \cos\alpha \cos^2\theta \sec^2\phi$$

But $\sec^2\phi = 1+\tan^2\phi$, and, substituting for $\tan\phi$ from (4.7):

$$\sec^2\phi = 1+\frac{\tan^2\theta}{\cos^2\alpha} = \frac{\cos^2\alpha\cos^2\theta+\sin^2\theta}{\cos^2\alpha\cos^2\theta}$$

$$= \frac{1-\cos^2\theta\sin^2\alpha}{\cos^2\alpha\cos^2\theta}$$

$$\therefore \frac{\omega}{\omega_1} = \frac{1-\cos^2\theta\sin^2\alpha}{\cos\alpha} \quad . \quad . \quad . \quad (4.8)$$

For a given value of α this expression is a maximum when $\cos\theta = 0$, i.e. when $\theta = \pi/2$, $3\pi/2$, etc.; and it is a minimum when $\cos\theta = \pm 1$, i.e. when $\theta = 0$, π, etc.

If the speed ω of the driving shaft is constant, the maximum speed ω_1 of the driven shaft is given by $\omega_1/\omega = 1/\cos\alpha$ and the minimum speed of the driven shaft by $\omega_1/\omega = \cos\alpha$.

The value of θ for which the speeds of the driving and the driven shafts are equal may be found by equating (4.8) to unity.

Then $$1-\cos^2\theta\sin^2\alpha = \cos\alpha$$

$$\therefore \cos^2\theta = \frac{1-\cos\alpha}{\sin^2\alpha} = \frac{1}{1+\cos\alpha}$$

$$\therefore \sin^2\theta = 1-\cos^2\theta = \frac{\cos\alpha}{1+\cos\alpha}$$

and $$\tan^2\theta = \cos\alpha \quad \text{or} \quad \tan\theta = \sqrt{\cos\alpha} \quad . \quad (4.9)$$

The angular acceleration of the driven shaft is given by $\mathrm{d}\omega_1/\mathrm{d}t$,

$$\therefore \frac{\mathrm{d}\omega_1}{\mathrm{d}t} = \frac{\mathrm{d}\omega_1}{\mathrm{d}\theta}\cdot\frac{\mathrm{d}\theta}{\mathrm{d}t} = \omega\frac{\mathrm{d}\omega_1}{\mathrm{d}\theta} = \frac{-\omega^2\cos\alpha\sin^2\alpha\sin 2\theta}{(1-\cos^2\theta\sin^2\alpha)^2} \quad (4.10)$$

The value of θ for which the acceleration is a maximum may be found by differentiating with respect to θ and equating to zero. The resulting expression is, however, very cumbersome, and it will be found that the following expression, which is derived from the exact expression by a simple approximation, gives results which are sufficiently close for most practical purposes:

$$\cos 2\theta \simeq \frac{2\sin^2\alpha}{2-\sin^2\alpha} \quad . \quad . \quad . \quad . \quad (4.11)$$

Even for a value of α as high as 30° this equation gives the value of θ accurate to within a few minutes. It should be noted that the angular acceleration of the driven shaft is a maximum when θ is approximately 45°, 135°, etc., i.e. when the arms of the cross are inclined at 45° to the plane containing the axes of the two shafts.

Since the universal joint provides a rigid connection between the two shafts so far as the transmission of torque is concerned, the moment of inertia of the masses attached to the driven shaft must be small. Otherwise very high alternating stresses may be set up in the parts of the joint owing to the alternate angular acceleration and retardation. The larger the shaft angle α the higher is the acceleration, so that this angle should always be kept as small as possible. In some drives a double universal joint is used. The power is transmitted from the driving shaft to the driven shaft through an intermediate shaft, at each end of which there is a universal joint. If the driving shaft and the driven shaft are equally inclined to the intermediate shaft and the two forks on the intermediate shaft lie in the same plane, it is easily seen that the speeds of the driving and the driven shafts are identical at every instant and fluctuations of speed are confined to the intermediate shaft which may be made short and light.

Example 1. A universal joint is used to connect two shafts which are inclined at 20° and the speed of the driving shaft is 1000 r.p.m. Find the extreme angular velocities of the driven shaft and its maximum acceleration.

The angular velocity of the driving shaft

$$= \omega = \pi . 1000/30 = 104{\cdot}7 \text{ rad/s}$$

But maximum value of $\omega_1 = \omega/\cos 20° = 111{\cdot}4$ rad/s and minimum value of $\omega_1 = \omega \cos 20° = 98{\cdot}3$ rad/s.

Using the approximate equation (4.11), the acceleration of the driven shaft is a maximum when:

$$\cos 2\theta \simeq \frac{2 \sin^2 20°}{2 - \sin^2 20°} = \frac{2.0{\cdot}1170}{1{\cdot}883}$$

$$\therefore\ 2\theta = 82°\ 52' \quad \text{or} \quad 277°\ 8'$$

$$\theta = 41°\ 26' \quad \text{or} \quad 138°\ 34'$$

Substituting for ω, α and θ in (4.10), the maximum angular acceleration of the driven shaft is given numerically by:

$$\left(\frac{100\pi}{3}\right)^2 \frac{0{\cdot}9397 . 0{\cdot}3420^2 . 0{\cdot}9923}{(1 - 0{\cdot}7497^2 . 0{\cdot}3420^2)^2} = 1373 \text{ rad/s}^2$$

The acceleration is in the opposite direction to the velocity, i.e. the driven shaft has maximum retardation, when $\theta = 41°\ 26'$ or $180° + 41°\ 26'$; and it is in the same direction as the velocity when $\theta = 138°\ 34'$ or $180° + 138°\ 34'$.

EXAMPLES IV

1. For what purpose is a pantograph used? Sketch one form of pantograph and show that it satisfies the required conditions.

2. Let OR be the diameter of a circle and Q a point on the circumference of the circle. Join O to Q and let P be a point on OQ produced. Show that, if OQ turns about O as centre and the product OQ.OP remains constant, the point P will move along a straight path perpendicular to the diameter OR.

3. Sketch the Peaucellier or the Hart straight-line motion and prove that the tracing point P describes a straight-line path.

4. Fig. 97 shows to scale the mechanism of the Crosby engine indicator. Show that the pencil point P traces a path which is approximately parallel to that of the indicator piston. What is the ratio of the displacement of the pencil to the displacement of the piston?

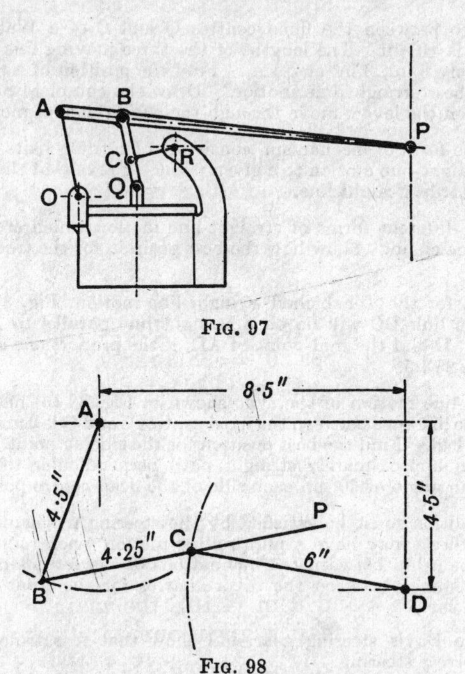

FIG. 97

FIG. 98

5. Fig. 98 shows part of the mechanism of a circuit breaker. A and D are fixed centres and the lengths of the links are: AB, 4·5 in.; BC, 4·25 in.; and CD, 6 in. Find the position of a point P on BC produced that will trace out an approximately straight vertical path 10 in. long.

6. The mechanism in Fig. 99 is a four-bar kinematic chain of which the centres A and B are fixed. The dimensions are AB = 2 ft, AC = CD = DB = 1 ft.

Find a point G on the centre line of the cross arm of which the locus is an approximately straight line even for considerable displacements from the position shown. L.U.A.

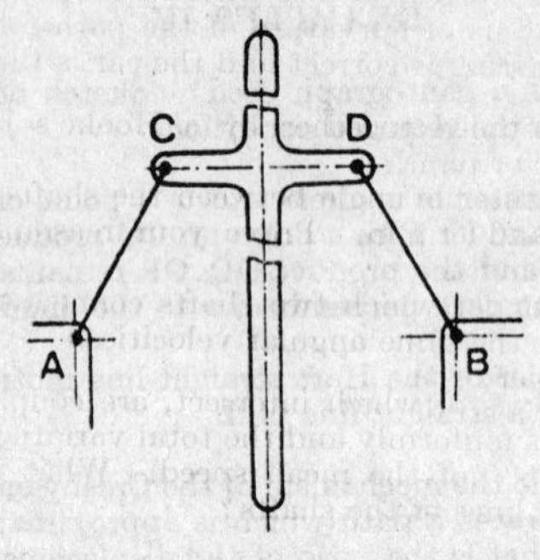

FIG. 99

7. The distance between the fixed centres O and C of a Watt straight-line motion, Fig. 85, is $10\frac{3}{8}$ in. The lengths of the three moving links OA, AD and DC are respectively 6 in., 3 in. and 4 in. Find the position of a point Q on AD which gives the best straight-line motion. Draw the complete path traced out by this point when the levers move through their full range of movement.

8. Describe one form of mechanism, consisting of turning pairs only, that will give an exact straight-line motion to a given point. Prove that the path followed by the point is a true straight line.

9. Sketch two different forms of straight-line motion which are based on the four-bar kinematic chain. Show how the best position for the tracing point may be found.

10. Show that for the Tchebicheff straight-line motion, Fig. 87, the point P which bisects the link BC will lie on a straight line parallel to AD when it is directly above A, D and the mid-point of AD, if the proportions of the links are BC : AD : AB : : 1 : 2 : 2·5.

11. A straight-line motion of the type shown in Fig. 89 (b) has the following dimensions. The distance between the fixed centres C and Q is 2 in. and the length of the link OC is 1 in. Find the best position for the tracing point P in order that it may follow an approximately straight path perpendicular to CQ while OC swings through an angle of 60° on each side of the dead-centre position.

12. What condition must be satisfied by the steering mechanism of a car in order that the wheels may have a pure rolling motion when rounding a curve? Deduce the relationship between the inclinations of the front stub axles to the rear axle, the distance between the pivot centres for the front axles and the wheelbase of the car.

13. Sketch the Davis steering gear and show that it satisfies the required conditions for correct steering.

If the distance between the pivots of the front axles is 3 ft 6 in. and the wheelbase is 8 ft 6 in., find the inclination of the track arms to the longitudinal axis of the car when the car is moving along a straight path.

14. The distance between the pivots of the front stub axles of a car is 51 in., the length of each track arm is 6 in. and the length of the track rod is 47·5 in. If the wheelbase of the car is 112 in. and the track is 56 in., find the radius of curvature of the path followed by the near-side front wheel at which correct steering is obtained when the car is turning to the right.

15. A car with a track of 4 ft 10 in. and a wheelbase of 9 ft has a steering mechanism of the Ackermann type, but with the track rod in front of the axle instead of behind it. The distance between the front stub axle pivots is 4 ft, the length of each track arm is 6 in. and the length of the track rod is 4 ft $4\frac{1}{2}$ in. Find the radius of curvature of the path followed by the near-side front wheel when the steering is correct and the car is turning to the right.

16. Two shafts are coupled together by a Hooke's joint, the driving shaft rotating uniformly at 600 r.p.m.

Find the greatest permissible angle between the shafts if the maximum speed of the follower shaft is 630 r.p.m. Prove your reasoning. What is then the minimum speed of this shaft?

State the conditions under which two shafts connected together by a double Hooke's joint shall have the same angular velocities. L.U.A.

17. Two shafts, the axes of which intersect, are coupled by a Hooke's joint, The driving shaft rotates uniformly and the total variation in speed of the driven shaft is not to exceed 8% of the mean speed. What is the greatest possible inclination of the centre lines of the shafts? L.U.A.

18. A Hooke's joint is to connect two shafts whose axes intersect at 150°. The driving shaft rotates uniformly at 120 r.p.m. Deduce a general expression for the angular velocity of the driven shaft.

The driven shaft operates against a steady torque of 100 lb ft and carries a flywheel whose weight is 100 lb and radius of gyration 0·5 ft. What is the maximum value of the torque which must be exerted by the driving shaft? L.U.A.

19. Two shafts A and B are connected together by a Hooke's joint with the axes inclined at 15°. The shaft A revolves at 2000 r.p.m. and the shaft B carries a flywheel of weight 20 lb and radius of gyration 3 in. Find the maximum torque in shaft B if it is assumed that the two shafts are torsionally rigid.

CHAPTER V

VALVE DIAGRAMS AND VALVE GEARS

62. The ideal form of the indicator diagram, which shows the cycle of operations in a steam-engine cylinder, is illustrated in Fig. 100. Admission of the live steam takes place at point A, just before the piston reaches the end of the return stroke, and continues until the piston reaches the point B on the outward stroke. Cut-off then takes place, followed by expansion of the steam to point C, when the valve opens to exhaust and release takes place. Exhaust continues during the return stroke until at point D the valve closes. The steam which remains in the cylinder is compressed from D to A and acts as a cushion for the reciprocating parts.

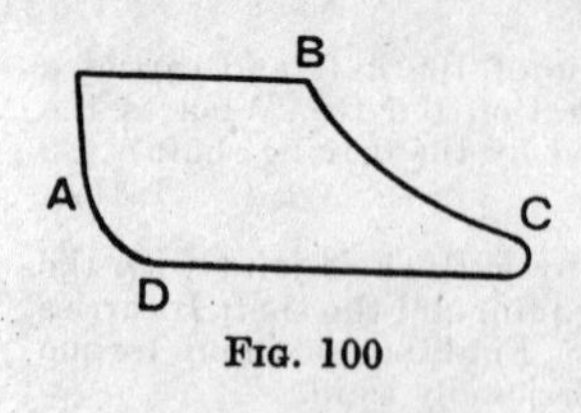

FIG. 100

In most steam engines a slide valve is used to determine the positions of the piston at which admission, cut-off, release and compression take place. Fig. 101 (a) shows diagrammatically the usual arrangement of valve, valve chest and cylinder for a double-acting engine. Steam from the boiler is admitted to the valve chest through O. The recess D in the valve is always in communication with the port E and this, in turn, is open either to atmosphere or to the condenser. Ports P serve to convey steam to and from the cylinder. The valve is driven from an eccentric keyed to the crankshaft. It reciprocates across the ports and opens them alternately to admit high-pressure steam from the valve chest and to exhaust the used steam through recess D to the exhaust port E.

In the figure the valve is shown in its mid-position relative to the ports. The outer edge of the valve overlaps the port P by the amount s, termed the *steam* lap. The inner edge of the valve overlaps the port P by the amount e, termed the *exhaust* lap.

In considering the displacement of the valve it will generally be sufficiently accurate to assume that it takes place with simple harmonic motion, since the obliquity of the eccentric rod is very small. On this assumption the eccentric centre line OE will be at right angles to the line of stroke when the valve is in its mid-position. This is shown to the right of Fig. 101 (a) for clockwise rotation of the crank.

An alternative type of valve is shown in Fig. 101 (b). This valve possesses certain advantages over the ordinary slide or D-valve, particularly for high-pressure, superheated steam. The valve consists essentially of two rigidly connected pistons which work in cylindrical liners and control the admission to, and exhaust from, the two ends of the cylinder. With this valve there is no unbalanced steam thrust between the valve and its seat, such as exists with the D-slide valve, and the power absorbed in operating

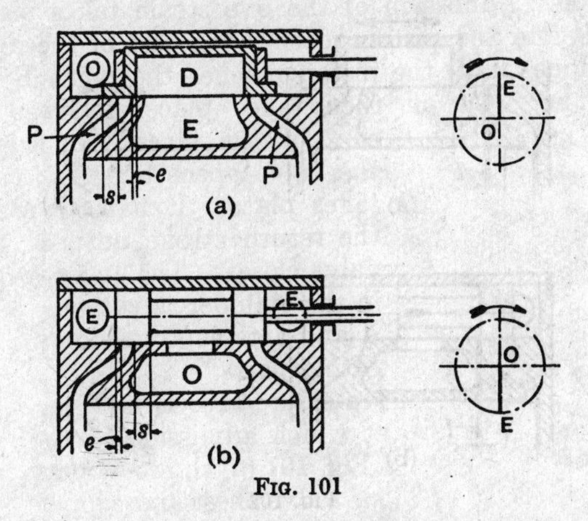

Fig. 101

the valve is therefore reduced. The live steam is usually admitted to the space between the two pistons and exhaust takes place from the ends of the valve chest. This has the advantage that the valve spindle packing, where the spindle passes through the end of the valve chest, is only subjected to the relatively low temperature and pressure of the exhaust steam and leakage is reduced. Most locomotives and the high-pressure cylinders of marine engines are fitted with piston valves. For clockwise rotation of the crank and with the valve in its mid-position, the eccentric position is shown to the right of Fig. 101 (b).

63. Relative Positions of Crank and Eccentric Centre Lines. (a) *Outside Admission.* At the beginning of the stroke of the piston from left to right, Fig. 102 (a), the crank will be on the dead centre and in order to admit steam the valve must be displaced from its mid-position towards the *right* by an amount at least equal to the steam lap s. In practice the displacement of the valve is greater than this by an amount l, termed the ***lead*** of the valve. The provision of lead ensures a larger opening of the port to steam during the early part of the stroke of the piston, and results in less wire-drawing or throttling of the steam during the admission period.

To give the above displacement of the valve from its mid-position, the eccentric centre line must be in advance of the 90° position by an angle α, such that $\sin \alpha = (s+l)/OE$.

The angle α is termed the *angle of advance* of the eccentric. The relative positions of the crank and eccentric centre lines remain unchanged during rotation of the crank. They are shown to the right of Fig. 102 (a).

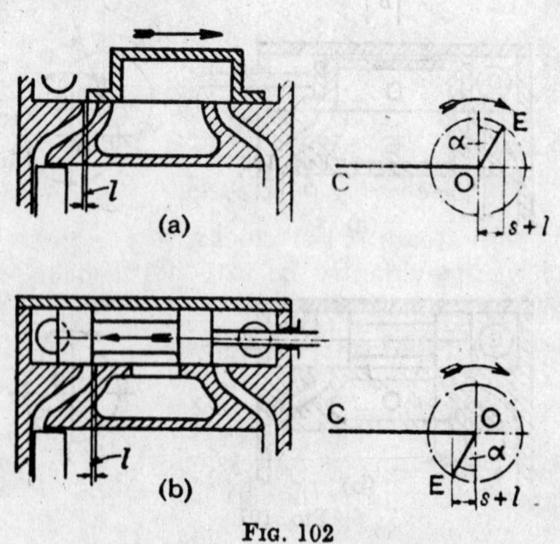

Fig. 102

(b) *Inside Admission.* At the beginning of the outward stroke of the piston, from left to right, Fig. 102 (b), the crank is on the inner dead centre and, in order to admit steam, the valve must be displaced from its mid-position towards the *left* by an amount at least equal to the steam lap s. As in the case of outside admission, the valve is given lead and the relative positions of the crank and eccentric centre lines are as shown at the right of Fig. 102 (b). As before, $\sin \alpha = (s+l)/OE$, but the angle of advance is now $180°+\alpha$.

64. Crank Positions for Admission, Cut-off, Release and Compression. In this Article the case of a valve with outside admission is considered. The same principles may be applied to a valve with inside admission.

At admission and cut-off for the cover end of the cylinder the outer edge of the valve coincides with the outer edge of the port. The valve is therefore displaced from its mid-position towards the right by an amount equal to the steam lap s, Fig. 103. At admission the valve is moving towards the right, as shown by the arrow A, so as to open the port to steam, whereas at cut-off the valve is moving towards the left, as shown by the arrow B, so as

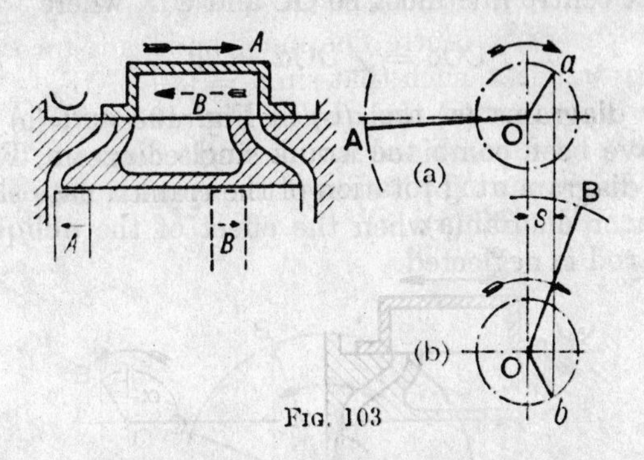

Fig. 103

to close the port to steam. For clockwise rotation of the crank these conditions can evidently be satisfied only if the eccentric centre line occupies the position Oa at admission and Ob at cut-off. The corresponding crank positions are then given by OA and OB where

$$\angle AOa = \angle BOb = 90° + \alpha$$

Similarly, at release and compression the valve is displaced from its mid-position towards the left by an amount equal to the exhaust lap e, Fig. 104, so that the inner edge of the valve coin-

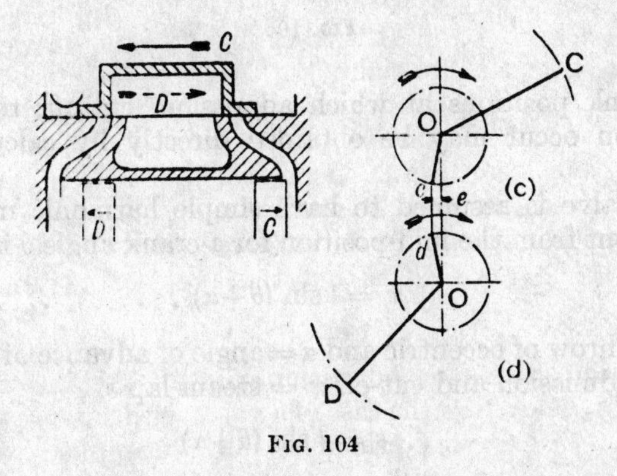

Fig. 104

cides with the inner edge of the port. At release the valve is moving towards the left, as shown by arrow C, so as to open the port to exhaust, while at compression the valve is moving towards the right, as shown by arrow D, so as to close the port to exhaust. To satisfy these conditions for clockwise rotation of the crank, the eccentric centre line must occupy the position Oc at release and Od at compression and the corresponding positions of the crank centre line must be OC and OD, where

$$\angle COc = \angle DOd = 90° + \alpha$$

The four diagrams (a) and (b) of Fig. 103 and (c) and (d) of Fig. 104 have been combined into a single diagram, Fig. 105 (a). Below this diagram at (b) is shown the approximate shape of the ideal indicator diagram, when the effect of the obliquity of the connecting rod is neglected.

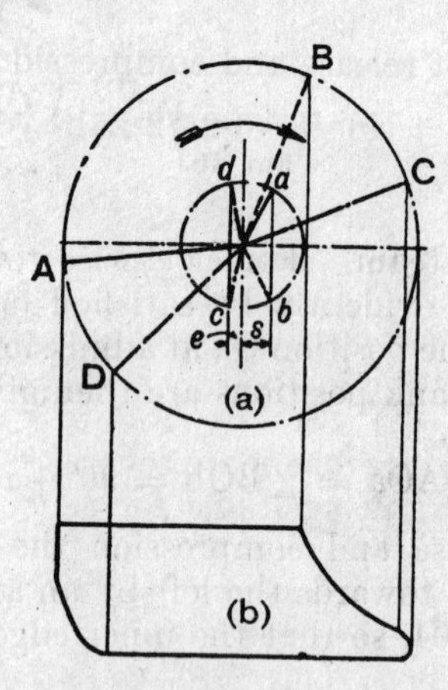

FIG. 105

The crank positions at which admission, cut-off, release and compression occur may be obtained directly by calculation as follows:

If the valve is assumed to have simple harmonic motion, its displacement from the mid-position for a crank angle θ is given by

$$x = t \sin(\theta + \alpha) \quad . \quad . \quad . \quad . \quad (5.1)$$

where t = throw of eccentric and α = angle of advance of eccentric.

But at admission and cut-off x = steam lap s.

$$\therefore \; s = t \sin(\theta + \alpha)$$

or

$$\theta + \alpha = \sin^{-1}(s/t) \quad . \quad . \quad . \quad . \quad . \quad (5.2)$$

The two values of θ which satisfy this equation give the crank positions for admission and cut-off.

Similarly, at release and compression, x = exhaust lap e and is negative.

$$\therefore \quad -e = t \sin(\theta+\alpha)$$

or
$$\theta+\alpha = \sin^{-1}(-e/t) \quad . \quad . \quad . \quad (5.3)$$

The two values of θ which satisfy this equation give the crank positions for release and compression.

Example 1. Let $s = 1\frac{1}{8}$ in., $e = \frac{1}{4}$ in., $t = 2\frac{1}{4}$ in., $\alpha = 35°$. Then, from (5.2), at admission and cut-off,

$$\theta+\alpha = \sin^{-1}(s/t) = \sin^{-1}(1{\cdot}125/2{\cdot}25) = 30° \quad \text{or} \quad 150°$$

$\therefore$ at admission $\theta = 30°-35° = -5°$ and at cut-off $\theta = 150°-35° = 115°$.

And from (5.3), at release and compression:

$$\theta+\alpha = \sin^{-1}(-e/t) = \sin^{-1}(-0{\cdot}25/2{\cdot}25) = 180°+6°\ 23' \quad \text{or} \quad 360°-6°\ 23'$$

$\therefore$ at release: $\theta = 186°\ 23'-35° = 151°\ 23'$

and at compression: $\theta = 353°\ 37'-35° = 318°\ 37'$

Fig. 105 is drawn to scale for the particulars given in this example.

65. Valve Diagrams. There are several graphical constructions which enable the crank positions for admission, cut-off, release

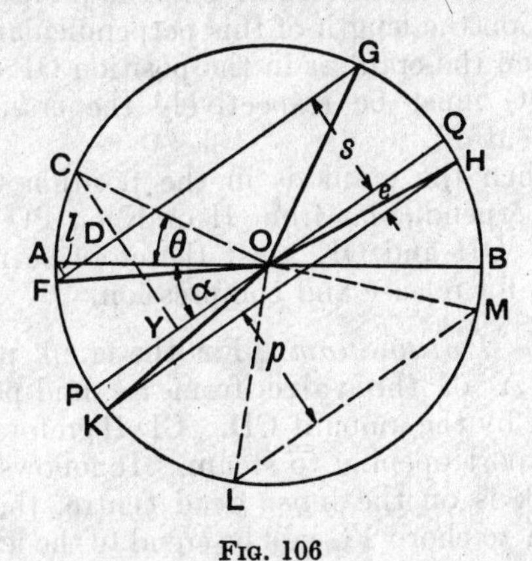

FIG. 106

and compression to be found directly without first determining the eccentric positions. Two of these constructions follow.

(a) *The Reuleaux Diagram*: Fig. 106. This diagram is simple to draw and for most problems on slide valves is the best one to use. A circle is drawn on AB as diameter, where AB is equal to the valve travel. For clockwise rotation of the crank a diameter PQ is drawn making an angle α (the angle of advance) with the diameter AB. Chords FG and KH are drawn parallel to diameter PQ and at distances from it respectively equal to the steam lap s and the exhaust lap e. The extremities F, G, H and K of the two chords are joined to the centre O of the circle in order to give the crank positions required.

Thus OF is the crank position for admission,
OG ,, ,, cut-off,
OH ,, ,, release,
OK ,, ,, compression.

Proof of the Construction. Take any crank position OC making an angle θ with the i.d.c. Draw CY perpendicular to PQ.

Then $CY = OC \sin COY = OC \sin(\theta+\alpha) = t \sin(\theta+\alpha)$.

But $t \sin(\theta+\alpha) = x =$ the displacement of the valve from its mid-position corresponding to crank angle θ, equation (5.1).

$$\therefore CY = x$$

With the above construction the length of the perpendicular from C to the diameter PQ is equal to the valve displacement from mid-position when the crank is in the position OC.

By construction the length of this perpendicular is equal to the steam lap s when the crank is in the position OF or OG. Therefor OF and OG must be respectively the crank positions for admission and cut-off.

Similarly, when the crank is in the position OH or OK the length of the perpendicular from H or K on PQ is equal to the exhaust lap e. OH and OK must therefore be respectively the crank positions for release and compression.

Opening of the Port to Steam. For the crank position OC, the displacement CY of the valve from its mid-position exceeds the steam lap s by the amount CD. CD therefore represents the amount of the port opening to steam. It follows from this that when the crank is on the inner dead centre, the perpendicular distance from A to chord FG will be equal to the lead of the valve.

The maximum possible opening of the port to steam is evidently the difference between the half-travel t of the valve and the steam lap s.

Similarly, the maximum possible opening of the port to exhaust is the difference between the half-travel t of the valve and the exhaust lap e. This difference may exceed the width of the actual port P, Fig. 101, through which the steam is admitted to and exhausted from the cylinder. In that case the port P will remain fully open for a certain period of crank rotation. The duration of this period may be found by drawing a chord LM parallel to KH at a distance from it equal to the width p of the steam port. The port will remain fully open to exhaust during the rotation of the crank from OM to OL.

(b) *The Bilgram Valve Diagram*: Fig. 107. As before, a circle is drawn on a diameter AB equal to the valve travel and diameter PQ is drawn inclined at the angle of advance α to AB. With centre P (or Q) two circles are drawn with radii respectively equal to the steam and exhaust laps.

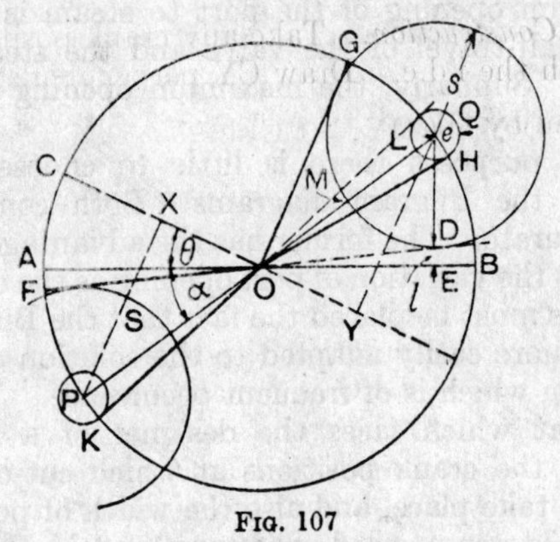

FIG. 107

FO produced and OG are tangential to the steam lap circle. They represent respectively the crank positions for cut-off and admission. Similarly KO produced and OH are tangential to the exhaust lap circle and they represent respectively the crank positions for compression and release.

Proof of the Construction. Take any crank position OC. From P and Q draw lines PX and QY perpendicular to OC to meet OC and CO produced respectively at X and Y. Then triangles POX, QOY are equal in all respects.

$$\therefore \text{ QY} = \text{PX} = \text{PO} \sin \text{POX}$$
$$= t \sin (\theta + \alpha)$$

$\therefore$ PY = PX = displacement of the valve from its mid-position corresponding to crank position OC. Hence, with this construction, the displacement of the valve from its mid-position for a given crank position is equal to the perpendicular distance of P or Q from that crank position.

Since, by construction, the perpendicular distance of Q from crank positions OF and OG is equal to the steam lap, these two crank positions must correspond respectively to admission and cut-off. In a similar way the distances of Q from OH and OK are each equal to the exhaust lap. Therefore, OH and OK must be respectively the crank positions for release and compression.

The port is open to steam while the crank turns from OF to OG, and is open to exhaust while the crank turns from OK to OH. The amount of port opening for any crank position such as OC is the difference between PX and the steam lap, i.e. SX. For the dead-centre position the port opening or lead is given by DE, where QE is perpendicular to AB.

The maximum opening of the port to steam is the difference between the half-travel of the valve and the steam lap. It is given by OM. Similarly, the maximum opening of the port to exhaust is given by OL.

For general purposes there is little to choose between the Reuleaux and the Bilgram diagrams. Both constructions are simple and accurate. The former has the advantage that it shows up more clearly the variation of port opening as the crank revolves, but against this must be placed the fact that the Bilgram diagram is very much more easily adapted to the solution of a particular type of problem which is of frequent occurrence. The particular problem is that which faces the designer of a steam engine, namely: Given the crank positions at which cut-off and release are required to take place, and also the width of port required to give a reasonable steam speed, what must be the throw and angle of advance of the eccentric and the steam and exhaust laps of the valve?

This problem cannot be solved directly by the Reuleaux diagram, but the Bilgram diagram may be used as follows:

Referring to Fig. 107, it will be seen that the point Q is:

(a) at a distance from O equal to the steam lap plus the maximum opening to steam,
(b) at a distance from OG equal to the steam lap,
(c) at a distance from AB equal to the steam lap plus the lead.

Draw a line AB, Fig. 108, to represent the line of stroke. Through any point O on AB draw OG to indicate the crank position at which cut-off takes place. Draw a circle with centre O and radius equal to the maximum opening to steam. Parallel to and above AB draw a line VW at a distance from AB equal to the lead of the valve. (N.B.—The amount of the lead would have to be assumed, say, $\frac{1}{8}$ in. to $\frac{1}{4}$ in.) The problem then resolves itself into finding the position of point Q such that Q is equidistant from OG, VW and the circle already drawn with centre O. The point Q may be found by trial, particularly if the line bisecting the angle between OG and VW is first drawn, as Q must lie on this bisecting line.

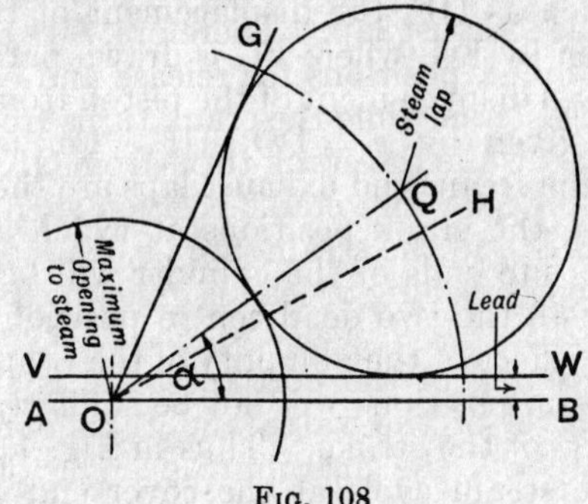

Fig. 108

When the position of Q has been found, the inclination of OQ to AB will give the angle of advance and the length OQ will give the throw of the eccentric. Moreover, the distance of Q from OG will be the steam lap required for the valve and the distance of Q from the dotted line OH, which is the crank position for release, will give the exhaust lap. The valve diagram may then be completed as shown in Fig. 107 and the crank positions for admission and compression determined.

66. The Piston Positions for Admission, etc. The Reuleaux and the Bilgram diagrams determine the crank positions at which the chief events in the cycle take place. The corresponding piston positions obviously depend on the ratio n of the length of the

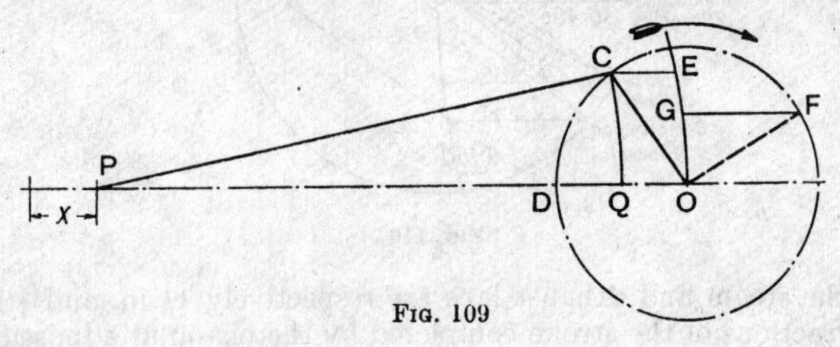

Fig. 109

connecting rod to the length of the crank. To find the piston position for a given crank position, the construction shown in Fig. 109 may be used. Let x be the displacement of the piston from the beginning of the stroke when the crank is in the position OC. Then, obviously, if the arc CQ is drawn with P as centre and

PC as radius, $DQ = x$. With centre on the line of stroke and radius PC draw an arc through O; through C draw CE parallel to the line of stroke. Then CE = QO = distance of the piston from mid-stroke when the crank is in the position OC. Hence, for any crank position such as OF, the displacement of the piston from mid-stroke is given by FG, where FG is drawn parallel to the line of stroke. The total displacement of the piston from the beginning of the stroke is given by $x = DO + GF$. For a double-acting engine, in which the steam and exhaust laps are the same for both ends of the valve, the crank positions at which admission, etc., take place at opposite ends of the cylinder will be similarly situated with respect to the two dead-centre positions of the crank. But, owing to the effect of the obliquity of the connecting rod, the corresponding piston positions will not be similarly situated with respect to the ends of the stroke. Thus in Fig. 110 the full lines correspond to the steam cycle at the cover end of the cylinder and the dotted lines to the steam cycle at the crank end of the cylinder, the steam and exhaust laps at the cover end of the valve having the same values as those at the crank end.

Example 2. Fig. 110 is drawn to scale for an engine with a connecting rod to crank ratio of 4; the valve is driven from an eccentric with a throw of $2\frac{1}{4}$ in. and an angle of advance of 35°

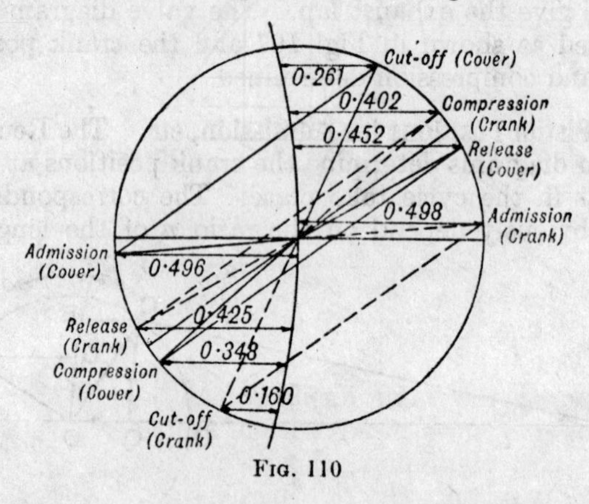

Fig. 110

and the steam and exhaust laps are respectively $1\frac{1}{8}$ in. and $\frac{1}{4}$ in. The fractions of the stroke completed by the piston at admission, etc., are given in the following table:

	Cover end	Crank end
Admission	0·996	0·998
Cut-off	0·761	0·660
Release	0·952	0·925
Compression	0·848	0·902

It will be seen from these figures that admission and compression take place later and cut-off and release take place earlier at the crank end than at the cover end.

It is generally desirable to equalise as far as possible the work done at the two ends of the cylinder. With this object the steam lap at the crank end may be made smaller than that at the cover end. For instance, in order to obtain cut-off at 0·761 of the stroke, the steam lap at the crank end would have to be reduced to 0·69 in. Admission would then take place at 0·982 of the stroke.

67. The Rectangular Valve Diagram: Fig. 111. This diagram enables the relative positions of the piston and the valve at any point of the stroke to be determined. The valve and piston displacements are plotted, not necessarily to the same scale, on a crank angle base. For convenience the two displacement curves are drawn with the line OX representing mid-stroke. In plotting the two curves the actual values of the piston and valve displacements for given values of the crank angles may be obtained either graphically or by calculation. If the angle of advance of the eccentric is α, then the eccentric leads the crank by the angle $90+\alpha$

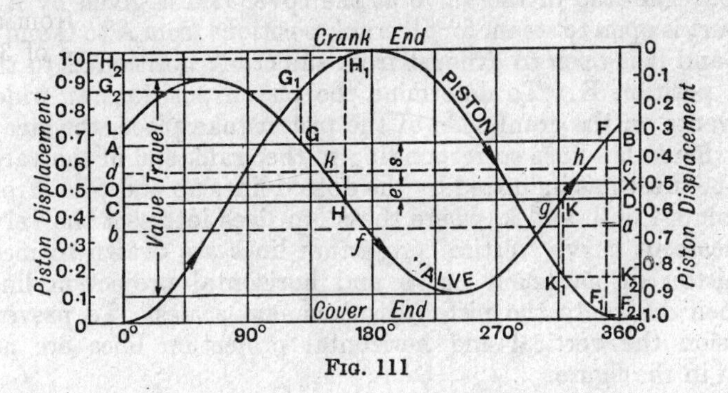

FIG. 111

and consequently the valve displacement reaches its maximum value for an angle $90+\alpha$ in advance of that for which the piston displacement is a maximum. To determine the events on the cover side of the piston, the full lines AB and CD are drawn as shown parallel to OX. The distances of AB and CD from OX are equal respectively to the steam and exhaust laps for the cover end of the valve and are set off to the same scale as the valve displacement curve. The line AB cuts the valve displacement curve at F and G. At G the valve is closing and cut-off is therefore taking place. At F the valve is opening and admission is taking place. The corresponding fractions of the stroke may be obtained by projecting vertically from the valve displacement curve to the piston displacement curve and then horizontally to the piston displacement scale, as shown by the dotted lines FF_1, F_1F_2 and GG_1, G_1G_2. Since G_1 is on the outward half of the piston displacement curve, the fraction of the stroke at which cut-off occurs is obtained by projecting from G_1 horizontally to G_2, when the fraction is read from the left-hand scale as 0·86. Similarly, since F_1 is on the return half of the piston displacement curve, F_1 is projected horizontally to F_2 on the right-hand scale and the fraction of the stroke at which admission takes place is read off as 0·99.

In the same way the points of release and compression may be found. At the point of intersection H of the line CD and the valve displacement curve, the port is opening to exhaust and release is taking place, while at the point of intersection K the port is closing to exhaust and compression is taking place. The corresponding piston positions for release and compression are shown by H_2 and K_2.

Reading from the scales, release takes place at 0·98 of the outward stroke and compression begins at 0·85 of the return stroke. The lead of the valve at the cover end is given by AL. The port is open to steam for all crank positions from A to G and F to B, and it is open to exhaust from the crank position H to the crank position K. To determine the piston positions at which the events on the crank side of the piston take place, the steam and exhaust lap lines corresponding to the crank end of the valve must be drawn as indicated by the dotted lines ab and cd. From the points f, g, h and k, where these lap lines intersect the valve displacement curve, vertical projection lines are drawn to meet the piston displacement curve and horizontal projection lines are then drawn to the piston displacement scales. To prevent confusion the vertical and horizontal projection lines are not shown in the figure.

68. The Oval Valve Diagram: Fig. 112. If the displacements of the valve are plotted as ordinates against the displacements of the piston as abscissæ, the oval valve diagram is obtained. The arrows serve to indicate the direction of the relative displacements as the piston moves on the outward and return strokes. Owing to the effect of the obliquity of the connecting rod and of the eccentric rod the curve is flatter on the return stroke than on the outward stroke. This diagram may be used to give directly the piston positions at which the various events in the steam cycle take place. The line OX is drawn across the diagram to correspond to the mid-stroke of the valve. Lines AB and CD are drawn parallel to OX and at distances from OX respectively equal to the steam lap and the exhaust lap at the cover end of the valve. At the points G and F, where AB intersects the oval curve, the displacement of the valve from its mid-position is equal to the steam lap. The displacement of the valve is increasing at the point F, so that this point corresponds to admission of the steam to the cylinder, while at point G the displacement is decreasing so that the supply of steam to the cylinder is cut-off. The point G lies on the outward half of the displacement curve and the fraction of the stroke at which cut-off occurs is found by projecting vertically from G to the piston displacement scale at the top of the diagram. For the given diagram this fraction is 0·86 of the stroke. Similarly, the point F lies on the return half of the displacement curve and, by projecting vertically downwards to the

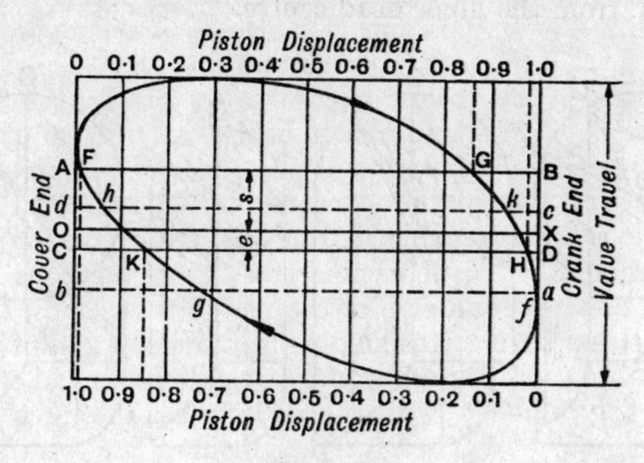

Fig. 112

piston displacement scale at the bottom of the diagram, the fraction of the stroke at which admission takes place is read off as 0·99. In the same way the point at which release takes place is found to be 0·98 of the outward stroke and the point at which compression begins is found to be 0·85 of the return stroke. The timing of the events on the crank side of the piston may be determined in exactly the same way by drawing the dotted lap lines ab and cd, which correspond to the steam and exhaust laps at the crank end of the valve.

The oval valve diagram, Fig. 112, is drawn to scale for the same particulars as the rectangular valve diagram, Fig. 111.

The rectangular and the oval valve diagrams are particularly useful for examining the timing of the events in the steam cycle of an engine which is fitted with a reversing gear, such as the Stephenson or Walschaert gear. In such gears the valve is not driven directly by a single eccentric, so that the Reuleaux or Bilgram diagrams cannot be used except for an approximate analysis.

69. Alteration of the Point of Cut-off with a Simple Slide Valve. It is of interest to consider how the point of cut-off for a given engine may be altered and what will be the effect of this alteration on the points of admission, release and compression.

Four valve diagrams with the corresponding ideal indicator diagrams are shown at (a), (b), (c) and (d) in Fig. 113. The first of these, (a), is for a slide valve of normal proportions with a normal angle of advance for the eccentric. It will be seen that cut-off occurs when the crank is in the position OG approximately 115° from the inner dead centre. The other diagrams (b), (c) and (d) relate to three different methods of obtaining cut-off at a crank position 90° from the inner dead centre.

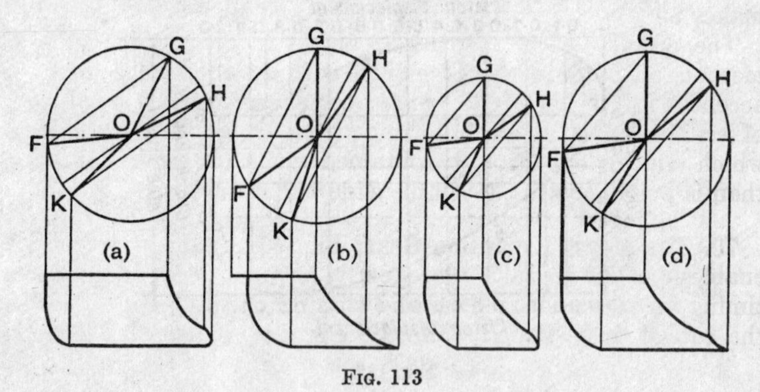

FIG. 113

In diagram (b) the earlier cut-off is obtained in the simplest possible way by increasing the angle of advance of the eccentric, while the throw of the eccentric and the steam and exhaust laps of the valve remain as in diagram (a). The effect of this alteration is to advance the timing of release, compression and admission, as well as that of cut-off, and the indicator diagram is modified as shown. Any considerable change of the point of cut-off from the normal by this method would involve objectionably early admission and release, and, in consequence, a shorter effective stroke.

In diagram (c) the earlier cut-off is again obtained by increasing the angle of advance of the eccentric, but in addition the travel of the valve is reduced by shortening the throw of the eccentric, in order to retain approximately the same timing for admission as in the normal diagram (a). The steam and exhaust laps are untalered. Release and compression take place earlier than in the normal diagram, but not so early as in diagram (b). The objection to this method is that the maximum opening of the port to steam and exhaust will be reduced by shortening the valve travel and wiredrawing or throttling of the steam will therefore take place.

It will be shown later, in Articles 74 *et seq.*, that *linking up* or *notching up* a steam-engine reversing gear is equivalent in its effect on the steam distribution to increasing the angle of advance and shortening the throw of the eccentric.

The fourth diagram (d) illustrates a method of obtaining earlier cut-off to which there are practical objections. The travel and the lead of the valve are the same as in diagram (a), but the steam lap of the valve and the angle of advance of the eccentric are both increased in order to give the earlier cut-off. The advantages of this method would be a normal timing of the admission and a smaller reduction in the maximum opening of the port to steam. However, the necessity for increasing the steam lap of the valve makes it unsuitable from a practical standpoint.

The disadvantages mentioned above in connection with the methods of obtaining early cut-off with a simple slide valve are accentuated, if cut-off is required to take place at a crank angle of less than 90°. In the following article a valve gear is described which enables cut-off to be obtained much earlier in the stroke than is possible with the simple slide valve.

70. The Meyer Expansion Gear: Fig. 114. This gear not only enables cut-off to take place early in the stroke with normal timing for admission, release and compression, but it also enables the cut-off to be varied while the engine is running. There are

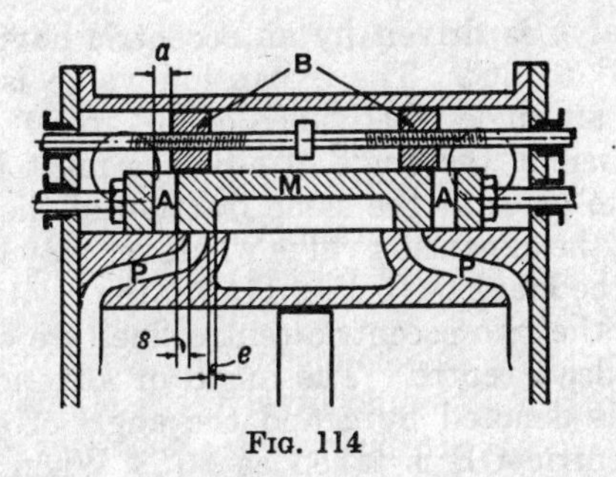

FIG. 114

two valves driven by separate eccentrics. The main valve M is similar to the ordinary slide valve, except that it is provided with extensions and the steam passes from the steam chest through the ports A, A. The back of the main valve is machined and a second valve, which generally takes the form of two separate plates B, B, reciprocates across it and alternately admits and cuts off the supply of steam to the ports A, A. Steam will not be admitted to the cylinder, even when the main valve is in such a position that ports A and P are in communication, unless at the same time the expansion plate has uncovered port A.

In Fig. 114 both the valves are shown in mid-position. The steam and exhaust laps of the main valve are denoted by s and e and the steam lap of the expansion valve by a. Note that the latter is negative, i.e. the port A is open to steam when the valve is in mid-position. The main valve determines the points of admission, release and compression and also the latest point of cut-off. The expansion valve merely serves to cut off the steam from the port A while this is still open to port P. The expansion valve or plate will cut-off or admit steam to port A at those crank positions in which its displacement relative to the main valve is towards the left (if we consider the steam cycle on the left of the piston) and equal to the steam lap a.

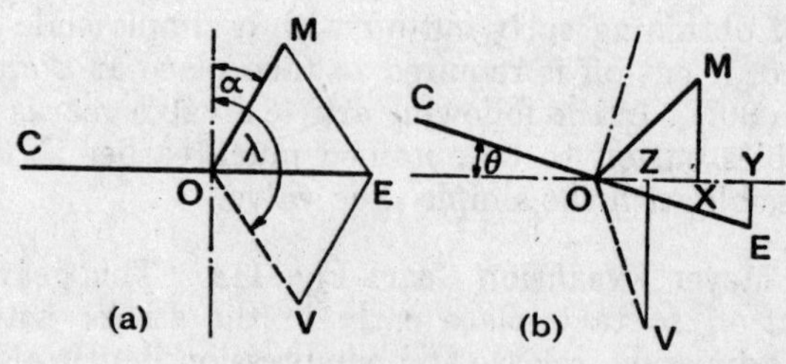

FIG. 115

The main valve is driven by an eccentric having an angle of advance of 25° to 30°. The expansion valve is driven by an eccentric with an angle of advance of 80° to 90°. If an engine has to be reversible, the angle of advance must be 90°, so that cut-off will take place at the same fraction of the stroke for the same setting of the expansion valve whatever may be the direction of rotation of the crank. In Fig. 115 (a) the relative positions of the crank and the two eccentric centre lines are shown with the crank on the dead centre. The angle of advance of the main eccentric OM is denoted by α and the angle of advance of the expansion eccentric OE is taken as 90°. When the crank has turned through an angle θ from the dead-centre position, as shown in Fig. 115 (b), the displacement of the main valve from its mid-position will be represented by OX, the projected length of OM. Similarly, the displacement of the expansion valve from its mid-position will be represented by OY, the projected length of

OE. In both cases it is assumed that the obliquity of the eccentric rods may be neglected. XY, the difference between OY and OX, will be the displacement of the expansion valve relative to the main valve, in this case towards the right. If through O a line OV is drawn parallel to and equal in length to ME, then OZ, the projected length of OV, will be equal to XY and therefore to the displacement of the expansion valve relative to the main valve. OV is termed the "virtual" or "equivalent" eccentric. Its angle of advance is λ. Cut-off will take place for the crank position in which Z lies at a distance a to the left of O. This position is most easily found by applying the Reuleaux construction to the "virtual eccentric" OV. In Fig. 116 (a) the Reuleaux valve

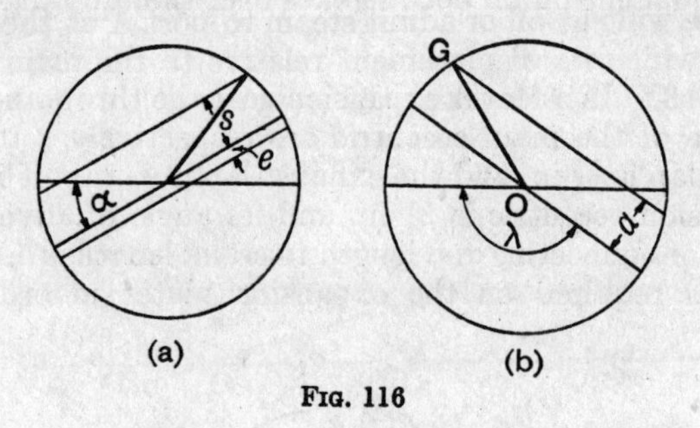

Fig. 116

diagram has been drawn for the main eccentric OM, in order to determine the crank positions for admission, release and compression and for the latest possible cut-off. In Fig. 116 (b) the Reuleaux valve diagram has been drawn for the virtual eccentric OV, bearing in mind that the steam lap for the expansion valve is negative. In this figure OG is the crank position at which cut-off takes place when the steam lap of the expansion valve is a. An increase or decrease of the steam lap a gives respectively a later or an earlier cut-off. One method of altering the steam lap a is illustrated in Fig. 114. The valve spindle is screwed with a right- and a left-hand thread so that, by rotating the spindle, the distance between the two expansion plates may be increased or decreased at will.

71. Minimum Width of the Expansion Plate. The Best Setting of the Plates. The maximum displacement, from the mid-position, of the expansion valve relative to the main valve is equal to the throw of the equivalent eccentric OV. The maximum overlap of the expansion valve and the port will be $OV-a$. Hence, the minimum width of the expansion plate $= OV-a+p$, where p is the width of the port A in the main valve. This width of expansion plate will just be sufficient to prevent steam from being re-admitted past the inner edge of the plate.

If cut-off is required to take place at the same fraction of the stroke for both strokes, then the steam lap a must be different for the two expansion plates, owing to the effect of the obliquity of the connecting rod. For instance, the difference between the two steam laps will not be the same for cut-off at 0·3 of the stroke as for cut-off at 0·4, or any other fraction of the stroke. When the expansion plates are assembled on the valve spindle, they may be given different laps, but the difference, once fixed, will remain constant for all values of the steam lap. It is therefore necessary to adopt that difference which will give as nearly as possible equal cut-off on both strokes over the full range of cut-off required.

Example 3. In a Meyer expansion gear the throw and the angle of advance of the main eccentric are respectively 2 in. and 30°, the steam lap is $\frac{13}{16}$ in. and the exhaust lap is $\frac{3}{16}$ in. The throw of the expansion eccentric is $2\frac{1}{4}$ in. and its angle of advance is 90°. The ratio of connecting-rod length to crank length is 5. Find the steam laps required on the expansion plates in order to give

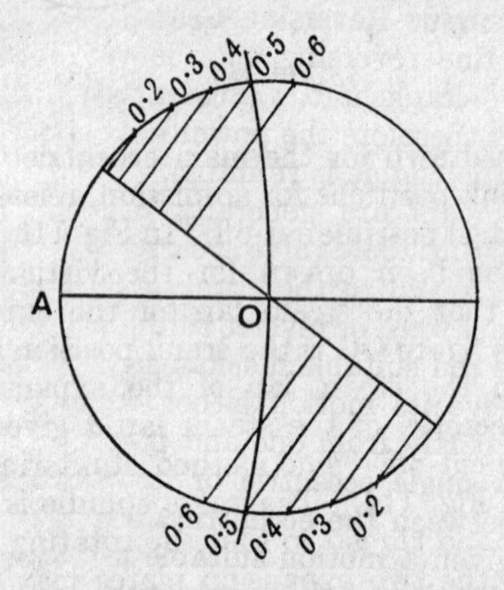

FIG. 117

cut-off at 0·2, 0·3, 0·4, 0·5 and 0·6 of the stroke on both strokes. What is the best setting of the plates? If the width of the steam port in the back of the main valve is $1\frac{1}{8}$ in., what is the minimum width which the expansion plates may have?

From Fig. 115, which is drawn to scale for the above gear, the equivalent eccentric has a throw OV of 2·14 in. and an angle of advance λ of 144·2°.

In Fig. 117 the Reuleaux diagram is drawn for the virtual eccentric OV. A circular arc is drawn through O with centre on OA produced and with radius equal to 5.OA. The crankpin positions corresponding to the fractions of the stroke at which cut-off is required to take place are then marked off as shown. The required values of the steam lap *a* are measured from the diagram and are entered in the table below.

Cut-off	0·2	0·3	0·4	0·5	0·6
Steam lap (cover) . .	0·49	0·93	1·29	1·61	1·85
,, ,, (crank) . .	0·81	1·26	1·59	1·84	2·02
Difference	0·32	0·33	0·30	0·23	0·17

From this table it appears that, if the expansion plates are set with the steam lap at the crank end 0·3 in. greater than that at the cover end, the cut-off will occur at approximately the same fraction of the stroke for both ends of the cylinder.

The minimum width of the expansion plates $= \text{OV}-a+p$ $= 2{\cdot}14-0{\cdot}49+1{\cdot}125 = 2{\cdot}78$ in.

N.B.—This is actually the minimum width of the expansion plate at the cover end. The minimum width of the expansion plate at the crank end need be only $2{\cdot}78-0{\cdot}3 = 2{\cdot}48$ in.

72. Steam-engine Reversing Gears. Primarily the function of the steam-engine reversing gear is to reverse the direction of rotation of the crankshaft. Incidentally it also enables the point of cut-off, and therefore the power developed by the engine, to be varied while the engine is running.

Reversing gears are generally classified as either (a) link motions or (b) radial valve gears. In the former class two eccentrics are keyed to the crankshaft, one for forward running and one for backward or reverse running. Between the eccentrics and the valve rod suitable mechanism is introduced to enable the valve to receive its motion either wholly from one of the two eccentrics or partly from one and partly from the other. In the latter class a single eccentric or its equivalent is used and the mechanism between the eccentric and the valve rod enables the valve to be given a motion suitable for either forward or reverse direction of rotation of the crank, and for early or late cut-off.

To determine accurately the piston positions at which admission, cut-off, release and compression take place for a given setting of the gear, the displacement curves for the valve and piston may be determined, either graphically or by calculation, and the rectangular valve diagram as described in Article 67 used. In most reversing gears the mechanism is so complicated that the calculation of the valve displacement curve would be an exceedingly laborious process. To obtain the required degree of accuracy by the use of graphical methods, the outline of the gear must be drawn to a large scale and for a large number of different positions. The graphical work may be simplified by the use of suitable templates.

In order to determine the approximate piston positions at which the various events take place, a simplified graphical construction may be used. The method consists in finding the throw and angle of advance of a single eccentric which, if driving the valve directly, would give to it a motion as nearly as possible identical with that which it receives from the reversing gear; this eccentric is termed a *virtual* or *equivalent* eccentric. The method of finding the throw and angle of advance of the virtual eccentric differs for the two types of reversing gear, and is dealt with in the following pages.

73. The Virtual Eccentric for a Valve with an Offset Line of Stroke. In Fig. 118, OC is the crank centre line, OE the eccentric centre line and EA the centre line of the eccentric rod. As the crank revolves the end A of the eccentric rod reciprocates along the line PA. It is required to find the throw and the angle of advance of an eccentric with axis at P, which will give to A the same motion as that which it derives from the actual eccentric OE.

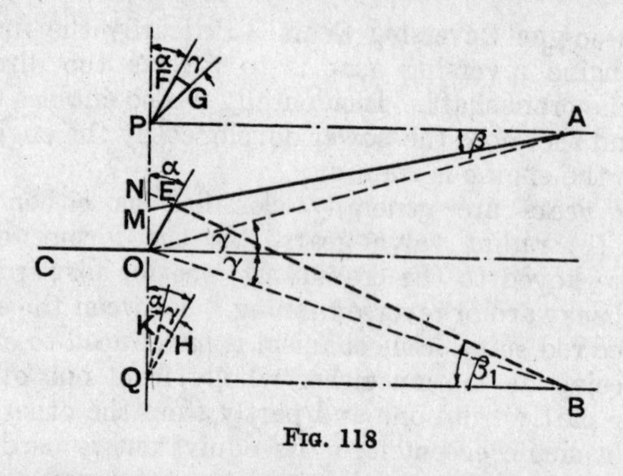

Fig. 118

Let α be the angle of advance of the actual eccentric, θ the angle through which the crank has turned from the dead centre.

Produce AE to cut at M a line through O perpendicular to PA.

If v_a is the velocity of the point A and v_e is the velocity of the point E, then it follows from Article 43 that:

$$\frac{v_a}{v_e}=\frac{OM}{OE}=\frac{\sin OEM}{\sin OME}=\frac{\sin\{(90-(\theta+\alpha+\beta)\}}{\sin(90-\beta)}$$

$$=\frac{\cos(\theta+\alpha+\beta)}{\cos\beta}$$

But $v_e = \omega . \text{OE}$, where ω is the angular velocity of the crank.

$$\therefore\ v_a = \omega . \frac{\text{OE}}{\cos \beta} . \cos (\theta + \alpha + \beta)$$

This is evidently the same velocity as A would have, if driven directly by an eccentric with centre P, throw OE/cos β and angle of advance $\alpha + \beta$. Although the angle β is not constant, its variation is small, as the eccentric rod EA is usually from 10 to 20 times as long as the eccentric throw OE. The angle γ represents approximately the mean value of β, so that the virtual eccentric may be taken to have a throw OE/cos γ and an angle of advance $\alpha + \gamma$.

If the line of stroke lies below the axis of rotation of the eccentric as for point B in Fig. 118, it can be shown in the same way that the virtual eccentric has the same throw OE/cos γ, but the angle of advance is $\alpha - \gamma$.

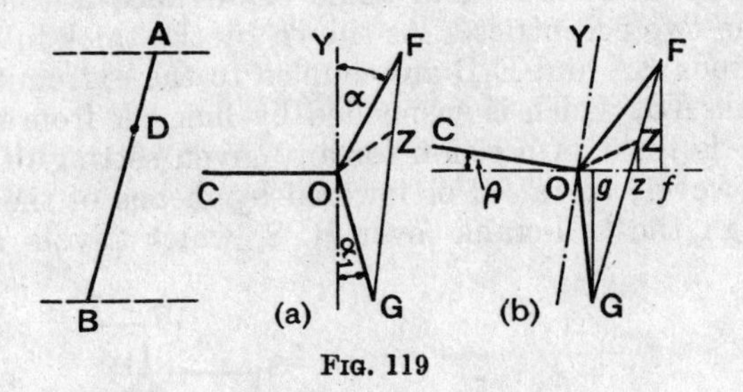

Fig. 119

The throw and angle of advance of the virtual eccentric may be determined graphically by the simple construction shown in the figure. Thus for A, the virtual eccentric is given by PG, where PF is equal and parallel to OE, FG is perpendicular to PF and the angle FPG is equal to γ. Similarly, for B, the virtual eccentric is given by QK, where QH is equal and parallel to OE, HK is perpendicular to QH and the angle HQK is equal to γ.

The usual arrangement in a link motion is for the ends A and B of a straight or curved link to be driven by separate eccentrics, which are keyed to the crankshaft and therefore revolve at the same speed and always in the same relative positions. The virtual eccentrics for the motion of the ends A and B may be determined as indicated above and it remains to show how the virtual eccentric for the motion of a given point D on the link AB may be found. Referring to Fig. 119, the ends A and B of the link AB are reciprocated along the paths shown by dotted lines. These two paths are assumed to be straight parallel lines. The virtual eccentrics for the motion of A and B are denoted at (a) by OF and OG respectively, the angles of advance being α and α_1. If the crank OC turns through the angle θ, Fig. 119 (b), the displacement of A from its mid-position is given by Of and the displacement of B from its mid-position by Og, so that the displacement of D from its mid-position will be given by Oz, where Oz is the projected length of OZ and Z divides FG in the same proportion as D divides AB. It follows that the virtual eccentric for the motion of D has a throw equal to OZ and an angle of advance equal to angle YOZ. The above principles may be applied to find the virtual eccentric for a given setting of any link motion.

74. The Stephenson Link Motion: Fig. 120. This is the most commonly used reversing gear of its class. It is simple in construction and gives a good steam distribution. The figure shows diagrammatically the arrangement of the gear in mid-position. OC is the crank centre line and OE and OE_1 the two eccentric centre lines. The throw and angle of advance is the same for each of the two eccentrics. As shown by the thick full lines, the eccentric rods EA and E_1B are coupled to the extremities of the curved link AB, which is suspended by link SA from the pin S. The pin S is fixed in position for any given setting of the gear. It can, however, be raised or lowered by means of the reversing rod through the bell-crank lever RPS, which pivots about the

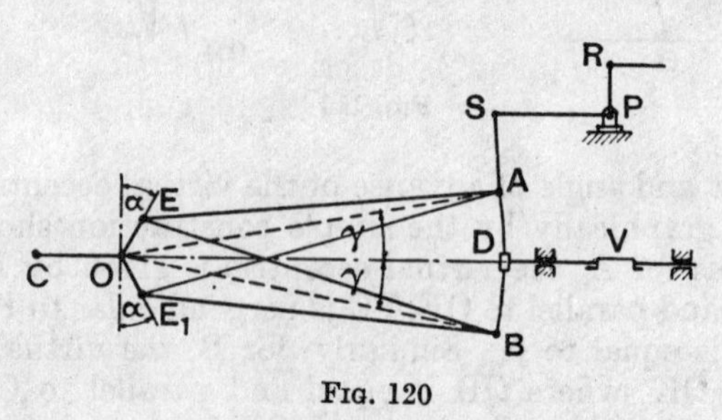

FIG. 120

fixed fulcrum P. This causes the curved link AB to slide through the block D and enables the latter to derive its motion from either B or A, or from any intermediate point on AB. In this way the point of cut-off may be altered and the direction of rotation of the crank may be reversed. The valve receives its motion from the block D and the valve rod is guided horizontally.

The Stephenson link motion described above, in which the eccentrics OE and OE_1 drive respectively the ends A and B of the curved link, is said to have *open* rods. If OE drives B and OE_1 drives A, as shown by the thin full lines, the gear is said to have *crossed* rods. This latter arrangement gives a different steam distribution.

When the curved link is lowered so that A and D coincide or raised so that B and D coincide, the link motion is in the full-gear position. The valve receives its motion wholly from one eccentric, whilst the other eccentric merely oscillates link AB without affecting the movement of the valve. On the other hand, when the gear occupies an intermediate position, as shown in Fig. 121, the valve will derive its motion partly from one eccentric and

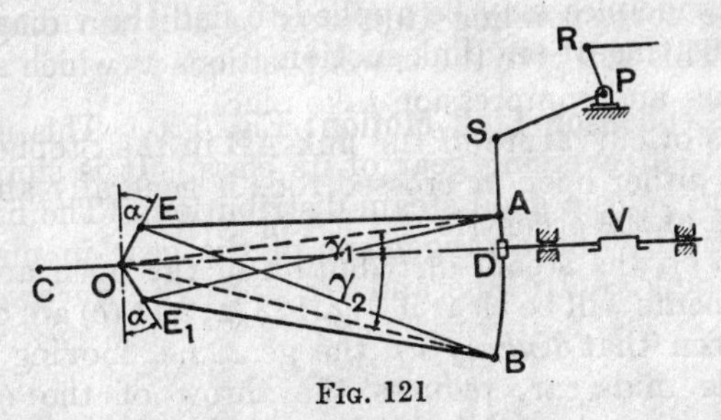

FIG. 121

partly from the other. The problem is to find the equivalent eccentric for the intermediate position.

If we assume that the ends A and B of the curved link move along straight paths which are parallel to the line of stroke of the valve, the equivalent eccentrics for the ends A and B and for the block D may be found as described in Article 73. This has been done for the two positions of the open rod gear, and the

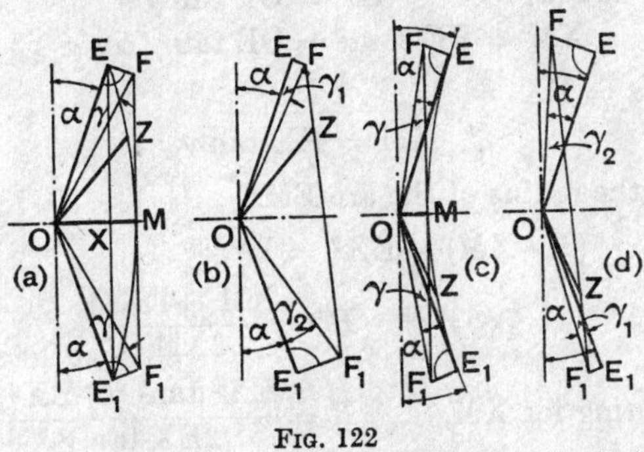

FIG. 122

construction is shown in Fig. 122 (a) and (b). In the former OM is the equivalent eccentric for mid-gear and in the latter OZ is the equivalent eccentric. If the construction is carried out for several different positions of the gear, a curve may be drawn through the various positions of Z. In practice a close approximation to this curve may be obtained by drawing a circular arc through E, M and E_1, as shown in Fig. 122 (a). The equivalent eccentric for the gear position of Fig. 121 may then be found by dividing the arc EME_1 at Z in the same proportion as D divides AB, and the result may be compared with that given by Fig. 122 (b).

Similarly, the equivalent eccentrics for the two positions of the crossed rod gear may be determined as shown in Fig. 122 (c) and (d).

Once the equivalent eccentric is known for a given setting of the gear, the corresponding Reuleaux or Bilgram diagram may be drawn. This will give the crank positions at which admission, cut-off, release and compression take place.

The radius of curvature of the link AB in the Stephenson link motion with either open or crossed rods is generally about equal to the length of the eccentric rod EA or E_1B.

The effect on the steam distribution of the open and crossed rod arrangements will be clear if Fig. 122 (a) and (c) are compared. It will be seen that *linking up* the gear, i.e. moving from full gear towards mid-gear, reduces the throw of the equivalent eccentric much more rapidly with crossed rods than with open rods. Since the displacement of the valve from its mid-position, when the crank is on the dead centre, is equal to the steam lap plus the lead, it follows that with open rods, the lead increases as the gear is linked up, while with crossed rods it decreases.

The radius of the curve EME_1 is approximately equal to $EE_1 . EA/2AB$. This can be shown as follows:

From Fig. 122 (a) $\quad EF = OE \tan \gamma$

and $\quad XM = EF \cos \alpha = OE \tan \gamma \cos \alpha$

But $OE \cos \alpha = EX$

$$\therefore XM = EX \tan \gamma$$

Let $R =$ the radius of the arc EME_1,

Then $\quad (R - XM)^2 + EX^2 = R^2$

Hence
$$R = \frac{XM^2 + EX^2}{2XM}$$

and, substituting for XM,
$$R = \frac{EX^2 \tan^2 \gamma + EX^2}{2EX \tan \gamma}$$
$$= \frac{EX \sec^2 \gamma}{2 \tan \gamma}$$
$$= \frac{EX}{\sin 2\gamma}$$

But $EX = EE_1/2$ and from Fig. 120, $\sin 2\gamma \simeq AB/EA$.

$$\therefore R \simeq EE_1 . EA/2AB$$

75. The Gooch Link Motion: Fig. 123. As already pointed out, the Stephenson link motion suffers from the disadvantage that the lead varies as the gear is *linked* or *notched* up. Admission, therefore, takes place earlier or later according as to whether open or crossed rods are used. The Gooch link motion was introduced in order to give constant lead for all gear positions.

It differs from the Stephenson link motion in that the curved link AB is convex towards the crankshaft instead of concave, and is suspended by link FG from a fixed fulcrum G on the engine frame. The valve rod is hinged at V and the portion DV is raised or lowered by the bell-crank lever KLM operated by the reversing rod. When D is raised so as to coincide with A, the valve receives its motion from eccentric OE and the crank revolves clockwise. When D coincides with B, the valve receives its motion from OE_1 and the crank revolves counter-clockwise. It is obvious that for the lead to be constant for all settings of the gear, the centre of curvature of

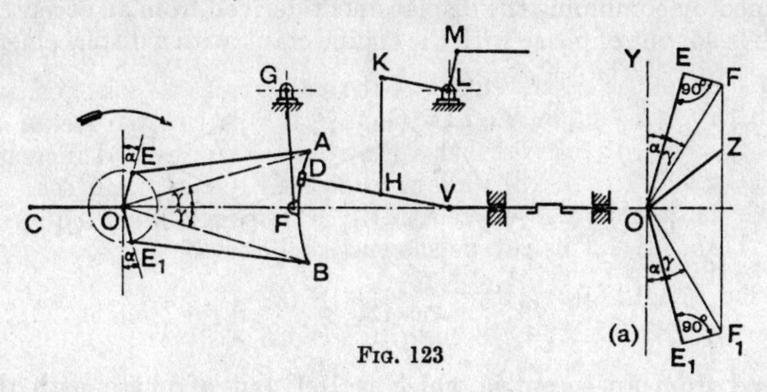

Fig. 123

AB must coincide with V when the crank is on either dead centre. It will then be possible to swing DV from one extreme position to the other without altering the position of V and therefore without displacing the valve relative to the ports.

The equivalent eccentric for any given setting may be found by a method similar to that given for the Stephenson link motion, if it is assumed that the motion of the valve is for all practical purposes the same as that of the block D. In the full gear positions D coincides with A or B and the equivalent eccentric is given by OF or OF_1, Fig. 123 (a), where OF and OF_1 are inclined at angle γ to OE and OE_1 and EF and E_1F_1 are respectively perpendicular to OE and OE_1. For any other position of block D along the curved link AB the equivalent eccentric is given by OZ, where Z divides FF_1 in the same proportion as D divides AB, and its angle of advance is given by angle YOZ.

The Gooch link motion has not been very widely adopted because the advantage of constant lead is outweighed by the disadvantages of greater complication and space requirement and less direct drive to the valve in the full gear positions.

76. Radial Valve Gears. The principles which underlie the operation of radial valve gears may be understood from Fig. 124. Let OC be the crank centre line and let OE be the eccentric centre line for a simple slide valve with outside admission; let OX and OY be the projected lengths of OE parallel and perpendicular to OC. When the crank has turned through the angle θ from the i.d.c., the displacement of the valve from its mid-position is given by OM. But OM $=$ ON$+$NM $=$ ON$+$OP since OX and YE are equal and parallel, so that the displacement which is given to the valve by the eccentric OE is equal to the sum of the displacements which would be given by the two component eccentrics OX and OY. It follows, therefore, that a suitable valve motion may be obtained by combining the displacement derived from an eccentric which is 90° out of phase with the engine crank with a displacement

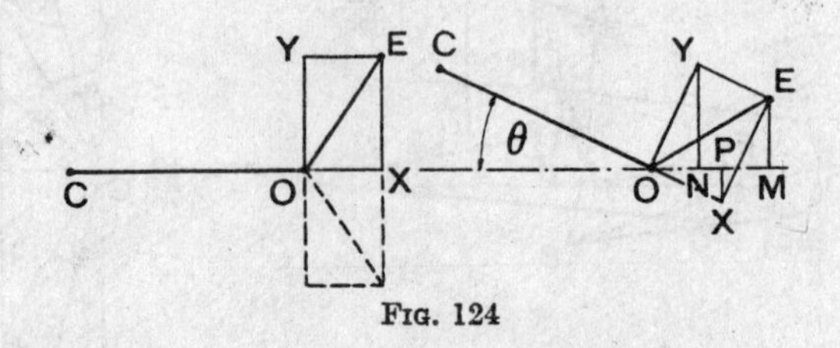

FIG. 124

derived from an eccentric which is 180° out of phase with the engine crank. If the valve has inside admission, the component eccentric, OX, must be in phase with the crank.

An examination of Fig. 124 shows that:

(a) The throw of the 180° component eccentric is equal to the displacement of the valve from its mid-position when the crank is on the dead centre, i.e. it is equal to the sum of the steam lap and the lead. If, therefore, the lead is to remain constant, the 180° component eccentric, OX, should have a constant throw for all settings of the gear.

(b) To reverse the direction of rotation of the crank all that is necessary is to change the direction of the 90° component eccentric as shown by the dotted line.

(c) If the throw of the 90° component eccentric is reduced, the resultant eccentric OE will have a larger angle of advance and a shorter throw. This, as pointed out in Article 69, will cause cut-off to take place earlier in the stroke of the piston.

77. The Walschaert Valve Gear: Fig. 125. This is probably the most extensively used of all reversing gears on modern locomotives. It shows very clearly the application of the above principles. A single eccentric OE is used and is set so that, when the crank is on the dead centre, F is in its mid-position. The eccentric rod EF oscillates the radius link FGH about the fulcrum G, which is fixed to the frame of the engine. The centre of curvature of the radius link should coincide as nearly as possible with pin K when the crank is on either of the dead centres. This condition must be satisfied in order that the lead of the valve shall remain constant for all settings of the gear. The position of the die-block H on the radius link FGH is determined by the reversing rod through the bell-crank UTS and the link SR. The link KLM is termed the combination lever. It is suspended from the pin L on the valve rod. Pin K derives its motion from the die-block H and ultimately from the eccentric OE, while pin M

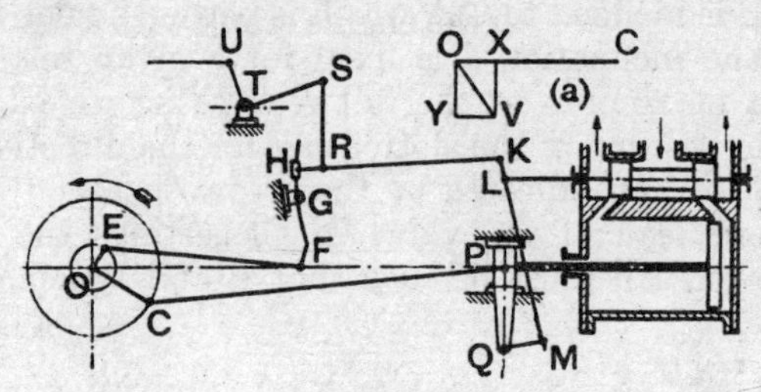

Fig. 125

derives its motion from the crosshead through the union link QM. It follows that the 90° component of the valve motion will be derived from the end K and the 0° component from the end M of the combination lever.

When the gear is in the mid-position, the die-block H coincides with the fulcrum G of the radius link and H remains stationary while the crank revolves. The motion of the valve is then in phase with that of the piston. The component OX therefore lies along the crank and its throw is given by OC.LK/MK. The position of the pin L on the link KM must be such that OX is equal to the steam lap plus the lead.

Since OX lies along the crank, the valve must have inside admission.

If the valve has outside admission, it is easily seen that the pin K must lie between the pins L and M.

When the die-block H is in the position shown in Fig. 121, the valve has a 90° component motion in addition to the 0° component motion. This component is received from the end K of the combination lever. The maximum displacement of K from its mid-position is given approximately by OE.HG/FG. The pin M acts as the fulcrum of the combination lever when transmitting the 90° component motion to L. Hence the maximum displacement of L from its mid-position is given approximately by OY = LM/KM.HG/FG.OE. The displacement of H, and therefore of L, is 180° out of phase with the displacement of F, since H and F lie on opposite sides of the fulcrum G of the radius link. The positions of the two component eccentrics OX and OY relative to the crank OC are shown at (a). The equivalent eccentric OV therefore corresponds to counter-clockwise rotation of the crank.

78. The Hackworth Valve Gear: Fig. 126. In this, the earliest of the radial valve gears, the eccentric OE is set directly opposite to the crank. The eccentric causes the die-block B to reciprocate along the slotted bar SS, which is pivoted to the frame at P. The slotted bar SS is inclined to OP, which is perpendicular to the line of stroke. The inclination θ is fixed for a given setting of the gear and is a maximum for the full gear positions. The dotted line shows the position of the slotted bar for the direction of rotation opposite to that indicated by the arrow.

For constant lead of the valve in all settings the length of the eccentric rod EB must be such that B and P coincide, when

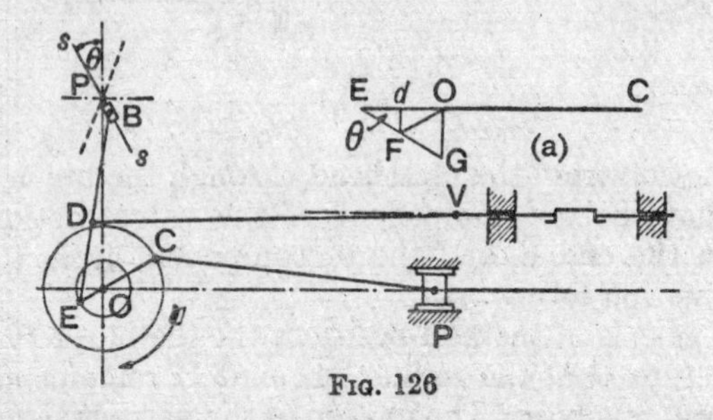

FIG. 126

the crank is in either of the dead-centre positions. The valve derives its motion from point D on EB, through the valve rod DV.

The approximate equivalent eccentric for the motion of the valve may be determined as follows:

With the gear in its mid-position, the bar SS will be vertical and the horizontal displacement of D will be to that of E as DB is to EB. The 180° component eccentric for the valve motion is therefore OE.DB/EB.

For the given setting of the gear the die-block B has a horizontal displacement, which is approximately a maximum when OE, and therefore the crank OC, is at right angles to the line of stroke. This maximum horizontal displacement of B is equal to OE tan θ.

The motion of the point D on the eccentric rod which drives the valve will be received partly from E and partly from B. The component of D's motion received from E will be equivalent to that of a 180° eccentric with a throw OE.DB/EB. The component received from B will be equivalent to a 90° eccentric with a throw (DE/BE)OE tan θ.

The equivalent eccentric for the valve motion will be the vector sum of these two components. Note that the 180° component has a throw independent of the setting of the gear. This throw is equal to the steam lap plus the lead. The magnitude of the 90° component will vary with the angle θ, i.e. with the particular setting of the gear.

The whole of the work required to determine the equivalent eccentric for a given setting may be done graphically as follows:

Set off OE, Fig. 126 (a), to represent the throw of the actual eccentric. Through E draw EG inclined at angle θ to OE, downwards for the full line position of SS and upwards for the dotted line position. Divide OE at d in the same proportion as D divides BE. Through d draw a line perpendicular to OE to meet EG at F and join O and F. Then OF is the equivalent eccentric for the motion of the valve V.

79. The Marshall Valve Gear. This is not illustrated, but is in general similar to the Hackworth gear. The point D divides EB externally instead of internally, and in consequence the eccentric is set parallel to the crank instead of opposite to it. The only other difference is that the pin B is attached to one end of a link which pivots about a fulcrum. This fulcrum is fixed for a given setting of the gear, but its position is altered in order to change the point of cut-off or to reverse the direction of rotation of the crank. Hence B moves along a circular arc instead of along a straight line. The approximate equivalent eccentric may be determined in the same way as for the Hackworth gear, θ being the mean inclination to the vertical of the curved path of the pin B.

80. The Joy Valve Gear: Fig. 127. This gear is unique in that the valve motion is derived ultimately from a pin A on the connecting rod, and no eccentric is used.

At A the link AB is pinned to the connecting rod and at B it is pinned to link BD which swings about the fixed fulcrum D. Link EFG receives its motion from pin E on AB, whilst the die-block, attached to EFG at F, slides along the curved slotted link which is pivoted at H to the frame of the engine.

Note that, if the lead is to remain approximately constant for all settings of the gear, the pin F should as far as possible coincide with the fulcrum H when the crank is on either dead centre.

The valve is driven from pin G through the valve rod GV. Reversal of the direction of rotation of the engine is effected by rotating the shaft, which carries the slotted link, about the axis through H normal to the plane of rotation of the crank, so that it is inclined in the opposite direction to the vertical. The mean inclination of the curved slotted link to the vertical is given by θ.

The method of determining the approximate equivalent eccentric for the given setting of the gear is similar to that already given for the Hackworth gear.

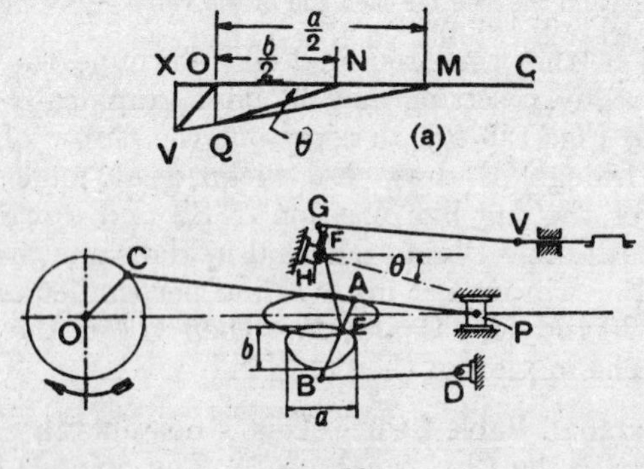

FIG. 127

The paths followed by the pins A and E during the crank rotation are indicated by thin full lines. The maximum horizontal and vertical displacements of E are denoted respectively by a and b.

The throw of the 180° component eccentric is then given approximately by $a/2 . \mathrm{GF}/\mathrm{FE}$ and the throw of the 90° component eccentric by $\mathrm{GE}/\mathrm{FE} . b/2 . \tan\theta$.

The complete graphical construction for the equivalent eccentric is shown at (a), Fig. 127. Set off OM equal to $a/2$ and divide OM externally at X in the same proportion as G divides FE. Through O and X draw lines perpendicular to OM and along OM mark off ON equal to $b/2$. Through N draw a line inclined at the angle θ to NO to cut the perpendicular through O at Q. Join M to Q and produce to cut the perpendicular through X at V. Then OV is the required virtual eccentric.

EXAMPLES V

1. Describe the action of a simple slide valve and show how the points of admission, cut-off, etc., may be determined, given the necessary particulars of the valve.

2. Show in section the arrangement of the steam chest with the valve in mid-position when (a) the valve has outside admission, (b) the valve has inside admission. What are the relative positions of the crank and the eccentric centre line in each case?

3. Show by means of sketches what is meant by the following terms: steam lap, exhaust lap, lead, angle of advance and throw of eccentric.

4. A slide valve has a travel of 5 in. The angle of advance of the eccentric is 35° and cut-off and release are required to take place at 0·75 and 0·95 of the stroke on both strokes. If the connecting rod is 4 cranks long, find the steam and exhaust laps and the lead for each end of the valve.

5. The travel of a slide valve is 4 in. and the lead at the crank end $\frac{1}{4}$ in. If the connecting rod is 4·5 cranks long, find the angle of advance of the eccentric, and the steam and exhaust laps to give cut-off at 0·65 of the stroke and release at 0·95 of the stroke.

6. The slide valve in a steam engine is required to cut-off the steam at 0·7 of the stroke at each end of the cylinder. The lead is $\frac{1}{4}$ in. at the cover end and the angle of advance of the eccentric is 40°. Find the valve travel and the steam lap at each end of the valve, taking the length of the connecting rod equal to $4\frac{1}{2}$ cranks.

7. A slide valve cuts off at 0·7 of the stroke for both ends of the cylinder. The steam lap and the lead at the cover end are $\frac{3}{4}$ in. and $\frac{1}{4}$ in. respectively. Find the travel of the valve and the angle of advance of the eccentric and the lead and steam lap at the crank end. The connecting rod is $4\frac{1}{2}$ cranks long. M.U.

8. A slide valve is driven by an eccentric having a throw of $2\frac{1}{2}$ in. and an angle of advance of 35°. The lead at the cover-end is $\frac{1}{4}$ in. and the connecting rod is 4 cranks long. Determine the maximum opening to steam and the fraction of the stroke at which cut-off occurs. What exhaust lap will be required to give release at 0·95 of the stroke?

9. A slide valve has a travel of 6 in. The lead at the cover end is 0·25 in. and cut-off is required to take place at 0·7 of the stroke on both strokes. If the connecting rod is 5 cranks long, determine the angle of advance of the eccentric, and the maximum opening to steam and the steam lap for the two ends of the valve.

10. A slide valve is required to give cut-off at 0·75 stroke for the cover end of the cylinder when the ratio of connecting rod length to crank length is 4, the maximum opening to steam is $1\frac{1}{4}$ in., and the lead of the valve is $\frac{1}{4}$ in. Find the throw and angle of advance of the eccentric and the steam lap of the valve. If the steam lap is the same for both ends of the valve, at what fraction of the return stroke will cut-off take place?

11. A slide valve is required to cut-off the steam at 0·7 of the stroke on both strokes, the maximum opening to steam and the lead at the cover end being $1\frac{1}{2}$ in. and $\frac{3}{16}$ in. respectively and the connecting rod 5 cranks long. Find the angle of advance and the throw of the eccentric and the steam laps for the two ends of the valve. M.U.

12. A piston valve admits steam from the inside. The maximum width of the port opening to steam is $1\frac{1}{2}$ in., the lead is $\frac{1}{4}$ in. Find the travel of the valve, the angle of advance of the eccentric and the steam lap to give cut-off at 0·72 of the stroke. The connecting rod is $4\frac{1}{2}$ cranks long. Sketch the relative positions of the crank and the eccentric.

13. A piston valve with inside admission has a travel of 5 in. and is required to give cut-off at 0·8 of the stroke for both strokes. Choosing a suitable lead for the cover end, find the angle of advance of the eccentric and the steam laps. The connecting rod is 4 cranks long. Sketch the relative positions of the crank and eccentric.

14. It is required to change the point of cut-off for an engine with a simple slide valve from 0·8 to 0·6 of the stroke without making any change in the valve dimensions. How may this be done and what effect will the change have on the points of admission, release and compression? The effect of the obliquity of the connecting rod may be neglected.

15. Explain carefully how to find the equivalent eccentric giving the relative displacement of the main and expansion valves of a Meyer valve gear, and prove that the construction is correct.

Then show how the steam lap of the expansion valve may be obtained to give cut-off at a given fraction of the stroke.

16. The main eccentric of a Meyer expansion valve gear has an angle of advance of 30° and a throw of $2\frac{1}{2}$ in. The expansion valve has an angle of advance of 90° and a throw of $2\frac{1}{4}$ in. The connecting rod is four cranks long and cut-off is to take place at 0·3 of the stroke on each stroke. Determine the steam lap required at each end of the expansion valve.

17. A Meyer expansion valve gear cuts off the steam at 0·4 of the stroke on both strokes. The main eccentric has a throw of $2\frac{3}{4}$ in. and an angle of advance of 35°. The expansion eccentric has a throw of $2\frac{1}{2}$ in. and an angle of advance of 90°. The connecting rod is 4 cranks long. Determine the steam laps required for the two ends of the expansion valve.

18. In a Meyer expansion valve gear the main eccentric has an angle of advance of 35° and a throw of 3 in. The expansion eccentric has an angle of advance of 90° and a throw of 3 in. The connecting rod is 5 cranks long. What laps will be required at the two ends of the expansion valve in order to give cut-off at 0·2 and 0·6 of the stroke on both strokes? M.U.

19. In a Meyer expansion gear, the main valve has a travel of 5 in.; steam lap 1 in.; exhaust lap $\frac{3}{16}$ in.; angle of advance of eccentric 30°. The expansion plate is driven from an eccentric of $2\frac{1}{2}$ in. throw and angle of advance 90°. Determine the lap of the expansion plate to cut-off at $\frac{1}{3}$ of the stroke and draw the probable indicator diagram. The connecting rod is 5 cranks long. M.U.

20. The following particulars refer to a Meyer expansion valve: travel of main valve 5 in.; lead at each end $\frac{1}{4}$ in.; angle of advance of main eccentric 35°; throw of expansion eccentric $2\frac{3}{4}$ in. and angle of advance 90°. Cut-off and release are required to take place at 0·35 and 0·93 of the stroke respectively for both strokes. The length of the connecting rod is $4\frac{1}{2}$ cranks. Find the steam and exhaust laps of the main valve and the steam lap of the expansion valve at each end. Sketch the arrangement showing the position of the main and expansion valves in relation to the cylinder ports when the piston is just about to commence the outward stroke.

21. In a Meyer expansion valve gear, the main eccentric has a throw of $2\frac{1}{2}$ in. and an angle of advance of 30°. The expansion eccentric has an equal throw, but an angle of advance of 90°. The connecting rod is 5 cranks long and it is required to vary the cut-off 0·2 to 0·6 of the stroke, keeping it as nearly as possible equal on both strokes. What will be the best setting of the expansion plates?

22. In a Meyer expansion valve gear the main valve has a travel of 6 in. and angle of advance 30°. The expansion valve has a travel of 5 in. and angle of advance 80°. The connecting rod is 4 cranks long and the steam and exhaust laps for the main valve are $1\frac{1}{8}$ in. and $\frac{3}{8}$ in. respectively. Determine, for the cover end only, the fractions of the stroke at which admission, release, and compression take place and the steam lap required on the expansion plate to give cut-off at 0·4 stroke.

23. The following data are taken from a horizontal steam engine fitted with a Meyer expansion valve: travel of main valve, 4 in.; angle of advance, $22\frac{1}{2}°$; lead, 0·25 in. at both ends of the cylinder. Travel of expansion valve, 4 in.; angle of advance, 90°; width of port through main valve, 1·5 in. Ratio of connecting-rod length to crank length 4 : 1, and release takes place at 170° on the out-stroke.

Find the latest point of cut-off possible for both ends of the cylinder, the point of release on the instroke and the points of admission and compression for both ends, the laps at both ends being equal.

Obtain the steam laps of the expansion valve to cut-off steam at 0·2, 0·3, 0·4, 0·5 and 0·6 stroke for both sides of the piston and deduce the best value for the difference of the laps as a compromise for this range of cut-offs.

What is the width of the narrowest possible expansion plate? L.U.A.

24. A variable cut-off valve of the Meyer type is required for a vertical engine in which the ratio connecting rod to crank is 4. The range of cut-off is from 0·1 to 0·75 for the out-stroke and 0·1 to 0·7 for the in-stroke. The greatest port opening on the main valve is $1\frac{3}{16}$ in., with a lead of $\frac{1}{8}$ in. for the out-stroke. The radii of the expansion and main eccentrics are the same and the angle of advance of the expansion eccentric is 90°.

Draw the necessary valve diagrams and sketch the valves and ports for mid-position for the two extreme cut-offs. Give the following particulars:

(1) Steam laps of main valve for both in and out-strokes and lead for up-stroke.
(2) Negative laps of expansion valve for earliest and latest cut-off.
(3) The cut-off on the in-stroke when cut-off on the out-stroke is 0·25.
(4) The exhaust laps on main valve for release at 0·95 on each stroke.

L.U.A.

25. The displacements in inches of the main and expansion valves of a Meyer gear are $x = 1{\cdot}5 \cos(\theta + \frac{2}{3}\pi)$ and $x = 1{\cdot}5 \cos(\theta + \pi)$. The outside lap of the main valve is 0·8 in. Find the point of cut-off by the main valve on the outward stroke and the negative laps for the expansion valve to give cut-off respectively at 0·2 and 0·3 of the outward stroke. The length of the connecting rod is 5·5 times the length of the crank. L.U.A.

26. Sketch and describe one form of valve gear for reversing a steam engine. Show how reversal is effected and how the points of cut-off and release may be determined given the steam and exhaust laps of the valve.

27. In a Stephenson link motion with open rods each eccentric has a throw of 3 in. and an angle of advance of 18°. The length of the curved slotted link is 16 in. and its radius of curvature is equal to the length of the eccentric rod, 45 in. Determine the throw and angle of advance of the equivalent eccentric when the motion is in the position mid-way between full-gear and mid-gear. Draw the valve diagram if the steam and exhaust laps are $\frac{3}{4}$ in. and $\frac{1}{4}$ in. respectively.

28. Sketch and describe a Stephenson link motion. Point out the different results obtained due to linking up for open and crossed rods respectively. The eccentrics in such a motion each have 4 in. throw. The eccentric rods are open and 6 ft long; the connecting link is 2 ft long and the angle between the eccentrics is 130°. Determine the equivalent eccentric when the block is 6 in. from one end of the link. If the steam lap is 1·5 in. and the connecting rod is 5 cranks long, at what point does cut-off take place?

29. Describe a method of obtaining the equivalent eccentric for a Stephenson link motion with open rods. How is the lead of the valve affected during the movement of the mechanism from full-gear to mid-gear?

If the throw of each eccentric is 3 in., the angle of advance 20°, the length of the slotted link 18 in. and of each eccentric rod 60 in., find the equivalent eccentric for (a) mid-gear and (b) half-way between full-gear and mid-gear.

30. What are the principles underlying the action of a radial valve gear? Show how these principles are applied in the case of a Walschaert valve gear and indicate how the equivalent eccentric for any given setting of the gear may be determined.

31. Referring to the Walschaert gear, Fig. 125, the line of stroke of the valve is 18 in. from the line of stroke of the piston. The dimensions of the various links are: OC, 13 in.; CP, 97 in.; FG, 12 in.; GH, 9 in.; KL, $4\frac{3}{8}$ in.; LM, $29\frac{5}{8}$ in.; OE, 6 in. Find the throw and the angle of advance of the equivalent eccentric for the given setting of the gear. If the steam lap is $1\frac{3}{8}$ in. and the exhaust lap $\frac{1}{2}$ in., draw the valve diagram and find the crank positions at which admission, cut-off, release and compression take place. Sketch the relative positions of the crank and the equivalent eccentric.

32. The dimensions of a Hackworth valve gear, Fig. 126, are as follows: OC, 12 in.; CP, 42 in.; OE $5\frac{7}{8}$ in.; EB, 32 in.; ED, 22 in. The pin B coincides with the fulcrum of the slotted link SS when the crank is on the dead centre. Find the throw and the angle of advance of the equivalent eccentric when the angle θ is (a) 30°, (b) −20°.

33. The dimensions of a Joy valve gear, Fig. 127, are as follows: OC, 13 in.; CP, 74 in.; CA, 49 in.; AB, 20 in.; BD, 28 in.; AE, 7 in.; EF, $20\frac{1}{2}$ in.; FG, 3 in.; GV, 42 in. The line of stroke of the valve is 16 in. from the line of stroke of the piston. The pin D is 16 in. from the line of stroke and 77 in. horizontally from the crankshaft centre. The fulcrum H of the slotted link is $13\frac{3}{8}$ in. from the line of stroke and 49 in. horizontally from the crankshaft centre. Find the angle of advance and the throw of the equivalent eccentric when the angle θ is 25°.

34. Choose a type of "radial" valve gear and determine the main dimensions to give the following distribution of steam in "full gear" to a locomotive: lead, 0·2 in.; maximum opening to steam, 1·55 in.; latest cut-off on out-stroke, 135°; crank radius, 13 in. Neglect the obliquities of all rods. What is the equivalent eccentric radius of the gear when cutting off steam at a crank angle of 90° on the out-stroke? L.U.A.

CHAPTER VI

FRICTION

81. The sliding of one solid body relative to a second solid body, with which it is in contact, is always resisted by a force called the force of *friction.* The force of friction acts in the opposite direction to that of the relative motion and is tangential to the surfaces of the two bodies at the point of contact. It follows that at every joint in a machine, owing to the relative motion between the two parts, friction forces arise and energy is absorbed. In order to reduce this waste of energy, it is clearly necessary that every effort should be made to reduce the magnitude of the friction forces. It is sometimes possible to alter the design of a joint so that a rolling motion between the parts is substituted for sliding motion, as, for instance, when a ball or roller bearing is used instead of a plain bearing in a turning pair. But in most practical joints it is not feasible to eliminate the sliding motion and the friction can be reduced only by the introduction of some form of lubricant between the surfaces, which will enable the surfaces to slide more easily. The ideal arrangement would be to have the contact surfaces completely separated by a layer or film of lubricant, so that fluid friction is substituted for solid friction. In considering the laws which govern the friction between two surfaces it is therefore necessary to distinguish between the three possible states of the surfaces: (a) dry; (b) greasy or partially lubricated; and (c) film or completely lubricated.

82. Friction between Dry Surfaces. Numerous experiments have shown that, when two solid bodies with smooth, dry surfaces are in contact, the least force required in order to cause the one body to slide over the other obeys approximately the following laws:

(1) The friction force is directly proportional to the normal load between the surfaces for a given pair of materials.

(2) The friction force depends upon the material of which the contact surfaces are made.

(3) The friction force is independent of the area of the contact surfaces for a given normal load.

(4) The friction force is independent of the velocity of sliding of the one body relative to the other body.

Referring to Fig. 128 (a), let R_n be the normal reaction between the surfaces and F the force required tangential to the contact surfaces in order to cause sliding with uniform velocity in the direction of F. Then F will be equal and opposite to the friction force and the first law states that F is directly proportional to R_n. The ratio of F to R_n is termed the *coefficient of friction* and is denoted by the Greek letter μ. The value of μ varies for different materials.

The third law is only approximately true. There is a limit to the intensity of pressure that can be allowed between a given pair of materials. When this limit is exceeded the phenomenon of "seizing" occurs; particles of material are torn from one, or both, of the surfaces and become welded to the other surface. For intensities of pressure below that at which seizing takes place, the value of μ is practically independent of the intensity of pressure.

The results of experiments also show that the fourth law is only approximately true. It is a well-known fact that in order to initiate sliding of one body over a second body a greater force is required than that which is necessary to maintain a uniform sliding motion. For this reason the ratio of the friction force, just before motion begins, to the normal load between the surfaces is referred to as the *coefficient of stiction* or the *coefficient of static friction.* Careful experiments have also shown that the coefficient of friction usually diminishes slowly but continuously as the velocity of sliding increases.

Although the laws of dry friction as given above are only approximately true, their use enables many problems into which friction enters to be solved with sufficient accuracy for most practical purposes.

Where a small amount of lubricant is introduced between the surfaces, so that they are in a greasy or partially lubricated condition, it is usual to assume the laws of dry friction hold and that the effect of the lubricant is simply to reduce the value of the coefficient of friction.

The friction of lubricated surfaces is dealt with in Articles 92 *et seq.*

83. The Limiting Angle of Friction. In Fig. 128 (b) a body of weight W is shown resting on a horizontal plane. If a horizontal force F is applied to the body, no relative motion will take place until F is equal to μW. But before motion begins the body will be in equilibrium under the three forces F, W, and the reaction R between the plane and the body. The reaction R must therefore be equal and opposite to the resultant of F and W and will be inclined at the angle α to the normal reaction R_n. As the force F is increased the angle α will increase until the body begins to slide along the plane. At this point the friction force has its maximum value and $F = \mu W$ so that $\tan \alpha = F/W = \mu$. This limiting value of the inclination to the normal of the reaction between the surfaces is termed the limiting angle of friction and is denoted by ϕ, so that $\tan \phi = \mu$. Hence, when one body slides over another body, the true reaction between them is always inclined at the angle ϕ to the normal to the contact surfaces. Further, the direction of the reaction is such as to oppose the sliding motion.

The limiting angle of friction may be shown in another way. If the plane on which the body rests is tilted, as shown in Fig. 129,

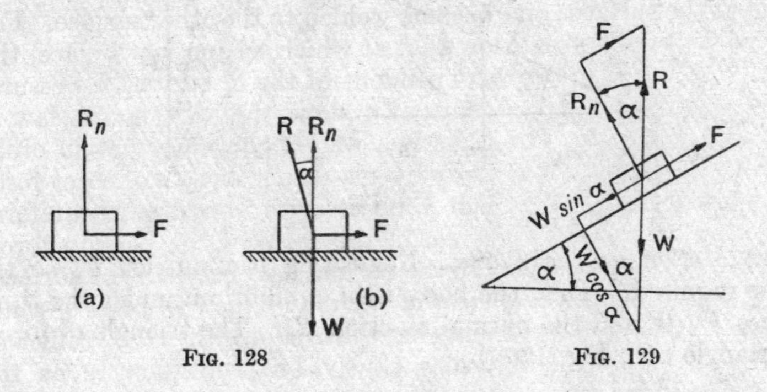

Fig. 128 Fig. 129

and the angle of inclination α of the plane to the horizontal is gradually increased, the body will remain at rest until $\alpha = \phi$, when it will begin to slide down the plane. Whatever the value of α, the weight W may be resolved into two components; one, $W \sin \alpha$, parallel to the plane, tends to cause the body to slide down the plane, the other, $W \cos \alpha$, perpendicular to the plane, is balanced by the normal reaction between the plane and the body. The body will only begin to slide when the component down the plane is equal to the limiting friction force between the body and the plane, i.e. when $W \sin \alpha = \mu W \cos \alpha$ or $\tan \alpha = \mu$. But when $\tan \alpha = \mu$, the inclination α must be equal to the limiting angle of friction ϕ. The reaction R of the plane on the body is equal and opposite to the weight W and it is obviously inclined at the angle ϕ to the normal to the plane at the instant the body begins to slide.

84. The Inclined Plane. In one form or another the inclined plane is very frequently used, and it is worth while to examine in some detail the relationship between the various forces which act on a body when it slides either up or down an inclined plane. Referring to Fig. 130 (a), let W be the weight of the body; α the inclination of the plane to the horizontal; ϕ the limiting angle of friction for the contact surfaces. Also let the force F be applied in the given direction in order to cause the body to slide with uniform velocity parallel to the slope. Then there are two different cases to consider, viz. (a) motion of the body up the plane and (b) motion of the body down the plane.

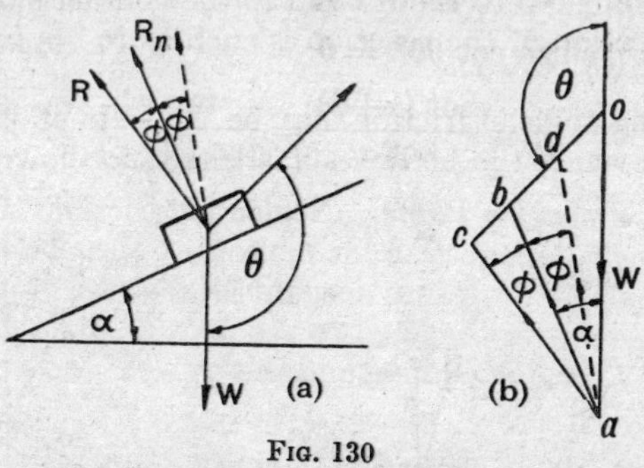

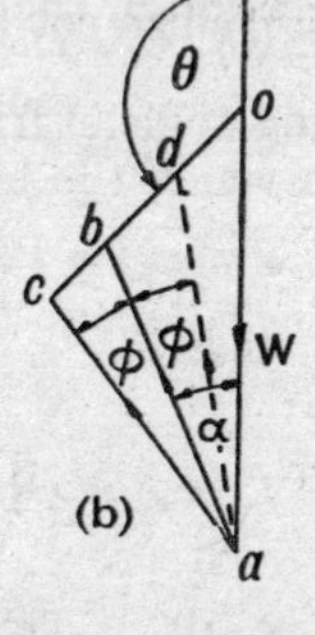

FIG. 130

(a) *Motion up the Plane.* Neglecting friction, let F' be the force required. Then the body is in equilibrium under the three forces F', W and the normal reaction R_n. The triangle of forces is triangle oab, Fig. 126 (b),

$$\therefore \frac{F'}{W} = \frac{\text{bo}}{\text{oa}} = \frac{\sin \text{oab}}{\sin \text{oba}} = \frac{\sin \alpha}{\sin (\theta - \alpha)} \quad . \quad . \quad (6.1)$$

When friction is taken into account let F be the force required Then the reaction R between the plane and the body is inclined to the normal at the friction angle ϕ and the triangle of forces is triangle oac, Fig. 130 (b),

$$\therefore \frac{F}{W} = \frac{\text{co}}{\text{oa}} = \frac{\sin \text{oac}}{\sin \text{oca}} = \frac{\sin (\alpha + \phi)}{\sin \{\theta - (\alpha + \phi)\}} \quad . \quad (6.2)$$

The ratio of the force required without friction to the force required with friction is given by:

$$\frac{F'}{F} = \frac{\sin \alpha}{\sin (\theta - \alpha)} \cdot \frac{\sin \{\theta - (\alpha + \phi)\}}{\sin (\alpha + \phi)} \quad . \quad . \quad . \quad . \quad . \quad . \quad (6.3)$$

$$= \frac{\sin \alpha}{\sin \theta \cos \alpha - \cos \theta \sin \alpha} \cdot \frac{\sin \theta \cos (\alpha + \phi) - \cos \theta \sin (\alpha + \phi)}{\sin (\alpha + \phi)}$$

$$= \frac{\cot (\alpha + \phi) - \cot \theta}{\cot \alpha - \cot \theta} \quad . \quad . \quad . \quad . \quad . \quad . \quad . \quad . \quad . \quad . \quad (6.4)$$

The ratio F'/F obviously represents the efficiency η of the inclined plane as a machine, so that:

$$\eta = \frac{\cot(\alpha+\phi)-\cot\theta}{\cot\alpha-\cot\theta} \quad . \quad . \quad . \quad . \quad (6.5)$$

If $\theta = 90°$, i.e. if the force is applied horizontally, then from (6.2):

$$\frac{F}{W} = \frac{\sin(\alpha+\phi)}{\cos(\alpha+\phi)} = \tan(\alpha+\phi) \quad . \quad . \quad (6.6)$$

and from (6.5), since $\cot 90° = 0$

$$\eta = \frac{\cot(\alpha+\phi)}{\cot\alpha} = \frac{\tan\alpha}{\tan(\alpha+\phi)} \quad . \quad . \quad . \quad (6.7)$$

(b) *Motion Down the Plane.* Neglecting friction, the triangle of forces is triangle oab just as for motion up the plane.

When friction is taken into account let F_1 be the force required. Then the reaction R is inclined to the normal at the friction angle ϕ, as shown by the dotted line, and the triangle of forces is triangle oad, Fig. 130 (b),

$$\therefore \frac{F_1}{W} = \frac{\text{do}}{\text{oa}} = \frac{\sin \text{oad}}{\sin \text{oda}} = \frac{\sin(\alpha-\phi)}{\sin\{\theta-(\alpha-\phi)\}} \quad . \quad . \quad . \quad (6.8)$$

and by analogy from (6.4)

$$\frac{F'}{F_1} = \frac{\cot(\alpha-\phi)-\cot\theta}{\cot\alpha-\cot\theta}$$

For motion down the plane the force W becomes the effort and the force F_1, or F', the resistance. The efficiency of the inclined plane as a machine is given by the ratio of the resistance F_1, which can be overcome with friction, to the resistance F', which can be overcome without friction,

$$\therefore \eta = \frac{F_1}{F'} = \frac{\cot\alpha-\cot\theta}{\cot(\alpha-\phi)-\cot\theta} \quad . \quad . \quad (6.9)$$

If $\theta = 90°$, i.e. if the force is horizontal, this equation reduces to:

$$\eta = \frac{\cot\alpha}{\cot(\alpha-\phi)} = \frac{\tan(\alpha-\phi)}{\tan\alpha} \quad . \quad . \quad (6.10)$$

85. Maximum Efficiency. For given values of θ and ϕ there is a value of α which gives maximum efficiency. When the direction of motion of the body is up the plane this value of α may be found most easily from (6.3):

$$\eta = \frac{F'}{F} = \frac{\sin\alpha}{(\sin\theta-\alpha)} \cdot \frac{\sin\{\theta-(\alpha+\phi)\}}{\sin(\alpha+\phi)}$$

Expressing the product of two sines in the numerator and in the denominator as the differences of two cosines, this may be written

$$\eta = \frac{\cos(\theta-\phi-2\alpha)-\cos(\theta-\phi)}{\cos(\theta-\phi-2\alpha)-\cos(\theta+\phi)} \quad . \quad . \quad (6.11)$$

This is clearly a maximum when $\cos(\theta-\phi-2\alpha)$ is a maximum, i.e. when $\theta-\phi-2\alpha = 0$

or when

$$\alpha = (\theta-\phi)/2 \quad . \quad . \quad . \quad . \quad . \quad (6.12)$$

Then substituting in (6.11),

$$\eta_{max} = \frac{1-\cos(\theta-\phi)}{1-\cos(\theta-\phi)} \quad . \quad . \quad . \quad (6.13)$$

If $\theta = 90°$, i.e. if the force is horizontal, then for maximum efficiency:

$$\alpha = (90-\phi)/2 = 45-\phi/2 \quad . \quad . \quad . \quad (6.14)$$

and

$$\eta_{max} = \frac{1-\cos(90-\phi)}{1-\cos(90+\phi)} = \frac{1-\sin\phi}{1+\sin\phi} \quad . \quad . \quad (6.15)$$

A graphical method of proving that the efficiency is a maximum when (6.12) is satisfied is shown in Fig. 131. From the triangle of forces, Fig. 130 (b), the efficiency is given by:

$$\eta = \frac{ob}{oc} = \frac{1}{1+bc/ob}$$

Hence the efficiency is a maximum when bc/ob is a minimum, i.e. when ob/bc is a maximum, and the problem resolves itself into that of finding the value of α which, with θ, ϕ and bc fixed, will make ob a maximum.

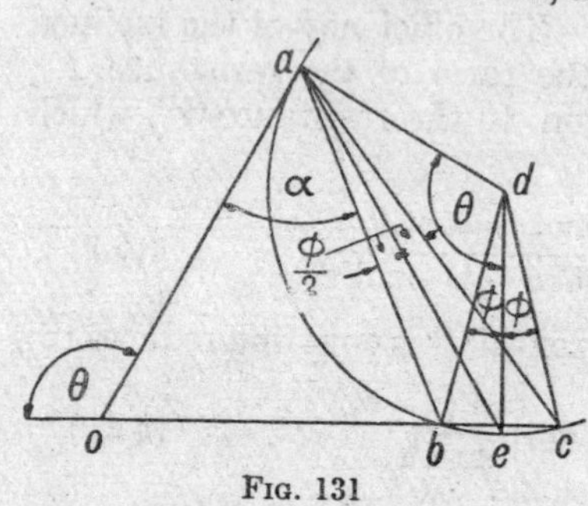

Fig. 131

Since angle bac has the fixed value ϕ, we know from a well-known theorem in geometry that the locus of a is a circle which passes through b and c. Draw this circle as shown in Fig. 131. The centre d is the point of intersection of the line ed which, bisects bc at right angles, and the line bd, which makes an angle $90-\phi$ with bc. Since the point a lies on the circumference of the circle and oa must make an angle θ with bc, it follows that the line oa must be tangential to the circle if ob is to be a maximum. Set off angle eda $= \theta$ and draw ao perpendicular to da. Then angle oab is the value of α which gives maximum efficiency.

Applying well-known principles of geometry, we have:

$$\angle \text{oae} = \tfrac{1}{2} \angle \text{ade} = \theta/2$$

Also
$$\angle \text{bae} = \angle \text{eac} = \phi/2$$

$$\therefore \alpha = \angle \text{oab} = \angle \text{oae} - \angle \text{bae} = \theta/2 - \phi/2 = (\theta - \phi)/2$$

For the sake of clearness the angle ϕ has been taken very much larger than its normal value in setting out Fig. 131.

86. The Inclined Plane with Guide Friction. As a further example of the inclined plane, consider the arrangement shown in Fig. 132. The elements A and B are free to slide through guides and they are in contact along the inclined plane surface CC. The angle between CC and the axis of sliding of A is $90-\alpha$ and that between CC and the axis of sliding of B is $90-\beta$.

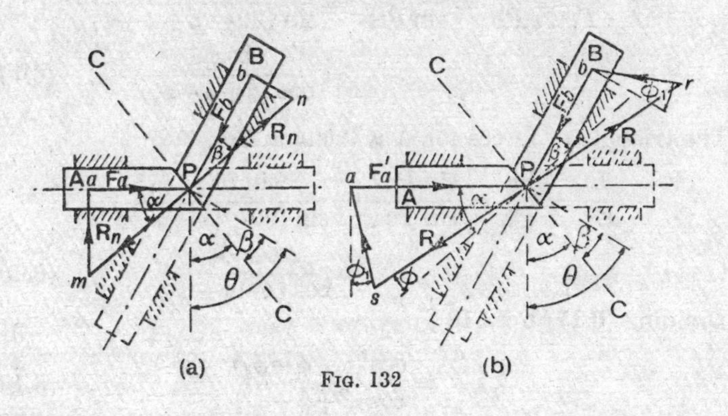

FIG. 132

A force F_a is applied to A in order to overcome the force F_b, which resists the sliding of B. It is required to find the relation between F_a and F_b and the efficiency of the arrangement as a machine when the limiting angle of friction for the contact surfaces of A and B is ϕ and that for the contact surfaces of each element A and B and the corresponding guides is ϕ_1.

(a) *Neglecting Friction.* The reactions between the various contact surfaces are along the normals to those surfaces. The element B is in equilibrium under the three forces F_b, the sum of the normal reactions of the two guides and the normal reaction R_n of A. The triangle of forces is triangle Pnb,

$$\therefore R_n = F_b/\cos\beta$$

Similarly, the element A is in equilibrium under the three forces

F_a, the sum of the normal reactions of the two guides and the normal reaction R_n of B. The triangle of forces is triangle Pma,

$$\therefore\ R_n = F_a/\cos\alpha$$

But the reaction of B on A is equal to the reaction of A on B, so that:

$$F_a/\cos\alpha = F_b/\cos\beta$$

or

$$F_a = F_b . \cos\alpha/\cos\beta \quad . \quad . \quad . \quad (6.16)$$

(b) *With Friction.* The true reactions between the contact surfaces are inclined to the normals at the friction angle and in a direction opposite to that of the relative sliding. Hence the true reactions are shown in Fig. 132 (b).

The triangle of forces for B is triangle Prb,

$$\therefore\ \frac{R}{F_b} = \frac{\text{Pr}}{\text{Pb}} = \frac{\sin \text{Pbr}}{\sin \text{Prb}} = \frac{\sin(90+\phi_1)}{\sin\{90-(\beta+\phi+\phi_1)\}}$$

$$= \frac{\cos\phi_1}{\cos(\beta+\phi+\phi_1)} \quad . \quad (6.17)$$

The triangle of forces for A is triangle Psa,

$$\therefore\ \frac{R}{F'_a} = \frac{\text{Ps}}{\text{Pa}} = \frac{\sin \text{Pas}}{\sin \text{Psa}} = \frac{\sin(90-\phi_1)}{\sin\{90-(\alpha-\phi-\phi_1)\}}$$

$$= \frac{\cos\phi_1}{\cos(\alpha-\phi-\phi_1)} \quad . \quad . \quad (6.18)$$

Dividing (6.17) by 6.18),

$$\frac{F'_a}{F_b} = \frac{\cos(\alpha-\phi-\phi_1)}{\cos(\beta+\phi+\phi_1)} \quad . \quad . \quad . \quad (6.19)$$

The efficiency of the arrangement as a machine is given by the ratio of the force F_a required without friction to that required with friction when F_b is constant, therefore, from (6.16) and (6.19),

$$\eta = \frac{\cos\alpha}{\cos\beta} . \frac{\cos(\beta+\phi+\phi_1)}{\cos(\alpha-\phi-\phi_1)}$$

Since $\beta = \theta - \alpha$, this may be written:

$$\eta = \frac{\cos\alpha}{\cos(\theta-\alpha)} . \frac{\cos(\theta-\alpha+\phi+\phi_1)}{\cos(\alpha-\phi-\phi_1)} \quad . \quad . \quad (6.20)$$

Expressing the product of the two cosines in the numerator and in the denominator as the sum of two cosines,

$$\eta = \frac{\cos(\theta+\phi+\phi_1)+\cos(2\alpha-\theta-\phi-\phi_1)}{\cos(\theta-\phi-\phi_1)+\cos(2\alpha-\theta-\phi-\phi_1)} \quad . \quad (6.21)$$

For given values of θ, ϕ and ϕ_1 the efficiency is clearly a maximum when $\cos(2\alpha-\theta-\phi-\phi_1)$ is a maximum, i.e. when $2\alpha-\theta-\phi-\phi_1 = 0$, or when

$$\alpha = \frac{\theta+\phi+\phi_1}{2} \quad . \quad . \quad . \quad . \quad (6.22)$$

Substituting for α in (6.21), we have:

$$\eta_{max} = \frac{\cos(\theta+\phi+\phi_1)+1}{\cos(\theta-\phi-\phi_1)+1} \quad . \quad . \quad . \quad (6.23)$$

If the friction at the guides is negligible, $\phi_1 = 0$ and the above equations may be simplified.

Example 1. If the angle θ is 60°, the coefficient of friction for the contact surfaces of A and B is 0·15 and that for the guides is 0·10, find the maximum efficiency.

$$\phi = \tan^{-1} 0{\cdot}15 = 8°\ 32' \quad \text{and} \quad \phi_1 = \tan^{-1} 0{\cdot}10 = 5°\ 43'$$

From (6.22), maximum efficiency is obtained when:

$$\alpha = \frac{60°+8°\ 32'+5°\ 43'}{2} = 37°\ 7{\cdot}5'$$

From (6.23), maximum efficiency is given by:

$$\eta_{max} = \frac{\cos 74°\ 15'+1}{\cos 45°\ 45'+1} = \frac{1{\cdot}2715}{1{\cdot}6978} = 0{\cdot}7489 \quad \text{or} \quad 74{\cdot}89\%$$

If the friction of the guides is negligible, the corresponding values are $\alpha = 34°\ 16'$ and $\eta_{max} = 84{\cdot}16\%$.

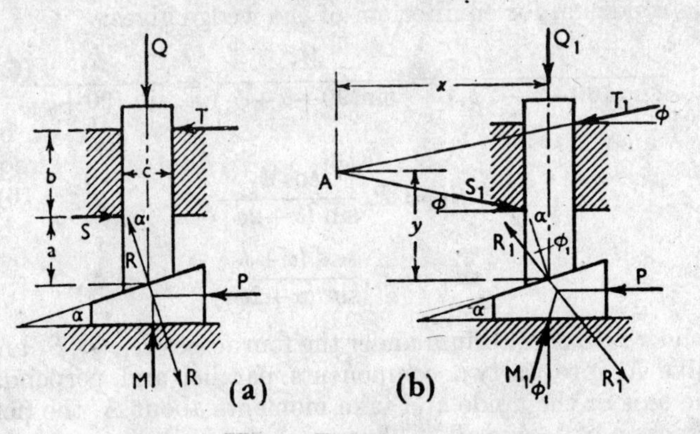

FIG. 13.

Example 2. Fig. 133 shows an arrangement in which a horizontal effort P is applied to a wedge in order to raise the slider against the vertical resistance Q. It is required to find the ratio of Q to P (a) when there is no friction at the contact surfaces, (b) when the friction angle at the contact surfaces of the wedge is ϕ_1 and at the contact surfaces of the guide is ϕ.

(a) *Without Friction.* The reactions at all the contact surfaces are normal to those surfaces as shown at (a).

The wedge is in equilibrium under the effort P, the reaction M of the base and the reaction R on the inclined face. These three forces therefore meet at a point and, from the triangle of forces,

$$P = R \sin \alpha, \quad M = R \cos \alpha \quad \text{and} \quad M = P \cot \alpha$$

The slider is in equilibrium under the force R applied by the wedge, the resistance Q and the reaction of the guide. Since R has a component perpendicular to the axis of the guide, the slider will tend to tilt in the guide and thus give rise to reactions S and T at the corners of the guide as shown, and

$$S - T = R \sin \alpha = P$$

$$Q = R \cos \alpha = M = P \cot \alpha \quad . \quad . \quad . \quad (1)$$

$$S.a = T(a+b)$$

$$\therefore S = (a+b)/b.P = (1+a/b)P$$

$$T = a/b.P$$

(b) *With Friction.* The reactions at the contact surfaces are inclined at the friction angle to the normals in such a way as to oppose the relative sliding, as shown at (b).

The condition for equilibrium of the wedge gives:

$$\frac{P}{\sin (180-\alpha-2\phi_1)} = \frac{M_1}{\sin(90+\alpha+\phi_1)} = \frac{R_1}{\sin (90+\phi_1)}$$

From which

$$R_1 = P.\frac{\cos \phi_1}{\sin (\alpha+2\phi_1)} \quad . \quad . \quad . \quad . \quad (2)$$

and

$$M_1 = P.\frac{\cos (\alpha+\phi_1)}{\sin (\alpha+2\phi_1)} \quad . \quad . \quad . \quad . \quad (3)$$

The slider is in equilibrium under the four forces R_1, Q_1, S_1 and T_1. Resolve R_1 into its two components parallel and perpendicular to the axis of the guide and take moments about A, the point of intersection of S_1 and T_1. Then:

$$\{R_1 \cos (\alpha+\phi_1) - Q_1\}x = \{R_1 \sin (\alpha+\phi_1)\}y$$

But

$$x = (b/2) \cot \phi = b/2\mu$$

and

$$y = b/2-(c/2) \tan \phi + a = b/2 + a - \mu c/2,$$

$$\therefore \{R_1 \cos(\alpha+\phi_1)-Q_1\}\frac{b}{2\mu}$$
$$= R_1\left(\frac{b}{2}+a-\frac{\mu c}{2}\right)\sin(\alpha+\phi_1)$$
$$Q_1 = R_1\left\{\cos(\alpha+\phi_1)-\mu\left(1+\frac{2a-\mu c}{b}\right)\sin(\alpha+\phi_1)\right\} \quad (4)$$

If a and c are small in comparison with b, this equation reduces to

$$Q_1 = R_1\{\cos(\alpha+\phi_1)-\mu \sin(\alpha+\phi_1)\}$$

and, since $\mu = \tan\phi$, this becomes

$$Q_1 = R_1 \frac{\cos(\alpha+\phi_1+\phi)}{\cos\phi} \quad . \quad . \quad . \quad . \quad (5)$$

Suppose $\alpha = 40°$, $\phi_1 = 10°$, $\phi = 6°$, $a = 0{\cdot}3b$, $c = 0{\cdot}3b$ and $P = 50$ lb.

Then, without friction, from (1):

$$Q = P\cot\alpha = 50.\cot 40° = 59{\cdot}6 \text{ lb}$$

and, with friction, from (2):

$$R_1 = P.\frac{\cos\phi_1}{\sin(\alpha+2\phi_1)} = 50.\frac{\cos 10°}{\sin 60°} = 56{\cdot}9 \text{ lb}$$

and, from (4), since $\mu = \tan 6° = 0{\cdot}1051$,

$$Q_1 = R_1\{\cos 50°-0{\cdot}1051(1+0{\cdot}6-0{\cdot}1051.0{\cdot}3)\sin 50°\}$$
$$= R_1\{0{\cdot}6428-0{\cdot}1051(1{\cdot}569)0{\cdot}7660\}$$
$$= R_1(0{\cdot}6428-0{\cdot}1263) = 0{\cdot}5165R_1$$
$$= 29{\cdot}4 \text{ lb}$$

The efficiency of the arrangement

$$= Q_1/Q = 29{\cdot}4/59{\cdot}6 = 0{\cdot}493 \quad \text{or} \quad 49{\cdot}3\%$$

If equation (5) is used, which is equivalent to neglecting the effect on the guide reactions of the tilting of the slider, then $Q_1 = 32{\cdot}0$ lb, and the efficiency is 53·7%.

87. Friction of a Screw and Nut. (a) *Square Thread.* The development of a screw thread when unwound from the body of the screw is an inclined plane, the inclination of the plane being equal to the helix angle of the thread, as shown in Fig. 134. Strictly speaking the helix angle decreases slightly from the root to the tip of the thread, but the depth of the thread is small in comparison with the radius of the screw and for practical purposes the helix angle at the mean radius of the thread is used. Let l be the lead of the thread or helix, i.e. the axial distance through which the nut would move if given one complete turn on a fixed screw; let r be the mean radius of the thread and α the lead angle or inclination of the equivalent inclined plane. Then:

$$\tan \alpha = l/2\pi r$$

It follows that rotation of the nut on the screw, or of the screw in the nut, is equivalent to sliding along an inclined plane.

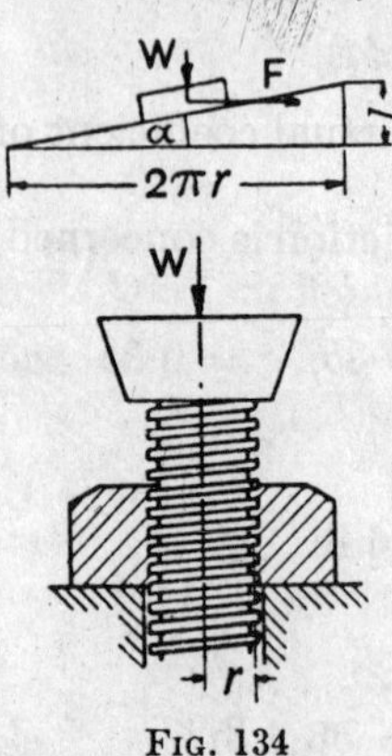

FIG. 134

The tangential force F required at the mean radius of the screw in a plane normal to the axis of the screw may therefore be expressed in terms of the axial load W, the inclination α of the developed plane and the angle of friction ϕ. The conditions are analogous to those already considered for an inclined plane with the force acting parallel to the base of the incline, i.e. $\theta = 90°$. If the nut is rotated so as to move the screw against the axial load, the latter has in effect been moved up the incline and the force F required may be calculated from equation (6.6). The turning moment which has to be exerted on the nut is therefore $F.r$. If an effort P is applied to a spanner at a distance L from the axis of the screw, then:

$$P.L = F.r = W.r \tan(\alpha+\phi) \quad . \quad . \quad (6.24)$$

If the nut is rotated in the opposite sense, the load in effect moves down the incline, and from equation (6.8) with $\theta = 90°$:

$$P_1.L = -W.r \tan(\alpha-\phi) \quad . \quad . \quad . \quad (6.25)$$

where P_1 is the effort which has to be exerted on the spanner *in the sense of rotation of the nut.* If $\alpha > \phi$, P_1 is negative, i.e. the nut will not remain at rest under the axial load W unless a torque is applied to it in order to prevent rotation. The efficiency of a screw and nut may be found from equation (6.7), and it follows from (6.14) that the efficiency is a maximum when the lead angle $\alpha = 45°-\phi/2$.

FIG. 135

(b) *Vee Thread.* In practice many screws are provided with vee threads and the normal reaction between the screw and the nut is therefore greater than when a square thread is used. The axial load W, Fig. 135, is assumed for simplicity to be concentrated at a single point on the thread. Since the axial component of the normal reaction R_n must be equal to W, we have:

$$R_n \cos\beta = W \quad \text{or} \quad R_n = W/\cos\beta$$

where 2β is the included angle between the sides of the thread.

But the friction force which acts tangentially to the surfaces of the threads is given by:

$$\mu R_n = \mu . W/\cos\beta = \mu_1 W$$

where $\mu_1 = \mu/\cos\beta$ and may be regarded as a virtual coefficient of friction.

The conditions for the vee thread, so far as friction is concerned, are identical with those for a square thread in which the coefficient of friction is μ_1. The corresponding friction angle is $\phi_1 = \tan^{-1}\mu_1$. This virtual friction angle may be substituted for the actual friction angle ϕ in the equations already given for the square-threaded screw.

88. Pivot and Collar Friction. It frequently happens that a rotating shaft is subjected to an axial thrust, and a bearing surface must be provided in order to take this thrust and to preserve the shaft in its correct axial position. Examples of shafts which carry an axial thrust are to be found in the propeller shafts of steamships, the shafts of steam turbines, vertical machine shafts, etc. The surface or surfaces on which the thrust is carried are usually plane surfaces at right angles to the axis of rotation, but occasionally conical surfaces, in which the axis of the cone coincides with the axis of rotation, may be used. Relative motion between the contact surfaces of a thrust bearing is resisted by the friction between the surfaces, and before such relative motion can take place a torque or couple must be applied to the shaft. The magnitude of the torque required may be determined approximately as follows:

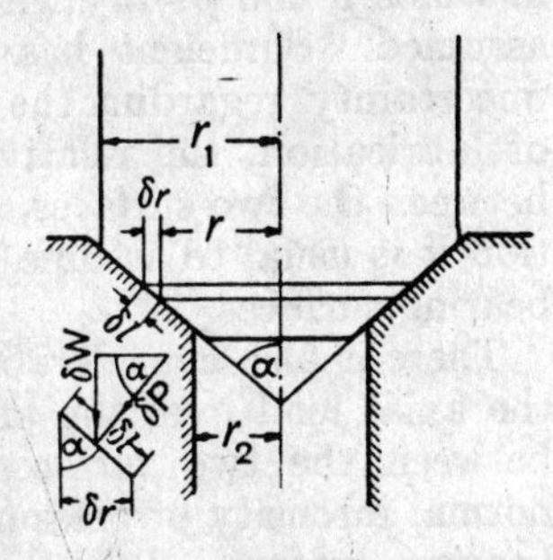

FIG. 136

Referring to Fig. 136, an axial load W is supported by a conical bearing surface with an apex angle 2α. The extreme radii of the actual area of contact are r_1 and r_2.

Consider a ring of bearing surface of radius r, radial width δr and width δl parallel to the conical surface.

Let p be the normal intensity of pressure between the surfaces at radius r and μ the coefficient of friction between the surfaces at radius r. Then the area of the ring of bearing surface $= \delta A = 2\pi r.\delta l$ and the normal load supported by the ring

$$= \delta P = p.\delta A = p.2\pi r.\delta l$$

The axial load supported by the ring

$$= \delta W = \delta P \sin \alpha = p.2\pi r.\delta l \sin \alpha$$

But $\delta l \sin \alpha = \delta r$,

$$\therefore \ \delta W = p.2\pi r.\delta r$$

and the total axial load

$$W = \int_{r_2}^{r_1} p.2\pi r.dr \quad . \quad . \quad . \quad . \quad (6.26)$$

The friction force on the ring of bearing surface

$$= \delta F = \mu.\delta P = \mu p.2\pi r.\delta l$$

and the friction moment about the axis of rotation of the shaft

$$= \delta M = \delta F.r = \mu p.2\pi r^2.\delta l = \mu p.2\pi r^2.\delta r/\sin \alpha$$

The total friction moment which resists the rotation of the shaft

$$= M = \int_{r_2}^{r_1} \mu p.2\pi r^2.dr/\sin \alpha \quad . \quad . \quad (6.27)$$

Before equations (6.26) and (6.27) can be integrated, the way in which μ and p vary with the radius r must be either known or assumed. Sufficient has already been said to emphasise the uncertainty regarding the value of μ, which varies with the degree of lubrication, the relative speed and the intensity of pressure between the two surfaces. In the absence of more exact information it is usual to assume that μ is constant for all points on the bearing surfaces.

There is also considerable uncertainty as to the distribution of the axial load over the area of the contact surfaces. If the fit between the two surfaces is assumed to be perfect, then the normal intensity of pressure will be the same at all points on the bearing surface. But the rate of wear of the surfaces must depend not only on the intensity of pressure but also on the velocity of rubbing between the surfaces, i.e. rate of wear $= f(p, v)$. There is very little information available as to the exact relationship between rate of wear, intensity of pressure and rubbing speed. If it is assumed that the rate of wear is directly proportional to the product pv, then, since v varies directly with r, the rate of wear $\propto p.r$.

Equations (6.26 and (6.27) will therefore be integrated on the assumptions that the coefficient of friction is the same at all points on the bearing surface and that either (a) the intensity of pressure is uniform or (b) the rate of wear is uniform, i.e. pr = constant.

(a) *Uniform Intensity of Pressure.* From equation (6.26):

$$W = 2\pi p \int_{r_2}^{r_1} r dr = p.\pi(r_1^2 - r_2^2) \quad . \quad . \quad (6.28)$$

and from equation (6.27):

$$M = 2\pi \frac{\mu}{\sin \alpha} p \int_{r_2}^{r_1} r^2 dr = \frac{2}{3} \frac{\mu}{\sin \alpha} p\pi(r_1^3 - r_2^3) \quad . \quad (6.29)$$

On substituting for p from (6.28) in (6.29):

$$M = \frac{2}{3} \frac{\mu}{\sin \alpha} W \frac{r_1^3 - r_2^3}{r_1^2 - r_2^2} \quad . \quad . \quad . \quad (6.30)$$

It is clear that, for a flat pivot or collar, equation (6.28) remains unchanged, but since $\alpha = 90°$, $\sin \alpha = 1$ and equation (6.30) reduces to:

$$M = \frac{2}{3} \mu W \frac{r_1^3 - r_2^3}{r_1^2 - r_2^2} \quad . \quad . \quad . \quad (6.31)$$

The friction moment for a conical pivot is therefore identical with that for a flat pivot which has a higher coefficient of friction $\mu_1 = \mu/\sin \alpha$.

(b) *Uniform Rate of Wear*: ($pr = C$). Substituting C for pr in equation (6.26):

$$W = 2\pi C \int_{r_2}^{r_1} dr = 2\pi C(r_1 - r_2) \quad . \quad . \quad (6.32)$$

and from equation (6.27):

$$M = 2\pi \frac{\mu C}{\sin \alpha} \int_{r_2}^{r_1} r dr = \pi \frac{\mu C}{\sin \alpha}(r_1^2 - r_2^2) \quad . \quad (6.33)$$

and substituting for C from equation (6.32):

$$M = \frac{\mu W}{\sin \alpha} \frac{r_1 + r_2}{2} \quad . \quad . \quad . \quad . \quad (6.34)$$

For a flat pivot equation (6.32) remains unchanged, while equation (6.34) reduces to

$$M = \mu W \frac{r_1 + r_2}{2} \quad . \quad . \quad . \quad . \quad (6.35)$$

Which of the two assumptions, uniform intensity of pressure or uniform rate of wear, should be used in any given problem is a matter of opinion. It would seem better to use that assumption which will give a result on the safe side. For instance, a comparison of equations (6.31) and (6.35) shows that the calculated friction moment is higher if uniform intensity of pressure is assumed than if uniform rate of wear is assumed. If, therefore, the problem is concerned with the power absorbed by friction in a thrust bearing, use the former assumption. If, on the other hand, the problem is to find the power which can be transmitted by friction between the surfaces, use the latter assumption.

89. The Thrust Bearing. The general arrangement of a thrust bearing of the ordinary type is shown diagrammatically in Fig. 137. A number of collars C is turned integrally with the shaft. Between each pair of collars there is a horse-shoe shaped bearing pad B, which is held in position by two or more lugs L. Long screwed

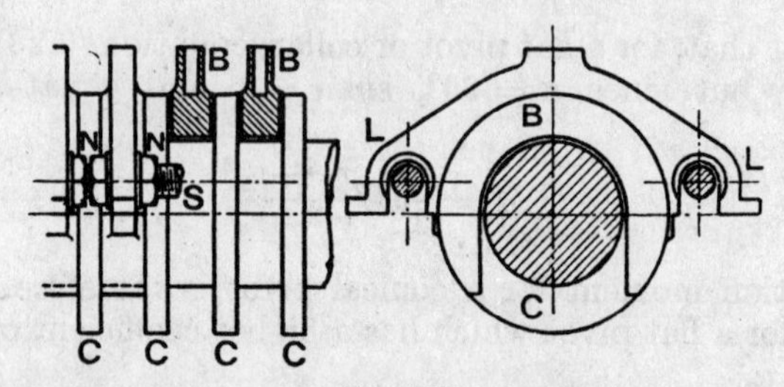

Fig. 137

bolts S pass through the lugs, and nuts N on either side of the lugs locate the bearing pads in the axial direction. When the pads are correctly adjusted, each one should carry an equal share of the total axial thrust W.

If n is the number of collars, then the friction moment for each collar from equation (6.31) is equal to:

$$\frac{2}{3}\mu\frac{W}{n}\frac{r_1^3-r_2^3}{r_1^2-r_2^2}$$

and the total friction moment for n collars:

$$M=\frac{2}{3}\mu W\frac{r_1^3-r_2^3}{r_1^2-r_2^2} \quad . \quad . \quad . \quad . \quad (6.36)$$

Thus the number of collars does not affect the friction moment.

The sole reason for providing a large number of collars is to reduce the intensity of pressure on the bearing surfaces, so that they may be effectively lubricated. In practice the bearing pressure is limited to about 50–60 lb/in^2.

According to Tower's experiments[1] the coefficient of friction for this type of bearing is about 0·035 to 0·040 when well lubricated.

Example 3. The thrust along a propeller shaft is 16 tons. The internal diameter of the bearing pads is 12 in. and the external diameter is 16 in. If the intensity of pressure is limited to 50 lb/in^2, the r.p.m. are 120 and μ is 0·04, find the power absorbed in overcoming friction and the number of collars required.

The total friction moment from equation (6.36) is:

$$M = \frac{2}{3}.0{\cdot}04.16.2240.\frac{8^3-6^3}{8^2-6^2}$$

$$= 10\,100 \text{ lb in.}$$

$$\therefore \text{ h.p. absorbed} = \frac{M.2\pi N}{12.33\,000} = \frac{10\,100.2\pi.120}{12.33\,000}$$

$$= 19{\cdot}3$$

If the bearing pads are shaped as in Fig. 137, the effective bearing surface per pad is easily calculated from the dimensions to be 58·5 in^2.

$$\therefore \text{ number of collars required} = \frac{16.2240}{58{\cdot}5.50}$$

$$= 12{\cdot}25 \text{ (say 12)}$$

If the bearing pressure is a nominal bearing pressure based on the annular area of the collar, then the number of collars required is:

$$n = \frac{16.2240}{\pi(8^2-6^2).50} = 8{\cdot}15 \text{ (say 8)}$$

90. Plate and Disc Clutches. Two types of friction clutch that are very widely used and that operate on the same principle are shown diagrammatically in Fig. 138. The clutch shown at (a) is a single plate clutch. The flywheel A is bolted to a flange on the driving shaft B. The plate C is fixed to a boss, which is free to slide axially along the driven shaft D but by means of splines is compelled to revolve with the shaft D. Two rings G of special friction material are riveted to A and E or, alternatively, to the plate C. The presser plate E is bushed internally, so as to revolve freely on the driven shaft D, and is integral with the

[1] *Proc. Inst. Mech. Eng.*, 1886.

withdrawal sleeve F. A number of spiral springs are arranged round the clutch as shown at S, so as to provide the axial thrust between the friction surfaces. The action of the clutch is as follows. When the withdrawal sleeve is displaced to the right, there is no axial pressure between the friction surfaces and the flywheel A revolves freely, while the plate C and the shaft D remain at rest. When the withdrawal force is removed from the sleeve F, the springs S force the presser plate E against the rings G and the friction between the contact surfaces of the rings G and the plate C transmits a torque to the shaft D. Provided that the friction torque on C is greater than the resisting torque on D, the shaft D will revolve. This type of clutch therefore enables the driven shaft to be started or stopped at will. To preserve the alignment of the driven shaft, a small spigot bearing is provided in the end of the driving shaft.

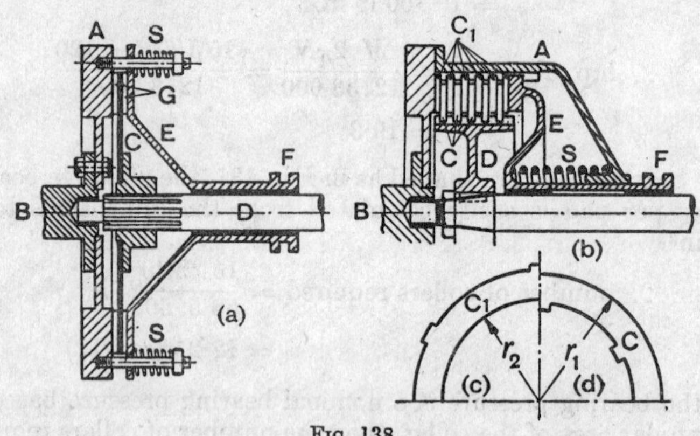

Fig. 138

The clutch shown at (b) is similar to that shown at (a) except that the number of surfaces at which slip can take place is increased. The outer casing A is bolted as before to the driving shaft B, but has a number of axial grooves cut on the internal surface. Alternate discs, marked C_1, have tongues or projections on the outer edge, as shown at (c). These are a sliding fit in the internal axial grooves on the casing A. The internal diameter of each disc C_1 is greater than the diameter of the driven member D. The other discs C have an external diameter less than the internal diameter of the casing A and are provided with tongues on their inner edges, as shown at (d). These are a sliding fit in the axial grooves on the driven member D. It follows that the discs C_1 must revolve with the casing A and the discs C must revolve with the driven member D. The action of the clutch is the same as that already described for the single plate clutch. The axial thrust exerted by the spring S forces the discs C_1 and C into contact and the friction between the contact surfaces enables a torque to be transmitted from the casing A to the shaft D. Displacement of the withdrawal sleeve F to the right removes the axial load from the discs and allows the outer casing to revolve freely, while the shaft D remains at rest. As already pointed out, the friction torque between each pair of contact surfaces should be determined from equation (6.35). It is given by:

$$M = \mu W \frac{r_1 + r_2}{2}$$

where W is the axial load and r_1 and r_2 are the external and internal radii of the contact surfaces.

It should be noted that, since the discs are free to slide axially under the spring pressure, each pair of contact surfaces is subjected to the full axial load W. Hence the total torque transmitted from the driving to the driven shaft is equal to n times the torque given by the above equation, where n is the number of pairs of surfaces between which sliding can take place. For the single plate clutch, $n = 2$, since there are two pairs of contact surfaces corresponding to the two sides of the plate C and the adjacent surfaces of the rings G. For the disc clutch shown it is easily seen that $n = 10$, since only one face of each of the outer discs C_1 is effective.

Example 4. A car engine rated at 12 h.p. gives a maximum torque of 65 lb ft. The clutch is of the single-plate type, both sides of the plate being effective. If the coefficient of friction is 0·3, the mean axial pressure is limited to 12 lb/in², and the external radius of the friction surface is 1·25 times the internal radius, find the dimensions of the clutch plate and the total axial pressure which must be exerted by the springs.

Let r_1, r_2 be the external and internal radii of the friction surfaces.

Then the total axial thrust

$$\begin{aligned} W &= p.\pi(r_1^2 - r_2^2) \\ &= 12.\pi.r_2^2(1{\cdot}25^2 - 1) \\ &= 21{\cdot}2 r_2^2 \text{ lb} \end{aligned}$$

Friction moment for each surface, from (6.35) $= \frac{1}{2}\mu W(r_1 + r_2)$. Total friction moment for the two sides of the clutch plate

$$= M = \mu W(r_1 + r_2) = 0{\cdot}3 . 2{\cdot}25 W r_2 = 0{\cdot}675 W r_2$$

Substituting for W, $M = 0{\cdot}675.21{\cdot}2r_2^3$

But the required friction moment $= 65$ lb ft $= 780$ lb in.

$$\therefore\ 0{\cdot}675.21{\cdot}2r_2^3 = 780$$

$$r_2 = \sqrt[3]{\frac{780}{0{\cdot}675.21{\cdot}2}} = 3{\cdot}79 \text{ in.}$$

$$r_1 = 4{\cdot}74 \text{ in.}$$

The dimensions of the friction surfaces are: outside dia. 9·5 in., inside dia. 7·6 in.

Example 5. A multi-plate friction clutch is required to transmit 100 h.p. at 3600 r.p.m. The plates are alternately steel and phosphor bronze and they run in oil. The coefficient of friction is 0·07, the axial pressure is 20 lb/in.^{2} and the internal radius of the friction surfaces is 0·8 of the external radius, which is 5 in. Find the number of plates required and sketch the arrangement.

The area of each friction surface

$$= \pi(r_1^2 - r_2^2)$$
$$= \pi.25(1 - 0{\cdot}8^2)$$
$$= 28{\cdot}3 \text{ in}^2$$

$$\therefore \text{ total axial thrust on the plates} = W = 20.28{\cdot}3$$
$$= 566 \text{ lb}$$

The friction moment for each pair of contact surfaces is given by (6.35):

$$M = \tfrac{1}{2}\mu W(r_1 + r_2) = \tfrac{1}{2}.0{\cdot}07.566.5(1 + 0{\cdot}8)$$
$$= 178 \text{ lb in.}$$

But total torque to be transmitted:

$$T = \frac{H.33\,000.12}{2\pi N}$$

where H is the horse-power and N is the r.p.m.

$$\therefore\ T = \frac{100.33\,000.12}{2\pi.3600} = 1752 \text{ lb in.}$$

$\therefore$ no. of effective friction surfaces required $= 1752/178 \simeq 10$

Hence, the plates must be arranged so that there are ten surfaces at which slip can take place. The arrangement will be similar to that shown diagrammatically in Fig. 138 (b), that is there must be eleven plates altogether, six revolving with the driver and five with the driven shaft.

91. Friction Circle. Friction Axis. The two members of a turning pair are shown in Fig. 139 (a) with the necessary mechanical clearance between the contact surfaces much exaggerated. When the pin is at rest in the bearing and the contact surfaces are frictionless, the reaction of the bearing on the pin will lie along the radial line AO through the point of contact A. But if the coefficient of friction is μ, corresponding to a friction angle ϕ, the direction of the reaction may lie anywhere between the two extreme positions AR and AS, which are inclined to the radial line AO at the angle ϕ. The limiting position of the reaction will be given by AR when the point of contact on the pin slides towards the right relative to the point of contact on the bearing, i.e. when the pin turns in the counter-clockwise sense relative to the bearing. Similarly, the limiting position of the reaction will be given by AS when the pin turns in the clockwise sense relative to the bearing. From O drop perpendiculars on each of the lines AR and AS. Then each of these perpendiculars

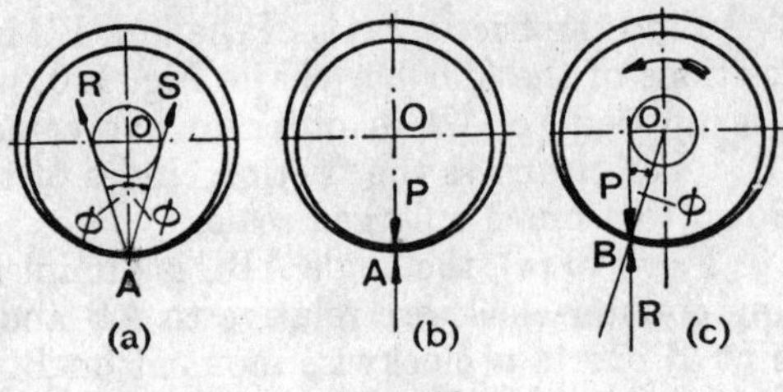

FIG. 139

will have a length, OA sin ϕ. But ϕ is usually small, so that sin ϕ may be assumed to be equal to tan ϕ or μ without sensible error. It follows that, if a circle is drawn with centre O and radius μr, the limiting positions AR and AS of the reaction of the bearing on the pin will be tangential to this circle. The circle is known as the *friction circle*, and so long as the pin is rotating in the bearing the reaction of the bearing on the pin must act along a line which is tangential to the friction circle. If, for example, the resultant force on the pin is vertical and the point of contact is at A, Fig. 139 (b), when the pin is at rest, then, when the pin rotates counter-clockwise relative to the bearing, the point of contact will move round to B, Fig. 139 (c), such that the line of action of P is tangential to the friction circle.

Note that the pin tends to climb up the bearing in the opposite sense to that of the rotation of the pin.

When a link of a mechanism is coupled by pin-joints to two adjacent links and the pin-joints are frictionless, the line of action of the thrust or pull transmitted along the link must coincide with the line joining the centres of the pins. But we have just seen that, when friction is taken into account, the reaction between a pin and its bearing no longer passes through the pin centre but is tangential to the friction circle of the pin. Hence the line of action of the thrust or pull along the link must coincide with the common tangent to the friction circles of the two pins. Since there are four common tangents to a given pair of friction circles, it is necessary to determine in any particular problem which of the four gives the true line of action of the thrust or pull. The line of action of the thrust or pull is known as the *friction axis* of the link. In order to fix the particular common tangent which determines the friction axis, it must be remembered that the friction at each pin opposes the relative motion between the pin and its bearing. Hence the thrust or pull along the friction axis must exert a moment about the pin centre which will overcome the friction moment. One or two examples will make the method clear.

Example 6. Find the friction axis of the link BC in each of the two given positions of the four-bar chain, Fig. 140, when a clockwise torque is applied to AB in order to overcome a resisting torque on CD. For clearness the friction circles of the pins at B and C are shown to a much enlarged scale.

Referring to Fig. 140 (a), the angle ABC is diminishing, so that BC is swinging counter-clockwise relative to AB and the friction of the pin-joint B exerts a clockwise moment on BC. Since BC

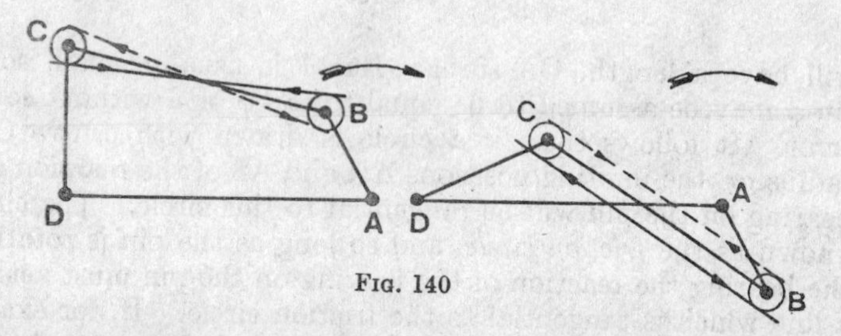

FIG. 140

is in tension, it follows that, to overcome this friction moment, the friction axis must be tangential to the friction circle of the pin B at a point above the centre of the pin. Similarly, the angle BCD is increasing so that CB is swinging counter-clockwise relative to CD and the friction of the pin-joint exerts a clockwise moment on CB. In order to overcome this friction moment, the friction axis must be tangential to the friction circle of pin C at a point below the centre of the pin. Hence, the friction axis must be as shown by the full line.

Again, referring to Fig. 140 (b), the angle ABC is increasing, so that the friction of the pin-joint B exerts a clockwise moment on BC. But BC is under compression and therefore the friction axis must be tangential to the friction circle of pin B at a point below the centre of the pin. Similarly, the angle BCD is decreasing, so that the friction of the pin-joint C exerts a counter-clockwise moment on CB. In order to overcome this friction moment the thrust along BC must be tangential to the friction circle of pin C at a point below the centre of the pin. Hence the friction axis must be as shown by the full line. It is left as an exercise for the student to show that if CD is the driving crank and AB the driven crank and the sense of rotation of AB and CD is unchanged, the friction axes will occupy the positions shown by dotted lines.

Example 7. The arrangement of the eccentric drive for the exhaust piston of a large two-stroke Diesel engine is shown in Fig. 141. The stroke is 16 in., the diameter of the eccentric is 30 in. and the diameters of the crankshaft journal and the crosshead pin are 14 in. and 8·8 in. The length of the eccentric rod between centres is 80 in. and the pull on the crosshead is 6 tons when the angle COP is 60°.

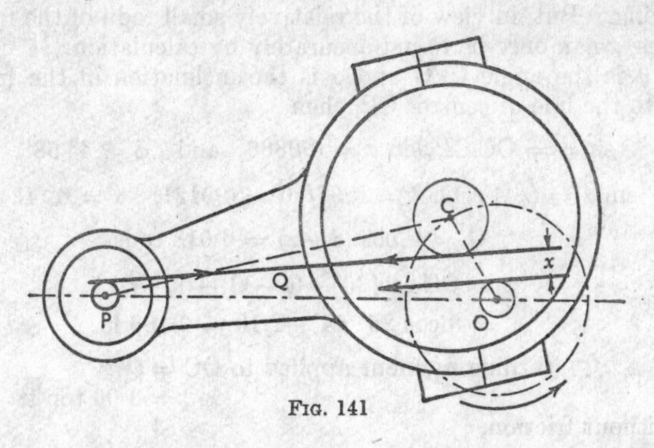

FIG. 141

Find the turning moment on the crankshaft (a) when friction at the bearings is neglected, (b) when the coefficient of friction at the bearings is 0·05.

The radii of the friction circles at O, C and P are: at O, $r_o = 0{\cdot}05 . 7 = 0{\cdot}35$ in., at C, $r_c = 0{\cdot}05 . 15 = 0{\cdot}75$ in. and at P, $r_p = 0{\cdot}05 . 4{\cdot}4 = 0{\cdot}22$ in.

These circles are shown, in the figure, which is not drawn to scale.

Let Q be the tensile force transmitted along the eccentric rod.

Then, since CP is swinging in the clockwise sense about the axis P, the force Q must exert a clockwise moment about P in order to overcome the friction between the pin and the bearing. Hence the line of action of Q must be tangential to the friction circle at a point above P.

Similarly, the angle OCP is increasing, so that the eccentric rod is swinging in the clockwise sense relative to the eccentric and in order to overcome the friction between the eccentric and the strap the line of action of Q must be tangential to the friction circle at a point below C.

The friction axis, or line of action of Q, is therefore as shown.

The force Q applied to the eccentric is equivalent to an equal and parallel force applied to the main bearing by the crankshaft journal. To overcome the friction between the journal and the bearing this force must be tangential to the friction circle at a point above O, as shown.

The effective turning moment applied to the crankshaft $= Q.x$.

The force Q may be found by drawing the triangle of forces at P, the third force being the side thrust of the guide bars on the crosshead. The distance x may be measured from a large-scale drawing. But, in view of the relatively small radii of the friction circles, x can only be found accurately by calculation.

If ϕ is the angle CPO and α is the inclination of the friction axis to the line of centres CP, then

$$\sin\phi = \text{OC/CP}.\sin\theta = 0{\cdot}0866 \quad \text{and} \quad \phi = 4^\circ\ 58'$$

and $\quad \sin\alpha = (r_c+r_p)/\text{CP} = 0{\cdot}97/80 = 0{\cdot}0121, \quad \alpha = 0^\circ\ 42$

$$\therefore\ Q = \text{P}/\cos(\phi-\alpha) = 6{\cdot}018 \text{ tons}$$

Also
$$x = \text{OC}\cos\{30^\circ-(\phi-\alpha)\}-(r_c+r_0)$$
$$= 8\cos 25^\circ\ 44'-1{\cdot}10 = 6{\cdot}106 \text{ in.}$$

$$\therefore \text{ turning moment applied to OC} = Q.x$$
$$= 3{\cdot}06 \text{ ton ft}$$

Without friction,

$$Q = F/\cos\phi = 6{\cdot}024 \text{ tons}$$

and
$$x = \text{perpendicular distance of CP from O}$$
$$= \text{OC}\cos(30^\circ-\phi) = 8.\cos 25^\circ\ 2' = 7{\cdot}248 \text{ in.}$$

$$\therefore \text{ turning moment applied to OC} = 3{\cdot}64 \text{ ton ft}$$

According to the above calculations, the efficiency of the mechanism is $3{\cdot}06/3{\cdot}64 = 0{\cdot}84$ or 84%.

In experiments quoted by C. C. Pounder[1] on an eccentric drive of approximately the same dimensions the coefficient of friction varied between 0·0035 and 0·0075. This would give a much higher efficiency.

Example 8. A horizontal pump, 4 in. stroke, is driven by means of an eccentric $11\frac{1}{2}$ in. dia., keyed to a shaft 4 in. dia. The shaft is driven by a vertical belt on a 14-in. dia. pulley. The belt embraces an arc of 180° and the coefficient of friction is 0·25. If the tension on the tight side of the belt is 350 lb, find the maximum horizontal force that can be transmitted to the pump when the radius of the eccentric is at 60° to the i.d.c. Assume the eccentric rod to be very long. The coefficient of friction between the eccentric strap and sheave and between the shaft and the bearing is 0·1.

From (7.4) $$T_1/T_2 = e^{\mu\theta} = e^{0 \cdot 25\pi} = 2 \cdot 193$$

But $T_1 = 350$ lb, so that

$$T_2 = 350/2 \cdot 193 = 159 \cdot 6 \text{ lb}$$

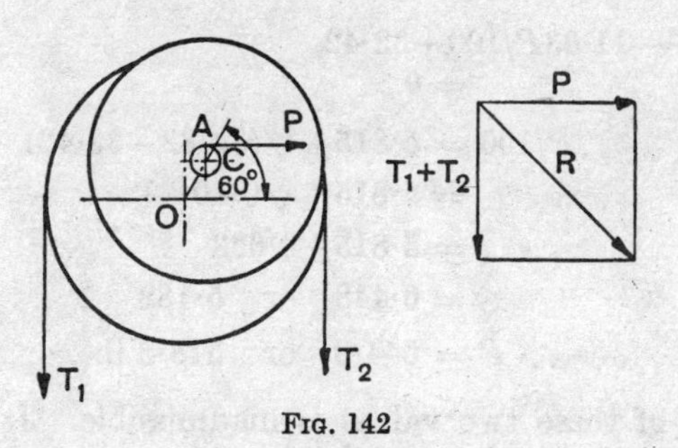

FIG. 142

Let P be the horizontal force transmitted to the pump and let R be the resultant load on the shaft bearings.

Then

$$R = \sqrt{\{P^2+(T_1+T_2)^2\}} = 100 . \sqrt{\{(P/100)^2+5 \cdot 096^2\}}$$

and the friction torque on the shaft

$$= \mu . R . \text{radius of shaft} = 0 \cdot 2R \text{ lb in.}$$

But the radius of the eccentric sheave is $5\frac{3}{4}$ in., so that the radius of the friction circle of the sheave is 0·575 in. and the line of action of the horizontal thrust along the eccentric rod must be

[1] C. C. Pounder, M.I.Mech.E. : "Some Current Types of Marine Diesel Engine," *Proc. Inst. Mech. Eng*, Vol. 160, 1949.

tangential to this friction circle. Hence the distance of the line of action of P from the axis of rotation of the shaft is

$$0{\cdot}575+2.0{\cdot}866 = 0{\cdot}575+1{\cdot}732 = 2{\cdot}307 \text{ in.}$$

The total resisting torque on the shaft is given by:

$$P.2{\cdot}307+0{\cdot}2R \text{ lb in.}$$

and this must be equal to the effective torque applied by the belt, i.e.

$$(T_1-T_2)7 = 190{\cdot}4.7 \text{ lb in.}$$

$$\therefore\ 1{\cdot}904.7 = 2{\cdot}307.P/100+0{\cdot}2\sqrt{\{(P/100)^2+5{\cdot}096^2\}} \quad . \quad . \quad (1)$$

$$\therefore\ 66{\cdot}64-11{\cdot}535.P/100 = \sqrt{\{(P/100)^2+5{\cdot}096^2\}}$$

$$\therefore\ 4441-1538P/100+133{\cdot}1(P/100)^2 = (P/100)^2+25{\cdot}97$$

$$\therefore\ 132{\cdot}1(P/100)^2-1538P/100+4415 = 0$$

$$\therefore\ (P/100)^2-11{\cdot}63P/100+33{\cdot}42 = 0$$

$$\begin{aligned}\therefore P/100 &= 5{\cdot}815\pm\sqrt{(33{\cdot}82-33{\cdot}42)}\\ &= 5{\cdot}815\pm\sqrt{0{\cdot}40}\\ &= 5{\cdot}815\pm0{\cdot}633\\ &= 6{\cdot}448 \quad\text{or}\quad 5{\cdot}182\\ \therefore\ P &= 644{\cdot}8 \quad\text{or}\quad 518{\cdot}2 \text{ lb}\end{aligned}$$

The larger of these two values is inadmissible. It corresponds to a negative sign in front of the second term on the right-hand side of equation (1).

92. Lubricated Surfaces. The results of experiments carried out by different observers on the friction of dry surfaces have frequently shown considerable discrepancies. There is little doubt that many of these discrepancies have been due to the presence of exceedingly minute traces of foreign matter on the surfaces. Recent research has shown that even a trace of lubricant is sufficient to modify appreciably the friction force required and that the bond between the lubricant and the metal surface is so strong that it is exceedingly difficult to ensure that no such trace remains.

The two most important properties of a lubricant, so far as the reduction of friction is concerned, are *viscosity* and *oiliness*.

The *viscosity* is a measure of the resistance offered to the sliding of one layer of lubricant over an adjacent layer. The absolute viscosity of a lubricant is defined as the force required to cause a plate of unit area to slide with unit velocity relative to a parallel plate when the two plates are separated by a layer of lubricant of unit thickness.

The layers of lubricant immediately adjacent to the surfaces have no motion relative to those surfaces. Thus, if one of the plates, say the lower plate, is fixed, the layer of lubricant adjacent to this surface is at rest while the layer adjacent to the surface of the upper plate moves with the same velocity as that plate. The intermediate layers move with velocities which are directly proportional to their distances from the surface of the fixed plate, Fig. 143.

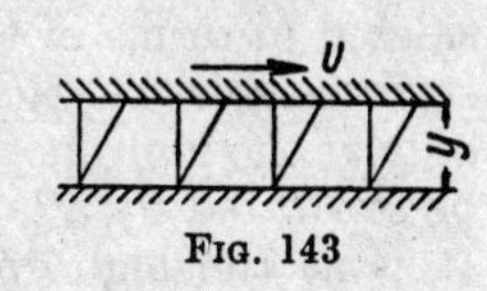

FIG. 143

Hence, it follows that the viscous force F required in order to cause a plate of area A to slide with velocity v relative to a parallel plate which is separated from it by a layer of lubricant of thickness y is given by:

$$F = \eta A . v/y \quad . \quad . \quad . \quad . \quad . \quad (6.37)$$

where η is the viscosity of the lubricant.

In the C.G.S. system of units, in which the unit of force is the dyne and the unit of length the centimetre, the unit of viscosity is termed the *poise*.

For practical methods of determining the absolute viscosity of a lubricant reference should be made to the specification of the British Standards Institution, B.S. 188:1937.

For commercial purposes the viscosities of liquids are measured by noting the time in seconds taken for a given volume of the liquid to flow through an orifice of given dimensions under specified standard conditions. The Redwood viscometer is an instrument of this type.

The viscosity of a liquid diminishes rapidly with an increase of temperature, but is only slightly affected by increase of pressure.

Oiliness is a property of the lubricant which it is more difficult to define. If two lubricants with identical viscosities are lightly smeared on two surfaces and the friction between those surfaces is tested under otherwise identical conditions, the friction force will be lower with one lubricant than with the other. The lubricant which gives the lower friction force is said to have the greater oiliness.

Viscosity and oiliness are entirely independent properties and which of the two properties exercises the controlling influence on the friction between two surfaces depends on the thickness of the layer of lubricant. If the layer is of finite thickness, so that no actual contact takes place between the surfaces, the friction is determined by the viscosity of the lubricant. If the layer is only a few molecules thick, the friction is determined by the oiliness of the lubricant.

93. Boundary Friction. However carefully the surfaces of a bearing are machined they are never perfectly smooth. Considered in terms of the dimensions of a molecule of lubricant the surfaces must be regarded as rough, so that contact will take place only at the high spots. The low areas between the high spots will be separated by a layer of lubricant of small but finite thickness. It is at the high spots that the condition known as boundary lubrication exists. The thickness of the film at these spots is exceedingly small, so small indeed that the viscosity of the lubricant plays no part in determining the friction between the surfaces, the oiliness of the lubricant being the important factor. But oiliness is a property which varies not only with different lubricants, but also with the same lubricant according to the material of the bearing surfaces. It has been established by the experiments of Hardy and others that there is some form of bond between the molecules of lubricant and the metal surface, a bond which is of the same nature as a chemical bond and the strength of which depends partly on the lubricant and partly on the metal. The lubricant is said to be adsorbed by the metal. The adsorbed film is exceedingly thin, but it is also exceedingly difficult to remove from the surface. The friction between two surfaces on which such a film has formed is essentially solid friction, i.e. there is no velocity gradient in the film, but the friction is considerably lower than for dry surfaces. The two surfaces are also much less liable to seize than are dry surfaces.

A feature of great practical importance is that once a film has been formed on the surfaces by running the bearing with a lubricant possessing a high degree of oiliness, it is possible to change to a lubricant with a much lower oiliness. The particular advantage of this lies in the fact that lubricants of high oiliness are liable to decompose or oxidise and are not suitable for general lubrication purposes. Vegetable oils such as castor oil, rape oil, and olive oil, particularly the first named, possess a high degree of oiliness, while mineral oils are relatively deficient in this property. Hence the practice has arisen of using an oil consisting of a mixture of mineral oil and a small percentage of vegetable oil. The vegetable oil has a greater affinity for the metal surface and is adsorbed on that surface. Perhaps the most striking results have been obtained with the solid lubricant graphite. The graphite is prepared in an extremely finely divided and chemically pure form known as colloidal graphite. In this form it can be mixed with a lubricating oil and the particles of graphite remain in suspension for an indefinite period. The graphite is deposited in a very thin layer on the surface of the metal and apparently increases the adsorption of oil on the surface. Tests carried out at the N.P.L. have shown that a bearing treated in this way is almost immune from seizure.

94. Film Lubrication. Film lubrication, in which the bearing surfaces are completely separated by a layer of lubricant, so that friction arises from the relative movement of the layers of oil and not from the rubbing of the actual surfaces, may be regarded as the ideal form of lubrication. Unfortunately special conditions of operation of the bearing are necessary in order to enable the film to be produced and maintained between the surfaces, and these conditions can only be satisfied in certain types of bearings.

The presence of a film of oil which completely separated the two bearing surfaces was first shown to exist in the course of experiments carried out on a journal bearing by Beauchamp Tower. These experiments are fully recorded in the *Proceedings of the Insitution of Mechanical Engineers*, 1885, to which reference should be made for a description of the apparatus used and the complete results. In the present connection the chief interest lies in the results of the experiments on a journal bearing with oil bath lubrication. In these experiments the bearing pad embraced rather less than one-half of the circumference of the journal and the bottom of the shaft dipped into a supply of oil which was contained in a reservoir. The oil adhered to the surface of the journal and was carried round to the bearing pad by the rotation of the journal. Tower found that the coefficient of friction was not constant, as it should be if the laws of dry friction held, but diminished with an increase of bearing pressure and increased with an increase of surface speed of the journal. Further, the actual value of the coefficient of friction was very small, being of the order 0·001 to 0·002 with mineral oil as the lubricant when the temperature of the bath was 90°F. Perhaps the most interesting result of the experiments was that relating to the pressure of the oil in the oil film. Small holes were drilled in the pad at various points along both its length and its width. Each of the holes was connected in turn to a Bourdon pressure gauge, so that the pressure of the oil in the film could be measured while the journal was rotating. Across the width of the pad the pressure was found to increase from zero at the inlet edge to a maximum value at a point a little to the outlet side of the centre line of the bearing and then to decrease to zero again at the outlet edge. Along the length of the pad the pressure remained very nearly constant, except near the two ends, where it rapidly fell down to zero, Fig. 144. The maximum pressure of the oil was much greater

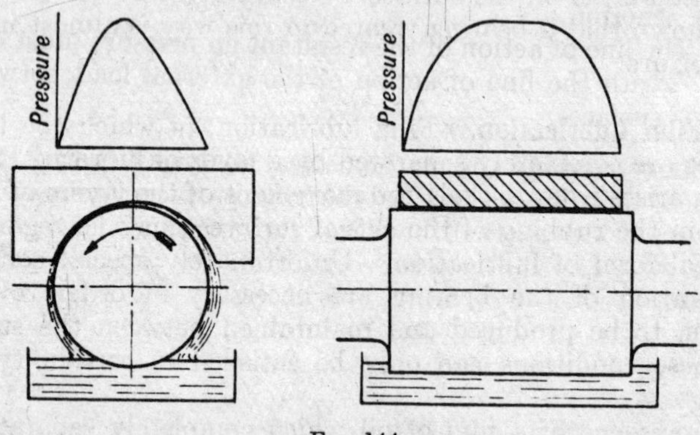

FIG. 144

than the nominal bearing pressure, but the total load corresponding to the oil pressures, as recorded on the pressure gauge, was substantially equal to the load carried by the bearing pad. This proved conclusively that the load was carried by the film of oil which separated the two surfaces and prevented metallic contact between them. Other results obtained by Tower in this series of experiments were:

(a) The tangential friction force per square inch of bearing surface was very nearly independent of the load.
(b) The tangential force diminished with an increase in the temperature of the lubricant in the oil bath.
(c) The tangential force was independent of the material used for the bearing pad.
(d) The tangential force was different for different lubricants.
(e) The thickness of the oil film was greater at the inlet or *on*-side of the bearing pad than at the outlet or *off*-side.

Since Tower's experiments had clearly shown that it was possible for a complete film of oil to separate the surfaces of the bearing and to transmit the bearing load, Osborne Reynolds was led to the conclusion that the conditions in the film of oil should be subject to hydrodynamic laws. In 1886 he read his classical paper on the mathematical theory of lubrication before the Royal

Society. In this paper he showed that the conditions essential to the formation of such a film are:

(a) A relative motion between the two surfaces in a direction approximately tangential to the surfaces.
(b) A continuous supply of oil to the surfaces.
(c) The ability of one of the surfaces to take up a small inclination to the other surface in the direction of the relative motion.
(d) The line of action of the resultant oil pressure must coincide with the line of action of the external load between the surfaces.

In order to simplify the mathematical analysis Osborne Reynolds assumed that the cylindrical bearing surfaces were of infinite transverse width, so that the flow of the lubricant took place wholly in the direction of the relative motion. Where, as in practical bearings, the width is finite, the analysis is complicated by the fact that flow also takes place in a direction at right angles to the relative motion, but the above conditions must still be fulfilled.

95. The Film Lubrication of Plane Surfaces. Although the existence of a film of oil which completely separated the two bearing surfaces was first demonstrated by Tower's experiments on a journal bearing, it will be simpler first of all to consider conditions as they apply to such a film between plane surfaces. Let Fig. 145 represent a section, taken parallel to the direction of relative motion, of the wedge-shaped film between two surfaces which are of infinite width normal to the section. The surface B moves with velocity v in the direction shown, while the surface A is at rest. The layer of oil immediately adjacent to the upper surface has the velocity v while the layer immediately adjacent to the surface A is at rest. The intermediate layers of oil have velocities which vary both across the thickness and along the length of the film. Since the surfaces are assumed to be of infinite width in the direction perpendicular to the section, no flow of oil is possible, except in the direction of v, and therefore the same quantity of oil must flow past all transverse sections of the film. Suppose that at the transverse section aa of the film the velocity gradient is constant, as it would be for a film between two parallel surfaces, then the mean velocity is $v/2$ and the

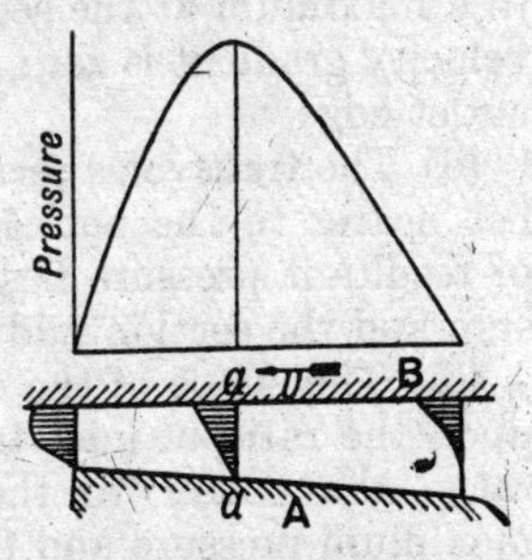

FIG. 145

quantity of oil flowing past this section in unit time is proportional to the area of the shaded triangle. It follows that for any other transverse section the shape of the velocity curve must be such that the corresponding shaded area will be equal to that of the triangle. For instance, at the inlet edge the film thickness is greater than at the section aa, so that the mean velocity is less than $v/2$ and therefore the velocity curve is convex when viewed in the direction of motion. On the other hand, at the outlet edge the film thickness is less than at aa, so that the mean velocity is greater than $v/2$ and therefore the velocity curve is concave when viewed in the direction of motion.

It is not possible to enter here into the theory of film lubrication, but mathematical analysis shows[1] that:

(a) For a thin film, in which the pressure may be assumed to be constant at all points of any one transverse section, the rate of change of velocity gradient across the film is constant.

(b) The slope of the curve of pressure variation along the film is, at any one transverse section, equal to the product of the viscosity of the oil and the rate of change of velocity gradient across the film.

From (a) it follows that the velocity curves are parabolic arcs, while from (b) the following facts may be deduced:

(i) The pressure in the film increases from zero at the inlet edge to a maximum at the section aa, where the rate of change of the velocity gradient is zero, and then diminishes again to zero at the outlet edge.

(ii) The transverse section at which the pressure is a maximum lies nearer to the outlet edge than the inlet edge, and the line of resultant pressure lies between the section of maximum pressure and the section midway along the film.

(iii) The shape of the curve of pressure variation depends only upon the ratio of inlet to outlet film thickness. The higher the ratio, the nearer to the outlet edge are both the section of maximum pressure and the line of resultant pressure.

(iv) The ordinates of the pressure curve for any one value of the ratio of inlet to outlet film thickness depend upon the values of the viscosity, the velocity, and the mean film thickness. The higher the viscosity or the velocity the greater is the pressure, but the thicker the film the lower the pressure.

It is assumed that the viscosity of the lubricant remains constant along the film.

[1] R. O. Boswall: *The Theory of Film Lubrication*, Chapter I.

Where, as in actual bearings, the viscosity diminishes along the film, owing to the rise of temperature caused by the heat generated in the film, this has the effect of shifting the line of resultant pressure towards the inlet edge.

Leakage of oil in the transverse direction, which is inevitable in bearings of finite transverse width, has the opposite effect.

The line of action of the resultant pressure may be shifted towards the inlet edge by extending the surface A beyond the inlet edge in the form of a rounded corner.

Since the speed v, the viscosity of the oil and the external load supported by the film may all vary, it is clearly necessary for the surface A to be able to alter its inclination to the surface B. This may be allowed for, in practice, by supporting on a pivot the pad which forms surface A. Although theoretically the pivot should be nearer to the outlet edge than to the inlet edge, since the lines of action of the resultant pressure and the external load must coincide, it is found that the shifting of the line of resultant pressure owing to the causes mentioned above is sufficient to enable a central pivot to be used quite satisfactorily.

96. The Michell and Kingsbury Bearings. Film lubricated plane surfaces were applied to thrust bearings in 1905 by A. G. M. Michell in England and A. Kingsbury in the United States. Working independently, they were the first to design bearings with the specific object of incorporating the principles which

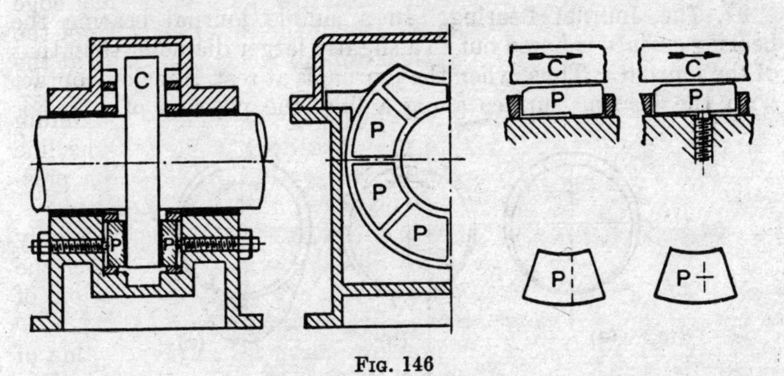

FIG. 146

had been laid down by Osborne Reynolds. A Michell bearing is shown in Fig. 146. The thrust is transmitted from the shaft to the bearing casing through a single collar C. The actual bearing surface is divided into a number of sector-shaped pads P, which may form a complete ring round the shaft or alternatively may extend only part way round. Each of the pads is supported in the casing in such a way that, although prevented from moving circumferentially, it is free to tilt and take up a small inclination to the surface of the collar in the direction of motion. Alternative methods of supporting the pads are shown in Fig. 146. Lubricant is supplied under pressure in large bearings, but the collar may simply dip into a well of oil in small bearings. The oil is carried round by the collar and a wedge-shaped film is formed between the surface of the collar and the surface of each pad. This film transmits the thrust from the collar, to the pads and thence to the casing.

For this type of bearing the friction is very much lower and the allowable bearing pressure is very much higher than for a bearing of the horse-shoe type, Fig. 137. Experiments have shown that the coefficient of friction is only about 0·003, whereas in Tower's experiments on a collar bearing the coefficient of friction was 0·035. The intensity of pressure reached the extremely high figure of 7000 lb/in^2 [1] in some experiments carred out on a Kingsbury bearing by the Westinghouse Co., U.S.A., and then the bearing failed not by rupture of the film but by the plastic flow of the white metal surface of the pads. In practice the bearing pressure used is about 300–400 lb/in^2. These pressures should be compared with the maximum pressure of 75 lb/in^2 at which the bearing seized in Tower's experiments on a collar bearing and the normal pressure of about 50 lb/in^2 used in the design of horseshoe thrust bearings.

97. The Journal Bearing. In a simple journal bearing the bearing surface is bored out to a slightly larger diameter than that of the journal. Thus, when the journal is at rest, it makes contact with the bearing surface along a line, the position of which is

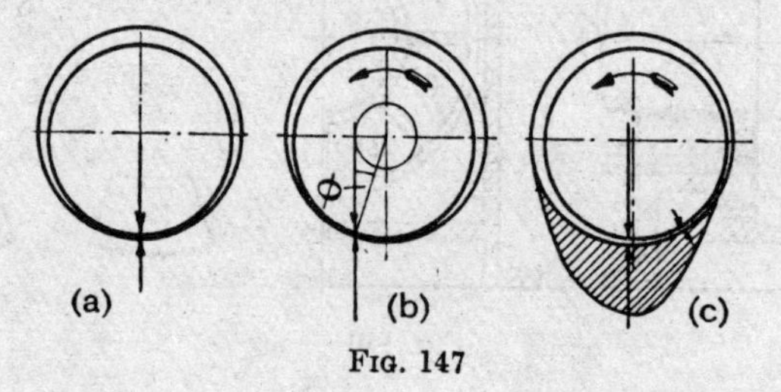

FIG. 147

determined by the line of action of the external load. If the load is vertical, as in Fig. 147 (a), the line of contact is parallel to the axis of the journal and directly below that axis. The crescent-shaped space between the journal and the bearing will be filled

[1] Hodgkinson: "Journal Bearing Practice", *Proc. Inst. Mech. Eng.*, 1929.

with lubricant. When rotation begins the first tendency is for the line of contact to move up the bearing surface in the opposite direction to that of rotation, as shown at (b), since when the journal slides over the brass, the true reaction of the brass on the journal is inclined to the normal to the two surfaces at the friction angle ϕ and this reaction must be in line with the load. The layer of lubricant immediately adjacent to the journal tends to be carried round with it, but is scraped off by the bearing, so that a condition of boundary lubrication exists between the high spots on the journal and bearing surfaces which are actually in contact.

As the speed of rotation of the journal increases, the viscous force which tends to drag the oil between the surfaces also increases, and more and more of the load is taken by the oil film in the convergent space between the journal and the brass. This gradually shifts the line of contact round the brass in the direction of motion of the journal Ultimately the film may break through, so that the two surfaces are completely separated and the load is transmitted from the journal to the brass by the oil. The film will only break through if it is possible for the resultant oil pressure to be equal to the load, and to have the same line of action. The pressure of the oil in the divergent part of the film may fall below that of the atmosphere, in which case air will leak in from the ends of the bearing. Assuming that the necessary conditions are fulfilled and that the complete film is formed, the point of nearest approach of the journal to the brass will by this time have moved to the position shown at (c) and the pressure distribution in the effective part of the oil film will be approximately as shown by the ordinates of the shaded polar diagram.

Let us consider how the tangential force F, which opposes the rotation of the journal, might be expected to vary as the speed of rotation increases.

At the instant rotation begins, there is solid friction between the surfaces and the force required is large. It is given by μW, where μ is the coefficient of friction, the value of which depends on the oiliness of the lubricant used. If there were no lubricant other than the greasy film on the surfaces, the force F would remain approximately constant at all speeds, as shown by the line AF, Fig. 148. On the other hand, if we imagine the surfaces to be separated by a film of lubricant and assume that the thickness of the film and the viscosity of the lubricant both remain constant, the force F will vary directly with the speed, equation (6.37), as shown by the line OD. But, as we have seen, the effect of an increase of speed in the bearing with a plentiful supply of lubricant is to cause more and more of the load to be supported by the convergent film and less and less to be supported by the small surface of the bearing which is actually in contact with the journal. Hence that part of the force F which is due to solid friction will diminish continuously, as shown by the line AB, and it will be zero for that value of the speed at which the film breaks through and separates the surfaces. The variation in the total value of F during this period will be approximately represented by the line AC. When the film has formed, the value of F will not follow the line CD for two reasons. First, the thickness of the film is not constant, but tends to increase as the speed increases, and, secondly, the viscosity of the lubricant diminishes as the speed increases, owing to the greater amount of heat generated in the film and the consequent rise of temperature of the lubricant. As shown by equation (6.37), an increase of film thickness and a decrease of viscosity both operate in the same direction to reduce the value of F for a given area A and velocity v.

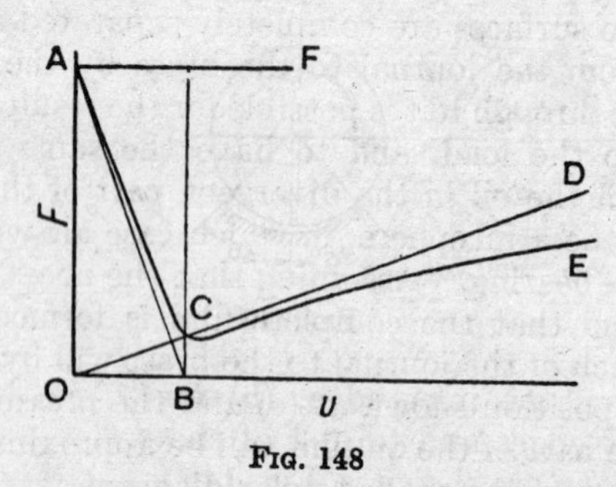

FIG. 148

The resulting variation of F for the journal bearing is shown by the line CE. That the curve ACE does indeed represent the way in which F (or μ) varies with the speed for a journal bearing is confirmed by experiment.

Both theory and experiment agree in showing that the simple cylindrical bearing surface which is shown in Fig. 147 is not the best for film lubrication. The oil pressure builds up in the convergent part of the film in a way similar to that already explained in connection with plane surfaces. But the maximum pressure is attained before the point of nearest approach of the two surfaces is reached, after which the pressure begins to fall. Theoretically, the oil pressure should become negative in the divergent part of the film and, in practice, the partial vacuum created will cause air to leak in from the ends of the bearing. It is therefore desirable to reduce the length of the arc over which the film has to be maintained and to provide a relatively large clearance between the journal and the brass over the remainder of the circumference. The arc of the brass over which the film is maintained may be either scraped so as to provide as nearly as possible a perfect fit between the bearing and the journal, in which case the brass is said to be "bedded", or, alternatively, it may be bored out to a slightly larger diameter than the journal, so as to provide a small though finite clearance between the journal and the brass, in which case the brass is said to be a "clearance" brass.

In a paper by Boswall and Brierley,[1] the authors give the results of experiments on both bedded and clearance brasses. They point out that with clearance brasses no difficulty was experienced in obtaining film lubrication until the length of the arc exceeded 105°, but that with bedded brasses the film refused to form if the length of the arc exceeded about 60°. The brasses were mounted in such a way that the line of action of the load occupied one of three positions in relation to the inlet and outlet edges of the brass. These three positions were (a) 0·6 of the arc

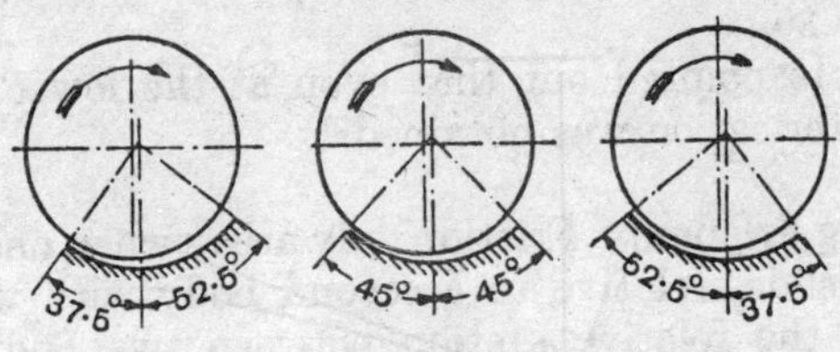

FIG. 149

of contact from the inlet edge, (b) midway between the inlet and the outlet edges and (c) 0·4 of the arc of contact from the inlet edge. They are shown in Fig. 149 for a clearance brass with an arc of contact of 90°.

For a given arc of contact and a given position of the load line, the coefficient of friction μ varied with the viscosity of the lubricant, the speed of rotation of the journal and the load. But, if values of μ were plotted against the function $\eta N/p$, a single curve was obtained, which conformed to the general equation:

$$\mu = K(\eta N/p)^n$$

where η is the inlet viscosity of the lubricant in poises, N is the r.p.m. of the journal and p is the nominal bearing pressure in lb/in², i.e. bearing load ÷ the product of diameter of journal and length of bearing. The index n had a value of approximately 0·58 in all cases and the coefficient K varied both with the arc of

[1] "The Film Lubrication of the Journal Bearing", R. O. Boswall and J. C. Brierley, *Proc. Inst. Mech. Eng.*, 1932.

contact and with the position of the load line. Values of K for clearance brasses under different conditions are shown in the table below.

VALUE OF K

Arc subtended	0·6 eccentric loading	Central loading	0·4 eccentric loading
45°	0·001 74	0·001 87	0·001 94
90°	0·001 60	0·001 87	0·002 14

In the above experiments the diameter of the journal was $2\frac{1}{2}$ in. and the length of the bearing surface was 4 in.; the clearance brasses were bored out to a diameter of 2·51 in., the inlet viscosity varied from about 0·22 to 0·56 poise and the speed of rotation of the journal from 2500 to 6500 r.p.m., while three values of the nominal bearing pressure were used, viz. 40, 80 and 120 lb/in^2. The range of values of $\eta N/p$ covered by the experiments was from about 2·5 to 30.

The authors pointed out that even at the lower limit of this range film lubrication was obtained.

98. Rolling Friction. Friction only arises when one body slides, or tends to slide, relative to a second body with which it is in contact. If the relative motion between two bodies is one of pure rolling, the two bodies make contact at a single point or along a line parallel to the axes of rotation of the two bodies. There is no relative sliding at the point or line of contact and therefore there is no friction. But in practice a pure rolling motion never exists. The surfaces of the two bodies are always more or less deformed by the reaction between them, so that the ideal point or line of contact degenerates into an area of contact. The harder the materials of the contact surfaces the less will be the deformation under a given reaction and the smaller will be the area of contact, but even with the hardest materials some deformation is inevitable. It is this deformation of one or both of the contact surfaces which gives rise to rolling friction.

Osborne Reynolds put forward a theory to account for rolling friction. This theory was based on an examination of two extreme cases in which the deformation of one of the contact surfaces is easily seen. In Fig. 150 (a) is shown the deformation produced in a thick sheet of some flexible material, such as rubber, when a hard steel roller rests on it. The lowest point of the roller sinks below the original surface level of the rubber sheet and squeezes out the rubber on each side of the roller as shown. The roller and the rubber are in contact over a surface and the surface fibres of the rubber are stretched where they make contact with the roller. If now the roller rolls on the rubber towards the left, the unstretched rubber fibres in advance of the roller must stretch when they come into contact with the roller, i.e. they must slip forwards relative to the roller surface thus introducing friction, while the stretched fibres under the roller must contract as the roller leaves them, i.e. they must slip forwards relative to the roller surface, thus again introducing friction. Both these frictional effects will tend to retard the rolling motion of the roller. The effect is obviously very similar if a flexible rubber roller rolls on a hard steel plate, the deformation of the roller being somewhat as shown in Fig. 150 (b) and the alternate extension and contraction of the rubber fibres setting up a scrubbing action which retards the motion of the roller. As already pointed out, in practice the deformation takes place on both surfaces. Where both the surfaces consist of hard materials the

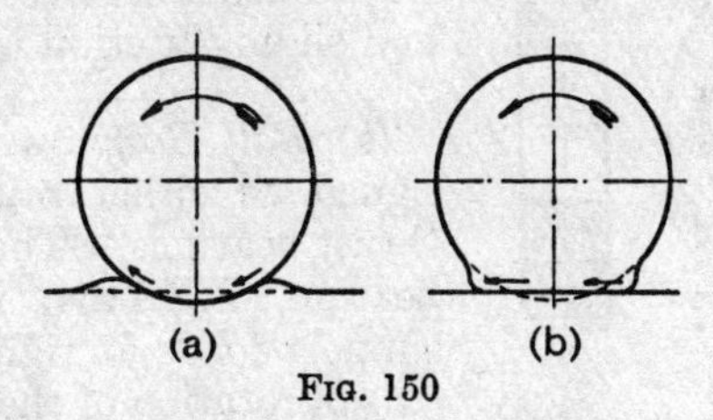

FIG. 150

deformation and therefore the magnitude of the rolling resistance is small, but where one or both of the surfaces consist of soft material, the deformation and therefore the magnitude of the rolling resistance may be large. Since the rolling resistance depends upon the extent of the deformation of the contact surfaces, it may be expected to vary in magnitude not only with the material of the contact surfaces and the load which acts between them but also with the relative curvature of the surfaces. In practice it is usual to assume that for a given pair of materials the rolling resistance is directly proportional to the load and is independent of the relative curvature of the contact surfaces.

Rolling resistance enters into all problems on the traction of wheeled vehicles and, as generally measured, it includes not merely the true rolling resistance between the wheel tyres and the surface of the track but also the frictional resistance at the bearings of the wheel axles. It is usually expressed as so many lb per ton of load supported by the wheels, and varies in extreme cases from a few lb per ton up to some hundreds of lb per ton. If the wheel or axle bearings are plain, the tractive resistance will be higher when starting from rest than when a steady slow speed has been reached. If ball or roller bearings are fitted the starting resistance will be no higher than the running resistance. For railway rolling stock, where both the tyres and rails are hard and smooth, the tractive resistance when in motion is about 5 to 10 lb per ton. For pneumatic-tyred vehicles on a concrete roadway it is about 25 to 35 lb per ton.

99. Ball and Roller Bearings. The substitution of rolling friction for the sliding friction, which is normally present between the two elements of a turning pair, results in a very considerable reduction in the frictional resistance which has to be overcome. This has led to the development of ball and roller bearings, in which the surfaces of the two elements of the turning pair are separated by a number of balls or rollers. Each ball or roller has rolling contact with the adjacent surfaces of the elements and sliding friction is eliminated.

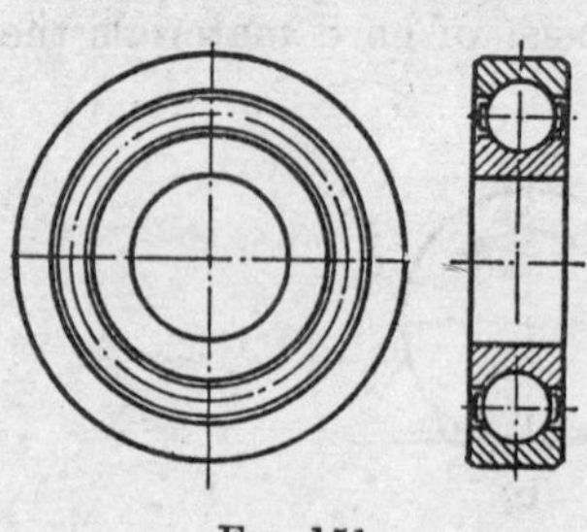

FIG. 151

(a) *Ball Bearings.* These are designed as either radial bearings or thrust bearings. The former type of bearing, Fig. 151, is intended to support loads at right angles to the axis of rotation of the shaft, although it will also sustain a certain amount of axial thrust. The inner and outer races on which the balls roll, as well as the balls themselves, are of hardened steel and are manufactured to extremely close limits, so that when the bearings are assembled there is no appreciable shake. The balls are separated from each other and spaced evenly round the races by means of a light metal cage. When mounting the races on the shaft and housing, the inner race should be a tight fit on the shaft and the outer race a tight fit in the housing so as to limit relative motion to that between the balls and the races. The ball races are grooved as shown, the radius of each groove being only slightly greater than the radius of the ball. Each ball then makes point contact with both the inner and the outer races. The double-row ball bearing shown in Fig. 152 (a) is manufactured by the Skefko Ball Bearing Co. Ltd., and the outer race is spherical with the centre on the axis of the shaft. By this construction the bearing allows the axis of the shaft to tilt relative to the housing and thus eliminates bending stresses in the neck of the shaft on which the inner race is mounted.

A single-row thrust bearing to support end thrust only is shown in Fig. 152 (b).

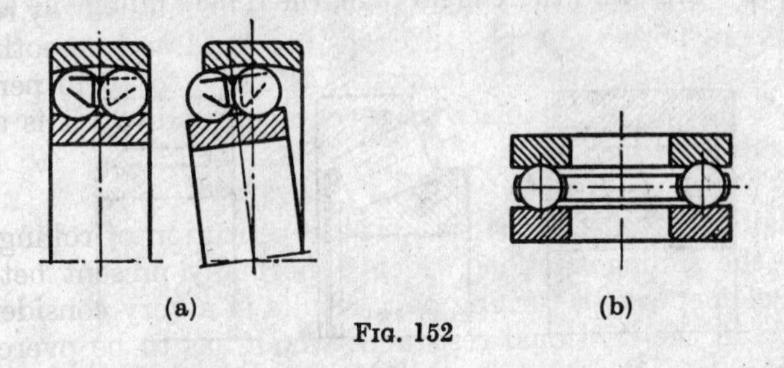

Fig. 152

(b) *Roller Bearings* Roller bearings are in general similar in construction to ball bearings, but the use of cylindrical rollers instead of balls enables greater journal loads to be carried for a given overall diameter of the bearing. Where the rollers are solid, Fig. 153 (a), they usually have a length equal to the diameter. The inner race is grooved, while the outer race is plane or, in some designs, slightly convex. The double-row Skefko roller bearing, Fig. 153 (b), has a spherical outer race, in order to allow self-alignment of the shaft. The rollers in this bearing obviously cannot be cylindrical but must be slightly barrelled. The Hyatt roller bearing, Fig. 154 (a), has long flexible rollers, which are in effect closely coiled helical springs, alternate rollers being wound with right- and left-hand spirals. The slight amount of flexibility provided in the roller ensures that the load is distributed over its full length.

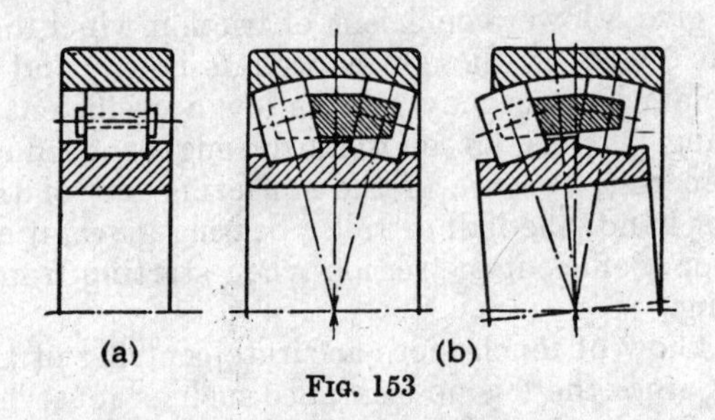

Fig. 153

In the bearing shown in Fig. 154 (b) tapered rollers are used and the bearing will support an end thrust as well as a radial load. The inner and outer races are conical and the inner race is grooved so as to prevent the rollers from running askew.

A comparatively recent development is the needle roller bearing, Fig. 154 (c). In this bearing the rollers are very small in diameter, from 2 to 4 mm, and there is no cage, the rollers filling the annular

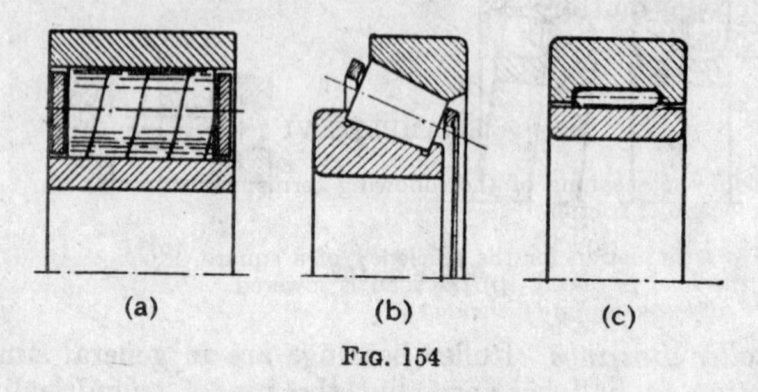

FIG. 154

space between the races. These bearings are particularly suitable for supporting heavy loads when the amount of relative motion between the races is small, as in the case of the small end bearings of internal-combustion-engine connecting rods.

The chief advantages of ball and roller bearings over plain journal bearings are:

(a) They are much shorter for a given load-carrying capacity.

(b) They give a lower coefficient of friction which lies between 0·0015 and 0·004. Although a well-designed and efficiently lubricated plain bearing may give as low a coefficient of friction when running, it has an enormously higher coefficient when starting from rest, because of the squeezing out of the oil film. On the other hand, the ball or roller bearing gives, if anything, a slightly lower coefficient of friction when starting from rest than when running.

(c) They allow of much more accurate centring of the shaft in the housing, since there is no clearance such as must be provided in the plain bearing in order to enable the oil film to be formed.

The chief disadvantages of ball and roller bearings are:

(a) They are generally more expensive than plain bearings.

(b) If overloaded or if an unsuitable lubricant is used, the hardened surfaces of the balls, rollers and races flake off. Once flaking begins the bearing becomes noisy and is rapidly destroyed.

So far as lubrication is concerned the chief purpose of the lubricant is to protect the surfaces of the balls or rollers and of the races from corrosion. It also serves to lubricate the contact surfaces of the balls or rollers and the cages. Very little lubricant is required and the bearings are usually packed with a high-quality grease on assembly and then run for long periods without further attention. Suitable covers must be fitted to the bearing housing in order to retain the lubricant and to prevent the ingress of any foreign matter.

EXAMPLES VI

1. Explain the meaning of the following terms: limiting friction; coefficient of friction; angle of friction.

2. Derive an equation for the efficiency of a square-threaded screw and nut when (a) the load is raised, (b) the load is lowered. What helix angle of the thread will give maximum efficiency?

3. A load of $\frac{1}{2}$ ton is raised by means of a screw jack. The thread is square with a pitch of $\frac{1}{2}$ in. and the mean diameter of the screw is 2 in. What force must be applied to the end of a lever 12 in. long which rotates the nut, if the coefficient of friction is 0·15?

4. A turnbuckle is used to tighten a wire rope. The threads are right and left-hand and are square in section. The pitch is $\frac{3}{8}$ in. and the mean diameter of the screw is $1\frac{1}{2}$ in. Assuming that the coefficient of friction between the screws and the nut is 0·15, determine the turning moment necessary (a) to tighten, (b) to slacken the wire, when the pull is 2000 lb. The wire rope is to be assumed not to twist. L.U.

5. A V-threaded screw passes through a nut which rests on a ball-thrust washer. The angle of the vee is 55°, the pitch of the screw is $\frac{1}{4}$ in. and the mean diameter of the screw is $2\frac{1}{4}$ in. If the axial load is 5 tons, the coefficient of friction for the screw and nut is 0·16 and the friction of the thrust washer is negligible, find the turning moment required on the nut in order to move the screw axially (a) against the load, (b) with the load.

What is the efficiency of the screw in case (a)?

6. A vertical screw of mean diameter $2\frac{1}{2}$ in. and with square threads of $\frac{1}{2}$ in. pitch supports a load of 5000 lb. It passes through the boss of a spur wheel of 70 teeth which acts as the nut. The axial thrust is taken on a collar bearing of 3 in. inside diameter and 5 in. outside diameter. The coefficient of friction for the screw and the collar bearing is 0·12. To raise the load the wheel is turned by means of a pinion of 18 teeth. If the efficiency of the wheel and pinion is 90%, what torque must be exerted on the pinion shaft?

7. A bevel gear is used for lifting a sluice gate. The gate, which weighs 5 tons, is subjected to a mean pressure of 50 lb/in^2 over a surface of diameter 10 ft. The vertical spindle which lifts the sluice has a square-threaded end which engages with a screwed bush fixed to the sluice. The mean diameter of the thread is $3\frac{1}{2}$ in. and the pitch of the thread is 1 in. The coefficient of friction between the sluice and the vertical facing on which it presses, and also between the screw and the bush, is 0·08. The bevel wheel keyed to the vertical spindle has 60 teeth and the bevel pinion with which it gears is driven direct by a constant-torque motor which develops 60 b.h.p. at a maximum speed of 600 r.p.m.

Assuming that frictional losses—other than losses due to friction at the screw thread and sluice facing—amount to 10% of the total power available, determine the maximum number of teeth for the bevel pinion. L.U.

8. The ends A and B of a link carry pins which fit into blocks, the blocks sliding in straight slots mutually at right angles and in the same plane. In a certain

position the link AB makes an angle of 60° with the path of A. A force acting on the block A in the direction of the slot overcomes a corresponding force on the block B. What is the efficiency of the mechanism if μ is the coefficient of friction between the blocks and the slots? Show how the friction on the pins may also be taken into account. W.S.S.

9. The slotted arm in Fig. 155 turns about a pin O and moves the guided link CD through the agency of the block A sliding in the slot. The block A is free to pivot about a pin fixed to CD. The coefficients of friction between the block and slot and between the link CD and its guides may be assumed equal. A torque of 30 lb ft is applied to the arm. What is the maximum value of μ if, when $\theta = 60°$, the torque is just insufficient to move the link CD?

Taking this value for μ, calculate the pull in the direction CD when $\theta = 45°$. What would its value be, for this position, if μ were zero? W.S.

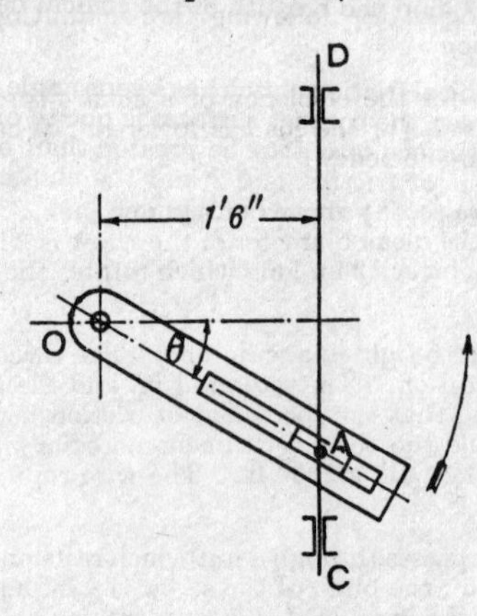

FIG. 155

10. A flat footstep bearing, 6 in. dia., supports a load of 1·5 tons. If the coefficient of friction is 0·05 and the r.p.m. 90, calculate the h.p. lost in overcoming friction.

11. Deduce an expression for the friction moment of a collar thrust bearing, and state what assumptions are made.

A thrust bearing has 12 collars, 16 in. external diameter and 12 in. internal diameter, and carries a load of 24 tons. If the coefficient of friction is 0·05, calculate the h.p. absorbed in overcoming friction at a speed of 105 r.p.m.

12. Sketch and describe a multiple-disc clutch.

A clutch has 9 discs, the contact faces of which are 6 in. external dia. and $4\frac{1}{2}$ in. internal dia. The coefficient of friction between the metal surfaces is 0·08 and the axial force is limited to 300 lb. What h.p. can be transmitted by the clutch at 1000 r.p.m.?

13. Describe, with sketches, the construction of a plate clutch.

A single-plate clutch has friction surfaces 9 in. internal and 12 in. external diameter. The intensity of pressure is limited to 10 lb/in^2 of contact surface and the coefficient of friction between the surfaces is 0·3. What is the maximum h.p. which can be transmitted by the clutch at 2500 r.p.m. if both sides of the plate are effective?

14. A motor-car clutch is required to transmit 45 h.p. at 3000 r.p.m. It is of the single-plate type, both sides of the plate being effective. If the coefficient of friction between the surfaces is 0·25, the axial pressure is limited to 10 lb/in^2

of plate area and the external diameter of the plate is 1·4 times the internal diameter, determine the dimensions of the plate. State what assumptions are made in the calculations and sketch the arrangement. M.U.

15. A conical pivot supports a load of 2 tons. The cone angle is 120° and the intensity of normal pressure is not to exceed 50 lb/in². The external diameter is three times the internal diameter. Find the dimensions of the bearing surface. If the coefficient of friction is 0·06 and the r.p.m. of the shaft 120, what h.p. is absorbed by friction?

16. A conical friction clutch with cast-iron contact surfaces transmits 130 h.p. at 1500 r.p.m. The cone angle is 20° and the coefficient of friction 0·20. If the mean diameter of the bearing surface is 15 in. and the intensity of normal pressure is not to exceed 40 lb/in², find the breadth of the conical bearing surface and the axial load required.

17. A leather-faced conical friction clutch has a cone angle of 25°. The intensity of normal pressure between the contact surfaces is not to exceed 8 lb/in² and the breadth of the conical surface is not to be greater than one-third of the mean radius. If the coefficient of friction is 0·2 and the clutch transmits 50 h.p. at 2000 r.p.m., find the dimensions of the contact surfaces.

18. The mean radius of the contact surfaces of a conical clutch is 8 in. and the breadth of the conical surface is 3 in. The clutch is lined with Ferodo and the coefficient of friction may be taken as 0·30. The cone angle is 35° and the intensity of normal pressure between the contact surfaces is 10 lb/in². What is the maximum h.p. that can be transmitted by the clutch at 1800 r.p.m. and what is the axial load required?

19. What is meant by (a) the friction circle of a pivot or journal, (b) the friction axis of a link? Show how the direction of the friction axis may be determined, given the nature of the force transmitted by the link, i.e. whether tension or compression, and the sense of rotation of the link relative to the adjacent links at the pin-joints.

20. The crank of a steam engine is 15 in. long and the connecting rod is 72 in. long. The journals at the crankshaft, crankpin and crosshead are respectively 7 in., $5\frac{1}{2}$ in. and 4 in. dia. If the effective thrust at the crosshead is 25 000 lb when the crank makes an angle of 60° with the i.d.c., find the reduction in the turning moment available at the crankshaft due to the friction of the journals The coefficient of friction is 0·05.

21. ABCD is a four-bar chain with AB as the driving link and AD as the fixed link. The lengths of the links are: AB, 3 in.; BC, 7 in.; CD, 6 in.; and AD, 9 in. The diameter of the pins at A, B, C and D is 1 in. and the coefficient of friction is 0·10. What torque must be applied to the driving shaft to which AB is keyed in order to overcome a resisting torque of 15 lb ft applied to the shaft to which CD is keyed, when the angle BAD is (a) 90°, (b) 180°, (c) 270°?

22. A body tied to a string is made to describe a circle 4 ft in diameter on a rough horizontal floor at a steady speed of 20 ft/s. In order to maintain this steady motion, the other end of the string must be constrained to move in a circle; the string is horizontal and tangential to this circle.

If the coefficient of friction between the body and the floor is 0·2, find the diameter of this circle and state the power exerted for each pound weight of the body. W.S.

23. Distinguish between the laws of friction for dry surfaces and for film-lubricated surfaces. What conditions must be satisfied in order that film-lubrication may be obtained?

24. Explain the difference between the properties *oiliness* and *viscosity* of a lubricant. In what circumstances will each of these properties exert the controlling influence on the friction between two lubricated surfaces?

25. Describe with sketches the Michell thrust bearing. Explain the principle on which it works and the advantages which it possesses over a thrust bearing of the horse-shoe type.

26. A journal under load begins to rotate in a bearing and the speed of rotation is gradually increased. If there is a plentiful supply of lubricant, explain how the position of the journal relative to the bearing and the friction on the journal vary from the instant at which rotation of the journal begins.

27. In an experiment to determine the variation of the coefficient of friction of a journal bearing with rubbing speed, a shaft which carried a heavy rotor was run up to speed, disconnected from the driving motor and allowed to run down under its own momentum. The following readings of the r.p.m. at different time intervals were taken:

Time, min. .	0	4	8	12	16	20	24	28	32	34	35
R.p.m. . .	300	261	223	188	156	125	96	69	43	27	0

The moment of inertia of the rotor was 5·75 ton ft^2, the combined weight of the rotor and shaft was 3·56 tons and the diameter of the journal bearings was 3 in. Assuming that the whole of the retardation arose from the friction of the bearings, plot a curve from the results to show the variation of μ with surface speed of the journal.

What conclusions do you draw from the shape of the curve?

28. Explain how rolling friction arises and why the coefficient of rolling friction is lower than the coefficient of sliding friction for a given pair of materials.

29. What are the advantages possessed by ball and roller bearings over plain bearings? Sketch and describe one type of self-aligning ball or roller bearing.

CHAPTER VII

BELT, ROPE AND CHAIN DRIVE

100. Ratio of the Tensions. Where power has to be transmitted between two shafts which are a considerable distance apart, a belt or rope drive is frequently used. In such drives the power transmitted depends upon the friction between the belt or rope and the pulley rim. Referring to Fig. 156 (a), let θ be the angle subtended at the pulley centre by that part of the belt or rope which is in contact with the pulley rim. Alternative forms for the cross-section of the pulley rim are shown at (b), (c) and (d). The cross-section shown at (b) is the one most commonly used;

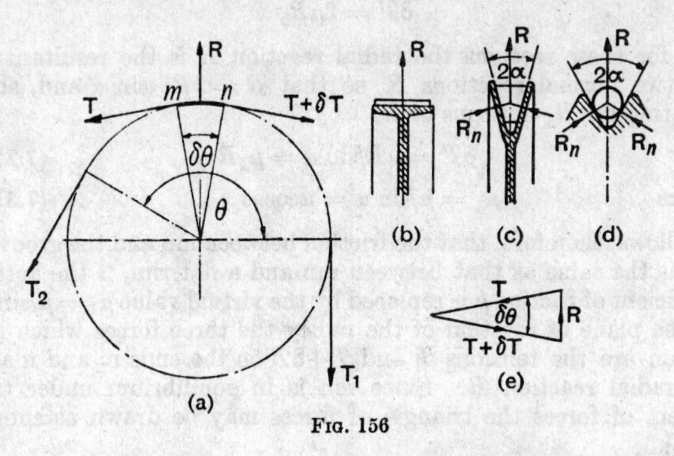

FIG. 156

the rim of the pulley is flat, or at least approximately so. In practice it is slightly crowned, since this is found to assist in maintaining the belt centrally on the rim. The grooved cross-sections (c) and (d) are used with wedge-section or circular section belts or ropes, and it should be noticed that the belt or rope does not rest on the bottom of the groove, but is wedged between the sides of the groove. The flat rim section may be regarded as the limiting case of a grooved section in which the groove angle $2\alpha = 180°$.

If the tension at one end of the belt is T_2 and the tension T_1 at the other end is increased gradually, the belt will ultimately begin to slip bodily round the pulley rim. The value of T_1 at which slip takes place will depend upon the values of T_2, θ and the coefficient of friction μ between the belt and the rim. Consider a short length mn of belt, which subtends an angle $\delta\theta$ at the pulley centre. Let T be the tension on the end m and $T+\delta T$ the tension on the end n. Then the difference of tension δT must be due to the friction between the length mn of the belt and the pulley rim, and it will depend upon the normal reaction between mn and the rim for the flat section (b), and upon the normal reactions between mn and the sides of the groove for the sections (c) and (d). Let R be the radial reaction between the pulley rim and the length mn of belt or rope and let R_n be the normal reaction between each side of the groove and the side of mn for the sections (c) and (d).

Then for section (b):

$$\delta T = \mu R \quad \ldots \ldots \quad (7.1)$$

and for sections (c) and (d):

$$\delta T = 2\mu R_n$$

But for these sections the radial reaction R is the resultant of the two normal reactions R_n so that $R = 2R_n \sin\alpha$ and, substituting for R_n in terms of R,

$$\delta T = \mu R/\sin\alpha = \mu_1 R \quad \ldots \quad (7.2)$$

where

$$\mu_1 = \mu/\sin\alpha = \mu \operatorname{cosec}\alpha \quad \ldots \quad (7.3)$$

It follows, therefore, that the friction between mn and the grooved rim is the same as that between mn and a flat rim, if the actual coefficient of friction μ is replaced by the virtual value $\mu_1 = \mu/\sin\alpha$. In the plane of rotation of the pulley the three forces which act on mn are the tensions T and $T+\delta T$ on the ends m and n and the radial reaction R. Since mn is in equilibrium under this system of forces the triangle of forces may be drawn as shown at (e).

From this triangle, since $\delta\theta$ and δT are small, $R \simeq T.\delta\theta$, and substituting this value of R in (7.1),

$$\delta T \simeq \mu T\delta\theta \quad \text{or} \quad \delta T/T \simeq \mu\delta\theta$$

If both sides of this equation are integrated between corresponding limits, then:

$$\int_{T_2}^{T_1} (dT/T) = \int_0^{\theta} \mu d\theta$$

$$\therefore \log_e(T_1/T_2) = \mu\theta$$

or

$$T_1/T_2 = e^{\mu\theta} \quad \ldots \quad (7.4)$$

As it stands this equation applies to the flat rim (b), but if μ_1 is substituted for μ, it will apply equally well to the grooved rims (c) and (d).

It must be emphasised that (7.4) gives the limiting ratio of the tensions T_1 and T_2 when the belt or rope is just about to slip bodily round the pulley rim. The actual ratio of the tensions may have a lower value, but cannot have a higher value than this limiting ratio.

The limiting ratio is very much increased, for given values of μ and θ, by using a grooved section. For instance, if θ is 165° and μ is 0·25, the limiting ratio for the flat rim is given by:

$$T_1/T_2 = e^{0\cdot25 \,.\, 11\pi/12} = 2\cdot054$$

If a wedge-section belt is used with a groove angle of 30°, then

$$\mu_1 = 0\cdot25/\sin 15° = 0\cdot966 \quad \text{and} \quad T_1/T_2 = e^{0\cdot966 \,.\, 11\pi/12} = 16\cdot15$$

Similarly, if a rope of circular section is used with a groove angle of 45°, then

$$\mu_1 = 0\cdot25/\sin 22\cdot5° = 0\cdot653 \quad \text{and} \quad T_1/T_2 = e^{0\cdot653 \,.\, 11\pi/12} = 6\cdot56$$

The maximum effective tangential pull exerted by the belt or rope on the pulley rim is, in each case, given by the difference between T_1 and T_2. It may be expressed in terms of the tension T_1 of the tight side, the magnitude of which is, of course, determined by the cross-section of the belt or rope and the allowable stress in the material.

For the flat belt under the above conditions the effective tension $T = T_1(1-T_2/T_1) = 0\cdot513T_1$; for the wedge-section belt, $T = 0\cdot938T_1$; and for the rope, $T = 0\cdot848T_1$. It is clear from these figures that the use of a grooved pulley rim with a suitable belt or rope section enables the materials to be employed more efficiently than where a flat rim is used.

So far it has been assumed that the pulley is stationary. If the pulley is mounted on a shaft, which is supported in bearings, then the effective tangential force exerted by the belt or rope on the pulley may be used to transmit power from the belt or rope to the pulley and thence to the shaft. The power transmitted may be determined when the effective tension and the speed of the belt or rope are known. But when the belt or rope is in motion the stresses in the material are not simply those which arise from the power transmitted. There is in addition the centrifugal stress due to the inertia of the belt or rope as it passes round the pulley rim. The magnitude of this stress may be determined as shown in the following article.

101. Centrifugal Stress in a Belt or Rope. Referring to Fig. 157, let r be the radius of the pulley, v the speed of the belt or rope, a the cross-sectional area and w the weight of the belt or rope per unit length. The weight of the short length mn which subtends the angle $\delta\theta$ at the pulley centre, is $w.r\delta\theta$ and the centripetal force which must be applied to mn is given by:

$$F_c = \frac{wr\delta\theta}{g}.\frac{v^2}{r} = \frac{wv^2}{g}.\delta\theta$$

So far as the tension in the moving belt is concerned, the conditions are the same as if the belt were at rest and a force equal to F_c were applied radially outward to mn as shown in Fig. 157. This force will give rise to tensile forces T_c at the ends of mn, and from the triangle of forces T_c may be expressed in terms of F_c.

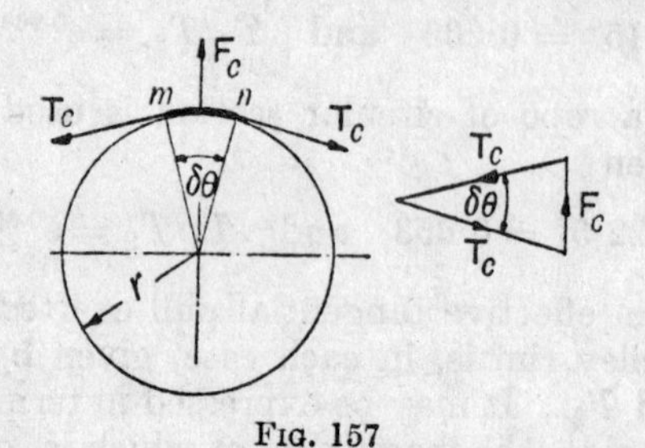

FIG. 157

Since $\delta\theta$ is small $F_c \simeq T_c\delta\theta$, and, substituting for F_c from the above equation:

$$wv^2/g.\delta\theta = T_c.\delta\theta$$

$$\therefore\ T_c = wv^2/g \quad . \quad . \quad . \quad . \quad (7.5)$$

The stress per unit area of the belt or rope material due to the inertia is given by:

$$f_c = T_c/a = w/a.v^2/g \quad . \quad . \quad . \quad . \quad (7.6)$$

It should be particularly noticed that the centrifugal stress is independent of the radius of curvature of the path.

It has been assumed so far that the rim of the pulley is flat and that the centrifugal inertia force therefore gives rise to a stress in the belt or rope material which is additional to the stresses caused by the tensions T_1 and T_2. If, however, the pulley rim is grooved as at (c) and (d), Fig. 156, it would appear at first sight that the centrifugal force may be either wholly or partly balanced by the friction between the sides of the belt or rope and the sides of the groove, in which case f_c will be either zero or will have a value less than that given by equation (7.6). But there are two other factors which have to be taken into account in this connection. First, if the power transmitted by the belt or rope is such that limiting friction exists in the tangential direction, i.e. if the belt or rope is just on the point of slipping bodily round the rim, there can be no friction force opposed to the centrifugal force. Since this condition of limiting friction rarely, if ever, exists in practice, there can be no doubt that the centrifugal stress in that part of the belt or rope, which is in contact with the rim, will be less than the stress calculated from equation (7.6). Secondly, and this is the more important factor, it has to be remembered that in any actual drive the part of the belt or rope between the pulleys is not straight but hangs in a curve. The free parts of the belt must therefore be subjected to the centrifugal stress given by equation (7.6). Hence, there is no justification for the assumption which is sometimes made that the centrifugal stress in a belt or rope running on a grooved pulley is less than that in the same belt or rope when running on a flat pulley.

102. The Power transmitted by a Belt or Rope. The power transmitted by a belt may be calculated, if the effective tension and the speed are known, from the equation:

$$H = Tv/550 \qquad (7.7)$$

where T is the effective tension in lb, v is the speed in ft/s and H is the horse-power.

For a given belt the value of H will be a maximum when the product Tv is a maximum. In what follows we shall refer to a belt drive, but it will be understood that similar conditions apply to a rope drive.

Let us consider, first of all, the ideal case in which the belt is made of a perfectly elastic material that is without mass. In these circumstances the free portions of the belt between the driving and the driven pulleys will be perfectly straight and there will be no centrifugal stress in the material. When the belt is fitted to the pulleys it will be given an initial tension T_0 and, under this tension, its total length will be L. As soon as power is supplied to one of the pulleys and transmitted to the other, the tensions in the two free lengths of belt will be changed; on the tight side the tension will increase from T_0 to T_1 and on the slack side the tension will decrease from T_0 to T_2. But the belt material is assumed to be perfectly elastic, so that, since the length L is unchanged, the mean tension must also be unchanged and $T_1 - T_0 = T_0 - T_2$. If the torque applied to the driving pulley is increased until the belt begins to slip, the two tensions T_1 and T_2

will then have their limiting ratio, and the actual values of T_1 and T_2 may be calculated from the two equations:

$$T_1 - T_0 = T_0 - T_2$$

and
$$\frac{T_1}{T_2} = e^{\mu\theta}$$

Writing k for T_1/T_2 and substituting for T_2 in terms of T_1 in the first of these equations, we have:

$$T_1 + T_1/k = 2T_0$$

$$\therefore\ T_1 = \frac{2k}{k+1} . T_0 \quad . \quad . \quad . \quad . \quad (7.8)$$

It will of course be understood that with an open drive the belt will begin to slip on the smaller pulley, since the angle of lap is smaller on this pulley than on the larger pulley. Hence the value of k must be calculated for the smaller pulley. In a drive with a crossed belt, the angle of lap is the same for both pulleys and slip should theoretically begin at the same instant on both pulleys.

Example 1. A flat belt is installed with an initial tension of 500 lb; the coefficient of friction between belt and pulley is 0·3; the angle of lap on the smaller pulley is 165° and the belt speed is 3600 ft/min. Determine the maximum horse-power which the belt can transmit, if it is assumed to be perfectly elastic and without mass.

$$\mu\theta = 0{\cdot}3 . 11\pi/12 = 0{\cdot}864; \quad \therefore\ k = T_1/T_2 = e^{0{\cdot}864} = 2{\cdot}37$$

From (7.8):

$$T_1 = \frac{2k}{k+1} T_0 = \frac{2 . 2{\cdot}37}{3{\cdot}37} . 500 = 703 \text{ lb}$$

$$\therefore\ T_2 = T_1/k = 297 \text{ lb}$$

the effective tension $\quad T = T_1 - T_2 = 406$ lb

and the horse-power transmitted,

$$H = \frac{406}{550} . \frac{3600}{60} = 44{\cdot}3$$

103. The Power transmitted by a Belt. Further Comments. Conditions in practice differ very widely from those assumed in the last article. Some of these differences and the effect which they have on the power transmitted will be briefly examined.

One important difference has been repeatedly demonstrated by experiment—the sum of the tensions T_1 and T_2 when the belt is transmitting power is always greater than twice the initial tension T_0. There are two reasons for this difference. In the first place the belt material is not perfectly elastic, so that the stress-strain curve is not a straight line. Instead the stress always increases at a greater rate than the strain. Consequently, for a given increase of strain a proportionately larger increase of stress is required at high values than at low values of the strain. But, if the length of the belt is the same when transmitting power as when at rest, the increase of strain on the tight side is equal to the decrease of strain on the slack side. It follows, therefore, that the increase of tension $T_1 - T_0$ on the tight side is greater than the decrease of tension $T_0 - T_2$ on the slack side and $T_1 + T_2 > 2T_0$.

The shape of the stress-strain curve is, however, not in itself sufficient to account for the large increase in the sum of the tensions which experiment shows. The second reason for the increase is to be found in the properties of the catenary curve in which the free lengths of the belt between the two pulleys hang, when the drive is horizontal or inclined. For a given span the length of the catenary curve varies with the tension at the supports. If s is the span, w the uniform load per unit span and T the tension at each support, Fig. 158, it can be shown that the length of the curve exceeds the span approximately by the amount $w^2s^3/24T^2$.[1] Hence, if it is assumed that the span is constant under all conditions, the length of the actual belt exceeds the length of a weightless belt by the amount $2w^2s^3/24T_0^2$, when the belt is running light, and by the amount $(w^2s^3/24)(1/T_1^2 + 1/T_2^2)$, when the belt is transmitting power. If the total length of the belt remains constant, then these two amounts must be equal and

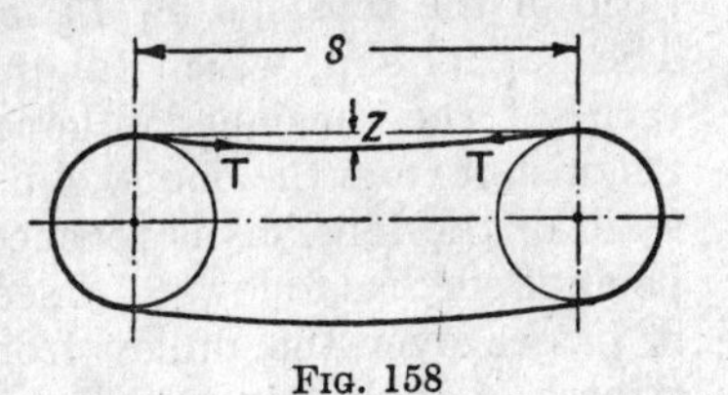

FIG. 158

$$1/T_1^2 + 1/T_2^2 = 2/T_0^2 \quad . \quad . \quad . \quad . \quad (7.9)$$

From this equation the values of $(T_1+T_2)/2T_0$ and of $(T_1-T_2)/T_0$ may be calculated for different values of T_1/T_0 as given in the table below.

T_1/T_0	1·2	1·4	1·6	1·8	2·0
$(T_1+T_2)/2T_0$.	1·038	1·110	1·194	1·285	1·378
$(T_1-T_2)/T_0$ . .	0·325	0·581	0·812	1·031	1·245

[1] Morley: *Strength of Materials*, Article 145.
The tension T at the supports $\simeq ws^2/8z$.
The difference between the length l of the catenary and the span $s \simeq 8z^2/3s$.

$\therefore$ eliminating z, $\quad l - s \simeq 8/3s \, . \, w^2s^4/64T^2 \simeq w^2s^3/24T^2$

It is clear from these figures that, as the power transmitted increases, i.e. as $(T_1-T_2)/T_0$ increases, the ratio $(T_1+T_2)/2T_0$ also increases. Of course, it has to be remembered that, as the sum of the tensions T_1+T_2 increases, so will the length of the belt increase, so that equation (7.9) is not strictly correct. Although the tightening effect of the catenary paths of the free parts of the belt is not so great as that shown in the table, nevertheless it is considerable. In order to determine its value with greater accuracy, it would be necessary to know the actual shape of the stress-strain curve for the belt material. For further information and for the results of experiments confirming the tightening effect, reference should be made to a paper by Dr. H. W. Swift.[1]

There is another difference between conditions as they exist in practice and the ideal conditions assumed in the last article. When power is transmitted by a belt or rope, there is always a difference between the peripheral speed of the driving pulley and that of the driven pulley. In a well-proportioned drive the difference is small. Nevertheless it is present, even though the ratio of the tensions T_1/T_2 is much less than the limiting ratio. The partial slip, which takes place with even low values of the ratio of the tensions, is generally referred to as *creep*, to distinguish it from the bodily slip which takes place when the limiting ratio of the tensions is reached. The phenomenon of creep may be explained as follows. Since the tension of the belt decreases as it passes over the pulley from the tight to the slack side, the stretch of the belt must also decrease. Hence it follows that the driving pulley must receive a longer length of belt than it pays out, and therefore the belt must creep back slightly relatively to the driving pulley rim. On the other hand, the driven pulley receives a shorter length of belt than it pays out, so that the belt must creep forward slightly relatively to the driven pulley rim. Obviously the difference between the peripheral speeds of the driving and the driven pulleys will depend upon the respective stresses in the tight and slack sides of the belt and the corresponding strains. Given a stress-strain curve for the belt material, the percentage loss of speed due to creep could be determined. In the paper already mentioned Swift suggests that creep does not take place over the complete arc of contact between the belt and the pulley, but is limited to an arc determined by the equation $T_1/T_2 = e^{\mu\beta}$. The effective angle of lap β extends backwards from the point at which contact between the belt and the pulley rim ceases. Over the remainder, $\theta-\beta$, of the actual angle of lap there is no change of tension and the inside surface of the belt and the pulley rim have

[1] H. W. Swift: "Power Transmission by Belts", *Proc. Inst. Mech. Eng.*, 1928.

identical speeds. In support of the above theory Swift points out that experiments show (a) that the peripheral speed of the pulley is identical with that of the oncoming belt and (b) that there is always a non-slipping or idle arc except when the limiting ratio $T_1/T_2 = e^{\mu\theta}$ is reached.

104. The Effect of Centrifugal Tension on the Power transmitted. The analysis of the effect of centrifugal tension on the power transmitted by a belt or rope, if made with the usual assumptions, gives results which are liable to be misleading. In what follows the usual analysis is first given and then the problem is examined in greater detail, separate consideration being given to vertical and horizontal drives.

The horse-power transmitted is given by (7.7):

$$H = Tv/550 = (T_1 - T_2)(v/550) = T_1(1-1/k)(v/550)$$

But the total tension on the tight side, which determines the maximum stress in the belt,

$$= T_t = T_1 + T_c$$

$$\therefore H = (T_t - T_c)(1-1/k)(v/550) \quad . \quad . \quad (7.10)$$

If T_t is assumed to be constant, the horse-power will first increase to a maximum and then decrease to zero as the speed is increased. But from (7.5) $T_c = wv^2/g$, so that H is a maximum when $(T_t - wv^2/g)v$ is a maximum.

Differentiating with respect to v and equating to zero, this gives:

$$T_t - 3wv^2/g = 0 \quad \text{or} \quad T_t = 3T_c$$

$$v = \sqrt{(T_t g/3w)} \quad . \quad . \quad . \quad . \quad (7.11)$$

also, from (7.10), $H = 0$, when $T_c = T_t$, i.e. when

$$v = \sqrt{(T_t g/w)} \quad . \quad . \quad . \quad . \quad (7.12)$$

If T_t is expressed in terms of the maximum safe stress f and the cross-sectional area a, $T_t = f.a$, and the speed for maximum H may be obtained from (7.11) and the corresponding value of H from (7.10).

In this analysis no reference is made to the initial tension T_0 with which the belt is assembled on the pulleys and it is tacitly assumed to be possible so to choose T_0 that, when the belt is transmitting power and the ratio of the driving tensions has its limiting value (7.4), the maximum tension T_t on the tight side remains constant at all speeds. In practice, however, it is impossible to satisfy this condition even for a vertical drive.

(a) *The Vertical Drive.* Referring to Fig. 159 (a), let T_0 be the initial tension in the belt when placed in position on the two pulleys. Then the total downward force on the bearings which support the upper shaft $= 2T_0 + W_b$, where W_b is the total weight of the belt, and the total upward force on the bearings which support the lower shaft $= 2T_0$. This, of course, assumes that the effect of the inclination of the sides of the belt to the vertical is negligible. Let us suppose that the speed of rotation of the upper shaft is gradually increased and that there is no resistance to the rotation of the lower shaft. The centrifugal tension T_c in the belt will increase with the speed according to equation (7.5). But if the portions of the belt between the pulleys remain straight, the total length of the belt remains constant and there is no change in the belt tension. What happens is simply that the tension arises more and more from the tendency of the two parts of the belt which are in contact with the pulley rims to fly outwards and less and less from the reactions of the pulley rims on the belt. The load on the bearings of the upper shaft decreases to $2(T_0 - T_c) + W_b$ and the load on the bearings of the lower shaft decreases to $2(T_0 - T_c)$. When the speed is such that $T_c = T_0$, the load on the bearings of the upper shaft is that due to W_b and the load on the lower shaft is zero. If the speed of the belt is still further increased, so that $T_c > T_0$, the belt stretches and hangs clear of the lower pulley, as shown in Fig. 159 (b).

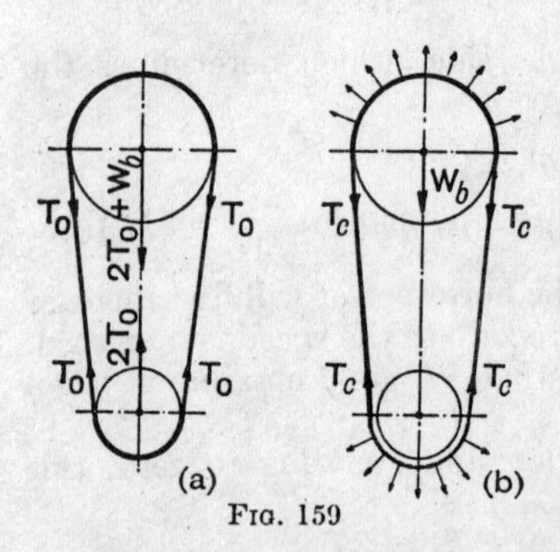

FIG. 159

Although in practice the portions of the belt between the pulleys will not remain perfectly straight, the above argument is in the main true, except that the load on the lower shaft will not be reduced to zero until the speed is somewhat greater than that for which $T_c = T_0$. For an actual photograph of a rubber belt running in the condition represented by Fig. 159 (b), the reader is referred to the late Professor Goodman's *Mechanics Applied to Engineering* (ninth edition), p. 357, Fig. 338.

Let us now suppose that the belt is running at a speed v, lower than that which makes $T_c = T_0$, and that at the same time power is transmitted from the upper to the lower shaft. Let T_t, T_s be the total tensions and T_1, T_2 be the driving tensions on the tight and slack sides respectively. Then $T_t = T_1 + T_c$, and $T_s = T_2 + T_c$. But, if the belt is perfectly elastic and its length remains unchanged, the mean tension also remains unchanged, so that $(T_t+T_s)/2 = T_0$.

$$\therefore\ T_1+T_2+2T_c = 2T_0$$

or
$$T_1+T_2 = 2(T_0-T_c) \quad . \quad . \quad (7.13)$$

Clearly, for a given initial tension T_0, the value of T_1+T_2 will decrease as the belt speed, and therefore T_c, increases. The limiting condition is reached when $T_c = T_0$. Under the assumed conditions, the belt is then incapable of transmitting any power. For a speed intermediate between zero and that for which $T_c = T_0$, the power transmitted may be found as follows:

Substitute $T_1 = kT_2$ in (7.12), then

$$T_2 = \frac{2}{k+1}(T_0-T_c)$$

and the effective tension:

$$T = T_1-T_2 = \frac{2(k-1)}{k+1}(T_0-T_c)$$

The horse-power transmitted is given by (7.7):

$$H = \frac{Tv}{550} = \frac{2(k-1)}{k+1}\,\frac{T_0-T_c}{550}.v \quad . \quad . \quad (7.14)$$

The value of H is proportional to $(T_0-wv^2/g)v$. Hence it follows that if T_0 is substituted for T_t in equations (7.11) and (7.12), the speeds for maximum and zero horse-power may be found.

(b) *The Horizontal Drive.* Let us assume, as in the case of the vertical drive, that the initial tension is T_0 when the belt is at rest, and that the speed of the belt is gradually increased, no resistance being offered to the rotation of the driven shaft.

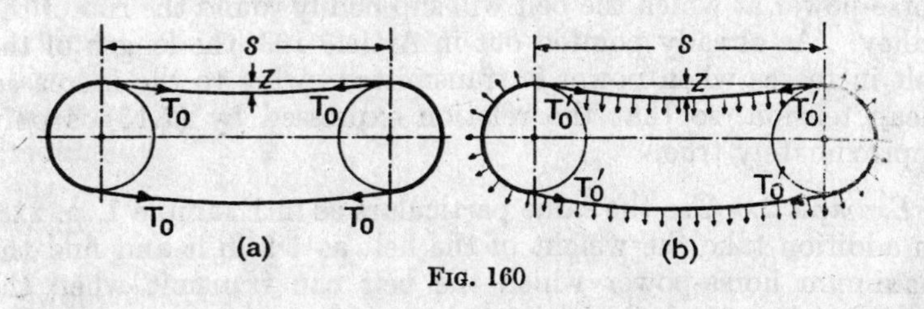

Fig. 160

Consider the conditions which exist so far as one of the hanging portions of the belt is concerned. When at rest, Fig. 160 (a), this portion of the belt has a span s approximately equal to the distance between the pulley centres, a weight w per unit length and a sag z. The tension T_0 is given approximately by $ws^2/8z$.

When running at the speed v, the load per unit length consists of the sum of the weight w and the inertia force wv^2/gR, where R is the radius of curvature of the path followed by the belt. The radius R, of course, varies from point to point along the path, but, since the path is not very different from a circular arc, R may be assumed to have the constant value $s^2/8z$. The tension under these conditions is therefore given by:

$$T_0' = \left(w+\frac{wv^2}{g}.\frac{8z}{s^2}\right)\frac{s^2}{8z} = T_0+\frac{wv^2}{g}$$

It should, however, be noted that the belt is elastic and therefore its length will be slightly greater when running than when at rest. The sag will also be greater and the tension will be slightly less than T_0+wv^2/g. It will be seen that the tension in the belt increases continuously as the speed increases. In this respect the horizontal drive is quite different from the vertical drive, in which the tension remains constant at the value T_0 until $T_c > T_0$. Hence it is impossible in the case of a horizontal drive for a condition ever to arise in which the belt ceases to make contact with the pulley rim.

If the belt is transmitting power, the total tension on the tight side is $T_t = T_1+T_c$ and the total tension on the slack side is $T_s = T_2+T_c$. Assuming the total length of the belt to remain unaltered, T_t, T_s and T_0' will be related by an expression similar to (7.9), which may be written:

$$2/T_0'^2 = 1/T_t^2+1/T_s^2 \quad . \quad . \quad . \quad (7.15)$$

This equation, taken in conjunction with (7.4), which gives the limiting ratio of the tensions, and (7.5), which gives the centrifugal tension, may be used to calculate the probable maximum horse-power at which the belt will slip bodily round the rim of the pulley. As already pointed out in Article 103, the length of the belt increases when power is transmitted owing to the increased mean tension, so that the relation expressed by (7.15) is only approximately true.

Example 2. Use the same particulars as in Example 1, p. 228. In addition take the weight of the belt as 1·2 lb/ft and find the maximum horse-power which the belt can transmit when the drive is (a) vertical, (b) horizontal.

(a) *Vertical drive*

From (7.5):

$$T_c = \frac{1{\cdot}2.60^2}{32{\cdot}2} = 134 \text{ lb}$$

From (7.14):

$$H = \frac{2.1{\cdot}37}{3{\cdot}37}.\frac{500-134}{550}.60 = 32{\cdot}5$$

N.B.—Maximum H occurs when $v = \sqrt{\dfrac{500.32{\cdot}2}{3.1{\cdot}2}} = 66{\cdot}8$ ft/s its value is $H = 32{\cdot}9$.

(b) *Horizontal drive*

$$T_0' = T_0 + T_c = 500+134 = 634 \text{ lb}$$

Also $T_t = T_1+T_c = T_1+134$ and $T_s = T_2+T_c = T_2+134$

From (7.15):

$$2/T_0'^2 = 1/T_t^2 + 1/T_s^2$$

But the limiting value of $T_1/T_2 = e^{\mu\theta} = 2{\cdot}37$,

$$\therefore \frac{2}{634^2} = \frac{1}{(2{\cdot}37T_2+134)^2} + \frac{1}{(T_2+134)^2}$$

By a process of trial and error T_2 is found to be 367 lb,

$\therefore T_1 = 2{\cdot}37.367 = 870$ lb, $T_t = 1004$ lb and $T_s = 501$ lb

$$\therefore \text{Horse-power transmitted} = \frac{(T_1-T_2)v}{550} = \frac{503.60}{550} = 54{\cdot}9$$

105. The Use of a Gravity Idler. In short belt drives, particularly where the velocity ratio between the pulleys is high, the maximum power which can be transmitted by the belt is much smaller than for a long drive operating under otherwise identical conditions. There are two reasons for this. First, the arc of contact on the smaller pulley is reduced when the centre distance is small and the limiting ratio of the tensions is correspondingly reduced. Secondly, the tightening effect of the catenary path followed by the free portions of belt between the pulleys is less marked, the shorter the centre distance. In order to enable the same amount of power to be transmitted with the short drive, it is therefore necessary to employ a much higher initial tension in the belt, and this is undesirable because of the greater bearing loads to which it gives rise. It is, however, possible to improve conditions in a short drive by adopting the arrangement shown in Fig. 161. A jockey pulley or idler is supported on an arm which

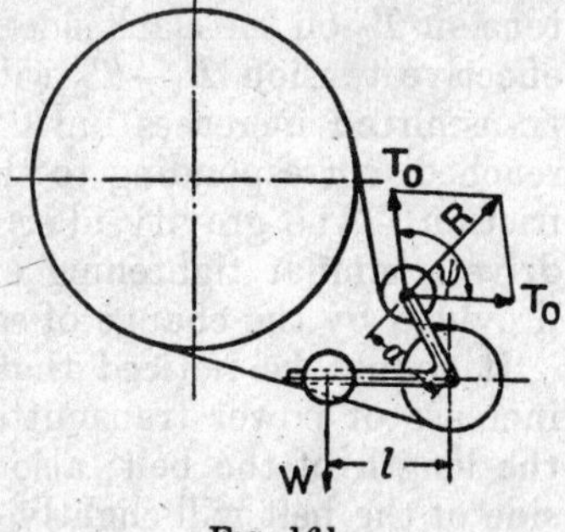

FIG. 161

is free to turn about pivots in a fixed frame. A deadweight is attached to the arm in such a way that it causes the idler to press against the slack side of the belt. This arrangement constitutes the well-known Lenix drive. The first and most obvious effect of the idler is to increase the arc of contact on the smaller pulley, and thus to increase the limiting ratio of the tensions at which the belt begins to slip bodily round the pulley. But a second and equally important effect of the idler is to increase the tension on the slack side of the belt, or perhaps it would be more correct to say that the second effect of the idler is to prevent the tension on the slack side from diminishing so rapidly, with an increase in the power transmitted, as it otherwise would do. This will be clear when it is remembered that the initial tension in a belt with a gravity idler is determined by the effective force exerted on the belt by the idler.

Thus $$R = 2T_0 \cos (\psi/2)$$

But $$R = Wl/a$$

where Wl is the total gravity torque on the arm to which the idler is attached and a is the perpendicular distance of the line of action of R from the pivot.

$$\therefore\ T_0 = \frac{Wl}{2a \cos (\psi/2)}$$

Suppose now that power is transmitted by the belt. Then, clearly if we neglect any change of length of the belt or of sag on the tight side of the belt, the position of the pivoted arm will remain unaltered and, in order to support the idler, the driving tension T_2 on the slack side must remain equal to T_0. Thus, the effective tension T_1-T_2 will increase continuously as the power transmitted increases until the limiting ratio of the tensions is reached, corresponding to the angle of lap and the coefficient of friction. The gravity idler therefore provides on a short-centre drive a similar tightening effect to that which is automatically provided by the change of sag on a long-centre drive.

It should be noticed that the increase of mean tension with increase of power transmitted will actually cause an increase of the length of the belt; also the increase of tension on the tight side of the belt will slightly decrease the sag. Both these effects will result in the pivoted arm of the idler turning through a small angle in the counter-clockwise sense in Fig. 161 and T_2 will be somewhat less than T_0.

Owing to the reverse bending of the belt as it passes over the idler pulley and the greater frequency of the stress changes due to bending, which is inseparable from a short-centre drive, the conditions are more severe than for the belt of a long-centre drive. It is therefore desirable to operate with a lower intensity of stress in such drives. For the same reasons an endless belt should be used.

106. Belt and Rope Materials. (a) *Flat Belts.* In addition to leather, belts are frequently made of other materials for which special advantages are claimed. There are several proprietary brands which consist of layers of woven camel hair or cotton, cemented together and impregnated with balata and other gums. These belts may be built up to any desired thickness by increasing the number of layers or plies of fabric. They are preferable to leather in warm climates, in damp atmospheres and in exposed positions. Rubber belts, consisting of layers of fabric impregnated with a rubber composition and having a thin layer of rubber on the faces, are very flexible but are quickly destroyed if allowed to come into contact with oil or grease. One of the principal advantages of these belts is that they may easily be made endless. The coefficient of friction between belt and pulley is a rather uncertain quantity under running conditions, probably owing to the fact that some allowance should be made for the pressure of the atmosphere on the outside of the belt. It is usual to assume a value of μ from 0·25 to 0·35 for leather and fabric belts on cast-iron pulleys, but the apparent coefficient of friction obtained from experiments on actual drives is often much higher.

The breaking strength of leather belting varies with the quality from 2000 to 8000 lb/in^2 and that of woven cotton belting from 6000 to 9000 lb/in^2. The thickness of a single leather belt is from $\frac{3}{16}$ in. to $\frac{1}{4}$ in. Double-thickness belts are obtained by cementing together two layers of single belt, and laminated belts of greater thickness by sewing together narrow strips of leather placed on edge. The usual thickness of woven cotton belts is from 0·25 to 0·35 in., but thicker belts may be obtained by increasing the number of layers or plies of fabric.

The maximum stress in belts is limited to from 200 to 400 lb/in^2, according to the quality, the type of joint used and the thickness of the belt. The thicker the belt the lower the stress used because of the additional stress caused by bending round the pulley rim. A rough rule sometimes used is that the *effective* tension per inch width should not exceed 40 lb for a belt of single thickness and 65 lb for a belt of double thickness.

Thin steel belts have been in use for some years, chiefly on the Continent. They are made from a specially prepared homogeneous steel in various thicknesses up to about 1 mm and are run on pulleys without camber, which are covered with a thin layer of cork. They have the advantages that there is no permanent stretch and the centrifugal tension is much lower for a given speed than in the case of a leather belt, so that higher belt speeds may be used. Their disadvantages are that shafts must be perfectly aligned and the pulley rims perfectly true, and that an absolutely satisfactory joint is difficult to devise. Recently this last difficulty has been overcome by producing endless belts which are rolled from seamless tubes.

(b) *Vee-belts.* Vee-belts may be built up of leather links, but, more often, they are made of fabric impregnated and covered with rubber and moulded to a trapezoidal shape. The groove angle of the pulley is generally 40°, but may, for some purposes, be as low as 30°. These belts are made endless and are particularly suitable for short drives. The wedging action of the belt in the groove gives a high value for the limiting ratio of the tensions, so that the material is efficiently used and it is therefore unnecessary to instal the belts with a high initial tension. The disadvantage is that the velocity ratio between the driving and the driven pulleys is liable to change as wear on the belt and groove takes place.

(c) *Ropes.* Ropes for transmitting power are usually made of cotton and are of circular, or sometimes of square, cross-section. In order to ensure long life, a large factor of safety is employed. It is usual to express the safe maximum tension on the rope in terms of the diameter, and a value of $160d^2$ lb, where d is the diameter in inches, is generally employed. The weight of the rope is approximately $0{\cdot}28d^2$ lb. Various diameters up to $2\frac{1}{2}$ in. are used, but the most common sizes for the transmission of large powers are $1\frac{5}{8}$ or $1\frac{3}{4}$ in. dia. In this country and on the Continent separate ropes connect corresponding grooves on the driving and the driven pulleys, but in the United States a continuous rope is threaded over all the grooves and then passed over a guide pulley in order to lead it from the last groove back to the first groove. The guide pulley shaft is mounted on slides and subjected to a gravity or spring pull which takes up any permanent stretch in the rope and maintains an approximately uniform tension.

One big advantage of rope drives is that a number of separate drives may be taken from the one driving pulley. For instance, in many spinning mills the lineshaft on each floor is driven by ropes passing directly from the main engine pulley on the ground floor.

The groove angle varies from 40° to 60°, but is generally 45°. As in the case of the vee-belt, the limiting ratio of the tensions is high, the actual coefficient of friction being about 0·3, but the wedging action of the rope in the groove gives a much higher virtual coefficient of friction.

(d) *Wire Ropes.* Where power has to be transmitted over long distances, as in mining, hauling, winding, etc., steel-wire ropes are used. The ropes run on grooved pulleys, but, contrary to the practice adopted with cotton ropes, they rest on the bottom of the grooves and are not wedged between the sides of the grooves. In order to increase the frictional grip, the bottom of the groove is provided with a wood or leather insertion. For the long spans used in wire rope power transmission, the tightening effect of the catenary path followed by the free portions of the rope is very marked. A sufficiently close approximation to the true magnitude of the tightening effect cannot always be obtained by assuming the curve in which the rope hangs to be a parabola, and it is necessary to take into account the true shape of the catenary curve.

The weight per foot and the working tension of wire ropes are generally expressed in terms of the circumference of the rope in inches. Average values are: weight per foot $0{\cdot}15C^2$ lb and working load $0{\cdot}3C^2$ tons, although there are wide variations, particularly in the working load, according to the design of the rope and the quality of the steel used.

107. Chain Drives. A chain may be regarded as a belt built up of rigid links, which are hinged together in order to provide the necessary flexibility for the wrapping action round the driving and driven wheels. These wheels have projecting teeth, which fit into suitable recesses in the links of the chain and thus enable a positive drive to be obtained. They are known as chain sprockets and bear a superficial resemblance to spur gears.

The *pitch* of the chain is the distance between a hinge centre of one link and the corresponding hinge centre of the adjacent link.

The *pitch circle diameter* of the chain sprocket is the diameter of the circle on which the hinge centres lie, when the chain is wrapped round the sprocket.

Referring to Fig. 162 (a), it will be seen that, since the chain links are rigid, the pitch becomes a chord, not an arc, of the pitch circle. The relation between the pitch circle diameter d, the pitch p and the number of teeth T on the sprocket, may be found as follows. The angle 2θ subtended at the sprocket centre by one pitch AB $= 360°/T$. But AC $=$ OA $\sin\theta$, so that

$$p/2 = (d/2) \sin (180°/T)$$

or

$$d = p \operatorname{cosec} (180°/T) \quad . \quad . \quad . \quad (7.16)$$

Because each link of the chain is rigid, the relation between the chain speed and the angular velocity of the sprocket will vary with the angular position of the sprocket. The extreme conditions are shown in Fig. 162 (a) and (b). Thus, if v is the speed of the chain and ω the angular velocity of the sprocket, then, for position (a) $v = \omega . OA$ and, for position (b), $v = \omega . OC = \omega . OA \cos \theta$.

In order to bring out clearly the kinematic conditions which obtain in a chain drive, an arrangement in which the sprockets have only 6 and 9 teeth is drawn in Fig. 163. Let the sense of rotation of the sprockets be counter-clockwise, with the smaller sprocket as the driver. Let 2θ, 2ϕ be the angles subtended by the chain pitch at the centres of the driving sprocket and the driven sprocket respectively. Then, obviously, the straight length of chain between the two sprockets must be an exact number of pitches and the positions in which its inclination to the line of centres is respectively a minimum and a maximum will be represented by AB and A_1B_1, where A, B_2 and B are in one straight

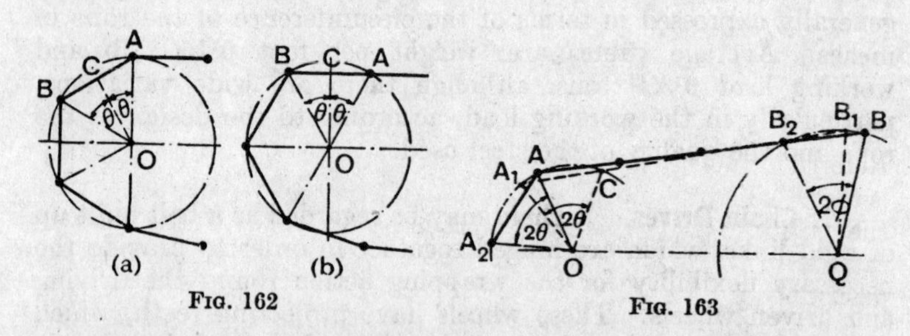

Fig. 162 Fig. 163

line, as also are A_1, C and B_1. Starting with the pin-centre on the driving sprocket in the position A, the length AB of chain will remain straight as the sprockets turn, until A reaches A_1 and B reaches B_1. As the driving sprocket continues to turn, the link A_1C of the chain will hinge about the pin-centre C and the straight length of chain between the two sprockets will be reduced to CB_1. While C moves to A, A_1 moves to A_2 and B_1 to B_2 and the length CB_1 of the chain remains straight. By this time each of the sprockets will have turned from its original position through an angle corresponding to one chain pitch. During the first part of the angular displacement, i.e. while OA moves to OA_1 and QB to QB_1, the arrangement is kinematically equivalent to the four-bar chain OABQ. Similarly, during the second part of the angular displacement, i.e. while OA_1 moves to OA_2 and QB_1 to QB_2, the arrangement is kinematically equivalent to the four-bar chain OCB_1Q. In these circumstances the ratio of the angular velocities of the two sprockets cannot be constant. This may perhaps most easily be shown by making use of the three centres in line theorem, Article 44, in order to find the instantaneous centre for the relative motion of the two links OA and QB. This centre lies at the point of intersection P of BA produced and QO produced, Fig. 164. Then, if ω, ω_1 are respectively the angular velocities of the driving and the driven sprockets, it follows that $\omega . OP = \omega_1 . QP$ or

$$\omega/\omega_1 = QP/OP = (QO+OP)/OP = 1+QO/OP \quad . \quad (7.17)$$

For a given chain drive, QO is constant but OP undergoes a periodic variation as the sprockets revolve, the period correspond-

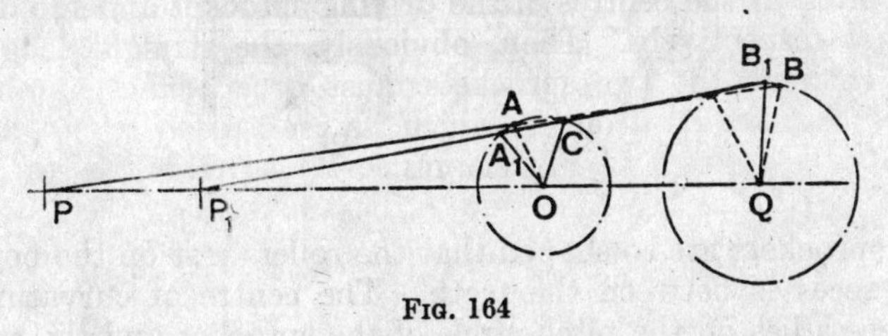

Fig. 164

ing to a rotation of the driving sprocket through an angle 2θ. Referring to Fig. 164, PO is a maximum, and the velocity ratio is therefore a minimum, when the straight length of chain occupies the position AB; whereas PO is a minimum, and the velocity ratio a maximum, when the straight length of chain occupies the position A_1B_1.

For the sake of clearness, Figs. 163 and 164 have been drawn for sprockets with impracticably small numbers of teeth. Normally, the smaller sprocket would have at least 17 teeth and the actual variation of velocity ratio would only amount to a few per cent. of the mean value.

108. Types of Chain. Two types of chain are described, namely (a) the roller chain and (b) the inverted tooth or silent chain. The former has practically superseded all other types for the transmission of power, so that the latter is chiefly of academic interest.

(a) *The Roller Chain.* The construction of this type of chain is shown in Fig. 165. The inner plates A are held together by steel bushes B, through which pass the pins C riveted to the outer links D. A roller R surrounds each bush B and the teeth of the sprocket bear on the rollers. The rollers turn freely on the bushes and the bushes turn freely on the pins. All the contact surfaces are hardened so as to resist wear and are lubricated so as to reduce friction. Fig. 165 (a) shows a simple roller chain, consisting of one strand only, but duplex and triplex chains, consisting of two or three strands, may be built up as shown in Fig. 165 (b), each pin passing right through the bushes in the two or three strands.

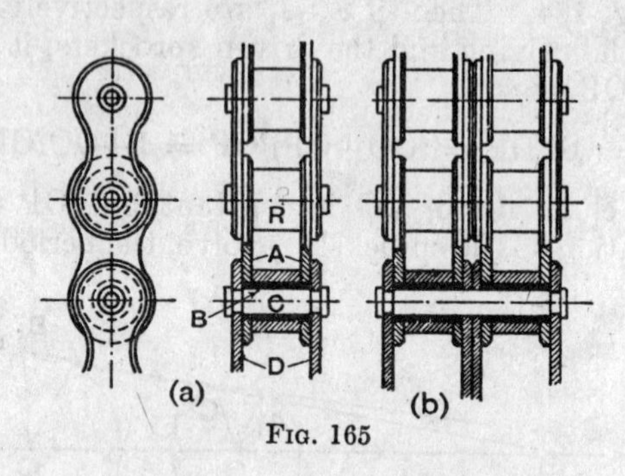

FIG. 165

The sprockets are so shaped that the rollers rest on the bottom of the recesses between the teeth. The centre of curvature of each recess lies on the pitch circle of the sprocket and the radius of curvature is a few thousandths of an inch larger than the radius of the roller. As each link of the chain enters or leaves the sprocket, the roller centre at one end will describe a circular arc relative to the roller centre at the other end, which is resting in a recess of the sprocket. Hence, if the chain is to enter the wheel without shock, the profile of the sprocket tooth should theoretically be a circular arc, concentric with the adjacent recess and of radius equal to the pitch of the chain less the radius of the roller, see Fig. 166 (a). There are, however, practical objections to this

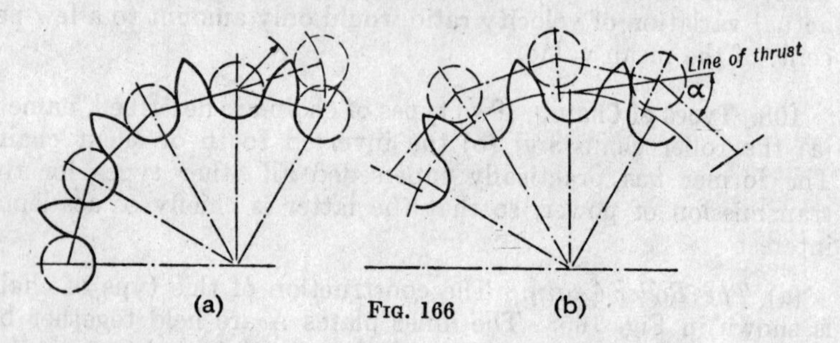

FIG. 166

shape of profile. For instance, it will not accommodate the increase of chain pitch which results from wear on the bearing surfaces without, at the same time, increasing undesirably the inclination α of the line of thrust between the tooth and the roller to the line of centres of the link. A more serious practical objection is that a different cutter would be required to machine the tooth spaces for each size of sprocket.

The British Standards Institution in B.S. 228:1934,[1] recommend a modified profile so that three cutters for each pitch will cover the complete range of numbers of teeth on the sprockets. No. 1 cutter is to be used for sprockets with from 9 to 12 teeth, No. 2 cutter for sprockets with from 13 to 19 teeth and No. 3 cutter for sprockets with 20 or more teeth. The shapes of the profiles of the three cutters are shown in Fig. 167. The corresponding tooth shapes are also shown for sprockets with the smallest number of teeth in each range. It will be seen that the root curve ab of the tooth extends in each case over an arc of 120° only. For the No. 1 cutter the remainder of the profile consists of a single arc, tangential to the root curve and of radius

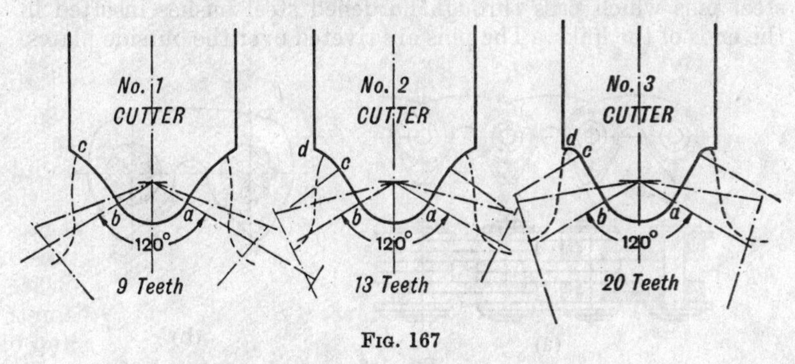

FIG. 167

1·2 times the chain pitch. For the No. 2 cutter the working part bc of the profile has a chord length of $0{\cdot}32p$ and a radius of $3p$, while the top cd of the tooth has a radius of $0{\cdot}48p$. Finally, for the No. 3 cutter the working part bc of the profile is straight and has a length of $0{\cdot}42p$, while the top cd of the tooth has a radius of $0{\cdot}10p$. In all cutters the different curves which form the profile meet tangentially and the radius of curvature for the root of the tooth is approximately 1% greater than the radius of the roller. It is not possible to express the root radius in terms of the pitch because the roller diameter is not a constant fraction of the pitch, but varies from $0{\cdot}575p$ to $0{\cdot}670p$, with a mean value of $0{\cdot}628p$. In the roller chain used for the drives of bicycles the pitch is 0·625 in. and the roller diameter is only 0·305 in. or $0{\cdot}488p$.

[1] This information is abstracted by permission from British Standard 228:1934 Steel Roller Chains and Chain Wheels, official copies of which can be obtained from the British Standards Institution, 2 Park St., London, W.1, price 3s. post free.

When wear takes place on the bearing surfaces of the chain, the pitch increases and the effective pitch circle diameter of the sprocket also increases, as shown in Fig. 166 (b), the rollers no longer resting on the bottom of the tooth spaces but riding up the working part of the tooth profiles.

(b) *The Inverted Tooth or Silent Chain.* The construction of this type of chain is shown in Fig. 168 (a). It is built up from a series of flat plates, each of which has two projections or teeth. The outer faces of the teeth are ground to give an included angle of 60° or, in some cases, 75°, and they bear against the working faces of the sprocket teeth. The inner faces of the link teeth take no part in the drive and are so shaped as to clear the sprocket teeth. The required width of chain is built up from a number of these plates, arranged alternately and connected together by hardened steel pins which pass through hardened steel bushes inserted in the ends of the links. The pins are riveted over the outside plates.

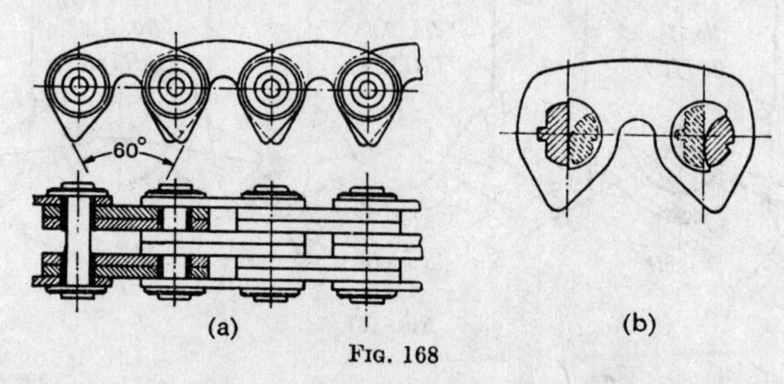

FIG. 168

The chain may be prevented from sliding axially across the face of the sprocket teeth by outside guide plates without teeth, or by a centre guide plate without teeth which fits into a recess turned in the sprocket.

Fig. 168 (b) shows the type of hinge used in the Morse silent chain. This reduces friction by substituting a hardened steel rocker on a hardened steel flat pivot for the pin and bush.

When the chain is new, the position which it takes up on the sprocket is shown in the upper part of Fig. 169. Each link as it enters the sprocket pivots about the pin on the adjacent link which is in contact with the sprocket. The working faces of the link are thus brought gradually into contact with the corresponding faces of the sprocket teeth. A similar action takes place as each link leaves the sprocket. Hence there is no relative sliding between the faces of the links and the faces of the sprocket teeth.

As wear takes place on the pins and bushes, the smooth action of the chain is not impaired, but the chain rides higher up the sprocket teeth and the effective pitch circle diameter of the sprocket is increased, as shown in the lower part of Fig. 169.

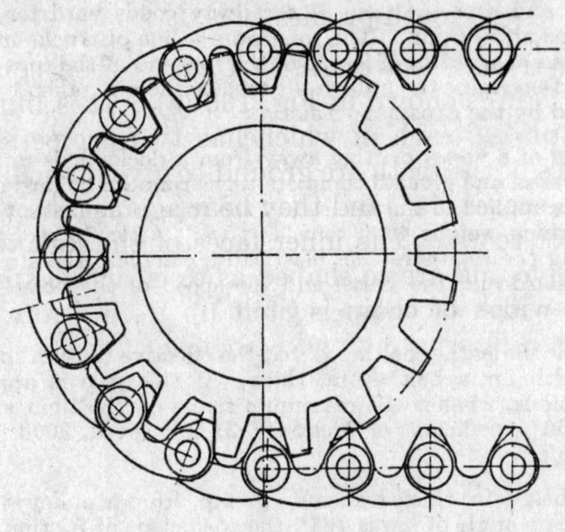

FIG. 169

It should be noticed that with this type of chain the pitch circle diameter of the sprocket may be greater than its outside diameter.

The number of teeth on the smaller sprocket of a chain drive should preferably be not less than 17 and the speed ratio should not exceed 6 or 7 to 1.

EXAMPLES VII

N.B.—Q7–Q13 are intended to be solved by applying the usual theory of the effect of centrifugal stress as given at the beginning of Article 104.

1. Deduce the equation for the limiting ratio of the tensions at the two ends of a belt or rope which is in contact with a pulley rim when (a) the pulley rim is flat, (b) the pulley rim is grooved and the belt or rope wedges between the sides of the groove.

2. Determine the limiting ratio of the tensions and express the difference of tension as a fraction of the maximum tension for the following cases:

(a) A flat belt in contact with a flat pulley rim. $\theta = 180°$, $\mu = 0{\cdot}4$.
(b) A circular section rope wedged in a grooved pulley rim. $\theta = 160°$, $\mu = 0{\cdot}3$, groove angle $= 50°$.

3. A rope is given three complete turns round a plain cylindrical post and a force of 50 lb is applied to one end. If the coefficient of friction between rope and post is 0·25, what force must be applied to the other end of the rope in order to cause it to slip round the post?

4. A hemp rope weighing $7\frac{1}{2}$ lb per fathom is wound $3\frac{1}{2}$ times round a fixed horizontal wooden beam of circular section. One end is hanging free and is 8 ft from the beam. Develop the theory to show whether it would be safe to attach to the other end an otherwise unsupported weight of 10 000 lb. The coefficient of friction between the rope and the beam is 0·4. W.S.

5. A capstan and rope are used in a railway goods yard for moving trucks. The capstan runs at 50 r.p.m. The rope from a line of trucks makes 2·75 turns round the capstan at a radius of 8 in. and the free end of the rope is pulled with a force of 30 lb. Determine the pull on the trucks, the h.p. taken by the trucks and the h.p. supplied by the capstan. Take $\mu = 0{\cdot}25$. L.U.

6. The motion of a vessel drifting away from a dockside is retarded by a rope secured to the vessel and given 3 complete turns round a bollard on the dockside. A pull of 80 lb is applied to the free end of the rope at an instant when the speed of the vessel, which weighs 4000 tons, is $\frac{1}{2}$ft/s. After 10 sec, the rope begins to slip; assuming the rope stretches elastically, calculate the stretch in the rope between the bollard and the vessel and the speed of the vessel when the rope slips. (μ between rope and bollard $= 0{\cdot}25$.)

7. A rough rule for leather belting is that the effective tension should not exceed 40 lb/in. of width for a belt $\frac{3}{16}$ in. thick. If this rule is applied under the following conditions, what is the maximum stress on the tight side of the belt? Angle of lap, 160°; coefficient of friction, 0·3; belt speed, 3000 ft/m; density of leather, 0·035 lb/in^3.

8. A leather belt $\frac{3}{8}$ in. thick transmits 50 h.p. from a pulley 4 ft dia. running at 240 r.p.m. The angle of lap is 165°, the coefficient of friction is 0·28 and the weight of 1 in^3 of leather is 0·035 lb. What width of belt will be required if the stress is limited to 350 lb/in^2?

9. Calculate the centrifugal tension in a belt which runs over two pulleys at a speed of [illegible]500 ft/min. The belt is 8 in. wide and $\frac{5}{16}$ in. thick, and weighs 0·035 lb/in^3.

If the belt embraces an angle of 165°, μ is 0·25 and the maximum permissible stress in the belt material is 350 lb/in^2, calculate the maximum h.p. transmitted at the above speed.

10. A belt is required to transmit 30 h.p. from a pulley 5 ft dia. running at 200 r.p.m. The angle embraced is 165° and the coefficient of friction 0·3. If the safe working stress for the leather is 350 lb/in^2, the weight of 1 in^3 of leather 0·035 lb and the thickness of the belt $\frac{3}{8}$ in., what width of belt will be required, taking into account the centrifugal force?

11. Deduce an expression for the centrifugal tension in a rope or belt passing round a pulley rim.

A rope pulley with 10 ropes and a peripheral speed of 4000 ft/min transmits 140 h.p. The angle embraced by each rope is 180°, the angle of the groove 40°, and the coefficient of friction between rope and groove 0·2. Assuming the ropes to be just on the point of slipping, calculate the tensions on the tight and slack sides of each rope, allowing for the centrifugal tension. The weight of each rope is 0·3 lb/ft.

12. A rope pulley with 15 ropes and a peripheral speed of 3500 ft/min transmits 400 h.p. The angle of the grooves is 60°, the angle embraced by the ropes 165°, the coefficient of friction 0·27, and the weight of the rope per foot of length 0·85 lb. Calculate the maximum tension in the rope, taking into account the effect of the centrifugal force.

13. A rope drive is required to transmit 1500 h.p. from a pulley 3·5 ft dia. running at 360 r.p.m. The safe pull in each rope is 490 lb and the weight of the rope per foot of length is 0·9 lb. The angle of lap is 150°, the groove angle 45°, and the coefficient of friction between rope and groove is 0·3. How many ropes will be required if allowance is made for the centrifugal stress?

14. A vertical belt drive is installed with an initial stress of 125 lb/in^2. The angle embraced on the smaller pulley is 165° and the coefficient of friction is 0·3. The density of the leather is 0·035 lb/in^3. Assuming the belt to be inextensible, plot a curve to show how the maximum h.p. which can be transmitted varies with the speed of the belt. Under these conditions what is the maximum stress in the belt and at what speed does it occur?

15. Explain briefly what factors influence the change in the sum of the tensions on the tight and slack sides of a belt as the power transmitted is increased. Why is a horizontal drive capable of transmitting more power than a similar vertical drive?

16. It is usually stated that the best speed at which to run a belt or rope is that at which the centrifugal tension is one-third of the maximum permissible tension. Explain the theoretical basis for this statement and criticise the assumptions made.

17. A horizontal belt drive is installed with an initial tension of 50 lb per in. of width. The angle embraced on the smaller pulley is 160° and μ is 0·3. The weight of a strip of belt 1 in. wide by 1 ft long is 0·1 lb. Assuming the belt to be inextensible, plot curves to show how (a) the maximum h.p. which can be transmitted per in. of width, (b) the maximum tension per in. of width, vary with the speed up to 120 ft/s.

18. If in the transmission of power between two shafts by means of a belt, the belt as a whole does not slip round either pulley, explain the nature of the "creep" of the belt on both pulleys.

Power is transmitted from a pulley 3 ft dia. running at 200 r.p.m. to a pulley 7 ft dia. by means of a belt 20 in. wide and 0·375 in. thick. Find the speed in r.p.m. lost by the driven pulley as a result of creep, if the tensions in the tight and slack sides of the belt are 62·5 lb. per in. width and 25 lb per in. width respectively, and Young's modulus for the material is 15 000 lb/ in^2. L.U.

19. What is the purpose of a gravity idler in a short-centre belt drive? Explain why a greater h.p. can be transmitted before slip begins when a gravity idler is used.

20. What are the relative advantages and disadvantages of (a) flat belts, (b) V-belts, (c) ropes and (d) chains for the transmission of power?

21. Describe the construction of (a) the roller chain, (b) the silent chain. How does wear affect the contact between the chain and sprocket in each case?

22. A chain connects two sprockets, the number of teeth being t on the smaller sprocket and T on the larger sprocket. Show that the instantaneous velocity ratio has a maximum possible value of $\dfrac{\tan(180/t)}{\sin(180/T)}$ and a minimum possible value of $\dfrac{\sin(180/t)}{\tan(180/T)}$. If $t = 12$ and $T = 60$, what is the maximum possible variation of velocity ratio?

CHAPTER VIII

BRAKES AND DYNAMOMETERS

109. The primary function of a brake is either to bring to rest a body which is in motion or to hold a body in a state of rest or of uniform motion against the action of external forces or couples. The primary function of a dynamometer is to measure the forces or couples which tend to change the state of rest or of uniform motion of a body. It will be seen, therefore, that the difference between a brake and a dynamometer is often one of purpose rather than of principle; an arrangement which is used as a brake may frequently be made to serve as a dynamometer with but the addition of a force-measuring device, such as a deadweight or spring balance. In the present chapter the principles of operation of representative types of brakes and dynamometers are discussed.

110. The Simple Block or Shoe Brake. A simple arrangement for applying a retarding or braking force to the rim of a drum or wheel is shown in Fig. 170. A block or shoe S of wood or metal, sometimes faced with a special friction material in order to give a high coefficient of friction, is forced into contact with the rim of the drum. The friction between the shoe and the drum causes a tangential force to act on the drum, which tends to prevent its rotation. The shoe is pressed against the drum by a force P applied to one end of a lever, generally called the brake hanger, to which the shoe is attached. The other end of the hanger is pivoted on a fixed fulcrum O. The shoe may be pivoted to the brake hanger as shown at (a) or rigidly attached to the hanger as shown at (b). The former arrangement is mechanically more complicated, but it enables the relation between the applied force P and the tangential braking force on the drum to be more accurately determined. Referring to Fig. 170 (a), the brake shoe is in equilibrium under two forces, the reaction of the drum and the reaction of the brake hanger. These two forces must therefore be in line. The reaction of the brake hanger on the shoe may be assumed to pass through the pin centre C, since the initial vibration of the shoe when the brake is applied will probably be sufficient to eliminate friction at the shoe pin. The reaction of the drum on the shoe may be assumed to be tangential to the friction circle of the drum. This is true for the reaction on each element of length of the shoe, but is not strictly true for the total reaction of the drum on the shoe because of the curvature of the contact surfaces. However, in view of the uncertainty which always exists as to the value of μ, the error involved in the assumption is negligible. From these three conditions the line of action of the resultant thrust may be completely determined. Assuming clockwise rotation of the drum, the line of thrust between the shoe and drum will therefore take up the position shown by the

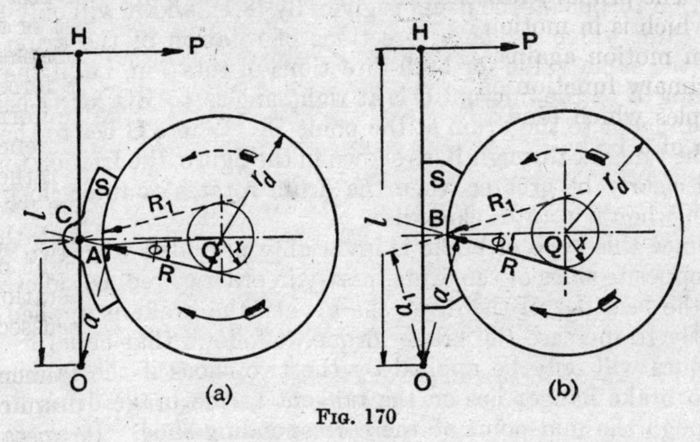

FIG. 170

full line. Similarly, when the drum rotates counter-clockwise, the line of thrust will take up the position shown by the dotted line.

The friction torque on the drum $= T_b = Rx$, where x is the perpendicular distance of the line of thrust from the drum centre Q, i.e. the radius of the friction circle of the drum $r_d . \sin \phi$. Also, for equilibrium of the brake hanger, $Pl = Ra$, where l and a are respectively the perpendicular distances of the lines of action of the effort P and the thrust R from the fulcrum O of the brake hanger. Substituting for R, the friction torque on the drum is given by:

$$T_b = Pl/a . x \quad . \quad . \quad . \quad . \quad . \quad (8.1)$$

In order that this torque shall be the same for both directions of rotation of the drum, the perpendicular distance a of the line of thrust from O must be the same for both directions of rotation, and this clearly requires that the line joining the centres of the two pins, O and C, shall be perpendicular to the line joining the centre of pin C to the drum center Q.

Where the brake shoe is rigidly attached to the hanger, as shown in Fig. 170 (b), the calculated value of the torque applied to the drum depends upon the assumptions made in determining the position of the line of action of the resultant thrust between the shoe and the drum. The simplest assumption is that the line of thrust passes through the mid-point B of the contact surface of the shoe. This is practically equivalent to the assumption of uniform intensity of pressure between the shoe and drum. Since the line of thrust may also be assumed to be tangential to the friction circle of the drum, it follows that for clockwise rotation, it is in the position shown by the full line and for counter-clockwise rotation in the position shown by the dotted line. As before, the friction torque on the drum is given by (8.1), and it will only have the same value for both directions of rotation of the drum if a has the same value for both directions of rotation, i.e. if the line joining B to the fulcrum O is at right angles to BQ, or O lies on the tangent to the drum at the point B. Where O lies to the left of the tangent through B, as shown in the figure, the friction torque will clearly be greater when the drum rotates counter-clockwise than when it rotates clockwise.

Since this type of brake is invariably provided with two shoes on opposite sides of the drum, partly in order to reduce side thrust on the bearings of the drum shaft when the brake is applied and partly to increase the brake torque, it follows that equal friction torques will only be applied by the two shoes if the fulcrum of each brake hanger lies on the tangent to the brake drum drawn through the mid-point of the corresponding shoe. It should be pointed out that this conclusion needs to be modified when other assumptions are made regarding the position of the line of thrust between shoe and drum. For instance, if, as is probable, the line of thrust intersects the contact surface of the shoe at some point other than the mid-point B, the fulcrum of the brake hanger should lie on the tangent drawn through that point instead of on the tangent drawn through B.

A closer approximation to the true point of intersection of the line of resultant thrust and the shoe surface may be obtained on the assumption that the material of the shoe obeys Hooke's law in compression.

Referring to Fig. 171 (a), a small angular displacement $\delta\theta$ of the hanger in the clockwise sense would cause the point C on the shoe to move to D, where CD is perpendicular to OC. Resolve the displacement CD into components CE and ED, respectively normal and tangential to the contact surface. Then the normal intensity of pressure at C will be directly proportional to the normal component CE. Join Q to C and produce to meet a line through O perpendicular to QC at F. From the similar triangles CDE, OCF, $CE/CD = OF/OC$, and therefore $CE = CD.OF/OC$. But CD/OC is the angular displacement $\delta\theta$ of the hanger. Therefore the normal intensity of pressure at C ∝ CE ∝ OF.$\delta\theta$ and, for a given value of $\delta\theta$, the normal intensity of pressure at C is directly proportional to OF.

Hence, for any other point on the contact surface of the shoe, the normal intensity of pressure is directly proportional to the length of the perpendicular from O to the line joining that point to the drum centre Q. If the total length of the arc of contact of the shoe is divided into a number of equal parts, the length of the perpendicular OF may be found corresponding to the mid-point of each part and a proportionate length may be marked off from the contact surface along the corresponding mid-ordinate. The length of each of the lines thus obtained will be proportional to the normal thrust on the corresponding part of the shoe and the line of resultant normal thrust may be found by combining the lengths vectorially and drawing a line through Q parallel to the

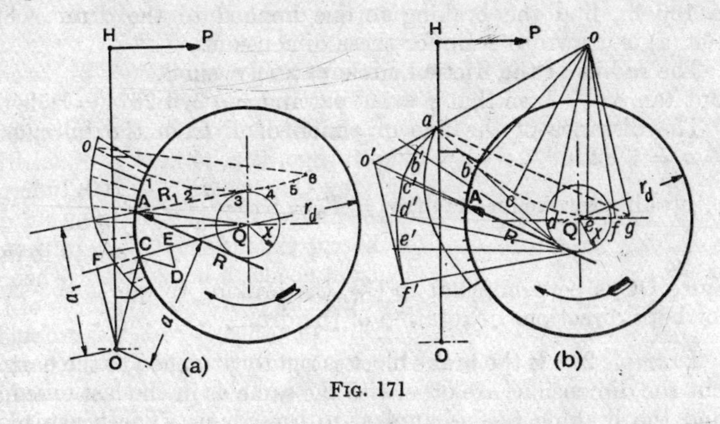

FIG. 171

resultant. The vector polygon is shown dotted in Fig. 171 (a). It will be seen that the line of thrust intersects the contact surface of the shoe at a point A above the mid-point. If the resultant friction force is assumed to be tangential to the drum at the point A, the resultant thrust between the shoe and the drum will pass through A and will be tangential to the friction circle. Although this assumption is not strictly correct, the line of action of the resultant thrust will be obtained with sufficient accuracy for all practical purposes. Strictly speaking, the resultant thrust ought to be found as shown in Fig. 171 (b). For each short length of the arc of contact, the true reaction is tangential to the friction circle and its magnitude is directly proportional to the normal intensity of pressure. The vector sum of the true reactions for the individual lengths into which the arc of contact is divided will be equal to the resultant thrust between the shoe and the drum. In order to fix the line of action of the resultant thrust, it is necessary to draw a link or funicular polygon. In Fig. 171 (b), o'ab'c'e'f' is the link polygon drawn in the usual way for the vector polygon abcdefg and any arbitrarily fixed pole o. The point o' is a point on the line of action of the resultant thrust and the direction of the resultant thrust is parallel to ag. It will be found that the line of thrust obtained by this construction is at a slightly greater distance from the axis of the drum than the radius of the friction circle. But, bearing in mind the uncertainty regarding the value of the coefficient of friction, it would appear that this refinement is hardly justified.

Example 1. A brake is arranged as in Fig. 170 (a). The drum is 12 in. dia., the distances CQ, OC and OH are respectively 7 in., 6 in. and 15 in. If the coefficient of friction is 0·3 and the effort P is 100 lb, find the braking torque applied to the drum. Fig. 166 (a) is drawn to scale for these dimensions.

The radius of the friction circle $= x = r_d \sin \phi$.

But $\tan \phi = 0{\cdot}3$, so that $\phi = 16° \, 42'$ and $x = 6 . 0{\cdot}2874 = 1{\cdot}724$ in.

The distance of the line of action of R from the fulcrum O $= a = 5{\cdot}82$ in.

$$\therefore \text{ braking torque on the drum} = T_b = Rx = \frac{Pl}{a} . x = \frac{100 . 15}{5{\cdot}82} . 1{\cdot}724$$
$$= 445 \text{ lb in.}$$

Since OC is perpendicular to CQ, the braking torque is the same for both directions of rotation of the drum.

Example 2. If the brake block is rigidly attached to the hanger, but the dimensions are otherwise the same as in the last example, find the braking torque applied to the drum. The brake block embraces 60° of arc. Fig. 170 (b) and Fig. 171 are drawn to scale for these dimensions.

From Fig. 170 (b), for clockwise rotation of the drum, a scales 6·0 in.

$$\therefore \text{ braking torque on the drum } T_b = Rx = \frac{Pl}{a} . x = \frac{100 . 15}{6} . 1{\cdot}724$$
$$= 431 \text{ lb in.}$$

Similarly, for counter-clockwise rotation of the drum, a_1 scales 5·5 in.

$$\therefore \text{ braking torque on the drum } T_b' = R_1 x \frac{Pl}{a_1} . x = \frac{100 . 15}{5{\cdot}5} . 1{\cdot}724$$
$$= 471 \text{ lb in.}$$

If the point A on the line of action of the resultant thrust between the block and the drum is found as already explained in connection with Fig. 171 (a), the distances a and a_1 scale respectively 6·48 in. and 6·38 in.

For clockwise rotation of the drum,

$$\text{braking torque on the drum} = T_b = Rx = \frac{Pl}{a}.x = \frac{100.15}{6 \cdot 48}.1 \cdot 724$$
$$= 400 \text{ lb in.}$$

and for counter-clockwise rotation,

$$\text{braking torque } T_b' = R_1 x = \frac{Pl}{a_1}.x = \frac{100.15}{6 \cdot 38}.1 \cdot 724 = 406 \text{ lb in.}$$

Example 3. Fig. 172 shows diagrammatically the arrangement of a brake. The two shoes are pivoted to the hangers and the brake is applied by means of a vertical force P, which acts through

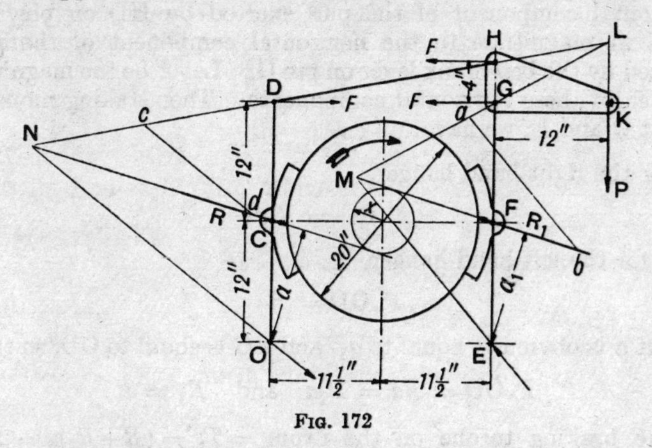

Fig. 172

the pin K on the bell-crank lever HGK. If the coefficient of friction is 0·35 and the braking torque required on the drum is 500 lb ft, find: (a) the magnitude of P; (b) the force transmitted along DH; and (c) the reactions at each of the pins O, E and G.

The radius of the friction circle $= x = r_d . \sin \phi$. But $\mu = \tan \phi = 0 \cdot 35$, so that $\phi = 19° 18'$ and $x = 10.0 \cdot 3303 = 3 \cdot 30$ in.

For clockwise rotation of the drum, the reactions between the two shoes and the drum are shown by R and R_1, which pass through C and F respectively and are tangential to the friction circle. The total braking torque T_b on the drum is therefore equal to $(R+R_1)x$.

Produce the line of action of the effort P to meet DH produced at L. Join L to G and produce to meet the line of action of R_1 at M. Join M to E. Then the bell-crank lever HGK is in equilibrium under the action of three forces which are applied at the pins H, G and K. The lines of action of the forces at H and K meet at L and therefore the reaction at the pin G must also pass through L.

Similarly, the right-hand brake hanger is in equilibrium under the three forces which act through E, F and G. But the lines of action of the forces through F and G intersect at M and therefore the reaction at the pin E must pass through M.

In the same way, the line of action of the reaction at pin O on the left-hand brake hanger may be found by joining O to the point of intersection of the line of action of R and HD produced.

Reverting to the bell-crank lever, it is clear that HGL is similar to the triangle of forces for this lever. Hence it follows that the horizontal component of the pull exerted by HD on pin D is equal in magnitude to the horizontal component of the push exerted by the bell-crank lever on pin G. Let F be the magnitude of each of these horizontal components. Then, taking moments about O and E, we have:

For the right-hand hanger,

$$F.\mathrm{EG} = R_1 a_1$$

and, for the left-hand hanger,

$$F.\mathrm{OD} = Ra$$

But a is obviously equal to a_1, and EG is equal to OD, so that:

$$F.\mathrm{OD} = Ra = R_1 a \quad \text{and} \quad R_1 = R$$

The braking torque on the drum $= T_b = (R+R_1)x = 2Rx$; and, substituting for R in the above equation:

$$F = \frac{R.a}{\mathrm{OD}} = \frac{T_b}{2x}.\frac{a}{\mathrm{OD}}$$

But T_b must be 500 lb ft, OD is 24 in., x is 3·30 in. and a scales 11·5 in.

$$\therefore \ F = \frac{500.12}{2.3{\cdot}30}.\frac{11{\cdot}5}{24} = 435 \text{ lb}$$

For the bell-crank lever we have, taking moments about G:

$$P.\mathrm{GK} = F.\mathrm{HG} \quad \text{or} \quad P = (4/12)435 = 145 \text{ lb}$$

Also, from the triangle of forces HGL, the tension in the link HD $= (\mathrm{HL/HG})P = 443$ lb and the reaction at the pin G $= (\mathrm{GL/HG})P = 488$ lb

Similarly, from the triangle of forces, Mab, for the right-hand brake hanger and the triangle of forces, Ncd, for the left-hand brake hanger, we find that the reactions at the pins O and E are respectively 550 lb and 640 lb.

111. The Internal-expanding Shoe Brake. A type of shoe-brake which is often used on self-propelled vehicles is shown in Fig. 173. Each shoe pivots at one end about a fixed fulcrum, while at the other end it rests against the face of a cam. The outer surfaces of the shoes are lined with Ferodo or other friction material which has a high coefficient of friction and good wearing properties. The shoes are normally held in the off position by a light spring S. To apply the brakes the cam is rotated by means of the braking force P applied to an arm keyed to the cam spindle. This forces the shoes into contact with the inside cylindrical surface of the brake drum, which is rigidly fastened to the road wheel. Friction between the shoes and the drum then applies the braking torque or couple to the drum.

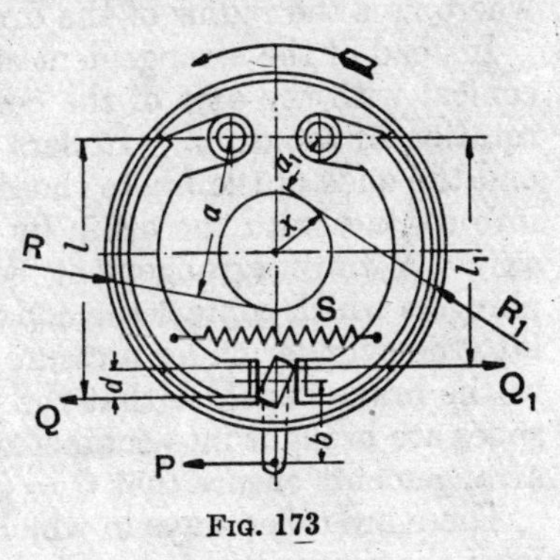

FIG. 173

The force analysis of this brake may be made as follows. Assume the material of the brake-shoe linings to obey Hooke's law in compression and find the line of action of the thrust on each shoe as already explained. Let R and R_1 represent the two thrusts for counter-clockwise rotation of the drum and a, a_1 the perpendicular distances of their lines of action from the respective shoe pivots. Also, let Q, Q_1 be the forces exerted by the cam on the ends of the respective shoes; l, l_1 the perpendicular distances of their lines of action from the shoe pivots; and d the perpendicular distance between Q and Q_1.

Even if the clearances between the shoes and the drum are correctly adjusted, so that when the brake is applied each shoe makes contact with the drum at the same instant, the forces Q and Q_1 will not be equal. This results from the fact that, for a given angular displacement of the cam, the angular displacement of the right-hand shoe is obviously slightly greater than the angular displacement of the left-hand shoe. The compression of the right-hand shoe lining is therefore greater than that of the left-hand shoe lining and Q_1 is greater than Q. In practice it is probably sufficiently accurate to assume that Q and Q_1 are equal.

Then, for equilibrium of the left-hand shoe, $Ql = Ra$, and, for equilibrium of the right-hand shoe, $Q_1 l_1 = R_1 a_1$.

But the torque on the drum

$$= T_b = (R+R_1)x$$

and, substituting for R and R_1

$$T_b = Q(l/a+l_1/a_1)x$$

For equilibrium of the cam, $Qd = Pb$ or $Q = Pb/d$,

so that
$$T_b = (Pb/d)(l/a+l_1/a_1)x \quad . \quad . \quad . \quad . \quad (8.2)$$

and, since x is the radius of the friction circle of the drum,

$$x = r_d \sin \phi \simeq \mu r_d$$

where r_d is the radius of the drum.

Instead of the arrangement shown in Fig. 173, the cam may be conical with the axis of the cone at right angles to the plane of rotation of the drum. Rollers are then placed between the cam and the ends of the brake shoes, so that the latter may be forced into contact with the drum by displacing the cam parallel to its axis. A third arrangement consists of a cylinder fitted with plungers which contact the ends of the brake shoes. Oil is forced into the cylinder under pressure from a master cylinder connected to the brake pedal, so that the plungers are forced apart and the shoes are brought into contact with the brake drum. Both these arrangements ensure that $Q = Q_1$ and also $l = l_1$.

There are other ways in which a modern braking system differs from the simple arrangement shown, but for information on these reference should be made to the appropriate literature.

112. The Band Brake. A flexible band may be used to apply a braking torque to a rotating drum. The flexible member may consist of a leather strap, one or more ropes, or a thin strip of steel lined with friction material. In order to apply the brake the band is tightened round the drum and the friction between the band and the drum provides the tangential braking force.

Referring to Fig. 174 (a), let θ be the angle of lap of the band on the drum, μ the coefficient of friction between the contact surfaces, and T_1 and T_2 the respective tensions on the tight and slack sides of the band. Then the limiting ratio of the tensions is given by $T_1/T_2 = e^{\mu\theta}$ and the tangential braking force on the drum by $Q = T_1 - T_2$.

A modification of the band brake is shown in Fig. 174 (b). The flexible steel band has a number of wood blocks fixed to the inside surface and the friction of the blocks on the drum provides the braking action. Each block embraces a short arc on the drum.

The ratio of the tensions on the tight and slack sides of the band may be found as follows. Let 2θ be the angle subtended at the drum centre by each brake block, T_0 the tension in the slack side of the band, T_1 the tension in the band between the first and second blocks, T_2 that between the second and third blocks, etc., T_n that in the tight side of the band, n the number of brake blocks and μ the coefficient of friction.

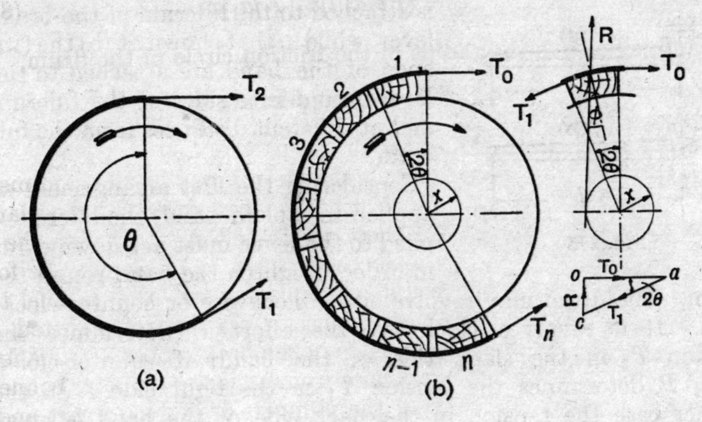

FIG. 174

Then the first block will be in equilibrium under three forces, the tensions T_0 and T_1 on the two ends of the band and the reaction R of the drum on the block. Since there is limiting friction between the block and the drum, the line of action of R will be tangential to the friction circle of the drum and it must also pass through the point of intersection of T_0 and T_1. Its inclination to the radial line drawn through the point of intersection of T_0 and T_1 may for all practical purposes be assumed equal to ϕ. The triangle of forces oac may then be drawn. From this triangle:

$$\frac{T_1}{T_0}=\frac{\sin aoc}{\sin aco}=\frac{\sin(90-\theta+\phi)}{\sin(90-\theta-\phi)}=\frac{\cos(\theta-\phi)}{\cos(\theta+\phi)}$$

$$=\frac{\cos\theta\cos\phi+\sin\theta\sin\phi}{\cos\theta\cos\phi-\sin\theta\sin\phi}=\frac{1+\tan\theta\tan\phi}{1-\tan\theta\tan\phi}=\frac{1+\mu\tan\theta}{1-\mu\tan\theta}$$

The triangle of forces is obviously similar for each block, and therefore the ratio of the tensions in the band is the same for each block. Hence the ratio of the tensions in the tight and slack sides is given by:

$$\frac{T_n}{T_0} = \frac{T_n}{T_{n-1}} \cdots \frac{T_3}{T_2} \cdot \frac{T_2}{T_1} \cdot \frac{T_1}{T_0} = \left(\frac{T_1}{T_0}\right)^n = \left(\frac{1+\mu \tan \theta}{1-\mu \tan \theta}\right)^n \qquad (8.3)$$

The ends of the plain band or of the band lined with wood blocks are attached to a brake lever, so that movement of the brake lever tightens the band round the drum and provides the braking action. Two possible arrangements are shown in Fig. 175. In the arrangement shown at (a), one end of the band is attached to the fulcrum of the brake lever, while in that shown at (b) the two ends of the band are attached to the lever on opposite sides of the fulcrum and at different distances from the fulcrum.

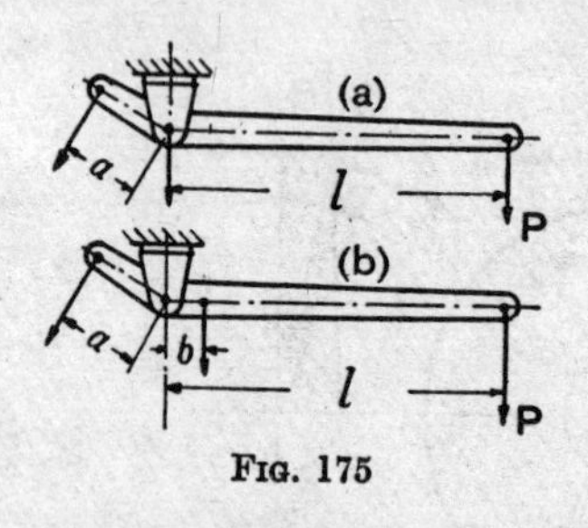

Fig. 175

Considering the first arrangement as applied to a plain band, the effort applied to the lever must act downwards in order to tighten the band round the drum. The drum may revolve either clockwise or counter-clockwise. If it revolves clockwise, the effort P determines the tension T_2 in the slack side of the band; if counter-clockwise, P determines the tension T_1 in the tight side. In the former case the tension in the slack side of the band is given by $T_2 = Pl/a$. But the ratio of the tensions $= T_1/T_2 = e^{\mu\theta}$ and the tangential braking force on the drum $Q = T_1 - T_2$, so that:

$$Q = T_1 - T_2 = T_2(e^{\mu\theta} - 1) = \frac{Pl}{a}(e^{\mu\theta} - 1) \quad . \qquad (8.4)$$

In the latter case, the tension in the tight side of the band is given by $T_1 = \frac{Pl}{a}$, so that:

$$Q = T_1 - T_2 = T_1 \frac{e^{\mu\theta} - 1}{e^{\mu\theta}} = \frac{Pl}{a} \frac{e^{\mu\theta} - 1}{e^{\mu\theta}} \quad . \qquad (8.5)$$

Hence, when the brake is so arranged that the brake lever applies the tension to the slack side of the band the braking force Q is $e^{\mu\theta}$ times as great as when it is so arranged that the brake lever applies the tension to the tight side of the band.

For the arrangement shown in Fig. 175 (b), the effort P must again act downwards, providing $a > b$. For clockwise rotation of the drum, the end of the band attached to A is the slack end, while that attached to B is the tight end. Hence, for a plain band,

$$Q = T_1 - T_2 = T_2(e^{\mu\theta} - 1)$$

But, for equilibrium of the brake lever,

$$T_2a = T_1b + Pl \quad \text{or} \quad T_2(a - be^{\mu\theta}) = Pl$$

and, substituting for T_2,

$$Q = Pl\frac{e^{\mu\theta}-1}{a-be^{\mu\theta}} \quad . \quad . \quad . \quad . \quad . \quad (8.6)$$

Similarly, for counter-clockwise rotation of the drum, we get:

$$Q = Pl\frac{e^{\mu\theta}-1}{ae^{\mu\theta}-b} \quad . \quad . \quad . \quad . \quad . \quad (8.7)$$

Since $a > b$ and $e^{\mu\theta} > 1$, the denominator of (8.7) is always positive, but it is possible for the denominator of (8.6) to be positive, zero or negative. The implication of a zero or negative value is that the brake is self-supporting. Once the band has been tightened round the drum the effort P may be reduced to zero. In practice a should always be greater than $be^{\mu\theta}$, otherwise gradual application of the brake will be impossible.

Example 4. The brake drum of a crane is 21 in. dia. and is keyed to the same shaft as the crane barrel, which is 15 in. dia. A band brake acts on the brake drum and is operated by a lever 18 in. long. One end of the band is attached to the fulcrum of the lever and the other end is attached to a pin on the lever at a distance of 4 in. from the fulcrum. The angle embraced by the band is 270° and μ is 0·3. What is the least force required at the end of the brake lever in order to support a load of ½ ton attached to a rope wound round the barrel? The arrangement of the brake lever is as shown in Fig. 175 (a).

In order to require the least effort P, the brake must be so arranged that rotation of the drum by the load tends to tighten that end of the band which is attached to the fulcrum of the brake lever.

The required tangential braking force on the drum

$$= Q = 1120.15/21 = 800 \text{ lb}$$

The ratio of the tensions in the two ends of the band

$$= T_1/T_2 = e^{\mu\theta} = e^{0\cdot3.3\pi/2} = 4\cdot111$$

Also $= 18$ in. and $a = 4$ in., so that, substituting in (8.4):

$$800 = \frac{P.18}{4}(4\cdot111-1)$$

from which $$P = \frac{800.2}{9.3\cdot111} = 55\cdot3 \text{ lb}$$

Example 5. A band-and-block brake is operated by a differential lever of the type shown in Fig. 175 (b). The distances a, b and l are respectively 4·5 in., 1 in. and 21 in. The brake drum is 30 in. dia. and there are 12 blocks, each of which subtends an angle of 20° at the drum centre. What is the least effort applied at the end of the lever which will provide a braking torque of 4000 lb ft on the drum if the coefficient of friction is 0·28?

For maximum braking torque, the brake must be so arranged that the tight side of the band is attached to the shorter arm b. The relation between Q and P is then given by (8.6).

From (8.3), the ratio of the tensions is given by:

$$k = \frac{T_n}{T_0} = \left(\frac{1+\mu \tan \theta}{1-\mu \tan \theta}\right)^n = \left(\frac{1+0{\cdot}28 . 0{\cdot}1763}{1-0{\cdot}28 . 0{\cdot}1763}\right)^{12} = 3{\cdot}274$$

The tangential braking force on the drum

$$= Q = T_n - T_0 = T_0(k-1)$$

Substituting k for $e^{\mu\theta}$ in (8.6), we get:

$$Q = \frac{P . l(k-1)}{a-b . k}$$

But $l = 21$ in., $a = 4{\cdot}5$ in., $b = 1$ in. and $k = 3{\cdot}274$,

$$\therefore\ Q = \frac{P . 21(3{\cdot}274-1)}{4{\cdot}5-3{\cdot}274} = 38{\cdot}9P$$

The required braking torque is 4000 lb ft and the radius of the drum is 15 in., so that $Q = 4000 . 12/15 = 3200$ lb.

$$\therefore\ 38{\cdot}9P = 3200 \quad \text{and} \quad P = 82{\cdot}2 \text{ lb}$$

113. The Braking of a Vehicle. It will be appropriate at this point to consider the retardation of a vehicle produced by the application of brakes. Fig. 176 shows diagrammatically a car moving up a plane surface inclined at the angle α to the horizontal. Let W be the weight of the car, h the height of the centre of gravity above the road surface, x its perpendicular distance from the rear axle, w the wheelbase of the car. It is required to find the retardation of the car when brakes are applied (a) to the rear wheels only, (b) to the front wheels only and (c) to all four wheels.

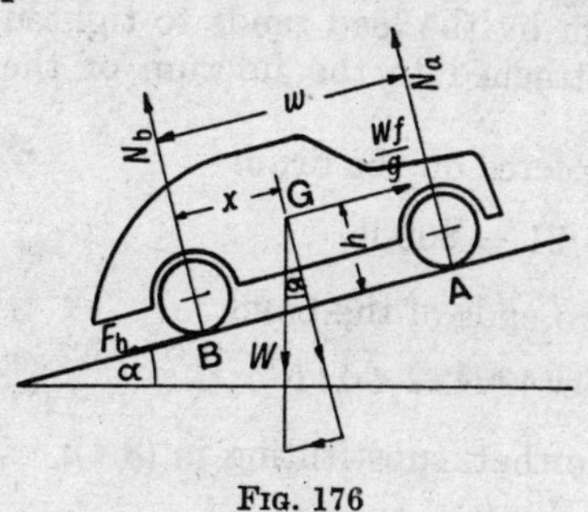

FIG. 176

In all cases the problem is reduced to the equivalent problem in statics by applying d'Alembert's principle. The reversed effective force, or the inertia force, is included with the system of forces which is actually applied to the car, so as to give a system of forces in equilibrium.

(a) *Brakes applied to the Rear Wheels.* Let F_b be the total braking force at the rims of the rear wheels due to the application of the brakes. Let N_a and N_b be the total normal reactions between the ground and the front and rear wheels respectively, f be the retardation and μ the coefficient of adhesion between the tyres and the road surface. Then the forces acting on the car may be reduced to the co-planar system shown and the car is in equilibrium under this system of forces.

Resolving parallel to the plane:

$$F_b + W \sin \alpha = Wf/g \quad . \quad . \quad . \quad . \quad (8.8)$$

Resolving perpendicular to the plane:

$$N_a + N_b = W \cos \alpha \quad . \quad . \quad . \quad . \quad (8.9)$$

A third equation is obtained by considering the equilibrium of the couples. For convenience take moments about G, then:

$$F_b h + N_b x = N_a(w - x)$$

$$\therefore \ F_b h = N_a w - (N_a + N_b)x$$

Substituting for $N_a + N_b$ from (8.9):

$$F_b h = N_a w - Wx \cos \alpha$$

$$\therefore \ N_a = \frac{Wx \cos \alpha + F_b h}{w} \quad . \quad . \quad . \quad (8.10)$$

From (8.9):

$$N_b = W \cos \alpha - N_a$$

$$= \frac{W(w-x) \cos \alpha - F_b h}{w} \quad . \quad . \quad . \quad (8.11)$$

Also, from (8.8):

$$f/g = F_b/W + \sin \alpha \quad . \quad . \quad . \quad (8.12)$$

The braking force F_b will depend upon the effort exerted by the driver on the brake pedal, but in no circumstances can it exceed the limiting value μN_b, since any attempt by the driver to increase F_b beyond this point will simply lock the wheels and cause them to skid. Hence the maximum possible retardation when the rear wheels only are braked is obtained from equation (8.12) by substituting μN_b for F_b:

$$f/g = \mu N_b/W + \sin \alpha$$

But from (8.11):

$$N_b = \frac{W(w-x)\cos\alpha - \mu N_b h}{w}$$

$$\therefore\ N_b = \frac{(w-x)\cos\alpha}{w+\mu h}W \quad . \quad . \quad . \quad (8.13)$$

so that

$$\frac{f}{g} = \frac{\mu(w-x)\cos\alpha}{w+\mu h} + \sin\alpha \quad . \quad . \quad . \quad (8.14)$$

Also, from (8.10):

$$N_a = \frac{(x+\mu h)\cos\alpha}{w+\mu h}W \quad . \quad . \quad . \quad (8.15)$$

(b) *Brakes applied to the Front Wheels.* In this case the braking force F_a acts through the point A. Apart from this the system of forces remains unchanged and equations (8.10), (8.11) and (8.12) apply if F_a is substituted for F_b. The maximum possible value of F_a is clearly equal to μN_a, and on substituting this value in (8.12) we get

$$f/g = \mu N_a/W + \sin\alpha$$

But from (8.10):

$$N_a = \frac{Wx\cos\alpha + \mu N_a h}{w}$$

so that

$$N_a = \frac{x\cos\alpha}{w-\mu h}.W \quad . \quad . \quad . \quad . \quad (8.16)$$

and

$$\frac{f}{g} = \frac{\mu x\cos\alpha}{w-\mu h} + \sin\alpha \quad . \quad . \quad . \quad (8.17)$$

Also, from (8.11):

$$N_b = \frac{(w-x-\mu h)\cos\alpha}{w-\mu h}.W \quad . \quad . \quad . \quad (8.18)$$

(c) *Brakes applied to the Four Wheels.* In this case there is a braking force F_a provided by the front wheels and a braking force F_b provided by the rear wheels. The only change required in equations (8.10), 8.11) and (8.12) is the substitution of F_a+F_b for F_b. The maximum possible braking force is $\mu N_a + \mu N_b$ or $\mu\ W\cos\alpha$.

From (8.12):

$$f/g = \mu\cos\alpha + \sin\alpha \quad . \quad . \quad . \quad . \quad (8.19)$$

From (8.10):

$$N_a = \frac{(x+\mu h)\cos\alpha}{w}W \quad . \quad . \quad . \quad (8.20)$$

From (8.11):

$$N_b = \frac{(w-x-\mu h)\cos\alpha}{w}W \quad . \quad . \quad . \quad (8.21)$$

If the car is moving down the plane, the term $W \sin \alpha$ must be moved from the left-hand side to the right-hand side of equation (8.8). This will not affect the equations for N_a and N_b, but it will alter the sign in front of the term $\sin \alpha$ in each of the equations (8.14), (8.17) and (8.19).

If the car is moving along a horizontal surface, then $\alpha = 0$ and the equations are simplified.

The application of the brakes to the car causes a couple to act which tends to rotate the car as a whole in a vertical plane about its centre of gravity. This couple increases the pressure between the front wheels and the road and decreases that between the rear wheels and the road, as may be seen if the reactions during braking are compared with those when the car is at rest or is moving with uniform speed. As the driver of the car gradually increases the pressure on the brake pedal, the friction force between the braked wheels and the road surface gradually builds up until the limiting condition is reached and the wheels are just on the point of becoming locked. Obviously, this limiting condition will be reached earlier with rear wheel brakes and later with front wheel brakes than if there were no transference of load from the rear to the front axle during braking.

Where four-wheel brakes are fitted, the relative magnitudes of the braking forces at the front and rear wheels are determined by the proportions of the mechanism between the brake pedal and the brake drums on the wheels. For most brake mechanisms the ratio F_a/F_b is constant. But, for maximum retardation, this ratio ought to be the same as N_a/N_b, and from equations (8.20) and (8.21):

$$\frac{N_a}{N_b} = \frac{x+\mu h}{w-x-\mu h} \quad . \quad . \quad . \quad . \quad (8.22)$$

Hence, to get maximum retardation under different conditions of road surface, i.e. for different values of μ, the ratio F_a/F_b ought to increase as the pressure on the brake pedal is increased. It is beyond the scope of this article to consider the problem in greater detail. Reference should be made to an article in the *Automobile Engineer*, January 1926, where a brake mechanism is described which is designed to provide for an increase in the ratio F_a/F_b as the pressure on the brake pedal is increased.

The conditions which govern the maximum acceleration of a self-propelled vehicle may be examined in the same way. The maximum possible tractive force is determined by the limiting friction between the driving wheels and the road surface. This force acts in the direction of motion and the couple caused by the combined effect of the tractive force and the inertia force increases the pressure between the rear wheels and the road and decreases that between the front wheels and the road.

Example 6. For a car w is 9·5 ft, h is 2 ft and x is 4 ft. If the car is moving along a level road at 30 m.p.h., find the minimum distance in which the car may be stopped when (a) the rear wheels are braked, (b) the front wheels are braked, (c) all the wheels are braked and the coefficient of friction between tyre and road is (i) 0·1, (ii) 0·6. What is the required ratio of N_a/N_b in each case when four-wheel brakes are used?

(i) Coefficient of friction $=0{\cdot}1$. (ii) Coefficient of friction $=0{\cdot}6$.

(a) *Rear wheels braked*

From (8.14):

$$\frac{f}{g} = \frac{0{\cdot}1 \,.\, 5{\cdot}5}{9{\cdot}5+0{\cdot}1 \,.\, 2} = 0{\cdot}0567 \qquad \frac{f}{g} = \frac{0{\cdot}6 \,.\, 5{\cdot}5}{9{\cdot}5+0{\cdot}6 \,.\, 2} = 0{\cdot}308$$

$$\therefore\ f = 1{\cdot}823 \text{ ft/s}^2 \qquad \therefore\ f = 9{\cdot}93 \text{ ft/s}^2$$

For uniform retardation $S = v^2/2f$,

$$\therefore\ S = \frac{44^2}{2 \,.\, 1{\cdot}823} = 531 \text{ ft} \qquad \therefore\ S = \frac{44^2}{2 \,.\, 9{\cdot}93} = 97{\cdot}5 \text{ ft}$$

(b) *Front wheels braked*

From (8.17):

$$\frac{f}{g} = \frac{0{\cdot}1 \,.\, 4}{9{\cdot}5-0{\cdot}1 \,.\, 2} = 0{\cdot}043 \qquad \frac{f}{g} = \frac{0{\cdot}6 \,.\, 4}{9{\cdot}5-0{\cdot}6 \,.\, 2} = 0{\cdot}289$$

$$\therefore\ f = 1{\cdot}383 \text{ ft/s}^2 \qquad \therefore\ f = 9{\cdot}31 \text{ ft/s}^2$$

$$\therefore\ S = \frac{44^2}{2 \,.\, 1{\cdot}383} = 700 \text{ ft} \qquad \therefore\ S = \frac{44^2}{2 \,.\, 9{\cdot}31} = 104 \text{ ft}$$

(c) *All wheels braked*

From (8.19):

$$f/g = \mu = 0{\cdot}1, \qquad f/g = \mu = 0{\cdot}6,$$

$$\therefore\ f = 3{\cdot}22 \text{ ft/s}^2 \qquad \therefore\ f = 19{\cdot}32 \text{ ft/s}^2$$

$$\therefore\ S = \frac{44^2}{2 \,.\, 3{\cdot}22} = 300 \text{ ft} \qquad \therefore\ S = \frac{44^2}{2 \,.\, 19{\cdot}32} = 50{\cdot}1 \text{ ft}$$

From (8.22):

$$\frac{N_a}{N_b} = \frac{4+0{\cdot}1 \,.\, 2}{9{\cdot}5-4-0{\cdot}1 \,.\, 2} = 0{\cdot}793 \qquad \frac{N_a}{N_b} = \frac{4+0{\cdot}6 \,.\, 2}{9{\cdot}5-4-0{\cdot}6 \,.\, 2} = 1{\cdot}21$$

114. Dynamometers. A dynamometer is essentially a device for measuring the forces or couples which tend to change the state of rest or of uniform motion of a body. There are many kinds of dynamometers, but reference can only be made to a few of the types used in measuring the power available from a uniformly revolving shaft. Broadly, two main types may be distinguished, namely, *absorption dynamometers* and *transmission dynamometers*. As the names imply, an absorption dynamometer absorbs the available power in doing work, usually against friction, whereas a transmission dynamometer transmits the available power unchanged, except for the small amount absorbed by friction at the joints of the dynamometer.

115. Absorption Dynamometers. These generally consist of some form of brake in which provision is made for measuring the friction torque on the drum.

(a) *The Prony Brake.* A simple type, known as the Prony brake, is shown in Fig. 177. It consists of two blocks of wood, each of which embraces rather less than one-half of the pulley rim. The two blocks can be drawn together by means of bolts, cushioned by springs, so as to increase the pressure on the pulley. One block carries an arm to the end of which a pull can be applied

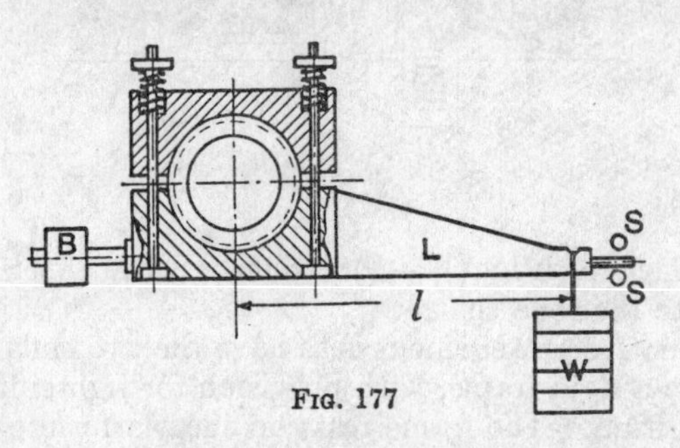

FIG. 177

by means of a deadweight or spring balance. A second arm projects from the block in the opposite direction and carries a balance weight B, which balances the brake when unloaded. The friction torque on the pulley may be increased by screwing up the bolts, until it balances the torque due to the available power. For counter-clockwise rotation of the drum, the arm L will float between the stops S with a weight W suspended from it. The torque on the drum is given by Wl and, knowing the speed of rotation of the pulley, the power absorbed may be calculated.

Wear of the blocks and variations in the coefficient of friction between the blocks and the pulley rim necessitate continual tightening of the bolts and render this type of brake unsuitable either for the absorption of large powers or for long-continued runs. In any case great care must be exercised to see that the lever L is always floating between the stops.

An alternative arrangement, which is much better for the absorption of larger powers, is so to arrange the brake that the end of the lever L rests on the platform of a weighing machine.

(b) *The Rope-brake Dynamometer.* This type of brake is generally much steadier in operation than the Prony brake. It is also suitable for the absorption of a wider range of powers. In general, two or more ropes rest on the pulley rim. They are spaced evenly across the width of the rim by means of three or four wooden blocks at different points round the rim. The total pull S on the slack ends of the ropes is registered on a spring balance, while the pull W on the right ends is provided by deadweights. The brake

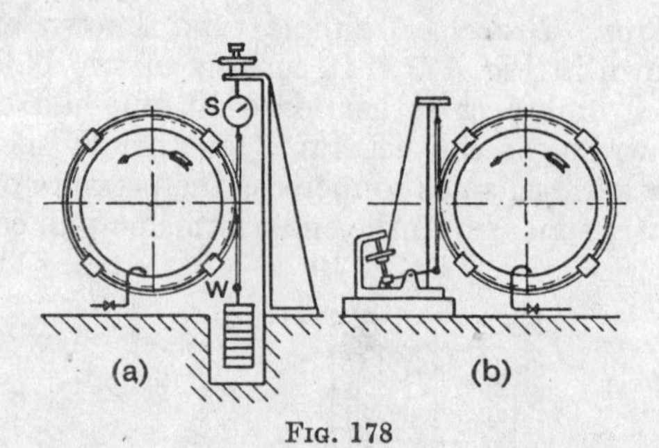

FIG. 178

torque is then given by $(W-S)r$, where r is the effective radius of the drum to the rope centre.

An alternative arrangement is to have the two ends of the ropes attached to a rigid frame, with provision for tightening the ropes round the drum. The frame rests on the platform of a weighing machine which registers the net downward force $W-S$ on the frame. The two arrangements are shown diagrammatically in Fig. 178.

Where power has to be absorbed continuously, as in engine testing, the brake drum should be separate from the flywheel. It may be either bolted to the flywheel rim or separately keyed to the crankshaft. In any case provision should be made for water cooling the rim in order to carry away the heat generated by the friction. The rim should be of channel section on the inside, so that cold water may be supplied at one point, carried round the rim and then removed by some form of scoop. Except for large powers it is unnecessary to provide a scoop; the supply of cold water may be adjusted so that it just makes good the loss of water by evaporation. Providing the inside surface of the rim is always covered by a layer of water, undue heating of the rim will not take place.

In the author's experience rope brakes seldom give trouble if the ropes are well greased with tallow, unless the ropes are too lightly loaded. It is better to use a higher stress than would be considered good practice in a rope drive, even though it means that the initial stretch is high.

(c) *The Heenan and Froude Dynamometer.* This dynamometer is very widely used for the absorption of a wide range of powers and is suitable for a wide range of speeds. It was invented by William Froude in 1877, and a section is shown in Fig. 179. A rotor A is keyed to the main shaft, to which the power to be

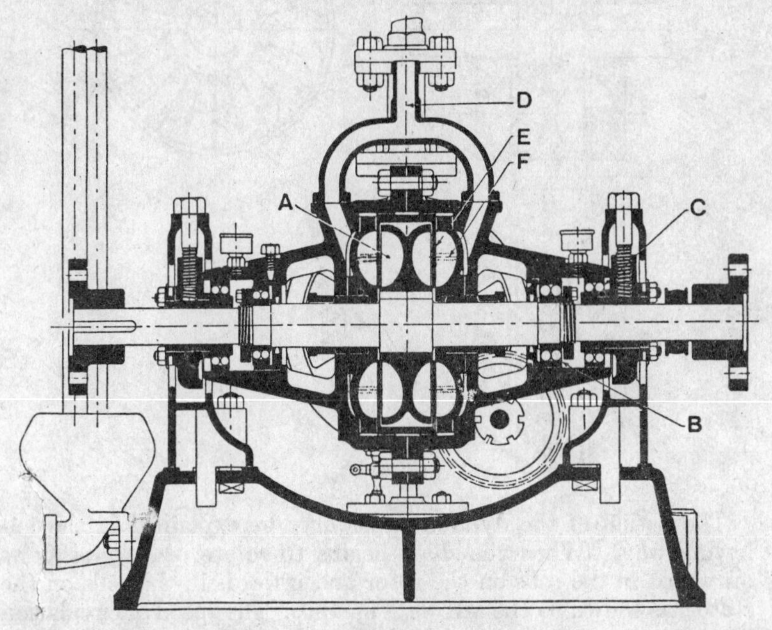

FIG. 179

measured is supplied. Surrounding the rotor is a stator fixed to the outer casing. The main shaft is supported on ball bearings B in the outer casing, and the outer casing is, in turn, supported on ball bearings C carried by brackets on the bed-plate. Water is supplied through a flexible pipe to the branch D. In each face of the rotor and in the adjacent faces of the stator there are semi-oval channels. Each channel is divided into a number of cells by semi-circular diaphragms, which are set obliquely at an angle of 45° to the plane of rotation so that the straight edge of the diaphragm coincides with the major axis of the oval channel. Water is conveyed to the cells through the passages F. The arrangement will be clear from the end elevation of the rotor and the developed circumferential section of the rotor and stator which are shown in Fig. 180.

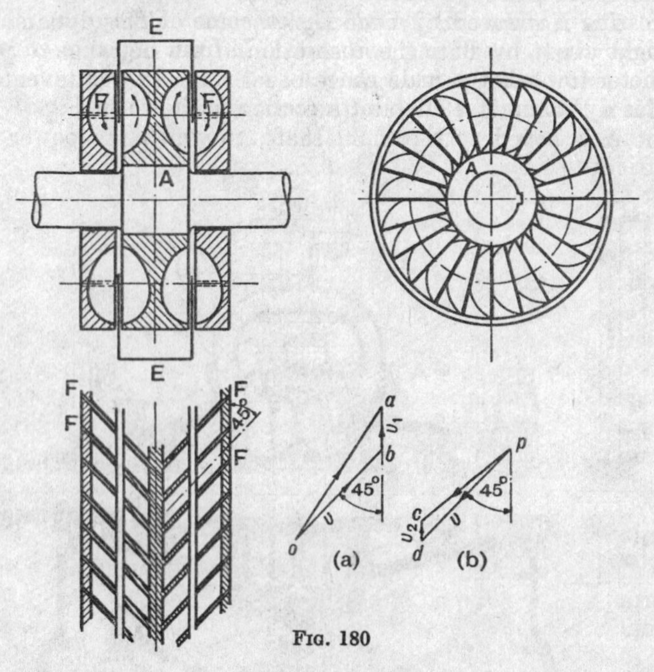

Fig. 180

The action of the dynamometer may be explained in the following way. When the shaft begins to rotate, the water flows outwards in the cells on the rotor and inwards in the cells on the stator, as shown by the arrows, Fig. 180. The speed of circulation in the vortex increases with the speed of rotation of the shaft. We may imagine the vortex to consist of a large number of filaments or rings of small cross-section, in each of which the speed of circulation remains constant. Let oab be the triangle of velocities at the point where one such filament crosses from the rotor to the stator. Then, if v is the velocity of circulation and v_1 is the tangential velocity of the rotor, the absolute velocity of the water represented by oa. This must be reduced to v as the water enters the stator and the reaction on the stator tends to revolve it in the same sense as the rotor. Similarly, from the triangle of velocities pcd at inlet to the rotor, it will be seen that the absolute velocity of the water is suddenly changed from v to that represented by pc. Again, the reaction on the stator tends to revolve it in the same sense as the rotor. The torque on the casing is measured by a deadweight attached to an arm fixed to the casing. Water is circulated continuously through the brake and the inside of the brake is always filled with water. The reaction torque on the casing is reduced by blanking off some of the cells. This is brought about by sliding two shields E from opposite ends of a diameter towards the main shaft.

This dynamometer enables the torque on the main shaft to be very accurately determined. The torque required to overcome the friction of the glands which prevent leakage between the main shaft and the outer casing, as well as that required to overcome the friction of the bearings B, is transmitted to the outer casing and included in the measured torque. The only torque not measured is the friction torque of the bearings C between the outer casing and the frame, and this is very small.

Other advantages are that the resisting torque on the main shaft may be varied at any time while the shaft is revolving, the heat generated is carried away by the continuous water circulation, the space occupied is small and there are no delicate parts or fine clearances to be maintained; moreover, the dynamometer is practically silent in operation.

(d) *The Swinging-field Dynamometer*. This dynamometer consists of an electric generator, the field system of which is mounted on trunnions so as to be able to revolve freely. The arrangement is similar to that of the outer casing of the hydraulic dynamometer. The rotor shaft is coupled to the source of power and when the rotor revolves the electromagnetic reaction on the field frame tends to cause rotation of the frame. This rotation is prevented by the application of a deadweight or a spring balance to an arm fixed to the field frame. From the measured reaction torque and the known speed of rotation of the shaft the power supplied can be calculated. The resisting torque is varied by altering the resistance in the armature circuit, thus altering the current generated.

Although strictly speaking this is an absorption dynamometer, the electrical energy generated can be fed back to the supply lines and usefully employed in lighting, etc. This constitutes one of the advantages of this type over the hydraulic dynamometer, but the chief advantage lies in the fact that the generator may be run as a motor. Thus, when testing high-speed internal-combustion engines, motoring tests provide the only satisfactory means of measuring the friction losses in the engine. With the swinging-field dynamometer it is possible to measure the power output of the engine at full throttle and full speed and then, with very little delay, reverse the electrical connections so that the engine is motored round at the same speed with the petrol supply cut-off. In this way, the friction torque is measured under the same conditions of engine temperature as those which obtained during the full-load test.

The disadvantages of the swinging-field dynamometer are that it is only suitable for comparatively high speeds and small powers, and it is much less robust than the hydraulic dynamometer.

For use in the routine testing of engines for motor cars, special automatic electric dynamometers have been developed. In these the swinging-field is dispensed with. The dynamometer is first used as a motor to run-in the engine, and the running-in period is continued until the friction torque is reduced to a predetermined figure. The petrol supply is then automatically turned on and the electrical connections to the dynamometer reversed, so that the engine is put under load. Finally, the load and speed are gradually built up until the desired output is obtained.

116. Transmission Dynamometers. This class of dynamometer is designed in order to allow of the measurement of the power which is usefully employed by a machine. The general principles which underlie the design of the various types will be briefly explained.

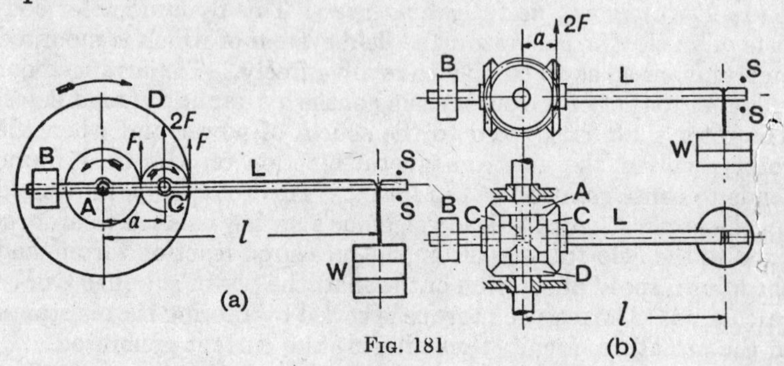

FIG. 181

(a) *The Epicyclic-train Dynamometer.* A simple epicyclic train of spur or bevel wheels, arranged as shown in Fig. 181, may be placed between the source of power and the machine and used to measure the power transmitted. Referring to Fig. 181 (a), the spur wheel A is keyed to the driving shaft and revolves in the counter-clockwise sense; the internal wheel D is keyed to the driven shaft and revolves in the clockwise sense. The power is transmitted from A to D through the intermediate wheel C. This wheel revolves freely on a pin fixed to the arm L and the latter can pivot freely about the common axis of the driving and the driven shafts. The tangential effort exerted by the wheel A on the wheel C and the tangential reaction of the internal wheel D on the wheel C are obviously equal, if the friction of the pin on which C revolves is neglected. Also these two forces both act in the upward direction, so that the total upward force on the arm L through the axis of wheel C is given by $2F$ and the corresponding torque on the arm by $2Fa$. This torque is balanced by suspending a deadweight W from the arm, which causes the arm to float between the stops S. The weight B balances the arm when the dynamometer is at rest. Then $2Fa = Wl$ or $F = Wl/2a$, and, given the radius and the speed of rotation of the wheel A, the power transmitted may be calculated.

The action of the bevel wheel epicyclic dynamometer, Fig. 181 (b), is similar. Wheels A and D are keyed to the driving and the driven shafts. The intermediate wheels C revolve freely on journals on the lever L and the latter pivots freely about the common axis of the driving and the driven shafts. If the driving shaft A revolves counter-clockwise and the driven shaft D clockwise and the total tangential force on the driving and the driven wheels is F, the counter-clockwise torque exerted on the lever is given by $2Fa$. This is balanced by the torque exerted by the deadweight W at the distance l from the axis of the lever.

Therefore $2Fa = Wl$ and the torque on the driving shaft $= Fa = Wl/2$. Given the speed of rotation of the driving shaft, the power transmitted may be calculated.

(b) *The Belt Transmission Dynamometer*. Two different types of belt transmission dynamometer are illustrated in Fig. 182, In each the design is such that while the belt is transmitting power. the difference between the tensions on the tight and slack sides may be measured. In the Tatham dynamometer, Fig. 182 (a), an endless belt passes from the driving pulley A over the intermediate pulleys C, C_1 to the driven pulley D. The driving and driven pulleys revolve about fixed axes, but the intermediate pulleys revolve on pins fixed to the lever L, which, in turn, pivots about the fulcrum E on the fixed frame. If the driving pulley A revolves counter-clockwise, the tight and slack sides of the belt are as shown. The total downward forces on the pins of wheels C and C_1 are respectively $2T_1$ and $2T_2$ and the net counter-clockwise moment on the lever L is $2(T_1-T_2)a$. This is balanced by suspending a known weight W from the lever at a distance l

from the fulcrum. Then $2(T_1-T_2)a = Wl$ and the effective tension is given by:

$$T_1-T_2 = Wl/2a$$

This, multiplied by the belt speed, gives the power transmitted from pulley A to pulley D. The power may of course be transmitted through the dynamometer in the opposite direction, i.e. from pulley D to pulley A.

The Von-Hefner Alteneck transmission dynamometer, Fig. 182 (b), is generally used on a horizontal drive. As before, the power is transmitted from A to D or from D to A. The driving belt passes over two jockey pulleys, C, C_1, and is arranged with the bottom side as the driving or tight side. The jockey pulleys ride loosely on pins fixed to a triangular-shaped frame, which is free to turn about a fixed axis through Q on the line of centres of the pulleys A and D. The net downward force on the jockey pulleys

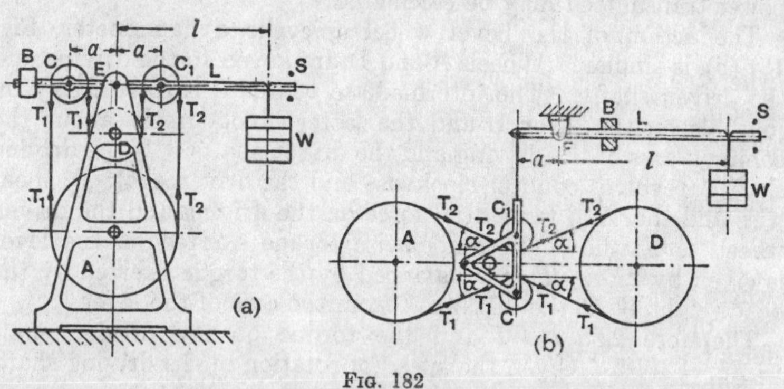

FIG. 182

caused by the difference between the belt tensions is transmitted to one end of the lever L, which pivots about the fixed fulcrum F, and is balanced by a deadweight W. The dynamometer is adjusted so that the lever floats midway between the stops S when the jockey pulley centres are equidistant from the line of centres of the main pulleys A and D. The balance weight B eliminates the effect of the deadweight of the frame, the jockey pulleys and the lever. The pulleys A and D are usually of equal size and the four straight portions of the belt are equally inclined at the angle α to the line of centres of A and D.

The downward force on the pulley C, due to the tension T_1 in the tight side of the belt $= 2T_1 \sin \alpha$.

Similarly, the upward force on the pulley C_1, due to the tension T_2 in the slack side of the belt $= 2T_2 \sin \alpha$.

Therefore the net downward force transmitted to the lever

$= P = 2(T_1 - T_2) \sin \alpha$, and, taking moments about the fulcrum of the lever, $Pa = Wl$, so that

$$T_1 - T_2 = Wl/2a \sin \alpha \quad . \quad . \quad . \quad (8.24)$$

This enables the effective tension on the belt to be determined and the power transmitted is then easily calculated.

(c) *Torsion Dynamometers.* A number of dynamometers make use of the elastic deformation of a steel shaft or spring in order to measure the torque transmitted. They have been developed principally in order to meet the need for measuring large powers, such as the power transmitted along the propeller shaft of a turbine or motor vessel.

When power is transmitted along a shaft, the driving end twists through a small angle relative to the driven end. The angle of twist in radians is given by:

$$\theta = f_s/C . l/r \quad . \quad . \quad . \quad . \quad . \quad . \quad (8.25)$$

where f_s is the shear stress at the surface of the shaft, C is the modulus of rigidity of the shaft material l is the length of the shaft and r is the external radius of the shaft.

But the shear stress f_s is directly proportional to the torque transmitted and is given by:

$$f_s = Tr/J \quad . \quad . \quad . \quad . \quad . \quad . \quad (8.26)$$

where T is the applied torque and J is the polar second moment of area of the shaft cross-section. For a solid shaft $J = \pi r^4/2$ and for a hollow shaft $J = \pi(r^4 - r_1^4)/2$, where r_1 is the internal radius.

Substituting for f_s from (8.25) in (8.26), we get

$$\theta = Tl/CJ \quad \text{or} \quad T = \theta . CJ/l \quad . \quad . \quad (8.27)$$

Hence, for a given shaft, the torque transmitted is directly proportional to the angle of twist and if the angle of twist can be measured the corresponding torque may be calculated.

For steel C is 5300 tons/in² and the mean shear stress at the surface of the shaft when transmitting full power will not usually exceed 2·5 to 3 tons/in². Hence, substituting in (8.25),

$$\theta = 2{\cdot}5/5300 . l/r \text{ rad} = 0{\cdot}027 . l/r \text{ deg.}$$

It will be seen that, in order to get an angle of twist of only one degree, the length of shaft between the measuring points must be approximately 40 times the radius or 20 times the diameter of the shaft. Hence, if the angle of twist is to be measured directly, a long length of shaft must be available. In practice the angle of twist usually has to be measured over a comparatively short length of shaft and some form of amplifying device must be incorporated in the dynamometer. The essential features of four different torsion dynamometers are briefly described below.

(i) *The Bevis-Gibson Flash-light Dynamometer*: Fig. 183. Two discs A and B are fixed to the shaft at points as far apart as possible. Each disc has a narrow radial slot and the two slots are in line when there is no torque transmitted along the shaft.

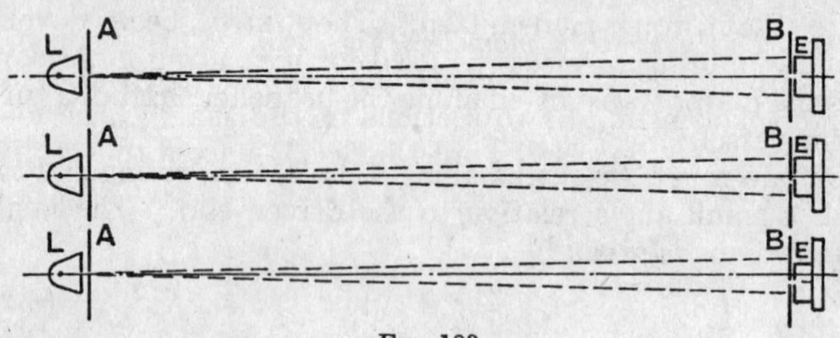

FIG. 183

Behind one disc A a powerful electric lamp is fixed to a bearing cap or other support. This lamp is masked so as to throw a narrow pencil of light parallel to the axis of the shaft and at the same distance from the axis as the radial slots in the discs A and B. Behind the disc B an eyepiece is supported on a fixed bracket, but by means of a vernier adjustment the eyepiece may be moved along an arc concentric with axis of the shaft.

With the shaft at rest the eyepiece is adjusted so as to receive the narrow pencil of light which passes from the lamp through the slots in the two discs. When the shaft revolves without transmitting torque a flash will be received in the eyepiece once per revolution; at high speeds, of course, the observer will not be able to distinguish the individual flashes. But, when the shaft is revolving and transmitting torque, the twist of the shaft between the discs A and B will cause one slot to lag behind the other and it will be necessary to displace the eyepiece along the circular arc, by means of the vernier, before the pencil of light which passes through the slots again enters the eyepiece. The vernier is provided with a scale so that the angular displacement of the eyepiece and therefore the angle of twist of the shaft may be measured.

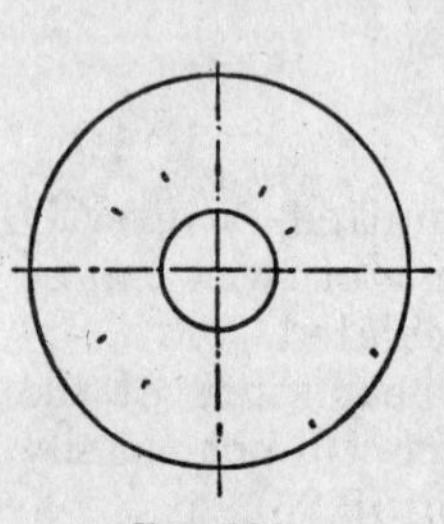

FIG. 184

Where a uniform torque is transmitted, it is sufficient to measure the angle of twist at a single angular position of the shaft. But where the torque varies it is necessary to measure the angle of twist at several different angular positions of the shaft. To do this, the discs A and B are provided with short radial slots arranged in spiral form as shown in Fig. 184. The lamp and eyepiece must then be moved radially so as to bring them in turn into line with the corresponding pair of slots in A and B. With twelve slots arranged as shown the angle of twist may be found for angular intervals of 30°.

(ii) *The Föttinger Dynamometer*: Fig. 185. This incorporates a purely mechanical contrivance for amplifying the relative angular displacement of two sections of the shaft. The disc A is fixed directly to the shaft, while the disc C is fixed to a stiff tube coaxial with the shaft and secured to the shaft at a point B distant l from A. When the shaft is transmitting power, the relative angular displacement of the discs A and C will be equal to the

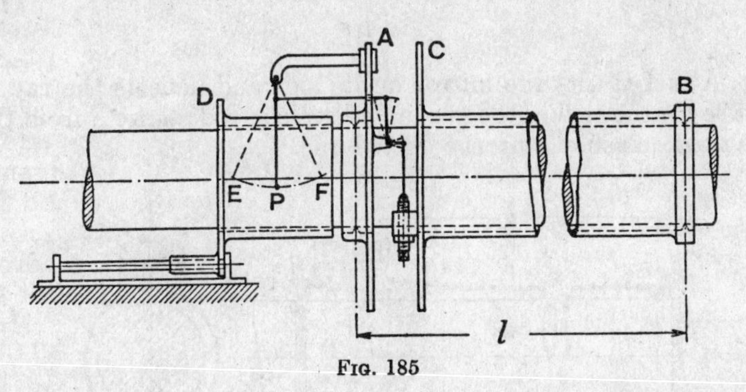

Fig. 185

angle of twist of the length l of the shaft. This angular displacement is amplified by the system of levers shown and the pencil P traces out a wavy line on a sheet of paper wrapped round the drum D which is coaxial with the shaft. The magnitude of the torque will be proportional to the curved ordinate EF of the wavy line, where F lies on the line traced by P when no power is transmitted by the shaft.

(iii) *The Hopkinson-Thring Dynamometer*: Fig 186. This dynamometer makes use of an optical method of amplifying the relative angular displacement of the discs A and C. A small mirror is supported on the disc A so as to pivot about a radial axis. A short arm fixed to the mirror is held lightly against a projection on the disc C by means of a spring. A beam of light from a lamp L is reflected from the mirror on to a graduated scale once per revolution of the shaft. When the shaft is at rest, the mirror is adjusted so that the hair line on the spot of reflected light coincides with the zero of the scale. Then, when the shaft is transmitting power, the relative angular displacement of the

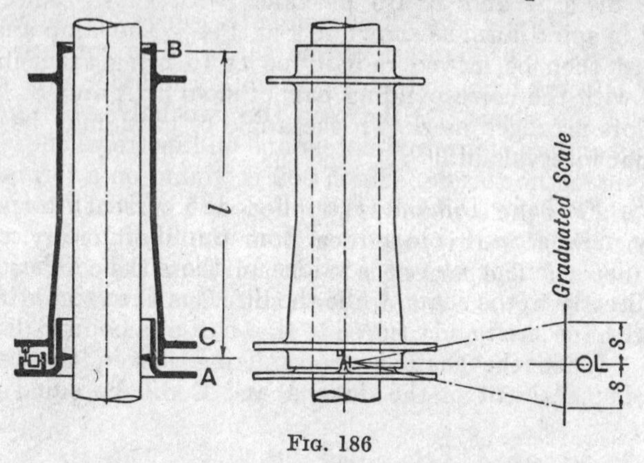

FIG. 186

discs A and C tilts the mirror on its axis and deflects the ray of light so that it strikes the graduated scale at a distance S from the zero mark, as shown by the dotted line.

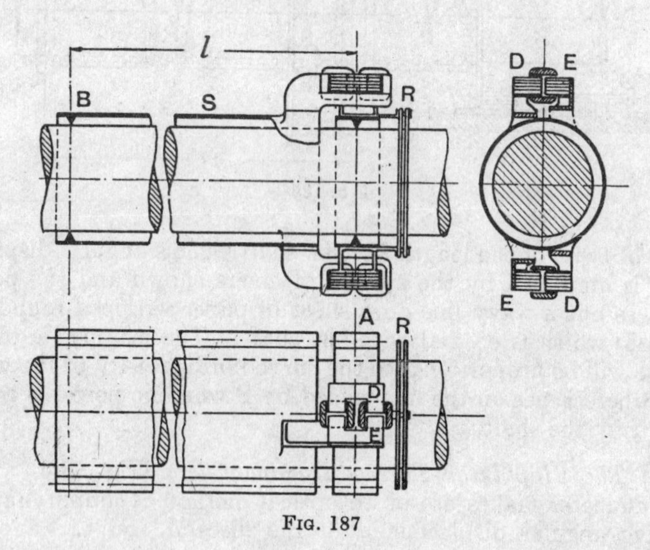

FIG. 187

(iv) *The Moullin Dynamometer*: Fig. 187. An electrical method of measuring the angle of twist is used in this dynamometer. This has the advantage that the recording instrument may be situated at a considerable distance from the dynamometer itself. The twist of the shaft varies the self-inductance of a coil which is supplied with alternating current. The strength of the current which flows through the coil thus varies with the twist of the shaft and therefore with the torque.

There are actually two laminated cores which are mounted at opposite ends of a diameter of the shaft. The cores are split, one half D being fixed to the disc A while the other E is fixed to the end of the sleeve S. Between the two halves of each core there is an air gap of approximately one millimetre when the shaft is transmitting no torque. Each coil is wound on a former fixed to the part D of the core and is supplied with current from a small alternator which is driven either from the shaft or by a small electric motor. The current is led to and from the coils by means of brushes which make contact with slip rings R mounted on, but insulated from, the disc A. When power is transmitted the twist of the shaft tends either to increase or to decrease the air gaps. This varies the self-inductance of the coils and the strength of the current which flows through them. The ammeter scale may be calibrated so as to indicate directly the torque transmitted.

EXAMPLES VIII

1. Show how the direction of the reaction between the shoe and the drum of a simple block brake may be found when (a) the shoe is rigidly attached to the hanger, (b) the shoe is pivoted to the hanger.

2. The dimensions of a block brake of the type shown in Fig. 170 (a) are as follows. The diameter of the brake drum is 27 in., the distance between the pins O and C is 15 in., and that between the pins O and H is 40 in.; the perpendicular distance of the brake hangar OCH from the axis of the drum is 15 in. and the coefficient of friction between the block and the drum is 0·35. What force P must be applied through the pin H in order to provide a braking torque of 100 lb ft on the drum?

3. If the brake shoe is rigidly attached to the hanger, as in Fig. 170 (b), and subtends an angle of 80° at the drum centre Q, but otherwise the dimensions are the same as in Question 2, find the applied force P. What is the value of P if the shoe material is assumed to obey Hooke's law?

4. Fig. 188 shows in outline a brake mechanism. The load W acting at A through the simple lever operates the upper bell-crank and applies the brake; an upward movement of A slackens the brake. The lengths in inches of certain of the links are given. The brake pulley is 16 ft in diameter and experiences a maximum torque of 408 000 lb ft when the brake is applied. Taking a value of 0·3 for μ, determine (a) a suitable value for the weight W, and (b) each of the side forces on the brake drum. Neglecting the effect of the tangential forces on the brake, calculate the force in the link BC. W.S.

5. The brake for a winding engine is arranged as shown in Fig. 189. The distance between the fixed pivots O and E of the brake hangers is equal to the diameter of the brake drum. The shoes are rigidly attached to the brake hangers and each subtends an angle of 70° at the drum centre. For the given dimensions and assuming $\mu = 0{\cdot}3$, find the load W required in order to provide a braking torque of 25 ton ft.

6. If the shoes of the brake in Question 5 are pivoted to the hangers at the points marked C and F in Fig. 189, find the value of W for the same braking torque. Also find the direction and magnitude of the reaction at each of the pin-joints O, C, D, H, G, F and E.

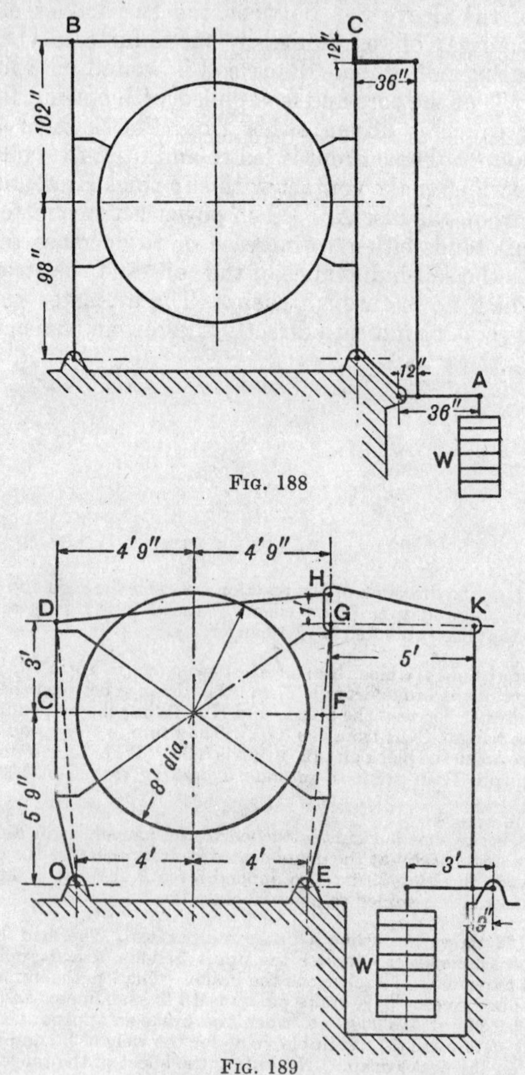

Fig. 188

Fig. 189

7. The arrangement of the transmission brake on a certain car is shown diagrammatically in Fig. 190. The shoes are pivoted at O and C and are brought into contact with the drum by rotating the shaft AB. The ends of this shaft are screwed right and left-handed and work in nuts in the ends of the shoes. The hand lever which rotates the shaft is 15 in. long from the point of application of the effort to the axis of the shaft. The mean diameter of the screwed ends of the shaft is $\frac{3}{4}$ in. and there are six threads with a lead of $2\frac{1}{4}$ in. If μ for the brake blocks is 0·30 and for the screws and nuts is 0·15, find the braking torque applied to the drum when the effort applied to the hand lever is 25 lb.

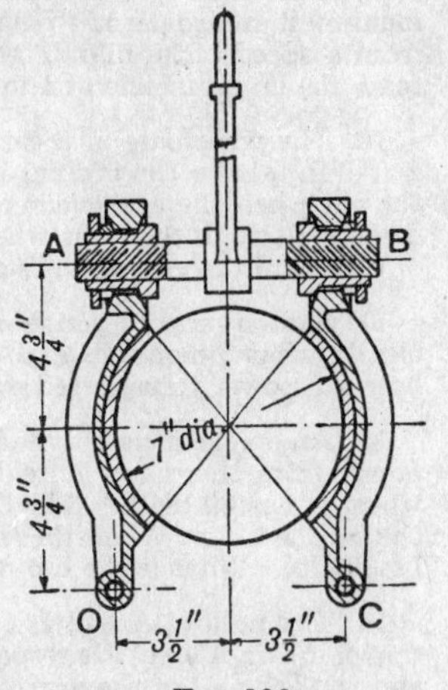

FIG. 190

8. The dimensions of an internal expanding brake, similar to that shown in Fig. 173, are: diameter of drum, 11 in.; the angle subtended by each shoe at the drum centre, 90°; the distance between the fulcrum centres, 3 in.; the distances of the fulcrum centres and of the cam axis from the drum centre $4\frac{1}{2}$ in.; the difference between l and l_1 1·25 in. and the distance of the line of action of P from the axis of the cam 4 in. Neglecting the pull of the releasing spring S, find the braking torque on the drum when P is 150 lb. $\mu = 0·3$.

9. A simple band brake is operated by a lever 20 in. long. The brake drum is 20 in. dia. and the brake band embraces five-eighths of its circumference. One end of the band is attached to the fulcrum of the lever, while the other is attached to a pin on the lever 4 in. from the fulcrum. If the coefficient of friction is 0·25 and the effort applied to the end of the brake lever is 100 lb, what is the maximum braking torque on the drum?

10. In a crab or winch the rope supports a load W and is wound round a barrel 18 in. dia. A differential band brake acts on a drum 30 in. dia. which is keyed to the same shaft as the barrel. The two ends of the bands are attached to pins on opposite sides of the fulcrum of the brake lever and at distances of 1 in. and 4 in. from the fulcrum. The angle of lap of the brake band is 240° and μ is 0·25. What is the maximum load W which can be supported by the brake when a force of 100 lb is applied to the lever at a distance of 36 in. from the fulcrum?

11. The drum of a band-and-block brake is 3 ft dia. and there are 14 blocks, each of which subtends an angle of 15° at the drum centre. One end of the band is attached to the fulcrum of the brake lever and the other to a pin 5 in. from the fulcrum. If the torque applied to the drum is to be 2000 lb ft, what effort must be applied to the brake lever at a point 30 in. from the fulcrum? Assume $\mu = 0·25$.

12. If in Question 11 the end of the band instead of being attached to the fulcrum is attached to a pin on the brake lever 1·5 in. from the fulcrum, what effort would be required in order to give the same torque on the drum?

13. A lorry has a 10-ft 6-in. wheelbase and the c.g. is 4 ft 3 in. in front of the rear axle and 3 ft above ground level. The coefficient of adhesion between tyre and road surface is 0·6 and brakes are applied to the rear wheels only. What is the minimum distance in which the lorry can be pulled up when travelling at 20 m.p.h.? What proportion of the total weight is carried by the front and the rear wheels during retardation?

14. An electric car travelling along a level track at 18 m.p.h. has the power cut off and brakes applied to bring it to rest. Its centre of gravity is midway between the wheels and 3 ft above the rail level. The wheelbase is 10 ft and the coefficient of friction between wheel and rail is 0·15. Find the minimum distance

travelled by the car before coming to rest if the brakes are applied (a) to the rear wheels only, (b) to all four wheels. M.U.

15. The wheelbase of a motor-cycle is 4 ft 3 in. and the centre of gravity is 2 ft above ground level and 2 ft 3 in. from the front axle. If the coefficient of friction between tyre and road is 0·6 (and the rear wheel only is braked) find the maximum retardation. What is the distance travelled before coming to rest from a speed of 30 m.p.h., when the motor-cycle is travelling (a) along a level road, (b) up an incline of 1 in 15, (c) down an incline of 1 in 15?

16. The wheelbase of a car is 9 ft and the c.g. is 4 ft from the rear axle and 2 ft 9 in. above the ground level. Find the maximum possible acceleration of the car when the coefficient of friction between tyre and road is 0·5 and (a) the drive is through the front wheels, (b) the drive is through the rear wheels.

What are the corresponding values of the acceleration when μ is only 0·2?

17. Sketch and describe one form of transmission dynamometer. State clearly what dimensions and measurements would have to be taken and explain how the power transmitted may be calculated.

18. An epicyclic gear dynamometer of the type shown in Fig. 181 (a) transmits power from the wheel A to the wheel D. The wheel A has 30 teeth and the wheel D has 80 teeth. The diametral pitch of the teeth is 5 and the length l of the arm is 3 ft. When the wheel A makes 500 r.p.m., it is found that W must be 160 lb. What is the h.p. transmitted?

19. The pulleys A and D of a belt transmission dynamometer similar to that shown in Fig. 182 (a) are respectively 40 in. and 15 in. dia. The belt is $\frac{3}{8}$ in. thick and the length l of the arm is 42 in. If the h.p. transmitted from D to A is 20 when the pulley D makes 600 r.p.m., what is the value of the load W?

20. The two pulleys A and D of a dynamometer of the type shown in Fig. 182 (b) have each a diameter of 3 ft and the distance between the shaft centres is 8 ft. The jockey pulleys have each a diameter of 1 ft. The distance between their centres is 18 in., and the thickness of the belt is $\frac{3}{8}$ in. The arms a and l of the lever are respectively 12 in. and 72 in. long. The load W required in order to balance the lever, when the belt transmits power from A to D and the speed of A is 300 r.p.m., is 77·5 lb. Calculate the h.p. transmitted.

21. Describe with sketches one form of torsion dynamometer and explain in detail the calculations involved in finding the h.p. transmitted. How would you proceed to calibrate the dynamometer?

CHAPTER IX

CAMS

117. Types of Cams. A *cam* is a reciprocating, oscillating, or rotating body which imparts reciprocating or oscillating motion to a second body, called the *follower*, with which it is in contact. The shape of the cam depends upon its own motion, the motion which is to be imparted to the follower, and the shape of the contact surface of the follower. There are many different types of cams, some of which are shown diagrammatically in Fig. 191. It will be seen that the follower usually has line contact with the cam, so that

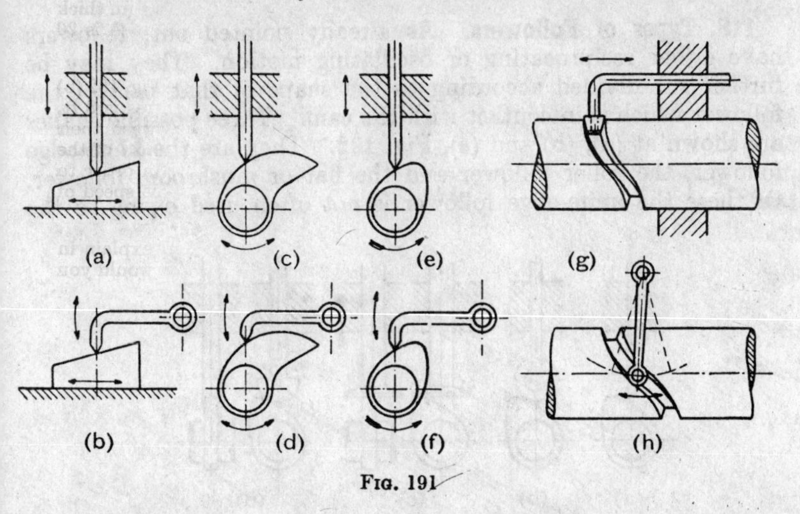

Fig. 191

the two bodies cam and follower constitute a *higher* pair. With few exceptions the motion of the follower is only determined positively by the cam during a part of each stroke, while during the rest of the stroke contact between the cam and the follower has to be maintained by an external force. The external force is frequently provided by a spring, although sometimes the weight of the follower itself is sufficient. In this connection it should be noticed that the cam does not, as would at first sight appear, determine positively the motion of the follower during the whole of its outstroke. Actually, owing to the inertia of the follower, it is during the first part of the outstroke and the latter part of the return or instroke that the motion of the follower is positively controlled by the cam.

Cams may be classified according to the direction of the displacement of the follower with respect to the axis of oscillation or of rotation of the cam. The two most important types are:

(a) Radial or disc cams, in which the working surface of the cam is so shaped that the follower reciprocates or oscillates in a plane at right angles to the axis of the cam. Examples (c) to (f) in Fig. 191 are radial cams.

(b) Cylindrical cams, in which the follower reciprocates or oscillates in a plane parallel to the axis of the cam, e.g. (g) and (h) in Fig. 191.

Since by far the greater number of cams used in practice belong to class (a), the following discussion will be limited to cams of this type.

118. Types of Followers. As already pointed out, followers have either reciprocating or oscillating motion. They may be further sub-divided according to the shape of that part of the follower which is in contact with the cam. Three possible shapes are shown at (a), (b) and (c), Fig. 192. They are the knife-edge follower, the roller follower and the flat or mushroom follower. Of these the knife-edge follower is not often used owing to the

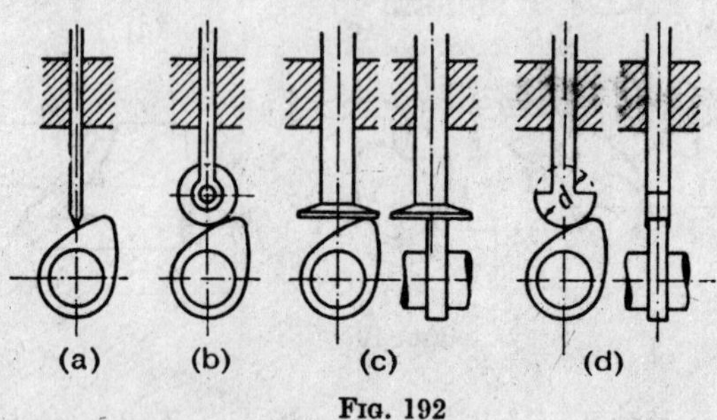

FIG. 192

rapid rate at which the knife-edge wears. The roller follower possesses the advantage that to a large extent a rolling motion between the contact surfaces is substituted for the sliding motion between the knife-edge and the cam. Note that sliding is not entirely eliminated, since the inertia of the roller prevents it from responding instantaneously to the changes of angular velocity required by the varying peripheral speed of the cam. Both the knife-edge and the roller followers give rise to considerable side thrust between the follower and the guide. The flat or mushroom follower has the advantage that the only side thrust on the guide is that due to friction between the contact surfaces of the cam and the follower. The relative motion between these surfaces is largely one of sliding, but wear may be diminished by offsetting the axis of the follower as shown in the figure, so that, as the cam rotates, the follower is also caused to rotate about its own axis. Where space is limited, as for instance in the cams which operate the valves of automobile engines, the flat-faced follower is generally used in preference to the roller follower because of the small diameter of the pin that would have to be used for the latter. In stationary gas and oil engines, however, where more space is available, the roller follower is preferred. Occasionally for automobile engines the end of the follower, instead of being flat, is machined to a curved surface as shown in Fig. 192 (d). It is then equivalent to a roller of diameter d, so far as the relative displacements of the cam and the follower are concerned. Theoretically, there is no limit imposed on the shape of the cam working surface when a knife-edged follower is used, but with a roller follower any concave portions of the working surface must have a radius at least equal to the radius of the roller, while with a flat follower it is clearly necessary for the working surface of the cam to be everywhere convex.

119. Displacement, Velocity and Acceleration-time Curves. The cam usually rotates at uniform speed, so that equal angular displacements take place in equal intervals of time. On the other

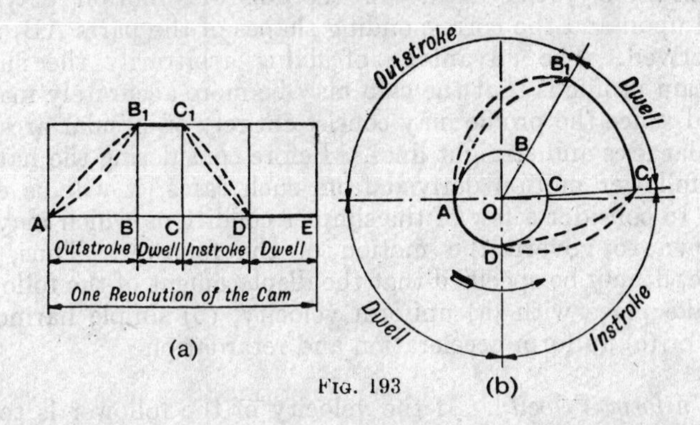

FIG. 193

hand, the follower is required to start from rest at the beginning of each stroke and to come to rest again at the end of each stroke, so that its velocity must be variable. It is also frequently necessary for the follower to remain at rest during part of the rotation of the cam. The periods of cam rotation during which the follower remains at rest are known as periods of dwell. A diagram may be drawn, as in Fig. 193 (a), to represent the relationship between the displacement of the follower and the angular displacement of the cam. In this diagram AE represents one revolution of the cam; it is divided at B, C and D, so that AB, BC, CD and DE represent in order the angular displacements of the cam during the outstroke of the follower, during the period of dwell at the end of the outstroke, during the instroke and during the period of dwell at the end of the instroke.

Suppose that a reciprocating knife-edge follower of the type shown in Fig. 192 (a) is used and that the line of stroke passes through the axis of the cam. Then the radial lines OA, OB, OC and OD may be set off as shown in Fig. 193 (b), where angle AOB is the angular displacement of the cam which corresponds to AB on the displacement diagram and angles BOC, COD and DOA are respectively the angular displacements which correspond to BC, CD and DE on the displacement diagram. The base circle radius OA of the cam is fixed from practical considerations, the difference between OB and OB_1 must clearly be equal to the follower stroke and the parts B_1C_1 and DA of the cam profile must be concentric arcs struck from the centre O. Each of the parts AB_1 and C_1D of the dam profile may have any one of an infinite number of different shapes, two of which are shown dotted in Fig. 193. For instance, the shapes may be arbitrarily fixed and the nature of the corresponding follower motion derived from them, or, alternatively, the nature of the follower motion may be decided upon and the corresponding shapes of the parts AB_1 and C_1D derived. The advantage of fixing arbitrarily the shape of the cam profile is that the cam may be more accurately manufactured, since the profile may consist entirely of circular arcs, or of circular arcs and straight lines. Before considering the nature of the follower motion derived from such cams, it will be convenient to consider a few of the simpler conditions which may be laid down to govern the motion of the follower. Thus, for instance, it may be specified that the displacement of the follower is to take place with (a) uniform velocity, (b) simple harmonic motion or (c) uniform acceleration and retardation.

(a) *Uniform Velocity.* If the velocity of the follower is to be uniform during the outstroke, the slope of the displacement curve must be constant, i.e. AB_1, Fig. 194 (a), must be a straight line. Similarly, if the velocity is to be uniform during the return stroke, the curve C_1D on the displacement diagram must be a straight line. The velocity diagram will then be as shown. These conditions are, however, impracticable, since the acceleration and retardation of the follower at the beginning and at the end of each stroke would require to be infinitely high. It is therefore necessary to modify the conditions which govern the follower motion, so that the acceleration and retardation are reduced to finite proportions. This may be done by rounding off the sharp corners at A, B_1, C_1 and D on the displacement diagram, so that the velocity of the follower increases gradually to its maximum value at the beginning of each stroke and decreases gradually to zero

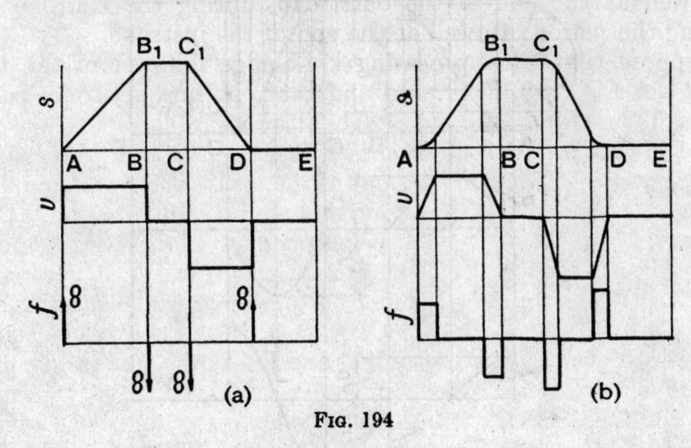

FIG. 194

at the end of each stroke. The modified displacement, velocity and acceleration diagrams are shown in Fig. 194 (b). In drawing these diagrams it has been assumed that the follower is accelerated or retarded uniformly. The follower motion takes place with uniform velocity, except for short periods at the beginning and at the end of each stroke. During these periods the acceleration of the follower is high, and will clearly be higher the shorter the period of cam rotation allowed for the increase or decrease of speed. The rounded corners of the displacement diagram are parabolic arcs.

(b) *Simple Harmonic Motion.* The displacement of the follower is identical with that which would be given by a uniformly rotating crank to which the follower is connected by an infinitely long connecting rod. The angles through which the cam rotates during the two strokes of the follower and during the periods of dwell are assumed to be the same as in Fig. 194. The displacement diagram, Fig. 195 (a), may be constructed by first drawing a semicircle on the follower stroke as diameter. This semicircle is divided into any convenient number of equal parts—eight in the present example. The angles through which the cam rotates during the outward and inward strokes of the follower are divided into the same number of equal parts. Points on the displacement diagram are then obtained by projecting across as indicated in the figure. The complete displacement curve is given by AB_1C_1DE. The velocity and acceleration diagrams corresponding to the displacement diagram of Fig. 195 (a) are shown at (b) and

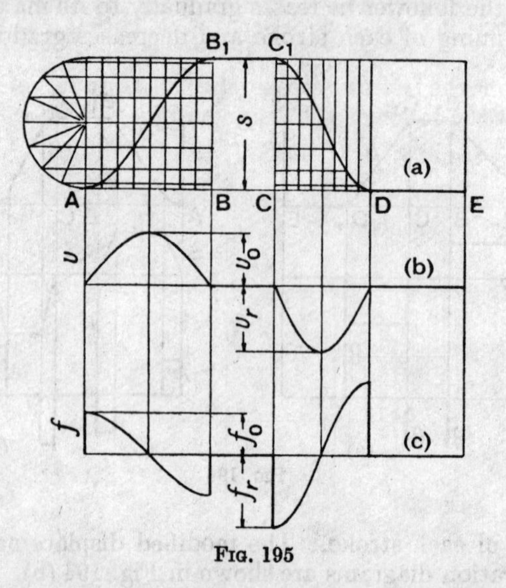

FIG. 195

(c) on the same figure. Since the follower motion is s mple-harmonic, the velocity curve is a sine curve and the acceleration curve is a cosine curve. The velocity of the follower is zero at the beginning and at the end of its stroke and increases gradually to a maximum at mid-stroke. On the other hand, the acceleration of the follower is a maximum at the ends of the stroke and diminishes to zero at mid-stroke.

Let S = stroke of the follower,

θ_o = the angle through which the cam rotates during the outstroke of the follower

and ω = the angular velocity of the cam.

Then the time required for the outstroke of the follower

$$= t_o = \theta_o/\omega$$

The point which defines the S.H.M. therefore moves at uniform speed round the circumference of a circle of diameter S in time $2t_0$ sec.

$$\therefore \text{ peripheral speed} = \pi S/2t_o = \pi\omega/\theta_o.S/2 \qquad (9.1)$$

$\therefore$ maximum velocity of the follower on the outward stroke

$$= v_o = \pi\omega/\theta_o.S/2 \quad . \quad . \quad . \quad . \quad (9.2)$$

Also the centripetal acceleration of the point which defines the S.H.M. $= v_o{}^2.2/S = \pi^2\omega^2/\theta_o{}^2.S/2$

$\therefore$ maximum acceleration of the follower on the outstroke

$$= f_o = \pi^2\omega^2/\theta_o{}^2.S/2 \quad . \quad . \quad . \quad . \quad (9.3)$$

Similarly, on the return stroke the maximum velocity and acceleration of the follower will be given by:

$$v_r = \pi\omega/\theta_r.S/2 \quad \text{and} \quad f_r = \pi^2\omega^2/\theta_r{}^2.S/2$$

where θ_r is the angle through which the cam turns during the return stroke of the follower.

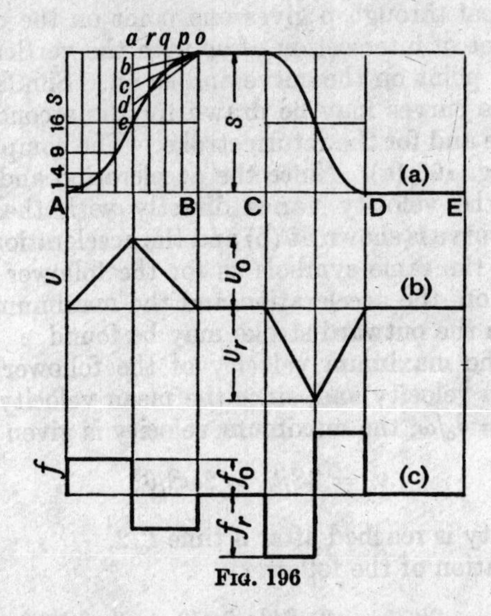

FIG. 196

(c) *Uniform Acceleration.* Since the displacement of the follower has to take place in a definite time, it is clear that the acceleration of the follower will have a minimum value when the first half of the displacement takes place with uniform acceleration and the second half of the displacement takes place with an equal uniform retardation. In these circumstances the maximum inertia force of the follower will have its lowest possible value. It is for this reason that the uniform acceleration curve is frequently used. Such a displacement curve is shown in Fig. 196 (a). It may be constructed as follows. The angle through which the cam rotates during the outward stroke of the follower is divided into an even number of equal parts—eight in the present example. These correspond to equal time intervals; during the first four intervals the follower is accelerated and during the second four it is retarded. For uniformly accelerated motion the displacement varies directly as the square of the time, so that the total displacements of the follower at the end of the first four intervals of time are in the proportion $1^2, 2^2, 3^2, 4^2$ or 1, 4, 9, 16. If, therefore, the half-stroke of the follower is divided into 16 equal parts, the points 1, 4, 9 and 16 may be projected across to the corresponding time ordinates. The points thus obtained may be joined by a smooth curve in order to give the displacement-time curve for the follower during the first half of the outward stroke. An alternative construction is given for the second half of the displacement curve on the outstroke. The distances oa and ae are each divided into four equal parts. The point of intersection of the line ob with the vertical through p gives one point on the displacement curve; the point of intersection of oc with the vertical through q gives a second point on the curve and so on. Similarly, the displacement-time curves may be drawn for the second half of the outward stroke and for the return stroke. The complete diagram is shown in Fig. 196 (a). Since the acceleration and retardation are uniform, the velocity varies directly with the time. The velocity-time curve is shown at (b) and the acceleration-time curve at (c). Using the same symbols as for the follower with simple harmonic motion, the acceleration and the maximum velocity of the follower on the outward stroke may be found.

Obviously the maximum velocity of the follower is equal to twice the mean velocity and, since the mean velocity is given by S/t_o, where $t_o = \theta_o/\omega$, the maximum velocity is given by:

$$v_o = 2S/t_o = 2\omega S/\theta_o \quad . \quad . \quad . \quad . \quad (9.4)$$

But this velocity is reached after a time $t_o/2$,
$\therefore$ the acceleration of the follower

$$f_o = 2v_o/t_o = 2\omega S/\theta_o . 2\omega/\theta_o = 4\omega^2 S/\theta_o^2 \quad . \quad (9.5)$$

Similar expressions will apply to the maximum velocity and the acceleration of the follower on the return stroke; the only change necessary is the substitution of θ_r for θ_o.

One modification of the uniform acceleration displacement curve may be noted. It is widely used in the design of cams for operating the valves of internal-combustion engines. Two desirable conditions for such cams are: (a) that they should open and close the valves as quickly as possible so as to provide a free flow of the gases to and from the cylinder, and (b) that they should require as small an external force as possible to maintain contact between the cam and the follower during the later part of the outward stroke and the early part of the return stroke. It is clear that, in order to satisfy the first condition, the acceleration of the follower on the outward stroke should be as high as possible; while, in order to satisfy the second condition, the retardation of the follower on the outward stroke should be as low as possible. On the return stroke the conditions are reversed—the acceleration should be low and the retardation high. But the acceleration of the follower on the outward stroke and the retardation on the return stroke are controlled positively by the cam and an increase

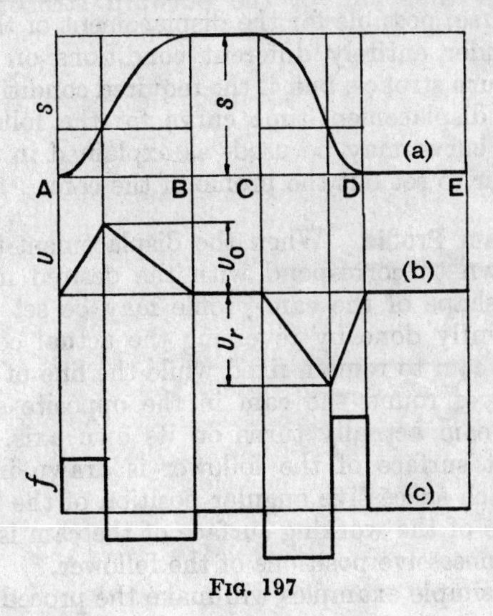

FIG. 197

of magnitude simply involves an increase in the pressure between the contact surfaces of the cam and the follower. On the other hand, the retardation of the follower on the outward stroke and the acceleration on the return stroke are only controlled by the cam, if the external force exerted on the follower by the spring, or by the deadweight, is greater than the inertia force of the follower. The cam may, therefore, with advantage be designed so as to provide an acceleration on the outward stroke that is considerably greater than the retardation on the outward stroke.

The corresponding displacement-time curve may be drawn in a similar way to that already described for the follower in which the acceleration and retardation are of equal magnitude, the only difference being that the periods of acceleration and of retardation are no longer equal. If, for instance, during the outward stroke the acceleration is k times the retardation, it is easily seen that the maximum velocity of the follower is twice the mean velocity whatever may be the value of k. But the duration of the acceleration period is clearly only $1/k$ times the duration of the retardation period and similarly the displacement of the follower during the acceleration period is only $1/k$ times the displacement of the follower during the retardation period. In Fig. 197 the displacement, velocity and acceleration-time curves have been drawn for a follower in which k is equal to 2.

It should be noted that Figs. 195, 196 and 197 are all drawn to the same scale and may therefore be directly compared.

It is, of course, possible for the displacement of the follower to take place under entirely different conditions on the outward and on the return strokes, but, if the required conditions are given, the complete displacement-time curve for the follower may be drawn. This curve may be used, as explained in the following article, in order to set out the profile of the cam.

120. The Cam Profile. When the displacement-time diagram has been drawn to correspond with the desired motion of the follower, the shape of the cam profile may be set out. This is most conveniently done by reversing the actual conditions and imagining the cam to remain fixed while the line of stroke of the follower revolves round the cam in the opposite sense to that in which the cam actually turns on its own axis. The profile of the contact surface of the follower is drawn in the correct position for each successive angular position of the line of stroke and the profile of the working surface of the cam is drawn so as to touch the successive positions of the follower.

One or two simple examples will make the procedure clear.

Example 1. *Cam with Knife-edge Reciprocating Follower.* A cam rotating clockwise at a uniform speed is required to give a knife-edge follower the motion defined below:

(i) Follower to move outwards through a distance of 1 in. during 120° of cam rotation.

(ii) Follower to dwell for 60° of cam rotation.

(iii) Follower to return to its initial position during 90° of cam rotation.

(iv) Follower to dwell for the remaining 90° of cam rotation.

The minimum radius of the cam is 2 in., the line of stroke of the follower is offset $\frac{3}{4}$ in. from the axis of the cam and the displacement of the follower is to take place with uniform and equal acceleration and retardation on both the outward and the return strokes.

The first step is to draw the displacement-time curve as shown in Fig. 198 (a). In this diagram it is convenient to represent the

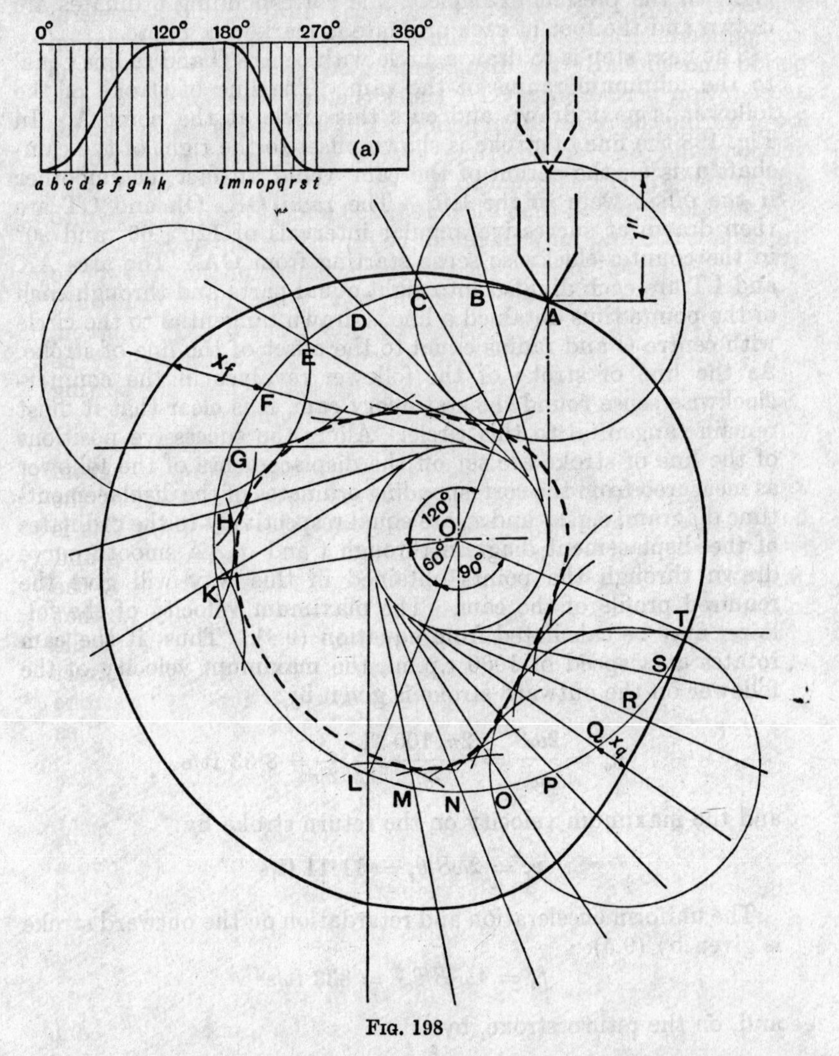

Fig. 198

displacement of the follower to the same scale as that to which the profile of the cam is to be drawn. The time scale or the scale for the angular displacement of the cam may be chosen quite arbitrarily. Since the acceleration and retardation are uniform, the displacement curves for the outward and return strokes consist of parabolic arcs. These may be drawn by the usual geometrical construction. The angles through which the cam rotates during each stroke are divided into an even number of equal parts—eight in the present example. The corresponding ordinates are drawn and the foot of each ordinate is marked a, b, etc.

The next step is to draw a circle with centre O and radius equal to the minimum radius of the cam. The line of stroke of the follower is next drawn and cuts this circle at the point A. In Fig. 198 the line of stroke is shown offset to the right of the cam-shaft axis O; the action of the cam would, in fact, be smoother if the offset were to the left. The radii OK, OL and OT are then drawn at successive angular intervals of 120°, 60° and 90° in the counter-clockwise sense starting from OA. The arcs AK and LT are each divided into eight equal parts and through each of the points thus obtained a line is drawn tangential to the circle with centre O and radius equal to the offset of the line of stroke. As the line of stroke of the follower revolves in the counter-clockwise sense round the stationary cam, it is clear that it must remain tangential to this circle. Along the successive positions of the line of stroke are set off the displacements of the follower as measured from the corresponding ordinates of the displacement-time diagram, e.g. x_f and x_q are equal respectively to the ordinates of the displacement diagram through f and q. A smooth curve drawn through the points obtained in this way will give the required profile of the cam. The maximum velocity of the follower may be calculated from equation (9.4). Thus, if the cam rotates at a speed of 1000 r.p.m., the maximum velocity of the follower on the outward stroke is given by:

$$v_o = \frac{2\omega S}{\theta_o} = \frac{2\pi . 100}{3} . \frac{1}{12} . \frac{3}{2\pi} = 8{\cdot}33 \text{ ft/s}$$

and the maximum velocity on the return stroke by:

$$v_r = 2\omega S/\theta_r = 11{\cdot}11 \text{ ft/s}$$

The uniform acceleration and retardation on the outward stroke is given by (9.5):

$$f_o = 4\omega^2 S/\theta_o^2 = 833 \text{ ft/s}^2$$

and, on the return stroke, by:

$$f_r = 4\omega^2 S/\theta_r^2 = 1482 \text{ ft/s}^2$$

Example 2. Cam with Roller Reciprocating Follower. When the reciprocating follower is fitted with a roller in order to reduce friction and wear, the displacement of the follower is determined by the displacement of the roller centre. The path of the roller centre relative to the cam may be determined in exactly the same way as for the knife-edge follower, except that OA must be the minimum distance of the roller centre from the axis of the cam and not the minimum radius of the cam. In other words, OA must be equal to the sum of the radius of the roller and the minimum radius of the cam. The successive positions of the roller centre having been determined, the corresponding roller profiles are drawn in and the required shape of the cam profile is obtained by drawing a smooth curve to touch the successive positions of the roller profile. If, for instance, the follower in the last example is fitted with a roller of $2\frac{1}{4}$ in. dia. and the minimum radius of the cam is $\frac{7}{8}$ in., the minimum distance of the roller centre from the axis of the cam is 2 in. and the path of the roller centre relative to the cam is therefore identical with that of the knife-edge follower. The shape of the cam profile is shown by the dotted line in Fig. 198.

Example 3. *Cam with Roller Oscillating Follower*. It is required to set out the profile of a cam to give the following motion to an oscillating follower:

(i) Follower to move outwards through an angular displacement of 15° during 90° of cam rotation.
(ii) Follower to dwell for 45° of cam rotation.
(iii) Follower to return to its initial position during 75° of cam rotation.
(iv) Follower to dwell during the remaining 150° of cam rotation.

The pivot of the oscillating follower is $4\frac{1}{2}$ in. from the axis of rotation of the cam, the distance between the pivot centre and the roller centre is 4 in., the roller is $1\frac{1}{2}$ in. dia. and the minimum radius of the cam is 2 in. The outward stroke of the follower is to take place with simple harmonic motion and the return stroke with uniform acceleration and retardation, the retardation being double the acceleration.

With an oscillating follower the displacement of the roller centre takes place along a circular arc and the length of the circular arc should strictly be used in setting out the displacement-time curve. However, where the angular displacement of the follower is small, the length of the chord may be used instead of the length of the arc without serious error. In the present example the length of the arc is 1·047 in. and the length of the chord is $2.4 \sin 7\frac{1}{2}°$ $= 1{\cdot}044$ in.

The complete displacement-time curve is shown in Fig. 199 (a). The angle through which the cam turns on the outward stroke is divided into eight equal parts and the displacement curve for simple harmonic motion is drawn in the usual way. During the return stroke, since the acceleration is to be one-half the retardation, two-thirds of the stroke must take place in two-thirds of the total time available with uniform acceleration and one-third of the stroke in one-third of the time with uniform retardation. In

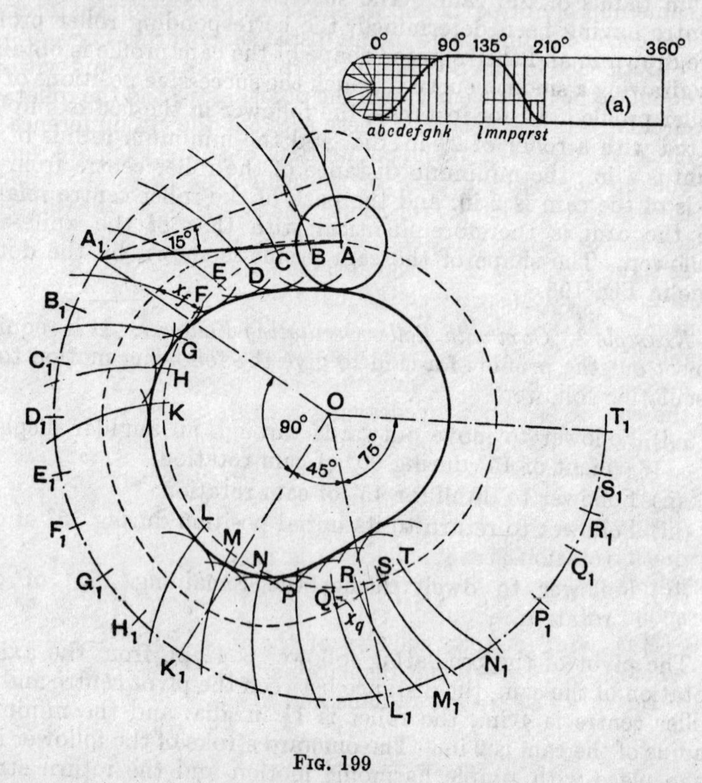

FIG. 199

the figure the angle of cam rotation during which acceleration of the follower takes place on the return stroke is divided into four equal parts, while the angle of cam rotation during which retardation takes place is divided into three equal parts. The parabolic arcs are drawn in the usual way.

In order to set out the profile of the cam, a circle is first drawn with centre O and radius equal to the minimum radius of the cam. The initial positions A and A_1 of the roller centre and the pivot centre are then marked off and circles are drawn with O as centre and OA and OA_1 as radii. Assuming that the cam is to rotate clockwise, the line which joins the follower pivot and the cam centre must be turned counter-clockwise about the fixed cam when setting out the profile. From OA_1 set off in succession angles of 90°, 45° and 75° to give the positions K_1,L_1 and T_1 of the follower pivot at the end of the outward stroke, at the beginning of the return stroke and at the end of the return stroke respectively. The arcs A_1K_1 and L_1T_1 are then divided into the same number of parts and at the same proportionate intervals as the corresponding parts of the displacement-time curve.

With each of the points thus obtained as centre and with radius equal to the distance between the pivot centre and the roller centre, circular arcs are drawn. These arcs cut the circumference of the circle of radius OA at the points B, C, etc., and each arc represents one position of the path of the roller centre as the follower is turned counter-clockwise round the fixed cam. Along each arc the corresponding displacement of the roller centre may be set off as measured from the displacement diagram. For example, the arcs x_f, x_q are respectively equal to the ordinates of the displacement diagram through f and q. The path of the roller centre relative to the cam is given by the smooth curve which joins the successive positions of the roller centre. The required profile of the cam is obtained by drawing a curve to touch the successive positions of the roller circumference as shown in Fig. 199.

If the cam rotates at a speed of 1000 r.p.m., the velocity and acceleration of the follower may be calculated as shown below.

On the outward stroke, when the follower has simple harmonic motion, the maximum velocity is given by an equation similar in form to equation (9.2), but with the symbols changed to correspond to angular motion.

Maximum angular velocity of the follower $= \omega_o = \pi\omega/\theta_0 . \beta/2$, where β = total angular displacement of the follower.

But $\theta_o = 90° = \pi/2$ radians and $\beta = 15° = \pi/12$ radians.

$$\therefore \ \omega_o = \frac{2\pi}{\pi} . \frac{\pi . 100}{3} . \frac{\pi}{24} = 27{\cdot}4 \text{ rad/s}$$

The maximum acceleration at the beginning of the outward stroke and the maximum retardation at the end of the outward stroke are given by an equation similar to (9.3):

Maximum angular acceleration of the follower

$$= \alpha_o = \frac{\pi^2\omega^2}{\theta_o{}^2} . \frac{\beta}{2} = 4\pi^2 . \frac{100^2}{9} . \frac{\pi}{24} = 5730 \text{ rad/s}^2$$

On the return stroke, when the follower moves with uniform

acceleration and retardation, the maximum velocity is twice the mean velocity.

Mean angular velocity of the follower $= \beta/t_r$.

But
$$t_r = \frac{\theta_r}{\omega} = \frac{75}{180}.\pi.\frac{3}{100\pi} = \frac{1}{80} \text{ sec}$$

$\therefore$ maximum angular velocity of the follower

$$= \omega_r = \frac{2\beta}{t_r} = \frac{2}{12}.\pi.80 = 41{\cdot}9 \text{ rad/s}$$

This velocity is reached in two-thirds of the time required for the cam to turn through the angle θ_r, i.e. in $\frac{2}{3}.\frac{1}{80}$ sec.

$\therefore$ acceleration of the follower $= 41{\cdot}9.3.80/2$

$$= 5030 \text{ rad/s}^2$$

During retardation this velocity is reduced to zero in one-third of the time required for the cam to turn through the angle θ_r, i.e. in $\frac{1}{3}.\frac{1}{80}$sec.

Retardation of the follower $= 41{\cdot}9.3.80$

$$= 10\,060 \text{ rad/s}^2$$

Example 4. Cam with Reciprocating Flat-faced Follower. It is required to set out the profile of a cam to give the following motion to a reciprocating follower with a flat, or mushroom, contact face.

(i) Follower to move outwards through a distance of 1 in. during 120° of cam rotation.
(ii) Follower to dwell for 30° of cam rotation.
(iii) Follower to return to its initial position during 80° of cam rotation.
(iv) Follower to dwell for the remaining 130° of cam rotation.

The minimum radius of the cam is 2 in. and the flat face of the follower is at right-angles to the line of stroke of the follower. Both the outward and the return strokes are to take place with simple harmonic motion.

The displacement-time diagram corresponding to the specified conditions is shown in Fig. 200 (a) and the times available for the outward and the return strokes are each divided into eight equal parts.

The same system of lettering is used as in the previous examples and the reader should have no difficulty in following the construction used in setting out the cam profile. As in the earlier examples, the profile of the contact face of the follower is set out in a number of positions, which conform to the desired follower motion, and then the cam profile is drawn so as to touch the successive positions of the contact face of the follower. Thus, when the line of stroke of the follower has turned through the angle AOE relative to the cam, the displacement x_e of the follower is given by the ordinate of the displacement diagram at the point e. The flat face of the follower therefore occupies the position shown dotted relative to the cam and the profile of the cam must be so

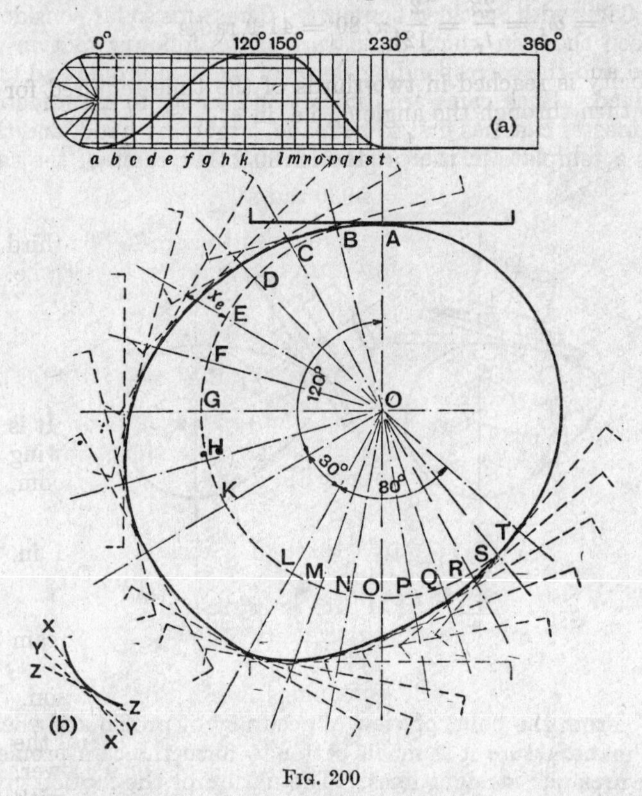

FIG. 200

shaped as to touch the dotted profile of the follower. It should be noted that, except at the beginning and at the end of each stroke, the point of contact of the cam profile and the follower profile does not lie on the line of stroke of the follower.

It may also be pointed out that with this type of follower it will sometimes be impossible to draw the cam profile so as to touch each of the desired positions of the follower. For instance, if three positions of the face of the follower which correspond to the desired motion of the follower are as shown by XX, YY and ZZ in Fig. 200 (b), it is clearly impossible to draw a curve which will touch YY as well as XX and ZZ, so that the desired follower motion could not be obtained. For this reason it is not usual to employ a flat-faced follower when the nature of the follower motion is arbitrarily specified. Flat-faced followers are, however, widely used when the cam profile is formed of circular arcs, and this type of cam and follower will be discussed in the next article.

121. Cams with Specified Contours. The cams so far considered have been those in which the nature of the follower motion was specified and the corresponding shape of the cam profile had to be determined. Such cams are difficult and costly to manufacture, since a master cam has first to be made, largely by hand, and then used as a template in the production of other cams of the same

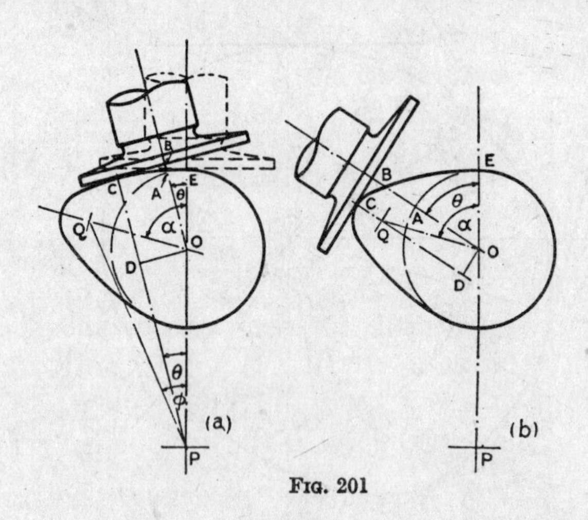

Fig. 201

shape. From the point of view of accuracy of profile and cheapness of manufacture it is much better to form the cam profile of circular arcs and straight lines. The nature of the motion given to the follower may then be determined. The valves of small internal-combustion engines are frequently operated by cams whose profiles consist entirely of circular arcs, and the followers either have flat faces or are fitted with rollers.

(a) *Circular Arc Cam with a Flat-faced Reciprocating Follower.* In Fig. 201 (a) the contour of a circular arc cam is shown. OE is the minimum radius of the cam and the flank consists of a circular arc with centre P. Let the cam remain fixed and the line of stroke of the follower turn in the opposite sense to that of the actual rotation of the cam, i.e. counter-clockwise in the figure. Then, when the line of stroke has turned through the angle θ relative to the cam, the flat face of the follower will be in contact with the cam profile at the point C, where PC is perpendicular to the face of the follower. The displacement of the follower is given by AB, where OB is perpendicular to BC and therefore parallel to PC. Draw OD perpendicular to PC.

Then, the displacement of the follower

$$= x = AB = BO - AO = CD - EO$$

But $\quad CD = PC - PD = PE - PO \cos\theta = EO + OP(1 - \cos\theta)$

$$\therefore\ x = OP(1 - \cos\theta) \quad . \quad . \quad . \quad . \quad (9.6)$$

Velocity of the follower,

$$v = dx/dt = dx/d\theta . d\theta/dt = \omega . dx/d\theta$$
$$= \omega . OP \sin\theta \quad . \quad . \quad . \quad . \quad . \quad . \quad . \quad (9.7)$$

Acceleration of the follower

$$= f = dv/dt = \omega . dv/d\theta = \omega^2 . OP \cos\theta \quad . \quad (9.8)$$

The three equations (9.6), 9.7) and (9.8) apply only while the follower is in contact with that part of the cam profile which has the centre of curvature P, i.e. for values of θ from 0 to ϕ, where ϕ = angle OPQ. When θ is greater than ϕ, the follower is in contact with the nose of the cam, the centre of curvature of which is Q.

Referring to Fig. 201 (b), the cam and follower are in contact at C on the nose of the cam. The displacement of the follower $= x = AB = OB - OA = CD - OA$.

But $\quad CD = CQ + QD = CQ + OQ \cos(\alpha - \theta)$

$$\therefore\ x = CQ - OA + OQ \cos(\alpha - \theta) \quad . \quad . \quad (9.9)$$

In this equation CQ, OA, OQ and α are constant for a given cam and, differentiating with respect to time, we have:

Velocity of the follower,

$$v = \omega . OQ \sin(\alpha - \theta) \quad . \quad . \quad . \quad . \quad (9.10)$$

Acceleration of the follower,

$$f = -\omega^2 . OQ \cos(\alpha - \theta) \quad . \quad . \quad . \quad (9.11)$$

The negative sign indicates that the follower is retarded while in contact with the nose of the cam. The three equations (9.9), (9.10) and (9.11) apply only while the follower is in contact with that part of the cam profile which has the centre of curvature Q, i.e. for values of θ between ϕ and α.

The velocity of the follower is a maximum when $\theta = \phi$ and its maximum value is given by $\omega.\text{OP}\sin\phi$.

The acceleration of the follower is a maximum when $\theta = 0$ and its value is given by $\omega^2.\text{OP}$.

The retardation of the follower is a maximum when $\alpha - \theta = 0$, i.e. at the end of the lift, and its value is given by $\omega^2.\text{OQ}$.

The motion of the follower is positively controlled by the cam so long as it is in contact with the flank of the cam, but contact between the follower and the nose of the cam is only maintained if the external spring force is greater than the inertia force. When the follower is in the full lift position, the spring must exert a force at least sufficient to give to the follower the acceleration $\omega^2.\text{OQ}$.

Cams of the above type are usually symmetrical. Occasionally the follower is given a short period of dwell at the end of the lift. The complete displacement, velocity and acceleration diagrams are shown in Fig. 202. These diagrams are drawn to scale for the cam of the following example.

Example 5. Lift $\frac{1}{2}$ in., minimum radius $1\frac{1}{8}$ in., nose radius $\frac{1}{8}$ in., no dwell; the angle α is 55° and the camshaft speed is 1200 r.p.m.

From the triangle OPQ, Fig. 201 (a):

$$\text{PQ}^2 = \text{OP}^2 + \text{OQ}^2 + 2.\text{OP}.\text{OQ}\cos\alpha$$

But
$$\text{PQ} = \text{PO} + 1\tfrac{1}{8} - \tfrac{1}{8} = \text{PO} + 1$$

∴ substituting and simplifying,

$$\text{OP} = \frac{\text{OQ}^2 - 1}{2(1 - \text{OQ}\cos\alpha)}$$

But $\text{OQ} = 1\frac{1}{8} + \frac{1}{2} - \frac{1}{8} = 1\frac{1}{2}$ in., so that $\text{OP} = 7{\cdot}477$ in.

Also
$$\sin\phi = (\text{OQ}/\text{PQ})\sin\alpha$$
$$\therefore \phi = 12^\circ\ 58'$$

While the follower is in contact with the flank of the cam, i.e. for values of θ from 0 to 12° 58′, we have:

Displacement of the follower, from (9.6),

$$x = 4{\cdot}477\,(1 - \cos\theta) \text{ in.}$$

Velocity of the follower, from (9.7),

$$v = \frac{\pi.120}{3}.\frac{4{\cdot}477}{12}\sin\theta = 46{\cdot}9\sin\theta \text{ ft/s}$$

Acceleration of the follower, from (9.8),

$$f = \left(\frac{\pi \, . \, 120}{3}\right)^2 . \frac{4 \cdot 477}{12} \cos\theta = 5892 \cos\theta \text{ ft/s}^2$$

While the follower is in contact with the nose of the cam, i.e. for values of θ from 12° 58′ to 55°, we have:

Displacement of the follower, from (9.9),

$$x = 1 \cdot 5 \cos(\alpha - \theta) - 1 \cdot 0 \text{ in.}$$

Velocity of the follower, from (9.10),

$$v = 15 \cdot 71 \sin(\alpha - \theta) \text{ ft/s}$$

Acceleration of the follower, from (9.11),

$$f = -1974 \cos(\alpha - \theta) \text{ ft/s}^2$$

From these equations the values given in the following table have been calculated and the curves of Fig. 202 have been plotted.

θ . . .	0°	5°	10°	12° 58′	20°	30°	40°	50°	55°
x in. .	0	0·0170	0·0680	0·1140	0·2288	0·3595	0·4489	0·4943	0·5000
v ft/s.	0	4·09	8·14	10·52	9·012	6·639	4·066	1·370	0
f ft/s²	5892	5869	5803	5742 −1467	−1617	−1789	−1906	−1966	−1974

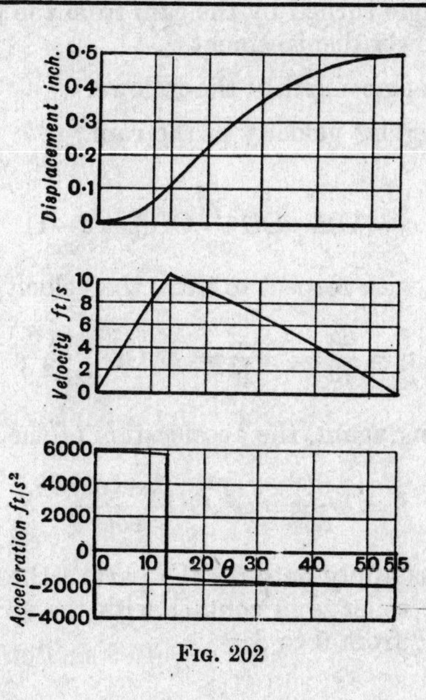

Fig. 202

(b) *Tangent Cam with Roller Reciprocating Follower.* When the reciprocating follower is fitted with a roller, the flank of the cam may be straight and tangential to the base circle. The cam is then known as a tangent cam and is shown in Fig. 203. The flanks AB and EF are straight lines tangential to the circular arc AF and the arcs BU and VE respectively. UV is a circular arc drawn with centre O. The path of the roller centre is shown by the dotted line. Since the whole of the cam profile is formed of circular arcs and straight lines, the working surface of the cam can be very accurately machined.

The outward displacement of the follower takes place partly while the roller is in contact with the straight flank AB and partly while it is in contact with the rounded corner BU. If θ is the angle through which the cam has turned from the beginning of the lift of the follower, then contact takes place between the follower and the straight flank AB for values of θ from 0 to ϕ, where $\tan \phi = GK/GO$, and between the follower and the rounded nose BU for values of θ from ϕ to α, where $\tan \alpha = CR/RO$.

The displacement, velocity and acceleration of the follower for a given value of θ may be found either analytically or graphically.

(a) *Analytical Solution.* (i) *Contact between the Roller and the Straight Flank* AB. Fig. 203 (a).

Let θ = the angle turned by the cam from the beginning of the follower displacement,

x = the displacement of the follower

and ω = the angular velocity of the cam.

Then

$$x = OH - OG = OG(\sec\theta - 1) \quad . \quad . \quad (9.12)$$

Differentiating with respect to time, the velocity of the follower

$$v = \frac{dx}{dt} = \omega\frac{dx}{d\theta} = \omega . OG \frac{\sin\theta}{\cos^2\theta} \quad . \quad . \quad (9.13)$$

and, differentiating again, the acceleration of the follower

$$f = \omega\frac{dv}{d\theta} = \omega^2 . OG\frac{2-\cos^2\theta}{\cos^3\theta} \quad . \quad . \quad . \quad (9.14)$$

These equations apply only for the part of the follower motion during which the roller is in contact with the straight flank AB, i.e. for values of θ from 0 to ϕ.

(ii) *Contact between the Roller and the Rounded Corner* BU. Fig. 203 (b). For values of θ from ϕ to α, the roller is in contact with the arc BU of the cam profile and the distance of the roller centre P from the centre C of the arc BU remains constant. The arrangement is, therefore, kinematically equivalent to a slider-crank chain with a crank of length OC, turning at uniform speed, and a connecting rod of length CP.

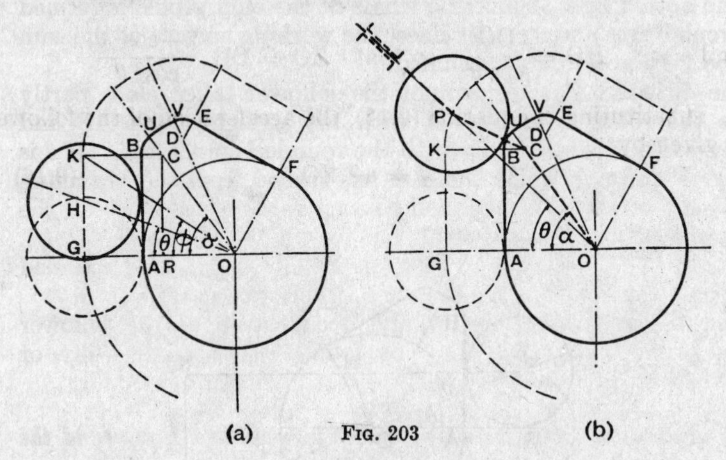

(a) FIG. 203 (b)

The displacement, velocity and acceleration of the follower may be found from the appropriate equations of Article 51, where $n = CP/OC$ and instead of θ the angle $\alpha - \theta$ is substituted. For a cam, the value of n is small, so that the approximate expressions of Article 51 cannot be used without introducing considerable errors in the calculated values of the velocity and acceleration of the follower. The exact expressions are, however, rather cumbersome and a graphical method of determining the velocity and acceleration will be found more expeditious.

(b) *Graphical Solution.* (i) *Contact between the Roller and the Straight Flank* AB. Referring to Fig. 204, draw through the roller centre H a line perpendicular to the straight flank AB to meet a line through O perpendicular to OH at N.

Then $ON = OH \tan\theta$ But $OH = OG/\cos\theta$, so that

$$ON = OG\frac{\tan\theta}{\cos\theta} = OG\frac{\sin\theta}{\cos^2\theta}$$

$\therefore$ substituting in (9.13), the velocity of the follower is given by:

$$v = \omega . ON \quad . \quad . \quad . \quad . \quad . \quad (9.15)$$

Similarly, in order to find the acceleration of the follower, set off HK = 2HN and draw a line through K perpendicular to HK to meet HO produced at Z.

Then $$HZ = \frac{HK}{\cos\theta} = \frac{2HN}{\cos\theta} = \frac{2HO}{\cos^2\theta}$$

But $$ZO = ZH - HO = HO\left(\frac{2}{\cos^2\theta} - 1\right) = HO.\frac{2-\cos^2\theta}{\cos^2\theta}$$

and $$HO = \frac{OG}{\cos\theta}, \text{ so that } ZO = OG.\frac{2-\cos^2\theta}{\cos^3\theta}$$

∴ substituting in equation (9.14), the acceleration of the follower is given by:

$$f = \omega^2.ZO \quad . \quad . \quad . \quad . \quad . \quad (9.16)$$

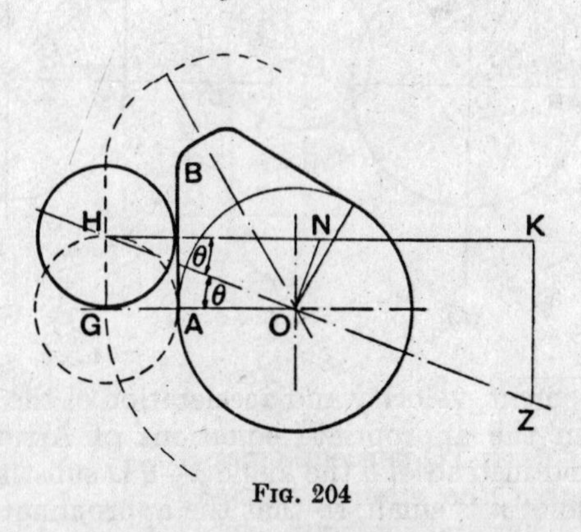

FIG. 204

It is left as an exercise for the student to show that these results could be obtained directly by imagining the cam to be extended so that a point H_1 on the cam coincides with the roller centre H. The velocity of H is then equal to the vector sum of the velocity of H_1 and the velocity of H relative to H_1. Similarly, the acceleration of H is the vector sum of the acceleration of H_1 and the acceleration of H relative to H_1.

(ii) *Contact between the Roller and the Rounded Corner* BU. Referring to Fig. 205, the instantaneous centre of the equivalent link CP is at I, the point of intersection of OC produced and the line through P normal to the line of stroke OS of the follower. If PC is produced to intersect at M a line through O perpendicular to OS, then triangles IPC, OMC are similar,

$$\therefore \; v_p/v_c = IP/IC = OM/OC$$

But $v_c = \omega . OC$, where ω is the angular velocity of the cam, so that

$$v_p = \omega . OM \quad . \quad . \quad . \quad . \quad . \quad (9.17)$$

The acceleration of the follower may be found by a modification of Klein's construction, Article 49. The purpose of Klein's construction is to find a point Q on CP such that $CQ = CM^2/CP$. Where CP is shorter than CM, as in Fig. 205, the point Q may be found as follows.

Draw through P a line perpendicular to CP; with centre C and radius CM draw an arc to intersect this perpendicular at T;

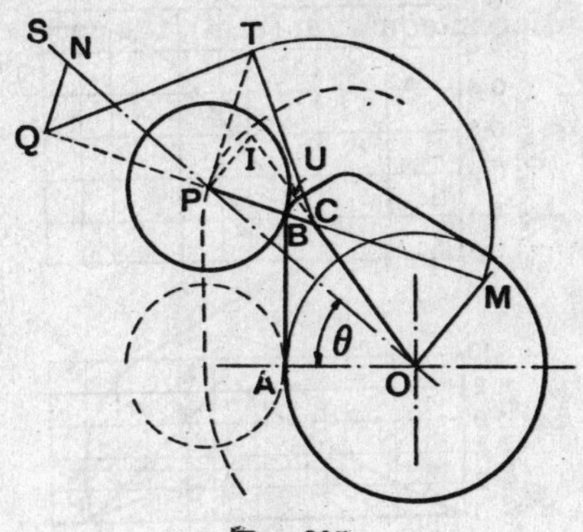

FIG. 205

through T draw TQ perpendicular to CT to intersect CP produced at Q. From the similar triangles PCT, TCQ it follows that $CQ = CT^2/CP$, and therefore, since $CT = CM$, $CQ = CM^2/CP$.

Hence, if QN is drawn perpendicular to CQ to intersect the line of stroke of the follower at N, it follows that the acceleration of P is represented by NO to the same scale as the acceleration of C is represented by CO.

But the acceleration of $C = \omega^2 . CO$,

$$\therefore \text{ the acceleration of } P = \omega^2 . NO \quad . \quad (9.18)$$

The displacement, velocity and acceleration diagrams for the outstroke of the follower when driven by a cam of this type are shown in Fig. 206. They are drawn to scale for the cam in the following example.

Example 6. Lift $\frac{9}{16}$ in., minimum radius 1 in., nose radius $\frac{7}{32}$ in., radius of the roller $\frac{5}{8}$ in., the total angle of cam rotation from the beginning of the outward stroke to the end of the return stroke is 120° and the camshaft speed is 800 r.p.m.

From Fig. 203 (a), $OC = 1+\frac{9}{16}-\frac{7}{32} = \frac{43}{32}$ in.,

$$\cos\alpha = \frac{OR}{OC} = \frac{OA-AR}{OC} = \frac{OA-BC}{OC} = \frac{25}{43}, \quad \therefore \alpha = 54^\circ\ 27'$$

and

$$\tan\phi = \frac{KG}{GO} = \frac{CR}{GO} = \frac{OC\sin\alpha}{GA+AO} = \frac{43}{52}\sin\alpha, \quad \therefore \phi = 33^\circ\ 56'$$

The angular velocity of the cam $= \pi.80/3 = 83{\cdot}8$ rad/s

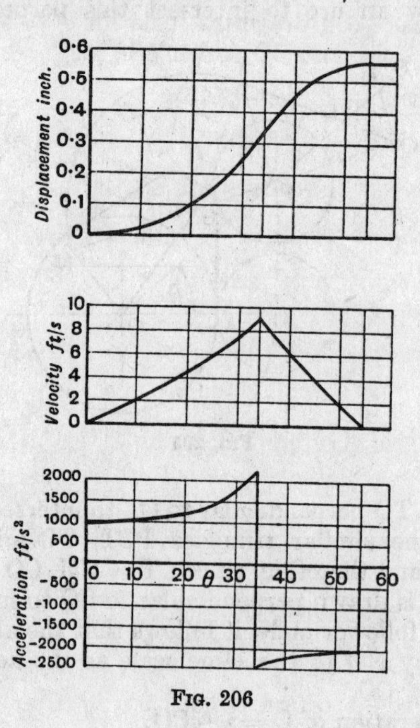

FIG. 206

While the follower is in contact with the straight flank of the cam, i.e. for values of θ from 0 to 33° 56′, the values of x, v and f are calculated from equations (9.12), 9.13) and (9.14).

While the roller is in contact with the nose of the cam, i.e. for values of θ from 33° 56′ to 54° 27′, we have

$$n = CP/OC = 27/32.32/43 = 27/43$$

and the values of x, v and f are calculated from equations (3.6), (3.9) and (3.10). Alternatively, they may be obtained graphically as explained above.

The values given in the following table have been calculated and the curves of Fig. 206 have been plotted from them.

θ . . .	0	10°	20°	30°	33° 56′	40°	50°	54° 27′
x in. . .	0	0·025	0·104	0·251	0·334	0·451	0·551	0·563
v ft/s . .	0	2·03	4·39	7·56	9·20	6·28	1·89	0
f ft/s² . .	950	1024	1278	1827	2183 −2466	−2190	−2045	−2034

N.B.—Fig. 204 has been drawn to scale for $\theta = 20°$. On this figure ON scales 0·63 in. and ZO scales 2·19 in.

∴ velocity of the follower, from (9.15),

$$= \omega . \text{ON} = 83{\cdot}8 . 0{\cdot}63/12 = 4{\cdot}4 \text{ ft/s}$$

and the acceleration of the follower, from (9.16),

$$= \omega^2 . \text{ZO} = 83{\cdot}8^2 . 2{\cdot}19/12 = 1280 \text{ ft/s}^2$$

Fig. 205 has been drawn to scale for $\theta = 40°$. On this figure OM scales 0·89 in. and NO scales 3·78 in.

∴ velocity of follower, from (9.17),

$$= \omega . \text{OM} = 83{\cdot}8 . 0{\cdot}89/12 = 6{\cdot}2 \text{ ft/s}$$

and the acceleration of the follower, from (9.18),

$$= \omega^2 . \text{NO} = 83{\cdot}8^2 . 3{\cdot}78/12 = 2210 \text{ ft/s}^2$$

The acceleration is towards O and is therefore negative.

(c) *Circular Arc Cam with Oscillating Roller Follower.* The contour of a circular arc cam with an oscillating follower is shown in Fig. 207. The cam turns about an axis through O and the follower about an axis through C. The roller turns about an axis through B, and P is the centre of curvature of the flank of the cam with which the roller makes contact. Since the distances between O and P, P and B, B and C, and C and O are constant, the arrangement is kinematically identical with a four-bar chain. The angular velocity and the angular acceleration of the oscillating follower cannot easily be calculated, but they may be found graphically by the methods given in Article 50. The link OP is in effect the driving crank of the four-bar chain and it turns at the uniform speed of the cam.

To Find the Angular Velocity of the Follower. The instantaneous centre of the link BP may be found. This lies at the point of intersection I of PO produced and CB produced. But, if a line is drawn through O parallel to BC to cut BP at M, then triangle OPM is similar to triangle IPB. Since the velocity of P $= v_p = \omega . OP$, where ω is the angular velocity of the cam, it follows that the velocity of B $= v_b = \omega . OM$ and the velocity of B relative to P $= v_{bp} = \omega . PM$.

$\therefore$ the angular velocity of the follower $= v_b/BC = \omega . OM/BC$

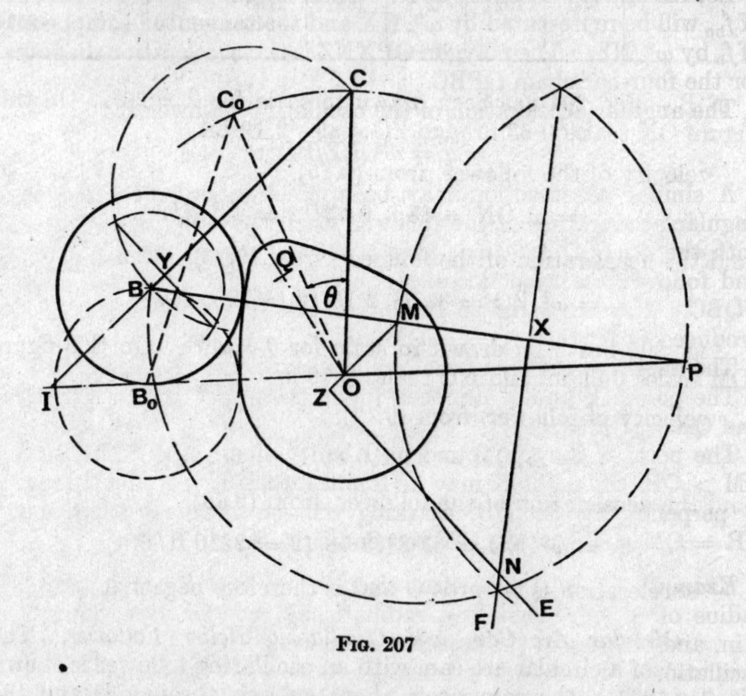

Fig. 207

To Find the Angular Acceleration of the Follower. An acceleration diagram may be drawn for the four-bar chain by applying Klein's construction as explained in Article 50.

The acceleration of P is wholly centripetal and is given by $f_p = \omega^2 . PO$.

On BC as diameter draw a circle, and with centre B and radius OM draw a second circle. Then the chord common to these two circles will cut BC at Y, such that the centripetal component of the acceleration, f_b, of B is given by $\omega^2 . BY$.

Mark off ZO equal to BY and parallel to BC and draw ZE perpendicular to BC to represent the direction of the tangential component of f_b.

On BP as diameter draw a circle, and with centre P and radius PM draw a second circle. Then the chord common to these two circles will cut BP at a point X such that the centripetal component of the acceleration of B relative to P is given by $\omega^2.XP$. Through X draw XF perpendicular to BP to represent the direction of the tangential component of the acceleration of B relative to P.

Let ZE and XF intersect at N. Then the tangential component of f_{bp} will be represented by $\omega^2.NX$ and the tangential component of f_b by $\omega^2.NZ$. The polygon OPXNZ is the acceleration diagram for the four-bar chain OPBC.

The angular acceleration of the oscillating follower BC

$$= \alpha = \omega^2.NZ/BC$$

A similar construction may be applied in order to find the angular acceleration of the follower when the roller is in contact with the nose of the cam. This is shown in Fig. 208. The cam and follower are kinematically equivalent to the four-bar chain OQBC. If a line is drawn through O parallel to BC to meet BQ produced at M, then $v_b = \omega.OM$ and $v_{bq} = \omega.QM$.

The angular velocity of the oscillating follower $= \omega.OM/BC$.

The points Y and X are then found such that $BY = OM^2/BC$ and $QX = QM^2/OB$.

The point Y may be found by Klein's construction, but, since $QM > QB$, the point X may be found by drawing a line through B perpendicular to QB, marking off the point R such that $QR = QM$ and drawing RX perpendicular to QR.

Example 7. A symmetrical circular arc cam has a minimum radius of $1\frac{1}{8}$ in. The flank radius is $4\frac{7}{8}$ in., the nose radius is $\frac{3}{8}$ in. and the rise of the cam is $\frac{1}{2}$ in. The distance of the axis of oscillation of the follower from the axis of the cam is 3 in., the length of the follower arm is 3 in. and the diameter of the roller is 2 in. The r.p.m. of the cam are 600.

Find the angular velocity and the angular acceleration of the follower when the cam has turned through angles of (a) 25° and (b) 45° from the beginning of the lift.

(a) When the cam has turned through 25° from the beginning of the lift, the roller is in contact with the flank of the cam, as shown in Fig. 207, which is drawn to scale for the above particulars.

The distance between the cam axis O and the centre of curvature P of the flank $= 4\frac{7}{8} - 1\frac{1}{8} = 3\frac{3}{4}$ in.

By measurement from the figure, OM = 0·79 in., NZ = 2·66 in.

$$\therefore\ v_b = \omega.OM = (\pi.600/30)0{\cdot}79 = 49{\cdot}6 \text{ in/s}$$

$\therefore$ angular velocity of the follower BC

$$= v_b/BC = 49{\cdot}6/3 = 16{\cdot}5 \text{ rad/s}$$

For counter-clockwise rotation of the cam, the angular velocity of BC is clockwise.

From the acceleration diagram OPXNZ, the tangential component of the acceleration of B

$$= \omega^2 . NZ = (\pi . 20)^2 . 2{\cdot}66 = 10\ 500 \text{ in./s}^2$$

$\therefore$ angular acceleration of BC $= 10\ 500/BC = 3500 \text{ rad/s}^2$

Since the tangential acceleration is in the direction NZ, the angular acceleration of BC is clockwise.

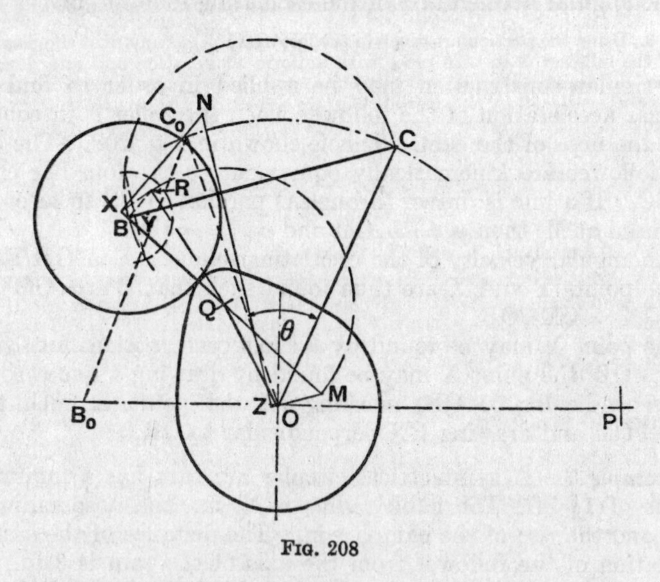

Fig. 208

(b) Fig. 208 is drawn to scale for this position of the cam and follower.

From the figure, OM scales 0·52 in. and NZ scales 3·24 in.

$\therefore$ angular velocity of the follower

$$= \omega . OM/BC = 20\pi . 0{\cdot}52/3 = 10{\cdot}9 \text{ rad/s}$$

For counter-clockwise rotation of the cam, the angular velocity of BC is clockwise.

The angular acceleration of the follower

$$= \omega^2 . NZ/BC = (20\pi)^2 . 3{\cdot}24/3 = 4260 \text{ rad/s}^2$$

Since the tangential acceleration of B is in the direction NZ, the angular acceleration of BC is counter-clockwise.

EXAMPLES IX

1. Sketch and describe the different types of followers which are used with radial or disc cams. Discuss the advantages and disadvantages of each type.

2. Draw the displacement-time, velocity-time and acceleration-time diagrams for a follower in order to satisfy the following conditions. The stroke of the follower is 1 in. The outward stroke takes place with simple harmonic motion during 90° of cam rotation, and the return stroke, also with simple harmonic motion, during 75° of cam rotation. The follower is to dwell in the out position for 45° of cam rotation and the cam turns at a uniform speed of 800 r.p.m.

3. Using the particulars given in the last example, except that the displacement of the follower is to take place with uniform acceleration and retardation, draw the displacement-time, velocity-time and acceleration-time diagrams when (a) the acceleration and retardation are equal, (b) the acceleration on the outstroke is twice the retardation on the outstroke and the retardation on the instroke is three times the acceleration on the instroke.

4. Draw the profile of a cam to give the following motion to a roller follower:

(a) Outward stroke during 50° of cam rotation.
(b) Dwell for 10° of cam rotation.
(c) Return stroke during 50° of cam rotation.
(d) Dwell for the remaining 250° of cam rotation.

The stroke of the follower is 1 in.; the diameter of the roller is $2\frac{1}{2}$ in.; the minimum radius of the cam is 2 in.; the line of stroke of the follower is radial and the outward and return strokes take place with uniform equal acceleration and retardation.

5. As Question 4, but the line of stroke of the follower is offset $\frac{1}{2}$ in. from the axis of the cam.

6. As Question 4, except that an oscillating roller follower is used. The distance of the follower pivot from the axis of the cam is $3\frac{3}{4}$ in. and the length of the follower arm from pivot centre to roller centre is 3 in.

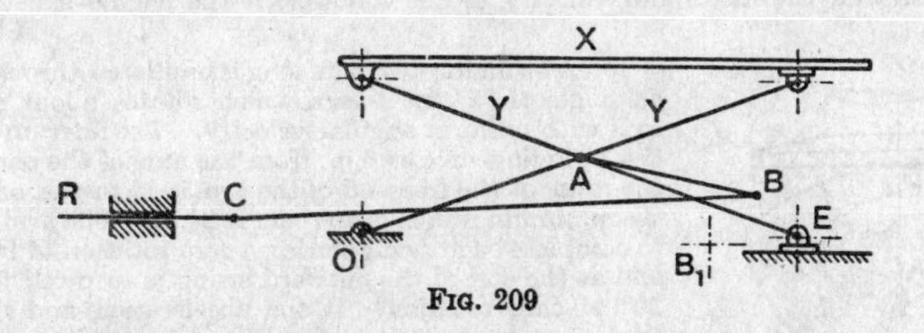

FIG. 209

7. The mechanism for raising a feed-table is shown in Fig. 209. The table X is supported on " lazy-tongs " levers YY, which are opened and closed by a rod R connected with the central pivot A by the links CB, BA. O is fixed and E moves along a horizontal guide. The joint B traverses a cam path which is of such a form that the horizontal velocity of R bears a constant ratio to the velocity of the table. Given the point B_1 at which the path of B is horizontal, construct the form of the cam path for a limited rise of the table from the given position.
W.S.S.

8. A cam is to give the following straight-line motion to a bar. The outstroke is to be made with simple harmonic motion during one-half of a revolution; the

return-stroke also takes one-half of a revolution, but during this stroke the acceleration is numerically equal to the deceleration. The stroke of the bar is 3 in. and the speed of the cam is 240 r.p.m. Find:

(a) the maximum value of the acceleration during the out-stroke;
(b) the numerical value of the acceleration during the return-stroke;
(c) the maximum values of the velocity during each stroke, all in in. sec units.

L.U.

9. A cam turning with uniform angular velocity operates a reciprocating follower through a roller 2 in. dia. The line of stroke of the follower is 1 in. from the axis of the cam, the stroke of the follower is 2 in. and the minimum radius of the cam is 2 in. The follower is required to move outwards and inwards with simple harmonic motion, each stroke occupying 75° of cam rotation. During the remainder of the cam rotation the follower is to rest at the bottom of its stroke. Draw the outline of the working surface of the cam.

10. The exhaust valve of a Diesel engine has a lift of $1\frac{1}{2}$ in. It is operated by a cam designed to give equal uniform acceleration and retardation during the opening and closing periods, each of which corresponds to 50° of cam rotation, and also to allow the valve to remain in the fully open position for 15° of cam rotation. The follower is provided with a roller 3 in. dia. and its line of stroke passes through the axis of the cam. If the minimum radius of the cam is $4\frac{1}{2}$ in., draw the outline of the cam.

11. A cam turning with uniform angular velocity operates an oscillating follower through a roller $\frac{3}{4}$ in. dia. The fulcrum of the follower is 2 in. from the axis of rotation of the cam and the distance from centre of roller to centre of fulcrum is $1\frac{3}{4}$ in.; the minimum radius of the cam is 1 in. and the angular displacement of the follower is 30°. The outward and inward displacements of the follower each occupy 60° of cam rotation and there is no dwell in the lifted position. If the displacement is simple harmonic, draw the outline of the cam and the velocity and acceleration diagrams of the follower for a camshaft speed of 200 r.p.m.

M.U.

12. The exhaust valve of a gas engine working on the Otto cyle has a lift of $\frac{3}{4}$ in. and is operated by a cam giving constant acceleration and retardation. The follower is provided with a roller $1\frac{1}{2}$ in. dia. and moves along a straight line passing through the axis of the cam. The valve opens 40° before the outer dead centre and closes 10° after the inner dead centre of the crank, and the minimum radius of the cam is $1\frac{5}{8}$ in. Draw the outline of the cam and determine the acceleration and the maximum velocity of the follower, if the engine runs at 250 r.p.m.

M.U.

13. A swinging arm 4 in. long is oscillated through an angle of 32° by a cam which rotates about its axis with uniform angular velocity. The fulcrum of the swinging arm is 6 in. from the axis of the cam, the roller at the free end of the arm is $1\frac{3}{4}$ in. dia. and the minimum radius of the cam is $2\frac{1}{2}$ in. The arm is to complete each swing during a cam rotation of 90° and at the end of the outward swing is to dwell for 30° of cam rotation. If the acceleration and retardation on each stroke are uniform and equal, draw the cam profile. What is the angular acceleration of the follower if the cam rotates at a speed of 300 r.p.m.?

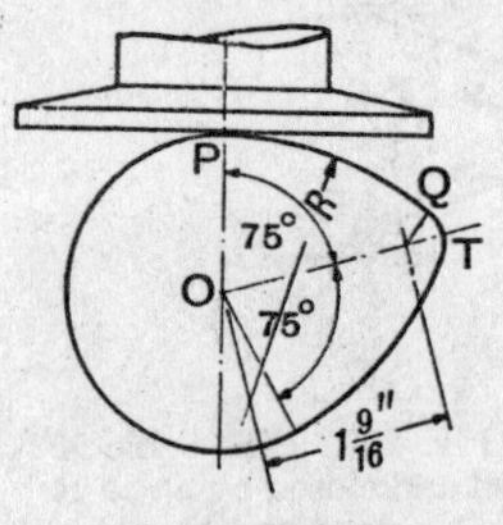

Fig. 210

14. Particulars of a cam with a flat-footed follower are given in Fig. 210. Calculate the value of the radius R and plot on a base of $3\frac{3}{4}$ in. = 150° a curve of displacement of the follower. If the speed of the cam is 500 r.p.m., calculate the accelerations of the follower at the points P, Q and T. If the combined mass of the follower and valve with which it is in contact is 10 lb, show how to calculate the stiffness of the spring which is employed to close the valve. Nose radius = $\frac{5}{16}$ in.; minimum diameter of cam = $2\frac{1}{2}$ in.

L.U.A.

15. A can operating a mushroom-ended follower has the following dimensions: minimum radius, 0·65 in.; lift of follower, 0·285 in.; radius of nose, 0·0625 in. If the total period of opening and closing of the valve corresponds to 110° of camshaft rotation, determine the radius of the flanks and the maximum velocity, acceleration and retardation of the follower. The camshaft speed is 900 r.p.m. M.U.

16. The valve timing for a four-stroke petrol engine is as follows: inlet opens 4° L.; inlet closes 50° L.; exhaust opens 50° E.; exhaust closes 10° L. Each valve has a lift of 0·4 in., the minimum radius of each cam is 0·8 in. and the nose radius is 0·1 in. The cams are of the circular-arc type with flat-faced followers. Set out the cam profiles and calculate the maximum velocity, acceleration and retardation of each follower, if the camshaft speed is 2000 r.p.m. What is the minimum force which must be exerted by the spring of each valve in order to overcome the inertia of the moving parts which weigh 0·4 lb for each valve?

17. A straight-sided cam with a circular tip gives a total lift of 1·25 in. to a valve, the stem of which carries a roller 1 in. in diameter. The reciprocating parts controlled by the cam weigh 1 lb. The cam has a base circle 1·75 in. in radius and acts on the roller through an angle of 120°. Find the pressure which must be exerted by the valve spring if the roller is to remain in contact with the cam when this rotates at 350 r.p.m. L.U

18. The following valve timing is used on a petrol engine: inlet opens 18° before t.d.c., closes 42° after b.d.c.; exhaust opens 62° before b.d.c., closes 13° after t.d.c. The lift of each valve is 7·5 mm, the diameter of the base circle of the cam is 30 mm and the nose radius of the cam is 3 mm. Flat-faced followers are to be used. Draw the cam profiles and show them in their correct relative angular positions on the camshaft. If the crankshaft speed is 5000 r.p.m. calculate the maximum velocity, the maximum acceleration and the maximum retardation of each follower on the outstroke.

19. A valve is actuated by a cam having the following dimensions: base circle 0·75 in., nose circle 0·546 in. diameter; distance between the centres of these circles 0·339 in.; the flanks are straight. The cam acts on a tappet roller 1 in. in diameter. The valve opens 25° before t.d.c. Determine:

(a) the lift of the valve;
(b) the crank angle at closure;
(c) the maximum velocity; and
(d) the maximum acceleration of the valve.

The camshaft, which runs at half engine speed, rotates at 1500 r.p.m. L.U.

20. The follower of a tangent cam is operated through a roller 2 in. dia. and its line of stroke intersects the axis of the cam. The minimum radius of the cam is $1\frac{1}{2}$ in., the nose radius is $\frac{1}{2}$ in. and the lift is 1 in. If the speed of rotation of the cam is 900 r.p.m., find the velocity and acceleration of the follower at the instant when the cam is 20° from the full-lift position. M.U.

21. A tangent cam for a Diesel engine has the following dimensions: minimum radius, $4\frac{5}{8}$ in.; lift, $1\frac{1}{2}$ in.; total angle through which the cam turns from the beginning of the outward stroke to the end of the inward stroke, 115°; nose radius, $2\frac{3}{8}$ in.; diameter of roller, 3 in.; camshaft speed, 115 r.p.m. Draw the complete displacement, velocity and acceleration-time curves and set out the cam profile. It may be assumed that the follower moves along a straight line which passes through the axis of the cam.

22. The tangent cam for a gas engine has a base circle $3\frac{1}{2}$ in. dia. and nose radius $\frac{3}{4}$ in. The lift is $\frac{7}{8}$ in., the roller 2 in. dia. and the r.p.m. of the cam are 160. Find the acceleration of the follower at the beginning and at the end of the lift and also the velocity and the acceleration at the instant the roller changes from contact with the straight flank to contact with the nose.

23. Use the same particulars as in Example 7, p. 309, and find: (a) the angular acceleration of the follower at the beginning and the end of both strokes; (b) the maximum angular velocity of the follower on both strokes; (c) the acceleration on both strokes at the instant the follower changes from contact with the curved flank to contact with the nose.

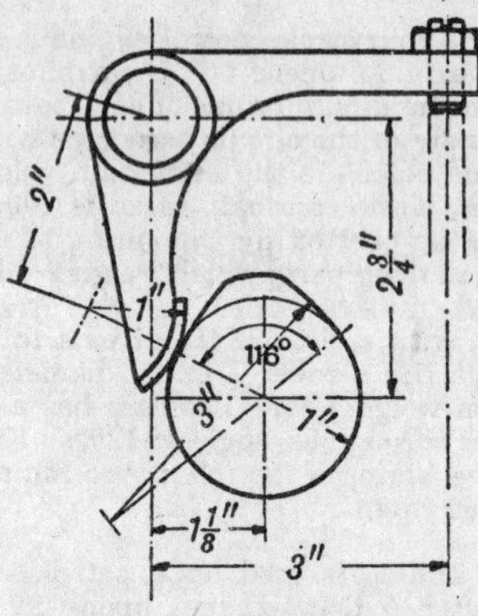

FIG. 211

24. The cam and rocker for an o.h.v. high-speed Diesel engine are shown in Fig. 211. The cam is symmetrical, the radius of curvature of the flanks being 3 in. and of the rounded corners $\frac{1}{8}$ in. The rise of the cam is $\frac{1}{4}$ in. and the angle through which the cam turns from the beginning of the outstroke to the end of the instroke of the follower is 116°. If the engine speed is 1200 r.p.m., find for both strokes: (a) the maximum velocity of the valve; (b) the acceleration of the valve at the beginning and at the end of each stroke; (c) the acceleration at the instant the follower changes from contact with the flank to contact with the rounded corner of the cam profile.

CHAPTER X

TOOTHED GEARING

122. Many different forms of toothed gearing are used for the transmission of motion or of power. These include *spur* gearing, in which the axes of the shafts connected by the gears are parallel; *bevel* gearing, in which the axes of the shafts intersect; and *skew* or *spiral* gearing, in which the axes of the shafts are non-parallel and non-intersecting. *Helical* gearing is the name given to a type of spur gearing in which, although the axes of the shafts are parallel, the teeth are cut on helices instead of straight across the wheels parallel to the axes. In most forms of toothed gearing the teeth are not theoretically necessary for the transmission of the required motion. From a practical point of view, however, the

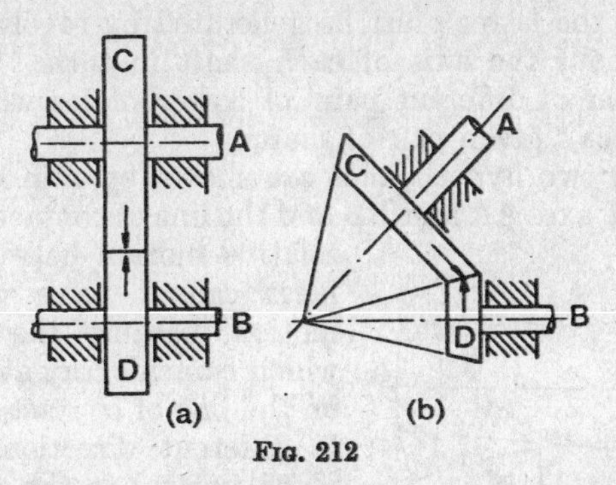

Fig. 212

teeth fulfil in all types of gearing the purpose of providing an invariable velocity ratio between the two shafts and of enabling a large torque to be transmitted from the driving shaft to the driven shaft.

Kinematically, a pair of spur gears is equivalent to a pair of cylindrical discs, keyed to parallel shafts, and having line contact as shown in Fig. 212 (a). If the two discs roll together *without slipping*, the peripheral speeds of the discs are equal and the ratio of the angular velocity of the driven shaft B to the angular velocity of the driving shaft A is equal to the ratio of the diameter of the disc C to the diameter of the disc D. Similarly, a pair of bevel wheels is kinematically equivalent to a pair of conical frusta, keyed to the intersecting shafts, and having line contact as shown in Fig. 212 (b). The apices of the complete cones, of which C and D are parts, obviously coincide with the point of intersection of the axes of the two shafts A and B. If C and D roll together *without slipping*, the peripheral speeds are equal at any particular point on the line of contact and the angular velocities of the two shafts A and B are inversely as the corresponding diameters of the frusta C and D. The cylindrical surfaces of C and D, Fig. 212 (a), and the conical surfaces of C and D, Fig. 212 (b), are known as *pitch surfaces*.

In the case of non-parallel, non-intersecting shafts the conditions are different and it is impossible to arrange for motion to be transmitted from one shaft to the other by surfaces which have a pure rolling action. The discs through which motion is transmitted are frusta of hyperboloids. A *hyperboloid* is the solid formed by revolving a straight line about an axis not in the same plane, such that every point on the line remains at a constant distance from the axis. Hence any two hyperboloids may be arranged to have line contact. This implies that, in order to connect two non-intersecting, non-parallel shafts by means of hyperboloids, the latter must be generated by revolving the line of contact about the axis of each shaft in turn. There is an infinite number of different pairs of hyperboloids which may be used to connect a given pair of shafts.

In Fig. 213 two hyperboloids are shown by thin lines. They turn about the axes AA and BB and the line of contact is CC. The relative motion between the surfaces cannot be a pure rolling motion, because the two points which coincide, say at the point P on the line of contact, are moving in different directions, one in a plane normal to the axis AA and the other in a plane normal to the axis BB. The coincident points will, however, have identical component velocities in a direction normal to the line of contact, if the relative sliding between the surfaces takes place wholly in a direction parallel to the line of contact. This is obviously what would happen if straight teeth were provided on the two hyperboloidal surfaces parallel to the line of contact so as to give a positive drive between the shafts. The two hyperboloidal frusta shown by thick lines in the figure may therefore be regarded as the pitch surfaces of a pair of skew

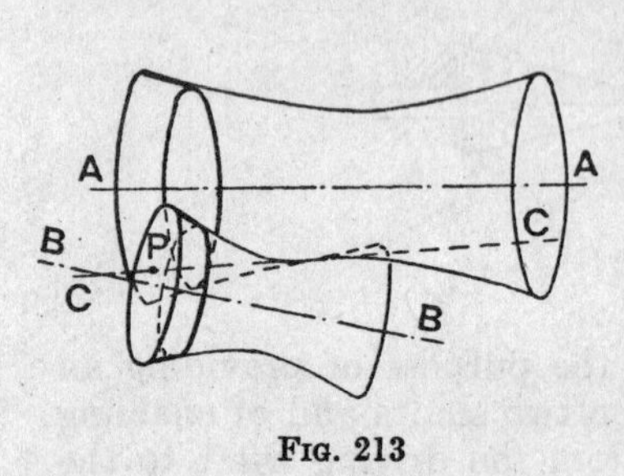

Fig. 213

gears, which would enable motion to be transmitted from shaft AA to shaft BB with line contact. Owing to the difficulty that would be experienced in manufacturing gearing with hyperboloidal pitch surfaces, this type of gearing is not used in practice and spiral gearing takes its place. Spiral gearing differs from true skew gearing in one very important respect. the contact between the pitch surfaces is point contact instead of line contact. Referring to Fig. 214 the two points which are in contact are moving in different directions and therefore the relative motion between the cylinders is partly rolling and partly sliding. In order to transmit uniform motion from one shaft to the other, it is necessary to specify that the relative sliding shall take place in a direction parallel to a tangential line which passes through the point of contact and occupies a fixed position relative to the axes of the two shafts. In practice the direction of this line is fixed by the direction of the helical teeth at the point of contact.

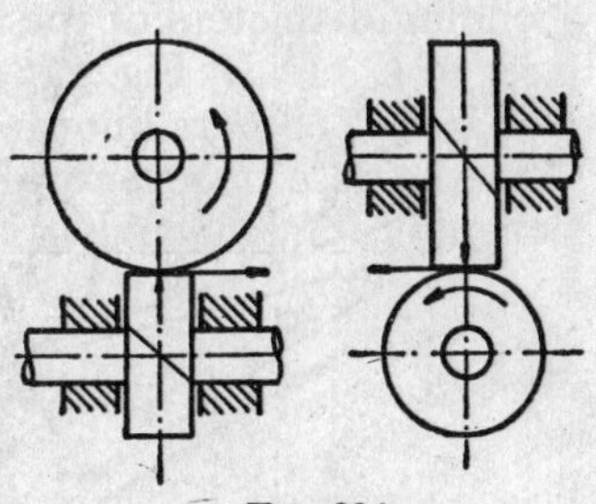

Fig. 214

The arrangements which are shown in Figs. 212 and 213 are obviously unsuitable where a positive drive is required. If the friction between the surfaces at the line or point of contact were relied upon in order to transmit power, not only would the transmitted power be small, but the velocity ratio between the shafts would be affected by slip. To make the drive positive it is necessary to provide teeth, which extend partly above and partly below the smooth cylindrical, conical or hyperboloidal pitch surfaces. The motion is then transmitted not by the rolling together of the pitch surfaces, but by the contact between the surfaces of these teeth. But, if the motion transmitted from the driving to the driven shaft is to be perfectly uniform and the velocity ratio is to be constant, the shapes of the contact surfaces of the teeth much satisfy certain conditions. These conditions are examined in the following article.

123. Motion Transmitted by two Curved Surfaces in Contact. Two rigid bodies, X and Y, are shown in Fig. 215. They turn about fixed axes through O and Q and the curved surfaces are in contact along a line through A. Let X be turning at a given instant with the angular velocity ω_x rad/s in the counter-clockwise sense. It is required to find the angular velocity ω_y of Y. At the given instant the point A, considered as a point on X, is moving in a direction perpendicular to OA with a velocity $v_{ax} = \omega_x . OA$. At the same instant the point A, considered as a point on Y, is moving in a direction perpendicular to QA with a velocity $v_{ay} = \omega_y . QA$. The direction of the common normal to the two surfaces at the point of contact is given by AB and the component

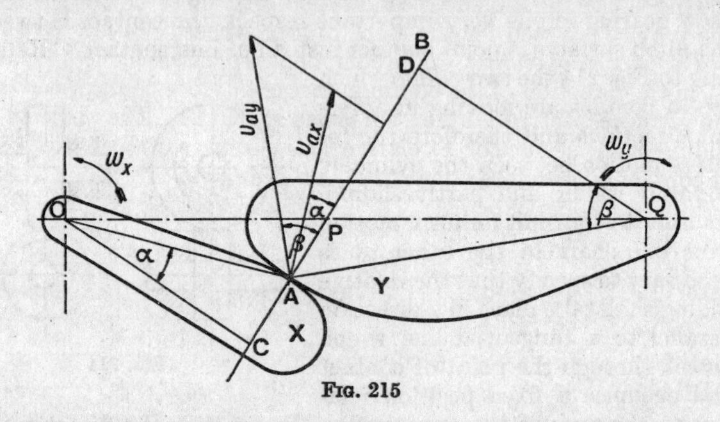

FIG. 215

of v_{ax} along AB is $v_{ax} \cos \alpha$ and the component of v_{ay} along AB is $v_{ay} \cos \beta$.

It is clear that if the two surfaces are to remain in contact these two component velocities must be equal, so that:

$$v_{ax} \cos \alpha = v_{ay} \cos \beta$$

Substituting for v_{ax} and v_{ay},

$$\omega_x . OA \cos \alpha = \omega_y . QA \cos \beta$$

Draw OC and QD perpendicular to AB. Then:

$$OC = OA \cos \alpha \quad \text{and} \quad QD = QA \cos \beta$$

so that

$$\omega_x . OC = \omega_y . QD$$

or

$$\omega_x / \omega_y = QD/OC$$

But, if P is the point of intersection of AB with the line of centres OQ, the triangles OCP, QDP are similar and $QD/OC = QP/OP$,

$$\therefore \omega_x / \omega_y = QP/OP \quad . \quad . \quad . \quad (10.1)$$

Hence the ratio of the angular velocities of X and Y is inversely as the ratio of the distances of P from the centres O and Q, or the common normal to the two surfaces at the point of contact A intersects the line of centres at a point P which divides the centre distance inversely as the ratio of the angular velocities. If, then, the velocity ratio is to be constant, the contact surfaces must be so shaped that the common normal intersects the line of centres at a fixed point P. The motion transmitted from X to Y will then be identical with that transmitted by pure rolling contact between two cylindrical surfaces with axes through O and Q and with a line of contact through P.

This is the fundamental condition which must be satisfied by the profiles adopted for the teeth of gear wheels.

The velocity of sliding. The components of the velocities v_{ax} and v_{ay} in the direction of the common tangent at the point of contact are given respectively by $v_{ax} \sin \alpha$ and $v_{ay} \sin \beta$. The velocity of sliding of the surface of Y relative to the surface of X at the point of contact is therefore equal to $v_{ay} \sin \beta - v_{ax} \sin \alpha$,

$$\begin{aligned} \therefore \text{ velocity of sliding} &= \omega_y . AQ \sin \beta - \omega_x . AO \sin \alpha \\ &= \omega_y . AD - \omega_x . AC \\ &= \omega_y(AP+PD) - \omega_x(CP-PA) \\ &= (\omega_x+\omega_y)AP + \omega_y . PD - \omega_x . CP \end{aligned}$$

But from (10.1) $\omega_y/\omega_x = OP/QP$ and from the triangles OCP, QDP, $OP/QP = CP/DP$, so that $\omega_y . PD = \omega_x . CP$,

$$\therefore \text{ velocity of sliding} = (\omega_x+\omega_y)AP$$

This result could be obtained more easily as follows. Since the point P is the instantaneous centre for the relative motion of X and Y and the angular velocity of Y relative to X is $\omega_x+\omega_y$ it follows at once that the velocity of sliding is equal to $(\omega_x+\omega_y)AP$.

124. Definitions. Before considering the theory of wheel teeth in detail, a few necessary definitions will be given of terms which are in general use in connection with toothed gearing. Some of these terms are illustrated in Fig. 216.

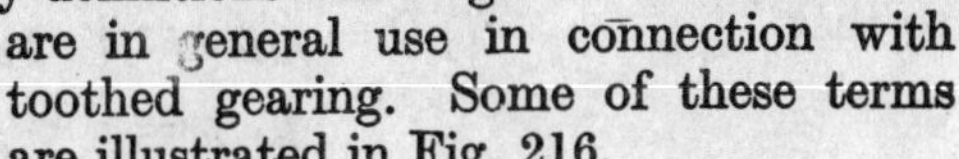

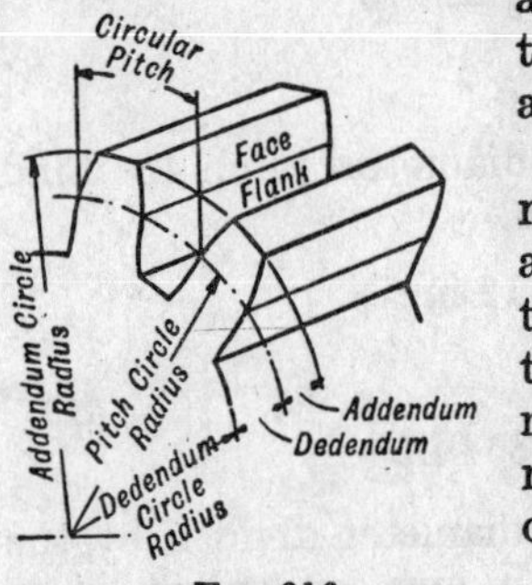

FIG. 216

The pitch circle diameter is the diameter of the circle which by a pure rolling action would transmit the same motion as the actual gear wheel. It should be noted that in the case of wheels which connect non-parallel shafts, the pitch circle diameter is different for each cross-section of the wheel normal to the axis of rotation.

The pitch point is the point of contact of two circles.

The pitch line is the line of contact of two pitch surfaces.

The circular pitch p is the distance measured along the pitch circle circumference from a point on one tooth to the corresponding point on the next tooth. Or, alternatively, the circular pitch is equal to the pitch circle circumference divided by the number of teeth T on the wheel, or if the diameter of the pitch circle is D, then

$$p = \pi D/T$$

Hence, if the number of teeth and the circular pitch are known,

$$D = pT/\pi$$

The use of the circular pitch is open to the very serious practical objection that D cannot be expressed exactly because of the presence of π in the denominator.

Example 1. A wheel has 48 teeth and the circular pitch is $\frac{3}{4}$ in. Find the pitch diameter.

$$D = 48.3/4\pi = 11{\cdot}459 \text{ in.}$$

This is not the exact diameter, but is only correct to five significant figures. Since the distance between the axes of a pair of mating wheels is equal to the sum of the pitch circle radii, it follows that the centre distance cannot be expressed exactly.

There are two different ways of stating the pitch which obviate this difficulty. They are given below.

The *diametral pitch* p_d is defined as the number of teeth per inch of pitch diameter:

$$\therefore\ p_d = T/D$$

But the definition of circular pitch

$$p = \pi D/T$$

so that $$T/D = \pi/p \quad \text{and} \quad p_d = \pi/p$$

or $$p_d \cdot p = \pi$$

Example 2. A wheel has 48 teeth of diametral pitch 4. Find the pitch diameter and the circular pitch.

$$D = T/p_d = 48/4 = 12 \text{ in.}$$

The circular pitch

$$p = \pi/p_d = \pi/4 = 0{\cdot}7854 \text{ in.}$$

The *module* m is defined as the pitch diameter divided by the number of teeth. It may be expressed in any convenient linear units, such as inches of pitch diameter per tooth, millimetres of pitch diameter per tooth, etc.

$$m = D/T \quad \text{or} \quad D = mT$$

But $$D/T = p/\pi, \quad \therefore\ m = p/\pi = 1/p_d$$

Example 3. A wheel has 48 teeth of module 6 mm. Find the pitch diameter and the circular pitch.

$$D = m.T = 6.48 = 288 \text{ mm} = 28{\cdot}8 \text{ cm}$$

$$p = \pi.m = 18{\cdot}85 \text{ mm} = 0{\cdot}7421 \text{ in.}$$

The advantage of stating the pitch of the teeth in terms of the diametral pitch or the module is shown in the above examples. The calculation of the pitch diameter of a wheel and of the distance between the centres of two mating wheels is simplified. Except in the case of certain diametral pitches, such as 3, 6, etc., which with certain combinations of teeth involve recurring decimals, the pitch diameter and centre distance can be expressed exactly.

It will of course be understood that for each of the three ways of stating the pitch, a series of standard cutters must be provided. A few of the standard cutters, which cover approximately the same range of circular pitches, are shown in the table below.[1]

Circular pitch in.	1	$\frac{3}{4}$	$\frac{1}{2}$	$\frac{1}{4}$	$\frac{1}{8}$
Diametral pitch	3	4	6	12	24
Module, mm.	8	6	4	2	1

The figures in the vertical columns of the table are *approximate* equivalents only. They give the nearest standard diametral pitches and modules for the circular pitches shown in the first line.

The *addendum* is the radial distance from the pitch circle to the top of the tooth.

The *dedendum* is the radial distance from the pitch circle to the bottom of the tooth space.

The *working surface* above the pitch surface is termed the *face* of the tooth and that below the pitch surface the *flank* of the tooth.

The *clearance* is the radial distance from the top of the tooth to the bottom of the tooth space in the mating gear.

A *pinion* is the smaller of two mating gear wheels.

A *rack* is a portion of a gear wheel which has an infinitely large number of teeth. Hence the pitch surface of a rack is a plane surface and the pitch circle circumference becomes a straight line.

The *pressure angle* or *angle of obliquity* ψ is the angle which the common normal to the two teeth at the point of contact makes with the common tangent to the two pitch circles at the pitch point.

The *path of contact* is the curve traced out by the point of contact of two teeth from the beginning to the end of engagement.

[1] Abstracted by permission from British Standard 436:1940, Machine Cut Gears. A. Helical and Straight Spur. Official copies can be obtained from the British Standards Institution, 2 Park St., London, W.1, price 10*s*. 6*d*. post free.

The *arc of contact* is the path followed by a point on either pitch line from the beginning to the end of engagement of a given pair of teeth. The arc of contact is divided into two parts. The part from the beginning of engagement until the two teeth are in contact at the pitch point is termed the *arc of approach*, and that from the pitch point to the end of engagement is termed the *arc of recess*.

125. Conjugate Teeth. It was shown in Article 123 that, in order to transmit a uniform velocity ratio, the shape of the profiles of the teeth must be such that the common normal at the point of contact passes through the pitch point. Within limits it is possible to choose an arbitrary shape for one of the profiles and then to determine the shape of the other profile to satisfy this condition. Teeth with profiles obtained in this way are spoken of as conjugate teeth.

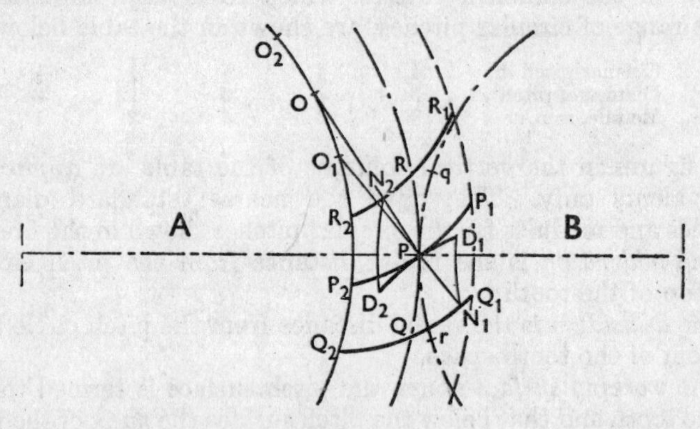

FIG. 217

In Fig. 217 part of a wheel A is shown and for simplicity the profile is assumed to be a circular arc P_1PP_2 of radius OP. It is required to find the profile of the tooth of a pinion B that will gear correctly with this wheel.

Let us suppose the wheel A turns on its axis until the given tooth profile occupies the position Q_1QQ_2, the centre of curvature of the profile moving at the same time to O_1. Then, in order to satisfy the fundamental condition, the point of contact of the pinion tooth and the wheel tooth must lie on the normal O_1N_1 which passes through the pitch point P. In other words the two teeth must be in contact at the point N_1. Hence the point D_1 on the profile of the pinion tooth may be found by setting off arc Pq along the pitch circle of B equal to arc PQ on the pitch circle of A, drawing the arc N_1D_1 concentric with B and marking off qD_1 equal to PN_1. Similarly, if A turns in the opposite sense until the given tooth profile occupies the position R_1RR_2, the corresponding position of the centre of curvature being O_2, then O_2N_2P is the normal to the profile which passes through P and N_2 is the point of contact of the two profiles. The point D_2 on the profile of the pinion tooth is found by a similar construction to that used for finding D_1. Thus arc Pr = arc PR, arc N_2D_2 is concentric with B and $rD_2 = PN_2$. It follows that the pinion tooth profile must pass through the points D_1, P and D_2. Obviously, the shape of the profile could only be determined accurately if the construction were repeated for a large number of positions of the tooth profile on A.

The above construction has been given in order to emphasise the fact that the choice of tooth profiles is theoretically a very wide one. Practical considerations, however, impose severe limitations on the choice of a suitable profile. In the past the cycloidal curve has been very widely used, but at the present time, for reasons that will appear later on, the involute curve is almost universally employed.

It will be seen later that gear cutting machines which generate the tooth profiles carry out mechanically the process which has just been described, i.e. the teeth of the cutter and the teeth of the wheel are conjugate teeth. The profiles of the cutter teeth are not, however, arbitrarily chosen but conform to the involute shape.

126. Cycloidal Teeth. A *cycloid* is the locus of a point on the circumference of a circle which rolls, without slipping, along a straight line.

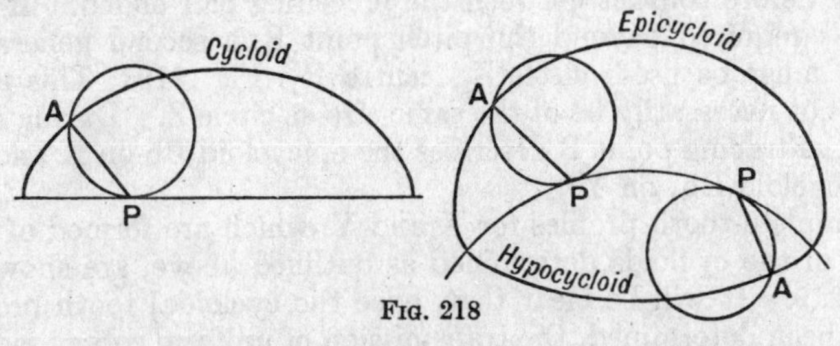

FIG. 218

If the circle rolls, without slipping, inside or outside a circular arc, the locus of a point on its circumference is termed respectively a *hypocycloid* or an *epicycloid*.

The three cycloidal curves are illustrated in Fig. 218. The important property of the cycloid which makes it suitable for use as a tooth profile is that the line joining the tracing point A to the point of contact P of the rolling circle and the line, straight or curved, along which it rolls is normal to the cycloid. This follows from the fact that the point of contact P is the instantaneous centre of the rolling circle and therefore the tracing point A is, at that instant, moving in a direction perpendicular to AP.

In Fig. 219 two circles X and Y with centres O and Q are in contact at point P. A smaller circle Z with centre R is also in contact with the other circles at point P. If the three circles now turn about their respective centres and roll together without slipping, a point A on the small circle will trace out a hypocycloid on X and an epicycloid on Y. Thus, if X turns in the clockwise sense through angle α, then Y and Z will turn respectively through angles β and γ. The three arcs PA, Pa and Pa_1 are equal in length, since there is pure rolling, and when motion began the three points A, a and a_1 all coincided at the pitch point P. During this rotation of the three circles the tracing point A on Z describes the portion aA of a hypocycloid on X and the portion a_1A of an epicycloid on Y. Point A is the point of contact of the two cycloids and AP is their common normal. This is always true whatever the length of the arc Pa so long as contact between the two cycloids is maintained. It follows therefore that Y could be driven positively by X if the portions a_1A and aA of the cycloids were used as the profiles of contact surfaces. Further, the velocity ratio would be constant and equivalent to that of pure rolling between the circumferences of X and Y. To transmit continuous motion from X to Y, it would be necessary to arrange for a number of similar pairs of cycloids to be so spaced round the circumferences of X and Y that contact between a second pair began before contact between the preceding pair ended. If contact is required beyond the pitch point P, a second generating circle must be used, circle Z_1, centre R_1, Fig. 219. This circle need not necessarily be of the same size as circle Z. During rotation the tracing point B describes the epicycloid Bb on X and the hypocycloid Bb_1 on Y.

Complete tooth profiles for X and Y, which are formed of portions of the cycloids determined as outlined above, are shown in Fig. 219. It will be clear that, once the cycloidal tooth profiles have been determined, the transmission of uniform rotary motion from X to Y requires that these two circles shall be in contact at point P. If the distance between the centres O and Q is either increased or decreased, the common normal to the two profiles at the point of contact will no longer intersect the line of centres OQ at a fixed point and therefore the motion transmitted will no longer be uniform.

In Fig. 220 two wheels with cycloidal teeth are shown in mesh. The point of contact of the profiles lies on the circumference of one or other of the rolling circles Z and Z_1, which are used to generate the profiles.

If X drives Y, the flank of the profile on X makes contact with the face of the profile on Y and the point of contact follows the arc AP until the pitch point P is reached. Beyond this point the face of the profile on X makes contact with the flank of the profile on Y and the point of contact follows the arc PB. The point A at which contact begins is given by the point of intersection of the

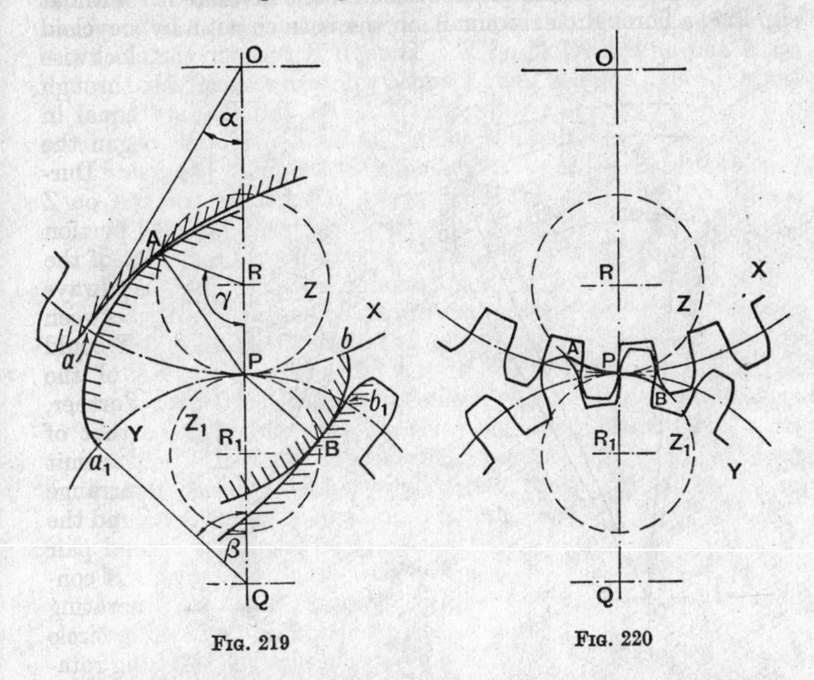

FIG. 219

FIG. 220

addendum circle of Y with the rolling circle Z and the point B at which contact ends by the point of intersection of the addendum circle of X with the rolling circle Z_1. The curved line APB is the path of contact.

It will be seen that the pressure angle varies with cycloidal teeth. It is a maximum at the beginning of engagement, when the point of contact between the teeth coincides with A, and diminishes in value as the point of contact moves along the arc AP. It is zero at the pitch point and then begins to increase again, reaching a maximum value at the end of engagement, when the point of contact coincides with B.

127. The Involute. An *involute* may be defined as the locus of a point on a straight line which rolls, without slipping, on the circumference of a circle, or, alternatively, it may be defined as the locus of a point on a cord which is held taut and unwound from a cylinder. The circle on which the straight line rolls or from which the cord is unwound is known as the *base* circle. The normal to the involute at a given point is obviously the tangent drawn from that point to the base circle and the radius of curvature of the involute at that point is equal to the length of the tangent. A geometrical construction for the involute is shown in Fig. 227. Through the point A on the base circle draw the tan-

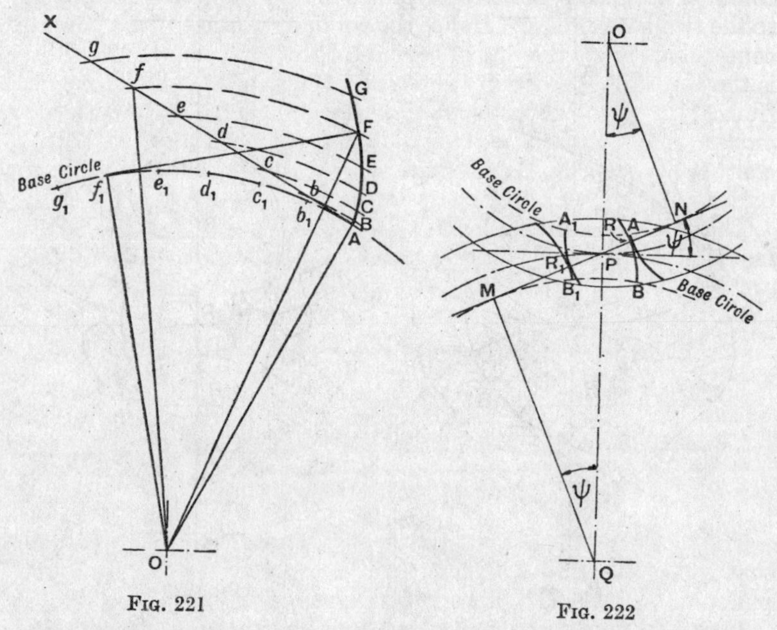

FIG. 221 FIG. 222

gent AX. Along AX and along the circumference of the base circle mark off a number of points b, c, d, e, f, etc., and b_1, c_1, d_1, e_1, f_1, etc., at equal distances apart. Let F be the position of the point on the tangent which originally coincided with A when the tangent has rolled on the base circle so that f coincides with f_1. Then F may be found as follows:

Through f draw a circular arc with O as centre; with centre f_1 and radius Af draw a second circular arc. The point of intersection of these two arcs will be the required point F on the involute curve.

Proof: By construction, $f_1F = Af$, $Of = OF$ and $Of_1 = OA$, so that the triangles Of_1F, OAf are equal in all respects.

Hence, angle Of_1F = angle $OAf = 90°$ and the line f_1F is tangential to the base circle.

But $f_1F = Af$ and by construction $Af = \text{arc } Af_1$, so that the length of the tangent Ff_1 = length of arc Af_1. The point F is therefore a point on the involute curve which passes through A.

Other points, such as B, C, D, etc., may be found in the same way and joined by a smooth curve to give the involute AF.

Let O and Q be the fixed centres of two base circles and let the corresponding involutes be in contact at the point R, Fig. 222. Since the normal of an involute at a given point is the tangent drawn from that point to the base circle, it follows that the common normal to the two involutes is also the common tangent to the two base circles. Hence the common normal at the point of contact intersects the line of centres at the fixed point P, and this is the necessary condition for the transmission of uniform motion from O to Q. The ratio of the angular velocities of the base circles, when motion is transmitted from one to the other by contact between the two involutes, is identical with that given by pure rolling between two circles of radii OP and QP.

The following alternative explanation may be given. Let A move to A_1 so that B moves to B_1 and R to R_1. Then, since arc NA = tangent NR and arc NA_1 = tangent NR_1, it follows that:

$$\text{arc } AA_1 = \text{arc } NA_1 - \text{arc } NA = NR_1 - NR = RR_1$$

Similarly,

$$\text{arc } BB_1 = \text{arc } MB - \text{arc } MB_1 = MR - MR_1 = RR_1$$

$$\therefore \text{arc } AA_1 = \text{arc } BB_1 = RR_1$$

The motion transmitted from O to Q is therefore identical with that which would be transmitted from a pulley of radius ON to a pulley of radius QM by means of a crossed belt.

It should be particularly noted that the shape of an involute curve is solely determined by the diameter of the base circle and that the distance between the centres of the base circles may be increased or decreased without affecting the motion transmitted. This constitutes one of the most valuable properties of the involute as a shape for the profiles of wheel teeth.

The common tangent NM to the base circles is inclined at the angle ψ to the common tangent to the pitch circles at the point P. For given base circles the angle ψ will clearly vary if the centre distance is varied. Since the point of contact of two involutes lies on the common tangent to the base circles, the angle ψ is the pressure angle.

128. Involute Teeth. In practice the involute profile extends from the base circle to the addendum circle and one pair of involutes is in contact for a small fraction only of a revolution of either wheel. If motion is to be transmitted continuously, it is necessary for a second pair of profiles to come into contact before engagement of the first pair ends. Actually there is always some overlapping during which two pairs of profiles are in contact at the same instant. In Fig. 222 the two positions for the profiles might in fact represent the profiles of two adjacent teeth. As we have already seen, arc AA_1 = arc BB_1, so that the distance between the profiles of two adjacent teeth on one wheel, measured along the base circle, must be the same as the distance between the profiles of the corresponding adjacent teeth on the other wheel. But for different base circle diameters a given profile on one wheel will not always make contact with the same profile on the other wheel, so that the necessary condition is that the profiles of all teeth on both wheels must be equally spaced round the circumference of the base circle. The distance between corresponding profiles of adjacent teeth, measured along the base circle, is known as the *base* pitch.

129. Standard Proportions for Interchangeable Gears. Where a set of gears is required any two of which will gear together correctly, it is necessary that the pitch, the addendum and the dedendum shall be the same for each wheel in the set. If the tooth profiles are cycloidal, it is also necessary that the rolling circle shall be of the same diameter for the faces and flanks of all wheels. In practice the diameter of this circle is made equal to one-half the pitch diameter of the smallest pinion in the set. The flanks of this pinion are then radial, since the hypocycloid becomes a diameter of the pitch circle.

If the tooth profiles are involute, the pressure angle ψ must be the same for all wheels in the set.

Standard proportions for the teeth originated with the Brown and Sharpe Manufacturing Co., and are as follows:

$$\text{Addendum} = p/\pi = 0{\cdot}3183p = 1/p_d = m$$
$$\text{Dedendum} = p/\pi + p/20 = 0{\cdot}3683p = 1{\cdot}1571/p_d = 1{\cdot}1571m$$
$$\text{For involute teeth } \psi = 14\tfrac{1}{2}°$$

There are serious objections to the use of a pressure angle of $14\frac{1}{2}°$, as the flanks of pinions with a small number of teeth are undercut so that correct tooth action is prevented and the teeth are weakened (see Article 130). These objections can be overcome to some extent by increasing the pressure angle and reducing the addendum of the tooth. The teeth are then known as stub teeth. Although there is a lack of uniformity in the proportions of stub teeth as adopted by different manufacturers, the pressure angle is generally 20° and the addendum is 0·8 of the B. and S. standard addendum, while the dedendum is equal to the B. and S. standard addendum.

In the Fellows stub-tooth system the tooth is shortened by using the addendum corresponding to a smaller pitch than that on which the pitch diameter is based. The proportions are then expressed in fractional form. Thus $\frac{5}{7}$ indicates that the diametral pitch is 5, but that the addendum corresponds to a diametral pitch of 7 and is therefore 0·1429 in. Other proportions are $\frac{3}{4}$, $\frac{4}{5}$, $\frac{6}{8}$, $\frac{7}{9}$, etc. In this system the clearance is larger than in the B. and S. system, as the dedendum is always 25% greater than the addendum.

The proportions recommended by the British Standards Institution in B.S.S. 436—1940[1] are:

$$\text{Addendum} = 0{\cdot}3183p = 1/p_d = m$$
$$\text{Dedendum} = 1{\cdot}25 \times \text{addendum} = 0{\cdot}3979p$$
$$\text{Pressure angle} = 20^\circ$$
$$\text{Working depth} = 0{\cdot}6366p = 2/p_d = 2m$$

But, where one or both wheels have a small number of teeth, these proportions are modified in order to obtain full involute action and teeth which are both stronger and more resistant to wear. The full recommendations of B.S.436 : 1940 are too lengthy to reproduce here, but the principles involved, in so far as they affect correct tooth action are covered in Articles 130–133.

For Class A1, precision-ground gears, a larger dedendum of 1·44×addendum is specified, but the working depth remains equal to the sum of the addenda of the mating teeth, i.e. $2m$ as given above.

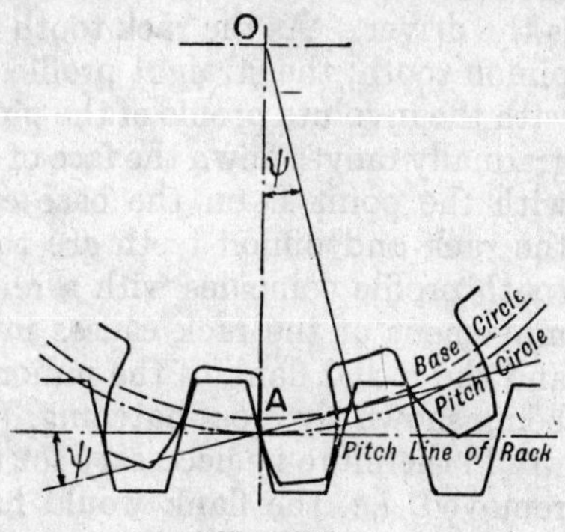

FIG. 223

130. The Involute Rack and Pinion. Interference. Since the path of contact for a pair of involute teeth is the tangent to the base circle which passes through the pitch point, it follows that the teeth of an involute rack are straight-sided, as shown in Fig. 223. The straight sides of the teeth are obviously at right angles to the path of contact. This simple shape is a great advantage and it has

[1] See footnote on page 321.

enabled very accurate methods of generating the teeth of involute gear wheels to be developed. These methods are described in a later section of the chapter. At present it will suffice to say that in modern methods of cutting gear wheels the accuracy with which the profiles of the gear teeth can be machined depends ultimately on the accuracy with which the profile of the rack tooth can be reproduced. Hence the simpler the shape of the rack tooth the greater the accuracy with which it can be reproduced and the greater the accuracy with which the profiles of the gear teeth can be machined.

In any involute system the shape of the rack tooth for unit pitch is taken as the standard and it is known as the *basic rack* tooth shape.

The involute profile of the tooth of the pinion extends from the base circle to the addendum circle. If the pinion has a small number of teeth, the dedendum circle is smaller in diameter than the base circle. At the point where the involute curve springs from the base circle, the curve is tangential to a radius of the base circle and theoretically the part of the tooth profile which lies between the base circle and the dedendum circle could be made radial, as shown in Fig. 223. This shape of profile would, however, give rise to interference between the tips of the rack teeth and the flanks of the pinion teeth.

Fig. 224 is drawn to scale for an 8-tooth pinion gearing with a rack and shows the locus of the corner A of the rack tooth, when the pitch surfaces of the rack and pinion roll together and the rack is the driver. As the rack tooth moves into engagement with the pinion tooth, the straight profile of the rack tooth makes contact with the involute profile of the pinion tooth. The point of contact gradually moves down the face of the pinion tooth until it coincides with the point D on the base circle. The relative positions of the rack and pinion teeth are then as shown at X and the rack tooth profile coincides with a radial line on the pinion. Further movement of the rack causes interference between the corner A and the radial flank of the pinion, the amount of the interference being shown by cross-hatching. In order to avoid interference it would therefore be necessary for this part of the pinion flank to be removed, i.e. the flank would have to be undercut. Note that during this part of the engagement between the two teeth only the extreme point A of the rack tooth makes contact with the undercut flank of the pinion tooth.

To illustrate the method of drawing the locus of the tip A of the rack tooth as the pitch line of the rack rolls on the pitch circle of the pinion, the construction for the two points c and c_1 is given below.

To find c mark off PT, PS and arc PR all equal and of any convenient length and draw TC parallel to the rack tooth profile PA. Along RO mark off Rb = PB = addendum of rack tooth. With centre O and radius OC draw arc Cc and mark point c such that bc = BC. Point c will then be the position of the tip of the rack tooth, when the pitch line of the rack is rolled on the pitch circle of the pinion until S coincides with R. Similarly, point c_1 is found by marking off $PT_1 = PS_1 =$ arc PR_1, drawing T_1C_1 parallel

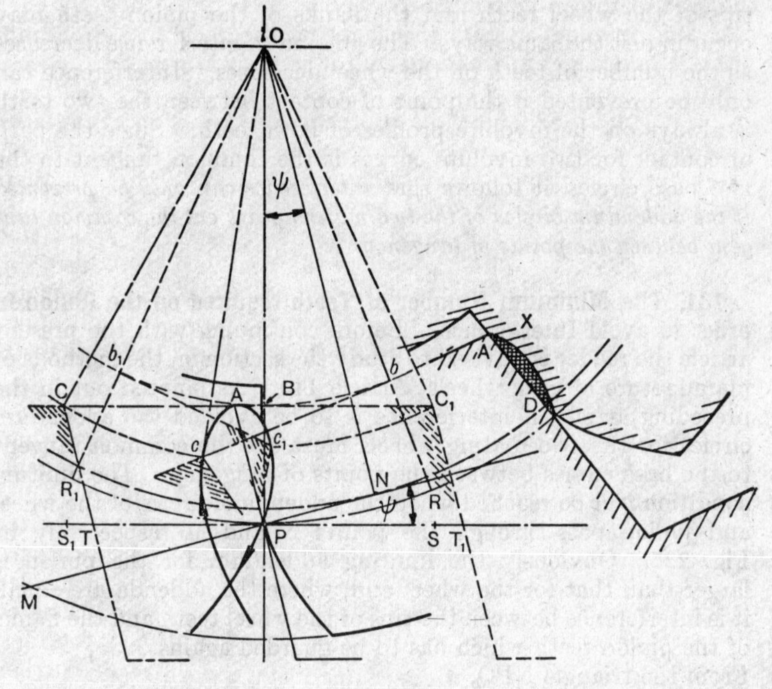

FIG. 224

to PA, describing arc C_1c_1 with centre O and radius OC_1 and marking c_1 such that $b_1c_1 = BC_1$. The tip of the rack tooth will be at point c_1 when the pitch line of the rack is rolled on the pitch circle of the pinion until S_1 coincides with R_1. If this construction is repeated for several values of PR and PR_1, a sufficient number of points will be obtained to enable the complete curve to be drawn.

It is clear that, if the flanks of the pinion are undercut as in Fig. 224, the path of contact will be discontinuous. As the rack tooth moves into engagement with the pinion tooth, contact begins at point M and moves along the path MPN. For that part of the period of engagement beyond N, the corner A of the rack tooth makes contact with the undercut flank of the pinion tooth and the path of contact lies along the addendum line of the rack. Actually, owing to the removal of a small part of the involute profile of the pinion tooth, it would not be possible to transmit uniform motion from the rack to the pinion.

When a pinion gears with a wheel, interference between the tips of the wheel teeth and the flanks of the pinion teeth may occur in just the same way. The amount of interference decreases as the number of teeth on the wheel decreases. Interference can only be prevented if the point of contact between the two teeth is always on the involute profiles of both teeth. Since the path of contact for two involute curves is the common tangent to the two base circles, it follows that *interference can only be prevented if the addendum circles of the two mating gears cut the common tangent between the points of tangency.*

131. The Minimum Number of Teeth required on the Pinion in order to avoid Interference. Before continuing with the present article the reader is advised to study the section on the methods of manufacture of gear wheels, Article 140. As pointed out in the preceding article, if interference is to be avoided the addendum circles for the two mating wheels must cut the common tangent to the base circles between the points of tangency. The limiting condition will be reached when the addendum circles of the wheel and pinion pass through the points N and M respectively in Fig. 225. Obviously the limiting addendum for the pinion is larger than that for the wheel and, where the addenda are equal, it is interference between the tips of the wheel teeth and the flanks of the pinion teeth which has to be guarded against.

From the triangle NPQ,

$$NQ^2 = PQ^2 + NP^2 - 2PQ.NP \cos NPQ$$

$$= PQ^2 + NP^2 + 2PQ.NP \sin \psi$$

But $NP = OP \sin \psi$

$$\therefore\ NQ^2 = PQ^2 + OP^2 \sin^2 \psi + 2PQ.OP \sin^2 \psi$$

$$= PQ^2\left\{1 + \frac{OP}{PQ}\left(\frac{OP}{PQ} + 2\right) \sin^2 \psi\right\}$$

$$\therefore\ NQ = PQ\sqrt{\left\{1 + \frac{OP}{PQ}\left(\frac{OP}{PQ} + 2\right) \sin^2 \psi\right\}}$$

and the addendum of the wheel = NQ−PQ

$$= PQ\left[\sqrt{\left\{1+\frac{OP}{PQ}\left(\frac{OP}{PQ}+2\right)\sin^2\psi\right\}}-1\right] \qquad (10.2)$$

Let t = number of teeth on the pinion

T number of teeth on the wheel

and $a_w.m$ = the addendum of the wheel, where a_w is a fraction and m is the module of the teeth.

Then $OP/PQ = t/T$ and $PQ = Tm/2$

Substituting in (10.2):

$$a_w m = \frac{Tm}{2}\left[\sqrt{\left\{1+\frac{t}{T}\left(\frac{t}{T}+2\right)\sin^2\psi\right\}}-1\right]$$

$$\therefore\ a_w = (T/2)\{\sqrt{(1+A\sin^2\psi)}-1\} \quad . \quad . \quad . \quad (10.3)$$

where $A = (t/T)(t/T+2)$, and its value depends only on the gear ratio.

From (10.3):

$$T = \frac{2a_w}{\sqrt{(1+A\sin^2\psi)}-1} \quad . \quad . \quad (10.4)$$

and therefore

$$t = \frac{2a_w.t/T}{\sqrt{(1+A\sin^2\psi)}-1} \quad . \quad . \quad . \quad (10.5)$$

This equation gives the minimum number of teeth required on the pinion in order to avoid interference between the flanks of the pinion teeth and the tips of the wheel teeth.

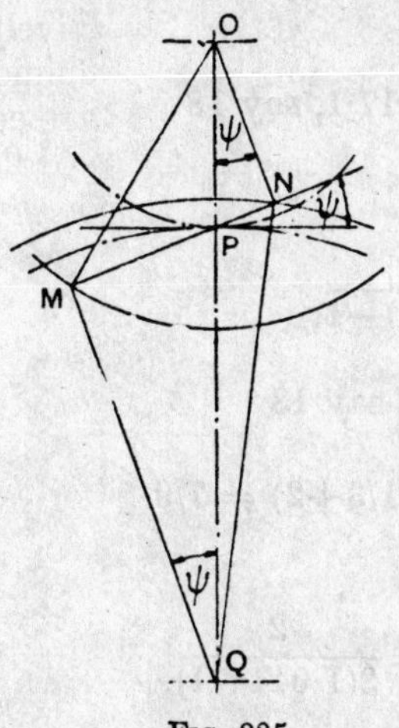

Fig. 225

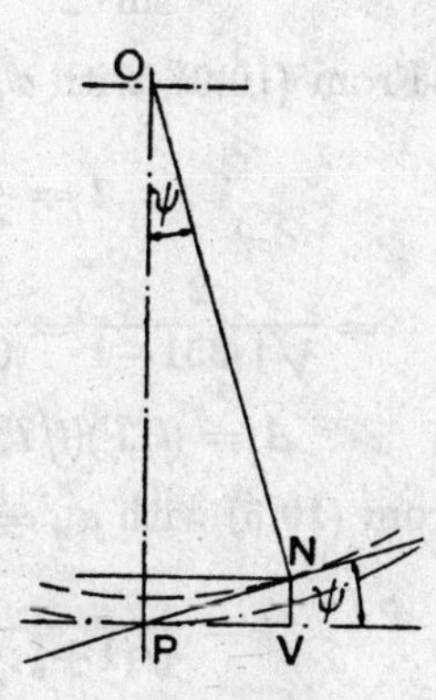

Fig. 226

For equal wheels $t = T$ and (10.5) reduces to:

$$t = \frac{2a_w}{\sqrt{(1+3\sin^2\psi)}-1} \quad . \quad . \quad . \quad (10.6)$$

While for a pinion gearing with a rack it is not difficult to show that (10.5) reduces to:

$$t = 2a_r/\sin^2\psi \quad . \quad . \quad . \quad . \quad (10.7)$$

where a_r is the addendum of the rack expressed as a fraction of the module.

Equation (10.7) could, of course, be obtained more easily from first principles. Thus, in Fig. 226, PN is the tangent drawn from the pitch point to the base circle and in order to avoid undercutting of the flanks of the pinion the addendum of the rack must not exceed NV.

But $\quad NV = PN \sin\psi \quad$ and $\quad PN = OP \sin\psi$

$$\therefore \; NV = OP \sin^2\psi$$

Also $NV = a_r m$ and $OP = tm/2$, so that $a_r = (t/2)\sin^2\psi$ which gives equation (10.7).

Example 4. What is the smallest number of teeth theoretically required in order to avoid interference on a pinion which is to gear with (a) a rack, (b) an equal pinion, (c) a wheel to give a ratio of 3 to 1?

The pressure angle is 20° and a standard addendum of one module is to be assumed.

(a) From (10.7), with $a_r = 1$,

$$t = \frac{2}{\sin^2 20^\circ} = \frac{2}{0{\cdot}1170} = 17{\cdot}1, \text{ say } 18$$

(b) From (10.6), with $a_w = 1$,

$$t = \frac{2}{\sqrt{(1+3\sin^2\psi)}-1}$$

$$= \frac{2}{\sqrt{1{\cdot}351}-1} = \frac{2}{0{\cdot}162} = 12{\cdot}4, \text{ say } 13$$

(c) $\qquad A = (t/T)(t/T+2) = (1/3)(1/3+2) = 7/9$

and from (10.5) with $a_w = 1$,

$$t = \frac{\frac{2}{3}}{\sqrt{(1+\frac{7}{9}.0{\cdot}1170)}-1} = \frac{2}{3(1{\cdot}044-1)}$$

$$= 2/0{\cdot}132 = 15{\cdot}1, \text{ say } 16$$

Although the values of t calculated in the above example represent the theoretical minimum numbers of teeth for the three cases, there are reasons why in practice larger numbers are used. One reason is that the tooth profiles are often generated with a rack cutter, or a hob. The limitation imposed by engagement with a rack must therefore necessarily apply during the generating process, independent of whether the finished pinion is to mate with a rack or a wheel.

There is another more important reason for using a larger number of teeth than the theoretical minimum. This is connected with the shape of the involute curve close to the base circle, which makes it unsuitable for inclusion in the working profile. The small and rapidly changing radius of curvature of this part of the involute would inevitably lead to high contact stresses and more rapid wear. In addition small inaccuracies in the finished profile close to the base circle would have a relatively much greater adverse effect on the smoothness of the motion transmitted than small inaccuracies in the profile further from the base circle.

It follows that the part of the involute which constitutes the working profile should never include that part of the involute which lies close to the base circle. Referring to Fig. 227, the involute profile CAG for the tooth of the wheel is shown in the extreme position when it makes contact with the extreme tip of the pinion tooth. As the wheel rotates clockwise and drives the pinion, the point of contact between the teeth moves along the profile CAG and ultimately reaches the extreme tip of the tooth when CAG has moved to DHB. The part CA of the profile, between the base circle and the point of intersection A of the addendum circle of the pinion and the common tangent to the base circles MN, does not form part of the working profile. Similarly, the part of the pinion tooth profile between its base circle and the point of intersection B of the addendum circle of the wheel and the common tangent MN does not form part of the working profile.

It is clear that the working profiles start further from the base circles the greater the distances MA and NB. From a practical point of view these distances should be as large as possible relative to the distance MN. Ideally, since all involute curves are geometrically similar, it would appear to be desirable that PA/PM = PB/PN. This necessarily means that for mating gears of unequal size the addendum of the wheel is smaller than the addendum of the pinion. So-called "corrected" gears are obtained by modifying the addenda in this way. (See Article 133.)

It is of interest to calculate the minimum total number of teeth in a gear pair with modified addenda, when the working depth is to be standard.

Equation (10.7) gives the relation between the maximum addendum of the rack cutter, the number of teeth on the pinion or wheel and the pressure angle. The addendum of the rack cutter when cutting the teeth of one wheel is also the addendum of the teeth on the mating gear.

In terms of the module, the maximum addendum of the rack cutter, when generating the tooth profiles, is given by

$$a_r m = (t/2)\, m \sin^2 \psi$$

$\therefore$ maximum addendum of wheel = maximum addendum of rack cutter when cutting the pinion

$$= (t/2) m \sin^2 \psi$$

maximum addendum of pinion = maximum addendum of rack cutter, when cutting the wheel

$$= (T/2) m \sin^2 \psi$$

and the maximum working depth = sum of maximum addenda

$$= \frac{t+T}{2}\, m \sin^2 \psi$$

But the standard working depth = $2m$.

$$\therefore \quad \frac{t+T}{2} m \sin^2 \psi = 2m \quad \text{or} \quad t+T = \frac{4}{\sin^2 \psi}$$

If $\psi = 14\tfrac{1}{2}°$,

$$t+T = 4/0{\cdot}0627 = 63{\cdot}8, \text{ say } 64$$

If $\psi = 20°$,

$$t+T = 4/0{\cdot}1170 = 34{\cdot}2, \text{ say } 35$$

Hence, if undercutting of the flanks is to be avoided and the standard working depth of 2 modules is required, the sum of the numbers of teeth on the mating wheels must not be less than 64 when ψ is $14\tfrac{1}{2}°$, nor less than 35 when ψ is 20°.

When stub-teeth are used, the standard addendum is less than the module and the sum of the numbers of teeth may be less than for teeth with an addendum of 1 module without giving rise to undercutting of the flanks. For instance, if the addendum is 0·8 module and the pressure angle is 20°, the working depth is 1·6 module and $t+T$ must be greater than 0·8 . 34·2 = 27·4 or, say, 28.

B.S. 436 : 1940 recommends that, for a pair of gear wheels which are to be mounted at standard centre distance with standard working depth, the total number of teeth shall be not less than 60.

This is considerably greater than the theoretical minimum of 35 calculated above. The purpose is to displace the working profiles along the involute curves so that the part of the involute close to the base circle is not included and thus contact conditions for the tooth profiles are improved.

132. The Length of the Arc of Contact. The length of the arc of contact must be at least equal to the circular pitch of the teeth, so that a second pair of teeth will begin to engage before engage-

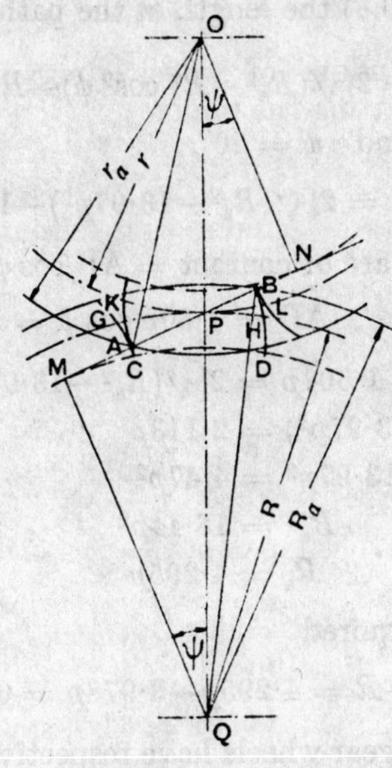

Fig. 227

ment between the preceding pair ends. Referring to Fig. 227, the addendum circles cut the common tangent MN at the points A and B. The length of the path of contact is therefore equal to AB.

$$AB = AP + PB$$

but $AP = AN - PN = \sqrt{(OA^2 - ON^2)} - OP \sin\psi$

$$= \sqrt{(OA^2 - OP^2 \cos^2\psi)} - OP \sin\psi$$

also $PB = BM - PM = \sqrt{(QB^2 - QM^2)} - QP \sin\psi$

$$= \sqrt{(QB^2 - QP^2 \cos^2\psi)} - QP \sin\psi$$

Since $OA = r_a$, $OP = r$, $QB = R_a$ and $QP = R$, we have:

$$AB = \sqrt{(r_a{}^2 - r^2 \cos^2\psi)} + \sqrt{(R_a{}^2 - R^2 \cos^2\psi)} - (r+R)\sin\psi \quad (10.8)$$

The arc of contact = arc GH = arc KL. But arc GH = arc CD/$\cos\psi$ and arc CD = path of contact AB,

$$\therefore \text{ arc of contact} = \frac{\text{path of contact}}{\cos\psi} = \frac{AB}{\cos\psi} \quad (10.9)$$

Example 5. Two gear wheels, each with 25 teeth of involute form and a pressure angle of 20°, are required to give an arc of contact equal to $1{\cdot}6p$, where p is the circular pitch. What is the addendum required?

From equation (10.8) the length of the path of contact is given by:

$$AB = 2\{\sqrt{(R_a^2 - R^2\cos^2\psi)} - R\sin\psi\}$$

But $R = 25p/2\pi$ and $\psi = 20°$,

$$\therefore AB = 2\{\sqrt{(R_a^2 - 13{\cdot}97p^2)} - 1{\cdot}361p\}$$

But from (10.9) the arc of contact = AB/$\cos\psi$,

$$\therefore AB = 1{\cdot}60p\cos\psi = 1{\cdot}504p$$

and

$$1{\cdot}504p = 2\{\sqrt{(R_a^2 - 13{\cdot}97p^2)} - 1{\cdot}361p\}$$

$$\therefore \sqrt{(R_a^2 - 13{\cdot}97p^2)} = 2{\cdot}113p$$

$$R_a^2 - 13{\cdot}97p^2 = 4{\cdot}47p^2$$

$$R_a^2 = 18{\cdot}44p^2$$

$$\therefore R_a = 4{\cdot}295p$$

The addendum required

$$= R_a - R = 4{\cdot}295p - 3{\cdot}978p = 0{\cdot}317p$$

Example 6. Two gear wheels have respectively 28 and 45 teeth and a standard addendum of one module. Find the length of the path of contact and the length of the arc of contact when the pressure angle is (a) $14\frac{1}{2}°$, (b) 20°.

The pitch radius of the smaller wheel,

$$r = 28m/2 = 14m$$

The addendum radius of the smaller wheel,

$$r_a = r + m = 15m$$

The pitch radius of the larger wheel,

$$R = 45m/2 = 22{\cdot}5m$$

The addendum radius of the larger wheel,

$$R_a = R + m = 23{\cdot}5m$$

(a) $\psi = 14\frac{1}{2}°$. Substituting in (10.8) the length of the path of contact:

$$AB = m\sqrt{(15^2-14^2.\cos^2 14\tfrac{1}{2}°)}+ m\sqrt{(23{\cdot}5^2-22{\cdot}5^2 \cos^2 14\tfrac{1}{2}°)}-m36{\cdot}5 \sin 14\tfrac{1}{2}°$$
$$= m\{\sqrt{(225-183{\cdot}7)}+\sqrt{(552{\cdot}3-474{\cdot}4)}-9{\cdot}139\}$$
$$= m(6{\cdot}426+8{\cdot}826-9{\cdot}139)$$
$$= 6{\cdot}113m$$

Or, since $p = \pi m$, $AB = 1{\cdot}945p$, the length of the arc of contact

$$= AB/\cos\psi = 6{\cdot}314m = 2{\cdot}010p$$

(b) $\psi = 20°$. Substituting in (10.8), the length of the path of contact:

$$AB = m\{\sqrt{(225-173{\cdot}1)}+\sqrt{(552{\cdot}3-447{\cdot}1)}-12{\cdot}48\}$$
$$= m(7{\cdot}204+10{\cdot}25-12{\cdot}48)$$
$$= 4{\cdot}97m = 1{\cdot}582p$$

The length of the arc of contact

$$= AB/\cos\psi = 5{\cdot}29m = 1{\cdot}683p$$

133. Methods of Reducing or Eliminating Interference. There are three ways by which undercutting of the flanks of pinions with small numbers of teeth may be avoided. They are as follows:

(a) The part of the flank of the pinion tooth which lies within the base circle and the part of the face of the gear tooth which engages with it may be made cycloidal instead of involute in shape.

(b) The addenda of the teeth on the wheel and pinion may be modified, the addendum of the wheel being reduced by the amount necessary to avoid interference and that of the pinion being correspondingly increased.

(c) The centre distance for two mating gears may be made larger than the standard centre distance, which has the effect of increasing the pressure angle and so avoiding interference.

(a) *Modified Involute or Composite System.* If a standard addendum and a pressure angle of $14\frac{1}{2}°$ are used, the smallest pinion that will gear with a rack without interference has 32 teeth. As it is desirable in practice to be able to use a pinion with only 12 teeth, the shape of the basic rack tooth has to be modified. Fig. 228 shows a rack gearing with a 12-tooth pinion. The profiles of the teeth consist partly of involute curves and partly of cycloidal curves. The cycloidal parts of the profiles are shown shaded. The rolling circles used each have a diameter equal to the pitch radius of the pinion. The involute parts of the profiles are shown unshaded. They are in contact while the point of contact travels along the common tangent MN to the base circles of two 12-tooth pinions. The shaded part of the flank of the pinion tooth makes contact with the shaded part of the face of the rack tooth, while the point of contact moves from A to M along the circumference of the rolling circle. Similarly, the shaded part of the face of the pinion tooth makes contact with the shaded part of the flank of the rack tooth while the point of contact moves from N to B along the circumference of the rolling circle.

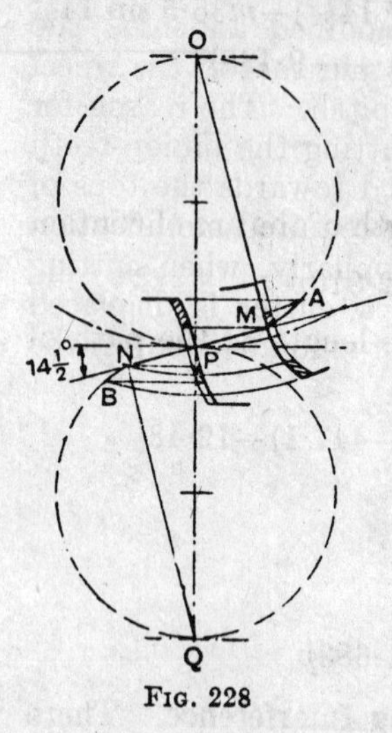

Fig. 228

Since the path of contact includes parts of the rolling circles, any two gears which are based on this system will only gear together correctly if the centre distance is exact, and one of the advantages of the pure involute profile is lost.

A more serious disadvantage of the composite system is that the profile of the rack tooth cannot be reproduced with the same degree of accuracy as the straight-sided profile of the involute rack.

(b) *Modified Addenda of the Wheel and Pinion.* The addendum of the wheel and the addendum of the pinion are generally made of equal lengths, but, as we have seen in Article 131, both addenda are limited by the condition that there shall be no undercutting of the flanks of the teeth. The permissible addendum of the pinion is, however, longer than the permissible addendum of the wheel. It is therefore possible, where wheels with small numbers of teeth are involved, to avoid undercutting of the flanks of the pinion by reducing the addendum of the wheel and yet to retain sufficient working depth for the teeth by increasing the addendum of the pinion. As far as possible the addendum of the pinion should be increased by exactly the same amount as the addendum of the wheel is reduced, so that the full standard working depth is retained.

The shape of the tooth profiles, when undercutting of the flanks is avoided by modifying the addenda in this way, is shown in Fig. 229, which is drawn to scale for a 10-tooth pinion meshing with a 30-tooth wheel, the pressure angle being 20°. For comparison the shapes of the teeth with standard addenda are also shown at X and Y. Note that with standard addenda the teeth of the pinion are undercut, but that with the modified addenda there is no undercutting of the flanks. The pitch-line thicknesses of the teeth are, however, different with the modified addenda, the thickness of the pinion tooth being greater and that of the wheel tooth being less than one-half the circular pitch. The reason for this difference of thickness is that when cutting the pinion teeth the pitch-line of the rack cutter is displaced towards the tops of the cutter teeth by an amount equal to the required modification of the addendum of the pinion teeth. Similarly, when cutting the wheel teeth the pitch line of the rack cutter is displaced towards the roots of the cutter teeth by an amount equal to the

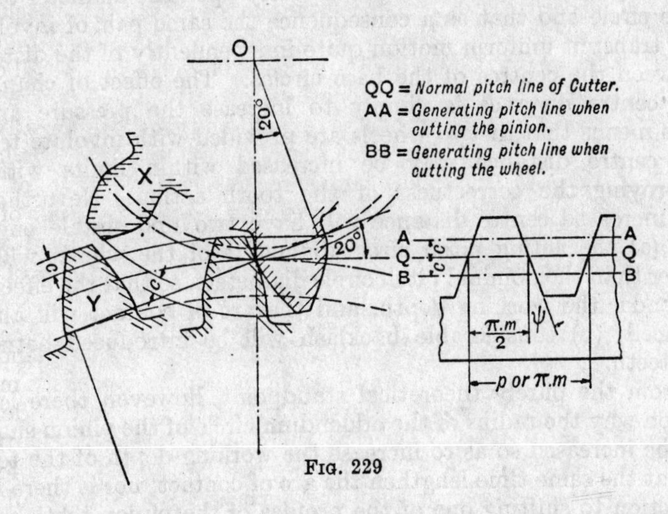

FIG. 229

required modification of the addendum of the wheel teeth. The sum of the pitch-line thicknesses of the wheel and pinion teeth is, of course, still equal to the circular pitch. The actual thicknesses of the teeth are equal to the widths of the tooth space on the rack cutter at the corresponding pitch line positions, and they are given by:

$$(\pi/2 \pm 2c \tan \psi)m$$

where c is the amount by which the addendum of the pinion is increased and the addendum of the wheel is reduced as compared with the standard addendum of one module.

For the 10-tooth pinion the addendum of the rack cutter to avoid interference is 0·585 module (from 10.7), and therefore $c = 1 - 0{\cdot}585 = 0{\cdot}415$.

The pitch-line thickness of the pinion tooth

$$= (\pi/2 + 0{\cdot}830 \tan 20°)m$$
$$= 1{\cdot}873m$$

and the pitch-line thickness of the wheel tooth

$$= (\pi/2 - 0{\cdot}830 \tan 30°)m$$
$$= 1{\cdot}269m$$

It will be seen from Fig. 229 that the general effect of modifying the addenda of the teeth so as to avoid interference is to strengthen the pinion teeth, but to weaken the wheel teeth.

(c) *Increased Centre Distance.* In Article 127 it was shown that the shape of the involute depends solely upon the diameter of the base circle and that as a consequence the same pair of involutes will transmit uniform motion quite independently of the distance between the centres of the base circles. The effect of changing the centre distance is simply to increase the pressure angle. This means that, if two wheels are provided with involute teeth, the centre distance may be increased within limits without destroying the correctness of the tooth action. Nevertheless the increased centre distance will have two undesirable effects, viz. (a) the acting pitch circle diameters of the wheels will be larger than the nominal pitch circle diameters, so that the effective addenda, the working depth, and the arc of contact will all be reduced; (b) considerable backlash will be introduced between the teeth.

From the purely theoretical standpoint, however, there is no reason why the radius of the addendum circle of the pinion should not be increased so as to increase the working depth of the teeth and at the same time lengthen the arc of contact; nor is there any objection to shifting one of the profiles of the pinion tooth so as to eliminate backlash. It would, of course, be possible to make the above changes on the wheel instead of on the pinion, or part of the change might be made on the pinion and part on the wheel. Fig. 230 shows part of a 10-tooth pinion and a 30-tooth wheel which have been separated so that the centre distance corresponds to 41 teeth, i.e. one more than the sum of the actual numbers of teeth on the wheel and pinion. The shapes of the teeth, as shown shaded, correspond to standard proportions and a pressure angle of $14\frac{1}{2}°$. The pinion tooth shape shown by the unshaded line has been obtained (a) by increasing the addendum and dedendum radii so as to give normal clearances, and (b) by shifting the right-hand profile, without altering its shape, until it touches the next wheel tooth profile and so eliminates backlash.

It will be seen that the pinion tooth is very much improved in shape by the above changes. The profile is for the most part above the base circle, so that undercutting is reduced, if not entirely eliminated, and the tooth is also stronger.

The above example is given in order to show that undercutting of the flanks of wheels with small numbers of teeth may be avoided by using a centre distance which is greater than the standard centre distance. It is not suggested that it would be practicable to make the necessary changes to the shape and dimensions of the pinion tooth, because, of course, a standard cutter would have to be used on the score of expense. But it would be quite practicable for the modifications to be made partly on the pinion and partly on the wheel and yet to use a standard cutter when forming the teeth. For instance, assume that a wheel blank for a 10-tooth pinion is of standard diameter, i.e. pitch diameter plus 2 modules, and has the teeth cut with a standard $14\frac{1}{2}°$ cutter, the distance of the nominal pitch line of the cutter from the axis of the blank being so adjusted that there is no undercutting of the flanks. Then the teeth obtained will be much shorter than the standard length and the shape will be as shown at X in Fig. 231. The outline of the rack cutter teeth is shown in its correct position relative to the axis of the blank and e is the distance of the actual pitch line PP of the cutter from the nominal pitch line QQ, when generating the teeth. For simplicity the additional length of cutter tooth required in order to give the clearance at the root of the pinion teeth is not shown, since it does not in any way affect the tooth action.

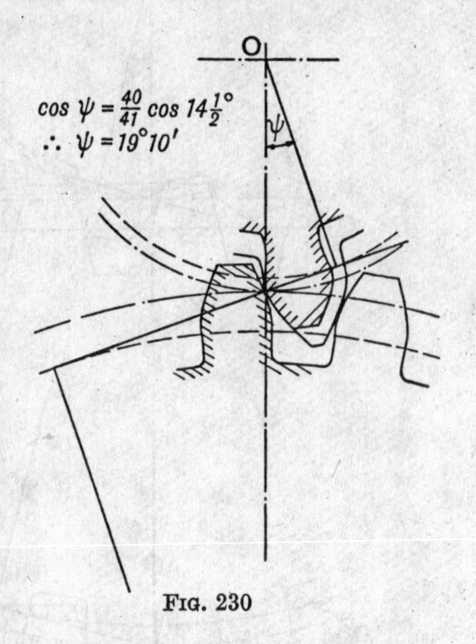

Fig. 230

Assume that a blank for a 30-tooth wheel of standard diameter also has the teeth cut in the same way with a standard $14\frac{1}{2}°$ cutter so as to avoid undercutting. The displacement e_1 of the generating pitch line P_1P_1 from the nominal pitch line Q_1Q_1 of the cutter will be much smaller than for the 10-tooth pinion, and the teeth will be only slightly shorter than the standard length. Their shape and the outline of the cutter teeth are shown at Y, Fig. 231.

The two wheels cut in the above way may be geared together, without backlash, if the centre distance is suitably chosen. Two of the teeth are shown in mesh at Z. The required centre distance will be greater than the standard centre distance and, since the diameters of the base circles are fixed by the numbers of teeth and the pressure angle and pitch of the cutter teeth, the working pressure angle will be larger than that of the cutter. The actual pitch diameters of the two wheels will also be larger than the nominal pitch diameters.

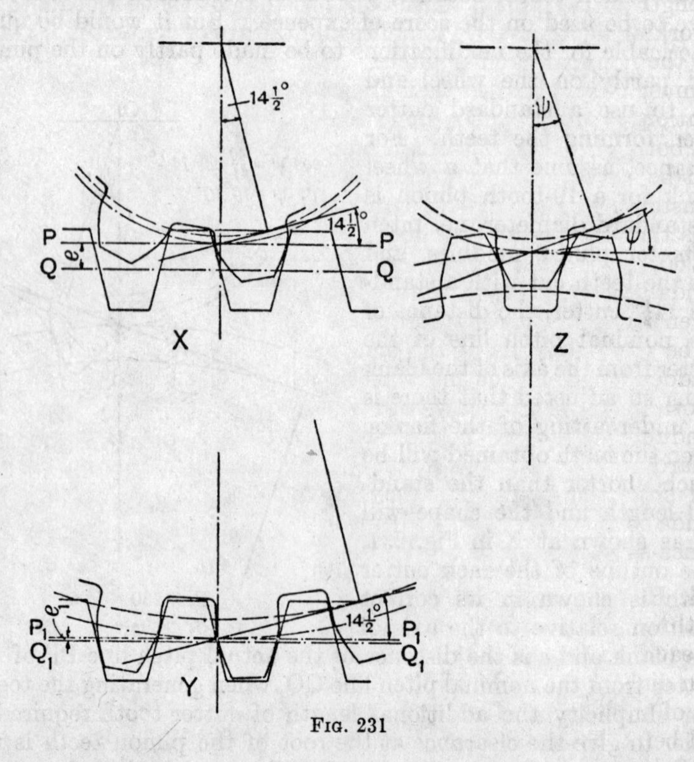

FIG. 231

It should be noted that in the particular combination shown the addendum radius of the wheel is slightly larger than the root radius of the pinion will allow, if clearance is neglected. This would not matter in practice, because the cutter teeth are actually extended so as to provide for clearance. A much more important point to notice is the very large clearance between the tops of the pinion teeth and the roots of the wheel teeth. This means that the working depth is small and the arc of contact short. But there is clearly no reason why the addendum diameter of the pinion should not be increased, so as to make the length of the pinion tooth equal to that of the wheel tooth. This would have the effect of lengthening the arc of contact.

The complete calculations involved in examining a combination similar to that of Fig. 231 are too lengthy to reproduce here. The reader is referred to *Spur Gears*, by Earle Buckingham, for the necessary details. It may perhaps be stated that the centre distance is increased in the above example from 20 to 20·62 modules, the working pressure angle is 20° 7′ and, if the outside diameter of the pinion is increased to 12·12 modules so as to make the length of the pinion tooth equal to that of the wheel tooth, the working depth is 1·874 modules and the arc of contact is 1·314 times the nominal circular pitch.

It must be emphasised that Fig. 231 is to be taken only as an indication of what can be done to obtain correct tooth action, when both wheels have a small number of teeth. In practice it would be undesirable to work so close to the theoretical limits. For instance B.S. 436:1940, standard pressure angle 20°, recommends that, if the total number of teeth in any gear pair is less than 60, both wheels shall, in general, be cut with modified addenda and an increased centre distance shall be used. A curve is given from which the appropriate increase of centre distance may be read in terms of the sum of the corrections applied to the addenda of the mating wheels. For the combination of a 10-tooth pinion and a 30-tooth wheel, it is specified that the addendum of the pinion shall be increased to 1·40 times the standard addendum of one module, but that the addendum of the wheel shall be standard. The sum of the corrections to the addenda is therefore 0·40 and the appropriate increase of centre distance as read from the curve is 0·375 module.

134. Helical Gears. In order to diminish noise and increase the smoothness of operation, the teeth of spur wheels may be cut so that each one forms part of a helix. A pair of helical gears is shown diagrammatically in Fig. 232 (a). It will be seen that the teeth of the two wheels are of opposite hand, those on the lower wheel being cut right-handed while those on the upper wheel are cut left-handed.

The pitch measured at right-angles to the teeth is termed the *normal* pitch, and differs from the circular pitch which determines the pitch circle diameters of the wheels.

Let p = the normal pitch,
p_c = the circular pitch
and α = the spiral angle of the teeth.

Then
$$p_c = p/\cos \alpha$$

The pitch circle diameter of a helical wheel with a given normal pitch is $1/\cos\alpha$ times the pitch circle diameter of a spur wheel with the same number of teeth and pitch.

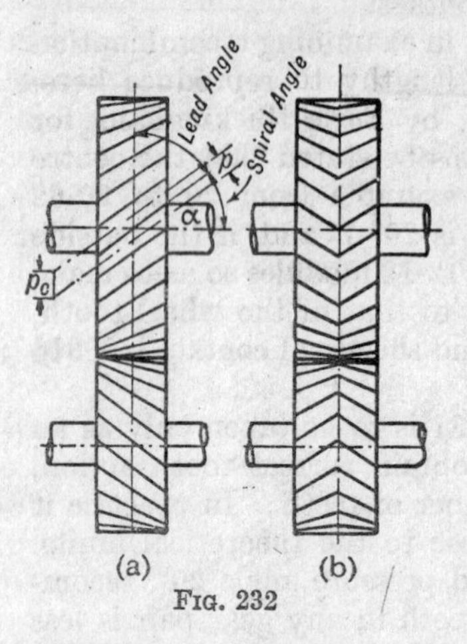

Fig. 232

Since the normal thrust between the teeth is inclined to the axis of rotation, there is an axial thrust on each shaft. To obviate this axial thrust the teeth are more often cut in the form of a double helix, as shown in Fig. 232 (b), when equal and opposite thrusts are produced on each wheel and no axial thrust is transmitted to the shafts.

The involute shape is invariably used for the profiles of single and double-helical gears and the action of the teeth is essentially the same as for normal spur gears. The chief advantage of helical gears lies in the gradual engagement of each pair of teeth, which contributes greatly to the smoothness and quietness of operation and at the same time increases the load-carrying capacity.

135. Bevel Gearing. When two shafts, the axes of which intersect, are to be connected by gearing, the wheels are known as bevel wheels. The pitch surfaces which by a pure rolling action will transmit the same motion as the bevel wheels are then frusta of cones, the apices of which coincide with the point of intersection of the axes of the two shafts: Fig. 212 (b). In order to provide a positive drive between the two shafts it is necessary to have interlocking projections or teeth on each conical frustum. The profiles of the teeth must be shaped, as in the case of spur gears, so as to satisfy the fundamental condition for the transmission of uniform motion. It is therefore clear that cycloidal or involute profiles may be adopted. The cycloidal tooth profile will be that generated by an element of a cone rolling on the pitch cone. If the generating cone rolls on the outside of the pitch cone it will sweep out the profile of the face of the tooth, and if it rolls on the inside of the pitch cone it will sweep out the profile of the flank of the tooth: Fig. 233. It is obvious that the apex of the generating cone must coincide with the apex of the pitch cone.

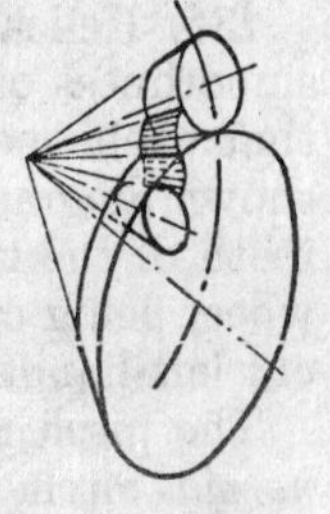
Fig. 233

It should also be noted that a point on the generating cone moves on a spherical surface, since the distance from the apex is constant. The cycloids are therefore spherical cycloids and the true shape of the path followed by a point on the generating cone cannot be represented by a curve on a plane surface. In practice a method known as Tredgold's approximation is employed. This approximation consists in substituting for the actual spherical surface the conical surface which is tangential to it. Thus in Fig. 234 the cycloidal profile of the outer edge of the tooth actually lies on the surface of a sphere of radius OA with centre O. The length

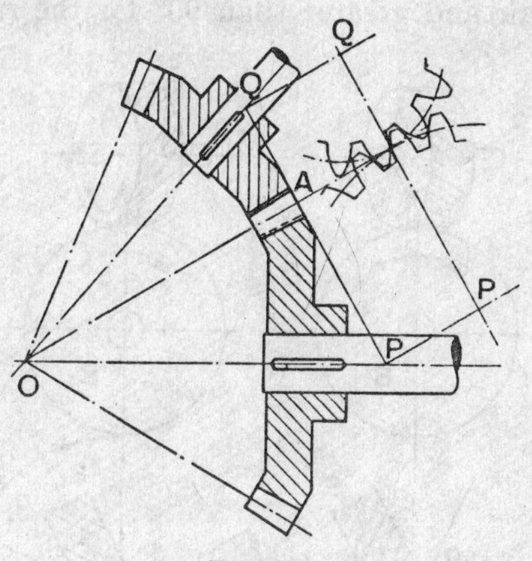

FIG. 234

of the side of the tangent cone is PA and the shape of the tooth profile is set out as for a spur gear of pitch radius PA, the diameter of the generating cone at A being used as the rolling circle.

In a similar way the approximate shape of the profile for involute teeth may be set out on the tangent cone, bearing in mind, of course, that the length of the side of the tangent cone is measured in this case from P to the base cone on which the involute is actually generated. As with spur gears, the teeth of bevel wheels are generally machined and the involute profile is almost universally employed.

136. Spiral Gears. As pointed out in Article 122 spiral gearing is used to connect non-parallel, non-intersecting shafts. The pitch surfaces are cylindrical and the teeth have point contact. Spiral gears are, therefore, only suitable for transmitting small powers. The shortest distance between the two shafts gives the centre distance for a pair of spiral gears. When the gears are viewed in the direction of this centre distance, the shaft angle is defined as the angle through which the axis of one shaft must be rotated in order to bring it parallel to the axis of the other shaft and with the two shafts revolving in opposite directions.

This definition of the shaft angle implies that a pair of helical spur gears may be regarded as the special case of a pair of spiral gears in which the shaft angle is zero.

Fig. 235 shows two pairs of spiral gears in which the relative positions of the shafts are identical, but for the same direction of rotation of the lower shaft the direction of rotation of the upper shaft is reversed. The shaft angle is less than 90° for the left-hand pair and greater than 90° for the right-hand pair.

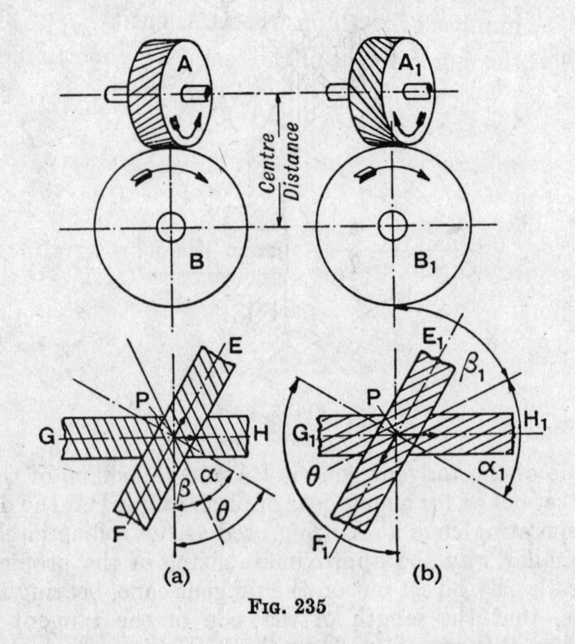

FIG. 235

In the plan views developments of the two pitch surfaces are shown in which P is the pitch point. To satisfy the assumed direction of rotation of the wheels, the line of the teeth must obviously lie between EP and PG for the left-hand pair and between E_1P and PH_1 for the right-hand pair. The small arrows indicate the directions of motion of the teeth at the pitch point.

It should be noticed that in both pairs of gears the shaft angle is equal to the sum of the spiral angles of the teeth of the mating wheels. Further, both the wheels A and B have left-hand helices, while both the wheels A_1 and B_1 have right-hand helices. If for each pair the positions of the shafts were interchanged, i.e. B above

A and B_1 above A_1 and all other conditions remained unchanged, all the wheels would have the directions of their helices reversed.

The normal pitch of the teeth must be the same for both wheels, but the circular pitches will be different unless the spiral angles for the teeth of both wheels are the same. Obviously the teeth should be designed with a standard normal pitch, so as to enable standard cutters to be used when cutting the wheels.

Let p = normal pitch,
p_a = circular pitch of wheel A,
p_b = circular pitch of wheel B,
T_a, T_b = number of teeth on wheels A and B,
α, β = the spiral angles of the teeth, i.e. the complements of the lead angles,
θ = the shaft angle,
C = centre distance, i.e. the shortest distance between the shafts,
D_a, D_b = pitch circle diameters of A and B
and G = gear ratio = T_b/T_a.

Then $$p_a = p/\cos\alpha \quad \text{and} \quad p_b = p/\cos\beta$$

Also $$D_a = \frac{T_a . p_a}{\pi} = \frac{T_a . p}{\pi\cos\alpha}$$

and $$D_b = \frac{T_b . p_b}{\pi} = \frac{T_b . p}{\pi\cos\beta}$$

he centre distance $$C = \frac{D_a + D_b}{2}$$

$$= \frac{T_a . p}{2\pi}\left(\frac{1}{\cos\alpha} + \frac{G}{\cos\beta}\right) \quad (10.10)$$

But $(T_a . p)/\pi$ is the pitch circle diameter of a spur gear with the same number of teeth and normal pitch as the spiral gear A.

Let $T_a . p/\pi = D'_a$,

then $$C = \frac{D'_a}{2}\left(\frac{1}{\cos\alpha} + \frac{G}{\cos\beta}\right) \quad . \quad . \quad . \quad (10.11)$$

137. The Efficiency of Spiral Gearing. It will be found on examination that the example on the inclined plane, which was dealt with in Article 86 and which is shown in Fig. 132, represents approximately the conditions that exist between the teeth of two spiral gears when transmitting power. Actually the normal reaction between the two teeth at the pitch point is inclined at the pressure angle ψ to the plane which is tangential to the pitch surfaces, whereas, of course, in Fig. 132 the normal reaction between the two surfaces lies in the plane of motion. The effect on the efficiency is, however, only small. It may be approximately allowed for by using a virtual coefficient of friction $\mu/\cos\psi$ instead of the actual coefficient of friction μ for the tooth surfaces. P is the pitch point, θ is the shaft angle, α and β are the spiral angles of the teeth on the two wheels, and the elements A and B are developments of the pitch surfaces. The axial thrust on the wheels is, of course, taken by thrust bearings fitted to the shafts.

A tangential effort will be required on each wheel in order to overcome the friction of the thrust and journal bearings, but, since the pitch radius will generally be large in comparison with the radius at which the friction force acts, the equivalent value of μ_1 will be small and to a first approximation ϕ_1 may be neglected. Hence the tangential effort F'_a exerted on the driver in order to overcome the tangential resistance F_b on the driven wheel may be found by substituting $\phi_1 = 0$ in equation (6.19),

$$\therefore\ F'_a = F_b \frac{\cos(\alpha-\phi)}{\cos(\beta+\phi)}$$

and the efficiency from equation (6.20) is given by:

$$\eta = \frac{\cos\alpha}{\cos(\theta-\alpha)} \cdot \frac{\cos(\theta-\alpha+\phi)}{\cos(\alpha-\phi)} \quad . \quad . \quad (10.12)$$

The efficiency is a maximum, from (6.22), when

$$\alpha = \frac{\theta+\phi}{2} \quad . \quad . \quad . \quad . \quad . \quad (10.13)$$

The corresponding maximum efficiency, from (6.23), is

$$\eta_{max} = \frac{\cos(\theta+\phi)+1}{\cos(\theta-\phi)+1} \quad . \quad . \quad . \quad (10.14)$$

It will be found that the maximum efficiency falls off rapidly as the shaft angle θ is increased. In practice this angle seldom exceeds 90°.

138. Design of Spiral Gears. The design of a pair of spiral gears is facilitated if the following facts are borne in mind:

(a) As in all other types of gearing, the velocity ratio depends solely on the numbers of teeth on the mating wheels.

(b) The efficiency is a maximum when the spiral angle of the teeth on the driver satisfies equation (10.13), i.e. the spiral angle should exceed the semi-angle of the shafts by some 2° to 4°.

(c) If the spiral angle of the driver is arbitrarily fixed, it will not usually be possible to specify arbitrarily an exact centre distance.

(d) If the exact centre distance is specified, there will in general be two values of α that will satisfy equation (10.11).

(e) If it is desired that one of the wheels should be as small as possible, that wheel should have a small spiral angle.

(f) A change in the value of the normal pitch will sometimes enable more efficient spiral angles to be used while still satisfying the specified conditions.

Example 7. A pair of spiral gears is required to connect two shafts with a shaft angle of 70°. The gear ratio is to be 2·5 to 1, the diametral pitch 8 and the centre distance approximately 5 in. Find the numbers of teeth on the two wheels and the spiral angles.

To find the number of teeth assume $\alpha = \beta$, then (10.11) reduces to

$$C = \frac{D'_a}{2}\left(\frac{1+G}{\cos\alpha}\right) \quad \text{or} \quad D'_a = \frac{2C}{1+G}\cos\alpha$$

But $D'_a = T_a/p_d$, and, on substituting,

$$T_a = \frac{2p_d.C}{1+G}\cos\alpha$$

Substituting the given data:

$$T_a = \frac{2.8.5}{1+2{\cdot}5}\cos 35° = 18{\cdot}7$$

But the gear ratio is 2·5 to 1, so that T_a must obviously be an even number, say 18. Then $T_b = 2{\cdot}5.T_a = 45$.

If the drive is to have maximum efficiency, $\alpha = (\theta+\phi)/2$ and, assuming $\phi = 6°$, $\alpha = 38°$ and $\beta = 32°$, therefore substituting in (10.11):

$$C = \frac{18}{2.8}\left(\frac{1}{\cos 38°}+\frac{2{\cdot}5}{\cos 32°}\right) = \frac{9}{8}(1{\cdot}269+2{\cdot}948) = 4{\cdot}744 \text{ in.}$$

If this is not sufficiently close to the desired centre distance of 5 in. and strength considerations will allow, a diametral pitch of 6 or 10 may be tried.

If $p_d = 6$, then $T_a = 14$, $T_b = 35$ and $C = 4{\cdot}920$ in.

If $p_d = 10$, then $T_a = 24$, $T_b = 60$ and $C = 5{\cdot}060$ in.

Where the centre distance has to be exactly 5 in., the spiral angles α and β must be altered so as to satisfy equation (10.11). This will, of course, mean some sacrifice of efficiency. As already pointed out, there are generally two values of α that will satisfy (10.11). These two values may be found approximately by means of the construction shown in Fig. 236. It is then a simple matter to find the exact values by a process of trial and error.

In Fig. 236, oa $= \frac{D'_a}{2}$, ob $= \frac{D'_b}{2}$, angle aob $= 180° - \theta$. The lines ap and bq are respectively perpendicular to oa and ob and the centre distance C is fitted in by trial, as shown by the two lines

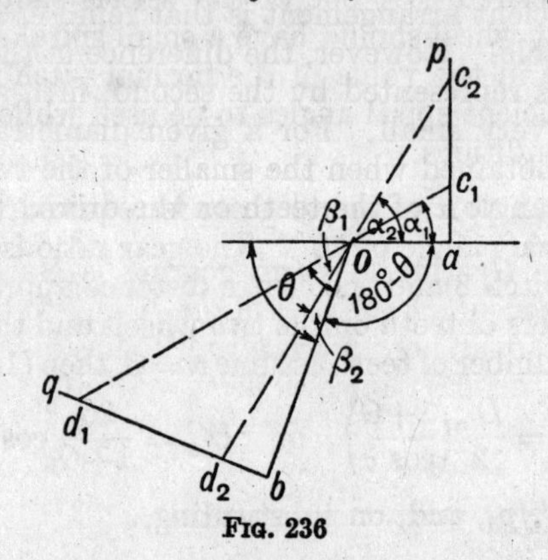

FIG. 236

c_1od_1 amd c_2od_2. The corresponding spiral angles are α_1, α_2 and β_1, β_2.

In order to enable a comparison to be made of the different arrangements which have been suggested for the spiral gear pair, the particulars are set down in the following table. The efficiency has been calculated for each case from the appropriate equation (10.12) or (10.14).

p_d	T_a	T_b	D_a	D_b	α	β	C	η%
6	14	35	2·961 2·828 3·913	6·878 7·172 6·087	38° 34° 26′ 53° 23′	32° 35° 34′ 16° 37′	4·920 5·000 5·000	86·36 86·26 84·84
8	18	45	2·855 2·581 4·250	6·632 7·419 5·750	38° 29° 18′ 58° 2′	32° 40° 42′ 11° 58′	4·744 5·000 5·000	86·36 85·90 83·68
10	24	60	3·046	7·074	38°	32°	5·060	86·36
	22	55	2·792 2·466 4·418	6·485 7·535 5·582	38° 26° 54′ 60° 8′	32° 43° 6′ 9° 52′	4·639 5·000 5·000	86·36 85·63 83·03

For each of the three diametral pitches, the first line gives the spiral angles and the centre distance which correspond to maximum efficiency. The next two lines give the spiral angles required and the efficiency of the drive when the centre distance is exactly 5 in. It should be noticed in this connection that it is impossible to obtain a centre distance of exactly 5 in. with wheels of 10 diametral pitch, unless T_a is reduced from 24 to 22. This accounts for the change in the numbers of teeth in the last three lines of the table.

It will be seen that, if the centre distance has to be exactly 5 in., the most efficient arrangement is that represented by the second line in the table. However, the difference in the efficiency of the arrangements represented by the second, fifth and ninth lines in the table is very small. For a given diametral pitch, a higher efficiency is obtained when the smaller of the two possible values of the spiral angle α of the teeth on the driving wheel is used.

139. Worm Gearing. Worm gearing is essentially a form of spiral gearing in which the axes of the driving and the driven shafts are usually, though not invariably, at right angles and the velocity ratio is high, the driving wheel being of small diameter. It differs from spiral gearing in one very important respect, namely, the teeth have line contact instead of point contact, so that the load-carrying capacity is much higher. The parallel type of worm, Fig.

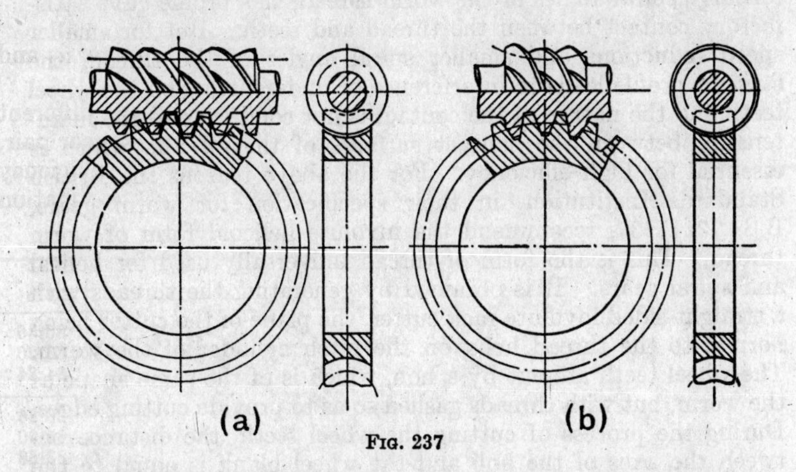

Fig. 237

237 (a), is most often used, but the enveloping worm, Fig. 237 (b), is claimed to have a greater load-carrying capacity. In appearance the parallel worm resembles a multiple-threaded screw and the wheel rim may be likened to a section, cut from a long nut by a plane parallel to the axis, and then bent so as to form an annular ring in a diametral plane of the screw or worm. As the worm rotates, the intersections of the worm threads with the plane of rotation of the wheel move parallel to the axis of the worm and

thus cause the wheel to rotate on its axis. The relative motion in the plane of the wheel is analogous to that of a rack and pinion and interference between the tips of the worm threads and the flanks of the wheel teeth may occur if the number of teeth on the wheel is small. It will be obvious, however, that, since the worm actually revolves, there is inevitably a considerable amount of sliding between the worm threads and the wheel teeth in a direction tangential to the thread surface and approximately normal to the plane of rotation of the wheel. In order to reduce this sliding to a minimum, the diameter of the worm should be as small as possible consistent with the provision of sufficient strength and stiffness.

It is not possible here to enter into a detailed discussion of either the different designs of worm or the nature of worm gear contacts. Reference should be made to articles and correspondence in the technical press and to papers read before the technical institutions. The threads on the earliest type of worm were straight-sided on an axial section, the profiles being identical with those of a $14\frac{1}{2}°$ or $15°$ involute rack. For large speed reductions and consequently for large spiral angles of the worm thread, this profile gave satisfactory contact between the thread and teeth. But for smaller speed reductions and smaller spiral angles of the thread, the straight profile leads to interference or undercutting on the wheel teeth and the nature of the contact is not conducive to the maintenance between the contact surfaces of the oil film which is essential for high efficiency. For the above reasons the British Standards Institution, in their specification for worm gears, B.S. 721:1937, recommend the involute helicoid form of worm thread. This is the form of thread universally used for helical and spiral gears. It is obtained by generating the threads with a straight-sided involute rack cutter, the plane of the cutter being normal to the thread helix on the pitch cylinder of the worm. The wheel teeth are cut by a hob, which is of the same shape as the worm, but with threads gashed so as to provide cutting edges. During the process of cutting the wheel teeth, the distance between the axes of the hob and the wheel blank is equal to the centre distance between the finished worm and wheel, and the hob and blank are given the correct relative angular displacements about their axes of rotation. With this method of cutting the blank, line contact is obtained between the worm threads and the wheel teeth.

On an axial section of the worm, the thread profiles are convex and the wheel tooth profiles are concave. This has the effect of distributing the load over a greater contact surface and also of assisting the formation of an oil film between the surfaces.

There has been a tendency in the past to regard a worm and wheel as a convenient but inefficient means of obtaining a large speed reduction between the driving and the driven shafts. Too much emphasis has also been placed on the need for a spiral angle of the worm thread of approximately 45° in order to obtain maximum efficiency. It is true that a worm and wheel can be an inefficient means of transmitting power and it is also true that maximum efficiency is obtained when the spiral angle of the worm threads is somewhat greater than 45°. But a correctly designed worm and wheel with proper lubrication has a maximum efficiency which may exceed 97 per cent and the falling off of efficiency is small over a wide variation in the spiral angle of the worm threads. So far as efficiency is concerned, the conditions are similar to those for a pair of spiral gears, but with this important difference: the line contact which exists between the threads and wheel teeth in a worm drive is much more favourable to efficient lubrication than the point contact which exists between the teeth of a spiral gear drive. Consequently the coefficient of friction between the contact surfaces in a worm drive will be low.

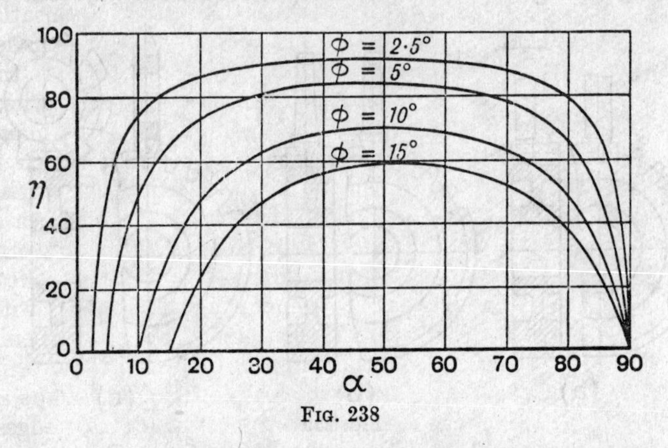

FIG. 238

In B.S. 721:1937 a curve is given showing the variation of coefficient of friction μ with rubbing speed v_s. A few figures taken from this curve are given in the table below:

v_s . . .	0	100	500	1200	3000
μ	0·150	0·054	0·030	0·020	0·014

N.B.—The rubbing speed v_s is equal to the pitch-line speed of the worm multiplied by sec λ where λ is the lead angle of the worm threads, i.e. the complement of the spiral angle α.

"The values of μ are based on the use of phosphor bronze wheels, case-hardened, ground and polished steel worms, and lubricated by a mineral oil having a viscosity of between 250 sec and 400 sec Redwood at 140°F and a specific gravity not exceeding 0·895 at 68°F."[1]

In order to show the way in which the efficiency varies with μ and α, values of the efficiency have been calculated from equation (10.12) and the curves of Fig. 238 plotted. It will be seen that, when the coefficient of friction is low, the efficiency curve is very flat in the region of maximum efficiency and also that, in order to get a high efficiency, a low coefficient of friction is much more important than a spiral angle of 45°.

140. Methods of Manufacturing Gear Wheels. The teeth on gear wheels may be either cast or machine cut. At the present time most teeth are machine cut, particularly where the pitch-line speed is high. Methods of cutting gear teeth generally involve either a milling or a planing process. Moreover, the teeth may be formed by using a suitably shaped tool or, where greater accuracy is required for the profile, they may be generated.

(a) *Formed Cutter Methods.* These are illustrated diagrammatically in Fig. 239. At (a) the spaces between adjacent teeth

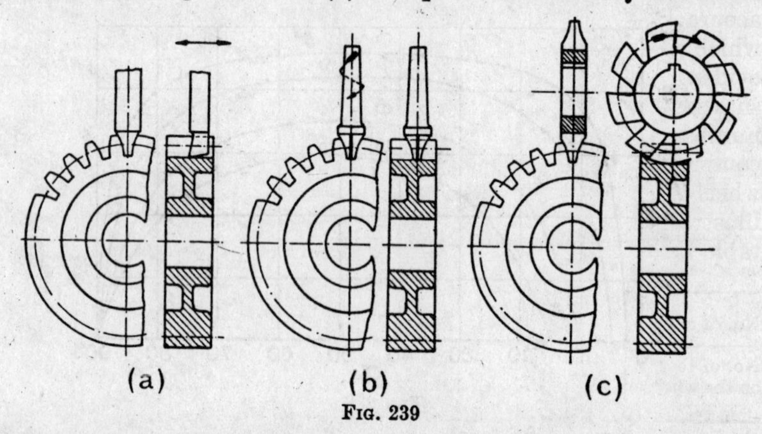

Fig. 239

are cut in the gear blank on a shaping machine. The tool is ground so that its cutting edge corresponds to the shape of the tooth space and it reciprocates parallel to the axis of the blank. When one space has been cut, the gear blank must be turned about its own axis through an angle corresponding to the pitch of the teeth in order to cut the second space. This process must be continued until all the tooth spaces have been cut.

[1] Abstracted by permission from British Standard 721:1937, Machine Cut Gears. C. Worm Gearing. Official copies can be obtained from the British Standards Institution, 2 Park St., London, W.1, price 5s. 4d. post free.

The second method, shown at (b), makes use of an end cutter or miller. The cutter rotates about an axis which is set radially with respect to the blank and at the same time the cutter is traversed parallel to the axis of the blank. The cutting edges lie on a surface of revolution, so that any axial cross-section of the cutter corresponds to the shape required for the space between two adjacent teeth on the finished wheel. There is less waste time with this method than with the first method. In practice it is only used for cutting wheels of very large pitch and wheels with double-helical teeth.

The most usual method of milling the teeth is that shown at (c). A flat circular cutter is used on the periphery of which the cutting teeth are formed. On a radial cross-section the shape of the cutting edge corresponds to the space between two adjacent teeth on the finished gear. The plane of rotation of the cutter is radial with respect to the blank. As the cutter rotates it is traversed parallel to the axis of the blank.

In each of the above methods most of the metal may be removed from the gear blank by means of straight-sided cutters. This will reduce the amount of wear on the finishing cutters. Even so, the accuracy of the tooth profiles depends upon the exactness with which the curved cutting edges on the tool are formed. It is evident too that the cutting edges will have to be differently shaped, not merely for each pitch but also for each change in the number of teeth on the wheel which is to be cut. In practice a compromise is effected by using one cutter to cover a range of wheel sizes. For example, in the case of flat circular cutters, as illustrated in Fig. 239 (c), a set is provided as indicated in the table below.

No. of cutter . .	1	2	3	4	5	6	7	8
No. of teeth on the wheel min.	135	55	35	26	21	17	14	12
No. of teeth on the wheel max.	∞	134	54	34	25	20	16	13

Each cutter has a cutting edge which is correct for the tooth space on a wheel which has the smallest number of teeth in the range.

(b) *Generating Methods.* The demand for greater accuracy in the shape of the tooth profiles has led to the development of machines for generating, rather than forming, the teeth on the gear blanks. Formed cutters such as those illustrated in Fig. 239 are suitable for producing either cycloidal or involute teeth. It is merely a question of obtaining as exactly as possible a contour for the cutting edge which corresponds to the shape of the space required between two adjacent teeth on the finished wheel. But, in the case of generated teeth, it would be much more difficult, if not actually impossible, to use the cycloidal system. For this reason all modern generating methods are based on the involute system, which has as one of its outstanding advantages the accuracy with which the profiles may be generated.

There are three methods of generating involute profiles. They are illustrated diagrammatically in Fig. 240. All three involve

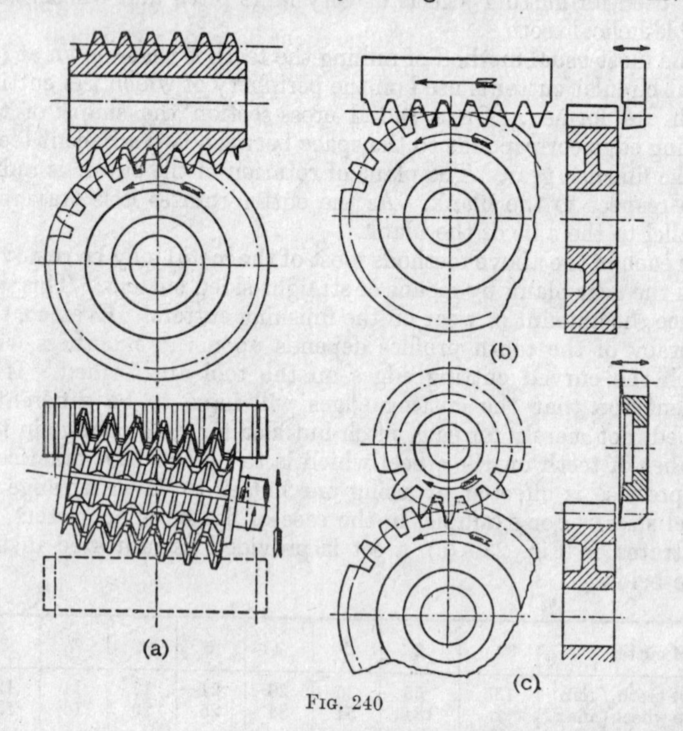

FIG. 240

the application of the same fundamental principles. Thus the cutter takes the form of a rack or pinion of the same pitch as that of the required wheel, and in the process of finishing the blank the pitch surface of the wheel rolls without slipping on the pitch surface of the cutter. The actual removal of the metal to give the spaces between adjacent teeth is brought about by moving the cutting edge either intermittently or continuously in a direction parallel to the axis of the blank.

The cutter in method (a) is termed a hob. It is essentially a screw or worm, and the normal cross-section of the thread corresponds in shape to the tooth of an involute rack such as would gear correctly with the wheel. A series of gashes runs axially across the threads, so as to provide a large number of cutting edges, and the teeth thus formed are relieved, so that there is clearance between the sides of the teeth and the sides of the groove cut in the gear blank. In operation the hob is set with its axis inclined to the axis of the blank at such an angle that the thread on the cutting side is parallel to the axis of the blank. The motion of the hob relative to the blank is of two kinds:

(i) The hob spindle and the work mandrel are connected by a gear train which causes the pitch surface of the wheel to roll without slipping on the pitch surface of the hob; in this way the profiles of the teeth on the wheel receive their correct shape.

(ii) The hob is traversed in the axial direction of the blank in order to remove the metal to form the tooth space across the full width of the blank.

The great advantage of this method of cutting gear wheels is that the process is continuous. There is no waste time from the beginning of the cut until the finished wheel is obtained. The hobbing process is therefore eminently suited to the mass-production of gear wheels.

In method (b), which is used on the Sunderland gear-shaper and machines of the same type, the cutting tool takes the form of a rack, some half-dozen or so teeth being provided. If the wheel blank were of plastic material, the teeth could be obtained by rolling the blank on the rack cutter in such a way that the two pitch surfaces had no relative sliding. In actual machines the cutting action is analogous to that of a shaping machine, only one stroke of the tool being effective, so that the cutting action is intermittent. Between each cutting stroke of the tool the relative positions of the blank and the tool are changed to correspond to a small amount of rolling of the one pitch surface on the other. To continue the cutting process round the full circumference of the blank, it is necessary to index the latter through one pitch relative to the cutter at appropriate intervals.

A slight variation of the above process is found in the Bilgram spur-gear planer. A single rack-tooth cutter is used and the gear blank is indexed, between each cutting stroke, through an angle which exceeds one pitch by the amount corresponding to the rolling together of the two pitch surfaces. The tool must be brought back to the starting position when the rolling together of the two pitch surfaces exceeds the arc of contact. The process is then repeated.

The chief advantage of the rack cutter lies in the accuracy with which the straight cutting edges may be produced. As the cutting process is intermittent the speed of production is not so high as with the hobbing machine.

In method (c), as used on the Fellows gear-cutting machine, the cutter takes the form of a pinion which has a pitch-circle diameter of from 3 to 4 in. The cutting action is intermittent, as in the case of the rack cutter, but some idle time is saved owing to the fact that the pitch surface of the cutter is cylindrical and therefore the rolling together of the pitch surfaces can take place continuously in the same direction. To obtain the same quality of finish on the surface of the teeth, more cuts will be required with a pinion cutter than with a rack cutter. The profiles of the pinion cutting teeth may be very accurately produced by means of a generating process in which a plane-face grinding wheel is used.

141. Methods of cutting Bevel Gears. In principle, the methods of cutting the teeth of bevel wheels are identical with those already described for spur gears. The teeth may be either *formed* or *generated*, but the process is, in any case, made more difficult by the change in the shape of the tooth across the face of the wheel. Not only does the width of the tooth space diminish from the back to the front of the wheel, but, in addition, the radius of curvature of the profile diminishes. Hence it is not possible to make either an end milling cutter or a circular milling cutter with a cutting profile which is correct for all normal sections of the tooth space.

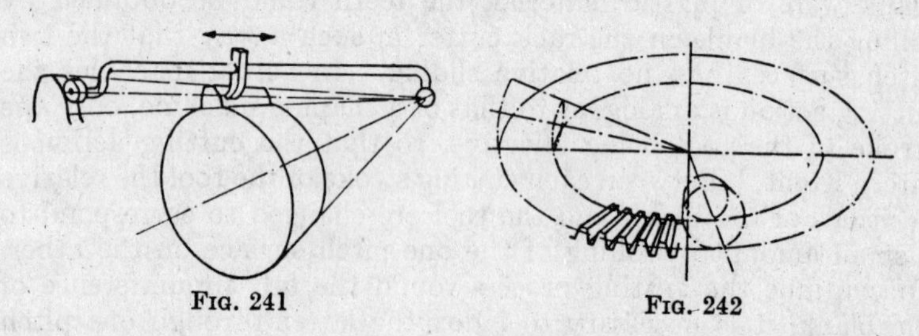

FIG. 241 FIG. 242

A more satisfactory method is to use a template to guide the tool and at the same time to cause it to reciprocate along a straight line which passes through the apex of the pitch cone. The arrangement is shown diagrammatically in Fig. 241. It is clear that only one side of the tooth space can be machined at a time.

The most usual method of machining bevel-gear teeth is essentially the same as the generating process for spur gears, in which a straight-sided rack cutter is used. The chief difference for bevel gears is that separate cutting tools must be used for the two sides of a tooth space. If the angle of the pitch cone of a bevel wheel is increased to 180°, the pitch cone becomes a plane surface and the teeth are radial with straight rack profiles. Such a wheel is termed a *crown* wheel, and, if we imagine the bevel gear blank to be made of plastic material, the straight-sided crown wheel teeth will form teeth of the correct shape in the gear blank when the pitch surfaces roll together without slipping, Fig. 242.

To cut the teeth of an actual bevel wheel one tooth of the crown wheel is, in effect, split and used as the cutting tool. The bevel-wheel spindle and the crown-wheel spindle are geared together so as to provide the correct rolling motion of the pitch surfaces. The two halves of the cutting tool are reciprocated along straight lines which pass through the apex of the pitch cone of the bevel wheel and thus remove the metal to form the tooth profiles. In

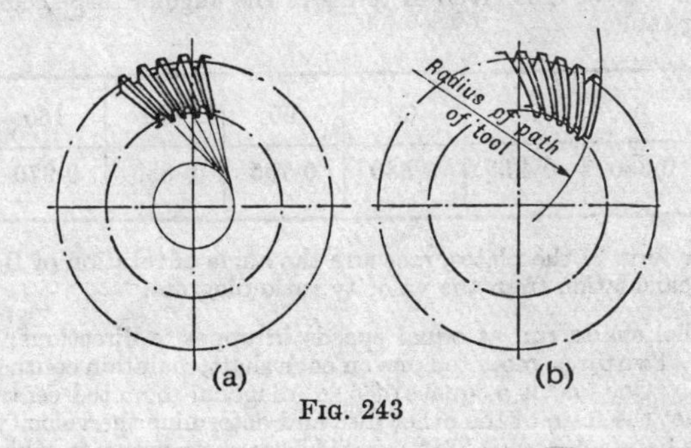

FIG. 243

practice the metal from the middle of the tooth space is removed in a preliminary operation so as to reduce the work of the finishing cutters.

In addition to bevels with straight teeth it is possible to produce bevels with so-called spiral and curved spiral teeth which offer certain practical advantages. Thus spiral teeth will be obtained if the straight line along which the cutting tool is reciprocated is tangential to a circle with the apex as centre, Fig. 243 (a).

Similarly, curved spiral teeth may be produced by causing the cutting tool to move along a circular instead of a straight path, Fig. 243 (b). This, incidentally, enables the motion of the tool during cutting to be made one of continuous rotation instead of one of reciprocation. Curved spiral bevels of this type are very largely used in the rear-axle drives of automobiles.

EXAMPLES X

1. Two rigid bodies A and B turn about fixed parallel axes and A drives B by direct contact. Show that the ratio of the angular velocity of B to that of A can only be constant if the contact surfaces are so shaped that the common normal at the point of contact intersects the line of centres at a fixed point.

2. Show that either the cycloidal or the involute shape for the profiles of wheel teeth satisfies the fundamental condition for the transmission of uniform motion. What are the principal advantages of the involute shape over the cycloidal shape?

3. Two points, 4 in. apart, are the centres of rotation of two non-circular plates A and B which maintain contact without slip through a complete revolution of A and the corresponding part of a revolution of B. The ratio r of the angular velocity of B to that of A is given in terms of the angular displacement θ of A in the following table:

θ deg. . .	0	30	60	90	120	150	180
r	0·390	0·450	0·580	0·765	0·885	0·970	1·000

Construct the form of the plates, measure the angle of rotation of B and check your result by calculation from the velocity ratio diagram. W.S.S.

4. Two parallel shafts run at equal speeds in opposite directions; their axes are 5 in. apart. Two discs, mounted one on each shaft, maintain contact throughout the motion. One disc is a square of 5 in. diagonal mounted centrally on its shaft. Construct the form of the other disc and determine the velocity of sliding when the shafts have rotated (i) $22\frac{1}{2}°$, (ii) 45° from the position in which a corner of the square is in contact with the other disc. W.S.S.

5. Explain the meaning of the following terms: circular pitch, diametral pitch, module, pressure angle, addendum, dedendum.

6. A wheel has 20 teeth, diametral pitch 3, addendum 1 module and dedendum 1·25 module. The profiles of the teeth are circular arcs of radius 1·5 in., the centres of curvature lying on a circle concentric with the pitch circle and of radius 3 in. Find graphically the required shape of the tooth profile on a 15-tooth pinion which is to gear correctly with the wheel.

7. Draw full size the profile of a rack tooth and the tooth of a 15-tooth pinion if the teeth are cycloidal, the diameter of the rolling circle for face and flank is equal to the pitch circle radius of a 12-tooth pinion and the circular pitch is 3 in. Use B. and S. proportions for the addendum and dedendum.

8. Draw full size the profiles of the teeth of a 15-tooth pinion and a 30-tooth wheel if the teeth are of involute shape. The diametral pitch is 1, the pressure angle is 20°, the addendum is 1 module and the dedendum 1·25 module.

9. What is meant by interference? Set out the profile of the tooth of a 6-tooth pinion to gear with a rack, if the pressure angle is 20°, the circular pitch is 3 in., the addendum is $0{\cdot}3183p$ and the dedendum $0{\cdot}3979p$.

10. Deduce an expression for the number of teeth in contact for a pair of gear wheels in terms of the pressure angle, the number of teeth on each of the wheels and the addendum.

What is the smallest number of teeth that can be used on each of two equal wheels in order to avoid interference if the pressure angle is $14\frac{1}{2}°$ and a standard

addendum of one module is used? Find also the length of the arc of contact in terms of the circular pitch.

11. The side thrust on the bearings of a pair of wheels with involute teeth is limited to one-third of the normal thrust between the teeth and, in addition, the addendum is 0·8 module. Find the least number of teeth that can be used on the pinion in order to avoid interference when (a) the pinion gears with an equal pinion; (b) the pinion gears with a wheel and gives a velocity ratio of 2·5 to 1; (c) the pinion gears with a rack. What is the length of the arc of contact in each case?

12. Two spur wheels each have 30 teeth of involute shape. The circular pitch is 1 in. and the pressure angle is 20°. If the arc of contact is to be equal to 2 × pitch, determine the minimum addendum of the teeth. L.U.

13. Two gear wheels of 4 in. and 6 in. pitch diameters have involute teeth of 8 diametral pitch and an angle of obliquity of 17°. The addenda are each $\frac{1}{8}$ in. Determine:

(a) the length of the path of contact;
(b) the number of pairs of teeth in contact;
(c) the angle turned through by the smaller wheel while any one pair of teeth are in contact. L.U.

14. If a standard addendum of one module is used, what is the smallest number of teeth that can be employed on a pinion to gear with a rack, the pressure angle being (a) $14\frac{1}{2}$°, (b) 20°? Prove the formula used.

15. A pair of spur wheels with 12 and 20 teeth is required with involute profiles and a pressure angle of 20°. What are the theoretical maximum addenda of the wheels if undercutting is to be avoided? Find also the working depth of the teeth and the length of the arc of contact, both expressed in terms of the diametral pitch, and the maximum velocity of sliding between the teeth in terms of the pitch line speed.

16. Two wheels with 42 and 19 teeth are cut with involute teeth of pressure angle 20° and diametral pitch 5. The addendum of each wheel is 0·2 in. Find: (a) the length of the arc of contact, (b) the number of pairs of teeth in contact, (c) the angle turned through by the smaller wheel, while any one pair of teeth is in contact.

17. A pair of spur wheels with involute teeth is to give a gear ratio of 3·5 to 1. If the arc of approach is to be not less than the circular pitch, the smaller wheel is the driver and the pressure angle is 20°, what is the least number of teeth that can be used on each wheel? What is the addendum of the wheel in terms of the circular pitch?

18. Deduce an expression for the length of the arc of contact for two involute wheels of unequal size in terms of the pressure angle ψ, the pitch circle radii and the addendum radii.

If the two wheels have 24 and 30 teeth, and a standard addendum of 1 module, and ψ is 20°, find the length of the arc of contact in terms of the circular pitch.

19. Two wheels with 15 and 24 teeth gear together. The addendum of each wheel is 0·8 module and ψ is $22\frac{1}{2}$°. What is the length of the arc of contact?

20. Involute pinions with a small number of teeth and B. and S. standard proportions cannot be cut without undercutting the flanks. Explain this statement. What modifications in the proportions may be introduced in order to avoid undercutting?

21. A gear wheel has involute teeth. The radius of the base circle is r_b, the radius of the pitch circle is r and the thickness of the tooth at the pitch circle is b. Show that at any other radius r_1, the thickness of the tooth is given by:

$$b_1 = r_1\{b/r - 2(\beta_1 - \beta)\}$$

where $\beta_1 = \tan\psi_1 - \psi_1$ and $\beta = \tan\psi - \psi$, in [which $\psi_1 = \cos^{-1} r_b/r_1$ and $\psi = \cos^{-1} r_b/r$

22. Two wheels with 15 and 25 teeth of involute shape are to be cut with a standard $14\frac{1}{2}°$ rack cutter, the generating pitch radii being so chosen that undercutting of the flanks of the teeth is avoided. Find the centre distance at which the two wheels will gear together without backlash and the corresponding pressure angle. What will be the working depth and the length of the arc of contact if the clearance is 10% of the working depth?

N.B.—The nominal addendum of the rack cutter is $0{\cdot}3683p$ and the tips of the teeth are to be assumed sharp-cornered.

23. In a spiral gear drive, the spiral angle of the teeth on the driving wheel has been fixed at 50°. The normal pitch of the teeth is $\frac{1}{2}$ in. and the driving wheel A runs at twice the speed of the driven wheel B. The shafts are at right angles and the shortest distance between their axes is approximately 7 in. Determine the dimensions of suitable gears for this drive, giving, for each wheel, (a) the number of teeth, (b) the spiral angle of the teeth, (c) the circular pitch and (d) the pitch diameter. Find also the exact distance between the axes.

If the friction angle is 5°, what is the efficiency of the wheels? L.U.

24. Determine the dimensions of a pair of wheels of a spiral gear drive connecting two shafts at right angles in order to satisfy the following particulars: approximate distance between shafts 8 in.; velocity ratio 2; normal pitch of teeth $\frac{1}{2}$ in.; slope of teeth of driver 50° with axis of shaft. State in your answer:

(a) the number of teeth in each wheel;
(b) the slope of teeth of the driven wheel;
(c) the circular pitch of each wheel; and
(d) the exact distance between the axes of the shafts. L.U.A.

25. A pair of screw wheels connects two shafts the centre lines of which are inclined at 80°, the velocity ratio is 2 and the driver has 25 teeth of a normal pitch of $\frac{1}{2}$ in. and a spiral angle of 30°. What is the least distance between the shafts? L. U.

26. A pair of screw wheels is to be designed to fulfil the following conditions: axes of the shafts at right angles; velocity ratio of driver to follower, 1 to 2; approximate centre distance, 4 in.; diametral pitch of cutter 8; the pitch line diameters of the two wheels are to be equal.

Find the spiral angles, the pitch line diameters, the numbers of teeth and the leads of the tooth helices. L.U.

27. In designing a pair of screw gears it is necessary to find the angles which the teeth of the wheels make with their respective axes.

If the data of the design permits, state what is the ideal ratio of these angles for maximum efficiency. Prove your statement. L.U.

28. A right-angled drive on a machine tool is to be made by two spiral gear wheels A and B. The wheels are to be equal in diameter with a normal pitch of $\frac{3}{8}$ in. and the axes are to be approximately 6 in. apart. The speed ratio of wheel A to wheel B is 5 to 2. Find:

(a) the apparent circular pitches of the teeth;
(b) the spiral angles of the teeth;
(c) the number of teeth on each wheel;
(d) the correct pitch-circle diameter.

If the friction angle is 6°, find the efficiency of the wheels. L.U.

29. For a right-angled spiral gear drive the normal pitch is $\frac{3}{8}$ in. the speed ratio is 5 to 2, the friction angle is 6°, the distance between the axes is approximately 6 in. and the efficiency is to be a maximum. Find:

(a) the apparent circular pitches;
(b) the spiral angles of the teeth;
(c) the numbers of teeth;
(d) the exact centre distance and the pitch-circle diameters of the two wheels;
(e) the efficiency of the drive.

30. Solve Question 29 if the distance between the axes is to be exactly 6 in. This will, of course, mean that maximum efficiency cannot be obtained.

31. A pair of spiral gears is required to connect two non-intersecting shafts which are inclined at 60°. The velocity ratio is 2 to 1, the diametral pitch of the teeth is 4 and the least distance between the axes of the shafts is approximately 10 in. Find the number of teeth on the wheels, the spiral angles and the efficiency, when (a) the least distance must be exactly 10 in., (b) the efficiency of the drive must be a maximum, (c) the pitch-circle diameters of the two wheels must be equal. The friction angle ϕ is 6° and the high-speed shaft is the driving shaft.

32. Deduce an expression for the efficiency of a worm and wheel on the assumption that the worm threads are square. Show that the efficiency is a maximum when the spiral angle of the worm threads is $45°+\phi/2$, where ϕ is the friction angle for the contact surfaces of the thread and wheel teeth.

33. In what respects does a worm-and-wheel drive differ from a spiral-gear drive? In what circumstances would the one type of drive be preferred to the other?

34. Distinguish between the methods of cutting wheel teeth in which (a) form cutters are used and (b) a generating process is used. What are the advantages and disadvantages of each method?

35. Describe the principles on which bevel-gear generating machines operate.

CHAPTER XI

GEAR TRAINS

142. Gear Trains. Any combination of gear wheels by means of which motion is transmitted from one shaft to another shaft is called a gear train. A single gear train may include any, or all, of the different kinds of gear wheels—spur, bevel, spiral, etc. —which were described in the preceding chapter.

The most usual type of gear train is that in which the wheels revolve about fixed axes. A second type of gear train is sometimes used, in which the axis of rotation of one or more of the wheels is carried on an arm which is free to revolve about the axis of rotation of one of the other wheels in the train. This type of gear train is known as an *epicyclic* train.

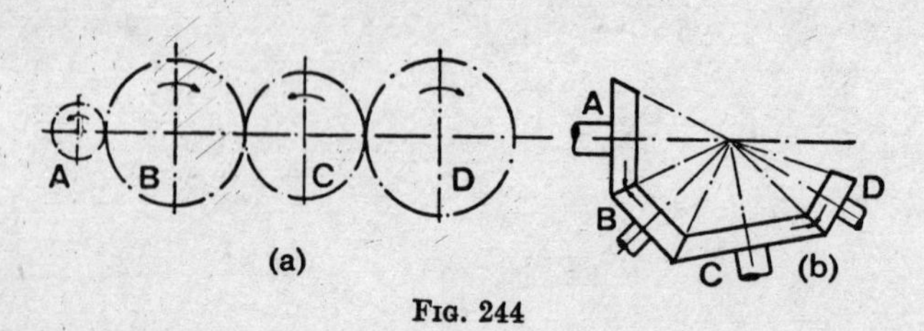

FIG. 244

A *simple* train is one in which each shaft carries one wheel only. With the exception of the first and last wheels in the train, each wheel acts both as a follower and as a driver.

Fig. 244 shows diagrammatically two examples of simple trains. It will be seen that in each of these examples the gear wheels are all of the same kind. The wheel A drives the wheel B, while B drives C and C drives D.

Let N_a, N_b, etc., be the r.p.m. of A, B, etc. Let T_a, T_b, etc., be the numbers of teeth on A, B, etc.

Then $$\frac{N_a}{N_b} = \frac{T_b}{T_a}, \quad \frac{N_b}{N_c} = \frac{T_c}{T_b}, \quad \frac{N_c}{N_d} = \frac{T_d}{T_c}$$

$$\therefore \frac{N_a}{N_d} = \frac{N_a}{N_b} \cdot \frac{N_b}{N_c} \cdot \frac{N_c}{N_d} = \frac{T_b}{T_a} \cdot \frac{T_c}{T_b} \cdot \frac{T_d}{T_c} = \frac{T_d}{T_a}$$

Hence the ratio of the speeds of A and D is inversely as the ratio of the numbers of teeth on the two wheels A and D. The intermediate wheels B and C have no effect on the velocity ratio given by the train, except in so far as they affect the sense of rotation of D for a given sense of rotation of A.

In the spur gear train, Fig. 244 (a), in which there are two intermediate wheels, the sense of rotation of D is opposite to that of A. This would be true for any even number of intermediate wheels. On the other hand, if the number of intermediate wheels is odd, the sense of rotation of D is the same as that of A.

Where the shafts intersect, as in bevel wheel trains, the above rule cannot be applied. It may happen that the first and last shafts of the train are in line, in which case a single intermediate wheel would cause the final wheel to rotate in the opposite direction to the first wheel, as in Fig. 245.

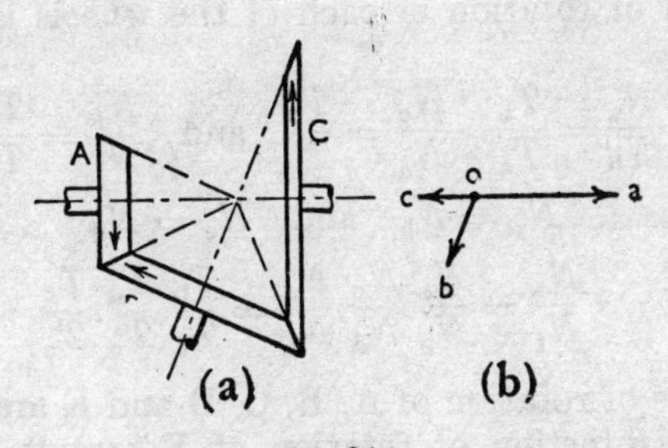

FIG. 245

One way of indicating the sense of rotation of bevel wheels is to use small arrows, as shown at (a). These arrows show the directions in which the various pitch surfaces are moving. A second way is to represent the plane and sense of rotation of each wheel by a vector drawn perpendicular to the plane of rotation. The arrow indicates the sense of rotation according to the right-handed screw rule. (See Article 17.) The three vectors shown at (b) are set off from the pole o to correspond to the three bevel wheels A, B and C of the train.

Intermediate wheels serve to bridge the gap between the first and last wheels of the train, and thus reduce the size of the individual wheels which would otherwise be required. They may also be used for driving auxiliaries incidental to the main drive. For instance, a train of gears may be used to transmit motion from the crankshaft to the camshaft of an internal-combustion engine, and the drive for an oil pump or a governor may be taken from one of the intermediate wheels.

A *compound* train is one in which each shaft, except the first and the last, carries two wheels, one of which acts as the follower and receives its motion from a wheel on a second shaft, while the other acts as a driver and transmits motion to a wheel on a third shaft. Examples of compound trains are shown in Fig. 246. The train (a) consists entirely of spur gears, while (b) consists of bevel gears and (c) includes spur, bevel and spiral gears.

It is clear that the two gears which are keyed to the same shaft must revolve at the same speed.

for (a) and (b) $$\frac{N_a}{N_b} = \frac{T_b}{T_a} \quad \text{and} \quad \frac{N_c}{N_d} = \frac{T_d}{T_c}$$

But $$N_c = N_b$$

$$\therefore \frac{N_a}{N_d} = \frac{N_a}{N_b} \cdot \frac{N_b}{N_d} = \frac{N_a}{N_b} \cdot \frac{N_c}{N_d} = \frac{T_b}{T_a} \cdot \frac{T_d}{T_c}$$

The direction of rotation of each of the wheels is shown by an arrow.

For (c) $$\frac{N_a}{N_b} = \frac{T_b}{T_a}, \quad \frac{N_c}{N_d} = \frac{T_d}{T_c} \quad \text{and} \quad \frac{N_e}{N_f} = \frac{T_f}{T_e}$$

But $$N_c = N_b \quad \text{and} \quad N_e = N_d$$

$$\therefore \frac{N_a}{N_f} = \frac{N_a}{N_b} \cdot \frac{N_c}{N_d} \cdot \frac{N_e}{N_f} = \frac{T_b}{T_a} \cdot \frac{T_d}{T_c} \cdot \frac{T_f}{T_e}$$

The directions of rotation of A, B, C, D and E are shown by the arrows. The direction of rotation of F cannot be determined unless the spiral angles of the teeth on E and F are given. If E has right-hand spiral teeth, the teeth on F must also be right-hand

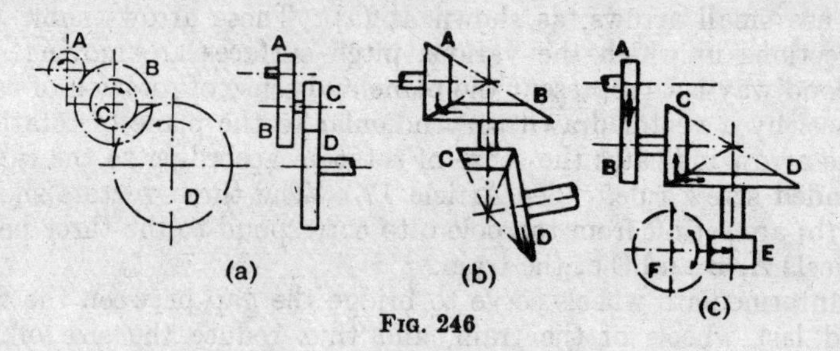

FIG. 246

and the direction of rotation of F is clockwise. On the other hand, if E has left-hand spiral teeth, the teeth on F must be left-hand and the direction of rotation of F is counter-clockwise.

The advantages of a compound train are that a large speed ratio may be obtained with wheels of small diameter and that the drive may be transmitted round bends or corners, and even, if necessary, returned upon itself so that the driving and driven shafts are co-axial.

As regards the arrangement of gear trains, the factors which have to be taken into account vary with the nature of the problem. and to bring out the differences of procedure one or two problems of different types will be considered.

143. Screw Cutting. In order to enable right- and left-hand screws of different pitches to be cut on a lathe, a set of change wheels is provided, each of which has a different number of teeth. The lead screw of the lathe, through which the saddle receives its traversing motion, has a definite pitch. Suitable change wheels must be arranged between the spindle and the lead screw so that the relative speeds of rotation of the work and the lead screw will result in the cutting of a screw of the desired pitch and of the desired hand. The problem is a straightforward one, and whether a simple or a compound train is used will depend upon the speed ratio required.

Example 1. The lead screw of a lathe has a right-hand single thread, pitch $\frac{1}{4}$ in. The smallest change wheel has 20 teeth, the largest 120 teeth and the numbers of teeth on intermediate sizes increase in steps of 5. Find a gear train suitable for connecting the spindle and the lead screw, when (a) a right-hand screw with 26 threads per inch has to be cut, and (b) a left-hand screw with 35 threads per inch has to be cut.

(a) Since the required screw is right-hand and has 26 threads per inch, it follows that the spindle must make 26 revolutions while the saddle moves 1 in. towards the headstock. But the saddle will move 1 in. towards the headstock while the lead screw makes 4 revolutions. Hence the spindle must make 26 revolutions while the lead screw makes 4 revolutions.

Let N_s = r.p.m. of spindle, N_l = r.p.m. of lead screw.

Then
$$\frac{N_s}{N_l} = \frac{26}{4} = \frac{13}{2}$$

The only wheel in the set of which 13 is a factor is that with 65 teeth.

$$\therefore \frac{N_s}{N_l} = \frac{13}{2} = \frac{65}{5} \cdot \frac{1}{2} = \frac{65}{25} \cdot \frac{5}{2} = \frac{65}{25} \cdot \frac{75}{30}$$

The wheels would be arranged as shown in Fig. 247 (a), the spindle and the lead screw revolving in the same direction.

(b) Since the required screw is left-hand and has 35 threads per inch, the spindle must make 35 revolutions while the saddle traverses 1 in. away from the headstock. The lead screw must therefore make 4 revolutions while the spindle makes 35 revolutions in the opposite direction.

$$\therefore \frac{N_s}{N_l} = \frac{35}{4} = \frac{7}{2} \cdot \frac{5}{2} = \frac{105}{30} \cdot \frac{100}{40}$$

The wheels would be arranged as shown in Fig. 247 (b).

The intermediate or idler wheel I is required in order to make the lead screw rotate in the opposite direction to the spindle. As we have already seen, the intermediate wheel does not affect the speed ratio, so that a wheel of any convenient size may be used.

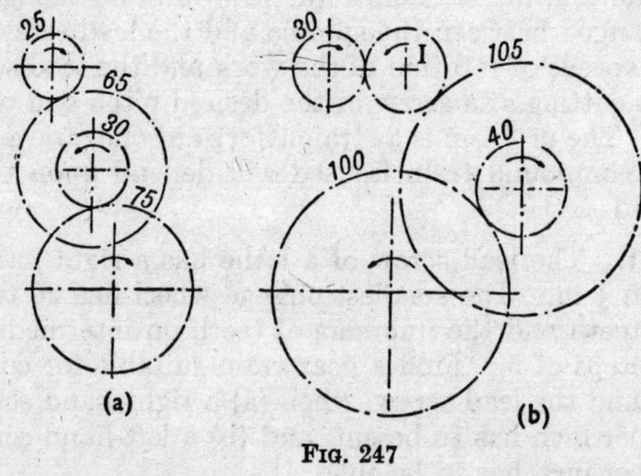

FIG. 247

144. Compound Trains for Large Speed Reductions. There are many machines in which the power is supplied through a high-speed motor. Such motors are both smaller and cheaper than slow-speed motors. The speed of rotation of the main shaft, at which the useful work is done by the machine, is often comparatively low and a train of gears is provided between the driving motor and the main shaft. The following factors have to be taken into account when arranging a gear train of this kind:

(a) The total speed reduction required and the largest speed reduction that can be allowed in one step.

(b) The space occupied by the gearing. This should be a minimum and therefore requires that the wheels shall be as small as possible. Hence the minimum allowable number of teeth on the pinions is a deciding factor.

(c) The pitch of the teeth. Since, apart from frictional losses, the power transmitted by each pair of gears in the train is the same, the pitch of the teeth must increase progressively from the high-speed to the low-speed shafts.

As an illustration the spur gear drive for the hoisting motion of an electric crane will be considered.

Example 2. A crane is required to hoist the load at a speed of 15 ft/min. The barrel on which the rope is wound is 2 ft diameter and the hoisting motor runs at 450 r.p.m. The rope is so arranged that the speed at which the rope is wound on the barrel is twice the speed of lift of the load.

Then r.p.m. of the barrel $= 2.15/2\pi = 4{\cdot}78$.

$$\therefore \text{ Total reduction of speed required} = 450/4{\cdot}78 = 94{\cdot}3/1$$

The speed ratio for a pair of spur gears in a drive of this kind will not usually exceed about 5 or 6 to 1 and therefore three reductions will be required.

If each pair of gears gives the same speed ratio, this ratio will be:

$$\sqrt[3]{(94{\cdot}3/1)} = 4{\cdot}55/1$$

The number of teeth on the various wheels in the train might therefore be chosen so as to give a ratio for each pair of approximately 4·55 to 1. Thus, if the minimum number of teeth is fixed at 20, the train might be as follows:

$$\frac{91}{20}\cdot\frac{91}{20}\cdot\frac{91}{20} = \frac{94{\cdot}2}{1}$$

This is sufficiently close to the required ratio.

But, before deciding definitely on the numbers of teeth, it is desirable to know what pitches are to be used for each pair of mating wheels. It will be assumed that considerations of strength require the diametral pitch to be 5 for the first pair of wheels, 3·5 for the second pair of wheels and 2·5 for the third pair of wheels. When choosing the numbers of teeth for the intermediate pair it is advisable, if possible, to make the total number of teeth a multiple of 7, so that the exact centre distance can be specified. If the pinion has 20 teeth and the wheel 92, the total number of teeth is 112, which is a multiple of 7 and would give a centre distance of exactly 16 in. Probably it would be better to increase the number of teeth on the pinion to 21, and retain a wheel with 91 teeth.

The minimum centre distance for the low-speed pair of mating wheels is influenced by the necessity for providing clearance between the tops of the teeth of the wheel of the intermediate pair and the drum, or barrel, on which the rope is wound. The addendum radius of a wheel with 91 teeth and a diametral pitch of 3·5 is:

$$\frac{91+2}{2.3{\cdot}5} = \frac{93}{7} = 13{\cdot}29 \text{ in.}$$

and the effective radius of the drum to the rope centre is 12 in., so that the minimum centre distance for the low-speed pair cannot well be less than about 26 in. This means that the total number of teeth for the low-speed pair must not be less than $26.2.2{\cdot}5 = 130$. The speed ratio required for this pair of wheels is:

$$\frac{94{\cdot}3}{1}\cdot\frac{21}{91}\cdot\frac{20}{91} = \frac{4{\cdot}79}{1}$$

$$\therefore \text{ number of teeth on the pinion} = \frac{130}{4{\cdot}79+1} = 22{\cdot}4 \text{ or, say, } 23$$

The number of teeth on the wheel $= 4{\cdot}79.23 = 110$.

The amended train consists of a high-speed pair with 20 and 91 teeth, an intermediate pair with 21 and 91 teeth, and a low-speed pair with 23 and 110 teeth.

The overall speed ratio or gear ratio is:

$$\frac{91}{20}\cdot\frac{91}{21}\cdot\frac{110}{23} = \frac{94{\cdot}3}{1}$$

The centre distances are:

for the high-speed pair $\dfrac{91+20}{2.5} = 11{\cdot}1$ in.

for the intermediate pair $\dfrac{91+21}{2.3{\cdot}5} = 16{\cdot}0$ in.

and for the low-speed pair $\dfrac{110+23}{2.2{\cdot}5} = 26{\cdot}6$ in.

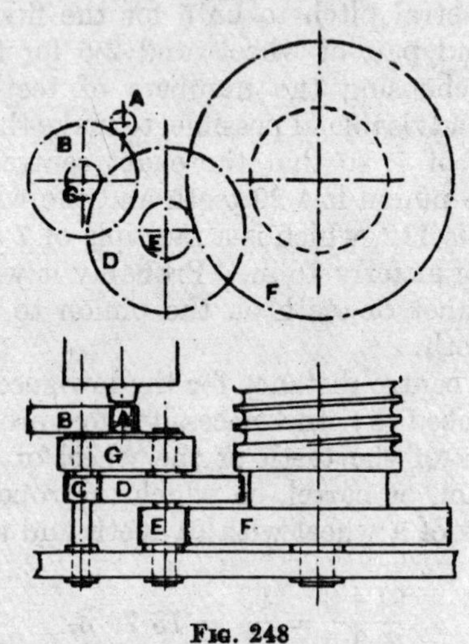

FIG. 248

So long as these centre distances are adhered to, there is considerable latitude in the actual positioning of the shaft centres. As a rule the centres would not be arranged in one straight line because of the large amount of space required. Instead they might be arranged as shown in Fig. 248. A plan view is given to show the relative positions of the various pairs of wheels. In this view G is a bracket which is designed to support one end of the driving shaft, one end of the shaft which carries the wheels B and C and one end of the shaft which carries the wheels D and E.

145. Compound Gear Trains with Co-axial Driving and Driven Shafts. In order to reduce the amount of space occupied, compound gear trains are frequently arranged with the driving and driven shafts co-axial. A simple example of this class of gear train is provided by the back gear of a lathe. The three-step cone pulley of a lathe may be connected directly to the spindle when the diameter of the work to be turned is small or the cut to be taken is light. For larger diameter work or for heavier cuts, a back gear is brought into operation. The drive is then taken from a pinion, keyed to the cone pulley, to a wheel on the back shaft, and a return drive conveys the power from a pinion on the back shaft to a wheel on the lathe spindle. By this means the spindle is made to run at a much lower speed than the cone pulley. Besides the speed ratio, the pitch, etc., an additional factor which has to be taken into account in arranging a gear train of this type is the necessity for the centre distance to be the same for both pairs of gears. Referring to Fig. 249, let G_1, G_2 be the speed ratios for the two pairs of mating wheels, r_1, r_2 the radii of the two pinions and C the centre distance.

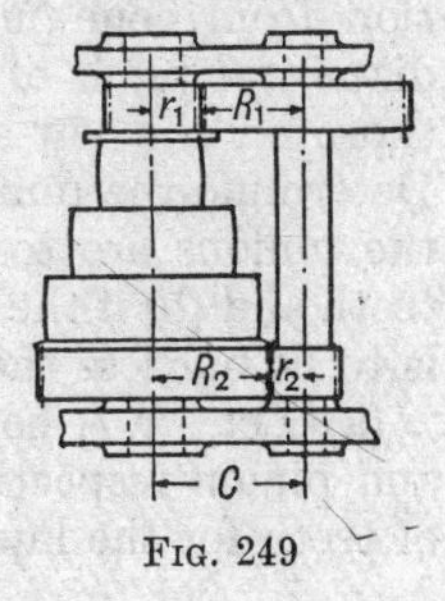

FIG. 249

Then $$C = r_1+G_1r_1 \quad \text{or} \quad r_1 = \frac{C}{G_1+1}$$

also $$C = r_2+G_2r_2 \quad \text{or} \quad r_2 = \frac{C}{G_2+1}$$

The total speed ratio from cone pulley to spindle $= G = G_1 . G_2$.

As already pointed out, the pinions in any gear train should always have as large a number of teeth as circumstances allow. This requires that the two pinions should have as nearly as possible equal numbers of teeth, since, if one of the two pinions is increased in size, the other must obviously be correspondingly diminished in size so as to give the same overall speed ratio G.

Hence it follows that, if the pitch of the teeth is the same for each pair of mating wheels, G_1 should equal G_2 and each should be equal to $\sqrt{G}$. If the pitch of the teeth is not the same for each pair, then r_1 and r_2, and consequently G_1 and G_2, will not be equal and the speed ratio for the pair of wheels with the smaller pitch will be greater than that for the pair with the larger pitch. In actual practice other factors may intervene. For instance, the minimum size of the pinion keyed to the cone pulley is largely determined by the diameter of the spindle on which the cone pulley rides. It is possible that, if the gears were designed so as to give equal numbers of teeth on the two pinions, the radius r_1 would be too small to allow of the requisite thickness of metal under the teeth. An example will make the point clear.

Example 3. A back gear is required for a lathe to give a reduction from cone-pulley speed to spindle speed of 9 to 1. The diametral pitch of the teeth on the high-speed pair is 7 and of those on the low-speed pair is 5. The centre distance is 7 in. Determine the number of teeth on each of the four wheels, (a) if the pinions are to have as nearly as possible equal numbers of teeth and (b) if the pitch diameter of the pinion on the cone pulley is to be not less than $3\frac{1}{4}$ in.

(a) Let T_1, t_1 be the numbers of teeth on the high-speed wheel and pinion respectively, let T_2, t_2 be the corresponding numbers of teeth for the low-speed pair.

$$\text{Then centre distance} = \frac{T_1+t_1}{2.7} = 7, \text{ so that } T_1+t_1 = 98.$$

$$\text{Also} \qquad T_2+t_2 = 7.2.5 = 70$$

$$\text{But} \qquad G = \frac{T_1}{t_1}.\frac{T_2}{t_2} = \frac{9}{1}, \quad \therefore \frac{98-t_1}{t_1}.\frac{70-t_2}{t_2} = \frac{9}{1}$$

$$\text{Let } t_1 = t_2, \text{ then} \qquad (98-t_1)(70-t_1) = 9t_1^2$$

This is a quadratic in t_1, the positive root of which is $t_1 = 20{\cdot}61$, the negative root being inadmissible. The nearest whole number is 21.

If each of the pinions has 21 teeth, then $T_1 = 77$ and $T_2 = 49$, so that

$$G = 77/21.49/21 = 8{\cdot}57/1$$

If $t_2 = 20$, then $T_2 = 50$ and the over-all speed ratio is:

$$G = 77/21.50/20 = 9{\cdot}17/1$$

The pitch circle diameter of the pinion on the cone pulley $= 21/7 = 3$ in.

(b) If the pitch diameter of the pinion on the cone pulley is to be not less than 3¼ in., the number of teeth t_1 must be not less than 22·8, say 23.

Then $T_1 = 98-23 = 75$ and $G_1 = 75/23 = 3{\cdot}26/1$

$\therefore\ G_2 = G/G_1 = 9/3{\cdot}26 = 2{\cdot}76/1$

But $t_2(G_2+1) = 70$

$\therefore\ t_2 = 70/3{\cdot}76 = 18{\cdot}62$, say 19

Then $T_2 = 70-19 = 51$

$G_2 = 51/19 = 2{\cdot}68/1$

and $G = 3{\cdot}26/1 . 2{\cdot}68/1 = 8{\cdot}75/1$

If this is not sufficiently close to the desired value, a change in the number of teeth, t_1 or t_2 or both, may be tried. For instance, if $t_1 = 24$ and $t_2 = 18$, then $T_1 = 74$ and $T_2 = 52$, so that

$$G = \frac{74}{24}.\frac{52}{18} = \frac{8{\cdot}91}{1}$$

146. The Motor-car Gear Box. The arrangement of a four-speed gear box of the sliding-gear type for a motor-car is shown diagrammatically in Fig. 250. The wheel A is keyed to the driving shaft and is in constant mesh with wheel B. The wheels B, D, F and H are rigidly fastened together and revolve freely on the lay-shaft. The mainshaft or driven shaft is splined, so that the wheel C and the compound wheel E-G must revolve with the shaft, but may also slide along it. The wheel C may engage directly with the wheel A through the dog clutch shown. The power is then transmitted directly to the mainshaft, and the sleeve on which the wheels B, D, F and H are mounted revolves idly. This corresponds to top gear and the driven shaft runs at the same speed as the driving shaft. Third gear is engaged by sliding wheel C along the mainshaft until it meshes with wheel D. The drive is then from A to B and back from D to C. First and second gears are provided by sliding the compound wheel E-G along the driven shaft, until either G meshes with H, which gives first gear, or E meshes with F, which gives second gear. The wheel C and the compound wheel E-G are shown in the neutral position in Fig. 250. The problem is to find the numbers of teeth

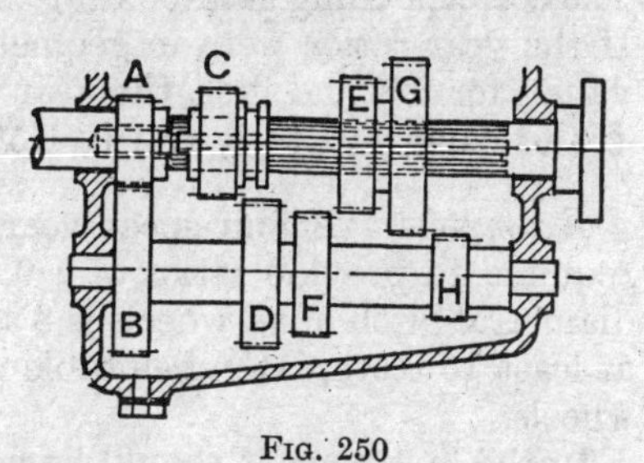

Fig. 250

on the different wheels that will give the required ratios of speed of driving shaft to speed of driven shaft. The choice of numbers of teeth is obviously limited by the fact that the centre distance must be identical for each pair of mating wheels. There is also a lower limit to the number of teeth on any one wheel. This limit may be 14 or 15, but it is desirable to adopt a larger number when space allows. The pitch of the teeth will be determined from considerations of strength. It is generally either 6 or 8 diametral pitch for a private car. The gear ratios of a car are always expressed in the form r.p.m. of engine divided by r.p.m. of driving wheels. Part of the gear ratio, indeed the whole of it in the case of top gear, is provided by the bevel or worm drive in the back axle. This ratio is usually about 5 or 5·5 to 1 except in the larger cars, where a somewhat smaller ratio is used.

The gear ratio in bottom gear is about four times the top gear ratio. The intermediate gears should theoretically be arranged so that the four ratios are in geometrical progression. In practice better results are obtained in the general performance of the car if the gap between the ratios decreases progressively from first to fourth or top. From an examination of the specifications of a large number of cars the author found that the mean ratios were:

Top	Third	Second	First
5·34	7·85	12·1	20·9

The corresponding gear-box ratios are 1·0, 1·47, 2·27 and 3·91 to 1. If the gear ratios were in geometrical progression with the same values for top and first, the gear-box ratios would be 1·0, $\sqrt[3]{3 \cdot 91}$, $\sqrt[3]{3 \cdot 91^2}$ and 3·91 to 1, or 1·0, 1·575, 2·48 and 3·91 to 1.

Example 4. A four-speed gear box, arranged as in Fig. 250, is required to provide ratios of 1·0, 1·47, 2·27 and 3·91 to 1. The diametral pitch of all wheels is 8 and the smallest pinion is to have at least 15 teeth. Find suitable numbers of teeth for the various wheels.

In the first place it should be noticed that it may be impossible to provide ratios exactly equal to those given. Since the pitch is the same for all wheels and the centre distance is the same for all pairs of mating wheels, the total number of teeth must be identical for each pair.

$$\therefore\ T_a + T_b = T_c + T_d = T_e + T_f = T_g + T_h$$

First Gear. To engage first gear the compound wheel is moved along the mainshaft until G meshes with H. The drive then takes place from A to B and from H to G. All the other gears revolve idly. This gives the biggest reduction of speed from the driving shaft to the mainshaft.

$$\therefore \frac{N_a}{N_g} = \frac{N_a}{N_b} \cdot \frac{N_h}{N_g} = \frac{T_b}{T_a} \cdot \frac{T_g}{T_h}$$

As already pointed out in connection with the last example, the two speed ratios T_b/T_a and T_g/T_h should be as nearly as possible equal, so that:

$$T_b/T_a = T_g/T_h = \sqrt{(3{\cdot}91/1)} = 1{\cdot}98/1$$

Let $T_a = T_h = 15$, then $T_b = T_g = 1{\cdot}98.15 \simeq 30$. This would mean that the actual ratio $N_a/N_g = 4{\cdot}0/1$ instead of $3{\cdot}91/1$.

A closer approximation to the value 3·91 to 1 could be obtained, if desired, by adopting the following numbers of teeth: $T_a = 16$, $T_b = 30$, $T_h = 15$ and $T_g = 31$. This would give

$$\frac{N_a}{N_g} = \frac{30}{16} \cdot \frac{31}{15} = \frac{3{\cdot}88}{1}$$

Note that $T_a + T_b = T_g + T_h = 46$.

Second Gear. To engage second gear, the compound wheel is moved along the mainshaft until E meshes with F. The drive is from A to B and from F to E.

$$\frac{N_a}{N_e} = \frac{N_a}{N_b} \cdot \frac{N_f}{N_e} = \frac{T_b}{T_a} \cdot \frac{T_e}{T_f} = \frac{2{\cdot}27}{1}$$

Then, if $T_a = 15$ and $T_b = 30$, $T_e/T_f = 2{\cdot}27/2 = 1{\cdot}135$.

But $$T_e + T_f = T_a + T_b = 45$$

$$\therefore \; 1{\cdot}135 T_f + T_f = 45$$

$$\therefore \; T_f = 45/2{\cdot}135 = 21{\cdot}1, \text{ say } 21$$

and $$T_e = 24$$

The actual ratio $N_a/N_e = 2.24/21 = 2{\cdot}286/1$.

Similarly, if $T_a = 16$ and $T_b = 30$, $T_e/T_f = 2{\cdot}27.16/30 = 1{\cdot}21$.

$$\therefore \; T_f = \frac{46}{2{\cdot}21} = 20{\cdot}8, \text{ say } 21$$

and $$T_e = 25$$

The actual ratio $N_a/N_e = 30/16.25/21 = 2{\cdot}23/1$.

Third Gear. To engage third gear, C is meshed with D and the drive is from A to B and from D to C,

$$\therefore \frac{N_a}{N_c} = \frac{N_a}{N_b} \cdot \frac{N_d}{N_c} = \frac{T_b}{T_a} \cdot \frac{T_c}{T_d} = \frac{1{\cdot}47}{1}$$

If $T_a = 15$ and $T_b = 30$, $T_c/T_d = 1{\cdot}47/2 = 0{\cdot}735$.

$$\therefore\ 1{\cdot}735 T_d = 45$$

$$T_d = 25{\cdot}9, \text{ say } 26$$

and $$T_c = 19$$

The actual ratio $N_a/N_c = 2.19/26 = 1{\cdot}46/1$.

Similarly, if $T_a = 16$ and $T_b = 30$, $T_c/T_d = 1{\cdot}47.16/30 = 0{\cdot}784$.

$$\therefore\ T_d = \frac{46}{1{\cdot}784} = 25{\cdot}8, \text{ say } 26$$

and $$T_c = 20$$

The actual ratio $N_a/N_c = 30/16.20/26 = 1{\cdot}44/1$.

The two alternative sets of gears and ratios are:

	A	B	C	D	E	F	G	H
(a)	15	30	19	26	24	21	30	15
(b)	16	30	20	26	25	21	31	15

	1st	2nd	3rd	4th
(a)	4·0	2·29	1·46	1·0
(b)	3·88	2·23	1·44	1·0

The distance between the shaft centres for wheels of 8 diametral pitch is (a) $45/(2.8) = 2{\cdot}8125$ in., (b) $46/(2.8) = 2{\cdot}875$ in.

147. The Speed Ratio of Epicyclic Trains. Where, as in epicyclic trains, the axes of rotation of the wheels are not all fixed, it is sometimes difficult to visualise exactly what happens when motion takes place. The important fact to bear in mind is that the relative motion between a pair of mating gear wheels is always the same, whether the axes of rotation are fixed or not. This relative motion is identical with the rolling together, without slip, of the pitch surfaces of the two wheels. For example, referring to Fig. 251 (a), B and C are the pitch circles of two wheels which are in contact at the pitch point P. The wheel B is fixed and the arm A, to which the axis of wheel C is attached, is free to revolve about the axis of B. Let the arm turn clockwise through the angle θ. Then, since C rolls without slipping on the circumference of B, the point on C which originally coincided with the pitch point P will now occupy the position T, such that arc RT = arc RP. Hence the radial line on C which originally coincided with QP now occupies the position ST. The angle through which C has turned relative to the fixed wheel is given by angle UST.

But $$\angle UST = \angle USR + \angle RST = \theta + \phi$$

and $$\phi = \frac{RT}{RS} = \frac{RP}{RS} = \frac{OR.\theta}{RS} = \frac{T_b}{T_c}.\theta$$

where T_b and T_c are the numbers of teeth on B and C respectively.

$$\therefore\ \angle \text{UST} = \theta+\phi = \theta(1+T_b/T_c)$$

Wheel C therefore turns in the same direction as the arm and $1+T_b/T_c$ times as fast as the arm. The speed ratio for any other epicyclic train may be found in the way just described.

Two more examples will be considered.

The epicyclic train shown in Fig. 251 (b) is similar to that shown in Fig. 251 (a), except that the intermediate wheel D is placed between the fixed wheel B and the wheel C. The pitch points are P and Q. Let the arm turn clockwise through the angle θ, then the point on D which originally coincided with the pitch point P

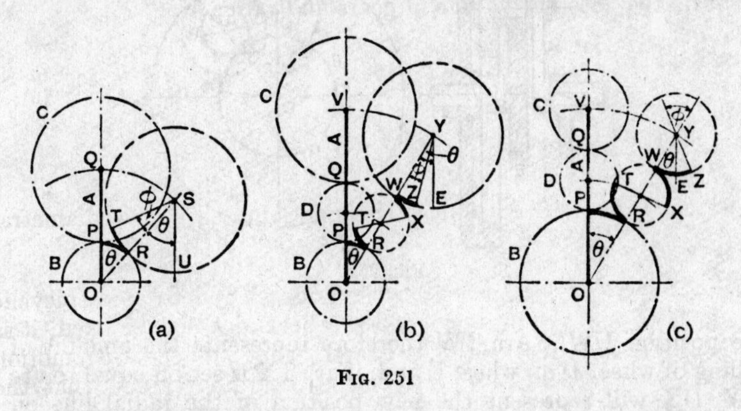

FIG. 251

will occupy the position T such that arc RT = arc RP. Similarly the points on D and C which originally coincided at the pitch point Q will occupy respectively the positions X and Z, such that arc WZ = arc WX = arc RP. This obviously means that the radial line on C, which originally coincided with VQ, now occupies the position YZ. The wheel C has therefore turned clockwise relative to the fixed wheel through an angle EYZ.

Let ϕ be the angle through which C turns relative to the arm and T_b, T_c be the numbers of teeth on B and C.

Then $$\phi = \frac{\text{WZ}}{\text{WY}} = \frac{\text{PR}}{\text{WY}} = \frac{\text{OP}.\theta}{\text{WY}} = \frac{T_b}{T_c}.\theta$$

But $\angle \text{EYZ} = \angle \text{EYW} - \angle \text{WYZ} = \theta-\phi = \theta(1-T_b/T_c)$

Hence wheel C turns $1-T_b/T_c$ times as fast as the arm and in the same direction as the arm. If $T_b > T_c$, $1-T_b/T_c$ is negative. The negative sign indicates that C turns in the opposite direction to the arm. See Fig. 251 (c).

A slightly more complicated epicyclic train is shown in Fig. 252. It is a compound train in which the internal wheel B is fixed, the compound wheel C-D revolves on a pin fixed to the arm A and the wheel E is co-axial with the fixed wheel B. Wheel B meshes with C and wheel E with D. Let the arm turn clockwise through the angle θ. Then wheel C will roll on the fixed wheel B and the radial line RP will move to the position UV such that arc SV = arc SP. Since wheel D is integral with wheel C, it must turn through the same angle as C, so that the radial line RQ moves to

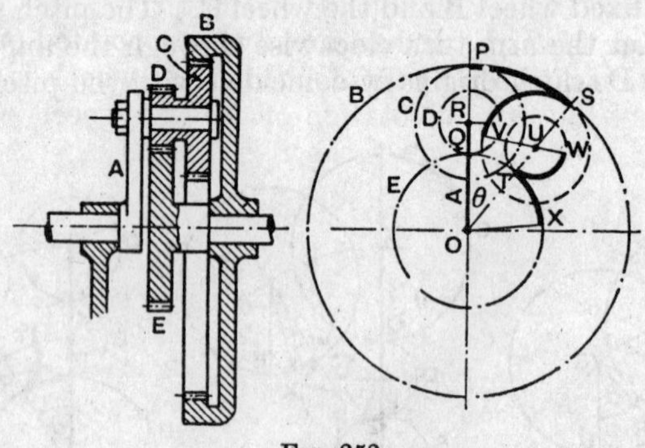

FIG. 252

the position UW. Arc TW therefore represents the amount of rolling of wheel D on wheel E and, if arc TX is set off equal to arc TW, OX will represent the new position of the radial line on wheel E which originally coincided with OQ. It follows that wheel E turns through angle QOX while the arm turns through the angle θ. Hence, the rotation of E is in the same direction as that of the arm but the speed of rotation is higher.

The relation between the speeds of rotation of wheel E and the arm A may be expressed in terms of the numbers of teeth on the various wheels as follows:

$$\angle QOX = \angle QOT + \angle TOX = \theta + \angle TOX$$

But $\angle TOX = TX/OT$,

and $$TX = TW = SV.\frac{UW}{UV} = SP.\frac{UW}{UV} = OP.\theta.\frac{UW}{UV}$$

$$\therefore \angle TOX = \frac{OP}{UV}\cdot\frac{UW}{OT}.\theta = \frac{T_b}{T_c}\cdot\frac{T_d}{T_e}.\theta$$

$$\therefore \angle QOX = \theta\left(1+\frac{T_b}{T_c}\cdot\frac{T_d}{T_e}\right)$$

This means that wheel E turns in the same direction as the arm and $1+T_b/T_c.T_d/T_e$ times as fast as the arm.

148. Alternative Ways of finding the Speed Ratio of Epicyclic Trains. The method of the last Article is too cumbersome to be used in the solution of any but the simplest epicyclic trains, and several more convenient methods of solving such problems have been devised. Some of these will be described as applied to the train shown in Fig. 252.

1. *Algebraic Method.* It has already been pointed out that the relative motion between a pair of mating gear wheels is always the same whether the axes of rotation of the two wheels are fixed or not. This relative motion corresponds to rolling of the two pitch surfaces without slip and is therefore determined by the number of teeth on the mating wheels. For the gear train shown the arm A, the spur gear E and the annulus B have a common axis of rotation, while the wheel C-D rotates about a parallel axis fixed to A. It is possible for the three members A, B, E of the train to be rotating about the common axis at different speeds N_a, N_b and N_e, but only two of these speeds can be arbitrarily assigned, since the relative speeds of rotation must conform to the numbers of teeth on the mating wheels.

The speed of E relative to the arm $= N_e - N_a$, and the speed of B relative to the arm $= N_b - N_a$. It follows that

$$\frac{N_e - N_a}{N_b - N_a} = -\frac{T_d}{T_e} \cdot \frac{T_b}{T_c} \quad . \quad . \quad . \quad (11.1)$$

where the negative sign is required because the relative speeds are opposite in sense.

It is clear that any one of the speeds may be zero, i.e. the corresponding member may be fixed, and the equation will then give the speed ratio for the other two members. Alternatively, any two of the members may have arbitrarily assigned speeds, in which case the equation will give the resulting speed of the third member.

With A fixed, $N_a = 0$,

$$\therefore \frac{N_e}{N_b} = -\frac{T_d}{T_e} \cdot \frac{T_b}{T_c}$$

With B fixed, $N_b = 0$,

$$\therefore \frac{N_e - N_a}{0 - N_a} = -\frac{T_d}{T_e} \cdot \frac{T_b}{T_c}, \quad \frac{N_e}{N_a} = 1 + \frac{T_d}{T_e} \cdot \frac{T_b}{T_c}$$

With E fixed, $N_e = 0$

$$\therefore \quad \frac{0-N_a}{N_b-N_a} = -\frac{T_d}{T_e}\cdot\frac{T_b}{T_c}, \quad \frac{N_b}{N_a} = 1+\frac{T_e}{T_d}\cdot\frac{T_c}{T_b}$$

If the speed of rotation N_c of the compound planet C-D is required, this may be obtained by considering the speeds of C-D and B, or of C-D and E, relative to the arm A. Thus,

$$\frac{N_c-N_a}{N_b-N_a} = \frac{T_b}{T_c} \quad . \quad . \quad . \quad . \quad (11.2)$$

and

$$\frac{N_c-N_a}{N_e-N_a} = -\frac{T_e}{T_d} \quad . \quad . \quad . \quad (11.3)$$

2. *Tabular Method.* Referring to Fig. 253 (i), let the arm A be fixed and the annulus B be turned counter-clockwise through the angle θ = angle ROJ. Then the planet C-D turns in the same sense through the angle α = angle RQH and the wheel E turns clockwise through the angle ϕ = angle POF. With the wheels locked so that no relative motion is possible, rotate the train about the axis of B in the clockwise sense through the angle θ. The net result of these two operations is shown at (ii) and is clearly the same as if, with the wheel B fixed, the arm were turned clockwise through the angle θ. The resulting rotation of E is $\theta+\phi$ clockwise.

If, after the first operation, the wheels are locked and the train is rotated counter-clockwise through the angle ϕ, the net result is shown at (iii) and corresponds to a counter-clockwise rotation of the arm through angle ϕ with the wheel E fixed. The resulting rotation of B is $\theta+\phi$ counter-clockwise.

Since $\quad \alpha = -\frac{T_b}{T_c}.\theta \quad$ and $\quad \phi = \frac{T_b}{T_c}\cdot\frac{T_d}{T_e}.\theta$

we get in the first case, with the annulus B fixed,

$$\frac{\text{Rotation of E}}{\text{Rotation of A}} = \frac{\theta+\phi}{\theta} = 1+\frac{T_b}{T_c}\cdot\frac{T_d}{T_e}$$

and, in the second case, with E fixed

$$\frac{\text{Rotation of B}}{\text{Rotation of A}} = \frac{\theta+\phi}{\phi} = 1+\frac{T_c}{T_b}\cdot\frac{T_e}{T_d}$$

In applying this method it is usual to assume θ to be one revolution. The steps required may then be tabulated as shown below: In this table line (a) corresponds to the first operation, i.e. with A fixed, B is turned through one revolution counter-clockwise; line (b) corresponds to the second operation, i.e. with the wheels locked the train is turned through one revolution clockwise; line (c) is obtained by adding together corresponding entries in lines (a) and (b) and is equivalent to a clockwise rotation of the arm A through one revolution with the annulus B fixed.

	Revolutions of A	Revolutions of B	Revolutions of C-D	Revolutions of E
(a)	0	-1	$-\frac{T_b}{T_c}$	$+\frac{T_b}{T_c}\cdot\frac{T_d}{T_e}$
(b)	1	1	1	1
(c)	1	0	$-\frac{T_d}{T_c}$	$1+\frac{T_b}{T_c}\cdot\frac{T_d}{T_e}$

If a solution is required for a fixed sun wheel E, line (a) would be set down for -1 revolutions of E with the arm fixed; line (b) would remain unchanged and line (c) would be found by adding together corresponding entries in lines (a) and (b).

3. *Graphical Method.* The relevant diagrams are shown in Fig. 253 below the corresponding diagrams used in explaining the tabular method. The same system of lettering is used in both sets of diagrams. Consider the two diagrams (iv) and (i), the points O_1, P_1, Q_1, R_1 in the former correspond to O, P, Q, R in the latter. The straight line P_1F_1, perpendicular to O_1P_1, is set off equal to the arc PF (or PG) and the point J_1 is obtained by joining F_1 to Q_1 and producing. It follows from the similar triangles $J_1R_1Q_1$, $F_1P_1Q_1$ that $R_1J_1/P_1F_1 = R_1Q_1/P_1Q_1$, so that R_1J_1 = arc RJ (or RH). But in Fig. 253 (i),

$$\phi = \text{arc PF/PO}, \quad \alpha = \text{arc PG/PQ}, \quad \theta = \text{arc RJ/RO}$$

$\therefore$ in (iv),

$$\phi = P_1F_1/P_1O_1 = \tan P_1O_1F_1, \quad \alpha = P_1F_1/P_1Q_1 = \tan P_1Q_1F_1$$

$$\theta = R_1J_1/R_1O_1 = \tan R_1O_1J_1$$

Hence, the angular displacements ϕ, α and θ are represented in (iv) by the tangents of the angles of inclination to the vertical of the appropriate lines, and a change in sense of the angular displacement is represented by a change in sense of the inclination to the vertical line O_1R_1.

If, now, a line is drawn perpendicular to O_1R_1 through a point a at any convenient distance h from O_1, F_1O_1 and J_1O_1 may be produced to intersect this line at e and b and a line may also be drawn from O_1 parallel to J_1F_1 to intersect it at d. It follows that:

$$\phi = \tan eO_1a, \quad \alpha = \tan dO_1a, \quad \theta = \tan bO_1a$$

and, therefore, the angular displacements ϕ, α and θ are represented to scale and in sense by the intercepts ae, ad and ab.

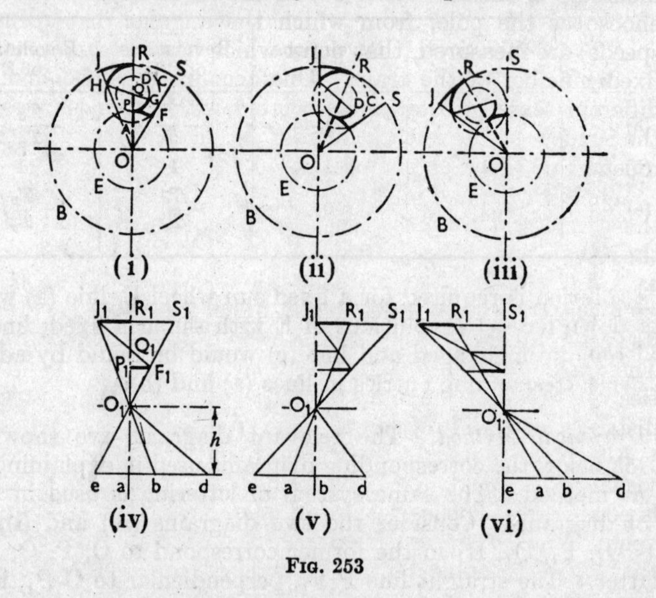

FIG. 253

The same procedure is followed in deriving the diagrams (v) and (vi) corresponding to (ii) and (iii). The distances between the points e, a, b and d will be exactly the same in all three diagrams (iv), (v) and (vi), the only difference being that the line ed is shifted to the left or to the right relative to the vertical line which passes through O_1.

With arm A fixed, diagram (iv), the point a is the pole from which angular displacements (or velocities) are measured. The displacements (or velocities) of C-D and B are opposite in sense to that of E and the relative magnitudes are given by the distances of the points d, b and e from a.

With the annulus B fixed, diagram (v), the point b is the pole, the displacements (or velocities) of E and A are of opposite sense to that of D and the relative magnitudes are given by be, ba and bd.

Similarly, with the gear E fixed, diagram (vi), the point e is the pole, and the displacements (or velocities) of A, B and D are all in the same sense, their relative magnitudes being given by ea, eb and ed.

Since the points e, a, b and d have the same relative positions in all three diagrams (iv), (v) and (vi), it is clear that (iv) contains all the information necessary to solve a problem on the particular gear train, whatever the fixed member or even if none of the members is fixed. Once (iv) is drawn it is only necessary to choose as the pole, from which the angular displacements (or speeds) are measured, that point which has the same letter as the fixed member of the train. Due account must be taken of the different sense of rotation associated with distances measured to the right or to the left of the pole, i.e. the distances from the pole must be regarded as vectors.

Where two of the members have arbitrarily assigned velocities, the position of a pole o must be found which is consistent with these velocities. For instance, suppose E turns in the same sense as A but at twice its speed, the pole o must divide ea externally, such that oe = 2oa; if E turns in the opposite sense to A at twice its speed, the pole o must divide ea internally such that oe = 2oa. The position of the pole having been fixed in this way, the velocities of the other members of the train are given by the vector distances of b and d from o.

Although the full construction shown in (iv) helps one to visualise the motions of the individual members of the train, it is possible to dispense with most of it and to set down directly the vector diagram of angular displacements eabd. It has been shown that the vectors ae, ab and ad represent completely the angular displacements ϕ, α and θ of the members E, C-D and B when the arm A is fixed.

But $$\alpha = -\frac{T_e}{T_d} \cdot \phi \quad \text{and} \quad \theta = -\frac{T_e}{T_d} \cdot \frac{T_c}{T_b} \phi$$

It therefore follows that:

$$\text{ae} : \text{ad} : \text{ab} :: 1 : -\frac{T_e}{T_d} : -\frac{T_e}{T_d} \cdot \frac{T_c}{T_b}$$

and the vector diagram eabd can be set off directly to satisfy this relation.

4. *Vector Method.* In some respects this is the simplest of all the methods described. It gives the vector diagram of angular displacements or velocities directly. Since angular displacement is a vector quantity, it can be represented according to the usual convention (Article 17, p. 16) by a line parallel to the axis of

rotation with an arrow to indicate the sense of rotation in conformity with the right-handed screw rule.

For the gear train shown in Fig. 254, there are four members each of which may rotate about parallel axes. If the arm A is fixed, all the axes of rotation are fixed and only the three members B, C-D and E can rotate. The relative angular velocities are determined by the numbers of teeth on the mating gear wheels.

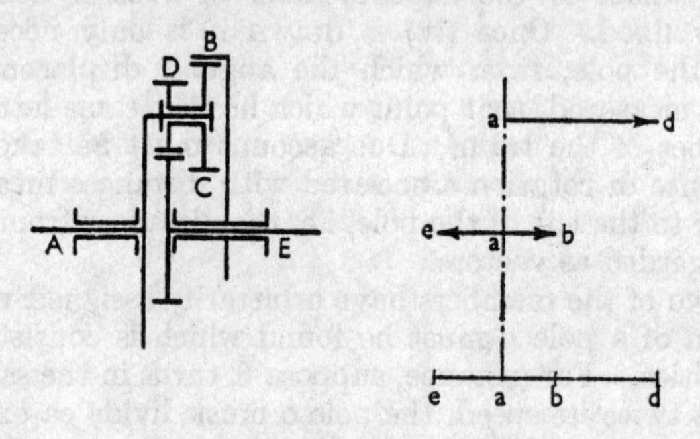

FIG. 254

Thus if B is given one revolution, C-D will make T_b/T_c revolutions in the same sense and E will make $T_b/T_c \cdot T_d/T_e$ revolutions in the opposite sense. These three angular displacements may be represented by three vectors perpendicular to the planes of rotation of the wheels. All the wheels of this particular train revolve in parallel planes, so that the three vectors are parallel as shown, where:

ab is the angular displacement of B with the arrow pointing to the right to indicate clockwise rotation of wheel B as viewed from the left,

ad is the corresponding angular displacement of the compound wheel C-D, also with the arrow pointing to the right, and

ae is the angular displacement of E with the arrow pointing to the left.

The relative lengths of the vectors are:

$$\text{ab}:\text{ad}:\text{ae}::1:\frac{T_b}{T_c}:\frac{T_b}{T_c}\cdot\frac{T_d}{T_e}$$

For convenience the vectors may be set off from a common pole as shown. The relative positions of the points e, a, b and d are then exactly the same in Fig. 254 as in Fig. 253 (iv), and the effect of changing the fixed member of the train, or, alternatively, of assigning arbitrary velocities to two members of the train, may be found by changing the position of the pole from which the displacements (or velocities) are measured as explained earlier in connection with Fig. 253 (iv).

Which of the four methods described above should be used in solving a particular problem is largely a matter of personal preference. Arguments can be advanced in favour of each, but for a simple train such as that used in explaining the different methods, there is probably little to choose between them.

Graphical methods are undoubtedly of assistance in visualising the motions of the individual members of the train, but are not always suitable except as a rough check. This applies particularly where, as often happens, a large speed ratio is provided. It is of course true that, even in such trains, the speed ratio can be obtained accurately by calculation based on the graphical construction, but there will then be no saving of effort. In a complex train, the number of lines required in the graphical construction is also liable to be confusing.

The vector method has two advantages. In the first place, each vector indicates the plane of rotation as well as the magnitude and sense of rotation. This is particularly valuable when the epicyclic train includes bevel wheels or other gears with non-parallel axes. In the second place, it simplifies the consideration of the relation between the externally applied torques, and thus enables the tooth loads to be found without difficulty.

In the examples which are given later, the solution will be found by at least two different methods. The reader is advised to use methods other than those used by the author, and then decide for himself which of the various methods he prefers.

149. Tooth Loads and Torques in Epicyclic Gear Trains. If the parts of an epicyclic gear train are all moving at uniform speeds, so that no angular accelerations are involved, the algebraic sum of all the external torques applied to the train must be zero, or

$$\Sigma(M) = 0$$

There are at least three external torques for every train, and in many cases there are only three. These are:

M_1 the input torque on the driving member,
M_r the resisting, or load, torque on the driven member,
M_b the holding, or braking, torque on the fixed member.

If there is no acceleration,

$$M_1 + M_r + M_b = 0 \quad . \quad . \quad . \quad . \quad (11.4)$$

or

$$F_1 r_1 + F_r r_r + F_b r_b = 0 \quad . \quad . \quad . \quad (11.5)$$

where F_1, F_r and F_b are the corresponding externally applied forces at radii r_1, r_r and r_b.

Further, if there are no internal friction losses at the bearings and at the contact surfaces of the wheel teeth, the net energy dissipated by the train must be zero, or

$$\Sigma(M\omega) = 0$$

$$\therefore\ M_1\omega_1 + M_r\omega_r + M_b\omega_b = 0 \quad . \quad . \quad (11.6)$$

where ω_1, ω_r and ω_b are the angular velocities of the three members to which the external torques are applied.

But for the fixed member, $\omega_b = 0$, so that

$$M_1\omega_1 + M_r\omega_r = 0 \quad . \quad . \quad . \quad . \quad (11.7)$$

From (11.7) the resisting, or load, torque:

$$M_r = -M_1 . \omega_1/w_r \quad . \quad . \quad . \quad . \quad (11.8)$$

and, from (11.4):

$$M_b = -(M_1 + M_r)$$

$$= M_1(\omega_1/\omega_r - 1) \quad . \quad . \quad . \quad . \quad (11.9)$$

These equations may be used to find the values of M_r and M_b (or F_r and F_b) when the input torque M_1 applied to the driving member is known. In addition, for complex trains, they may be used to find the tooth loads or torques on all the intermediate members through which power is transmitted.

Example 5. An epicyclic speed-reduction gear is shown in Fig. 255. The driving shaft carries on the arm A a pin, on which the compound wheel B-C is free to revolve. Wheel C meshes

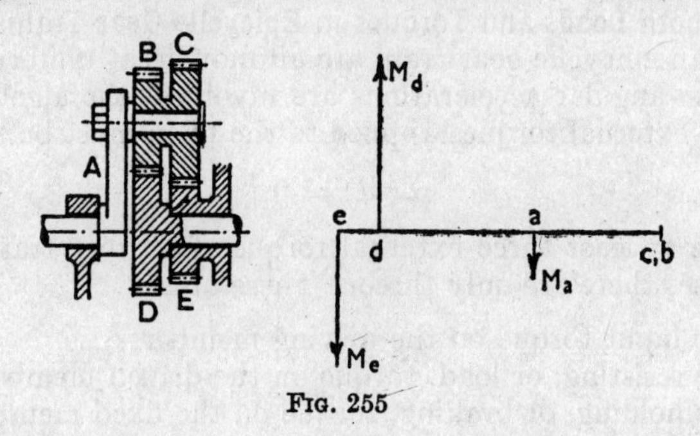

FIG. 255

with the fixed wheel E and wheel B meshes with a wheel D keyed to the driven shaft. The numbers of teeth on the wheels are: $T_b = 27$, $T_c = 30$, $T_d = 24$ and $T_e = 21$. Find the ratio of the speed of the driving shaft to the speed of the driven shaft.

If the input torque to the driving shaft is 20 lb ft, what are the load torque on D and the holding torque on E?

(1) *Algebraic Method.* With the arm fixed:

$$\frac{N_d}{N_e} = \frac{T_e}{T_c} \cdot \frac{T_b}{T_d}$$

and therefore when the arm is not fixed:

$$\frac{N_d - N_a}{N_e - N_a} = \frac{T_e}{T_c} \cdot \frac{T_b}{T_d} = \frac{21}{30} \cdot \frac{27}{24} = \frac{63}{80}$$

But wheel E is fixed, so that $N_e = 0$,

$$\frac{N_d - N_a}{0 - N_a} = \frac{63}{80} \quad \text{or} \quad N_d = N_a\left(1 - \frac{63}{80}\right) = N_a \cdot \frac{17}{80}$$

$$\therefore\ N_a/N_d = 80/17 = 4 \cdot 71/1$$

(2) *Tabular Method.* The table is filled in as shown below. The wheel E is to be the fixed wheel, so that in the first operation, when the arm is fixed, wheel E is given one revolution counter-clockwise as viewed from the left, say.

	Revolutions of A	Revolutions of E	Revolutions of B-C	Revolutions of D
(a) . . .	0	−1	$\frac{T_e}{T_c} = \frac{21}{30}$	$-\frac{T_e}{T_c} \cdot \frac{T_b}{T_d} = -\frac{21}{30} \cdot \frac{27}{24}$
(b) . . .	1	1	1	1
(a)+(b) .	1	0	$1 + \frac{21}{30} = \frac{17}{10}$	$1 - \frac{21}{30} \cdot \frac{27}{24} = \frac{17}{80}$

$$\therefore\ N_a/N_d = 80/17 = 4 \cdot 71/1$$

The driving shaft therefore runs at 4·71 times the speed of the driven shaft and the two shafts revolve in the same direction.

Assuming that there are no friction losses and that the members are revolving at uniform speeds,

$$M_a + M_d + M_e = 0 \quad . \quad . \quad . \quad . \quad . \quad (1)$$

and

$$M_a\omega_a + M_d\omega_d + M_e\omega_e = 0 \quad . \quad . \quad . \quad . \quad (2)$$

But E is fixed, so that $\omega_e = 0$ and

$$M_a\omega_a + M_d\omega_d = 0$$

$$\therefore\ M_d = -M_a \cdot \omega_a/\omega_d = -4 \cdot 71 M_a$$

$$= -94 \cdot 2 \text{ lb ft}$$

Also from (1):

$$M_e = -(M_a + M_d)$$
$$= -(20 - 94{\cdot}2)$$
$$= 74{\cdot}2 \text{ lb ft}$$

The torque applied by D to B

$$= -M_d . T_b/T_d = 94{\cdot}2 . 27/24 = 106 \text{ lb ft}$$

and that applied by E to C

$$= -M_e . T_c/T_e = -74{\cdot}2 . 30/21 = -106 \text{ lb ft}$$

These must of necessity be equal and opposite. The holding torque on E is in the same sense as the input torque on A, while the load torque on D is opposite in sense to the input torque.

The corresponding tooth loads on D and E may be found when the pitch of the teeth is given.

(3) *Vector Method.* With the arm fixed, give D one revolution. Then the compound wheel B-C will make $-T_d/T_b$ revolutions and the wheel E will make $+T_d/T_b . T_c/T_e$ revolutions. Set off from pole a vectors to represent these angular displacements. Thus,

$$\text{ad} : \text{ac} : \text{ae} :: 1 : -\frac{T_d}{T_b} : \frac{T_d}{T_b} . \frac{T_c}{T_e}$$

With the wheel E fixed, the point e becomes the pole from which the angular displacements are measured, and

$$N_a/N_d = \text{ea}/\text{ed} = 4{\cdot}71$$

The three external torques are applied to A, D and E, and equation (1) implies that the angular velocity diagram edac can be treated as a lever to which the external torques are applied. For equilibrium the torques M_e and M_a must have the same sense and the torque M_d must have the opposite sense, as shown.

Then:

$$M_a . \text{ad} = M_e . \text{ed}$$
$$\therefore \; M_e = M_a . \text{ad}/\text{ed} = 3{\cdot}71 M_a = 74{\cdot}2 \text{ lb ft}$$

and

$$M_d . \text{de} = M_a . \text{ae}$$
$$\therefore \; M_d = M_a . \text{ae}/\text{de} = 4{\cdot}71 M_a = 94{\cdot}2 \text{ lb ft}$$

In an epicyclic train of this type the speed reduction is greater if the wheels D and E are made more nearly equal in size. When D and E are equal in size, the speed reduction is infinitely large, the wheel D remaining at rest while the arm A revolves. If the fixed wheel E is larger than D, the sense of rotation of the latter is reversed for the same sense of rotation of the arm. Thus, if T_e is 26 and T_c is 25, T_b and T_d remaining unchanged, we have:

$$\frac{N_d}{N_a} = 1-\frac{T_e}{T_c}\cdot\frac{T_b}{T_b} = 1-\frac{26}{25}\cdot\frac{27}{24} = -\frac{17}{100}$$

$$\therefore \frac{N_a}{N_d} = -\frac{100}{17} = -\frac{5{\cdot}89}{1}$$

The driving shaft therefore turns at 5·89 times the speed of the driven shaft and the two shafts revolve in opposite directions.

The Trojan epicyclic gear box, Fig. 261, makes use of epicyclic trains of the same type as that in the above example.

Example 6. An epicyclic bevel gear train is shown in Fig. 256. The fixed wheel B meshes with the pinion C. The wheel E on the driven shaft meshes with the pinion D. The pinions C and D are keyed to a shaft, which revolves in bearings on the arm A.

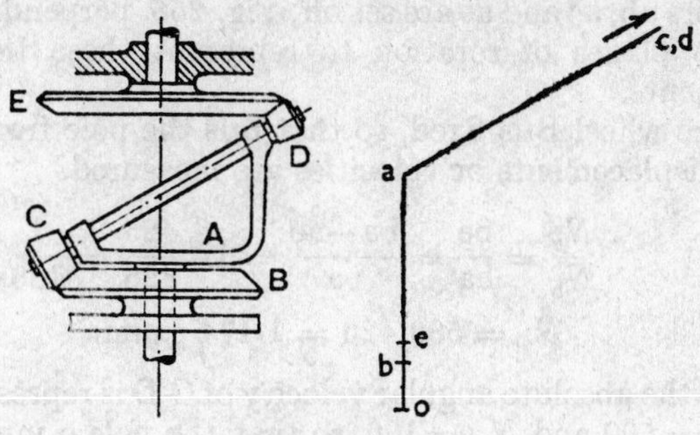

FIG. 256

The arm A is keyed to the driving shaft. The numbers of teeth are: $T_b = 75$, $T_c = 18$, $T_d = 17$ and $T_e = 71$. Find the speed of the driven shaft if (a) the driving shaft makes 500 r.p.m., (b) the wheel B turns in the same sense as the driving shaft at 100 r.p.m., the driving shaft still making 500 r.p.m.

(1) *Algebraic Method.* When the arm is fixed

$$\frac{N_e}{N_b} = \frac{T_b}{T_c}\cdot\frac{T_d}{T_e}$$

$$\therefore \frac{N_e-N_a}{N_b-N_a} = \frac{T_b}{T_c}\cdot\frac{T_d}{T_e} = \frac{75}{18}\cdot\frac{17}{71} = \frac{425}{426}$$

(a) Wheel B is fixed, so that $N_b = 0$, and

$$\frac{N_e - N_a}{0 - N_a} = \frac{425}{426}$$

$$\therefore \; N_e/N_a = 1 - 425/426 = 1/426$$

and $$N_e = 500/426 = 1{\cdot}174 \text{ r.p.m.}$$

(b) $N_a = 500$, $N_b = 100$,

$$\therefore \; \frac{N_e - 500}{100 - 500} = \frac{425}{426}$$

$$\therefore \; N_e = -\frac{425}{426}.400 + 500$$

$$= 100{\cdot}9 \text{ r.p.m.}$$

(2) *Vector Method.* If A is fixed and B is given one revolution clockwise as seen in plan, C-D will make $T_b/T_c = 75/18$ revolutions clockwise as viewed in the direction C-D, and E will make $T_b/T_c \,.\, T_d/T_e = 425/426$ of a revolution clockwise as seen in plan. The vectors ab, ac and ae are set off, Fig. 256, perpendicular to the respective planes of rotation to represent these three angular displacements.

(a) Here wheel B is fixed, so that b is the pole from which the angular displacements or velocities are measured.

$$\therefore \; \frac{N_e}{N_a} = \frac{\text{be}}{\text{ba}} = \frac{\text{ba}-\text{ae}}{\text{ba}} = 1 - \frac{425}{426} = \frac{1}{426}$$

and $$N_e = 500/426 = 1{\cdot}174 \text{ r.p.m.}$$

Note that the absolute angular velocity of C-D is represented by bc.

(b) $N_a = 500$ and $N_b = 100$, so that the pole o must divide ab externally, such that

$$\text{ao/bo} = N_a/N_b = 5 \quad \text{or} \quad \text{ao/ab} = 1{\cdot}25$$

Then

$$\frac{N_e}{N_a} = \frac{\text{eo}}{\text{ao}} = \frac{\text{ao}-\text{ae}}{\text{oa}}$$

$$= \frac{1{\cdot}25 - 425/426}{1{\cdot}25}$$

$$= 1 - 0{\cdot}7982 = 0{\cdot}2018$$

$$\therefore \; N_e = 100{\cdot}9 \text{ r.p.m.}$$

The angular velocity of C-D is now represented by oc. It is impracticable in this problem to draw the vectors so that eb is shown to scale.

Example 7. Referring to Fig. 257, the two wheels S_1 and S_2 are integral with the driving shaft. The wheel P_1 revolves on a pin attached to the arm A, which is integral with the driven shaft, and P_1 gears with the internal wheel I_1, which is co-axial with the driving shaft. The wheel P_2 meshes with S_2 and the fixed internal wheel I_2 and revolves on a pin fixed to the internal wheel I_1. The numbers of teeth are $T_{s_1} = 31$, $T_{s_2} = 26$, $T_{i_1} = 83$ and $T_{i_2} = 88$.

If the input to the driving shaft is 30 h.p. at 3000 r.p.m., find:

(a) the output speed and torque,

(b) the holding torque on I_2,

(c) the tooth loads on all the wheels, which have a diametral pitch of 8.

N.B.—There are usually three planet wheels P_1 spaced round the sun wheel S_1 at equal angular intervals of 120°, and similarly three planet wheels P_2 spaced round S_2. The three planets serve both to distribute the load on the teeth of the sun and of the internal wheel and to balance the centrifugal forces. Kinematically one planet is sufficient for each sun.

This train corresponds to second gear in the Wilson gear box, Fig. 262

(1) *Tabular Method.* Since the internal wheel I_2 is fixed, and the pin on which P_2 revolves is fixed to the internal wheel I_1, the first step in solving the problem is to find the speed ratio of S_2 to I_1. Then, knowing the speeds of I_1 and S_1, the next step is to find the speed of A.

In filling up the table (page 439), line (a) gives the revolutions of S_2 for one revolution of I_2, when the arm, i.e. I_1, is fixed. Line (b) corresponds to the rotation of the locked train through one revolution in a direction opposite to that of I_2 in line (a). Line (c) is the sum of lines (a) and (b) and gives the revolutions of S_2 for revolution of the arm I_1, when I_2 is fixed. Line (d) gives the revolutions of I_1 for one revolution of S_2, when I_2 is fixed.

The remaining lines are for the train S_1, I_1, A. Line (e) is put in by inspection. It is obtained from line (d) by substituting the numbers of teeth on S_1, I_1 for those on S_2, I_2. It is at this point that the student may find difficulty in deciding on the next step to take. The simplest plan to follow is to work backwards from line (h). Since the two suns S_1 and S_2 are integral with the driving shaft, they must revolve at the same speed. But it has already been found in line (d) that for one revolution of S_2 the annulus I_1 makes 13/57 of a revolution. These two figures are therefore entered in the appropriate columns of line (h).

The annulus I_1 is fixed for line (f), so that we may enter 0 in column three of this line. Since (h) is the sum of lines (f) and (g), it follows that the figure to be entered in column three of line (g) is 13/57 and the same figure must be entered in each of the other two columns in this line. In order to get the required figure of unity in the second column of line (h), we must have 1—13/57, i.e. 44/57, in the corresponding column of line (f). The revolutions of the arm in column one of line (f) automatically follow and, by adding (f) and (g), we get the required revolutions of the arm A and therefore of the driven shaft.

The driven shaft revolves in the same direction as the driving shaft and at a speed of $3000 \times 0{\cdot}438 = 1314$ r.p.m.

	Revolutions of arm	Revolutions of sun	Revolutions of internal gear or annulus
For Train S_2, I_2, I_1.			
(a)	0	$\frac{88}{26}$	−1
(b)	1	1	1
(c) = (a)+(b) . .	1	$1+\frac{88}{26}=\frac{57}{13}$	0
(d)	$\frac{13}{57}$	1	0
For Train S_1, I_1, A.			
(e)	$\frac{31}{114}$	1	0
(f)	$\frac{44}{57}\cdot\frac{31}{114}$	$\frac{44}{57}$	0
(g)	$\frac{13}{57}$	$\frac{13}{57}$	$\frac{13}{57}$
(h) = (f)+(g) . .	$\frac{44}{57}\cdot\frac{31}{114}+\frac{13}{57}=0{\cdot}438$	1	$\frac{13}{57}$

(2) *Graphical Method.* Since the pitch-circle radii of the wheels are directly proportional to the number of teeth, a scale of tooth numbers may be set off and used as a scale of pitch-circle radii, as shown in Fig. 257. For the train S_2, P_2, I_2, the tooth numbers are $S_2 = 26$ and $I_2 = 88$, marked B and D on the scale. The axis of the planet P_2 is fixed to the annulus I_1 and the radius of rotation of this axis is marked C, equivalent to 57 teeth. Pitch line velocities are represented by horizontal vectors set off from the vertical axis OY. Since the annulus I_2 is fixed, its pitch line velocity is nil and is represented by point D.

Let BS_2' represent the pitch line velocity of S_2. Then the tangential velocity of the axis of planet P_2 will be given by CI_1'', where I_1'' lies on the straight line DS_2'. But the planet axis is fixed to the annulus I_1, so that the pitch line velocity of I_1 must be given by $83/57 . CI_1'' = FI_1'$, where I_1' is found by joining O to I_1'' and producing to intersect the horizontal line through F (at 83 the number of teeth on annulus I_1).

The angular velocities of S_1 and S_2 are equal so that the pitch line speed of S_1 is found by drawing a horizontal line through E (31 teeth) to intersect at S_1' the line OS_2' produced.

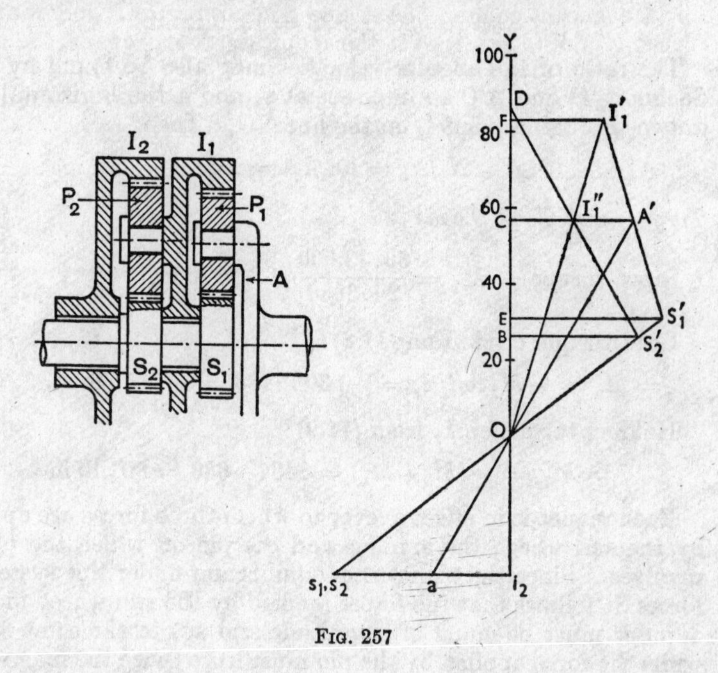

Fig. 257

The top point of the planet P_1 has the same tangential velocity as the annulus I_1 and the bottom point of the planet has the same tangential velocity as the sun S_1. Hence the axis of P_1 must have the tangential velocity CA′ where A′ lies on the straight line $I_1'S_1'$. The point C is the same as for the first train because the sum of the teeth on sun and planet is the same for both trains.

The angular velocity of the driving shaft

$$= N_s = BS_2'/BO \quad \text{or} \quad ES_1'/EO$$

and the angular velocity of the driven shaft

$$= N_a = CA'/CO$$

$$\therefore \frac{N_s}{N_a} = \frac{CO}{EO} \cdot \frac{ES_1'}{CA'} = \frac{57}{31} \cdot \frac{ES_1'}{CA'}$$

and, if ES_1' and CA' are scaled from the velocity diagram and substituted it will be found that

$$N_s/N_a = 2{\cdot}28$$

The two shafts turn in the same sense and the velocity of the driven shaft

$$= N_a = 3000/2{\cdot}28 = 1310 \text{ r.p.m.}$$

The ratio of the angular velocities may also be found by producing $S_1'O$ and $A'O$ to intersect at s_1 and a the horizontal line drawn through a point i_2 on the line YO. Then

$$N_s/N_a = i_2s_1/i_2a = 2{\cdot}28$$

Torques and Tooth Loads

$$\text{Input torque} = M_1 = \frac{30.33{,}000}{2\pi.3000} = 630 \text{ lb in.}$$

Load torque on A from (11.8)

$$= M_a = -M_1.\omega_1/\omega_a = -630/0{\cdot}438 = -1437 \text{ lb in.}$$

Holding torque on I_2 from (11.9)

$$= M_{i_2} = -(M_1+M_a) = 1437-630 = 807 \text{ lb in.}$$

Each planet is in effect a lever to which three forces are applied by the sun wheel, the annulus and the pin on which the planet revolves. Since the planet is in equilibrium under this system of forces, it follows that the forces applied by the sun wheel and the annulus must be equal in magnitude and act in the same sense, while the force applied by the pin must have twice the magnitude and act in the opposite sense.

Let F_1, F_2 be the tangential forces applied by the sun wheels S_1, S_2 to their respective planets P_1, P_2.

Then,
$$M_1 = F_1r_1 + F_2r_2$$
$$M_a = -2F_1.r_a$$
and
$$M_{i_2} = F_2.r_{i_2}$$

Since the wheels are all 8 diametral pitch,

$$r_1 = \frac{31}{2.8} = \frac{31}{16} \text{ in.,} \quad r_2 = \frac{26}{16} = \frac{13}{8} \text{ in.,} \quad r_{i_1} = \frac{83}{16} \text{ in.}$$

$$r_{i_2} = \frac{88}{16} = \frac{11}{2} \text{ in.,} \quad r_a = \frac{57}{16} \text{ in.}$$

$$\therefore\ F_1 = -M_a/2r_a = 1437.8/57 = 201{\cdot}6 \text{ lb}$$

and

$$F_2 = M_{1_2}/r_{i_2} = 807.2/11 = 146{\cdot}7 \text{ lb}$$

As a check,

$$M_1 = F_1r_1 + F_2r_2 = 201{\cdot}6.31/16 + 146{\cdot}7.13/8$$
$$= 391 + 239 = 630 \text{ lb in.}$$

The suns S_1 and S_2 therefore transmit 62·1% and 37·9% of the input torque.

The annulus I_1 has two forces applied to it, one by the planet P_2 through its pin and the other by the planet P_1. Since these are the only forces which tend to rotate I_1 about its axis and I_1 is in equilibrium, it follows that the algebraic sum of their moments must be zero.

$$\therefore\ 2F_2.r_a - F_1.r_{i_1} = 0$$

This equation is satisfied by the values of F_1 and F_2 as determined above, the torque transmitted from one train to the other through I_1

$$= F_1r_{i_1} = 201{\cdot}6.83/16 = 1047 \text{ lb in.}$$

If there are three planets in each train, the tooth loads are theoretically reduced to one-third of the above values.

150. Some Applications of Epicyclic Trains. Epicyclic gear trains are frequently used where it is necessary to obtain a large speed reduction and the available space is limited. They have also come into prominence in connection with the gear boxes of motor-cars and other self-propelled vehicles. A few typical examples will be described.

(a) *The Cyclometer Mechanism.* This is shown diagrammatically in Fig. 258. There are two co-axial internal wheels C and D, of

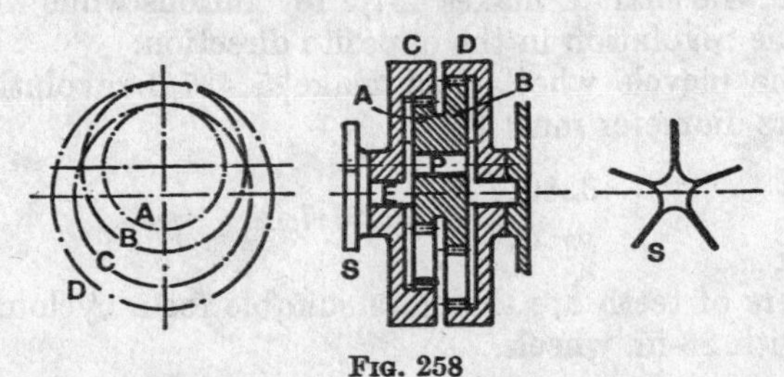

Fig. 258

which C is fixed. The compound wheel A-B is free to revolve on a pin, which is attached to the arm on the driving shaft E. The wheel B meshes with D and the wheel A with C. The driving shaft carries the star wheel S which is operated by a striker fixed to the bicycle wheel and makes one-fifth of a revolution for each revolution of the wheel. The wheel D makes one revolution while the bicycle travels a distance of one mile.

Example 8. The numbers of teeth are: A, 19; B, 20; C, 22; and D, 23. What must be the diameter of the bicycle wheel if D makes one revolution per mile?

(1) *Tabular Method.* The table may be filled in as follows:

	Revolutions of E	Revolution of C	Revolutions of D
(a) E fixed . .	0	−1	$-\frac{22}{19}\cdot\frac{20}{23}=-\frac{440}{437}$
(b)	1	1	1
(c) = (a)+(b) .	1	0	$1-\frac{440}{437}=-\frac{3}{437}$

Therefore the shaft E and the star wheel S make 437/3 revolutions while D makes one revolution in the opposite direction.

(2) *Algebraic Method.*

$$\frac{N_c-N_e}{N_d-N_e}=\frac{T_d}{T_b}\cdot\frac{T_a}{T_c}=\frac{23}{20}\cdot\frac{19}{22}=\frac{437}{440}$$

When C is fixed, $N_c=0$,

$$\therefore\ \frac{0-N_e}{N_d-N_e}=\frac{437}{440}\quad\text{and}\quad\frac{N_d}{N_e}=-\frac{3}{437}$$

As before, the shaft E makes 437/3 revolutions while the wheel D makes one revolution in the opposite direction.

Hence the bicycle wheel must make 5.437/3 revolutions per mile, and its diameter must be

$$\frac{5280.12.3}{\pi.5.437}=27{\cdot}7\text{ in.}$$

The numbers of teeth are therefore suitable for a cyclometer for a bicycle with 28-in. wheels.

(b) *Humpage's Gear.* This is a bevel-wheel epicyclic and is shown diagrammatically in Fig. 259. The driving and driven shafts A and B are co-axial and each carries a bevel wheel C, D. Wheel C gears with E and E with the fixed wheel G; wheel D gears with F, which is compound with E. The compound wheel E-F

revolves freely on the arm attached to H, and H revolves about the same axis as the shafts A and B. In the actual gear there are either two or three arms on H, each of which carries a compound wheel identical with E-F.

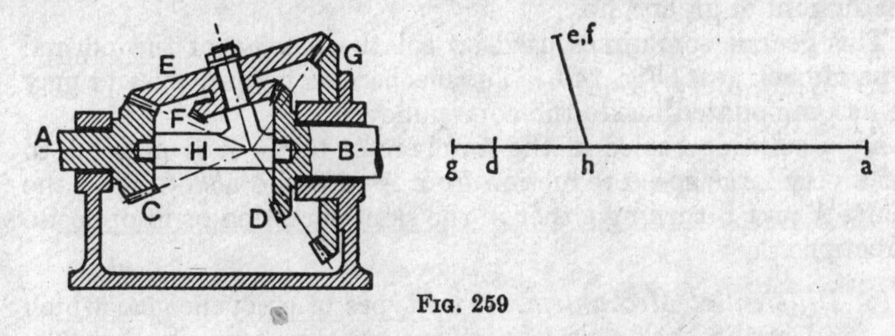

FIG. 259

Example 9. Let the number of teeth on C, D, E, F and G be respectively 15, 20, 39, 16 and 32. Then by the tabular method, we have:

	Revolutions of H	Revolutions of G	Revolutions of A	Revolutions of B
(a) H fixed . .	0	−1	$+\frac{32}{15}$	$-\frac{32}{39}\cdot\frac{16}{20}$
(b)	1	1	1	1
(c) = (a)+(b) .	1	0	$1+\frac{32}{15}=\frac{47}{15}$	$1-\frac{32}{39}\cdot\frac{16}{20}=\frac{67}{195}$
Multiply by $\frac{15}{47}$	$\frac{15}{47}$	0	1	$\frac{67}{195}\cdot\frac{15}{47}=\frac{1}{9{\cdot}118}$

The shaft A turns in the same direction as the shaft B and 9·118 times as fast.

Vector Method. With H fixed, give G one revolution clockwise as viewed from the right. Then the compound wheel E-F will make $T_g/T_e = 32/39$ revolutions counter-clockwise as seen in plan, A will make $T_g/T_a = 32/15$ revolutions in the opposite sense to G, and D will make $T_g/T_e . T_f/T_d = 32/39 . 16/20$ $= 128/195$ revolutions in the same sense as G.

Set off the vectors hg, he, ha and hd, as shown to represent the angular displacements relative to H. The directions of the vectors are perpendicular to the respective planes of rotation and conform to the right-handed screw convention.

In the actual gear, G is the fixed wheel, so that g becomes the pole from which the angular displacements are measured, and

$$N_a/N_b = N_a/N_d = \text{ga/gd} = 9{\cdot}12/1$$

The displacements of A and B are in the same sense.

The angular displacement of H is represented by gh and is in the same sense as that of A and D.

The angular displacement of E-F is represented by ge, the vector sum of gh and he.

This gear is sometimes used on a lathe instead of the normal type of back gear, Fig. 249. The mechanism is compact and may be accommodated inside the cone-pulley.

By a suitable choice of the numbers of teeth, it is possible to get a very large speed reduction from A to B and also to have the shafts A and B turning either in the same direction or in opposite directions.

(o) *Differential Mechanism.* Two types of epicyclic gear which are used in the back axle of a motor-car are shown in Fig. 260. The purpose of the differential mechanism is to enable the back wheels to revolve at different speeds when the car is rounding a corner. In the arrangement shown at (a) the wheels in the epicyclic train are bevel wheels, while in that shown at (b) they are spur wheels. The two equal wheels A and B are keyed to the two halves of the rear axle, and the bevel wheels C, Fig. 260 (a),

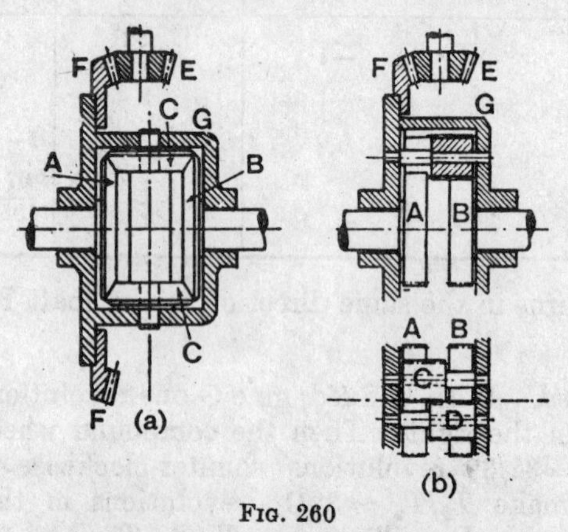

FIG. 260

or the spur wheels C and D, Fig. 260 (b), revolve on pins which are carried by the casing G. The drive is transmitted from the propeller shaft to the casing through the bevel wheels E and F. The bevel pinions C gear with the wheels A and B, so that if the casing G is stationary the wheels A and B revolve in opposite directions. Similarly, the spur pinions C and D gear respectively with the wheels A and B and also with each other, as shown in the plan view, so that when the casing is stationary A and B revolve in opposite directions. It is easily seen that if the car is moving along a straight path the wheels A and B and also the casing will revolve at the same speed, and the bevel pinions C, or the spur pinions C and D, will remain stationary relative to the casing. If, however, the car follows a curved path, the wheels A and B will revolve at different speeds and the casing will revolve at a speed which is the arithmetic mean of the speeds of A and B. At the same time the bevel pinions C, or the spur pinions C and D, will revolve on their pins.

(d) *Three-speed and Reverse Gear Box.* The essential features of the Trojan epicyclic gear box are shown in Fig. 261. The casing Q rotates at engine speed and carries three studs, one of which is shown at P. The three studs are arranged symmetrically about the axis of rotation of the casing and on each stud a cluster of four wheels, D, E, F and G, is free to revolve. Wheel G on the cluster gears with wheel H which is keyed to the propeller shaft. D, E and F gear with K, L and M, which are respectively keyed to sleeves integral with the drums A, B and C. Each of the three sleeves is free to revolve about the axis of the propeller shaft, but rotation of any one of them may be prevented by tightening the brake band on the corresponding drum. In the figure, wheels M and L are smaller and wheel K larger than wheel H. If M or L is prevented from rotating by tightening the brake band on C or B, the propeller shaft turns in the same direction as the casing Q, but at a lower speed (see Example 5); whereas, if K is held stationary, the propeller shaft turns in the opposite direction to the casing Q. First gear is engaged by preventing rotation of drum B, since this gives the biggest reduction in speed from Q to the propeller shaft. Tightening the brake band on drum C gives second gear. Top gear is engaged by locking the train, so that the propeller shaft rotates at the same speed as the casing. This is brought about by means of two semi-circular bands which are anchored to the drum C and which can be contracted round the drum B. The arrangement is such that, while B and C are locked together, both of them can revolve. The gear ratios may be calculated, as explained in Example 5, when the number of teeth on each wheel is known.

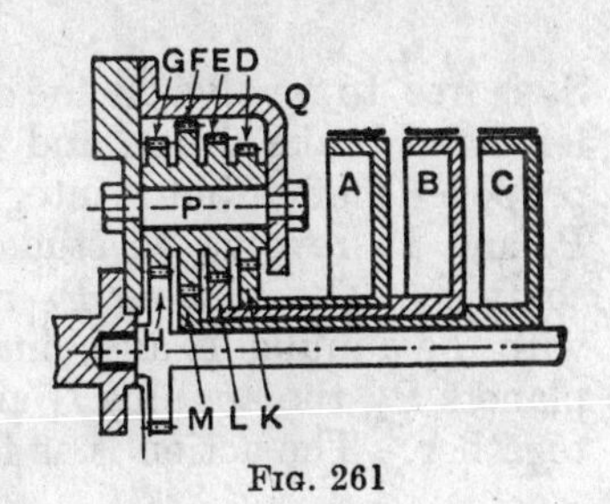

FIG. 261

(e) *The Wilson Gear Box.* The arrangement of this epicyclic gear box is shown diagrammatically in Fig. 262. It provides four forward ratios and a reverse ratio. It will be seen that there are four sun wheels, S_r, S_1, S_2 and S_3, and four internal gear wheels or annuli, I_r, I_1, I_2 and I_3. Each sun is geared to the corresponding annulus by means of three planet wheels, only one of which is shown. The planets are denoted by P with the appropriate suffix. There are four brake drums, which may be brought to rest by means of the contracting brake bands B_r, B_1, B_2 and B_3. The suns S_1 and S_2 are integral with the engine shaft E. The sun

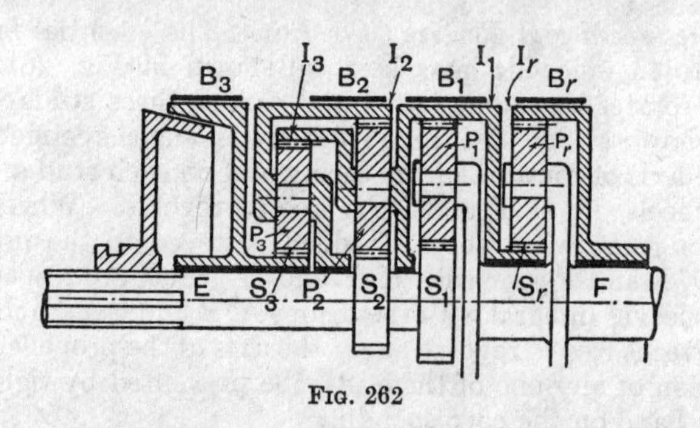

FIG. 262

S_3 is free to revolve on the engine shaft, but is integral with the left-hand brake drum, and the sun S_r is free to revolve on the propeller shaft F but is integral with the annulus I_1. The planets P_r and P_1 revolve on studs which are carried by the propeller shaft F. The planets P_3 revolve on studs which are integral with the annulus I_2 and, finally, the annulus I_3, the studs for the planets P_2, the annulus I_1 and the sun S_r are all rigidly fastened together. The action is as follows:

(a) To engage first gear the brake band B_1 is contracted and the annulus I_1 is thus brought to rest, so that motion is transmitted from E to F through the simple epicyclic train S_1, P_1 and I_1.

(b) To engage second gear the brake band B_2 is contracted so that the annulus I_2 is brought to rest. The motion is then transmitted through the compound epicyclic train S_2, P_2, I_2 and S_1, P_1, I_1.

(c) To engage third gear the brake band B_3 is contracted so that the sun S_3 is brought to rest. The motion is then transmitted from E to F through the three epicyclic trains.

(d) To engage reverse gear the brake band B_r is contracted so that the annulus I_r is brought to rest. The two trains S_1, P_1, I_1 and S_r, P_r, I_r then determine the motion transmitted from E to F.

(e) To engage top gear a small cone clutch shown on the extreme left of the figure is used. This clutch prevents all relative motion between the sun S_3 and the engine shaft E. Hence the epicyclic gears are locked together and the shaft F runs at the same speed as the shaft E.

It should be noted that only one of the brake bands is contracted for each gear ratio, all the others remaining free.

For details of the band-contracting mechanism and of the pre-selecting mechanism the reader should consult the paper by W. G. Wilson, on " Epicyclic Gearing " in the *Proceedings of the Institution of Automobile Engineers*, 1931–2.

Example 10. The numbers of teeth on the wheels of a gear box of the above type are: $S_1 = 26$, $S_2 = 24$, $S_3 = 23$, $S_r = 31$, $I_1 = 70$, $I_2 = 72$, $I_3 = 61$ and $I_r = 71$. The wheels are all 6 diametral pitch and the input torque is 1500 lb in. Find the ratio of engine speed to propeller-shaft speed and the tooth loads for third gear.

It is not proposed to give the solutions for first, second and reverse gear, which follow similar lines to that of Example 7, p. 393.

Third Gear. Brake band B_3 is contracted so that S_3 is at rest.

(1) *Algebraic Method.* For the train S_3, P_3, I_3 the axes of the planets are fixed to I_2.

$$\therefore \frac{N_{s_3}-N_{i_2}}{N_{i_3}-N_{i_2}} = -\frac{61}{23}$$

But S_3 is fixed,

$$\therefore \frac{0-N_{i_2}}{N_{i_3}-N_{i_2}} = -\frac{61}{23} \quad \text{and} \quad \frac{N_{i_3}}{N_{i_2}} = \frac{84}{61}$$

For the train S_2, P_2, I_2 the axes of the planets P_2 are fixed to I_3,

$$\therefore \frac{N_{s_2}-N_{i_3}}{N_{i_2}-N_{i_3}} = -\frac{72}{24}$$

Substituting for N_{i_2} in terms of N_{i_3}:

$$\frac{N_{s_2}-N_{i_3}}{(61/84)N_{i_3}-N_{i_3}} = -\frac{72}{24} \quad \text{and} \quad \frac{N_{s_2}}{N_{i_3}} = \frac{51}{28}$$

But $N_{i_3} = N_{i_1}$ and $N_{s_2} = N_{s_1}$, so that $N_{s_1}/N_{i_1} = 51/28$.

For the train S_1, P_1, I_1, the axes of rotation of the planets are fixed to and revolve with the propeller shaft F.

$$\therefore \frac{N_{s_1}-N_f}{N_{i_1}-N_f} = -\frac{T_{i_1}}{T_{s_1}} = -\frac{70}{26} = -\frac{35}{13}$$

and, substituting for N_{i_1} in terms of N_{s_1},

$$\frac{N_{s_1}-N_f}{(28/51)N_{s_1}-N_f}=-\frac{35}{13}$$

$$\therefore\ N_{s_1}-N_f=-\frac{35}{13}\cdot\frac{28}{51}\cdot N_{s_1}+\frac{35}{13}\cdot N_f$$

$$\therefore\ \frac{N_{s_1}}{N_f}=\frac{48}{13}\cdot\frac{1}{2\cdot479}=\frac{1\cdot49}{1}$$

(2) *Vector Method.* For each of the simple trains consisting of sun wheel, planet, planet carrier and annulus, a vector diagram of angular displacements or velocities of each member relative to the planet carrier may be drawn as explained on p. 385. The relevant information for drawing the diagrams is given in the following table

Train	Planet carrier fixed	Revs of sun	Revs. of planet	Revs. of annulus
1	0	1	$-\frac{26}{22}=-\frac{13}{11}$	$-\frac{26}{70}=-\frac{13}{35}$
2	0	1	$-\frac{24}{24}=-1$	$-\frac{24}{72}=-\frac{1}{3}$
3	0	1	$-\frac{23}{19}$	$-\frac{23}{61}$

The planet carrier for the first train is fixed to the driven shaft F, and the three vectors fs_1, fp_1 and fi_1 are set off from pole f as shown in Fig. 263 (i).

The planet carrier for the second train is the annulus I_1, and the corresponding vectors are set off from the pole i_1, Fig. 263 (ii).

The planet carrier for the third train is the annulus I_2, and the vectors are set off from the pole i_2, Fig. 263 (iii).

The speeds of S_1 and S_2 are necessarily equal, since both wheels are integral with the driving shaft. The speeds of the annulus I_1 and the annulus I_3 are also equal since they are rigidly fastened together by the pins on which the planets P_2 rotate. The three vector diagrams (i), (ii) and (iii) may therefore be combined to give the single diagram (iv), in which s_1 and s_2 coincide and also i_1 and i_3 coincide. This involves changing the scale of (ii) so that s_2i_1 is equal in length to s_1i_1 in (i), and changing the scale of (iii) so that i_2i_3 is equal in length to i_2i_1 in (ii).

The simplest way of obtaining the combined diagram (iv) is to use the graphical construction shown. Draw vertical lines through the points i_1 in (i) and (ii) and the point i_3 in (iii).

Through s_1 and s_2 draw lines of equal inclination θ, say 45°, to give the points q_1 and q_2, and through i_2 diagram (iii), draw a line parallel to q_2i_2, diagram (ii), to give the point q_3.

Start diagram (iv) by drawing the triangle $s_1i_1q_4$ similar to the triangle $s_1i_1q_1$ in (i). From q_4 draw lines parallel to q_1f and q_1p_1, also lines parallel to q_2i_2 and q_2p_2 and finally lines parallel to q_3s_3 and q_3p_3. Mark the points of intersection of these lines with s_1i_1 in (iv) with the appropriate letter f, p_1, i_2, p_2, s_3, p_3. The vector diagram s_1s_3 will then give the relative angular velocities of all the members of the train, as may readily be proved

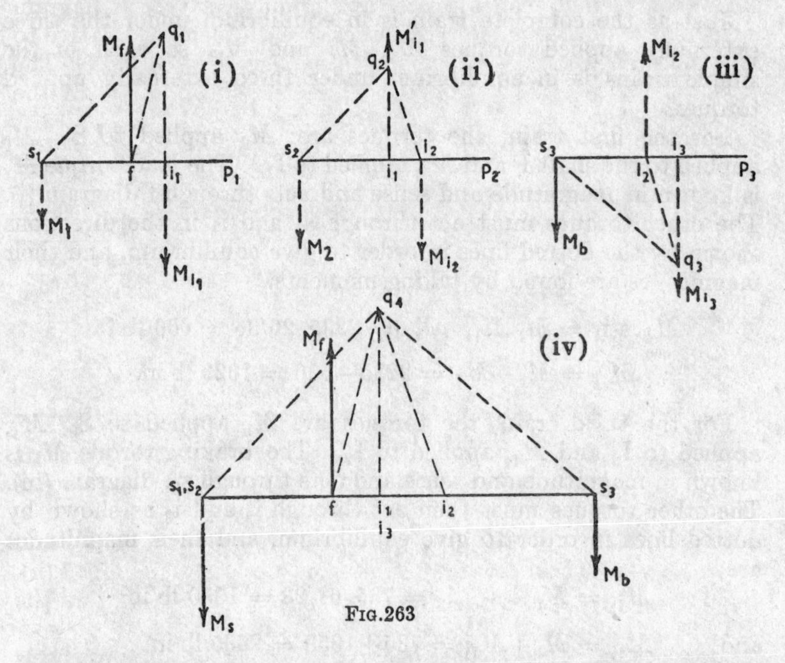

FIG. 263

by similar triangles. Note that to prevent confusion the lines from q_4 which fix the positions of p_1, p_2 and p_3 are not actually shown.

Since the sun wheel S_3 is the fixed wheel in the actual train, this is the pole from which the angular velocities are to be measured. The driving shaft speed is that of the sun wheels S_1 and S_2 and the driven shaft speed is that of F.

$$\therefore\ N_s/N_f = s_3s_1/s_3f = 1{\cdot}49/1$$

The external torques are the input torque M_1 applied to the shaft which carries the two wheels S_1 and S_2, the load torque M_f

applied to the driven shaft F, and the holding, or brake torque M_b applied to S_3. The vector diagram s_1s_3 may therefore be treated as a lever with the three "forces" M_s, M_f and M_b applied at s_1, f and s_3. It follows that for equilibrium M_s and M_b will act in the same sense and M_f in the opposite sense.

$$\therefore \; M_f/M_s = s_3s_1/s_3f = 1.49/1$$

$$M_f = 1{\cdot}49M_s$$

$$= 2235 \text{ lb in.}$$

and $$M_b = M_f - M_s = 0{\cdot}49M_s = 735 \text{ lb in.}$$

Just as the complete train is in equilibrium under the three externally applied torques M_s, M_f and M_b, so each of the simple trains is in equilibrium under three externally applied torques.

For the first train, the torques are: M_1 applied to S_1, M_f applied to the shaft F and M_{i_1} applied to I_1. The load torque M_f is known in magnitude and sense and acts through f diagram (i). The other torques must act through s_1 and i_1 in the directions shown by the dotted lines in order to give equilibrium, and their magnitudes are found by taking moments.

$$M_1.s_1i_1 = M_f.fi_1, \quad M_1 = 2235.26/96 = 606 \text{ lb in.}$$

$$M_{i_1} = M_f - M_1 = 2235-606 = 1629 \text{ lb in.}$$

For the third train, the torques are M_b applied to S_3, M_{i_3} applied to I_3 and M_{i_2} applied to I_2. The braking torque M_b is known in magnitude and sense and acts through s_3, diagram (iii). The other torques must then act through i_2 and i_3 as shown by dotted lines in order to give equilibrium, and their magnitudes are:

$$M_{i_3} = M_b.i_2s_3/i_2i_3 = 735.61/23 = 1950 \text{ lb in.}$$

and $$M_{i_2} = M_b + M_{i_3} = 735+1950 = 2685 \text{ lb in.}$$

For the second train, torques are applied to S_2, and I_2 and the planet carrier which is integral with I_1 and I_3. The torque applied through i_2 in (ii) is equal and opposite to the torque applied through i_2 in (iii), since I_2 is in equilibrium under the torques applied by the reaction of the planet P_3 on its pin and the reaction of the teeth of planet P_2 on the teeth of I_2. Similarly, the compound member I_3-I_1 is in equilibrium under the three torques applied to it by the reaction on the teeth of I_1, the reaction on the teeth of I_3 and the reaction of the planet P_2 on its pin. Hence the torque M'_{i_1} which is shown acting through i_1 in (ii) must be equal and opposite to the algebraic sum of the torques M_{i_1} and M_{i_3} in (i) and (iii).

Taking moments about i_1,

$$M_2 = M_{i_2}.i_1i_2/i_1s_2 = 2685.24/72 = 895 \text{ lb in.}$$

and $$M'_{i_1} = M_2+M_{i_2} = 895+2685 = 3580 \text{ lb in.}$$

As a check,

$$M_1+M_2 = \text{total input torque}$$
$$= 606+895 = 1501 \text{ lb in.}$$

and $$M'_{i_1} = M_{i_1}+M_{i_3} = 1629+1950 = 3579 \text{ lb in.}$$

The magnitudes of the torques applied to the individual members having been found as above, the corresponding tooth loads may be calculated. The results are given in the following table:

Member	Torque lb in.	Teeth	Radius in.	F (total) lb	F (3 planets) lb
S_1	606	26	$2\frac{1}{6}$	280	93·3
S_2	895	24	2	447·5	149·2
S_3	735	23	$1\frac{11}{12}$	381	127
I_1	1629	70	$5\frac{5}{6}$	279	93
I_2	2685	72	6	447·5	149·2
I_3	1950	61	$5\frac{1}{12}$	383	127·7
F	2235	48	4	558·9	186·3

A check on the accuracy of the working is provided by the entries in the last column, since the tangential force applied to each sun wheel S should be equal to that applied to its annulus I.

EXAMPLES XI

1. Two parallel shafts are connected by spur gearing. The diametral pitch of the teeth is 4, the distance between the shaft axes is approximately 8·5 in. and one wheel is to turn at four times the speed of the other. Find the numbers of teeth on the two wheels and the exact centre distance. If the centre distance is to be exactly 8·5 in., what is the velocity ratio nearest to 4 that can be obtained?

2. As Question 1, except that the diametral pitch is 5, the approximate centre distance 7 in. and the velocity ratio 2·70.

3. A train of spur gears is required to give a total reduction of 250 to 1 in four steps. No pinion is to have less than 20 teeth and the diametral pitches are to be 5 for the first step, 3·5 for the second, 2·5 for the third and 1·5 for the fourth. The centre distances must not involve recurring decimals. Find the numbers of teeth, the pitch-circle diameters and the centre distances for a suitable train of gears.

4. A total reduction of approximately 12 to 1 is required in two steps by means of spur gears. The driving and driven shafts are co-axial and the diametral pitch of the teeth is to be 5 for the high-speed pair and 3 for the low-speed pair. If no wheel is to have less than 18 teeth, find the numbers of teeth, the pitch-circle diameters and the centre distance for a suitable train of gears.

5. A gear box of the sliding-gear type has two indirect speeds and the constant mesh wheels are equal. The distance between the centre line of the secondary shaft and that of the driving and driven shafts, which are in line, is to be as nearly as possible $6\frac{1}{4}$ in. The gear ratios are to be approximately 2 to 1 and $4\frac{1}{2}$ to 1 and the diametral pitch of all wheels is to be 5.

What arrangement of wheels will satisfy the required conditions and what are the resulting gear ratios? L.U.

6. As Question 5, except that the centre distance is to be $4\frac{3}{8}$ in., the gear ratios approximately 1·75 to 1 and 3·4 to 1 and the diametral pitch 8.

7. A four-speed gear box with three indirect speeds, similar to that shown in Fig. 250, is required to give ratios of approximately 1·5 to 1, 2·5 to 1 and 4 to 1. The diametral pitch of all the teeth is 8 and the centre distance is $2\frac{3}{4}$ in. Find the numbers of teeth on the wheels and sketch the arrangement.

8. As Question 7, except that the centre distance is to be 5 in., the gear ratios approximately $1\frac{3}{4}$ to 1, $2\frac{3}{4}$ to 1 and 5 to 1 and the diametral pitch 6.

9. The fixed internal wheel B, Fig. 252, has 92 teeth; the wheels C and D have respectively 25 and 15 teeth; and the wheel E has 52 teeth. If the arm A makes 130 r.p.m. and the input torque is 20 lb ft, what are the speed of E, the resisting torque on E and the holding torque on B?

10. Referring to Fig. 255, the wheels B, C, D and E have respectively 30, 27, 21 and 24 teeth. If the arm A makes 2400 r.p.m. and the input torque is 130 lb in., find the speed and load torque on D and the holding torque on E.

11. Referring to Fig. 255, the wheels B, C, D and E have respectively 25, 26, 75 and 74 teeth. If the arm A makes 470 r.p.m. and the wheel D 100 r.p.m. in the same sense, what is the speed of wheel E? If the input torque to D is 35 lb ft, what torques must be applied to A and E?

12. An internal wheel B with 80 teeth is keyed to a shaft F. A fixed internal wheel C with 82 teeth is concentric with B. A compound wheel DE gears with the two internal wheels; D has 28 teeth and gears with C, while E gears with B. The compound wheel revolves freely on a pin which projects from a disc keyed to a shaft A co-axial with F. If the wheels all have the same pitch and the shaft A makes 800 r.p.m., what is the speed of shaft F?

If the torque input to shaft A is 40 lb ft, what is the load torque on shaft F and the holding torque on wheel C?

13. For an epicyclic gear of the type shown in Fig. 255, the numbers of teeth are: wheel B, 30; wheel C, 33; wheel E, 24. If the teeth are all of the same pitch, find the speed of A in terms of that of D. If A is required to revolve in the opposite direction to D at approximately four times the speed of D, what change must be made in the numbers of teeth on wheels C and E and what is the actual speed ratio?

14. The internal wheels I_1, I_2 of a compound epicyclic gear, Fig. 257, have respectively 79 and 83 teeth, the sun wheels S_1 and S_2 have respectively 23 and 19 teeth. If the speed of the driving shaft is 2500 r.p.m., what is the speed of the driven shaft? If the power input is 40 h.p. and all the wheels are 5 diametral pitch, find:

(a) the load torque on A and the holding torque on I_2;
(b) the fractions of the input torque transmitted by the sun wheels S_1 and S_2;
(c) the tooth loads.

15. In the compound epicyclic gear of Fig. 264, A is the driving shaft and B the driven shaft. The internal wheel or annulus I_2 is fixed, the pinions P_1 and P_2 revolve on pins fixed to the arms C and D, which are in turn fixed to the driven shaft B, and the internal wheel I_1 and the sun wheel S_2 form a compound wheel which revolves about the common axis of shafts A and B. The sun wheel S_1 is keyed to the driving shaft A. The numbers of teeth are S_1, 24; I_1, 66; S_2, 28; I_2, 62. If the shaft A turns at 1500 r.p.m., the input torque is 60 lb ft, and all the wheels are 6 diametral pitch, find:

(a) the speed of the driven shaft;
(b) the load torque on B and the holding torque on I_2;
(c) the fractions of the load torque transmitted by C and D;
(d) the tooth loads.

16. The numbers of teeth on the wheels of a Wilson epicyclic gear box, Fig. 262, are: S_1, 23; S_2, 23; S_3, 20; S_r, 31; I_1, 67; I_2, 67; I_3, 58; and I_r, 65. Find the gear ratios provided by the gear box.

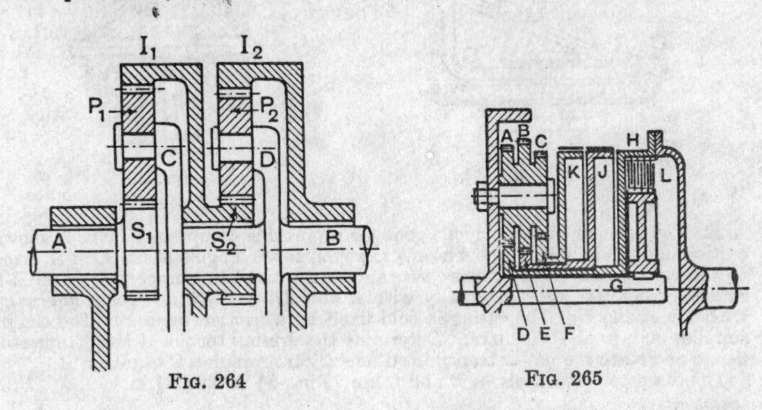

Fig. 264

Fig. 265

17. Fig. 265 shows diagrammatically the arrangement of the Ford model T epicyclic gear box. The cluster A, B, C revolves on a pin attached to the engine flywheel. The wheels D, E, F are attached to concentric sleeves, which surround an extension G of the engine crankshaft. Brake drums H, J, K are integral with the wheels D, E and F respectively. The propeller shaft and rear axle are driven from the drum H. Bottom gear is engaged by contracting a brake band round drum J; reverse gear by contracting a brake band round drum K; and top gear by bringing into operation the plate clutch L, which locks together the extension shaft G and the brake drum H. The numbers of teeth on the various wheels are A, 27; B, 33; C, 24; D, 27; E, 21; and F, 30. Find the gear ratios in bottom and reverse gears.

18. The numbers of teeth on the wheels of a Humpage epicyclic gear box, Fig. 259, are: C, 12; D, 34; E, 40; F, 16; and G, 46. If the shaft A makes 1500 r.p.m., what is the speed of shaft B?

19. A bevel gear epicyclic is shown in Fig. 266. The wheel A is keyed to the driving shaft, the wheel F to the driven shaft and the wheel E is fixed. The arm G which supports the inclined shaft is free to turn about the common axis of the driving and the driven shafts, and the wheels B, C and D are keyed to the inclined shaft. The wheels A and B are equal in size; C has 19 teeth; D, 18; E, 7[illegible], and F, 74 teeth. Find the speed of F in terms of the speed of A. What is the speed ratio when E has 81 teeth?

20. In the epicyclic pulley block shown in Fig. 267, A is a fixed wheel of 48 teeth; B and C are intermediate wheels carried on an eccentric which is keyed to the

shaft. B has 30 and C 31 teeth. The wheel D with the sprocket wheel F runs loose on the shaft. D has 50 teeth. The sprocket wheel E is keyed to the shaft and is 10 in. diameter. The wheel F is 7 in. diameter. Find the velocity ratio of the block. L.U.

21. A rotating arm A, making 100 r.p.m. in the clockwise direction, carries two wheels B and C, which are in gear. The wheel B has 60 teeth and the same axis of rotation as the arm. A wheel D with 120 internal teeth and mounted on the axis of rotation of the arm gears with C. Find the speed and direction of rotation of D when B is fixed and of B when D is fixed. Find in each case the r.p.m. of C on its own pin. L.U.

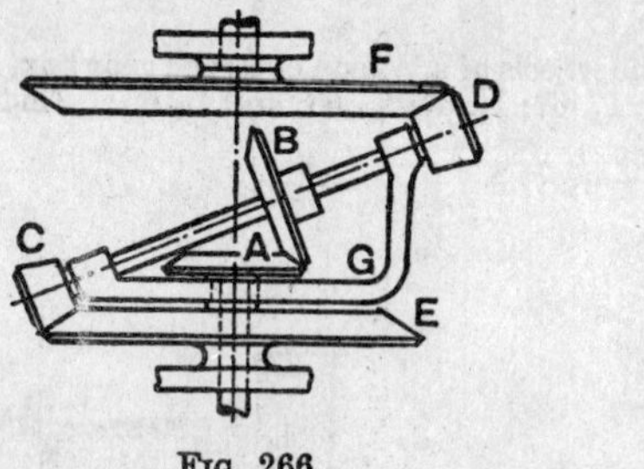

Fig. 266

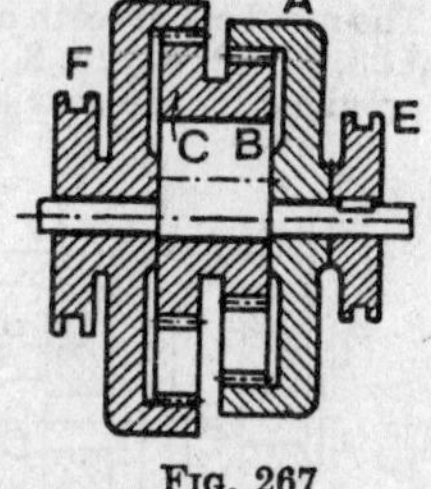

Fig. 267

22. The scheme of an epicyclic gear for a portable compressor drive is shown sectionally in Fig. 268. The wheel A is keyed to the engine shaft. Arm D, fixed to the compressor shaft, carries two axles on which the compound wheel B-C rotates freely. C gears externally with A and drives B, which gears internally with the casing E. The casing is held fixed by a friction band F. The engine supplies 30 h.p. at 1000 r.p.m. Determine the driving torque at the compressor flange and state the power transmitted directly from pinion C to pinion B.

The diameters of wheels A, B and C are $7\frac{1}{2}$ in., $3\frac{1}{2}$ in. and $2\frac{3}{4}$ in. L.U.

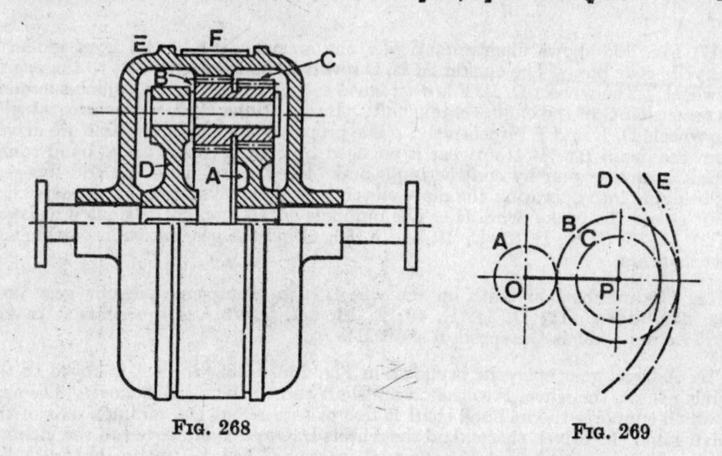

Fig. 268

Fig. 269

23. Fig. 269 shows diagrammatically a compound epicyclic gear. Wheels A, D and E are free to rotate independently on spindle O, while B and C are compound and rotate together on spindle P, on the end of arm OP. All the wheels have teeth of the same pitch. A has 12 teeth, B has 30 and C has 14 teeth cut externally. Find the number of teeth on wheels D and E which are cut internally.

If the wheel A is driven clockwise at 1 r.p.s., while D is driven counter-clockwise at 5 r.p.s., determine the magnitude and direction of the angular velocities of arm OP and wheel E. L.U.

24. In the gear shown in Fig. 270 the spindle M is driven from the pinion J keyed to the mainshaft. The sleeve L is driven partly by the disc N keyed to

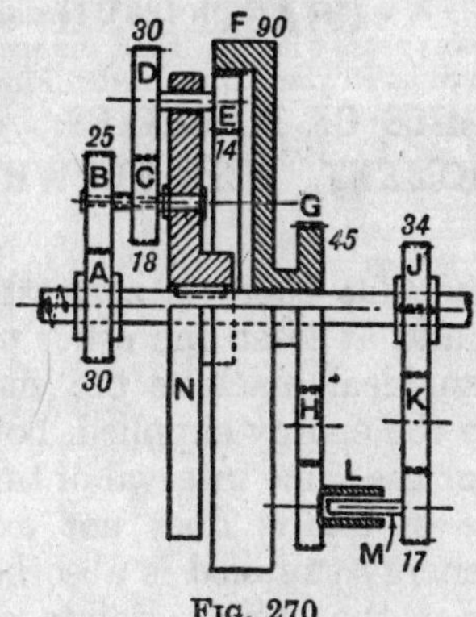

Fig. 270

the mainshaft and partly by the pinion A, which can be rotated independently. BC rotates on a pin fixed in N, DE rotates freely in N and FG is free on the mainshaft. The mainshaft is driven at 350 r.p.m. in the given direction. What must be the speed of A in order to make L, which has 20 teeth, rotate at a speed 80 r.p.m. slower than M? (The number of teeth in each wheel is indicated on the diagram; those without numbers are idle wheels.) L.U.

CHAPTER XII

DYNAMICS OF MACHINES. TURNING MOMENT. THE FLYWHEEL

151. In every machine there is at least one point at which energy is supplied and at least one other point at which energy is delivered. In an ideal machine the energy delivered would be exactly equal to the energy supplied, both amounts of energy being measured over the same interval of time. But in an actual machine this state of affairs does not exist for two reasons: first, some of the energy supplied is absorbed in overcoming the inevitable friction at the various joints or couplings, so that the energy delivered is less than that supplied, the ratio of the two amounts of energy being the mechanical efficiency of the machine; secondly, during the given interval of time, the kinetic and potential energies of each link will, in general, change, so that either some of the energy supplied will be absorbed in increasing the total energy, kinetic plus potential, of the moving parts, or alternatively the energy supplied will be supplemented by the decrease in total energy of the moving parts. In this connection it must be emphasised that during the interval of time required for the machine to complete a full cycle of operations, the net change of kinetic or potential energy for each moving part is nil, since at the end of the cycle each moving part occupies exactly the same position and is moving at exactly the same speed as at the beginning of the cycle. Hence, taken over a complete cycle, the energy delivered by the machine is not affected by the inertia of the moving parts and only differs from the energy supplied to the machine by the amount of the energy absorbed in overcoming friction at the joints.

When analysing the forces which act on the various parts, or links, of a machine, it is often convenient to consider separately the forces which would act if the moving parts were without mass and those which arise from the inertia of the moving parts. It is then easier to follow the effect of a change in the mean speed of movement of the parts on the total force to which any one part is subjected.

152. Effort and Resistance. Referring to Fig. 271, let M represent any ideal machine to which an effort F is applied and by means of which a resistance F_r is overcome. During a short interval of time δt, let the effort move through a distance δx and the resistance through a distance δx_r. Then the work done by the effort must be equal to the work done against the resistance, or $F.\delta x = F_r.\delta x_r$.

If both sides of this equation are divided by δt we have, in the limit when δt becomes infinitesimal,

$$F.v = F_r.v_r \quad . \quad . \quad . \quad . \quad . \quad . \quad (12.1)$$

where v and v_r are respectively the velocities at the given instant of the effort and the resistance.

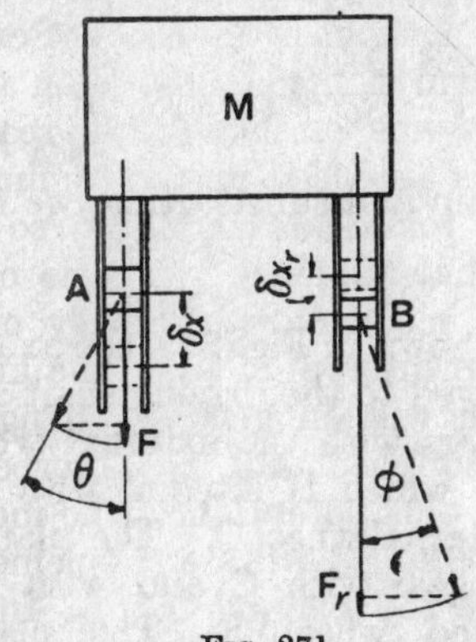

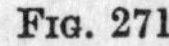

FIG. 271

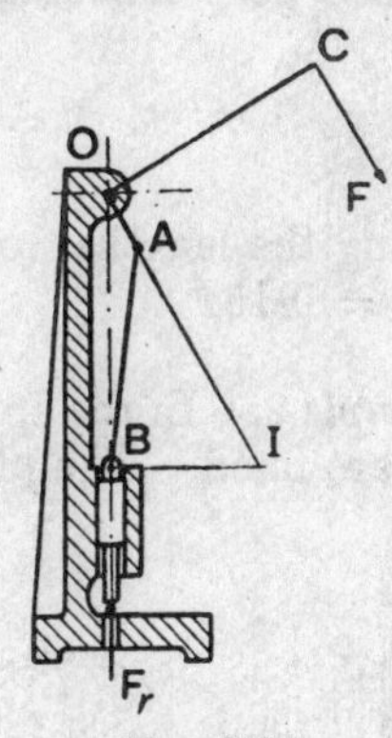

FIG. 272

If the effort and the resistance are applied in directions other than those in which the points A and B of the machine are constrained to move, as shown by the dotted lines, then clearly the work done by the effort is given by the product of δx and the component F' of F in the direction of δx; similarly, the work done against the resistance is given by the product of δx_r and the component F_r' of F_r in the direction of δx_r. Equation (12.1) then applies if F' and F_r' are substituted for F and F_r.

Should the energy be supplied to the machine by means of a torque T applied to one link and this energy be used to overcome a resisting torque T_r applied to another link, then by similar reasoning, we get

$$T\omega = T_r\omega_r \quad . \quad . \quad . \quad . \quad . \quad (12.2)$$

where ω and ω_r are respectively the angular velocities at the given instant of the link to which the energy is supplied and the link from which the energy is delivered.

Example 1. A toggle mechanism such as might be used in a small hand punch is shown in Fig. 272. The effort F is applied to the point C on the bell-crank lever COA and the resistance F_r to the block B, which is guided along a vertical path. What is the least value of F in terms of F_r for the given position of the mechanism?

If friction is neglected, the rate of doing work at C must be equal to the rate of doing work at B.

$$\therefore\ F.v_c = F_r.v_b \quad \text{or} \quad F = v_b/v_c.F_r$$

But $v_c = \text{OC}/\text{OA}.v_a$; also, the instantaneous centre of link AB is at I, where I is the point of intersection of OA produced and a line through B perpendicular to v_b. So that, $v_a/v_b = \text{IA}/\text{IB}$.

$$\therefore\ F = \frac{v_b}{v_a}.\frac{v_a}{v_c}.F_r = \frac{\text{IB}}{\text{IA}}.\frac{\text{OA}}{\text{OC}}.F_r$$

Scaling the lengths from the diagram and substituting, we find that $F = 0{\cdot}15F_r$.

Example 2. In the epicyclic gear shown in Fig. 273, the axes O and P are fixed. The pinion A is keyed to the driving shaft and gears with the external wheel B, which is integral with the internal wheel C. The planets D gear with C and with the fixed wheel E. The planet carrier is keyed to a driven shaft F coaxial with the fixed wheel E. The numbers of teeth are: A, 25; B, 90; C, 83; E, 31. Find the torque exerted on the driven shaft if the torque on the driving shaft is 10 lb ft.

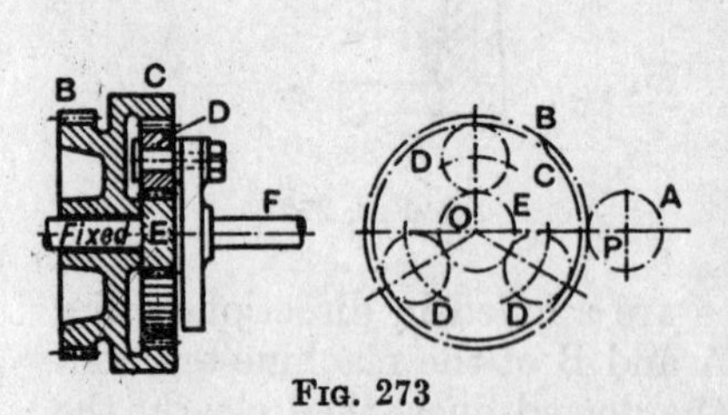

FIG. 273

Let N with the appropriate suffix represent the speed of rotation of the corresponding wheel or shaft. Then, for the epicyclic train C, D, E:

$$\frac{N_e - N_f}{N_c - N_f} = -\frac{83}{31}$$

But the wheel E is fixed, so that $N_e = 0$,

$$\therefore\ \frac{0-N_f}{N_c-N_f} = -\frac{83}{31}$$

from which $$N_c/N_f = 114/83$$

Since B and C are rigidly fastened together, $N_b = N_c$.

But $\dfrac{N_a}{N_b} = -\dfrac{90}{25} = -\dfrac{18}{5}$, so that $\dfrac{N_a}{N_c} = -\dfrac{18}{5}$

and $$\frac{N_a}{N_f} = \frac{N_a}{N_c}\cdot\frac{N_c}{N_f} = -\frac{18}{5}\cdot\frac{114}{83} = -4{\cdot}94$$

If friction is neglected, the rate of doing work on the driven shaft is equal to the rate of doing work on the driving shaft.

$$\therefore\ T_f . N_f = T_a . N_a \quad \text{or} \quad T_f = \frac{N_a}{N_f} . T_a = -4{\cdot}94 . 10 = -49{\cdot}4 \text{ lb ft}$$

The negative sign merely indicates that the torque exerted on the driven shaft is of opposite sense to that exerted on the driving shaft.

Example 3. The high-pressure cylinder of a horizontal cross-compound steam engine has a diameter of 9 in. and a stroke of 24 in. The diameter of the piston rod is 2 in., the length of the connecting rod is 60 in. and the r.p.m. are 120. When the crank has turned through 40° from the i.d.c., the steam pressures on the cover and crank sides of the piston are respectively 160 and 32 lb/in² abs. Neglecting friction and the inertia of the moving parts, find the torque exerted on the crankshaft.

The net steam thrust on the piston

$$= F = \pi/4 . 9^2 . 160 - (\pi/4)(9^2 - 2^2)32$$
$$= 63{\cdot}6 . 160 - 60{\cdot}5 . 32$$
$$= 8240 \text{ lb}$$

Let F_t be the equivalent resisting force, which acts through the crankpin C tangential to the crank circle.

Then $$F_t v_c = F . v_p$$

where v_c and v_p are respectively the velocities of the crankpin and the piston.

Referring to Fig. 274, I is the instantaneous centre of the connecting rod CP, so that v_p/v_c = IP/IC.

But, from the similar triangles IPC, OMC, we have OM/OC = IP/IC and therefore v_p/v_c = OM/OC. Substituting in the above equation:

$$F_t . \text{OC} = \text{F} . \text{OM}$$

But F_t.OC is equal to the torque T exerted on the crankshaft,

$$\therefore\ T = F . \text{OM} \quad . \quad . \quad . \quad . \quad (12.3)$$

The length OM may be found either from a scale drawing or by calculation. Since $OM/OC = v_p/v_c$, it follows from equation (3.11) that:

$$OM/OC \simeq \sin\theta + (\sin 2\theta)/2n$$

But $n = CP/OC = 5$ and $\theta = 40°$,

$$\therefore\ OM/OC \simeq 0{\cdot}6428 + 0{\cdot}9848/10 \simeq 0{\cdot}7413$$

and the length of OM $\simeq 0{\cdot}7413$ ft

$\therefore$ the torque exerted on the crankshaft $= F.OM \simeq 8240.0{\cdot}7413$
$\simeq 6108$ lb ft

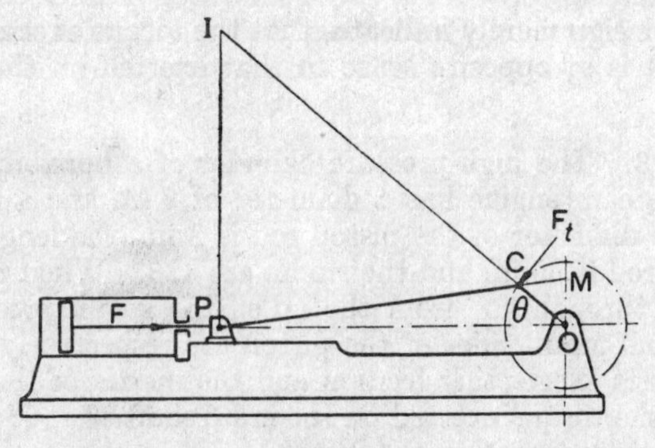

FIG. 274

153. The Effective Force and the Inertia Force of a Link. Each part or link of a machine has a more or less complicated motion which changes from instant to instant. In general the motion consists of rotation about an axis, the instantaneous axis of the link, with varying angular velocity. This is equivalent to a motion of translation of the c.g. of the link together with rotation about an axis through the c.g. Both the linear velocity of the c.g. and the angular velocity of rotation vary from instant to instant. It was shown in Article 28 that motion of this kind is produced when the line of action of the force applied to a rigid body does not pass through the c.g. The magnitude and direction of the linear acceleration of the c.g. depend upon the magnitude and direction of the applied force and the mass of the body; the magnitude and sense of the angular acceleration in turn depend on the magnitude and sense of the moment of the applied force about an axis through the c.g. and on the moment of inertia of the body about this axis. Given the linear acceleration of the c.g. and the mass, the magnitude and direction of the applied force may be calculated; given the angular acceleration and the moment of inertia, the moment of the applied force and therefore the distance of its line of action from the c.g. may be calculated.

A link of a mechanism is constrained to move in a definite way by the adjacent links to which it is connected. The resultant of all the forces applied to the link through those connections, together with any other external forces, such as the force of gravity, must be equal to the force required to accelerate the link. This force is termed the effective force and its magnitude is given by the product of the mass of the link and the linear acceleration of its centre of mass.

The direct way of finding the magnitude and the line of action of the effective force for a given link of a mechanism is to find first the acceleration f_g of the c.g. and the angular acceleration α, as explained in Articles 46–50. The magnitude F of the effective force is then given by $W/g.f_g$, where W is the weight of the link, and its line of action is parallel to f_g. The distance h of the line of action from the c.g. and the side of the c.g. on which the line of action lies must then conform to the magnitude and sense of α. Thus $h = W/g.k^2\alpha/F$, where k is the radius of gyration of the link about an axis through the c.g. perpendicular to the plane of motion. Referring to Fig. 275 (b) G is the position of the c.g. of the link AB, f_a and f_b the accelerations of the points A and B, represented to scale in (a) by oa and ob, and ab is the acceleration of B relative to A. The vector ab may be resolved into two components ap and pb perpendicular and parallel to the link AB. Then ap is the tangential component $f_{ba}{}^t$ of the acceleration of B relative to A, and the angular acceleration of AB is clockwise and its magnitude is $\alpha = f_{ab}{}^t/\text{AB}$. The acceleration f_g of G is given by og, where g divides the acceleration image ab in the same proportion as G divides AB.

The effective force

$$F = W/g.f_g$$

and

$$h = W/g.k^2\alpha/F$$

Since α is clockwise and F acts in the sense og, the line of action of F is at the distance h above G as shown.

The line of action may however be obtained graphically without the need for any calculation, if the link is replaced by a dynamically equivalent two-mass system.

In Article 27 it was shown that any two-mass system which is to be dynamically equivalent to a given link must have the two masses rigidly fastened together at distances a and b on opposite sides of the c.g. of the link, such that $a.b = k^2$, where k is the radius of gyration of the link about an axis through the c.g. Further, the sum of the two masses must be equal to the mass of the link and the c.g. of the two masses must coincide with the c.g. of the link.

Referring to Fig. 275 (c), the position of one of the two masses which form an equivalent dynamical system may be fixed arbitrarily. The position of the other mass must then be found so that the product of the distances of the two masses from G is equal to k^2. Let one of the masses be placed at A, then the position C of the other mass may be found graphically.

Erect GK perpendicular to AB and equal in length to k; join A to K and draw KC perpendicular to AK to intersect AB at C. Then, from the similar triangles AGK, KGC, we have AG/GK = GK/CG or AG.CG = $KG^2 = k^2$. The link AB is then dynamically equivalent to a system of two masses, one of which

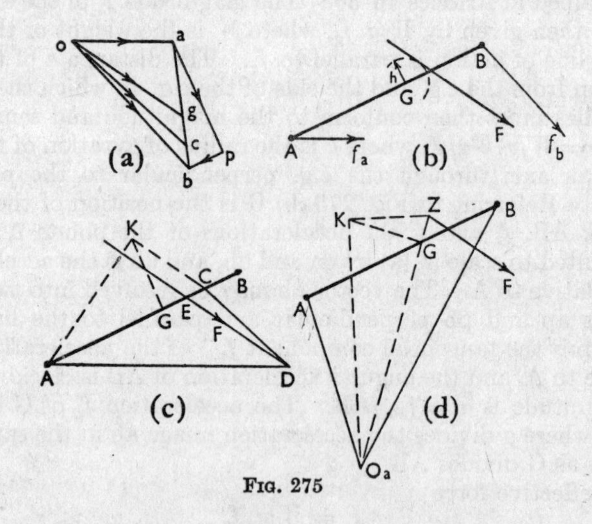

FIG. 275

is placed at A and the other at C. The resultant accelerating force on the link must therefore coincide with the resultant accelefating force on the masses at A and C. It must therefore pass through the point of intersection of the accelerating force on C with the accelerating force on A. The accelerations of C and G are given respectively in magnitude and direction by oc and og, where c and g divide the acceleration image ab in the same proportions as C and G divide the link AB. Hence the line of action of the accelerating force on AB passes through D, which is the point of intersection of f_a produced with a line drawn through C parallel to oc. The magnitude of the accelerating force F is given by $W/g . f_g$, where f_g is the acceleration of the c.g. of the link and is represented to scale by og.

It is often more convenient to place one of the two masses which form the equivalent dynamical system at the acceleration centre of the link. As explained in Article 47, the acceleration centre O_a may be found by drawing triangle O_aAB similar to the acceleration triangle oab. Then, referring to Fig. 275 (d), the other mass must be placed at Z on O_aG produced, such that $GZ = k^2/O_aG$. The point Z may be found by a graphical construction similar to that used for finding C in Fig. 275 (c). But the mass at O_a has zero acceleration, so that the resultant accelerating force on the two-mass system must pass through Z. In other words, Z must be a point on the line of action of the accelerating force on link AB.

Each of the individual links of a machine may be examined in this way in order to find the magnitude and the line of action of the effective force applied to the link. D'Alembert's principle,

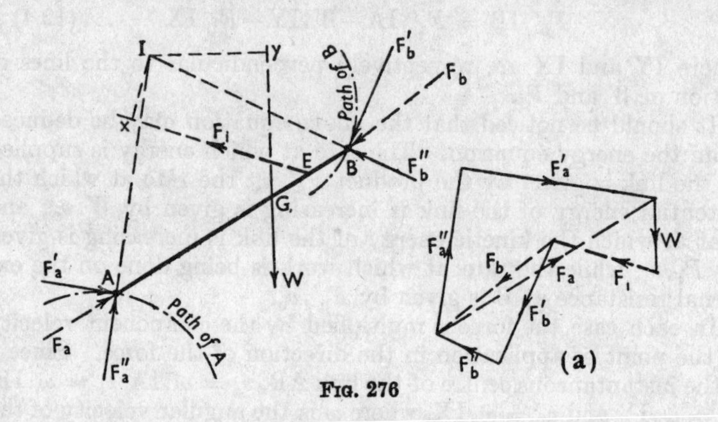

FIG. 276

Article 28, may then be used and for each link the inertia force F_i equal and opposite to the effective force, may be included with the external forces applied to the link in order to give a system of forces which is in equilibrium.

Referring to Fig. 276, AB is a link of a machine with the pins A and B constrained to move along the paths shown. The weight W acts through the centre of gravity G and the inertia force F_i through E. The magnitude and line of action of F_i are found as explained above.

The force F_a which is applied to the link AB at pin A by the adjacent link will usually have a component F_a' tangential to the path of A and also a component F_a'' perpendicular to the path of A. The former component does the useful work on the link, while the latter constrains the pin A to follow the given path. Similarly, the force F_b which is applied to the link AB at pin B by the adjacent link will have a component F_b' tangential to the path of B and a component F_b'' perpendicular to the path of B.

If the component F_a' in the tangential direction is given, then the component in the normal direction and the two components of F_b may be found from the necessary conditions for the equilibrium of the link. These conditions are that (a) the vector sum of all the forces which act on the link shall be zero and (b) the algebraic sum of the moments of the forces about any point in their plane shall be zero.

Since the components of F_a and F_b normal to the paths of A and B are both unknown, we must obviously take moments about the point of intersection I of the lines of action of these two components. We are then able to find F_b', the tangential component of F_b, by equating moments:

$$F_b'.\mathrm{IB} = F_a'.\mathrm{IA} - W.\mathrm{IY} - F_i.\mathrm{IX} \quad . \quad (12.4)$$

where IY and IX are respectively perpendicular to the lines of action of W and F_i.

It should be noticed that the above equation may be deduced from the energy equation. The rate at which energy is supplied to the link is given by the product $F_a'.v_a$; the rate at which the potential energy of the link is increasing is given by $W.v_w$; and that at which the kinetic energy of the link is increasing is given by $F_i.v_i$, while the rate at which work is being done on the external resistance at B is given by $F_b'.v_b$.

In each case the force is multiplied by the component velocity of the point of application in the direction of the force. Since I is the instantaneous centre of the link AB, $v_a = \omega.\mathrm{IA}$, $v_b = \omega.\mathrm{IB}$, $v_w = \omega.\mathrm{IY}$ and $v_i = \omega.\mathrm{IX}$, where ω is the angular velocity of the link AB.

But $$F_a'.v_a = F_b'.v_b + F_i.v_i + W.v_w$$

$$\therefore \; F_a'\omega.\mathrm{IA} = F_b'.\omega.\mathrm{IB} + F_i.\omega.\mathrm{IX} + W.\omega.\mathrm{IY}$$

or $$F_b'.\mathrm{IB} = F_a'.\mathrm{IA} - W.\mathrm{IY} - F_i.\mathrm{IX} \quad \text{as before}$$

When the magnitude of F_b' has been found, the magnitudes of F_a'' and F_b'' may be obtained by drawing the polygon of forces as shown at (a) in Fig. 276.

It is customary, when solving problems of this type, to ignore the effect of the pull of gravity on the link. Obviously this should only be done when the weight of the link is small in comparison with the other forces which act on the link, otherwise the results obtained may involve considerable error.

Example 4. Fig. 276 is drawn to scale for a link AB 12·5 in. long. The c.g. is 4 in. from B and the inertia force F_i passes through the point E, 2 in. from B, in the given direction. The weight of the link is 3 lb, the inertia force F_i is 5 lb and the component F_a' is 9 lb. Scaled from the diagram, the lengths are: IA, 10·75 in.; IB, 10·15 in.; IX, 2·92 in.; and IY, 5·5 in.

Substituting in (12.4):

$$F_b'.10{\cdot}15 = 9.10{\cdot}75 - 3.5{\cdot}5 - 5.2{\cdot}92$$

and
$$F_b' = 6{\cdot}47 \text{ lb}$$

From the polygon of forces: $F_a'' = 7{\cdot}66$ lb and $F_b'' = 3{\cdot}0$ lb, the total force applied to the link AB at the pin A

$$= F_a = \sqrt{(F_a'^2 + F_a''^2)} = 11{\cdot}82 \text{ lb}$$

and the total force applied to the link AB at the pin B

$$= F_b = \sqrt{(F_b'^2 + F_b''^2)} = 7{\cdot}13 \text{ lb}$$

154. Inertia Forces in the Reciprocating Engine. As already pointed out, it is usually convenient to consider the forces which are brought into play owing to the inertia of the links separately from those which would arise if the links were without mass. So far as the reciprocating engine is concerned, the problem to be solved is that of finding the torque which must be exerted on the crankshaft in order to overcome the inertia of the connecting rod and of the reciprocating parts. Because of the importance of this problem a method of calculating the torque required will be given in addition to a method of finding the torque graphically.

(a) *Graphical Method*: Fig. 277. The first step is to draw the acceleration diagram for the mechanism. This may be done by using Klein's construction, Article 49, which gives a triangle OCN, such that the sides CO, NO and NC represent to scale the acceleration of the crankpin C, the acceleration of the piston P and the acceleration of P relative to C. The construction is not actually shown but the triangle is shown to an enlarged scale at (b). The side CN is the acceleration image of the connecting rod CP.

The acceleration of P is $f_p = \omega^2.\text{NO}$ in the sense from N to O. The inertia force F_p of the reciprocating parts therefore acts in the opposite sense as shown.

The positions of two masses which form a system dynamically equivalent to the rod may then be found as explained in Article 27. One of the masses is arbitrarily placed at the small end centre P and the position D of the other mass must then satisfy the relation $\text{GD} = k^2/\text{GP}$. The accelerations of D and G are given by

$$f_d = \omega^2.\text{dO} \quad \text{and} \quad f_g = \omega^2.\text{gO}$$

where d and g divide CN in the same proportion as D and G divide CP.

Since the accelerating forces on the masses at P and D intersect at E, where DE is drawn parallel to dO, their resultant must also pass through E. But their resultant is equal to the accelerating force on the rod, so that the line of action of the accelerating force on the rod is given by a line drawn through E parallel to gO and acting in the sense from g to O. The inertia force F_i of the connecting rod therefore acts through E in the opposite sense as shown.

The forces which act on the rod are the inertia force F_p of the reciprocating parts, which acts along the line of stroke through P, the side thrust F_n between the crosshead and the guide bars, which

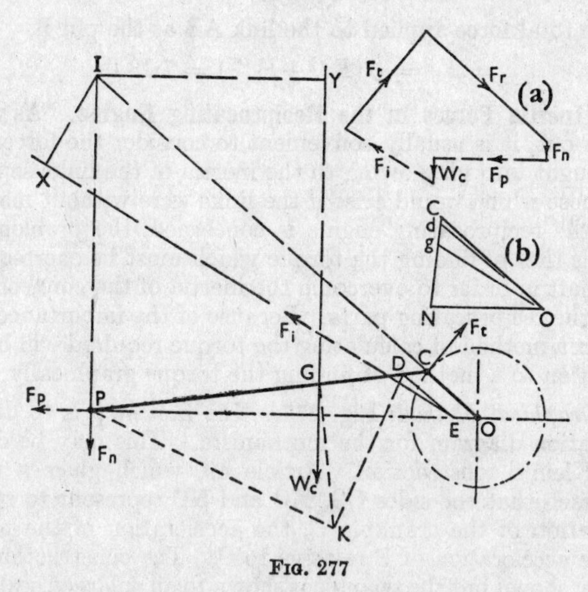

Fig. 277

also acts through P at right angles to the line of stroke, the inertia force F_i of the rod, the weight of the rod W_c and the forces F_r and F_t which act through C parallel and perpendicular respectively to the crank OC. Taking moments about I, the point of intersection of F_n and F_r, we have:

$$F_t.IC = F_p.IP + F_i.IX + W_c.IY \quad . \quad . \quad (12.5)$$

where IX and IY are respectively perpendicular to F_i and W_c.

From the above equation F_t may be calculated and, by drawing the polygon of forces as shown at (a), F_n and F_r may be obtained.

The torque which must be exerted on the crankshaft in order to overcome the inertia of the moving parts is then given by $F_t.OC$.

(b) *Calculation of the Inertia Torque.* The effect of the inertia of the connecting rod on the crankshaft torque may be obtained approximately by dividing its mass into two parts, one part being placed at the crankpin C and the other at the crosshead pin P, while the c.g. of the two masses coincides with G, the c.g. of the rod (Fig. 278). The mass at C clearly has no effect on the crank-

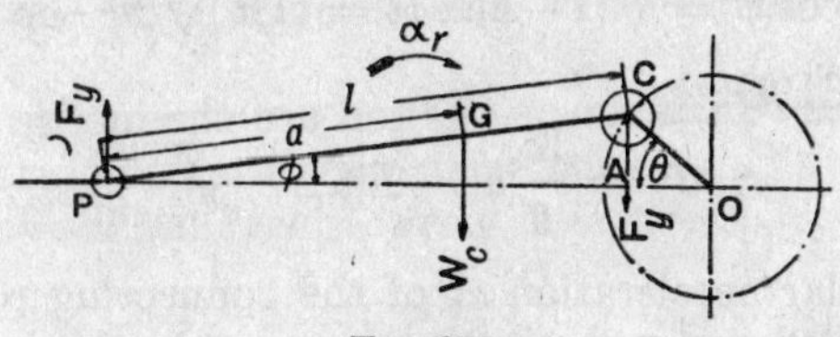

FIG. 278

shaft torque, since its inertia force acts radially outwards along the crank. The effect of the mass at P may be found by including it with the reciprocating masses. Let W_c be the weight of the rod, R the weight of the reciprocating parts, l the length of the rod and a the distance of G from P. Then the total equivalent reciprocating mass weighs

$$R+\frac{l-a}{l}.W_c$$

The inertia force F_p of this mass is

$$-\left(R+\frac{l-a}{l}.W_c\right)\frac{f_p}{g}$$

where f_p is the acceleration of the reciprocating parts, and the corresponding torque exerted on the crankshaft, by analogy with equation (12.3), is given by:

$$T_p = F_p.\text{OM} \quad . \quad . \quad . \quad . \quad . \quad (12.6)$$

This equation only gives the approximate inertia torque because it assumes that two masses, placed one at C and the other at P, are dynamically equivalent to the rod. As explained in Article 27, this assumption is not correct. It is necessary to apply a correction couple T' to the two-mass system. The correction couple is given by equation (2.33) and for the connecting rod this may be written $T' = (W_c/g)a(l-L)\alpha_r$, where L is the length of the equivalent simple pendulum, when swung about an axis through P, and α_r is the angular acceleration of the rod. The couple T' may be applied to the system by two equal and opposite vertical forces F_y acting through P and C. Then

$$F_y.\text{AP} = T' = \frac{W_c}{g}a(l-L)\alpha_r \quad . \quad . \quad (12.7)$$

The corresponding torque on the crankshaft $= T_c = F_y.AO$ and, substituting for F_y,

$$T_c = \frac{W_c}{g} a(l-L)\alpha_r \frac{AO}{AP}$$

But $\quad AO = OC\cos\theta \quad$ and $\quad AP = CP\cos\phi$

Also $\quad \cos\phi = \sqrt{(1-\sin^2\phi)} = (1/n)\sqrt{(n^2-\sin^2\theta)}$

where $n = CP/OC$.

$$\therefore\ T_c = \frac{W_c}{g} a(l-L)\alpha_r \frac{\cos\theta}{\sqrt{(n^2-\sin^2\theta)}} \quad . \quad (12.8)$$

The angular acceleration α_r of the connecting rod is given by equation (3.12) and, substituting:

$$T_c = -\frac{W_c}{g} a(l-L) \frac{\omega^2(n^2-1)\sin\theta}{(n^2-\sin^2\theta)^{3/2}} \cdot \frac{\cos\theta}{\sqrt{(n^2-\sin^2\theta)}}$$

$$= -\frac{W_c}{g} a(l-L) \frac{\omega^2(n^2-1)\sin 2\theta}{2(n^2-\sin^2\theta)^2} \quad . \quad . \quad . \quad (12.9)$$

Since n^2 is usually large in comparison with both $\sin^2\theta$ and 1, this may be written:

$$T_c \simeq -\frac{W_c}{g} a(l-L) \frac{\omega^2}{2n^2} \sin 2\theta \quad . \quad . \quad (12.10)$$

There is, of course, in addition the torque due to the pull of gravity on the rod. It is easily seen that the vertical force through C is given by $W_c.PG/PC$, or $W_c.a/l$, and therefore the torque exerted on the crankshaft by gravity is given by

$$T_w = -W_c.a/l.AO$$

Or, since $AO = OC\cos\theta$,

$$T_w = -W_c(a/n)\cos\theta \quad . \quad . \quad . \quad (12.11)$$

The total torque exerted on the crankshaft by the inertia of the moving parts is the algebraic sum of T_p, T_c and T_w.

Example 5. The reciprocating parts of the engine referred to in Example 3, p. 415, weigh 580 lb, the connecting rod weighs 500 lb, the c.g. of the connecting rod is 19 in. from the crankpin centre and the radius of gyration of the rod about an axis through the c.g. is 25 in. Find the inertia torque on the crankshaft when $\theta = 40°$ and the r.p.m. are 120.

(a) *Graphically*. Fig. 277 is drawn to scale to correspond with the above particulars.

The acceleration of the piston f_p

$$= \omega^2.\text{NO} = (\pi.120/30)^2.9{\cdot}6/12 = 126{\cdot}4 \text{ ft/s}^2$$

The inertia force due to the reciprocating parts

$$= F_p = R/g.f_p = 580/32{\cdot}2.126{\cdot}4 = 2275 \text{ lb}$$

The acceleration of the c.g. of the rod

$$= f_g = \omega^2.\text{gO} = 16\pi^2.10{\cdot}7/12 = 140{\cdot}8 \text{ ft/s}^2$$

The inertia force due to the rod

$$= F_i = W_c/g.f_g = 500/32{\cdot}2.140{\cdot}8 = 2185 \text{ lb}$$

To find D, GK is set off perpendicular to CP and equal in length to the radius of gyration of the rod; P is joined to K and KD drawn perpendicular to PK. Then from the similar triangles PGK, KGD, it follows that GD = GK^2/GP.

Scaled from the figure, IP = 57·5 in., IC = 77·5 in., IX = 18·6 in and IY = 40·5 in.

Substituting in equation (12.5):

$$\begin{aligned} F_t &= (2275.57{\cdot}5 + 2185.18{\cdot}6 + 500.40{\cdot}5)/77{\cdot}5 \\ &= (130\,800 + 40\,700 + 20\,250)/77{\cdot}5 \\ &= 191\,750/77{\cdot}5 = 2470 \text{ lb} \end{aligned}$$

The torque exerted on the crankshaft due to the inertia of the moving parts = 2470 lb ft in the counter-clockwise sense.

From the force polygon, the forces F_n and F_r scale respectively 3380 lb and 280 lb.

(b) *By Calculation.*

The total equivalent reciprocating mass

$$= (580 + 19/60.500) = 738{\cdot}3 \text{ lb}$$

and the acceleration of the reciprocating parts, from equation (3.12),

$$\begin{aligned} = f_p &= 16\pi^2.1.(\cos 40° + \cos 80°/5) \\ &= 157{\cdot}9(0{\cdot}7660 + 0{\cdot}0347) \\ &= 126{\cdot}4 \text{ ft/s}^2 \end{aligned}$$

The total equivalent inertia force

$$= -738{\cdot}3/32{\cdot}2.126{\cdot}4 = -2895 \text{ lb}$$

Also, from Example 3, p. 415, OM = 0·7413 ft, so that, substituting in equation (12.6), $T_p = -2895.0{\cdot}7413 = -2146$ lb ft.

The length of the equivalent simple pendulum when the connecting rod is swung about the small-end centre

$$= L = a + k^2/a = 41 + 25^2/41 = 56{\cdot}24 \text{ in.} \quad \text{or} \quad 4{\cdot}687 \text{ ft}$$

Substituting in equation (12.10):

$$T_c = -\frac{500}{32{\cdot}2}\cdot\frac{41}{12}(5-4{\cdot}687)\frac{16\pi^2}{2.25}0{\cdot}9848 = -51{\cdot}6 \text{ lb ft}$$

N.B.—If the exact equation (12.9) is used instead of (12.10), the value of T_c is $-51{\cdot}2$ lb ft. The error involved in using the approximate expression for T_c is therefore negligible.

The torque on the crankshaft due to the pull of gravity on the rod, from equation (12.11), is:

$$T_w = -500.\frac{41}{12.5}.0{\cdot}7660 = -261{\cdot}7 \text{ lb ft}$$

The total torque exerted on the crankshaft due to the inertia of the moving parts

$$= T_p + T_c + T_w = -2146 - 51{\cdot}6 - 261{\cdot}7 = -2459 \text{ lb ft}$$

Note that in this example the magnitude of the inertia torque is underestimated when the usual assumption is made that the mass of the rod may be divided into two parts, one concentrated at the crankpin and the other at the crosshead pin. So far as the inertia effect of the rod itself is concerned the approximate method gives a crankshaft torque of $158{\cdot}3/738{\cdot}3.2146 = 460$ lb ft, whereas the exact method gives $460+51{\cdot}6$, or $511{\cdot}6$ lb ft. The error in the result given by the approximate method is therefore $-51{\cdot}6/511{\cdot}6.100$, or $-10{\cdot}3\%$. The error in the total inertia torque is of course much smaller. Note too the relatively large magnitude of the torque due to the pull of gravity on the rod.

155. Inertia Forces in the Four-bar Chain. ABCD, Fig. 279, is a four-bar chain with AD as the fixed link. It is required to find the effect of the inertia of the links BC and CD when the driving crank AB turns with uniform angular velocity and the mass, the centre of gravity and the radius of gyration of both BC and CD are given.

The first step is to draw the acceleration diagram for the mechanism. This is most easily done by the method of Article 50, which gives ABXQTA as the acceleration diagram, Fig. 279 (a):

The acceleration of B $= f_b = \omega^2$.BA.
The acceleration of C $= f_c = \omega^2$.QA.
The acceleration of C relative to B $= f_{cb} = \omega^2$.QB.

The centripetal and tangential components of f_c are given by $f^c_c = \omega^2$.TA and $f^t_c = \omega^2$.QT, and the centripetal and tangential components of f_{cb} are given by $f^c_{cb} = \omega^2$.XB and $f^t_{cb} = \omega^2$.QX.

The next step is to consider the forces applied to the link CD, Fig. 279 (b).

The angular acceleration of CD

$$= \alpha = f^t_c/CD = \omega^2.QT/CD \text{ (clockwise)}$$

and the clockwise torque applied to accelerate CD

$$= W_h k_d^2/g.\alpha = W_h k_d^2/g.\omega^2.QT/CD$$

where k_d = radius of gyration of CD about D.

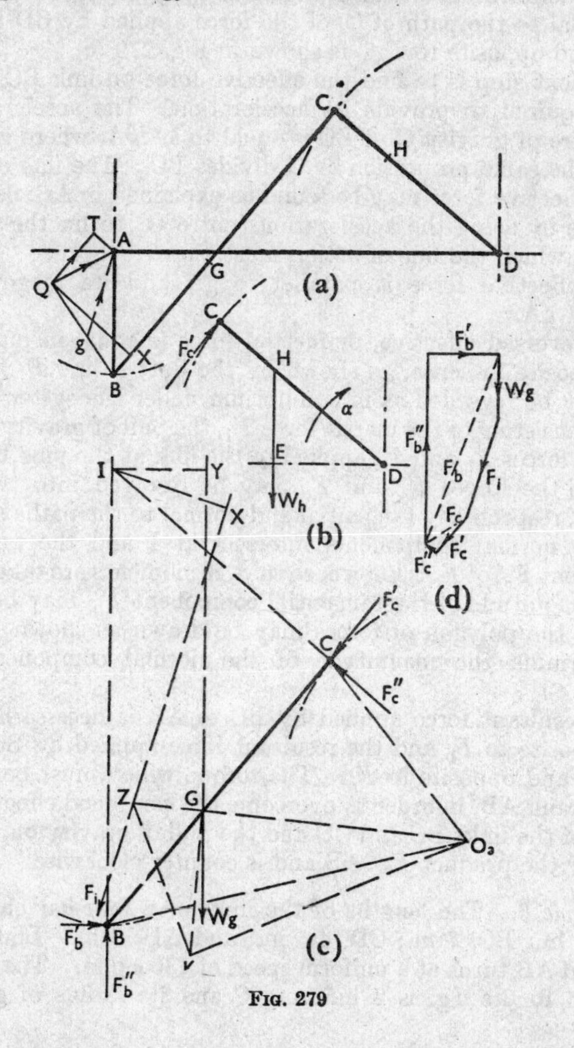

FIG. 279

Hence the total clockwise torque which must be applied to overcome the pull of gravity and accelerate CD

$$T_c = W_h.\mathrm{ED} + W_h k_d^2/g.\omega^2.\mathrm{QT/CD}$$

The corresponding component, tangential to the path of C, of the force applied by BC to CD

$$= F_c' = T_c/\mathrm{CD}$$

This force acts in the direction shown in Fig. 279 (b).

Since action and reaction are equal and opposite, the component, tangential to the path of C, of the force applied by CD to BC is equal and opposite to F_c', as shown in Fig. 279 (c).

The next step is to find the effective force on link BC, i.e. the force required to provide its acceleration. The acceleration of the centre of gravity G of BC is equal to ω^2.gA, where g divides BQ in the same proportion as G divides BC. The line of action of the effective force may be found as explained in Article 153, in this case by using the acceleration centre O_a to fix the point Z through which the line of action must pass.

The effective force is parallel to gA and its magnitude is $W_g/g.\omega^2$.gA.

The reversed effective, or inertia, force is equal in magnitude and opposite in sense, as shown by the dotted line F_i, and the link may be regarded as in equilibrium under the system of four forces consisting of the inertia force F_i, the pull of gravity W_g and the two forces F_b and F_c applied to the link at the pins B and C. Each of the forces F_b and F_c may be resolved into two components, respectively tangential and normal to the paths of B and C. The normal components intersect at I and the tangential component F_c' of F_c is known, so that, if moments are taken about I, the magnitude of the tangential component F_b' may be found. Finally, the polygon of forces may be drawn as shown in order to determine the magnitude of the normal components F_c'' and F_b''.

The resultant force applied by BC to AB is necessarily equal and opposite to F_b and the resultant force applied by BC to CD is equal and opposite to F_c. The torque which must be applied to the crank AB, in order to overcome the combined effects of the inertia of the links BC and CD and the pull of gravity on them, is given by the product F_b'.AB and is counter-clockwise.

Example 6. The lengths of the links of a four-bar chain are: AB, 2·5 in.; BC, 7 in.; CD, 4·5 in.; and AD, 8 in. Link AD is fixed and AB turns at a uniform speed of 180 r.p.m. The link BC weighs 5 lb, its c.g. is 4 in. from C and its radius of gyration about an axis through the c.g. is 2·9 in. The link CD weighs 3 lb, its c.g. is 1·5 in. from C and its radius of gyration about an axis through D is 3·5 in. When BA is at right angles to AD and B and C lie on opposite sides of AD, find the torque exerted on AB to overcome the inertia of the links and the forces which act on the pins B and C.

Fig. 279 is drawn to scale for the particulars of the above example.

Link CD. The angular acceleration of CD

$$= \alpha = \omega^2 . \mathrm{QT/CD} \text{ (clockwise)}$$

But $\omega = 6\pi$ rad/s and QT scales 1·35 in.

$$\therefore \ \alpha = (6\pi)^2 . 1{\cdot}35/4{\cdot}5 \doteq 106{\cdot}6 \text{ rad/s}^2$$

and the torque applied to accelerate CD

$$= 3/32{\cdot}2 . (3{\cdot}5/12)^2 . 106{\cdot}6$$
$$= 0{\cdot}845 \text{ lb ft}$$
$$= 10{\cdot}14 \text{ lb in.}$$

The additional clockwise torque required to overcome gravity

$$= W_{\mathrm{h}} . \mathrm{ED} = 3 . 2{\cdot}3 = 6{\cdot}9 \text{ lb in.}$$

$\therefore$ total clockwise torque to be applied to CD

$$= T_{\mathrm{c}} = 10{\cdot}14 + 6{\cdot}9 = 17{\cdot}0 \text{ lb in.}$$

The corresponding component, tangential to the path of C, of the force applied by BC to CD

$$= F_{\mathrm{c}}' = T_{\mathrm{c}}/\mathrm{CD} = 17{\cdot}0/4{\cdot}5 = 3{\cdot}78 \text{ lb}$$

N.B.—If the pull of gravity on the link is neglected, then:

$$F_{\mathrm{c}}' = 10{\cdot}14/4{\cdot}5 = 2{\cdot}25 \text{ lb}$$

Link BC. The acceleration of G $= f_{\mathrm{g}} = \omega^2 . \mathrm{gA}$ and gA scales 1·8 in., so that $f_{\mathrm{g}} = 36\pi^2 . 1{\cdot}8/12 = 53{\cdot}3$ ft/s^2.

The effective force applied to BC is parallel to gA and its magnitude is 5/32·2 . 53·3 = 8·28 lb. Since $\mathrm{O_aG}$ scales 5·6 in. and k_{g} is 2·9 in.,

$$\mathrm{GZ} = k_{\mathrm{g}}^2/\mathrm{O_aG} = 2{\cdot}9^2/5{\cdot}6 = 1{\cdot}50 \text{ in.}$$

The inertia force F_{i}, equal and opposite to the effective force, passes through the point Z and is shown dotted.

Moments are then taken about I of the four forces F_{b}', F_{c}', F_{i}

and W_g to find F_b'. Scaled from Fig. 279 (c) IB = 9 in., IC = 5·8 in., IX = 2·49 in. and IY = 1·93 in.

$$\therefore\ F_b' = (3{\cdot}78\,.\,5{\cdot}8 + 8{\cdot}28\,.\,2{\cdot}49 + 5\,.\,1{\cdot}93)/9{\cdot}0$$
$$= (21{\cdot}9 + 20{\cdot}6 + 9{\cdot}65)/9{\cdot}0 = 5{\cdot}78 \text{ lb}$$

N.B.—If the pull of gravity on the links is neglected,

$$F_b' = \frac{2{\cdot}25\,.\,5{\cdot}8 + 8{\cdot}28\,.\,2{\cdot}49}{9{\cdot}0} = \frac{13{\cdot}0 + 20{\cdot}6}{9{\cdot}0} = 3{\cdot}73 \text{ lb}$$

The torque which must be exerted on AB in order to overcome the inertia of the links = F_b'.AB = 5·78.2·5 = 14·5 lb in. If the pull of gravity is neglected the torque = 3·73.2·5 = 9·4 lb in. From the force polygon F_c'' is found to be 1·0 lb and F_b'' 15·2 lb. Neglecting gravity, the corresponding values are F_c'' 0·4 lb and F_b'' 10·0 lb, but F_c'' now acts in the opposite direction to that shown. The total force applied to the link BC at pin C is the vector sum of F_c' and F_c'' and is given by F_c = 3·96 lb. Similarly, the total force applied at pin B is the vector sum of F_b' and F_b'' and is given by F_b = 16·3 lb.

Neglecting gravity, the values of F_c and F_b are respectively 2·32 lb and 10·7 lb, but the force polygon for this case is not shown.

156. Turning Moment Diagram for a Steam Engine. Figure 280 shows diagrammatically the arrangement of a single-cylinder, double-acting, horizontal steam engine. Above the cylinder the indicator diagrams are drawn, that for the cover end in full lines and that for the crank end in dotted lines. For the sake of clear-

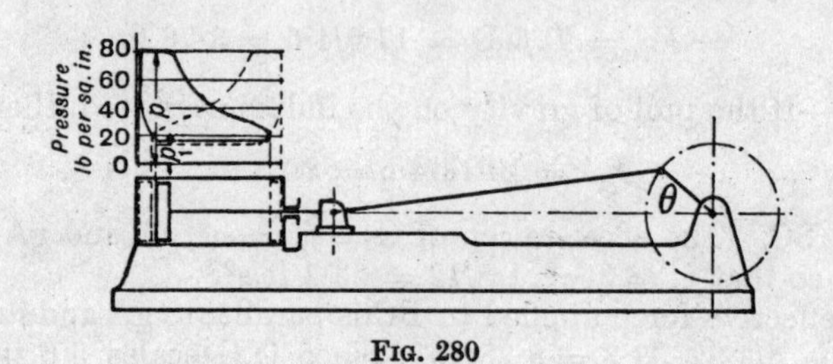

FIG. 280

ness the exhaust pressure line for the cover end is drawn slightly above that for the crank end, although it will be understood that actually the exhaust pressure is the same for both ends of the cylinder. With the piston in the given position and the crank turning clockwise, the pressure of the steam on the cover side of the piston is given by the ordinate p; similarly, the pressure on the crank side of the piston is given by the ordinate p_1. The net steam thrust which tends to move the piston towards the right is therefore given by $F_s = p.A - p_1.A_1$, where A, A_1 are respectively the areas of the cover and crank sides of the piston. If, for the present, the effect of the inertia of the moving parts is ignored, F_s will produce a torque on the crankshaft the magnitude of which may be calculated from equation (12.3). The variation of the turning moment due to the steam pressure, calculated in this way, for a complete revolution of the crankshaft is shown in Fig. 281. The mean height of the curve represents the mean steam torque on the crankshaft and is shown in the figure. When

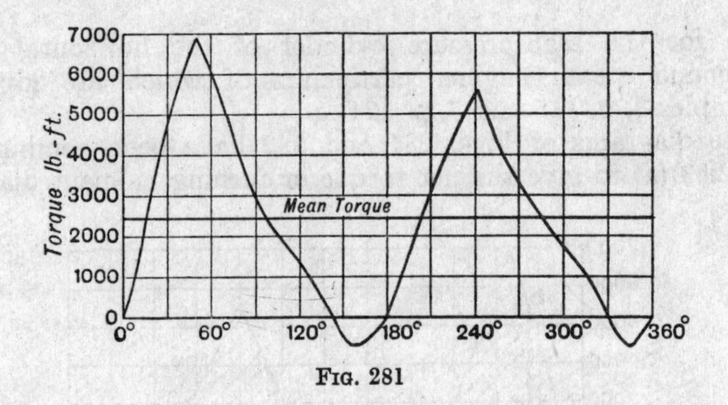

FIG. 281

the mean speed of rotation of the crankshaft is constant, the mean height also represents the mean resisting moment, which is due partly to the frictional resistances between the moving parts of the engine, but chiefly to the power taken from the engine crankshaft. Fig. 281 is drawn to scale for the high-pressure cylinder of a horizontal, cross-compound steam engine, particulars of which are given in Example 3, p. 415.

For a vertical engine the effect of the deadweight of the reciprocating parts must be taken into account. Obviously, when the piston is moving downwards the weight of the parts must be added to the effective steam thrust, and, conversely, when the piston is moving upwards the weight of the parts must be subtracted from the effective steam thrust.

157. Inertia Torque. The torque required on the crankshaft in order to overcome the inertia of the reciprocating parts and of the connecting rod may be calculated as explained in Article 154. The torque required varies for a complete revolution of the crankshaft, as shown in Fig. 282.

For each half-revolution of the crank the diagram has a positive and a negative loop. The two loops are different in shape because of the effect of the obliquity of the connecting rod, but they are of equal area since the energy absorbed in accelerating the parts is returned during subsequent retardation. Fig. 282 is drawn to

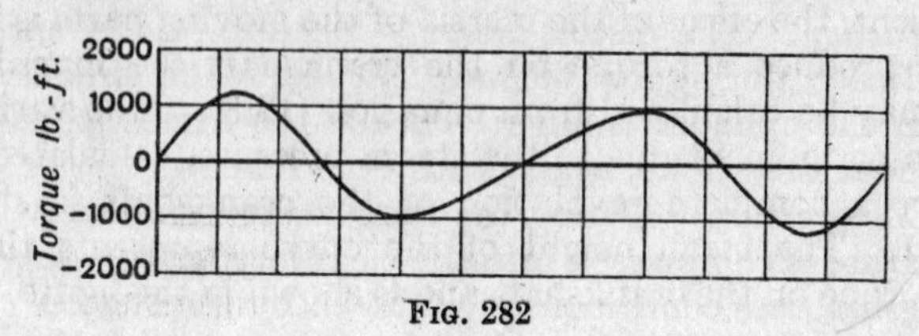

FIG. 282

scale for the high-pressure cylinder of the horizontal cross-compound steam engine, particulars of which are given in Examples 3, p. 415 and 5, p. 424.

The diagrams of Figs. 281 and 282 have been combined in Fig. 283 (a) to give the net torque or turning moment diagram.

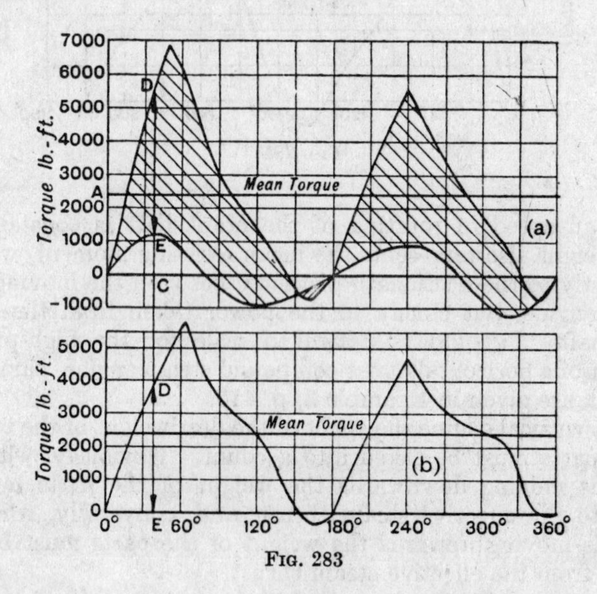

FIG. 283

The net torque for a given crank position is represented by the ordinate of the shaded portion of the diagram. Thus for the crank position represented by C the steam torque is given by CD, the torque required to accelerate the parts by CE and the net torque by ED. The variation of net torque is shown in Fig. 283 (b).

158. Turning Moment Diagram for an Internal-combustion Engine. Fig. 284 shows diagrammatically the arrangement of a single-cylinder, single-acting internal-combustion engine operating on the four-stroke cycle.

Taking a gas or petrol engine as typical the four strokes are:

(1) *Suction.* The piston draws a mixture of fuel and air into the cylinder during its outward stroke. The pressure of the gases in the cylinder is slightly below the atmospheric pressure throughout the stroke, as shown by the line ab.

(2) *Compression.* During the next inward stroke of the piston the mixture is compressed into the clearance space. The pressure of the gases rises continuously as shown by the line bc.

(3) *Expansion.* At the end of the compression stroke the mixture is fired by an electric spark and combustion takes place practically at constant volume, as shown by the line cd. The

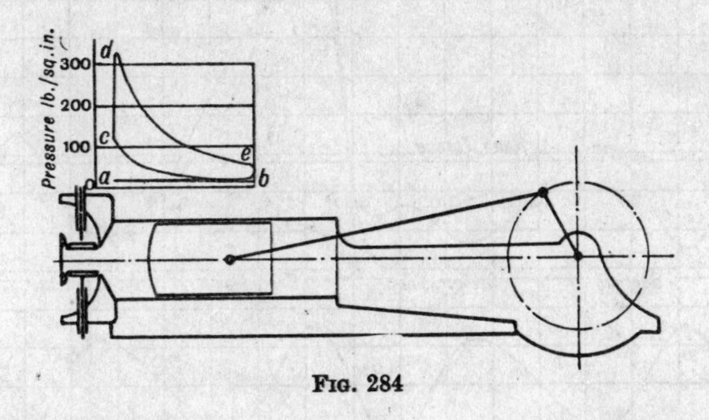

FIG. 284

products of combustion then expand, driving out the piston and doing useful work on the crankshaft. This expansion is shown by the line de.

(4) *Exhaust.* On the next return stroke of the piston the products of combustion are pushed out of the cylinder.

The cycle of operations differs slightly from the above for most oil engines. Instead of a mixture of air and fuel, air only is drawn into the cylinder during the suction stroke and compressed into the clearance space on the return stroke. The temperature of the compressed air, combined in some cases with the temperature of an uncooled portion of the cylinder head, is sufficiently high to ignite the oil as it is sprayed into the cylinder. Combustion, expansion and exhaust follow as in the gas or petrol engine.

The turning moment on the crankshaft for different crank positions may be calculated in exactly the same way as for the steam engine, and the turning moment diagrams for a complete cycle may be drawn as shown in Fig. 285. The gas torque diagram is shown by the dotted lines in Fig. 285 (a). The pressure of the gases on the working side of the piston is slightly less than the pressure of the atmosphere on the crank side of the piston during suction and there is a small negative gas torque loop. During compression a large negative gas torque loop is obtained owing to the work done by the piston on the gases. During expansion the expanding gases do work on the piston and a large positive gas torque loop is obtained, while during exhaust work is done by the piston on the gases, since the pressure of the gases in the

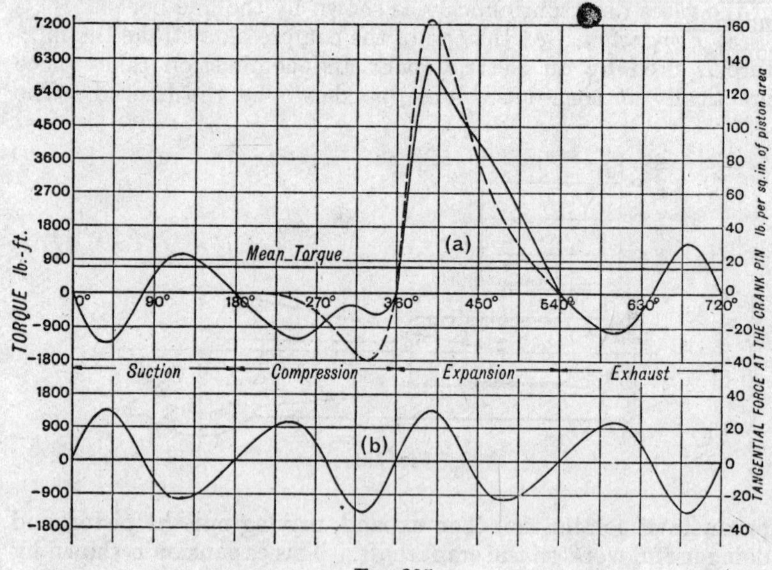

Fig. 285

cylinder is slightly higher than the external atmospheric pressure, and the gas torque diagram again shows a small negative loop. In Fig. 285, which is drawn to scale, the negative loops on the suction and exhaust strokes are too small to be shown.

The variation of the torque required to accelerate the moving parts is shown in Fig. 285 (b). For each half-revolution of the crank the curve is, of course, similar in shape to the corresponding diagram for a steam engine, Fig. 282.

The net turning moment on the crankshaft for a given crank position is given by the difference between the gas torque and the torque required to accelerate the moving parts. It is shown by the full line in Fig. 285 (a). The torque diagrams of Fig. 285

have been drawn to scale for a single-cylinder gas engine. The indicator diagram is shown to scale in Fig. 284. The cylinder is 9 in. dia., the stroke is 17 in. and the r.p.m. are 270. The reciprocating parts weigh 110 lb and the connecting rod weighs 150 lb. The length of the rod is 38·25 in. between centres, its c.g. is 13·75 in. from the crankpin centre and the radius of gyration about the c.g. is 15·25 in.

159. Turning Moment Diagram for Multi-cylinder Engines. So far we have dealt only with single-cylinder engines. In order to determine the shape of the total turning moment diagram for a multi-cylinder engine, the turning moment diagrams for the individual cylinders must first be determined. These are then combined so that the total torque at the instant one of the cranks

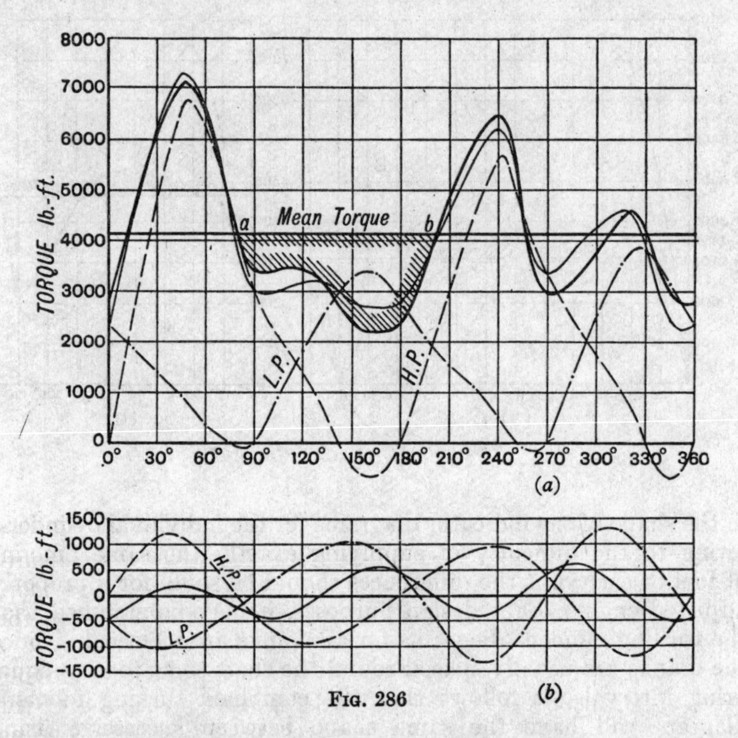

FIG. 286

occupies a given position is the algebraic sum of the torques exerted on the individual cranks at the same instant. The turning moment diagrams for a horizontal cross-compound steam engine are shown in Fig. 286. The low-pressure crank leads the high-pressure crank by 90°, so that the turning moment diagram for the low-pressure cylinder must be displaced 90° to the left relative to the diagram for the high-pressure cylinder. This brings the torques exerted on the two cranks at the same instant along the same ordinate so that they may be combined directly. Referring to Fig. 286, the dotted line shows the turning moment diagram for the high-pressure cylinder and the chain-dotted line that for the low-pressure cylinder. The combined diagrams are shown by the full line. Fig. 286 (a) is drawn for the steam torque only and Fig. 286 (b) for the accelerating torque only. The thick line in Fig. 286 (a) shows the net turning moment diagram when the steam torque and the inertia torque are combined.

With multi-cylinder internal-combustion engines the indicator diagram and the weight of the reciprocating parts should be identical for all cylinders. In practice there may be small differences

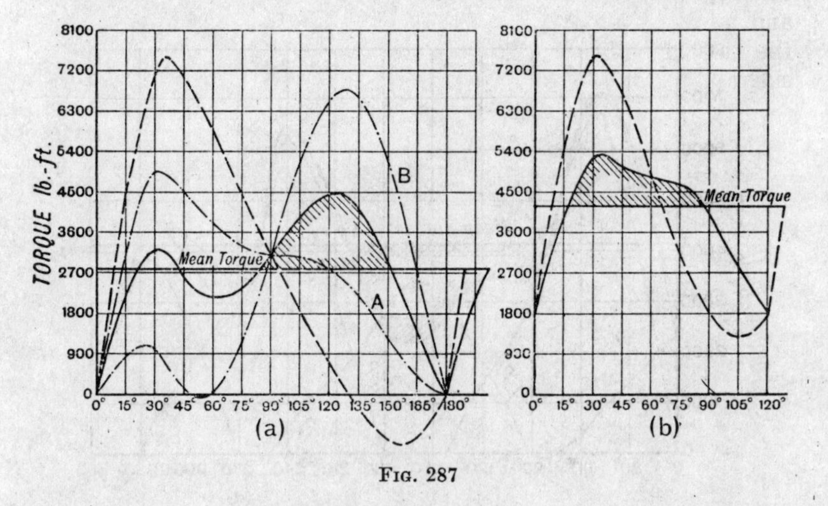

Fig. 287

in the shape of the indicator diagrams for the individual cylinders, owing to the difficulty of supplying exactly the same amount of fuel to each, but the differences should be small for a properly adjusted engine. For design purposes, it may be assumed that the turning moment diagrams for all cranks are identical. Since the cranks are usually spaced round the shaft so as to give equal firing intervals, it follows that the combined turning moment diagram will have the same shape between successive firing intervals. In Fig. 287 (a) and (b) the turning moment diagrams are shown respectively for a four-cylinder and a six-cylinder gas engine based on the single-cylinder diagrams of Fig. 285. The gas torque diagram is shown by dotted lines, and the combined diagram, gas torque plus inertia torque, by full lines in each case.

Incidentally the effect of engine speed on the shape of the combined diagram is shown by the chain-dotted lines in the case of the four-cylinder engine. Curve A corresponds to a running speed 25% below and curve B to a running speed 25% above the normal speed of rotation.

It should be pointed out that the combined turning moment diagrams of Fig. 287 have been drawn on the assumption that the shape of the gas torque diagram is independent of the speed of rotation of the crank. This will not be true for any actual engine, but the change of shape with speed will be relatively small.

160. The Fluctuation of the Crankshaft Speed. The couple or torque, which resists rotation of the crankshaft, is to all intents and purposes uniform for most engines, while, as we have seen, the turning moment exerted on the crankshaft fluctuates considerably. This means that, except for isolated crank positions,

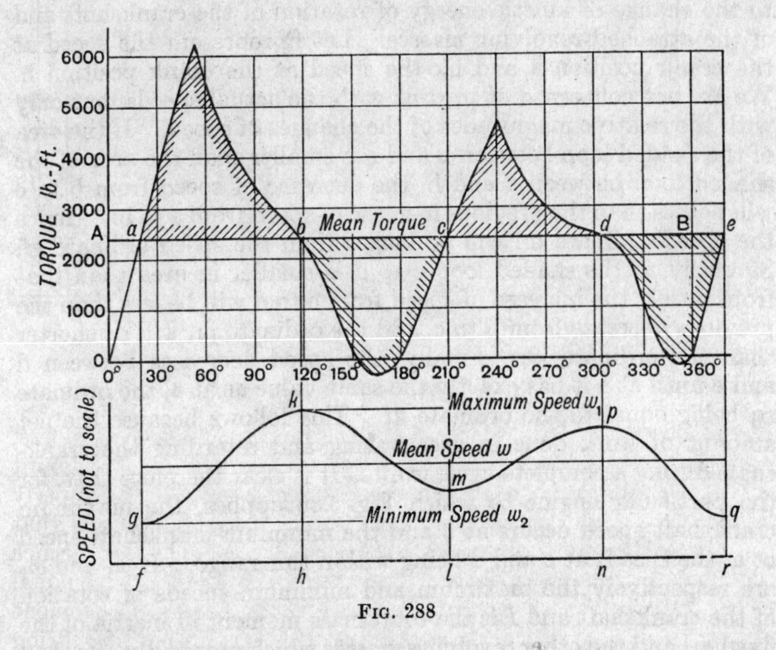

FIG. 288

there is an unbalanced torque which tends either to increase or to decrease the speed of rotation of the crankshaft.

The net turning moment diagram for a single-cylinder double-acting steam engine, Fig. 283 (b), is repeated in Fig. 288, and below the turning moment diagram is shown the approximate shape of the curve of fluctuation of speed of the crankshaft. This curve is obtained as follows: assuming that the resistance to rotation of the crankshaft is uniform, the resisting moment line will coincide with the mean turning moment line AB and will therefore intersect the net turning moment line at the points a, b, c, d and e, the point e being the point a for the following cycle. At these points of intersection the turning moment and the resisting moment are equal and consequently the slope of the speed curve is zero. Between a and b the turning moment is greater than the resisting moment and the unbalanced torque accelerates the crankshaft; conversely, between b and c the turning moment is less than the resisting moment and the unbalanced torque therefore retards the crankshaft. Similarly, between c and d the unbalanced torque accelerates the crankshaft, while between d and e it again retards the crankshaft. Obviously the area of the shaded loop between a and b represents the work done by the unbalanced accelerating torque and is equal to the change of kinetic energy of rotation of the crankshaft and of the attached revolving masses. Let fg represent the speed at the crank position a and hk the speed at the crank position b. We are not concerned at present with the actual speeds, but only with the relative magnitudes of the changes of speed. If the area of the shaded loop between b and c is smaller than the area of the shaded loop between a and b, the decrease of speed from b to c will be less than the previous increase of speed from a to b. Hence the speed ordinate ml will be longer than the speed ordinate gf. Similarly, if the shaded loop c to d is smaller in area than that from b to c, the increase of speed from c to d will be less than the previous decrease from b to c, and the ordinate pn will be shorter than the ordinate kh. Finally, the speed decreases between d and e until at e it has exactly the same value as at a, the ordinate rq being equal to the ordinate gf. This follows because the net amount of work done in accelerating and retarding the crankshaft during a complete cycle is nil. It is clear therefore that, for the particular engine to which Fig. 288 applies, the maximum crankshaft speed occurs at b and the minimum crankshaft speed at a, the speeds at c and d being within this range. If ω_1 and ω_2 are respectively the maximum and minimum speeds of rotation of the crankshaft and I is the total mass moment of inertia of the flywheel and the other revolving masses which are rigidly attached to the crankshaft, then the change of kinetic energy of rotation is given by $\frac{1}{2}I(\omega_1^2-\omega_2^2)$. This is clearly equal to the energy represented by the area of the shaded loop between a and b. Alternatively, it is equal to the energy represented by the algebraic sum of the areas of the other shaded loops, b to c, c to d and d to e.

It is important to realise that the maximum and minimum speeds of rotation of the crankshaft need not necessarily occur at opposite ends of the same shaded loop, although this happens to be so in the case of the single-cylinder engine to which Fig. 288 applies.

The net turning moment diagram for a single-cylinder, single-acting, four-stroke-cycle gas engine, Fig. 285, is repeated in Fig. 289, and below it is shown the approximate shape of the curve of fluctuation of speed. Assuming, as before, that the resisting moment is uniform, the resisting moment line cuts the net turning

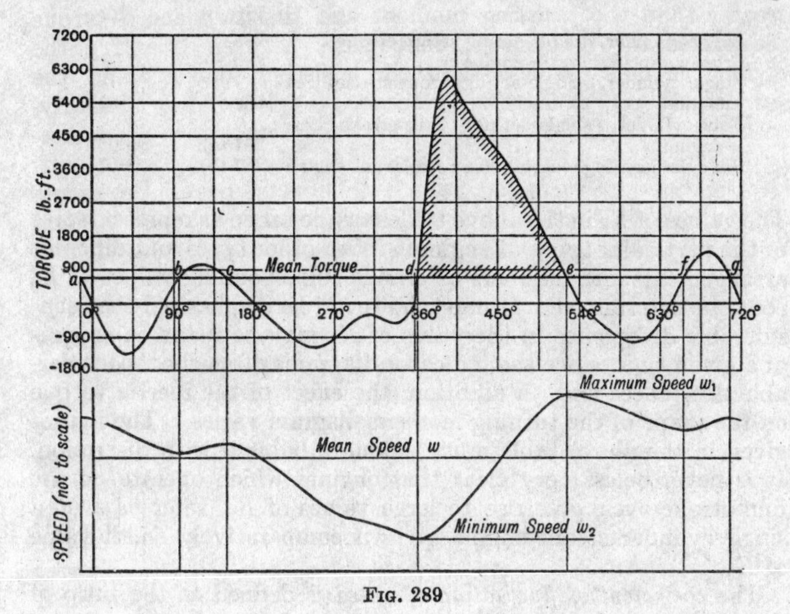

FIG. 289

moment line at the points a, b, c, d, e, f, g and at these points the slope of the speed curve is zero. Proceeding as described for the steam-engine diagram, it is found that the absolute minimum speed occurs at point d and the absolute maximum speed at point e, the other maxima and minima occurring within this range. The area of the shaded loop between the points d and e therefore represents the amount of energy which has to be absorbed by the revolving parts of the engine while the speed increases from ω_2 to ω_1.

161. The Coefficient of Fluctuation of Energy and the Coefficient of Fluctuation of Speed. In the preceding pages we have seen how the crank positions at which the angular velocity of the crankshaft has its minimum and maximum values may be determined from the shape of the turning moment diagram. We have also seen how the net amount of excess energy between these two crank positions may be determined from the same diagram. This excess energy is termed the fluctuation of energy and may be expressed as a fraction of the indicated work done by the engine during one revolution of the crank. The fraction is known as the *coefficient of fluctuation of energy* and will be denoted by K_e. The values of K_e, as determined from the turning moment diagrams given earlier are as follows:

Engine	Fig.	K_e
Single-cylinder, double-acting steam engine . .	288	0·21
Cross-compound steam engine	286	0·096
Single-cylinder, single-acting, four-stroke gas engine	289	1·93
Four-cylinder, single-acting, four-stroke gas engine	287 (a)	0·066[1]
Six-cylinder, single-acting, four-stroke gas engine	287 (b)	0·031

The values of K_e in the above table may be taken as representative of the particular types of engine. With other types and different arrangements and numbers of cranks corresponding values of K_e could be determined. It must, however, be emphasised that considerable divergence in the value of K_e may be found in engines of a given type, since the indicator diagrams often show considerable differences and, in addition, the effect of the inertia torque on the shape of the turning moment diagram varies. The figures given in the above table must therefore be used with discretion. It is nevertheless very clear that engines which operate on the four-stroke cycle give rise to large values of K_e, whereas even a single-cylinder steam engine gives a comparatively small value of K_e.

The *coefficient of fluctuation of speed* is defined as the ratio of the difference between the maximum and minimum angular velocities of the crankshaft to its mean angular velocity. It will be denoted by K_s. Thus, if ω_1, ω_2 and ω are respectively the maximum, the minimum and the mean angular velocities, then:

$$K_s = (\omega_1 - \omega_2)/\omega \quad . \quad . \quad . \quad . \quad (12.12)$$

Evidently the difference $\omega_1 - \omega_2$ may be made as small as desired by having revolving parts of sufficiently large rotary inertia keyed to the crankshaft, but there is no point in making $\omega_1 - \omega_2$ less than necessary. The value of K_s to be used in fixing the inertia of the revolving parts is largely determined by the purpose for which the particular engine is required. For general purposes as, for example, when driving lineshafts for machine tools, a much

[1] This value is for the normal speed of rotation as shown by the full-line diagram.

larger variation of speed may be allowed than when the engine is direct-coupled to an electric generator. Where engines are coupled to alternators, which are to work in parallel, the permissible fluctuation of speed is even smaller. The following table gives average values of K_s which may be adopted when the engine is used for the specified purpose:

	K_s
For engines driving agricultural machinery	0·05
For engines driving workshop shafting	0·03
For engines driving weaving and spinning machines . .	0·02–0·01
For engines driving direct-current generators	0·006

In order to enable alternators driven by different engines to be satisfactorily operated in parallel, the angular deviation should not exceed 2·5 electrical degrees. To satisfy this requirement the B.E.A.M.A. recommend that the permissible value of K_s be determined from the equation:

$$K_s = n/6p \quad . \quad . \quad . \quad . \quad (12.13)$$

where n = number of impulses per revolution of the crankshaft,
and p = number of poles on the alternator.

The number of poles on the alternator depends on the speed of rotation and the frequency of the current generated. In this country the frequency is usually 50 cycles per second, and, since the number of cycles per revolution of the alternator is equal to the number of pairs of poles on the alternator, the frequency per second is given by:

$$F = \frac{N}{60} \cdot \frac{p}{2} = \frac{Np}{120}$$

where N is the speed of rotation in r.p.m.

Example 7. A six-cylinder, single-acting, four-stroke cycle oil engine is required to generate alternating current at a frequency of 50 cycles per second at a speed of approximately 210 r.p.m. Find the actual speed, the number of poles on the alternator and the required value of K_s.

Since $F = 50$, $Np = 50.120$ and since $N = 210$, $p/2 = 14{\cdot}3$. But $p/2$ must be a whole number, say 14, so that the number of poles $p = 28$ and consequently $N = 214{\cdot}3$.

The number of impulses per revolution is three, since each cylinder gives one impulse in two revolutions of the crankshaft.

Therefore, from equation (12.13):

$$K_s = 3/(6.28) = 1/56 = 0{\cdot}0179$$

For a single-cylinder engine of the same type and running at the same speed, the permissible value of K_s would obviously be only one-sixth as great, or $K_s = 0{\cdot}003$.

162. The Flywheel. As already explained, the fluctuations of the turning moment exerted on the crankshaft are reflected in fluctuations of the speed of rotation. The fluctuations of turning moment depend upon the type of engine—steam, gas, oil, etc.—and the number and arrangement of the cranks; while the permissible fluctuation of speed depends upon the purpose for which the engine is used. The function of the flywheel is to act as a reservoir which will absorb energy during those periods of crank rotation when the turning moment is greater than the resisting moment and will restore the energy during those periods when the turning moment is less than the resisting moment. Absorption of energy is necessarily accompanied by an increase of speed and restoration of energy by a decrease of speed. The flywheel must be so proportioned that these changes of speed do not exceed the permissible limits. When calculating the dimensions of the flywheel, it is customary to ignore the effect of the rotary inertia of the other revolving parts of the engine.

Let I = mass moment of inertia of the flywheel = Wk^2/g,
ω = mean speed of rotation,
ω_1, ω_2 = maximum and minimum speeds of rotation respectively,
E = indicated work per revolution of the crankshaft
and K_e, K_s = coefficients of fluctuation of energy and speed respectively.

The amount of energy, which has to be absorbed by the flywheel with an increase of speed from ω_2 to ω_1, is found, as already explained, from the turning moment diagram. Where the data for the construction of the turning moment diagram are not available, a value of K_e must be assumed, based on the turning moment diagram for a similar type of engine.

The fluctuation of energy is then given by:

$$E_f = K_e . E$$

The change of kinetic energy of the flywheel

$$= \tfrac{1}{2}I(\omega_1^2 - \omega_2^2)$$
$$= \tfrac{1}{2}I(\omega_1 - \omega_2)(\omega_1 + \omega_2)$$

But from the definition of the coefficient of fluctuation of speed,

$$K_s = (\omega_1 - \omega_2)/\omega \quad \text{or} \quad \omega_1 - \omega_2 = K_s . \omega$$

Also, although the mean speed is not necessarily, or usually, the arithmetic mean between the maximum and minimum speeds, the difference $\omega_1-\omega_2$ is generally so small a fraction of ω as to justify the assumption that $\omega_1+\omega_2 = 2\omega$, so that, substituting for $\omega_1-\omega_2$ and $\omega_1+\omega_2$, the change of kinetic energy of the fly-wheel may be written $IK_s\omega^2$.

Equating this to the fluctuation of energy,

$$IK_s\omega^2 = E_f = K_eE$$

$$\therefore \tfrac{1}{2}I\omega^2 = \frac{E_f}{2K_s} = \frac{K_eE}{2K_s} \quad . \quad . \quad . \quad (12.14)$$

The term on the left-hand side of the above equation is the mean kinetic energy of the flywheel.

Example 8. The speed of rotation of the compound steam engine for which Fig. 286 is the turning moment diagram is 120 r.p.m. and the i.h.p. is 94·5. If the fluctuation of speed is not to exceed 1% of the mean speed and the radius of gyration of the flywheel is 3·5 ft, what weight of flywheel will be required?

The fluctuation of energy is represented by the area of the shaded loop a to b. Taking into account the scales to which the diagram is drawn, it is found to amount to 2500 ft lb.

Substituting in equation (12.14):

$$\tfrac{1}{2}I\omega^2 = \frac{E_f}{2K_s} = \frac{2500.100}{2}$$

$$\therefore\ Wk^2 = 32{\cdot}2\left(\frac{30}{\pi.120}\right)^2 \frac{2500.100}{2240} = 22{\cdot}8 \text{ ton ft}^2$$

Since k is 3·5 ft,

$$\therefore\ W = 22{\cdot}8/3{\cdot}5^2 = 1{\cdot}86 \text{ tons}$$

Example 9. The gas engine for which Fig. 289 is the turning moment diagram has two flywheels each of which weighs 1000 lb and has a radius of gyration of 2·25 ft. The speed of rotation is 270 r.p.m. and the i.h.p. is 35·8. Find the fluctuation of speed.

Fluctuation of energy = area of shaded loop d to e = 8450 ft lb. Or, since

$$E = 35{\cdot}8.33\,000/270 = 4380 \text{ ft lb}$$

and, from the table on p. 440, $K_e = 1{\cdot}93$, the fluctuation of energy

$$E_f = K_eE = 1{\cdot}93.4380 = 8450 \text{ ft lb}$$

From equation (12.14):

$$K_s = K_eE/I\omega^2$$

But the moment of inertia of each flywheel

$$= 1000 . 2{\cdot}25^2 = 5063 \text{ lb ft}^2$$

and

$$\omega = \pi N/30 = 9\pi \text{ rad/s}$$

$$\therefore\ K_s = \frac{32{\cdot}2}{10\,125} . \frac{8450}{81\pi^2} = 0{\cdot}0337 \quad \text{or} \quad 3{\cdot}4\%$$

The fluctuation of speed is therefore 3·4% or 1·7% on either side of the mean speed.

Example 10. A single-cylinder, single-acting, four-stroke oil engine develops 25 i.h.p. at 300 r.p.m. The work done by the gases during the expansion stroke is 2·3 times the work done on the gases during the compression stroke and the work done during the suction and exhaust strokes is negligible. If the turning moment diagram during expansion is assumed to be triangular in shape and the speed is to be maintained within 1% of the mean speed, find the moment of inertia of the flywheel.

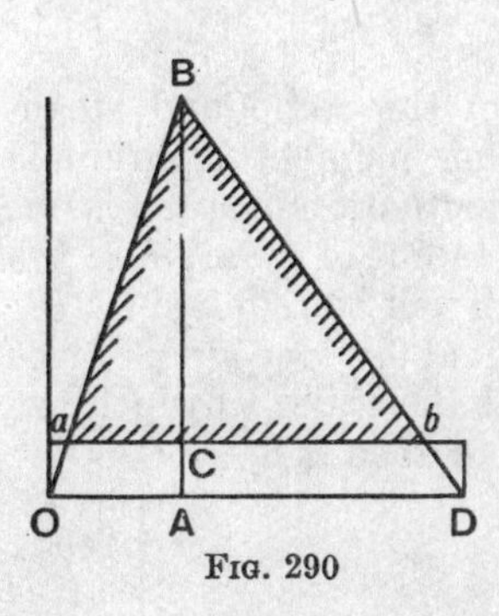

FIG. 290

The indicated work done per revolution E

$$= 25 . 33\,000/300 = 2750 \text{ ft lb}$$

$$\therefore \text{ indicated work done per cycle} = 5500 \text{ ft lb}$$

Let $\qquad W_e$ = work done by the gases during expansion.

Then $\qquad W_e/2{\cdot}3$ = work done on the gases during compression

and $W_e(1-1/2{\cdot}3)$ = indicated work done per cycle = 5500 ft lb.

$$\therefore\ W_e = 5500 . 2{\cdot}3/1{\cdot}3 = 9740 \text{ ft lb}$$

The turning moment diagram for expansion is shown in Fig. 290. The maximum torque

$$= AB = 9740 . 2/\pi \text{ lb ft}$$

But the mean turning moment

$$= AC = 2750/2\pi \text{ lb ft}$$

and the maximum excess turning moment

$$= CB = AB - AC = (1/\pi)(19\,480 - 1375) = (1/\pi)18\,105 \text{ lb ft}$$

The fluctuation of energy = shaded area a to D and, from the similar triangles aBb, OBD, this is given by:

$$\left(\frac{BC}{BA}\right)^2 .\,\text{area OBD} = \left(\frac{BC}{BA}\right)^2 .\,W_e = \left(\frac{18\ 105}{19\ 480}\right)^2 .\,9740 = 8410 \text{ ft lb}$$

This corresponds to a value of $K_e = 8410/2750 = 3{\cdot}06$.

From equation (12.14):

$$I = E_f / K_e \omega^2$$

But $\omega = \pi . 300/30 = 10\pi$, $K_s = 2/100$, $I = (W/g)k^2$,

$$\therefore\ Wk^2 = \frac{32{\cdot}2}{100\pi^2} . \frac{8410}{2} . 100 = 13\ 720 \text{ lb ft}^2$$

A cast-iron flywheel to run at 300 r.p.m. would have an outside diameter of about 6 ft and a radius of gyration of about 2·5 ft.

$$\therefore \text{ weight of flywheel required } W = 13\ 720/2{\cdot}5^2 = 2190 \text{ lb}$$

Example 11. The turning moment diagram for a quadruple-expansion marine steam engine is drawn to scale in Fig. 291. The original diagram of which this is a reproduction was drawn to the following scales: 1 in. = 15 ton ft and 1 in. = 40°. The areas of the loops above and below the mean turning moment line, taken in order, are 0·12, 0·34, 0·91, 0·81, 0·15, 0·18, 1·86 and 1·71 sq. in. The total moment of inertia of the engine and propeller masses is 150 ton ft². Estimate the percentage variation from the mean speed which is 100 r.p.m.

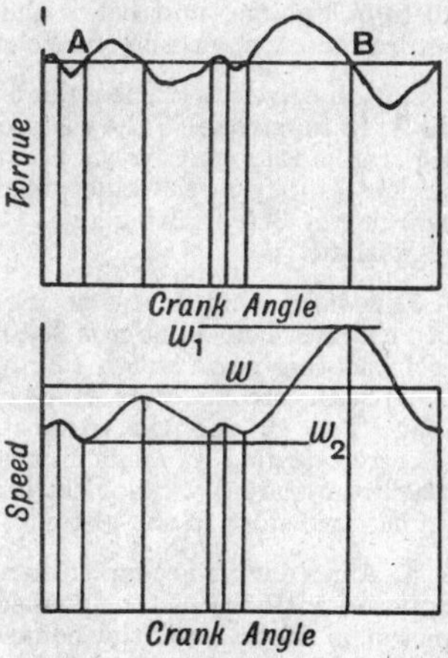

FIG. 291

Proceeding as described in Article 160, it is found that the minimum engine speed occurs at the crank position corresponding to point A and the maximum engine speed at the crank position corresponding to point B.

The fluctuation of energy is represented by the algebraic sum of the areas of the loops between the points A and B or between the point B and the point A on the following cycle, i.e. the fluctuation of energy is represented by:

$$-1{\cdot}71 + 0{\cdot}12 - 0{\cdot}34 = -1{\cdot}93 \text{ sq. in.}$$

But 1 sq. in. of turning moment diagram area

$$= 15.\pi/180.40 = 10{\cdot}47 \text{ ft tons of energy}$$

$$\therefore \text{ fluctuation of energy} = 1{\cdot}93.10{\cdot}47 = 20{\cdot}2 \text{ ft tons}$$

From equation (12.14):

$$\tfrac{1}{2}I\omega^2 = 20{\cdot}2/2K_s$$

$$\therefore K_s = \frac{20{\cdot}2.32{\cdot}2.30^2}{150.\pi^2.100^2} = 0{\cdot}0396$$

The total fluctuation of speed is 3·96% and the variation of speed is 1·98% on either side of the mean speed.

EXAMPLES XII

1. A vertical double-acting steam engine has a cylinder 12 in. dia. by 18 in. stroke, and runs at 200 r.p.m. The reciprocating parts weigh 400 lb, the piston rod is $2\frac{1}{2}$ in. dia. and the connecting rod is 45 in. long. When the crank has turned through 120° from the top dead centre, the steam pressure above the piston is 40 lb/in.^{2} gauge, and below the piston 2 lb/in^2 gauge. Calculate the effective turning moment on the crankshaft.

2. A horizontal single-cylinder single-acting oil engine has a cylinder 10 in. dia. × 18 in. stroke. The reciprocating parts weigh 360 lb, the connecting rod is 4·5 cranks long and the r.p.m. are 300. When the crank has turned through an angle of 40° from the inner dead centre on the firing stroke, the pressure in the cylinder is 320 lb/in^2 gauge. Calculate the effective turning moment on the crankshaft.

3. A horizontal double-acting steam engine has a stroke of 12 in. and runs at 250 r.p.m. The cylinder is $8\frac{1}{2}$ in. dia., the connecting rod is 5 cranks long and the reciprocating parts weigh 150 lb. Steam is admitted at 80 lb/in^2 gauge for one-third of the stroke, after which expansion takes place according to the hyperbolic law $PV = C$, and the exhaust pressure is -12 lb/in^2 gauge. Calculate the effective turning moment on the crankshaft when the crank has turned through 120° from the i.d.c. Neglect the effect of clearance and of the difference of area on the two sides of the piston. M.U.

4. A gas engine is coupled to a compressor, the two cylinders being horizontally opposed with the pistons connected to a common crankpin. The stroke of each piston is 20 in. and the connecting rods are each 50 in. long. The cylinder diameters are 8·5 in. and 9·5 in., and the reciprocating parts weigh 290 lb and 360 lb for the engine and compressor respectively. When the crank has turned through 60° from the i.d.c. on the firing stroke, the pressure of the gas in the engine cylinder is 150 lb/in^2 gauge and the pressure in the compressor cylinder is 15 lb/in^2 gauge. If the crank makes 200 r.p.m. and the flywheel weighs 3000 lb and has a radius of gyration of 3 ft, calculate the angular acceleration of the flywheel at this instant. M.U.

5. A horizontal single-cylinder single-acting Otto cycle gas engine has a bore of 11·5 in. and a stroke of 21 in. The engine runs at 180 r.p.m., the ratio of compression is 5·5, the maximum explosion pressure is 450 lb/in^2 abs. and expansion follows the law $PV^{1{\cdot}3} = C$. If the weight of the reciprocating parts is 360 lb and the connecting rod is 48 in. long, calculate the turning moment on the crankshaft when the crank has turned through 60° from the i.d.c.

6. A vertical single-cylinder single-acting Diesel engine has a cylinder 12 in. diameter and a stroke of 18 in. The reciprocating parts weigh 500 lb, the connecting rod is 40 in. long and the r.p.m. are 240. The ratio of compression is 14 and the pressure remains constant during the injection of the oil for one-tenth of the stroke. If the index in the laws for compression and expansion is 1·35, find the effective turning moment on the crankshaft when the crank makes an angle of 75° with the i.d.c.

7. The mechanism of a small hand punch is shown in Fig. 292. OA is 6 in., AB is 3 in. and AC is 9 in. The angle BAC is 90° and B is guided along a vertical line $1\frac{1}{2}$ in. from the fixed fulcrum O. The points A, D, E and F are spaced equally along the path of A. For the given position of the mechanism and the positions in which the pin A coincides with D, E and F, find the resistance which can be overcome at B when a force of 40 lb is applied at C in a direction perpendicular to AC.

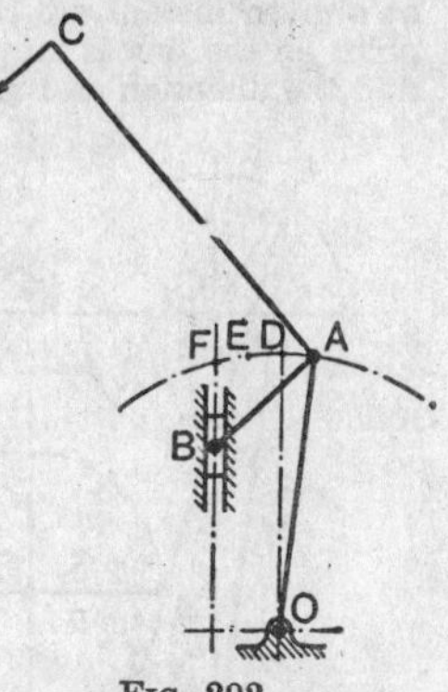

Fig. 292

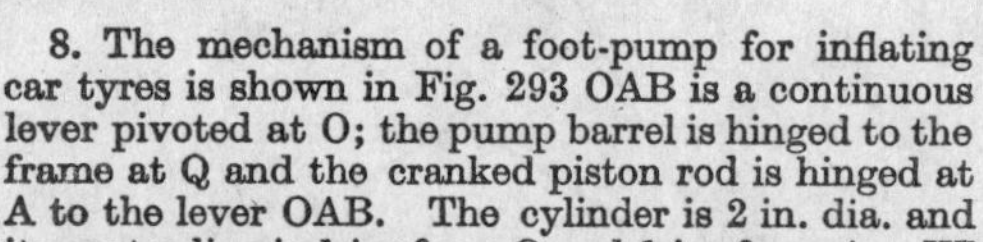

8. The mechanism of a foot-pump for inflating car tyres is shown in Fig. 293 OAB is a continuous lever pivoted at O; the pump barrel is hinged to the frame at Q and the cranked piston rod is hinged at A to the lever OAB. The cylinder is 2 in. dia. and its centre line is 1 in. from Q and 1 in. from A. When the angle AOQ is 45° the air pressure in the cylinder is 25 lb/in² gauge. Neglecting friction, what vertical force must be applied at B, if the compressive force in the spring is 15 lb?

9. Fig. 294 shows the mechanism of a moulding press used in plastics. The crank OA rotates counter-clockwise from the horizontal through about 150°, dragging with it the crank O_1B. Movement continues until OA and AB are in line. The crank O_1B operates, through the link BC, a ram at C constrained to move vertically.

Determine the velocity ratio of C to A (a) when OA is in its initial horizontal position, (b) when OA has moved through 75°. Neglecting friction, compare the

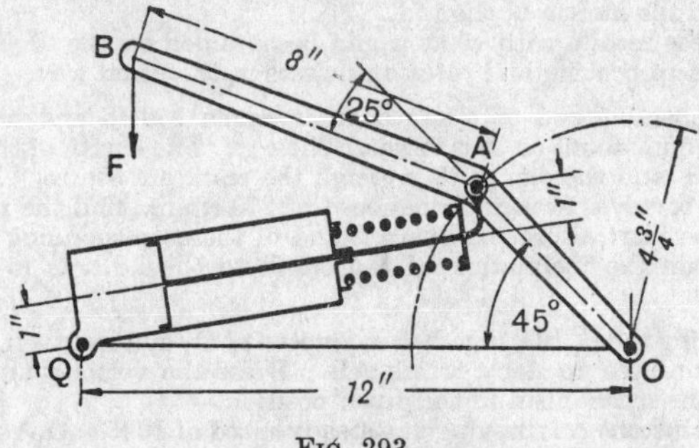

Fig. 293

ratios of the forces transmitted vertically at C and tangentially at A for the two positions (a) and (b). Comment on the velocity of the ram and the force exerted by it as OA approaches its extreme position. W.S.

10. A, B, O are the corners of a mechanism, Fig. 295. O is a fixed pivot and A can move along a horizontal guide. All the pivots are frictionless. Find the vertical velocities of B, G and H if A moves horizontally at a speed V. Hence, or otherwise, find the force P at A required to balance a vertical force Q applied successively at B, G and H. To get rid of the force P at A, a spring is fitted

between the pins E and F. What force must the spring apply to the pins when Q is applied successively at each of the three pins B, G and H? W.S.S.

11. The weight, the radius of gyration and the position of the c.g. of a link of a mechanism are given; the velocity and acceleration of one point on the link at a given instant are completely specified and the direction in which a second point on the link is moving at the same instant is also specified. Show how to find the direction and magnitude of the resultant accelerating force on the link.

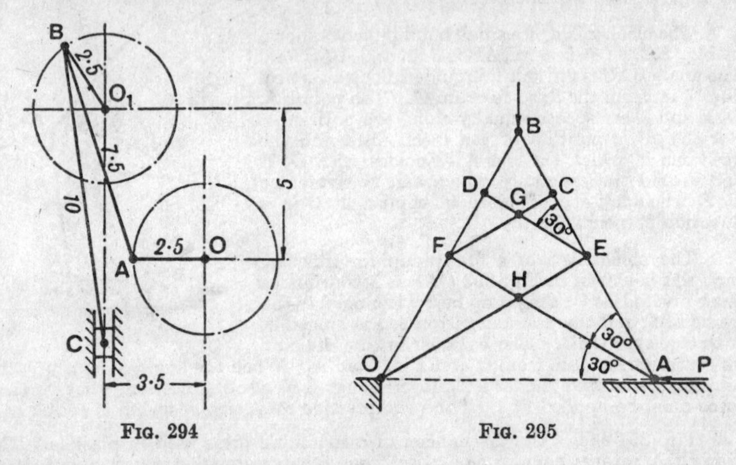

FIG. 294 FIG. 295

12. The connecting rod of a horizontal steam engine is 90 in. long and weighs 800 lb. The c.g. is 40 in. from the crankpin centre and the radius of gyration about the c.g. is 25 in. The stroke is 36 in. Reduce the rod to an equivalent dynamical system of two concentrated masses, one of which is at the crosshead pin. For an engine speed of 120 r.p.m. and a crank position 60° from i.d.c., determine the force and torque on the engine frame—referred to the crankshaft centre—due to the inertia of the rod.

Contrast these results with what would be obtained by the division of the rod between the reciprocating and rotating masses in the usual way. L.U.A.

13. The connecting rod of a gas engine weighs 150 lb and has a radius of gyration of 14·5 in. about an axis through the c.g. The length of the rod between centres is 40 in. and the c.g. is 13 in. from the crankpin centre. If the crank is 9 in. long and revolves at a uniform speed of 270 r.p.m., find the magnitude and direction of the inertia force on the rod and of the corresponding torque on the crankshaft when the inclination of the crank to the i.d.c. is (a) 30°, (b) 90°, (c) 135°.

14. A swinging rod O_2BC, Fig. 296, is pivoted at O_2 and is driven by a vibrating crank O_1A connected to it by a link AB. Draw the velocity and acceleration diagrams for the mechanism in the given position.

At the given instant A is moving at a steady speed of 10 ft/s: $O_1A = O_2B = 6$ in. What turning moment must be applied to O_1A to counterbalance the inertia effect of a mass of 10 lb concentrated at C? $O_2C = 18$ in. W.S.S.

15. The crank OC of the mechanism shown, Fig. 297, is driven through the given position at constant angular velocity ω rad/s. CQ is a uniform rod of mass M lb and mass-centre G. $OC = r$; $CQ = 3r$; $QP = 1{\cdot}5r$. Find (a) the acceleration of G and the angular acceleration of CQ, (b) the torque on OC and the forces at O and P due to the inertia of CQ. W.S.S.

16. The connecting rod of a vertical reciprocating engine is 8 ft long between centres and weighs 1000 lb. Its mass centre is 3 ft from the centre of the big-end bearing. When suspended from the crosshead pin and allowed to swing,

the period of oscillation is found to be 2·93 sec. The crank is 2 ft long and rotates at 240 r.p.m.

When the crank has turned through 45° from the top-dead-centre position, find, due to the inertia of the connecting rod:

(a) the magnitude and the line of action of the resultant force acting upon the connecting rod;
(b) the reaction at the crosshead guide;
(c) the force on the main bearing;
(d) the torque on the crankshaft. L.U.A.

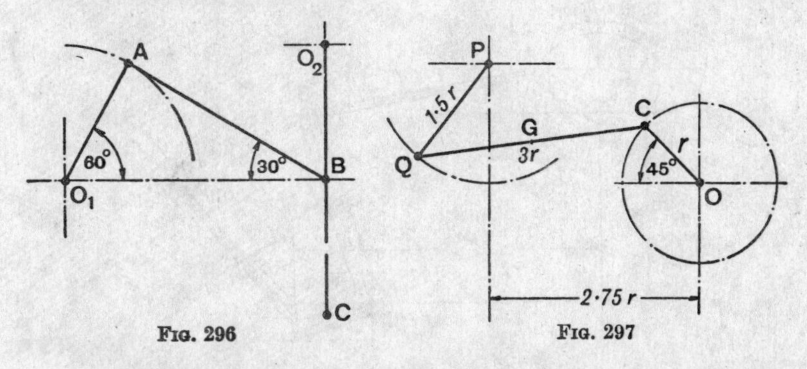

FIG. 296

FIG. 297

17. The connecting rod of a vertical high-speed Diesel engine weighs 5·16 lb and is 12 in. long between centres. Its c.g. is 3·92 in. from the centre of the big-end bearing. When suspended in a vertical plane and allowed to swing about the axis of the small-end bearing, it makes 50 complete oscillations in 53·5 sec. The stroke of the piston is 5 in. and the reciprocating parts of the engine weigh 2·35 lb.

Find the torque exerted on the crankshaft due to the inertia of the moving parts when the crank makes angles of (a) 35° and (b) 140° with the t.d.c. and the speed of rotation is 1600 r.p.m.

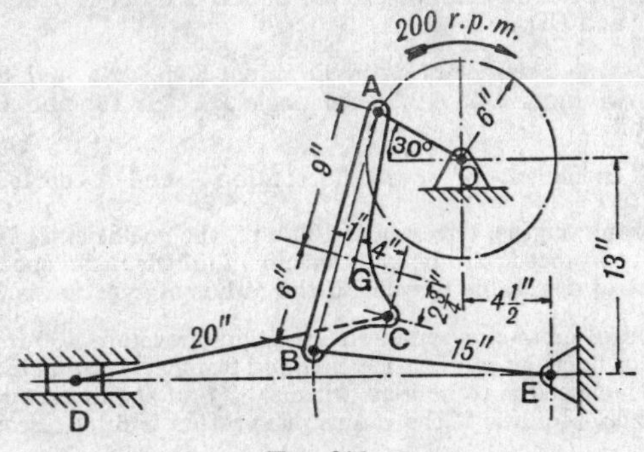

FIG. 298

18. In the mechanism shown in Fig. 298, the swinging link ABC has a weight of 90 lb and its c.g. is at the point marked G. Its radius of gyration about the c.g. is 4 in. Construct a velocity diagram for the mechanism in the configuration shown and determine (a) the angular velocity and kinetic energy of ABC and (b) the turning moment on the crank OA for a force of 1000 lb acting on the block D in the direction of motion. L.U.A.

19. Fig. 299 shows a reciprocating element E, driven by a crank OA, through a bell-crank lever BCD. The crank speed is 200 r.p.m. in the direction indicated. The bell-crank with allowance for attached rods weighs 12 lb and the radius of gyration about C is $5\frac{1}{4}$ in. The reciprocating element and its connections weigh 15 lb. The motion of E is resisted by a force P of 100 lb.

Determine, for the position of the mechanism given in the figure, (a) the linear acceleration of E and the angular acceleration of BCD; (b) the torque required at crank OA to overcome the specified resistance and inertias. L.U.A.

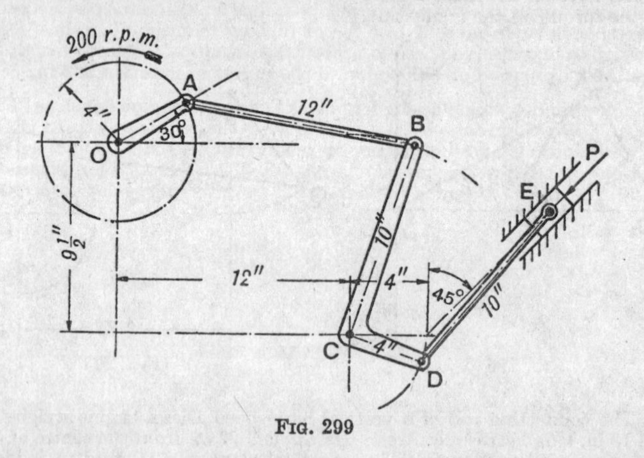

FIG. 299

20. ABCD is a four-bar chain with AD as the fixed link. The lengths of the links are; AB 3 in., BC 13·5 in., CD 6 in., and DA 15 in., and AB turns at a uniform speed of 120 r.p.m. The link CD weighs 20 lb and has a radius of gyration of 3·5 in. about an axis through D. The link BC weighs 10 lb, its c.g. is 4·5 in. from B and its radius of gyration about an axis through the c.g. is 5·4 in. For the configuration in which the angle BAD is 30° and B and C lie on opposite sides of AD, find the torque which must be exerted on AB in order to overcome the inertia of the links BC and CD.

21. Use the same particulars as in Example 6, p. 428, and find the inertia torque on the driving crank AB, when angle BAD is 15° and B and C lie on opposite sides of AD.

22. Define "coefficient of speed fluctuation" and "coefficient of energy fluctuation".

In a single-crank engine, developing 100 h.p., the coefficient of energy fluctuation is 0·20 and the speed is to be kept within 1% of the mean speed of 120 r.p.m. Find the weight of a suitable flywheel if the radius of gyration is 3·5 ft.

23. A horizontal cross-compound steam engine develops 400 h.p. at 90 r.p.m. The coefficient of fluctuation of energy is found from the turning moment diagram to be 0·1 and the speed is to be kept within 0·5% of the mean speed. Find the weight of flywheel required if the radius of gyration is 6 ft. Prove the formula you use.

24. Deduce the equation connecting the mean kinetic energy of the flywheel, the coefficients of energy and speed fluctuation and the indicated work done per cycle.

A gas engine working on the Otto cycle develops 25 i.h.p. at 240 r.p.m. The coefficient of fluctuation of energy is 1·90; the flywheel weighs 1 ton and has a radius of gyration of 3 ft. What is the cyclical speed variation from the mean?

25. A gas engine working on the Otto cycle develops 200 h.p. at 110 r.p.m. The coefficient of fluctuation of energy is 2·0; the speed has to be kept within ½% of the mean speed; and the centrifugal stress in the flywheel rim is not to exceed 800 lb/in². Calculate the maximum diameter of the flywheel and the minimum weight of the rim. The weight of 1 in³ of cast-iron may be taken as 0·26 lb.

26. The turning moment diagram for a four-cyclinder petrol engine is drawn to the following scales: turning moment, 1 in. = 100 lb ft; crank angle, 1 in. = 25°. The fluctuation of energy between the crank positions for minimum and maximum speeds corresponds to an area of 1·05 in² of turning moment diagram. Determine the weight of flywheel required to prevent a fluctuation of speed greater than 0·5% from the mean speed of 1500 r.p.m. if the radius of gyration is 5 in.

27. A single-cylinder, single-acting four-stroke cycle gas engine develops 25 h.p. at 260 r.p.m. The work done by the gases during the explosion stroke is three times the work done on the gases during the compression stroke. The work done on the suction and exhaust strokes may be neglected. Determine the fluctuation of energy. If the flywheel weighs 1·5 tons and has a radius of gyration of 2 ft 3 in., calculate the cyclical fluctuation of speed.

28. The turning moment diagram for a four-cylinder petrol engine is drawn to the following scales: turning moment, 1 in. = 1000 lb in.; crank angle, 1 in. = 25°. The curves are identical for each half-revolution of the crankshaft and the areas below and above the mean turning moment line taken in order are 0·22, 0·93, 0·06, 0·38 and 1·03 in². The engine runs at 1500 r.p.m. and the flywheel weighs 40 lb and has a radius of gyration of 5 in. Calculate the coefficient of fluctuation of speed.

29. A machine punching 1½ in. holes in a 1¼ in. plate does 15 in. tons of work per square inch of sheared area. The punch has a stroke of 4 in. and punches one hole every 10 seconds. The maximum speed of the flywheel at the radius of gyration is 90 ft/s. Find the weight of the wheel if the speed at this radius is not to fall below 80 ft/s during each punch. L.U.

30. Neglecting the effect of the arms and boss show that the minimum weight W of flywheel required may be expressed in the form

$$W = (12K_e\rho/K_s f_c)E$$

where K_e and K_s are the coefficients of fluctuation of energy and speed, ρ is the density of the rim material in lb/in³, f_c is the hoop stress in the rim in lb/in² and E is the indicated work done per revolution.

A gas engine develops 30 h.p. at 275 r.p.m. If K_e is 2·2, K_s is 1/75, f_c is 750 and ρ, for cast-iron, is 0·26, find the minimum weight of flywheel and the mean diameter of the rim.

31. During the outward stroke of the piston of a double-acting steam engine the turning moment has a maximum value of 13 000 lb ft when the crank makes an angle of 60° with the i.d.c. During the inward stroke the maximum turning moment is 10 000 lb ft when the crank makes an angle of 280° with the i.d.c. The turning moment diagrams on a crank angle base may be assumed triangular for both strokes. Find the crank angles at which the speed has its maximum and minimum values and the coefficient of fluctuation of energy.

If the crank makes 80 r.p.m., the radius of gyration of the flywheel is 5 ft and the speed is to be kept within 0·5% of the mean speed, what weight of flywheel will be required?

32. The turning moment diagram for a three-cylinder engine is drawn to the following scales: crank displacement, 1 in. = 40°; turning moment, 1 in. = 5000 lb ft.

During one revolution of the crank the areas above and below the mean turning moment line taken in order are: 0·60, 0·69, 0·64, 0·75, 0·78 and 0·58 in². If the speed is to be kept within 1% of the mean speed, which is 90 r.p.m. and the radius of gyration of the flywheel is 3·5 ft, what weight of flywheel will be required? M.U.

33. A steam engine develops 50 b.h.p. at 250 r.p.m. against a steady load. The mechanical efficiency is 88% and the frictional losses may be assumed constant. The flywheel weighs 0·6 ton and has a radius of gyration of 2·25 ft. If the load is suddenly diminished to one-quarter of the full load and no change in the steam supply takes place for two complete revolutions after the reduction of load, calculate the speed of the engine at the end of this period. M.U.

34. A single-cylinder double-acting pump is driven through gearing at 40 r.p.m. by an electric motor which gives a uniform torque. The resisting torque for each half-revolution of the pump shaft may be assumed to follow a sine curve with a maximum value at 90° and 270° of 3090 lb ft. Find what weight of flywheel will be required on the pump shaft to keep the speed within $1\frac{1}{2}$% of the mean speed, if the radius of gyration of the flywheel is 4 ft. The flywheel effect of the motor armature and gear wheels is equivalent to 1000 lb at a radius of 3 ft on the pump shaft. M.U.

35. A machine shaft running at a mean speed of 200 r.p.m. requires a torque which increases uniformly from 1000 lb ft to 3000 lb ft during the first half revolution, remains constant for the following revolution, decreases uniformly to 1000 lb ft during the next half revolution and then remains constant again for the next two revolutions, this cycle being repeated. It is driven by a motor which exerts a constant torque and has a rotor weighing 1000 lb with a radius of gyration of 10 in. If in addition a flywheel with a weight of 2 tons and a radius of gyration of 2 ft is fitted to the shaft, what will be the percentage fluctuation of speed during the cycle and the required h.p. of the motor? M.U.

36. The turning moment diagram for a quadruple-expansion marine engine is drawn to the following scales: 1 in. = 50 ton ft and 1 in. = 40°. The areas of the loops above and below the mean turning moment line taken in order are: 0·12, 0·34, 0·91, 0·81, 0·15, 0·18, 1·86 and 1·71 in². If the moment of inertia of the propeller and entrained water is 1000 ton ft² and the mean speed of rotation is 100 r.p.m., what is the value of the coefficient of fluctuation of speed ? M.U.

37. The turning moment diagram for a four-stroke, single-acting gas engine may be assumed for simplicity to be represented by four triangles, the areas of which measured from the line of zero pressure are as follows: expansion stroke, 8·50 in²; exhaust stroke, 0·80 in²; suction stroke, 0·56 in²; and compression stroke, 2·14 in². Each square inch represents 1000 ft lb.

Assuming the resisting torque to be uniform, find the weight of the rim of a flywheel required to keep the speed between 98 and 102 r.p.m.; the mean radius of the rim is 3 ft. L.U.

38. A four-stroke-cycle internal-combustion engine indicates 16 h.p. at 160 r.p.m. with 79 explosions or power strokes per minute. The work done on the gases during the compression stroke is 0·30 of the effective work as represented by the area of the indicator diagram.

If the total fluctuation of speed is limited to $2\frac{1}{2}$% of the normal speed, find the mass in tons of each of the two flywheels, the radius of gyration of each wheel being 3 ft. L.U.

39. A gas engine has a cylinder diameter of 11 in. and a stroke of 20 in. and runs at a mean speed of 200 r.p.m. Reckoned from atmosphere, the mean pressure in the cylinder is: during the working stroke +100 lb/in²; during suction −1; during compression +30; during exhaust +2.

Assuming that the instants of lowest and highest speed respectively coincide with the beginning and the end of the working stroke, find the moment of inertia of the flywheel required to keep the speed during a working cycle within 1% of the mean speed.

Also find the drop of speed which will then occur during a cycle in which there is no admission. L.U.

40. A small single-cylinder, four-stroke cycle oil engine of 5 in. stroke develops 7 h.p. at 1000 r.p.m. The excess energy delivered during the power stroke is

78% of the energy per cycle. The engine is fitted with a combined flywheel and belt pulley of weight 160 lb and radius of gyration $7\frac{1}{4}$ in. The rotating parts of the engine—part connecting rod, crankpin, etc.—are equivalent to 9 lb concentrated at crank radius and are balanced by weights fixed to the crankwebs, the c.g. and radius of gyration of which are respectively 3 in. and $3\frac{3}{4}$ in.

Estimate the range of speed fluctuation of the engine and state the percentage error that would be incurred in this estimate by considering the flywheel only.
L.U.

41. The crank effort diagram for an engine is given by T lb ft $= 5000 + 8000 \sin \theta + 1200 \sin 2\theta + 100 \sin 3\theta$, where θ is the crank angle. The resisting torque is uniform and the r.p.m. are 120. Determine: (a) the h.p. of the engine, (b) the weight of the flywheel rim, of mean radius 4 ft, so that the fluctuation of speed shall not exceed $\pm 1\%$, (c) the angular acceleration of the flywheel when $\theta = 0°$ and $\theta = 45°$.

42. The torque exerted on the crankshaft of an engine when corrected for balance is given by the expression T ton ft $= 10 + 5 \sin 2\theta - 7 \cos 4\theta$. Assuming that the resistance is uniform, find the moment of inertia of the flywheel, if the speed variation is not to exceed 0·3% above or below the mean speed, which is 120 r.p.m.
W.S.S.

43. A twin-cylinder engine is single-acting with its cranks set at right angles and it runs at 1500 r.p.m. The torque–crank angle diagram is practically a triangle for the power stroke with a maximum torque of 120 lb ft at 60° after dead centre of the corresponding crank. The torque on the return stroke is negligible.

Find (a) the h.p. developed; (b) the weight of flywheel, concentrated at 8 in. radius, to keep the speed within $\pm 3\%$ of the mean speed; (c) the angle turned through by the crank while it is being speeded up.
L.U.

CHAPTER XIII

GOVERNORS

163. The function of the governor must be carefully distinguished from that of the flywheel. The former is required to maintain, as closely as possible, a constant mean speed of rotation of the crankshaft over long periods during which the load on the engine may vary. The latter, as pointed out in Article 162, serves to limit the inevitable fluctuations of speed during each cycle which arise from the fluctuations of turning moment on the crankshaft. On the one hand the governor exercises no control over the cyclical fluctuations of speed, while on the other hand the flywheel has no effect on the mean speed of rotation.

If the load on the engine is constant, the mean speed of rotation will be constant from cycle to cycle. But if the load changes, the mean speed will also change, unless the output of the engine is adjusted to the new demand. It is the purpose of the governor to make this adjustment automatically.

It is, of course, desirable that the energy supplied to the engine should be altered by exactly the right amount immediately the change of load takes place. But if the adjustment of the supply of energy to the engine is to be carried out automatically, use has to be made of the tendency for the mean speed of rotation to change as the load changes. In other words, a change of speed must take place before the energy supplied to the engine can be automatically adjusted to the new load. Hence it follows that the mean speed of rotation will tend to increase continuously as the load on the engine decreases. The governor and the mechanism which it operates should be so designed that this increase of the mean speed of rotation is as small as possible. In any actual engine the problem is complicated by the fact that the load may change immediately after a fresh charge has been supplied to the engine cylinder and, since the governor can only affect the charge admitted during the next and succeeding cycles, some lag between the change of load and the change of engine output is inevitable.

164. Types of Governors. Governors are generally of one of two types, either (a) centrifugal or (b) inertia.

In the first type two or more masses termed the governor balls are caused to revolve about the axis of a shaft, which is driven through suitable gearing from the engine crankshaft. Each ball is acted upon by a force which acts in the radially inward direction, and is provided by a deadweight, a spring or a combination of the two. This force is termed the controlling force and it must increase in magnitude as the distance of the ball from the axis of rotation increases. When the governor balls are revolving at a uniform speed, the radius of rotation will clearly be such that the outward inertial or centrifugal force is just balanced by the inward controlling force. If the speed of rotation now increases owing to a decrease of load on the engine, the governor balls will move outwards until the centrifugal force is again balanced by the controlling force. Conversely, if the speed of rotation decreases owing to an increase of load on the engine, the governor balls will move inwards until the centrifugal force is again balanced by the controlling force. This movement of the balls is transmitted by the governor mechanism to the valve which controls the amount of energy supplied to the engine, so that movement in the outward direction reduces the valve opening and movement in the inward direction increases the valve opening.

Governors of the second type operate on a different principle. The governor balls are so arranged that the inertia forces caused by an angular acceleration or retardation of the governor shaft tend to alter their positions. The amount of the displacement of the governor balls caused by the inertia forces is controlled by suitable springs and, through the governor mechanism, alters the amount of energy supplied to the engine. The obvious advantage of this type of governor lies in its more rapid response to the effect of a change of load, since the displacement of the balls is determined by the rate of change of speed of rotation, as distinct from an actual change of speed of rotation, such as is required in governors of the first type. This advantage is offset, however, by the practical difficulty of arranging for the complete balance of the revolving parts of the governor. For this reason centrifugal governors are much more frequently used than are inertia governors, and only the former type will be dealt with here.

165. Centrifugal Governors. It was pointed out above that the controlling force in this type of governor is provided by a deadweight, a spring or a combination of both deadweight and spring. There are many governors in which the control is wholly, or mainly by means of deadweights; there are many others in which the control is mainly by means of springs. One or two examples of each type will be described and analysed in detail.

Three representative governors of the former type are illustrated in Fig. 300. That shown at (a) is the simplest, and, although now obsolete, is interesting as being the forerunner of the later examples which are shown at (b) and (c). It is the original form of governor as used by Watt on some of his early steam engines. Each ball is attached to an arm which is pivoted on the axis of rotation. The sleeve is attached to the governor balls by arms, pin-jointed at both ends, and is free to slide along the governor shaft. The weight of the sleeve is balanced by the components, parallel to the axis of rotation, of the tension T_2 in the two lower arms. The inward centripetal force which is required in order to maintain each ball in a circular path of radius r, when the angular velocity of rotation is ω, is provided by the radial components of the tension T_2 in the lower arm and the tension T_1 in the upper arm. It is convenient to suppress the rotation of the governor and reduce the problem to one in statics. It is then necessary to apply a

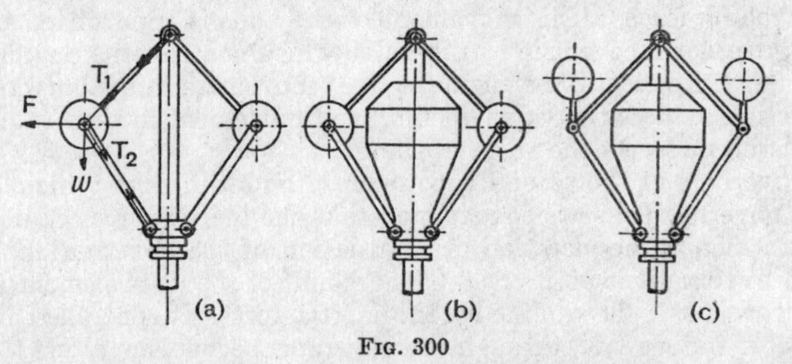

FIG. 300

force F acting radially outward through the ball centre, as shown in Fig. 300 (a), equal in magnitude to the force which the ball actually exerts on the governor when the latter is revolving. This force is the centrifugal force. In these circumstances, each ball will be in equilibrium under the four forces, T_1, T_2, the weight w and the centrifugal force F. Obviously a change in the radius of rotation of the governor balls will involve a movement of the sleeve along the shaft. This movement is transmitted by suitable mechanism to the valve which controls the energy supplied to the engine. An increase of speed raises the sleeve and is arranged to reduce the supply of energy. Conversely, a decrease of speed lowers the sleeve and increases the supply of energy.

The type of governor which is illustrated at (b) is known as the Porter governor. The only respect in which it differs from the Watt governor is in the use of a heavily weighted sleeve. The action is exactly the same as that of the simpler governor. The advantages of the loaded sleeve will be made clear later.

A third form of loaded governor is shown at (c). This is the Proell governor. It is similar to the Porter governor in that it has a heavily weighted sleeve, but differs from it in the arrangement of the balls. These are carried on extensions of the lower arms instead of at the junction of the upper and lower arms. The action of this governor is again similar to that of the Watt governor. An increase of the speed of rotation increases the radius of rotation and raises the sleeve, thus reducing the amount of energy supplied to the engine. Conversely, a decrease of speed results in a decrease in the radius of rotation, thus lowering the sleeve and increasing the amount of energy supplied to the engine.

166. The Porter Governor. In Fig. 301 one-half of a Porter governor is shown diagrammatically. If the weights of the upper and lower arms are assumed to be negligible in comparison with the weight of the ball, the forces acting through the pin joint B in the equivalent system with rotation suppressed consist of the force F, equal to the centrifugal force which the ball exerts, the weight w, and the tensions T_1 and T_2 in the upper and lower arms, all of which act through the ball centre, together with one-half the sleeve load W, which acts through the pin C. The equation connecting F, w and W is most simply derived by taking moments about the instantaneous centre I of the lower arm BC. Since B moves along the circular arc which has A as centre and AB as radius, and C moves parallel to the axis of the governor, the instantaneous centre I lies at the point of intersection of AB produced and a line drawn through C perpendicular to the governor axis.

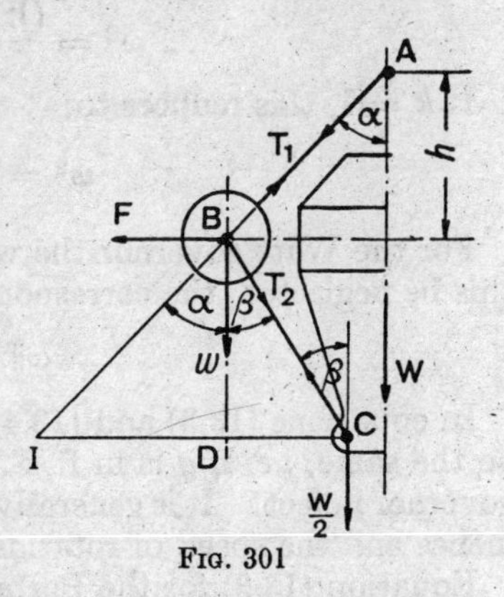

FIG. 301

Taking moments about I:

$$F.\text{BD} = w.\text{ID} + \frac{W}{2}.\text{IC},$$

$$\therefore\ F = w.\frac{\text{ID}}{\text{BD}} + \frac{W}{2}.\frac{\text{IC}}{\text{BD}}$$

$$= w\tan\alpha + \frac{W}{2}(\tan\alpha + \tan\beta)$$

$$= \left\{\frac{W}{2}(1+k) + w\right\}\tan\alpha \quad . \quad . \quad . \quad (13.1)$$

where $k = \tan\beta/\tan\alpha$.

Note that k will have a different value for each radius of rotation of the governor balls, unless the upper and lower arms are of equal length and the pins A and C are either on the governor axis or are at equal distances from it. In the latter event the angles α and β are equal for all radii of rotation and $k = 1$.

The distance of the plane of rotation of the governor balls from the point of intersection of the upper arms (produced if necessary) with the axis of rotation is termed the height of the governor and is denoted by h.

Then $\tan \alpha = r/h$.

Also
$$F = (w/g)\omega^2 r$$

where ω = angular velocity of the governor.

$\therefore$ substituting in equation (13.1):

$$\frac{w}{g}\omega^2 r = \left\{\frac{W}{2}(1+k)+w\right\}\frac{r}{h}$$

$$\therefore \omega^2 = \frac{(W/2)(1+k)+w}{w}\cdot\frac{g}{h} \quad \ldots \quad (13.2)$$

If $k = 1$, this reduces to:

$$\omega^2 = \frac{W+w}{w}\cdot\frac{g}{h} \quad \ldots\ldots \quad (13.3)$$

For the Watt governor the weight of the sleeve is small and, if this be neglected, the corresponding equation is:

$$\omega^2 = g/h \quad \ldots\ldots \quad (13.4)$$

In equations (13.3) and (13.4) the linear units for g and h must be the same, i.e. if g is in ft/s^2, then h must be the height of the governor in feet. It is generally more convenient to work with h in inches and the speed of rotation in r.p.m. instead of in rad/s.

Equation (13.3) for the Porter governor then becomes:

$$N^2 = \frac{W+w}{w}\cdot\frac{35\,230}{h''} \quad \ldots\ldots \quad (13.5)$$

or

$$h'' = \frac{W+w}{w}\cdot\frac{35\,230}{N^2} \quad \ldots\ldots \quad (13.6)$$

and for the Watt governor:

$$h'' = \frac{35\,230}{N^2} \quad \ldots\ldots \quad (13.7)$$

Corresponding values of h and N for the Watt governor are given in the following table:

N r.p.m.	40	60	80	100	150	200
h in.	22	9·78	5·5	3·52	1·56	0·88

It will be seen that the height diminishes very rapidly as the speed of rotation increases. This type of governor is therefore only suitable for low speeds of rotation, not exceeding, say, about 75 r.p.m.

It is clear from equations (13.6) and (13.7) that for a given speed N the height of the Porter governor may be made as much greater than the height of the Watt governor as desired by choosing a suitable value for the ratio W/w.

The following is an alternative way of deriving equation (13.1).

Referring to Fig. 301, the pin joint B between the upper and lower arms must be in equilibrium under the system of four forces shown. These are the weight w, the tensions T_1 and T_2 in the upper and lower arms and the centrifugal force F. They must form a closed polygon as shown in Fig. 302 (a). The sleeve is also in equilibrium under a system of three forces. These are the load W on the sleeve, and the tensions T_2 in the two lower arms. They must form a closed triangle as shown in Fig. 302 (b).

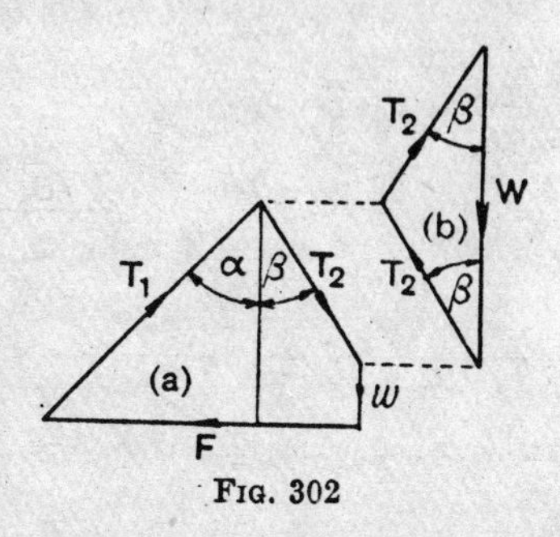

FIG. 302

From the triangle of forces on the sleeve we have:

$$W = 2T_2 \cos\beta \quad \text{or} \quad T_2 = W/2\cos\beta$$

From the polygon of forces on the ball we have, resolving vertically:

$$T_1 \cos\alpha = T_2 \cos\beta + w = W/2 + w$$

or

$$T_1 = \frac{W/2+w}{\cos\alpha}$$

and, resolving horizontally,

$$F = T_1 \sin\alpha + T_2 \sin\beta$$

so that, substituting for T_1 and T_2:

$$\begin{aligned} F &= (W/2+w)\tan\alpha + (W/2)\tan\beta \\ &= \{(W/2)(1+k)+w\}\tan\alpha \end{aligned}$$

where $k = \tan\beta/\tan\alpha$.

Example 1. A Porter governor has all four arms of length 12 in. The upper arms are pivoted on the axis of rotation and the lower arms are attached to the sleeve at distances of 1·5 in. from the axis. The weight of each ball is 15 lb and the load on the sleeve is 120 lb. If the extreme radii of rotation of the governor balls are 8 in. and 10 in., find the corresponding equilibrium speeds.

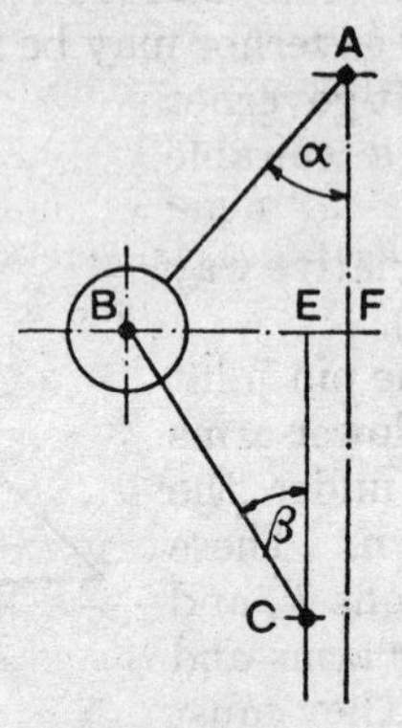

FIG. 303

Referring to Fig. 303 and using the suffixes 1 and 2 to distinguish between the maximum and minimum speeds of rotation, we have:

(a) *At Minimum Radius.*

$$AF = \sqrt{(AB^2 - BF^2)} = \sqrt{(12^2 - 8^2)} = 8{\cdot}944 \text{ in.}$$

$$CE = \sqrt{(12^2 - 6{\cdot}5^2)} = 10{\cdot}08 \text{ in.}$$

$$\therefore\ k_2 = \frac{\tan\beta_2}{\tan\alpha_2} = \frac{BE}{CE}\cdot\frac{AF}{BF} = \frac{6{\cdot}5}{10{\cdot}08}\cdot\frac{8{\cdot}944}{8{\cdot}0} = 0{\cdot}720$$

$\therefore$ substituting in equation (13.2):

$$N_2^2 = \frac{60\,.\,1{\cdot}720 + 15}{15}\cdot\frac{35\,230}{8{\cdot}944}$$

and $$N_2 = 176{\cdot}3 \text{ r.p.m.}$$

(b) *At Maximum Radius.*

$$AF = \sqrt{(12^2 - 10^2)} = 6{\cdot}633 \text{ in.};\ CE = \sqrt{(12^2 - 8{\cdot}5^2)} = 8{\cdot}47 \text{ in.}$$

$$\therefore\ k_1 = \frac{\tan\beta_1}{\tan\alpha_1} = \frac{BE}{CE}\cdot\frac{AF}{BF} = \frac{8{\cdot}5}{8{\cdot}47}\cdot\frac{6{\cdot}633}{10} = 0{\cdot}665$$

$\therefore$ substituting in equation (13.2):

$$N_1^2 = \frac{60\,.\,1{\cdot}665 + 15}{15}\cdot\frac{35\,230}{6{\cdot}633}$$

and $$N_1 = 201{\cdot}4 \text{ r.p.m.}$$

167. The Proell Governor. The Proell governor may be analysed in the same way as the Porter governor by considering the equilibrium of the lower arm. Thus, referring to Fig. 304 (a), the instantaneous centre I of the lower arm BC lies at the point of intersection of AB produced with a line drawn through C at right angles to the governor axis. Then, taking moments about I and assuming the extension BG of the lower arm to be vertical, we have:

$$F.\mathrm{DG} = w.\mathrm{ID}+(W/2)\mathrm{IC}$$

Divide both sides by BD. Then:

$$F.\frac{\mathrm{DG}}{\mathrm{BD}} = w.\frac{\mathrm{ID}}{\mathrm{BD}}+\frac{W}{2}\frac{\mathrm{IC}}{\mathrm{BD}} = w\tan\alpha+\frac{W}{2}(\tan\alpha+\tan\beta)$$

$$= \left(w+\frac{W}{2}\right)\tan\alpha+\frac{W}{2}\tan\beta$$

$$\therefore\ F = \frac{\mathrm{BD}}{\mathrm{DG}}\left\{\left(w+\frac{W}{2}\right)\tan\alpha+\frac{W}{2}\tan\beta\right\}$$

This may be written in the form:

$$F = \frac{\mathrm{BD}}{\mathrm{DG}}\left\{\frac{W}{2}(1+k)+w\right\}\tan\alpha \quad . \quad . \quad (13.8)$$

where $k = \tan\beta/\tan\alpha$.

If $\beta = \alpha$, then $k = 1$ and equation (13.8) becomes:

$$F = \frac{\mathrm{BD}}{\mathrm{DG}}(W+w)\tan\alpha \quad . \quad . \quad . \quad (13.9)$$

But $\tan\alpha = r/h$ and $F = (w/g)\omega^2 r$, so that, substituting in equation (13.8), we get:

$$\frac{w}{g}\omega^2 r = \frac{\mathrm{BD}}{\mathrm{DG}}\left\{\frac{W}{2}(1+k)+w\right\}\frac{r}{h}$$

$$\therefore\ \omega^2 = \frac{\mathrm{BD}}{\mathrm{DG}}\left\{\frac{W}{2}(1+k)+w\right\}.\frac{g}{w.h} \quad . \quad . \quad (13.10)$$

and, if $k = 1$, this reduces to

$$\omega^2 = \frac{\mathrm{BD}}{\mathrm{DG}}.\frac{W+w}{w}.\frac{g}{h} \quad . \quad . \quad . \quad . \quad (13.11)$$

On comparing equations (13.10) and (13.11) with (13.2) and (13.3), it will be at once obvious that the effect of placing the ball at G instead of at the pin-joint B is to reduce the equilibrium speed for given values of w, W and h. Hence in order to give the same equilibrium speed for given values of W and h, it is necessary to use a ball of smaller mass in the Proell governor than in the Porter governor.

A more important effect of the change in position of the ball is to reduce the increase of speed necessary in order to lift the sleeve by a given amount. This will be brought out more clearly by working through a numerical example.

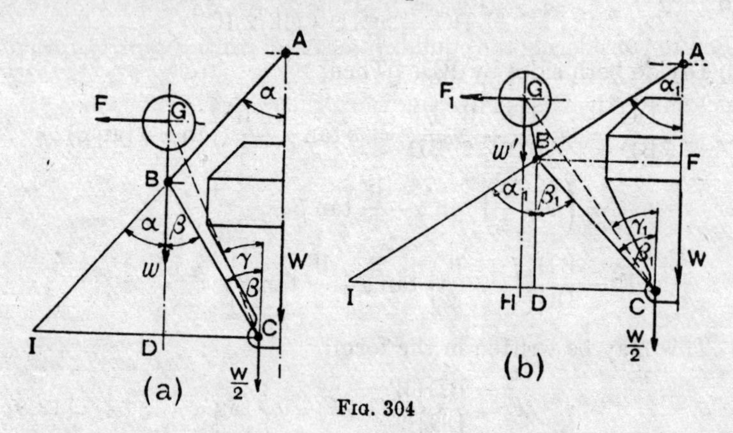

FIG. 304

Example 2. A Proell governor has the same dimensions as the Porter governor in Example 1, except that the ball is carried on an extension of the lower arm, such that BG = 4 in. It is required that this governor shall have the same equilibrium speed as the Porter governor, when the radius of rotation of the ball is 8 in. and the ball centre is vertically above the pin-joint B. Find: (a) the weight of ball required, and (b) the maximum equilibrium speed, if the sleeve lift is equal to that of the Porter governor.

(a) *At Minimum Radius.* Let w' be the weight of ball required in the Proell governor in order to give the same equilibrium speed at the radius of 8 in.

Then, from equation (13.10),

$$\omega^2 = \frac{BD}{DG}\left\{\frac{W}{2}(1+k)+w'\right\}\frac{g}{w'.h}$$

But for the Porter governor, from (13.2),

$$\omega^2 = \frac{(W/2)(1+k)+w}{w}\cdot\frac{g}{h}$$

∴ equating:

$$\frac{(W/2)(1+k)+w}{w} = \frac{BD}{DG}\cdot\frac{(W/2)(1+k)+w'}{w'}$$

But, for Example 1, $w = 15$ lb, $W = 120$ lb, $k = 0{\cdot}720$, BD = CE = 10·08 in. Also DG = BD+BG = 14·08 in.

$\therefore$ substituting:

$$\frac{60.1{\cdot}720+15}{15} = \frac{10{\cdot}08}{14{\cdot}08}.\frac{60.1{\cdot}720+w'}{w'}$$

and $$w' = 10{\cdot}32 \text{ lb}$$

Before considering the equilibrium speed at maximum radius, it will be convenient to find the values of CG, β and γ in Fig. 304 (a), which shows the Proell governor when the radius of rotation is 8 in.

$$CG = \sqrt{(DG^2+CD^2)} = \sqrt{(14{\cdot}08^2+6{\cdot}5^2)} = 15{\cdot}51 \text{ in.}$$

$$\tan\gamma = \frac{CD}{DG} = \frac{6{\cdot}5}{14{\cdot}08}; \quad \therefore \gamma = 24^\circ\ 47'$$

$$\sin\beta = \frac{CD}{BC} = \frac{6{\cdot}5}{12}; \quad \therefore \beta = 32^\circ\ 48'$$

(b) *At Maximum Radius*: Fig. 304 (b).

$$\sin\alpha_1 = \frac{BF}{AB} = \frac{10}{12}; \quad \therefore \alpha_1 = 56^\circ\ 27' \text{ and } \tan\alpha_1 = 1{\cdot}5080$$

$$\sin\beta_1 = \frac{CD}{BC} = \frac{8{\cdot}5}{12}; \quad \therefore \beta_1 = 45^\circ\ 6' \text{ and } \tan\beta_1 = 1{\cdot}0035$$

Also $\gamma_1 = \gamma+\beta_1-\beta = 24^\circ\ 47'+45^\circ\ 6'-32^\circ\ 48' = 37^\circ\ 5'$.

Taking moments about I, the instantaneous centre of the lower arm, we have:

$$F_1.GH = w'.IH+(W/2)IC$$
$$= w'(IC-CH)+(W/2)IC$$

$$\therefore F_1.GC\cos\gamma_1 = (w'+W/2)BD(\tan\alpha_1+\tan\beta_1)-w'.CG\sin\gamma_1$$

$$\therefore F_1 = \frac{(w'+W/2)BD(\tan\alpha_1+\tan\beta_1)}{GC\cos\gamma_1} - w'.\tan\gamma_1$$

$$= \frac{70{\cdot}32.8{\cdot}471(1{\cdot}5080+1{\cdot}0035)}{15{\cdot}51.0{\cdot}7977} - 10{\cdot}32.0{\cdot}7559$$

$$= 120{\cdot}9-7{\cdot}79 = 113{\cdot}1 \text{ lb}$$

The radius of rotation of the ball $= r_1 = CG\sin\gamma_1+1{\cdot}5$
$= 10{\cdot}86$ in.

$$\therefore \text{the equilibrium speed} = N_1 = \frac{30}{\pi}\sqrt{\left(\frac{F_1}{w'}.\frac{32{\cdot}2.12}{r_1}\right)}$$

$$= \frac{30}{\pi}\sqrt{\frac{113{\cdot}1.386{\cdot}4}{10{\cdot}32.10{\cdot}86}} = 188{\cdot}6 \text{ r.p.m.}$$

The equilibrium speeds in this example should be compared with those for the Porter governor in the previous example. It will be seen how much smaller is the change of speed for the same lift of the sleeve.

168. Spring-loaded Governors. In addition to the types of governors so far considered there are many others in which the control is effected, either wholly or in part, by means of springs. Representative spring-loaded governors are shown diagrammatically in Fig. 305. That shown at (a) is an example of the well-known Hartnell governor. The bracket A is keyed to, and revolves with, the governor shaft. The thrust exerted by the spring that surrounds the governor shaft is transmitted by means of three struts T, which pass through holes drilled in the bracket A, and the centrifugal force on each ball causes an upward thrust on the sleeve S, which is balanced by the downward thrust of the spring. The lock-nuts N enable the spring thrust for a given

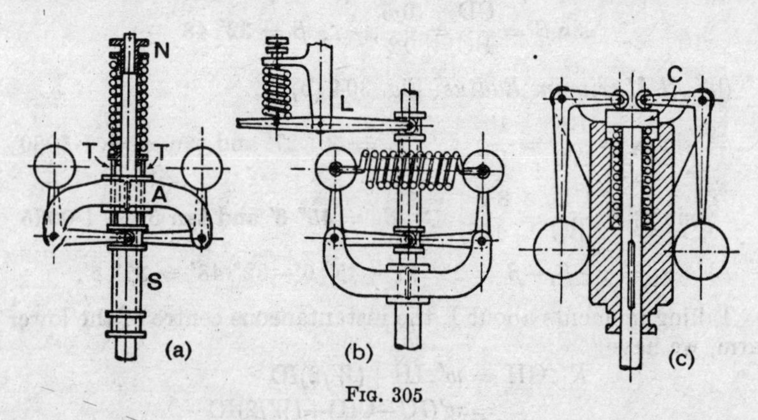

FIG. 305

sleeve position to be altered while the engine is running, so that the equilibrium speed may be adjusted between certain limits. In the type of governor shown at (b), two springs, which straddle the sleeve, are connected directly between the governor balls. These two springs are in tension, and it is evident that the load on the bell-crank pivots will be much smaller than in the case of the types (a) and (c), where the spring thrust has to be transmitted through the bell-crank levers. This type of governor, however, invariably requires an auxiliary spring which is also in tension and which produces a load on the sleeve that practically balances the centrifugal forces on the governor balls. One end of the auxiliary spring is attached to the lever L, through which the movement of the sleeve is transmitted to the valve which controls the supply of energy to the engine. The other end is attached to an adjusting screw, which passes through a hole in a fixed bracket and carries two lock-nuts for varying the spring tension. This auxiliary spring enables the equilibrium speed corresponding to a given sleeve position to be varied while the engine is running.

In the type of governor shown at (c) the pivots for the bell-crank levers are carried by the moving sleeve. The spring is compressed between the sleeve and the cap C, which is fixed to the end of the governor shaft. The rollers on the ends of the horizontal arms of the bell-crank levers press on the cap C, so that the sleeve is lifted against the compression of the spring. It is customary for a small auxiliary spring to be attached to the governor lever in a similar way to that shown for type (b), in order to provide for some adjustment of the equilibrium speed.

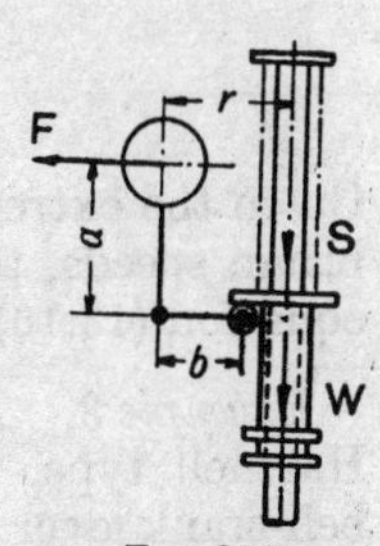

FIG. 306

169. The Hartnell Governor: Fig. 305 (a). The relation between the dimensions of the governor, the equilibrium speed and the spring load may be found by considering the equilibrium of one of the bell-crank levers, Fig. 306.

Let w = weight of each ball,
W = weight of the sleeve,
S = the force exerted on the sleeve by the spring which surrounds the governor spindle,
p = stiffness of the spring,
ω = the speed of rotation,
a, b = the vertical and horizontal arms of the bell-crank levers
and r = the radius of rotation.

Then, taking moments about the fulcrum of the bell-crank lever and neglecting the effect of the pull of gravity on the governor balls and arms:

$$F.a = \frac{W+S}{2}.b \quad \text{or} \quad W+S = 2F.\frac{a}{b} \quad . \quad (13.12)$$

Let the suffixes 1 and 2 be used to denote the values at maximum and minimum radius respectively.

Then, at maximum radius:

$$W+S_1 = 2F_1.a/b \quad . \quad . \quad . \quad . \quad (13.13)$$

and, at minimum radius:

$$W+S_2 = 2F_2.a/b \quad . \quad . \quad . \quad . \quad (13.14)$$

Subtracting (13.14) from (13.13):

$$S_1-S_2 = 2(a/b)(F_1-F_2)$$

The lift of the sleeve $x = (b/a)(r_1-r_2)$, so that

$$S_1-S_2 = p(b/a)(r_1-r_2) \quad . \quad . \quad (13.15)$$

and

$$p(b/a)(r_1-r_2) = 2(a/b)(F_1-F_2)$$

$$\therefore\ p = 2\left(\frac{a}{b}\right)^2 \cdot \frac{F_1-F_2}{r_1-r_2} \quad . \quad . \quad (13.16)$$

Given the extreme radii of rotation and the corresponding equilibrium speeds, the stiffness of the spring may be calculated from equation (13.16).

Example 3. The following particulars refer to a governor of the Hartnell type, Fig. 306: weight of each ball, 3 lb; lengths of bell-crank lever arms, $a = 4$ in., $b = 2$ in.; distance of the fulcrum of each bell-crank lever from the axis of rotation, $3\frac{1}{2}$ in.; maximum and minimum radii of rotation of the governor balls, $4\frac{1}{2}$ in. and 3 in. The minimum equilibrium speed is to be 300 r.p.m. and the maximum equilibrium speed is to be 6% greater than this. Find the rate or stiffness of the spring and the equilibrium speed when the radius of rotation of the balls is $3\frac{1}{2}$ in.

$$F_2 = \frac{w}{g}\omega_2^2 r_2 = \frac{3}{32 \cdot 2}\left(\frac{\pi . 300}{30}\right)^2 \frac{3}{12} = 23 \cdot 0 \text{ lb}$$

$$F_1 = F_2 \cdot \left(\frac{N_1}{N_2}\right)^2 \cdot \frac{r_1}{r_2} = 23 \cdot 0 . 1 \cdot 06^2 . \frac{4 \cdot 5}{3} = 38 \cdot 8 \text{ lb}$$

From equation (13.16):

$$p = 2\left(\frac{a}{b}\right)^2 \cdot \frac{F_1-F_2}{r_1-r_2}$$

$$= 2\left(\frac{4}{2}\right)^2 \cdot \frac{38 \cdot 8 - 23 \cdot 0}{4 \cdot 5 - 3}$$

$$= 84 \cdot 3 \text{ lb/in.}$$

Since the effect of gravity is neglected, F will vary directly with the sleeve load and therefore with the radius.

$\therefore$ when the radius is $3\frac{1}{2}$ in., the centrifugal force will be:

$$F_2 + (F_1 - F_2)\frac{0 \cdot 5}{1 \cdot 5} = 23 \cdot 0 + \frac{38 \cdot 8 - 23 \cdot 0}{3} = 28 \cdot 27 \text{ lb}$$

$$\therefore\ \frac{3}{32 \cdot 2} \cdot \left(\frac{\pi N}{30}\right)^2 \cdot \frac{3 \cdot 5}{12} = 28 \cdot 27$$

$$\therefore\ N = 308 \cdot 3 \text{ r.p.m.}$$

170. Spring-controlled Governor of the Type shown in Fig. 305 (b).

Let w = weight of each ball,
W = weight of sleeve,
P = combined pull of the ball springs,
S = pull of auxiliary spring,
p_b = stiffness of each ball spring,
p_a = stiffness of auxiliary spring,
F = centrifugal force of each ball
and r = radius of rotation of the balls.

Referring to Fig. 307, the total downward force on the sleeve $= W+Sy/x$.

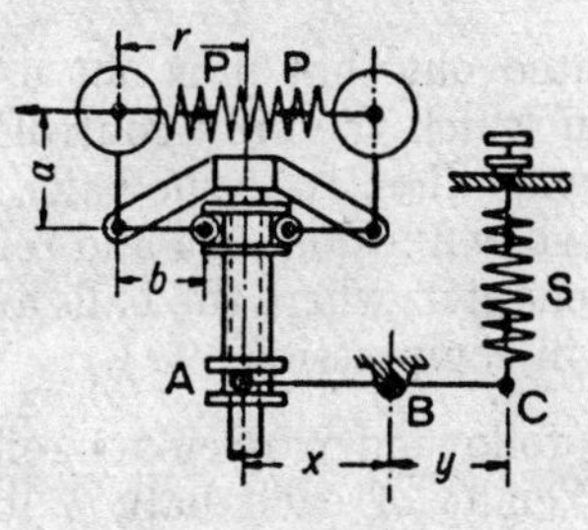

FIG. 307

Taking moments about the fulcrum of the bell-crank lever and neglecting the effect of the pull of gravity on the ball, we have:

$$(F-P)a = \frac{W+Sy/x}{2}.b$$

At minimum equilibrium speed:

$$(F_2-P_2)a = \frac{W+S_2y/x}{2}.b \quad . \quad . \quad (13.17)$$

At maximum equilibrium speed:

$$(F_1-P_1)a = \frac{W+S_1y/x}{2}.b \quad . \quad . \quad (13.18)$$

Subtracting equation (13.17) from equation (13.18):

$$\{F_1-F_2-(P_1-P_2)\}a = (S_1-S_2)y/x.b/2 \quad (13.19)$$

But, if the radius increases from r_2 to r_1, the ball springs extend by the amount $2(r_1-r_2)$ and the auxiliary spring extends by the amount $(r_1-r_2)b/a.y/x$.

$$\therefore\ P_1-P_2 = 2p_b.2(r_1-r_2) = 4p_b(r_1-r_2)$$

and

$$S_1-S_2 = p_a.b/a.(y/x)(r_1-r_2)$$

$\therefore$ substituting in equation (13.19) and transposing:

$$F_1 - F_2 = 4p_b(r_1 - r_2) + \frac{p_a}{2}\left(\frac{b}{a}\cdot\frac{y}{x}\right)^2 (r_1 - r_2)$$

$$\therefore\ 4p_b + \frac{p_a}{2}\left(\frac{b}{a}\cdot\frac{y}{x}\right)^2 = \frac{F_1 - F_2}{r_1 - r_2} \quad . \quad . \quad (13.20)$$

It will be clear from equation (13.20) that either p_a or p_b may be fixed arbitrarily and the value of the other stiffness then calculated to suit.

If no auxiliary spring is used, i.e. if $p_a = 0$, then equation (13.20) reduces to:

$$p_b = \frac{F_1 - F_2}{4(r_1 - r_2)} \quad . \quad . \quad . \quad . \quad (13.21)$$

It should be pointed out that it is not usually practicable to design a governor in which the only control is that provided by tension springs attached directly to the balls. A more practicable arrangement is that shown in Fig. 315 and referred to in Question 13 at the end of the chapter, where the balls are directly controlled by separate springs in compression.

Example 4. The following particulars refer to a governor of the above type. Weight of each ball, 5 lb; minimum radius, 5 in.; maximum radius, 7 in.; minimum speed, 240 r.p.m.; maximum speed, 5% greater than the minimum; lengths of bell-crank lever arms, $a = 6$ in., $b = 4$ in. The combined stiffness of the two ball springs is 3 lb/in. Find the equivalent stiffness of the auxiliary spring referred to the sleeve.

$$F_2 = \frac{w}{g}\omega_2^2 r_2 = \frac{5}{32{\cdot}2}\cdot\left(\frac{\pi . 240}{30}\right)^2 \cdot \frac{5}{12} = 40{\cdot}9 \text{ lb}$$

$$F_1 = F_2\left(\frac{N_1}{N_2}\right)^2 \cdot \frac{r_1}{r_2} = 40{\cdot}9 . 1{\cdot}05^2 . \frac{7}{5} = 63{\cdot}1 \text{ lb}$$

The equivalent stiffness of the auxiliary spring referred to the sleeve

$$= p = p_a(y/x)^2$$

$\therefore$ from equation (13.20):

$$p = p_a\left(\frac{y}{x}\right)^2 = 2\left(\frac{a}{b}\right)^2\left(\frac{F_1 - F_2}{r_1 - r_2} - 4p_b\right)$$

$$= 2\left(\frac{6}{4}\right)^2\left(\frac{63{\cdot}1 - 40{\cdot}9}{7 - 5} - 2.3\right)$$

$$= 23{\cdot}0 \text{ lb/in.}$$

If no auxiliary spring were fitted, so that $p_a = 0$, the **combined** stiffness of the ball springs would have to be $p_b = 11{\cdot}1/2 = 5{\cdot}55$ lb/in.

To provide a pull of 40·9 lb on each ball when the distance between the ball centres is 10 in., the free length of the springs would be only $10-40{\cdot}9/5{\cdot}55 = 2{\cdot}63$ in.

It is quite evident that it would be impracticable, if not actually impossible, to design a spring to satisfy these conditions. It is for this reason, as well as to enable the equilibrium speed to be adjusted, that the auxiliary spring is provided.

Let the free length of each ball spring be 8 in. Then at the minimum radius of 5 in. the combined pull of the two ball springs is 6 lb and the force to be exerted at the sleeve by the auxiliary spring may be found as follows:

Let Q_2 be the force on the sleeve due to the auxiliary spring. Then $Q_2 = S_2.y/x$ and substituting in equation (13.17),

$$(F_2-P_2)a = \frac{W+Q_2}{2}.b$$

But $F_2 = 40{\cdot}9$ lb, $P_2 = 6$ lb, $a = 6$ in., $b = 4$ in.

$$\therefore\ W+Q_2 = (40{\cdot}9-6)(2.6/4)$$
$$= 104{\cdot}7 \text{ lb}$$

If the axis of the governor is horizontal, the weight W of the sleeve will not affect the equilibrium and the whole of the force on the sleeve must be provided by the auxiliary spring.

171. Spring-controlled Governor of the Type shown in Fig. 305. (c). This type of governor may be analysed most simply by taking moments about the instantaneous centre of all the forces which act on one of the bell-crank levers. Referring to Fig. 308, the instantaneous centre of the bell-crank lever is at I, since the path of the fulcrum B is parallel to the governor axis and the path of the roller centre C is at right angles to the governor axis.

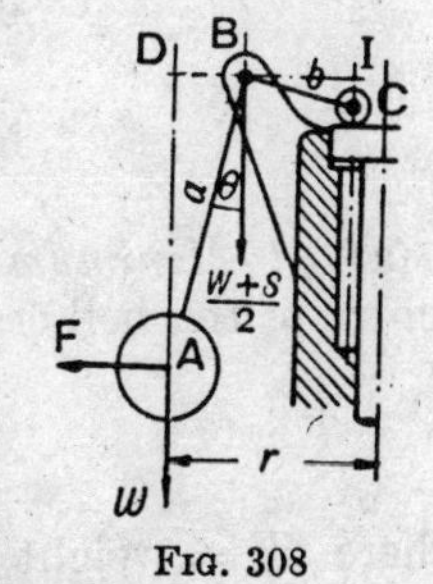

Fig. 308

There are two forces, w and F, which act through the ball centre A, and in addition there is one-half the total sleeve load, which acts through the fulcrum B. The total sleeve load will consist of the deadweight W together with the load due to the main compression spring, which surrounds the spindle, and the auxiliary spring, if one is fitted. The total equivalent spring load on the sleeve is denoted by S.

Taking moments about I:

$$F.\mathrm{AD} = w.\mathrm{DI} + \frac{W+S}{2}.\mathrm{BI}$$

$$\therefore\ W+S = \frac{2}{\mathrm{BI}}(F.\mathrm{AD} - w.\mathrm{DI}) \quad . \quad . \quad (13.22)$$

Example 5. The following particulars refer to a governor of the above type. The weight of each ball is 3 lb, the weight of the sleeve is 15 lb, the lengths of the arms a and b of the bell-crank levers are $4\frac{1}{2}$ in. and 2 in., the distance of the fulcrum of each bell-crank lever from the axis of rotation is $2\frac{1}{2}$ in., the minimum radius of rotation is $2\frac{1}{2}$ in., the corresponding equilibrium speed is 240 r.p.m. and the sleeve is required to lift $\frac{1}{2}$ in. for an increase of speed of 5%. Find the stiffness of the governor spring.

$$F_2 = \frac{w}{g}\omega_2{}^2 r_2 = \frac{3}{32{\cdot}2}.\left(\frac{\pi.240}{30}\right)^2.\frac{2{\cdot}5}{12} = 12{\cdot}25 \text{ lb}$$

If we assume that the two arms of each bell-crank lever are at right angles, the increase of radius for a sleeve lift of 0·5 in. is evidently:

$$a/b.0{\cdot}5 = 4{\cdot}5/2.0{\cdot}5 = 1{\cdot}125 \text{ in.};\ \therefore\ r_1 = r_2 + 1{\cdot}125 = 3{\cdot}625 \text{ in.}$$

$$\therefore\ F_1 = (N_1/N_2)^2.r_1/r_2.F_2 = 1{\cdot}05^2.3{\cdot}625/2{\cdot}5.12{\cdot}25 = 19{\cdot}6 \text{ lb}$$

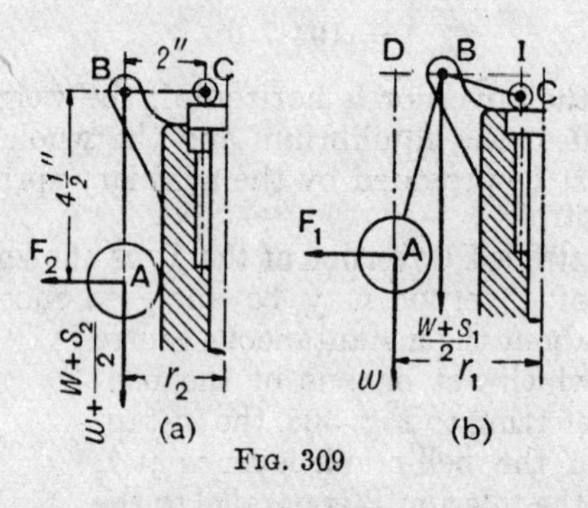

FIG. 309

(a) *At Minimum Radius*: Fig. 309 (a). The instantaneous centre of the bell-crank lever coincides with the roller-centre C,

$$\therefore\ F_2.\mathrm{AB} = \left(w + \frac{W+S_2}{2}\right)\mathrm{BC}$$

where W = weight of sleeve, S_2 = spring load on sleeve and w = weight of ball.

$$\therefore\ 12{\cdot}25.4{\cdot}5 = \left(3 + \frac{15+S_2}{2}\right)2 \quad \text{and} \quad S_2 = 34{\cdot}2 \text{ lb}$$

(b) *At Maximum Radius*: Fig. 309 (b). I is the instantaneous centre of the bell-crank lever,

$$\therefore\ F_1.\mathrm{AD} = w.\mathrm{DI} + \frac{W+S_1}{2}.\mathrm{BI}$$

$$\therefore\ W+S_1 = 2\left(F_1.\frac{\mathrm{AD}}{\mathrm{BI}} - w.\frac{\mathrm{DB}+\mathrm{BI}}{\mathrm{BI}}\right)$$

But $\mathrm{DB} = r_1 - r_2 = 1{\cdot}125$ in. and $\mathrm{BI} = \sqrt{(\mathrm{BC}^2 - \mathrm{IC}^2)} = 1{\cdot}936$ in.

$$\therefore\ W+S_1 = 2\left(19{\cdot}6.\frac{4{\cdot}5}{2} - 3.\frac{1{\cdot}125+1{\cdot}936}{1{\cdot}936}\right)$$

$$= 2(44{\cdot}1 - 4{\cdot}74)$$

$$= 78{\cdot}8 \text{ lb}$$

$$\therefore\ S_1 = 78{\cdot}8 - 15 = 63{\cdot}8 \text{ lb}$$

$$\therefore\ \text{stiffness of the spring} = S_1 - S_2 \div \text{sleeve lift}$$

$$= \frac{63{\cdot}8 - 34{\cdot}2}{0{\cdot}5}$$

$$= 59{\cdot}2 \text{ lb/in.}$$

172. Definitions. Before proceeding further it is desirable to give one or two definitions of terms which are used in connection with governors.

(a) *Sensitiveness.* In order to maintain as closely as possible a constant mean speed of rotation, whatever may be the load on the engine, it is clearly desirable that the movement of the sleeve should be as large as possible and the corresponding change of equilibrium speed as small as possible. The smaller the fractional change of speed for a given displacement of the sleeve, or the bigger the displacement of the sleeve for a given fractional change of speed, the more sensitive is the governor said to be.

This definition of sensitiveness is quite satisfactory when the governor is considered as an independent mechanism, but when the governor is fitted to an engine the practical requirement is simply that the change of equilibrium speed from the full load to the no load position of the sleeve should be as small a fraction as possible of the mean equilibrium speed. The actual displacement of the sleeve is immaterial, provided that it is sufficient to change the energy supplied to the engine by the required amount. For this reason sensitiveness is more correctly defined as the ratio of the difference between the maximum and minimum equilibrium speeds to the mean equilibrium speed.

(b) *Stability.* A governor is said to be stable when for each speed within the working range there is only one radius of rotation of the governor balls at which the governor is in equilibrium.

(c) *Isochronism.* A governor is said to be isochronous when the equilibrium speed is constant for all radii of rotation of the balls within the working range.

It follows from the definition of sensitiveness given above that an isochronous governor will be infinitely sensitive. The sleeve will remain in the lowest position until the equilibrium speed is reached, when the slightest further increase of speed will cause the governor balls to fly out to their maximum radius and thus lift the sleeve to its highest position.

(d) *Hunting.* This is the name given to a condition in which the speed of the engine controlled by the governor fluctuates continually above and below the mean speed. It is caused by the use of a governor which is too sensitive and which, therefore, changes by too large an amount the supply of energy to the engine when a small change in the speed of rotation takes place. Thus, to take an extreme case, let us suppose that an isochronous governor is fitted to an engine and that the engine is running under a steady load. If a slight increase of the load takes place, the speed of rotation will fall and the governor sleeve will immediately fall to its lowest position. This will open the control valve wide and the supply of energy to the engine will now be in excess of its requirements, so that the speed will rapidly increase again and the sleeve will rise to its highest position. As a result of this movement of the sleeve, the control valve will cut off the supply of energy to the engine and the speed will once more begin to fall, the cycle being repeated indefinitely. Such a governor would admit either the maximum or the minimum amount of energy and could not possibly admit an amount of energy between these two extremes. The effect, as we have seen, will be to cause wide fluctuations in the speed of rotation of the engine. In other words, the engine will hunt.

173. Governor Effort and Power. The effort of a governor is the force which the governor can exert at the sleeve on the mechanism which controls the supply of energy to the engine. When the speed is constant the effort is zero, but if a sudden change of speed takes place, the sleeve tends to move to its new equilibrium position and a force is exerted on the mechanism. This force gradually diminishes to zero as the sleeve moves to the equilibrium position corresponding to the new speed. The mean force exerted during the given change of speed is termed the effort. For convenience in comparing different types of governors, it is usual to define the effort as that which can be exerted for a 1% change of speed.

The power of a governor is defined as the work done at the sleeve for a given percentage change of speed. It is the product of the governor effort and the displacement of the sleeve. The power required will obviously depend on the form of the controlling mechanism which the governor is called upon to operate. In a gas engine governed on the hit-and-miss principle, for instance, very little work has to be done by the governor, whereas in a steam engine governed by alteration of the cut-off, the governor is required to do more work and must, therefore, be more powerful. Where the power required is large, it is usual to employ compressed air or oil under pressure in order to change the position of the valve or valves that control the supply of energy to the engine. The governor is then simply called upon to move a small pilot valve, which admits the compressed air or oil to a cylinder in which moves a piston connected by suitable linkage to the energy supply valves.

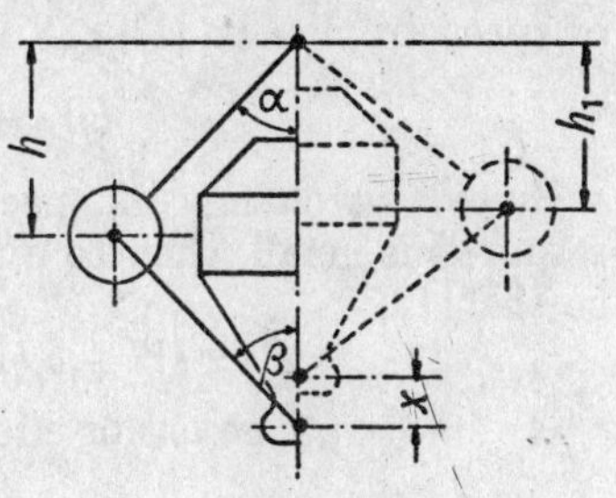

FIG. 310

The governor effort and power may be determined in the following way. The Porter governor is used to illustrate the method, but the same principle is adopted for any other type of governor. Referring to Fig. 310, let N be the equilibrium speed corresponding to the configuration shown by the full lines and let a sudden increase of speed, $c.N$, take place, for which the equilibrium position is shown by the dotted lines. In order to prevent the sleeve from rising when the increase of speed takes place, a downward force will have to be exerted on the sleeve. The magnitude of this force may be determined by finding what increase is required in the load W on the sleeve in order to cause the governor to revolve in equilibrium in the full line position at the increased speed $(1+c)N$.

When the speed is N r.p.m. and the angles α and β are equal so that $k = 1$, the height h is given by equation (13.6):

$$h = \frac{W+w}{w} \cdot \frac{35\,230}{N^2}$$

If the speed increases to $(1+c)N$ and the height remains the same, we have:

$$h = \frac{W_1+w}{w} \cdot \frac{35\,230}{(1+c)^2 N^2} \qquad . \quad . \quad (13.23)$$

where W_1 is the required sleeve load.

On equating the two expressions for h and reducing,

$$W_1+w = (W+w)(1+c)^2,$$

$$\therefore\ W_1 = (W+w)(1+c)^2-w$$

and

$$W_1-W = (W+w)\{(1+c)^2-1\} \quad . \quad . \quad (13.24)$$

W_1-W is evidently the downward force P which must be applied in order to prevent the sleeve from rising when the increase of speed takes place. It is therefore equal to the force exerted by the governor sleeve on the controlling mechanism, immediately after the increase of speed has taken place and before the sleeve begins to move. As the sleeve rises to the new equilibrium position shown by the dotted lines, this force gradually diminishes to zero. The mean force Q exerted by the sleeve during the change of speed from N to $(1+c)N$ is therefore equal to

$$(W_1-W)/2 = P/2$$

Since the change of speed for which the governor effort is required is small, then to a first approximation $(1+c)^2 = 1+2c$, so that:

$$P \simeq (W+w)(1+2c-1) \simeq (W+w)2c$$

$$\therefore \text{ governor effort } Q \simeq P/2 \simeq c(W+w) \quad (13.25)$$

The governor power is the product of the effort Q and the sleeve displacement x:

But $x = 2(h-h_1)$, where h_1 = height corresponding to the increased speed $(1+c)N$, i.e. $h_1 = h/(1+c)^2$.

$$\therefore\ x = 2h\left\{1-\frac{1}{(1+c)^2}\right\} \simeq 2h.\frac{2c}{1+2c} \quad . \quad (13.26)$$

$$\therefore \text{ governor power} = Q.x \simeq \frac{4c^2}{1+2c}(W+w)h \quad (13.27)$$

If α and β are not equal, then equation (13.25) for the governor effort becomes:

$$Q \simeq c\left(W+\frac{2w}{1+k}\right) \quad . \quad . \quad . \quad . \quad (13.28)$$

where $k = \tan\beta/\tan\alpha$.

Similarly, equation (13.26) for the lift of the sleeve will no longer apply, since x is not equal to $2(h-h_1)$. It can be shown that x is approximately equal to $(1+k)(h-h_1)$.

But $h_1 = h/(1+c)^2$, so that

$$x \simeq (1+k)h\left\{1-\frac{1}{(1+c)^2}\right\} \simeq (1+k)h.\frac{2c}{1+2c}$$

$$\therefore \text{ governor power} \simeq \frac{2c^2}{1+2c}\left\{W(1+k)+2w\right\}h$$

$$\simeq \frac{4c^2}{1+2c}\left\{\frac{W}{2}(1+k)+w\right\}h \quad (13.29)$$

Example 6. Determine the governor effort and power for the Porter governor of Example 1, p. 459, when the sleeve is in its lowest position. The fractional increase of speed is 1%.

Here $W = 120$ lb, $w = 15$ lb, $k = 0{\cdot}720$, $c = 0{\cdot}01$ and $h = 8{\cdot}944$ in.

From equation (13.28),

$$Q = 0{\cdot}01\left(120+\frac{2.15}{1{\cdot}720}\right) = 1{\cdot}374 \text{ lb}$$

Also $$x = \frac{2c}{1+2c}(1+k)h = \frac{0{\cdot}02}{1{\cdot}02}.1{\cdot}720.8{\cdot}944$$

$$= 0{\cdot}302 \text{ in.}$$

$$\therefore \text{ governor power} = Q.x = 1{\cdot}374.0{\cdot}302$$

$$= 0{\cdot}415 \text{ in. lb}$$

Example 7. Determine the effort and power of the spring-loaded governor of Example 4, p. 468, when the radius of rotation is 6 in. and the increase of speed is 1%

Since the deflections of the ball springs and of the auxiliary spring are directly proportional to the change of radius of rotation of the balls, the centrifugal force will also vary directly with the radius, and we may write:

$$F = Ar+B$$

where A and B are constants which depend on the stiffness of the springs.

It has already been determined that $F = 40{\cdot}9$ lb when $r = 5$ in., and $F = 63{\cdot}1$ lb when $r = 7$ in.

To satisfy these two conditions, the equation must be:

$$F = 11{\cdot}1r-14{\cdot}6 \quad . \quad . \quad . \quad . \quad . \quad (1)$$

so that when $r = 6$ in. $F = 52{\cdot}0$ lb.

If the speed increases by 1% while the radius remains unchanged, the centrifugal force on each ball will increase to $1{\cdot}01^2.52{\cdot}0$.

The increase is approximately $2.0{\cdot}01.52{\cdot}0 = 1{\cdot}04$ lb, and the force required to prevent the sleeve from rising is $2a/b.1{\cdot}04 = 3{\cdot}12$ lb, therefore the mean force exerted as the sleeve moves to the new equilibrium position $= 3{\cdot}12/2 = 1{\cdot}56$ lb.

The radius of rotation corresponding to the new speed may be found as follows:

From equation (1),

$$F/r = 11{\cdot}1 - 14{\cdot}6/r$$

But $F = 53{\cdot}04$ lb, when $r = 6$ in.

$$\therefore \frac{53{\cdot}04}{6} = 11{\cdot}1 - \frac{14{\cdot}6}{r}$$

from which $r = 6{\cdot}46$ in.

If the radius of rotation increases by $0{\cdot}46$ in., the sleeve will lift $b/a . 0{\cdot}46 = 0{\cdot}307$ in.

$$\therefore \text{ governor power} = 1{\cdot}56 . 0{\cdot}307 = 0{\cdot}479 \text{ in lb}$$

174. Controlling Force. When the speed of rotation is uniform, each governor ball is subjected either directly or indirectly to an inward pull which is equal and opposite to the outward centrifugal reaction. This inward pull is termed the controlling force, and a

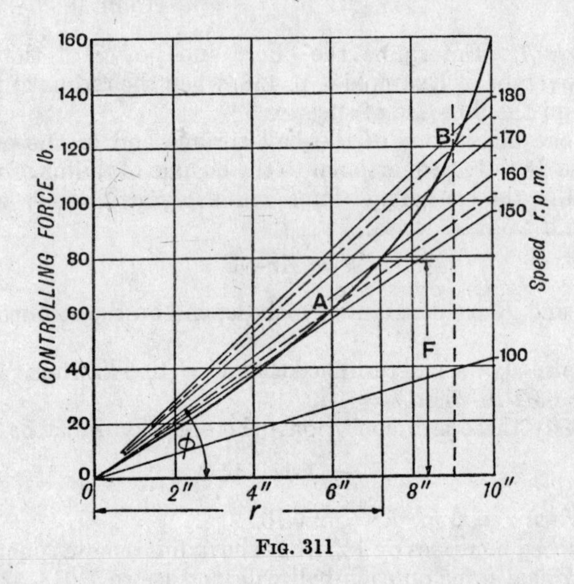

FIG. 311

curve drawn to show how the pull varies with the radius of rotation of the ball is called a controlling force curve, although in some types of governor it may be a straight line. The curve enables the stability and sensitiveness of the governor to be examined and also shows clearly the effect of friction. Such a curve is shown in Fig. 311.

Since the controlling force is equal and opposite to the centrifugal force, we have $F = (w/g)\omega^2 r$,

or $$\omega = \sqrt{(g/w . F/r)} = \sqrt{\{(g/w) \tan \phi\}} \quad . \quad (13.30)$$

where ϕ is the inclination to the r axis of the line joining a given point on the curve to the origin.

If the governor is to satisfy the conditions for stability, the equilibrium speed must increase as the radius of rotation of the governor balls increases, i.e. the shape of the controlling force curve must be such that the angle ϕ increases continuously as r increases. It also follows from the definition of sensitiveness that the change in the value of ϕ over the range of radius of rotation should be as small as possible in order to provide the most sensitive governor. Further, if the controlling force curve is a straight line which passes through the origin, the angle ϕ will be constant for all values of the radius and the governor will be isochronous.

From equation (13.30):

$$\tan \phi = (w/g)\omega^2 = C . N^2$$

Using this relation, values of ϕ may be calculated for different values of N and lines may be drawn radiating from the origin, as shown in Fig. 311. These enable the equilibrium speed corresponding to a given radius of rotation to be determined.

Alternatively, the same result may be obtained more simply by setting-off a speed scale along any arbitrarily chosen ordinate. The controlling force is calculated for a constant radius and for different speeds to cover the full working range. The values thus obtained are then set off along the ordinate that corresponds to the chosen radius and marked with the appropriate speeds as shown in Fig. 311. This figure is drawn to scale for the Porter governor, particulars of which are given in the following example.

Example 8. The dimensions of a Porter governor are $w = 15$ lb, $W = 90$ lb, length of each arm $= 12$ in. and all arms are pivoted on the axis of rotation. The extreme radii of rotation are 6 in. and 9 in. Draw the controlling force curve and set off a speed scale along the ordinate corresponding to a radius of 10 in.

The controlling force exerted on each governor ball may be expressed in terms of the weight of the ball and the deadweight on the sleeve. It is equal to the sum of the components of the tensions T_1 and T_2 in the upper and lower arms.

From Fig. 302 (a),

$$F = \left(\frac{W}{2}+w\right)\tan\alpha+\frac{W}{2}\tan\beta$$

$$= \left\{\frac{W}{2}(1+k)+w\right\}\tan\alpha$$

But $\tan\alpha = r/h$ and in the example $\alpha = \beta$,

$$\therefore\ F = (W+w)\frac{r}{h} = (W+w)\frac{r}{\sqrt{(l^2-r^2)}}$$

where l = length of the governor arm.

$\therefore$ substituting the given values:

$$F = 105.r/\sqrt{(144-r^2)}$$

From this equation the following table of values is obtained:

r in.	2	3	4	5	6	7	8	9	10
F lb	17·7	27·1	37·2	48·2	60·6	75·4	93·9	119	158·4

These are plotted to give the controlling force curve of Fig. 311. To set off the speed scale along the ordinate through r = 10 in., we have:

$$F = \frac{w}{g}\omega^2 r = \frac{15}{32{\cdot}2}\cdot\left(\frac{\pi N}{30}\right)^2\cdot\frac{10}{12} = 0{\cdot}004\ 25N^2$$

Corresponding values of F and N are given in the table below:

N r.p.m.	100	150	155	160	165	170	175	180
F lb	42·5	95·6	102·1	108·8	115·7	122·8	130·1	137·7

The speed scale is then marked off as shown on the figure.

The range of equilibrium speeds for the governor is found by drawing lines from the origin through the two points A and B on the controlling force curve that correspond to the extreme radii of rotation. These two lines are shown dotted and they intersect the speed scale at approximately 154 and 176 r.p.m. To avoid confusion radiating lines are drawn for speeds of 100, 150, 160, 170 and 180 r.p.m. only.

Example 9. The following particulars refer to a governor of the type shown in Fig. 305 (c). The weight of each ball is 3 lb and the weight of the sleeve is 15 lb. The arms, a and b, of the bell-crank levers are $4\frac{1}{2}$ in. and 2 in. long respectively, and are at right angles to each other. The extreme radii of rotation are $2\frac{1}{2}$ in. and $4\frac{1}{2}$ in. At the minimum radius the ball centres are vertically below the pivots of the bell-crank levers and the spring load is

34·2 lb. The rate of the spring is 59 lb/in. Draw the controlling force curve and erect a speed scale along the ordinate through $r = 5$ in.

Referring to Fig. 308, it is simpler to take the angle θ as the variable and to determine the controlling force and the radius of rotation of the ball for different values of θ. Let a be the length of AB and b the length of BC.

Taking moments about I, the instantaneous centre of the bell-crank lever, we have:

$$F.\text{AD} = w.\text{DI} + \frac{W+S}{2}.\text{BI}$$

$$\therefore \; F.a\cos\theta = w(a\sin\theta + b\cos\theta) + \frac{W+S}{2} b\cos\theta$$

$$\therefore \; F = w\left(\tan\theta + \frac{b}{a}\right) + \frac{W+S}{2}.\frac{b}{a} \quad . \quad . \quad . \quad (1)$$

But S = spring load on the sleeve
= spring load at minimum radius + rate of spring × sleeve lift
$= 34{\cdot}2 + 59.x = 34{\cdot}2 + 118\sin\theta$

Substituting in equation (1):

$$F = 3\left(\tan\theta + \frac{2}{4{\cdot}5}\right) + \frac{15+34{\cdot}2+118\sin\theta}{2}.\frac{2}{4{\cdot}5}$$

$$= 3\tan\theta + 26{\cdot}2\sin\theta + 12{\cdot}25 \quad . \quad . \quad . \quad . \quad . \quad (2)$$

Also $$r = 2{\cdot}5 + \text{BD} = 2{\cdot}5 + 4{\cdot}5\sin\theta \quad . \quad . \quad . \quad (3)$$

From equations (2) and (3), the following values of F and r are calculated for different values of θ.

$\theta°$	0	5	10	15	20	25	30
F lb	12·25	14·80	17·33	19·84	22·30	24·73	27·08
r in.	2·5	2·89	3·28	3·67	4·04	4·40	4·75

The controlling force curve is plotted in Fig. 312. It will be seen that the curve is very nearly a straight line.

To erect a speed scale along the ordinate through $r = 5$ in., we have:

$$F = \frac{w}{g}\omega^2 r = \frac{3}{32{\cdot}2}.\left(\frac{\pi N}{30}\right)^2.\frac{5}{12} = 0{\cdot}000\,425N^2$$

From this equation corresponding values of F and N are calculated as given in the table below.

N r.p.m.	100	200	240	245	250	255	260	265
F lb	4·25	17·0	24·48	25·51	26·6	27·6	28·7	29·9

These values are used to mark off the speed scale as shown on the figure. The extreme equilibrium speeds are then determined by drawing dotted lines from the origin through the points A and B on the controlling force curve at which the radii are 2·5 in. and 4·5 in. respectively. These intersect the speed scale at 240 r.p.m. and at 257·5 r.p.m. respectively.

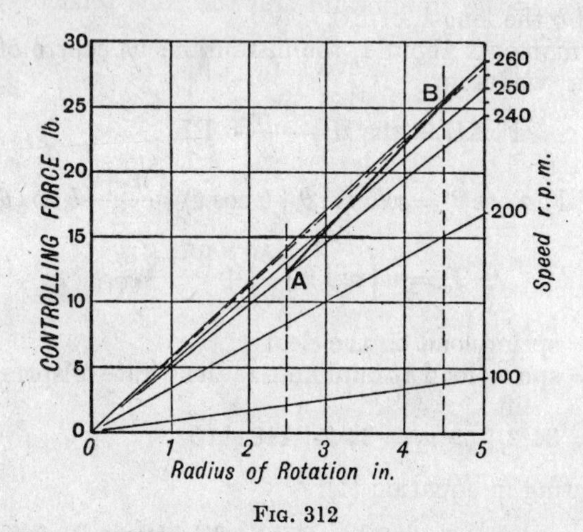

Fig. 312

175. The Stability of Spring-controlled Governors. The controlling force curve for a spring-controlled governor is usually, as in the last example, a straight or an approximately straight line. It has already been shown that for the governor to be stable the ratio F/r must increase as r increases. Hence the controlling force curve when produced must intersect the F axis below the origin, so that the equation to the curve is of the form $F = ar - b$. By increasing the initial tension of the spring, the curve may be raised parallel to itself so as to make b either zero or positive. If b is zero, the controlling force curve will pass through the origin, and the ratio F/r will be constant for all radii, so that the governor becomes isochronous. If, on the other hand, b is positive, then the ratio F/r decreases as r increases, so that the equilibrium speed of the governor decreases with an increase of the radius of rotation of the balls. Such a governor will be unstable, as the following considerations show. When the governor is at rest, the balls are at their minimum distances from the axis of rotation and the sleeve is in its lowest position. As the speed of rotation increases no movement of the ball or sleeve takes place until the equilibrium speed which corresponds to the minimum radius of rotation is reached. The slightest further increase of speed then upsets the equilibrium and immediately causes the governor balls to fly out to their maximum radius. This has the effect of cutting off the supply of energy to the engine and the speed begins to fall. The speed continues to fall until the equilibrium speed which corresponds to the maximum radius of rotation is reached. The slightest further decrease of speed once more disturbs the equilibrium and the balls immediately return to their minimum radius. It follows that the governor sleeve can only occupy one or other of the extreme positions and the control valve on the engine must be either wide open or closed. The conditions are, in fact, similar to those referred to in Article 172 under the definition of *hunting*, but with this difference: with an isochronous governor the slightest change of speed above or below the constant equilibrium speed of the governor causes the governor sleeve to move from the lowest to the highest position, or vice versa, whereas, with the unstable governor, once the governor sleeve has moved from one extreme position to the other, a *finite* change of speed is required in order to cause it to move back again. Hence the degree of hunting with an unstable governor will be much greater than with an isochronous governor.

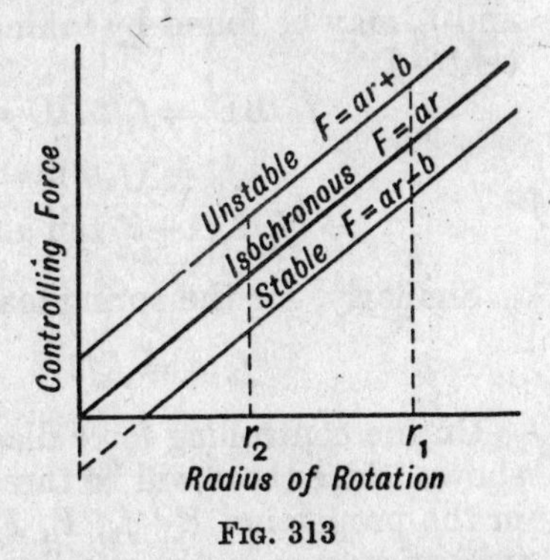

FIG. 313

Controlling force curves for stable, isochronous and unstable governors are shown in Fig. 313.

176. Friction and Insensitiveness. So far the governor has been assumed to be frictionless. In actual fact there is always friction in the joints of the governor and of the mechanism which it operates. Since a friction force always acts in the opposite direction to that of motion, it is clear that friction tends to prevent the upward movement of the sleeve and the outward movement of the balls when the speed of rotation increases, and conversely it tends to prevent the downward movement of the sleeve and the inward movement of the balls when the speed of rotation decreases.

Let f_s = the force required at the sleeve to overcome the friction of the governor and its mechanism,

f_b = the corresponding radial force required at each ball,

W_s = the total load on the sleeve

and F = the controlling force on each ball.

Then, for a given configuration of the governor, the sleeve load will be W_s+f_s if the speed is increasing, and W_s-f_s if the speed is decreasing. In the same way the controlling force will be $F+f_b$ if the speed is increasing, and $F-f_b$ if the speed is decreasing.

For a given type of governor there is a simple relation between f_s and f_b.

Thus, for the Porter governor, Fig. 301, the relation between f_s and f_b may be found by taking moments about the instantaneous centre I:

$$f_b.BD = f_s/2.IC = (f_s/2)(ID+DC)$$

$$f_b = (f_s/2)(\tan\alpha+\tan\beta)$$

$$= (f_s/2)(1+k)\tan\alpha = (f_s/2)(1+k)(r/h) \quad . \quad . \quad (13.31)$$

Similarly, for the spring-loaded governors, Figs. 306, 307, 308,

$$f_b.a = f_s/2.b \quad \therefore \ f_b = f_s.b/2a \quad . \quad . \quad (13.32)$$

On the controlling force diagram the effect of friction is clearly shown, since there will be three curves, the ordinates of which are in the proportion $F+f_b$, F, $F-f_b$. The upper curve is that for increasing speeds, the middle curve is that obtained when friction is neglected, and the lower curve is that for decreasing speeds. These three curves are shown in Fig. 314. For the radius OA the controlling force neglecting friction is AB and the equilibrium

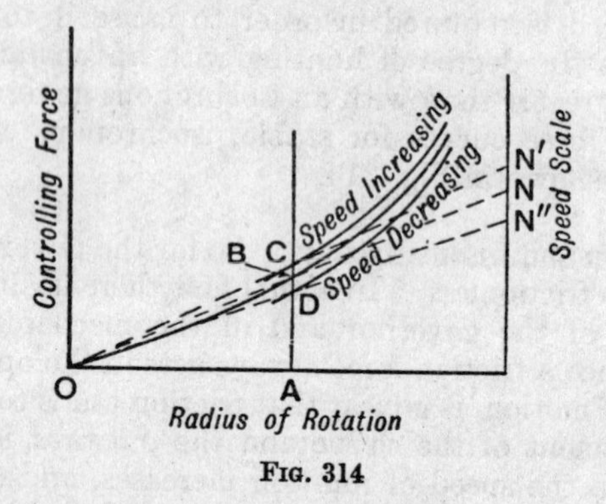

FIG. 314

speed from the speed scale is N. If the speed is increasing, the effect of friction is to increase the controlling force to AC and the speed to N', while, if the speed is decreasing, the controlling force is reduced by friction to AD and the corresponding speed to N''. This means that, when the radius is OA, the speed of rotation may vary between the limits N'' and N' without causing any displacement of the governor sleeve. The governor is said to be insensitive over this range of speed. The ratio $(N'-N'')/N$ is termed the *coefficient of insensitiveness* of the governor. It may be expressed in terms of the friction force f_b and the controlling force F as follows.

Since the controlling force at a given radius is proportional to the square of the speed, we have:

$$F = k.N^2 \quad . \quad . \quad . \quad . \quad (13.33)$$

$$F+f_b = k(N')^2 \quad . \quad . \quad . \quad . \quad (13.34)$$

$$F-f_b = k(N'')^2 \quad . \quad . \quad . \quad . \quad (13.35)$$

Subtracting (13.35) from (13.34):

$$2f_b = k\{(N')^2-(N'')^2\}$$

Dividing by (13.33):

$$\frac{2f_b}{F} = \frac{N'^2-N''^2}{N^2} = \frac{N'-N''}{N} \cdot \frac{N'+N''}{N}$$

But $N'+N''$ is approximately equal to $2N$:

$$\therefore \frac{2f_b}{F} \simeq \frac{N'-N''}{N}.2$$

and the

$$\text{coefficient of insensitiveness} = \frac{N'-N''}{N} \simeq \frac{f_b}{F}$$

Example 10. The friction of the Porter governor, particulars of which are given in Example 8, p. 477, is equivalent to a force of 3 lb at the sleeve. Find the coefficient of insensitiveness at the extreme radii of rotation.

From equation (13.31):

$$f_b = (f_s/2)(1+k)(r/h)$$

But in the present example, since all the arms are of equal length and are pivoted on the axis of rotation, $k = 1$ and the above equation reduces to:

$$f_b = f_s.r/h$$

It was shown in Example 8 that:

$$F = (W+w)r/h$$

$$\therefore \text{coefficient of insensitiveness} = \frac{N'-N''}{N} = \frac{f_b}{F} = \frac{f_s}{W+w}$$

i.e. it is independent of the radius of rotation.

Substituting $f_s = 3$ lb, $W = 90$ lb, $w = 15$ lb,

$$\frac{N'-N''}{N} = \frac{3}{105} = 0{\cdot}029 \quad \text{or} \quad 2{\cdot}9\%$$

Example 11. For the spring-loaded governor of Example 9, p. 478, friction in the governor and mechanism is equivalent to a force of 1·5 lb at the sleeve. Find the coefficient of insensitiveness at the extreme radii of rotation.

The lengths of the arms, a and b, of the bell-crank levers are 4·5 in. and 2 in., and at the extreme radii of 2·5 in. and 4·5 in. the values of the controlling force, read from Fig. 312, are respectively 12·25 lb and 25·4 lb.

From Fig. 308 (a) it is easily seen that the force f_b at the ball centre is equivalent to a force f_s at the sleeve, if:

$$f_b . AD = f_s/2 . BI$$

or

$$f_b = f_s/2 . BI/AD = f_s/2 . b/a$$

$$= 1{\cdot}5/2 . 2/4{\cdot}5 = \tfrac{1}{3} \text{ lb}$$

At the minimum radius:

$$\frac{N' - N''}{N} = \frac{f_b}{F_2} = \frac{1}{3 . 12{\cdot}25} = 0{\cdot}0272 \quad \text{or} \quad 2{\cdot}72\%$$

At the maximum radius:

$$\frac{N' - N''}{N} = \frac{f_b}{F_1} = \frac{1}{3 . 25{\cdot}4} = 0{\cdot}0131 \quad \text{or} \quad 1{\cdot}31\%$$

EXAMPLES XIII

1. Distinguish carefully between the function of (a) the flywheel, (b) the governor of an engine.

2. The arms of a Porter governor are 12·5 in. long and are pivoted on the axis of rotation. Each ball weighs 15 lb and the central load is 90 lb. Find the equilibrium speeds corresponding to radii of 8 in. and 10 in.

3. The arms of a Porter governor are 12 in. long. The upper arms are pivoted on the axis of rotation and the lower arms are attached to the sleeve at distances of 1½ in. from the axis of rotation. The load on the sleeve is 150 lb and each ball weighs 20 lb. Determine the equilibrium speed when the radius of rotation of the balls is 9 in.

4. The arms of a Proell governor are 11 in. long and are pivoted on the axis of rotation. Each ball is carried on an extension, 4 in. long, of the lower arm and weighs 10 lb. The central load on the sleeve is 150 lb. If the ball centres are vertically above the pin-joints connecting the upper and lower arms when the radius of rotation is 7·5 in., calculate the corresponding equilibrium speed.

5. The arms of a Proell governor are 12 in. long; the upper arms are pivoted on the axis of rotation, while the lower arms are pivoted at a radius of 1·5 in. Each ball weighs 10·5 lb. and is attached to an extension, 4 in. long, of the lower arm; the central load is 120 lb. At the minimum radius of 6·5 in. the extensions to which the balls are attached are parallel to the governor axis. Find the equilibrium speeds corresponding to radii of 6·5, 7·5, 8·5 and 9·5 in.

6. A spring-controlled governor has two balls, each weighing 5 lb and each attached to the arm of a bell-crank lever which pivots about a fixed fulcrum. The other arms of the bell-crank levers carry rollers which lift the sleeve against the pressure exerted by a spring surrounding the governor spindle. The two arms of each bell-crank lever are of equal length and the minimum and maximum radii of rotation of the governor balls are 3 in. and $4\frac{1}{2}$ in. If the sleeve is to begin to lift at 240 r.p.m. and the increase of speed allowed is 7%, find the initial load on the sleeve and the required stiffness of the spring. M.U.

7. Each ball of a governor weighs 3 lb and is attached to one arm of a bell-crank lever. The other arms of the bell-crank levers lift the sleeve against the force exerted by a spring under compression, which surrounds the governor spindle. The lengths of the ball and sleeve arms of the bell-crank levers are respectively 5 in. and 3 in. The fulcrum of each bell-crank lever is $3\frac{1}{2}$ in. from the axis of the governor spindle. The maximum and minimum radii of rotation of the governor balls are respectively $4\frac{1}{2}$ in. and 3 in. The sleeve is to begin to lift at a speed of 300 r.p.m. and the maximum speed is to be 6% greater. Find the rate or stiffness of the spring and the equilibrium speed for a radius of $3\frac{1}{2}$ in.

8. In a governor of the type shown in Fig. 308 the weight of each ball is 3 lb; the weight of the sleeve is 15 lb; the lengths of the arms a and b of the bell-crank levers are $4\frac{1}{2}$ in. and $1\frac{1}{2}$ in.; the distance of the fulcrum of each bell-crank lever from the axis of rotation is 2 in.; and the minimum radius of rotation of the governor balls is $2\frac{1}{2}$ in. At this radius the arm b is horizontal. Find the initial thrust in the spring and the rate or stiffness of the spring in order that the sleeve may begin to lift at 240 r.p.m. and may rise 0·3 in. for an increase of speed of 5%.

9. A spring-controlled governor of the Hartnell type with a central spring under compression has balls each of which weighs 5 lb. The ball and sleeve arms of the bell-crank levers are at right angles and are respectively 5 in. and 3 in. long. For the lowest position of the governor sleeve, the radius of rotation of the balls is 4 in. and the ball arms are parallel to the governor axis. Find the initial load on the spring in order that the sleeve may begin to lift at 300 r.p.m. If the stiffness of the spring is 125 lb/in., what is the equilibrium speed corresponding to a sleeve lift of 0·5 in.? M.U.

10. Each ball of a spring-loaded governor is attached to one arm of a bell-crank lever. The other arm of each lever presses against the sleeve and lifts the sleeve against the force exerted by a spring under compression, which surrounds the governor spindle. The pivots of the bell-crank levers are fixed to the spindle at 3 in. radius. The length of the ball arm of each lever is 6 in., the length of the sleeve arm is 3 in. and the two arms are at right angles. The weight of each ball is 5 lb and the stiffness of the spring is 150 lb/in. When the ball arms are parallel to the governor spindle the equilibrium speed is 300 r.p.m. Neglecting friction, find the sleeve lift for an increase of speed of 6%.

11. For a governor of the type shown in Fig. 308 the weight of each ball is 3 lb, the weight of the sleeve 10 lb, the lengths of the arms a and b of the bell-crank levers 5 in. and 2 in., the distance of the fulcrum from the axis of rotation and the minimum radius of rotation of the governor balls are each $2\frac{1}{2}$ in. Find the initial thrust in the spring and the stiffness of the spring in order that the sleeve may begin to rise at 300 r.p.m., and may rise 0·4 in. for an increase of speed of 5%. M.U.

12. In a governor of the type shown in Fig. 308, the two arms of the bell-crank levers are at right angles and their lengths are a, 5 in., b 2·5 in. The distance of each pivot from the axis of rotation is 3·5 in. and the minimum radius of rotation is also 3·5 in. The weight of each ball is 5 lb and the weight of the sleeve is 20 lb. Find the initial thrust in the spring and the stiffness of the spring in order that the sleeve shall begin to lift at 180 r.p.m. and shall lift 0·25 in. for an increase of speed of 8%.

13. Fig. 315 shows a spring-controlled governor in its mid-position. The sleeve has a total travel of 1¼ in. Each revolving mass has a weight of 7 lb and is directly controlled by a spring of stiffness 50 lb/in. compression. In the mid-position the compression of each spring amounts to 2½ in. The weight of the gear operated by the governor—reduced to the sleeve—amounts to 25 lb. Calculate the range of speed and state the sensitiveness of the governor. How does this alter if there is a frictional effect to be taken into account? L.U.

14. The weight of each revolving mass of a spring-controlled governor of the type shown in Fig. 315 is 20 lb and the stiffness of each spring is 150 lb/in. If the length of each spring is 4½ in. when the radius of rotation is 2¾ in. and the equilibrium speed is 360 r.p.m., find the free length of each spring and state whether the governor is isochronous.

15. In a vertical spring-loaded governor, the bell-crank ball levers are pivoted at 6 in. radius. The ball arms are vertical and 6 in. long. The horizontal arms are 3 in. long. Each ball has an effective weight of 8 lb. The balls are connected by springs and the motion of the governor sleeve is transmitted through a lever to a spring S, which has a stiffness of 80 lb/in. of elongation. The length of this lever is 15·75 in. and it is pivoted at a point 7 in. from the governor sleeve.

The governor has a normal speed of 300 r.p.m., the rise in speed at no load is 3% and the sleeve moves 1·25 in. to cut off steam. Determine the necessary stiffness of the ball springs and find what additional extension of spring S will be required to raise the normal speed by 5%. L.U.A.

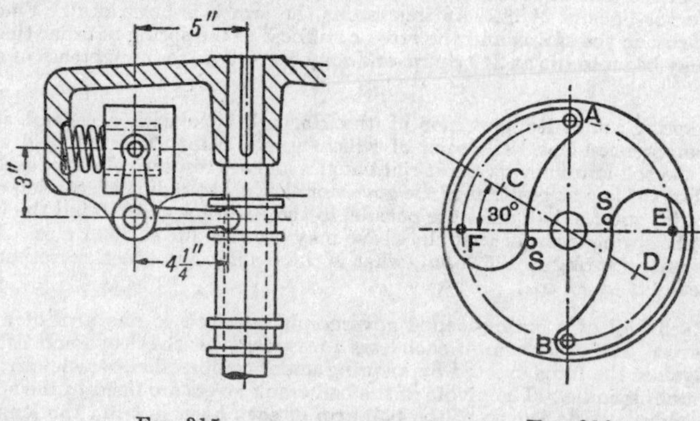

FIG. 315 FIG. 316

16. In a governor of the type shown in Fig. 307, the two springs attached directly to the balls each have a stiffness of 3 lb/in. and a free length of 5 in. The weight of each ball is 8 lb, the length of the ball arm of each bell-crank lever is 4 in. and that of the sleeve arm is 3 in.; the lever ABC is pivoted at its mid-point. When the radius of rotation of the balls is 4 in., the equilibrium speed is 240 r.p.m. If the sleeve is to lift 0·25 in. for an increase of speed of 5%, determine the required stiffness of the auxiliary spring. M.U.

17. Two masses AF and BE, Fig. 316, each weigh 12 lb and are pivoted at A and B to a disc which revolves about a fixed axis. The two masses are connected at E and F by a helical spring, the axis of the spring being at right angles to AB. The centres of gravity of the masses are on the line CD and each is 6 in. from the axis of rotation of the disc. A, B, E and F are all on a circle of radius 8 in. When the disc is at rest, the two masses are pulled by the spring on to the stops S. Determine:

(a) The pull in the spring so that the two masses will just float from the stops when the speed is 300 r.p.m.

(b) The stiffness of the spring so that the masses will revolve with C and D at a radius of 7 in. when the speed is 400 r.p.m. L.U.

18. Define the following terms: controlling force, sensitiveness, stability, isochronism, governor effort.

19. Sketch the controlling force curves for a spring-controlled governor to correspond to (a) a stable governor, (b) an isochronous governor, (c) an unstable governor. In the case of the unstable governor, explain clearly what will happen when the speed of rotation is first increased from zero up to the maximum equilibrium speed and is then diminished below the minimum equilibrium speed.

20. Show that the effect of friction in the governor mechanism is to make the governor insensitive over a certain range of speed at each radius of rotation of the governor balls. Deduce an expression for the coefficient of insensitiveness in terms of the controlling force and the equivalent friction force at each ball.

21. The arms of a Proell governor are 11 in. long. The upper arms are pivoted on the axis of rotation, while the lower arms are pivoted to the sleeve at a distance of 1·5 in. from the axis of rotation. Each ball weighs 12 lb and is carried on an extension, 4 in. long, of the lower arm. The central load is 200 lb. The extensions of the lower arms, to which the balls are attached, are parallel to the governor axis when the radius of rotation is 8 in. Find: (a) the equilibrium speed for the above configuration; (b) the equilibrium speed for a radius of rotation of 9 in.; (c) the mean force exerted at the sleeve during the above change of radius.

22. The weight of each ball of a Proell governor is 12 lb, the central load is 300 lb and the arms are all 10 in. long. The arms are open and are each pivoted at a distance of 2 in. from the axis of rotation. The extension of the lower arm to which each ball is attached is 5 in. long and the radius of rotation of the balls is 9 in. when the arms are inclined at 40° to the axis of rotation. Find: (a) the equilibrium speed for the above configuration; (b) the coefficient of insensitiveness, if the friction of the governor mechanism is equivalent to a force of 4 lb at the sleeve.

23. A Porter governor has all four arms of equal length, 12 in., and pivoted on the axis of rotation. Each ball weighs 16 lb and the weight on the sleeve is 85 lb. The extreme radii of rotation of the governor balls are 7 in. and 9·75 in. Draw the controlling force curve and erect a speed scale along the ordinate corresponding to a radius of 10 in. What are the extreme equilibrium speeds for the governor?

24. The arms of a Proell governor are all of equal length, l, and are pivoted on the axis of rotation. The extensions of the lower arms to which the balls are attached are of length a; each ball weighs w lb and the central load on the sleeve is W lb. At the minimum radius of rotation the extensions a are parallel to the governor axis. Show that the governor will be stable providing that the inclination α of the arms to the governor axis satisfies the relation:

$$\cos\alpha < \tfrac{1}{2}[-\beta\gamma+\sqrt{\{\beta^2\gamma^2-4\beta^2\gamma+4\}}]$$

where $\beta = a/l$ and $\gamma = (W+2w)/(W+w)$

25. A spring-loaded governor has the balls attached to the vertical arms of bell-crank levers, the horizontal arms of which lift the sleeve against the pressure exerted by a spring. The weight of each ball is 6 lb and the lengths of the vertical and horizontal arms of the bell-crank levers are 6 in. and $4\frac{1}{2}$ in. respectively. The extreme radii of rotation of the balls are 4 in. and 6 in. and the governor sleeve begins to lift at 240 r.p.m. and reaches its highest position with a $7\frac{1}{2}$% increase of speed. Determine the required stiffness of the spring and the average force exerted at the sleeve for an increase of speed of 2% above that corresponding to a radius of rotation of 5 in.

26. The controlling force curve of a spring-controlled governor is a straight line. The weight of each governor ball is 9 lb and the extreme radii of rotation are 4·5 in. and 7 in. If the values of the controlling force at the above radii are respectively 46·0 and 82·5 lb and friction of the mechanism is equivalent to 0·5 lb at each ball, find: (a) the extreme equilibrium speeds of the governor; (b) the equilibrium speed and the coefficient of insensitiveness at a radius of 6 in.

27. The controlling force in lb and the radius of rotation in in. for a spring-loaded governor are related by the expression $F = 15{\cdot}5r - 16{\cdot}7$. The weight of each ball is 10 lb and the extreme radii of rotation of the balls are 4 in. and 7 in. Find the maximum and minimum equilibrium speeds. If the friction of the governor mechanism is equivalent to a force of 1 lb at each ball, find the coefficient of insensitiveness of the governor at the extreme radii. M.U.

CHAPTER XIV

BALANCING

177. The high speeds of rotation at which engines and other machines are required to run at the present day have made it increasingly important that all revolving and reciprocating parts should be as completely balanced as possible. Not only are the bearing loads and the stresses in the members increased by the dynamic forces which arise from any lack of balance, but there is also the possibility that unpleasant and even dangerous vibrations may be set up by these forces. In this chapter the balancing of both revolving and reciprocating masses will be considered.

178. Balance of a Single Revolving Mass. There are two different cases to consider, viz: (a) that in which the balance weight may be arranged to revolve in the same plane as the disturbing weight, and (b) that in which the balance weight cannot revolve in the same plane as the disturbing weight.

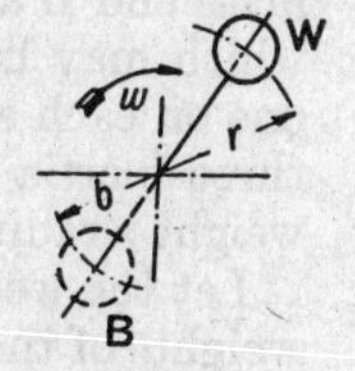

FIG. 317

(a) Referring to Fig. 317, let a mass, of weight W, be attached to a shaft which rotates with angular velocity ω and let r be the distance of the centre of gravity from the axis of rotation. Then a centrifugal force due to the inertia of the mass will act radially outwards and will produce a bending moment on the shaft. To counteract the effect of this inertia force a balance weight may be introduced in the plane of rotation of the disturbing mass, such that the inertia forces of the two masses are equal and opposite.

Let B = weight of the balancing mass and b = distance of the centre of gravity of the balance weight from the axis of rotation.

Then, for balance,

$$(B/g)\omega^2 b = (W/g)\omega^2 r$$

or

$$Bb = Wr \quad . \quad . \quad . \quad . \quad . \quad (14.1)$$

The product Bb may be split up in any convenient way. Generally the radius b is made as large as possible so as to keep down the value of B.

(b) If the balancing mass cannot be introduced in the same plane of rotation as the disturbing mass, it is not sufficient to use a single balancing mass. For, although the two inertia forces are equal in magnitude and opposite in direction, they have different lines of action and therefore give rise to a couple which tends to rock the shaft in its bearings. It is clear that to put the system in perfect balance two balancing masses must be used and the three masses must be so arranged that the resultant dynamic force and couple on the shaft are zero. This requires that the lines of action of the three inertia forces shall be parallel and that the algebraic sum of their moments about any point in the same plane shall be zero.

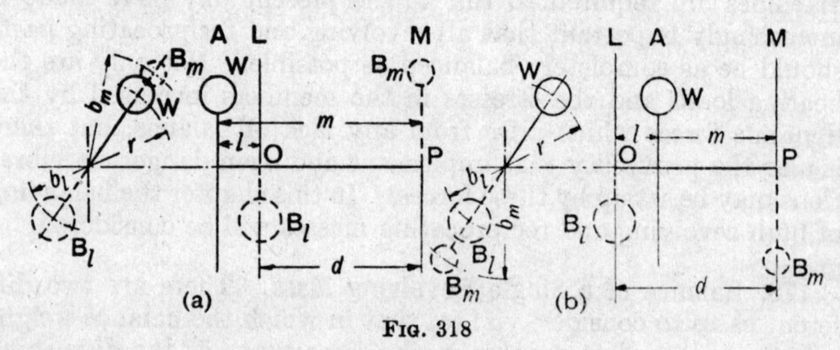

FIG. 318

In Fig. 318, A denotes the plane of rotation of the disturbing mass and L and M the planes of rotation of the balancing masses which may lie both on the same side of A, as in Fig. 318 (a), or one on each side of A, as in Fig. 318 (b). The magnitudes of the three inertia forces are proportional to the respective products, weight × radius, since ω^2/g is the same for each mass.

Let the weight of the disturbing mass in plane A be W and the weights of the balancing masses in planes L and M be respectively B_l and B_m; also let the distances of the centres of gravity from the axis of rotation be respectively r, b_l and b_m, the distances of planes L and M from plane A be respectively l and m and the distance between the planes L and M be d.

Then, taking moments about O, the point of intersection of plane L and the axis of rotation,

$$B_m b_m d = Wrl$$

or

$$B_m b_m = Wrl/d \quad . \quad . \quad . \quad . \quad (14.2)$$

Similarly, taking moments about P, the point of intersection of plane M and the axis of rotation:

$$B_l b_l d = Wrm$$

or

$$B_l b_l = Wrm/d \quad . \quad . \quad . \quad . \quad (14.3)$$

These two equations apply to both cases (a) and (b) of Fig. 318.

Example 1. $W = 200$ lb, $r = 9$ in., $b_l = b_m = 15$ in., $l = 10$ in., $d = 50$ in.

Then for case (a), $m = d+l = 60$ in.; for case (b), $m = d-l = 40$ in.

Substituting in equations (14.2) and (14.3 , the balance weights required are found to be: for case (a), $B_m = 24$ lb, $B_l = 144$ lb, and for case (b), $B_m = 24$ lb, $B_l = 96$ lb.

The relative positions of the disturbing mass and the balance masses are shown in the figure.

The moments of any two of the forces about a point on the line of action of the third force must be equal and opposite. Hence the sense of each of the outer forces must be opposite to that of the middle force.

179. Several Masses revolving in the Same Plane. A number of masses, of weight W_a, W_b, W_c, etc., is shown in Fig. 319 (a). Each mass is rigidly attached to a shaft which revolves about an axis through O perpendicular to the plane of the paper. The radii of rotation of the masses are r_a, r_b, r_c, etc., and their relative angular positions are as shown. When the shaft revolves, a centrifugal force acts on the shaft radially outwards through each

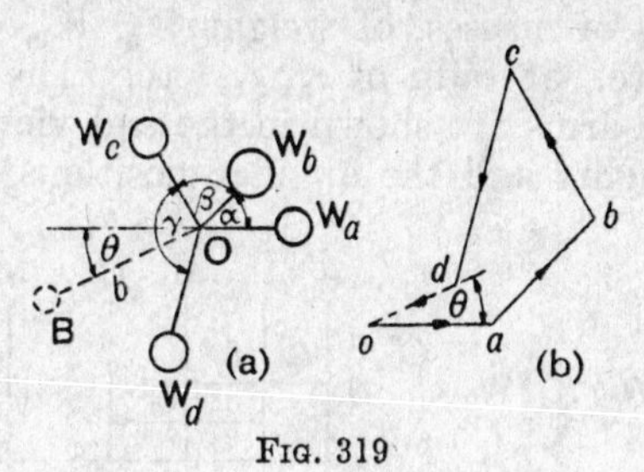

FIG. 319

mass centre. Each of these forces is proportional to the corresponding product, Wr, and their resultant may be found either by drawing the polygon of forces, as shown at (b), or by resolving each force along two mutually perpendicular directions. The product Bb for the balance mass is represented to scale by the closing line do of the force polygon and the direction of the radius of rotation of the balance mass must be parallel to do.

Fig. 319 is drawn to scale to correspond with the following example.

Example 2. The weights of the four masses W_a, W_b, W_c and W_l are respectively 200, 300, 240 and 260 lb, the corresponding radii of rotation are 9, 7, 10 and 12 in. and the angles α, β and γ

are respectively 45°, 75° and 135°. Find the position and magnitude of the balance weight required if the radius of rotation is 24 in.

Resolving each force horizontally and vertically, the horizontal component of the force due to the balance weight is given by:

$$H_b = 1800 + 2100 \cos 45° - 2400 \cos 60° - 3120 \cos 75°$$
$$= 20(90 + 74{\cdot}24 - 60 - 40{\cdot}37) = 20.63{\cdot}87$$

and the vertical component is given by:

$$V_b = 2100 \sin 45° + 2400 \sin 60° - 3120 \sin 75°$$
$$= 20(74{\cdot}24 + 103{\cdot}92 - 150{\cdot}69) = 20.27{\cdot}47$$

$$\therefore \; Bb = 20\sqrt{(63{\cdot}87^2 + 27{\cdot}47^2)} = 1391 \text{ lb in.}$$

and $$B = 1391/24 = 57{\cdot}9 \text{ lb}$$

Also $\tan \theta = V_b/H_b = 27{\cdot}47/63{\cdot}87$, so that $\theta = 23° \; 16'$.

From the force polygon, Fig. 319 (b), do scales 1390 lb in. and θ measures 23·5°.

180. Several Masses revolving in Different Planes. Fig. 320 shows a system of masses, of weight W_a, W_b, etc., revolving in planes A, B, etc., at radii of r_a, r_b, etc. The relative angular positions of the arms are shown in the end view. It is required to find the weights and the angular positions of two balancing

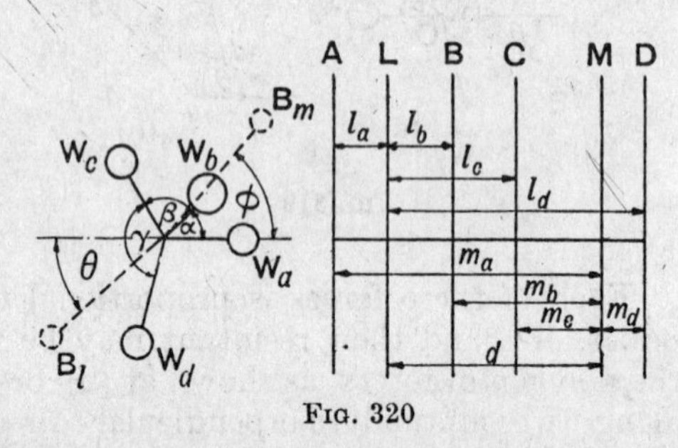

Fig. 320

masses in planes L and M that will give a complete balance of the system.

Two methods of solving this problem will be given. Although at first sight the second method may appear less simple than the first method, it does in fact entail rather less arithmetical work and is to be preferred for solving most problems on balancing.

First Method. This is an extension of that given for a single mass. It consists in finding the balancing masses required in planes L and M for each of the disturbing masses taken separately.

The separate balancing masses in each of the planes L and M are then combined into single resultant balancing masses.

The data may be conveniently set down in a table as shown below. In this table there will be one line for each disturbing mass, but only the line corresponding to the mass in plane A is entered:

Plane	Weight, W	Radius, r	Force $\div \frac{\omega^2}{g}$, Wr	Distance from		Balancing force $\div \frac{\omega^2}{g}$	
				Plane L	Plane M	Plane L	Plane M
A	W_a	r_a	$W_a r_a$	l_a	m_a	$-\frac{W_a r_a m_a}{d}$	$+\frac{W_a r_a l_a}{d}$

The signs in the last two columns are put in by inspection, remembering that, of three parallel forces which are in equilibrium, the middle force is opposite in direction to the two outer forces. For convenience, when the radius of the balancing mass is in the same direction as the radius of the disturbing mass, the positive sign is used, and when in the opposite direction the negative sign is used. The four separate forces in column seven are combined to give the single resultant force for plane L, and, similarly, the four separate forces in column eight are combined to give the single resultant force for plane M.

Example 3. Referring to Fig. 320, the particulars are as given in the first six columns of the following table:

Plane	W, lb	r in.	Force $\div \frac{\omega^2}{g}$, Wr lb in.	Distance from		Balancing force $\div \frac{\omega^2}{g}$	
				Plane L, l	Plane M, m	Plane L, $Wrm \div d$	Plane M, $Wrl \div d$
A	200	9	1800 (15)	12	62	−18·6	+ 3·6
B	300	7	2100 (17·5)	15	35	−12·25	− 5·25
C	240	10	2400 (20)	30	20	− 8·0	−12·0
D	260	12	3120 (26)	60	10	+ 5·2	−31·2

The angles α, β and γ are respectively 45°, 75° and 135° and the distance d between the planes L and M in which the balance weights are to be placed is 50 in.

To simplify the calculations, the actual values of Wr have been divided by 120 to give the figures in brackets in column four. These values have been used in calculating the forces for columns seven and eight. The separate forces for planes L and M are shown to scale at (a) and (b) in Fig. 321 and the resultant forces for planes L and M are found by drawing the force polygons as at (c) and (d). The resultants scaled from the drawings are as follows:

For plane L:

$$B_l b_l = 32{\cdot}1 . 120 = 3850 \text{ lb in.}$$

For plane M:

$$B_m b_m = 21{\cdot}3 . 120 = 2560 \text{ lb in.}$$

If $b_l = b_m = 24$ in., then

$$B_l = 161 \text{ lb} \quad \text{and} \quad B_m = 107 \text{ lb}$$

The angles θ and ϕ, which give the inclinations of the radii to the horizontal, are respectively 40° and 49° and the positions of these two radii in relation to the radii of the disturbing weights are shown dotted in the end view, Fig. 320.

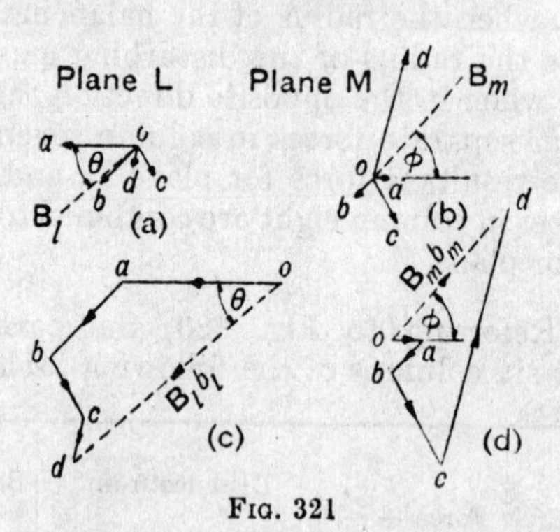

FIG. 321

Second Method. This method is due to the late Professor W. E. Dalby. It is based on the fact that a force acting on a rigid body, which is fixed at one point, is equivalent to an equal and parallel force acting through the fixed point, together with a couple which tends to cause rotation in the plane containing the line of action of the force and the fixed point.

If a number of forces act on a rigid body, each force may be replaced by an equal and parallel force acting through the fixed point, together with a couple. For the rigid body to be in equilibrium, the resultant force at the fixed point must be zero and the resultant couple on the body must also be zero. The construction of the force polygon is quite straightforward, but some difficulty may be found in drawing the couple polygon. According to the usual convention a couple may be represented by a vector at right angles to the plane of the couple. The length of the vector represents to a convenient scale the magnitude of the couple and the arrow-head points in the direction in which a right-handed screw would move if acted upon by the couple. Where the planes of all the couples are at right angles to a given plane, as in the present problem, it is possible to adopt a simpler convention, as the following considerations will show.

Referring to Fig. 322 (a), each of the centrifugal forces due to the masses revolving in planes A, B, C and D is equivalent to an equal and parallel force through point P on the shaft, together with a couple, the magnitude of which is given by the product of the force and its distance from the fixed point P. According to the usual convention, each of these couples may be represented

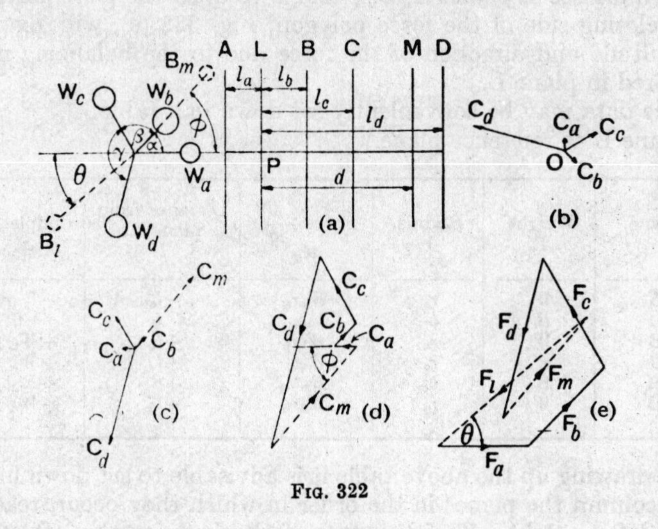

FIG. 322

by the vectors shown in the end view, Fig. 322 (b), i.e. OC_a is the couple vector for the revolving mass in plane A, OC_b the couple vector for the mass in plane B, and so on. It is clear that the planes of these couples are all normal to the plane of the paper in this view and the vectors lie in the plane of the paper and at right angles to the corresponding radii. If the couple vectors are all turned counter-clockwise through 90°, Fig. 322 (c), their relative positions will remain unchanged, but OC_b, OC_c and OC_d will now be parallel to, and in the same direction as the radii r_b, r_c and r_d respectively, while OC_a will be parallel to the radius r_a, but in the opposite direction. Hence the couple vectors may be fixed in their correct relative positions by drawing them radially outwards along the corresponding radii for all masses which lie on one side of the fixed point P and radially inwards along the corresponding radii for all masses which lie on the other side of the fixed point.

It should be noted that the fixed point P is taken as the point of intersection of the plane of rotation of one of the balancing masses, plane L, and the axis of rotation. This is necessary in order to eliminate the couple due to the mass in this plane. The plane L is known as the reference plane. The four known couples due to the masses in planes A, B, C and D may be represented in magnitude and direction by the sides of a polygon and the closing side of the polygon will represent the couple due to the other unknown balancing mass in plane M, Fig. 322 (d). Its magnitude and direction can thus be determined and then used with the known masses in planes A, B, C and D to draw the force polygon. The closing side of the force polygon, Fig. 322 (e), will give the magnitude and direction of the force due to the balancing mass required in plane L.

The data may be conveniently set down in a table.

Plane L as reference plane.

Plane	Weight, W	Radius, r	Force $\div \frac{\omega^2}{g}$, Wr	Distance from plane L, l	Couple $\div \frac{\omega^2}{g}$, Wrl
A	W_a	r_a	$W_a r_a$	$-l_a$	$-W_a r_a l_a$
L	B_l	b_l	$B_l b_l$	0	0
B	W_b	r_b	$W_b r_b$	$+l_b$	$+W_b r_b l$
C	W_c	r_c	$W_c r_c$	$+l_c$	$+W_c r_c l_c$
M	B_m	b_m	$B_m b_m$	$+d$	$+B_m b_m d$
D	W_d	r_d	$W_d r_d$	$+l_d$	$+W_d r_d l_d$

In drawing up the above table it is advisable to set down in the first column the planes in the order in which they occur, reading from left to right. The distances of all planes to the left of the reference plane may be regarded as negative and those to the right as positive. The negative sign indicates that, when drawing the couple polygon, the vector should be set off in the radially inward sense of the corresponding radius.

Example 4. The same particulars are used as for the problem worked out by the first method. From these particulars the following table is filled in.

Plane L as reference plane.

The couple polygon is drawn to scale in Fig. 322 (d). The closing line measures 1060 units,

$$\therefore B_m = 1060/10 = 106{\cdot}0 \text{ lb}; \quad \phi = 49^\circ$$

Plane	Weight, W lb	Radius, r in.	Force $\div \frac{\omega^2}{g}$, Wr	Distance from plane L, l	Couple $\div \frac{\omega^2}{g}$, Wrl
A	200	9	1800 (15)	−12	−180
L	B_l	24	$24B_l$ ($B_l \div 5$)	0	0
B	300	7	2100 (17·5)	+15	+262·5
C	240	10	2400 (20)	+30	+600
M	B_m	24	$24B_m$ ($B_m \div 5$)	+50	$+10B_m$
D	260	12	3120 (26)	+60	+1560

Using this value of B_m, the force polygon, column four in the above table, is shown to scale at (e), Fig. 322. The closing line measures 32·2 units.

$$\therefore\ B_l = 5.32{\cdot}2 = 161 \text{ lb}; \quad \theta = 40°$$

These results are, of course, identical with those obtained by the first method.

181. The Effect on the Engine Frame of the Inertia of a Reciprocating Mass. It was shown in Article 51 that the acceleration of the reciprocating parts of an engine could be expressed with sufficient accuracy for most practical purposes by equation (3.12):

$$f_p = f_c\{\cos\theta + (\cos 2\theta)/n\}$$

where f_p is the acceleration of the piston, f_c the acceleration of the crankpin, θ the inclination of the crank to the i.d.c. and n the ratio of length of connecting rod to length of crank.

If R is the weight of the reciprocating parts, then the force required to accelerate those parts is given by:

$$F = (R/g)f_c\{\cos\theta + (\cos 2\theta)/n\} \quad . \quad . \quad (14.4)$$

This force is provided by the pull of the connecting rod. Referring to Fig. 323 the connecting rod is in tension and the force Q

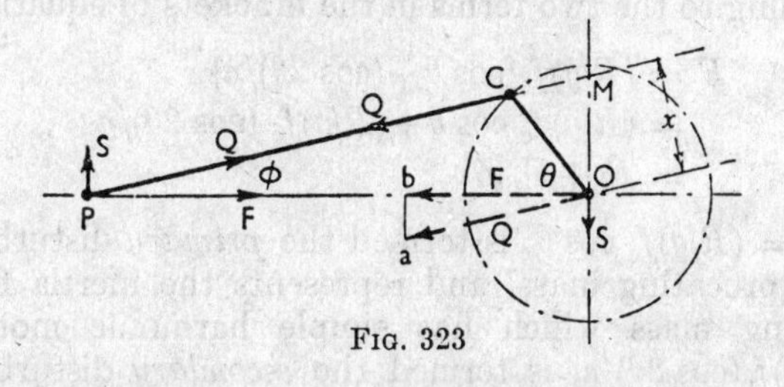

FIG. 323

applied by the connecting rod to the crankpin C is equivalent to an equal and parallel force through O together with the couple

$Q.x$. The couple $Q.x$ tends to retard the rotation of the crank-shaft and its effect is taken into account in finding the net turning moment exerted on the crankshaft. The force at O is transmitted from the crankshaft through the main bearings to the engine frame. Both the force at O and that at P may be resolved parallel and perpendicular to the line of stroke. The horizontal components are equal and opposite, the one acting through P accelerates the reciprocating parts and the other through O is an unbalanced force applied to the frame. The latter tends to cause the frame to slide backwards and forwards on its foundation as the crank rotates. The two vertical components are equal and opposite and constitute a couple applied to the frame which tends to rotate the frame in the clockwise sense.

The magnitude of the couple is $S.\mathrm{OP}$. But the triangles Oba, POM are similar

$$\therefore \mathrm{ba/Ob} = S/F = \mathrm{OM/OP} \quad \text{and} \quad S.\mathrm{OP} = F.\mathrm{OM}$$

Also, $\mathrm{ba/Oa} = S/Q = \mathrm{OM/PM}$

and $x/\mathrm{OM} = \cos\phi = \mathrm{OP/PM}$, so that $\mathrm{OM/PM} = x/\mathrm{OP}$

$$\therefore S/Q = x/\mathrm{OP} \quad \text{and} \quad S.\mathrm{OP} = Q.x$$

$$\therefore S.\mathrm{OP} = F.\mathrm{OM} = Q.x$$

The full effect on the engine frame of the inertia of the reciprocating mass is equivalent to the force F along the line of stroke at O and the clockwise couple of magnitude $S.\mathrm{OP}$.

N.B.—The force applied to the frame at O is equal and opposite to the force required to accelerate the reciprocating mass, i.e. it is equal to the reversed effective, or inertia, force. The couple applied to the frame has the same sense and magnitude as the external couple applied to the crankshaft in order to accelerate the reciprocating mass.

The inertia force may be conveniently divided into two parts corresponding to the two terms in the brackets of equation (14.4).

$$\begin{aligned} F &= (R/g)f_c\{\cos\theta + (\cos 2\theta)/n\} \\ &= (R/g)f_c \cos\theta + (R/g)f_c (\cos 2\theta)/n \\ &= F_p + F_s \end{aligned}$$

where $F_p = (R/g)f_c \cos\theta$, is termed the *primary* disturbing force of the reciprocating mass, and represents the inertia force of a reciprocating mass which has simple harmonic motion, and $F_s = (R/g)f_c(\cos 2\theta)/n$, is termed the *secondary* disturbing force of the reciprocating mass, and represents the correction which is required in order to allow approximately for the effect of the obliquity of the connecting rod.

It will be seen from these expressions that the maximum value of the secondary force is only $1/n$ times the maximum value of the primary force, but that this maximum value occurs four times per revolution of the crank, as compared with twice per revolution of the crank for the maximum primary force.

It is important to note the essential difference between the unbalanced force due to a reciprocating mass and the unbalanced force due to a revolving mass. The former varies in magnitude but is constant in direction, while the latter is constant in magnitude but varies in direction. In general, therefore, a *single* revolving mass cannot be used to balance a reciprocating mass, nor vice versa. There are, however, occasions in which it is desirable to obtain a partial balance of a reciprocating mass by means of a revolving balance weight. Before considering the wider problem of the complete balancing of reciprocating masses, one or two examples of partial balancing will be given.

182. Partial Primary Balance. Referring to Fig. 324, let R be the weight of a reciprocating mass driven from the crank OC, of length r, which turns with uniform angular velocity ω. Then, for the given crank position, the primary disturbing force is given by:

$$F_p = (R/g)\omega^2 r \cos\theta$$

This is clearly equal in magnitude to the component, parallel to the line of stroke, of the centrifugal force produced by an equal mass attached to, and revolving with, the crankpin. Let us suppose that a balance weight B is fixed, as shown, at radius b directly opposite to the crank. Then the component parallel to the line of stroke of the centrifugal force of this mass will be given by $(B/g)\omega^2 b \cos\theta$. This component is opposite in direction to the primary force of the reciprocating parts and the resultant disturbing force parallel to the line of stroke is therefore equal to $(R/g)\omega^2 r \cos\theta - (B/g)\omega^2 b \cos\theta$ or $(Rr - Bb)(\omega^2/g)\cos\theta$.

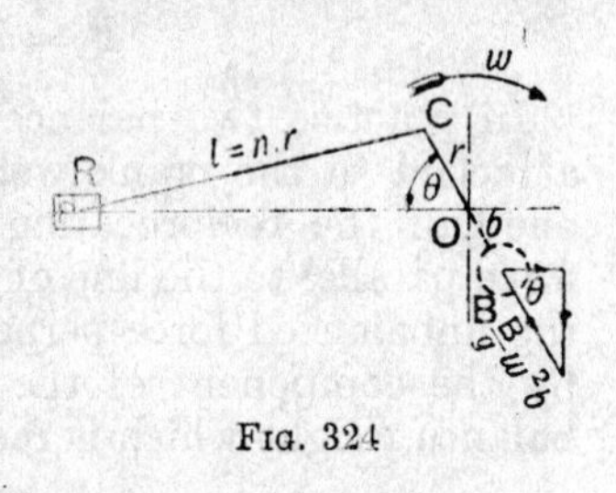

Fig. 324

Hence, if $Bb = Rr$, there is no disturbing force parallel to the line of stroke.

But the centrifugal force of the revolving mass has a component perpendicular to the line of stroke, the magnitude of which is $(B/g)\omega^2 b \sin\theta$. As the crank revolves, this component will go through the same variations of magnitude as the original primary force of the reciprocating parts, but its maximum values will occur when the crank is at right angles to the line of stroke. The introduction of the revolving balance weight has merely served to change the direction of the disturbing force. It is usually preferable to make $Bb = c.Rr$, where $c < 1$. This will have the effect of reducing the unbalanced force parallel to the line of stroke to $(1-c)Rr(\omega^2/g)\cos\theta$ and the unbalanced force at right angles to the line of stroke will be $cRr(\omega^2/g)\sin\theta$. Obviously, the magnitude of the unbalanced force on the engine frame is least when $c = 0{\cdot}5$. But, where a disturbing force parallel to the line of stroke is more harmful than one perpendicular to it, as, for instance, in locomotives, a higher value of c may be used.

If the balance weight B has to balance the revolving parts as well as give a partial balance of the reciprocating parts, then:

$$Bb = Wr + cRr = (W+cR)r$$

Example 5. A single-cylinder oil engine has a stroke of 15 in. and the crank makes 300 r.p.m. The reciprocating parts weigh 150 lb and the revolving parts are equivalent to 180 lb at crank radius. A revolving balance weight is introduced at a radius of 6 in. to balance the whole of the revolving parts and one-half of the reciprocating parts. Find the balance weight required and the residual unbalanced force on the crankshaft.

The total equivalent revolving weight at crank radius, which has to be balanced, is

$$W+cR = 180+0{\cdot}5.150 = 255 \text{ lb}$$

$$\therefore\ Bb = 255.7{\cdot}5$$

and, since $b = 6$ in.,

$$B = 255.7{\cdot}5/6 = 318{\cdot}8 \text{ lb}$$

In practice two balance weights, each of 159·4 lb, would be attached to the crank webs as shown in Fig. 325. Since only one-half the reciprocating parts are balanced, the unbalanced force parallel to the line of stroke is given by $\frac{1}{2}(R/g)\omega^2 r\cos\theta$, and the unbalanced force perpendicular to the line of stroke is given by the component of the centrifugal force of that part of the balance weight which is required for the reciprocating parts, i.e.

$$\tfrac{1}{2}(R/g)\omega^2 r\sin\theta$$

$\therefore$ the resultant unbalanced force

$$= \tfrac{1}{2}(R/g)\omega^2 r\sqrt{(\sin^2\theta+\cos^2\theta)} = \tfrac{1}{2}(R/g)\omega^2 r$$

$$= \frac{1}{2}\cdot\frac{150}{32{\cdot}2}\left(\frac{\pi.300}{30}\right)^2\frac{7{\cdot}5}{12} = 1435 \text{ lb}$$

Referring to Fig. 326, Oa is the primary disturbing force and Ob the centrifugal force due to the revolving balance weight: Oc is the residual unbalanced force parallel to the line of stroke and Oe the unbalanced force at right angles to the line of stroke. The resultant unbalanced force on the engine frame is given by Of, the vector sum of Oc and Oe. Since Od = Oa/2, therefore Oc = Od and angle cOf = θ.

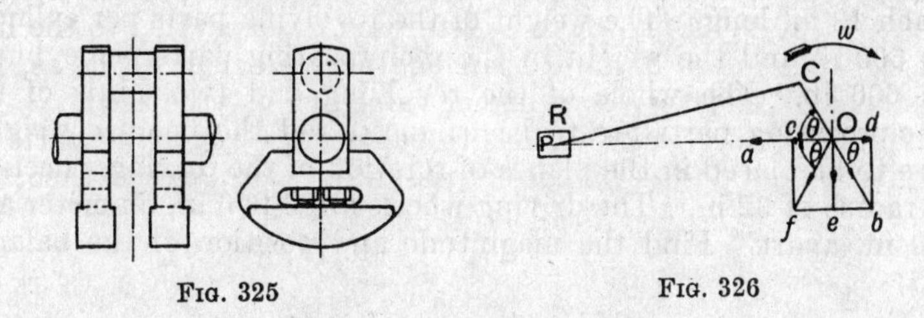

FIG. 325

FIG. 326

Hence, the force remains constant in magnitude as the crank revolves and its line of action is inclined to the i.d.c. at the same angle θ as the crank, but it revolves in the opposite sense to the crank.

183. Partial Balance of Locomotives. In many locomotives there are two cylinders of the same dimensions placed symmetrically either between the frames or outside the frames. The two cranks are invariably at right angles to each other, so that one crank, at least is away from the dead centre, and it is always possible to start the locomotive. The ratio of the length of the connecting rod to the length of the crank is generally large, so that the secondary forces are small. In any case, as we shall see later, since the cranks are at right angles, the secondary force for one set of reciprocating parts is always equal and opposite to that for the other set. The primary disturbing forces not only cause a variation of the tractive effort of the locomotive but, owing to the distance between the cylinder centre lines, they also give rise to a swaying couple. This acts in the plane of the cylinders and has to be resisted by the side pressure between the flanges of the wheel tyres and the inside of the rails. To reduce the magnitude of the swaying couple, revolving balance weights are introduced. But, as shown in the preceding article, the revolving balance weights cause unbalanced forces to act at right angles to the line of stroke. These forces vary the downward pressure of the wheels on the rails and cause oscillation of the locomotive in a vertical plane about a horizontal axis. Since a swaying couple is more harmful than an oscillating couple, it has been the practice to use a value of c from $\frac{2}{3}$ to $\frac{3}{4}$ in two-cylinder locomotives with two pairs of coupled wheels. But in large four-cylinder locomotives with three or more pairs of coupled wheels the value of c is frequently as low as $\frac{2}{5}$.

The following example will show how the unbalanced forces and couples may be calculated for a given engine.

Example 6. A two-cylinder, uncoupled locomotive has inside cylinders 27 in. apart; the cranks are at right angles and are each 13 in. long. The weight of the revolving parts per cylinder is 500 lb and the weight of the reciprocating parts per cylinder is 600 lb. The whole of the revolving and two-thirds of the reciprocating parts are to be balanced and the balance weights are to be placed in the planes of rotation of the driving wheels at a radius of 32 in. The driving wheels are 6 ft 6 in. diameter and 60 in. apart. Find the magnitude and position of the balance

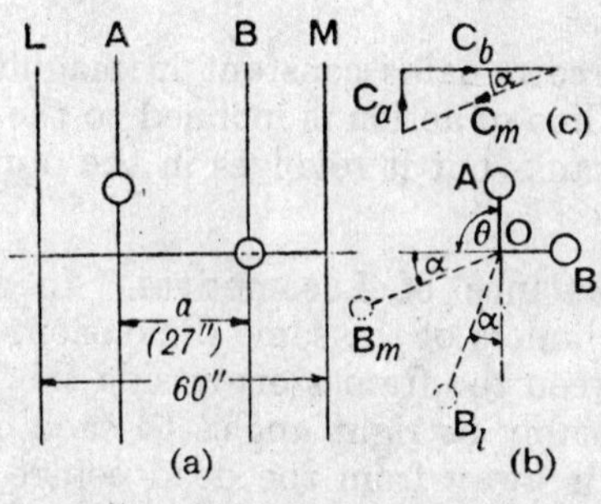

FIG. 327

weights and the maximum unbalanced forces and couples acting on the locomotive when the speed is 60 m.p.h.

The total equivalent weight of the revolving parts per cylinder at crank radius $= 500+\frac{2}{3}.600 = 900$ lb.

Referring to Fig. 327 (a), A and B are the planes of rotation of the cranks and L and M the planes of rotation of the wheels and balance weights. Taking plane L as reference plane, the data may be used to fill in the table below:

Plane	Weight, W	Radius, r	Force $\div \frac{\omega^2}{g}$, Wr	Distance from plane L, l	Couple $\div \frac{\omega^2}{g}$, Wrl
L	B_l	32	$32B_l$	0	0
A	900	13	900.13	16·5	900.13.16·5 (C_a)
B	900	13	900.13	43·5	900.13.43·5 (C_b)
M	B_m	32	$32B_m$	60	32.60.B_m (C_m)

From the last column the couple polygon may be drawn, Fig. 327 (c). The closing line C_m may be scaled, or, by calculation:

$$C_m = \sqrt{(C_a^2+C_b^2)} = 900.13\sqrt{(16{\cdot}5^2+43{\cdot}5^2)}$$
$$\therefore\ 32.60B_m = 900.13.46{\cdot}52$$
$$\therefore\ B_m = 283{\cdot}3 \text{ lb}$$

Also
$$\tan\alpha = C_a/C_b = 16{\cdot}5/43{\cdot}5$$
$$\therefore\ \alpha = 20^\circ\ 46'$$

Having found B_m, the force polygon may be drawn from column four in the table and B_l obtained in magnitude and direction. Since the engine is symmetrical, it is, however, obvious that $B_l = B_m$ and that the radii of the balance weights must be symmetrically placed as shown in the end view, Fig. 327 (b).

To find the unbalanced forces and couples, it is necessary to know what part of each balance weight is required for the reciprocating masses. This is clearly given by:

$$B_r = 400/900.283{\cdot}3 = 126 \text{ lb}$$

(a) *Hammer Blow.* The unbalanced force acting in the plane of each wheel will be the component perpendicular to the line of stroke of the centrifugal force of B_r. This will be a maximum when the c.g. of the balance weight is directly above or below the wheel centre. Its effect will be to cause a variation in the pressure between the wheel and the rail. This variation is shown for one revolution of the wheel in Fig. 328, where L is the static wheel load.

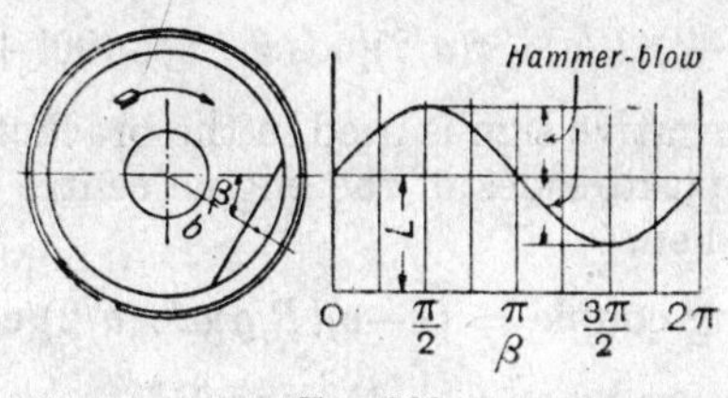

FIG. 328

The maximum variation of wheel load from the mean is $(B_r/g)\omega^2.b$ and is termed the *hammer-blow*. Its magnitude in the present example is:

$$\frac{126}{32{\cdot}2}\cdot\left(\frac{88.12}{39}\right)^2\cdot\frac{32}{12} = 7670 \text{ lb}$$

Since the hammer-blow varies as the square of the speed, it may, at high speeds, be greater than the static wheel load L and the wheel may actually lift from the rail.

(b) *Variation of Tractive Effort.* The variation of tractive effort is caused by the unbalanced part of the primary disturbing force of the reciprocating masses. If the cranks rotate clockwise in Fig. 327 (b) and θ is the inclination of OA to the i.d.c., then $90°+\theta$ is the inclination of OB to the i.d.c.

Then

$$\text{variation of tractive effort} = (1-c)(R/g)\omega^2 r\{\cos\theta+\cos(90°+\theta)\}$$
$$= (1-c)(R/g)\omega^2 r(\cos\theta-\sin\theta)$$

This is a maximum when $\cos\theta-\sin\theta$ is a maximum, i.e. when

$$d/d\theta(\cos\theta-\sin\theta) = 0,$$

or when $$-\sin\theta-\cos\theta = 0,$$

or when $$\tan\theta = -1,$$

i.e. when $$\theta = -45° \quad \text{or} \quad +135°$$

$\therefore$ maximum variation of tractive effort

$$= \pm\sqrt{2}(1-c)(R/g)\omega^2 r$$

$$= \pm\sqrt{2}.\frac{1}{3}.\frac{600}{32{\cdot}2}.\left(\frac{88.12}{39}\right)^2.\frac{13}{12} = \pm 6990 \text{ lb}$$

(c) *Swaying Couple.* The unbalanced parts of the primary disturbing forces cause a horizontal swaying couple to act on the locomotive owing to the distance a between the cylinder centre lines. Taking moments about the engine centre line the resultant unbalanced couple is equal to:

$$(1-c)(R/g)\omega^2 r(a/2)\{\cos\theta-\cos(90°+\theta)\}$$

Note that the negative sign is used in the bracket because the two forces lie on opposite sides of the engine centre line about which moments are taken.

$$\therefore \text{ swaying couple} = (1-c)(R/g)\omega^2 r(a/2)(\cos\theta+\sin\theta)$$

This is a maximum when $\theta = 45°$ or $225°$.

$$\therefore \text{ maximum swaying couple} = \pm\frac{1-c}{\sqrt{2}}.\frac{R}{g}\omega^2 ra$$

$$= \pm\frac{1}{3\sqrt{2}}.\frac{600}{32{\cdot}2}\left(\frac{88.12}{39}\right)^2.\frac{13}{12}.\frac{27}{12}$$

$$= \pm 7870 \text{ lb ft}$$

184. Coupled Locomotives. The example of the last article represents a type of locomotive which is not used at the present day. It is usual to have two or three pairs of wheels coupled together so as to increase the adhesive weight. In such locomotives the coupling rod cranks are set at 180° to the adjacent driving cranks, and in determining the position and magnitude of the balance weights required each coupled axle must be separately considered. For instance, for the driving axle there will be the two sets of coupling rod masses to consider, as well as the two sets of cylinder masses and the balancing masses in the planes of the wheels. In order to reduce the hammer-blow in coupled locomotives, the balance weights required for the reciprocating parts are distributed between the coupled wheels instead of being concentrated in the driving wheels. The effect of the separate balance weights on the engine frame is the same horizontally as that of single balance weights in the driving wheels, but the variation of wheel-load on the driving wheels is reduced, part of the variation being transferred to the coupled wheels.

With this method of balancing the reciprocating parts, the balance weights required for the coupled wheels are found most directly by considering the axles to have imaginary cranks, parallel to the actual cranks on the driving axle, and carrying the appropriate fraction of the reciprocating masses.

185. Primary Balance of Multi-cylinder In-line Engines. The usual arrangement with multi-cylinder engines is to have the cylinder centre lines all in the same plane and on the same side of the crankshaft centre line. This constitutes the *in-line* engine. Other types of engine will be referred to later.

The conditions which must be satisfied in order to give primary balance of the reciprocating parts of a multi-cylinder engine are that the algebraic sum of the forces shall be zero and that the algebraic sum of the couples about any point in the plane of the forces shall also be zero,

i.e. $$\Sigma(R/g)\omega^2 r \cos\theta = 0 \quad . \quad . \quad . \quad (14.5)$$

and $$\Sigma(R/g)\omega^2 ra \cos\theta = 0 \quad . \quad . \quad . \quad (14.5)$$

where a is the distance of the plane of rotation of the crank from a parallel reference plane. It is further necessary that these two equations shall be satisfied for all angular positions of the crankshaft relative to the dead centres. It has already been shown that the disturbing force due to a reciprocating mass is identical with the component parallel to the line of stroke of the centrifugal force produced by an equal mass attached to, and revolving with, the crankpin. Let Fig. 329 be an end view of the crankshaft of a four-cylinder engine with reciprocating masses attached to the cranks OA, OB, etc. Further, let oa, ab, bc, cd be vectors representing the centrifugal forces produced when revolving masses respectively equal to the reciprocating masses are attached to the crankpins. Then it follows that the primary forces for the individual cylinders are equal to the components of oa, etc., along the line of stroke. Hence, if PQ is parallel to the line of stroke, ef, fg, gh and he are the primary forces and, since the algebraic sum of these four forces is zero, the engine is balanced for primary forces when the crankshaft is in the position shown. But now, suppose the crankshaft turns clockwise through an angle γ, then the effect is the same as if the crankshaft remained fixed and the line of stroke turned counter-clockwise through an equal angle γ, as shown by PS. For this position of the line of stroke relative to the crankshaft, the primary forces are represented by kl, lm, mn and nk_1.

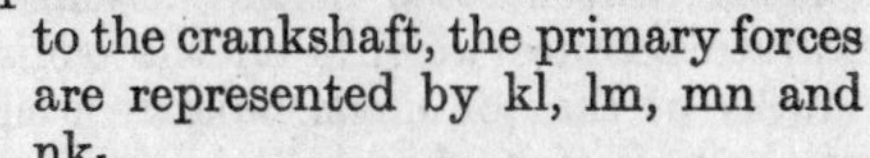

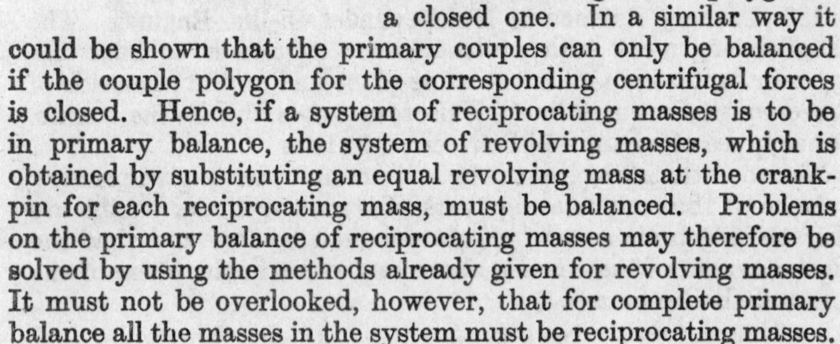

FIG. 329

The algebraic sum of these is not zero, but is equal to kk_1. Clearly, therefore, the primary forces can only balance for this new position of the crankshaft if k_1 coincides with k, i.e. if d coincides with o and the centrifugal force polygon is a closed one. In a similar way it could be shown that the primary couples can only be balanced if the couple polygon for the corresponding centrifugal forces is closed. Hence, if a system of reciprocating masses is to be in primary balance, the system of revolving masses, which is obtained by substituting an equal revolving mass at the crankpin for each reciprocating mass, must be balanced. Problems on the primary balance of reciprocating masses may therefore be solved by using the methods already given for revolving masses. It must not be overlooked, however, that for complete primary balance all the masses in the system must be reciprocating masses.

186. Secondary Balance of Multi-cylinder In-line Engines. The secondary disturbing force due to a reciprocating mass arises from the obliquity of the connecting rod. Its magnitude is $(R/g)\omega^2 r(\cos 2\theta)/n$. Just as it was found convenient to consider the primary force as equal to the component of a centrifugal force, so it is convenient to regard the secondary force as equal to the component of a centrifugal force. Thus the expression for secondary force may be written $(R/g)(2\omega)^2(r/4n)\cos 2\theta$, and this is identical with the component parallel to the line of stroke of the centrifugal force of a mass R, attached to a crank of length $r/4n$, revolving at twice the speed of the actual crank. The two cranks coincide in position at the i.d.c., so that the imaginary secondary crank always makes an angle 2θ with i.d.c.

The conditions for the complete secondary balance of an engine are that:

$$\Sigma(R/g)(2\omega)^2(r/4n)\cos 2\theta = 0 \quad . \quad . \quad (14.7)$$

and

$$\Sigma(R/g)(2\omega)^2(r/4n)a\cos 2\theta = 0 \quad . \quad . \quad (14.8)$$

for all angular positions of the crankshaft relative to the dead centres.

As for primary balance, these conditions can only be satisfied if the force and couple polygons are closed for the corresponding system of revolving masses. This means that, if to each imaginary secondary crank a revolving mass is attached, equal to the corresponding reciprocating mass, then the system thus obtained must be completely balanced. The following examples will serve to illustrate the application of the principles of primary and secondary balance which have been given in the last two articles.

Example 7. Fig. 330 shows diagrammatically a two-cylinder engine with the cranks at 180° and the cylinders on the same side of the crankshaft centre line. Find to what extent the engine is

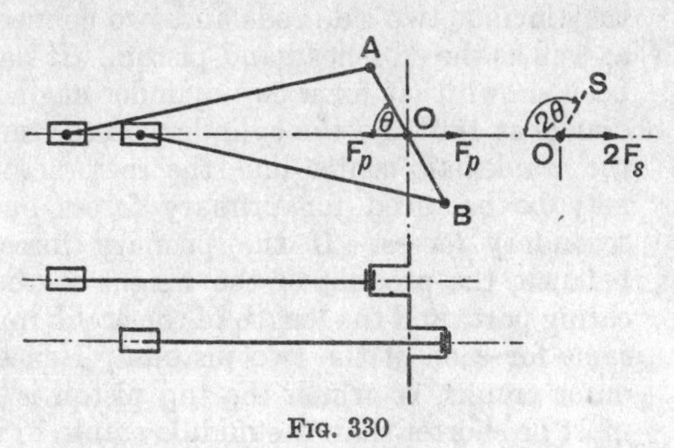

FIG. 330

balanced for primary and secondary forces and couples, if the reciprocating masses, the crank lengths and the connecting rod lengths are identical for each cylinder.

Primary Force. Since the actual cranks are directly opposite to each other, it follows at once that the primary forces are equal and opposite for all positions of the crankshaft relative to the i.d.c. Hence the primary forces are balanced.

Secondary Force. The secondary crank for each cylinder is inclined to the i.d.c. at an angle equal to twice that of the corresponding actual crank. Hence the two secondary cranks are parallel, as shown by the dotted line OS, and the resultant secondary force is equal to the sum of the individual secondary forces.

Primary and Secondary Couples. As regards the couples, it is clear that the two cranks must revolve in different planes, and therefore the primary forces, although equal and opposite, must give rise to an unbalanced couple. Also, since the resultant secondary force is not zero, it will have a moment about any point not on its own line of action, which is, of course, midway between the lines of stroke.

Example 8. *The Opposed Piston Engine.* Fig. 331 shows diagrammatically one cylinder of a type of internal-combustion engine which has been developed both for large slow-speed marine engines and for small high-speed aeroplane and automobile engines. Two pistons reciprocate along the same cylinder but in opposite directions. The front, or bottom, piston operates a central crank through a connecting rod in the usual way. The back, or top, piston operates two outer cranks which are set at 180° to the centre crank. Thus the two pistons move inwards and outwards together. Evidently the reciprocating parts for the top piston will be heavier than those for the bottom piston, since they include two side rods and two connecting rods as well as the crosshead and piston. It has already been shown that for a two-cylinder engine with the cranks at 180° and the cylinders on the same side of the crankshaft centre line, the reciprocating parts may be balanced for primary forces but not for secondary forces. If the primary forces are to balance, the product of the weight of the reciprocating parts and the length of the crank must be the same for each of the two pistons. Hence the two outer cranks, to which the top piston is attached, must be shorter than the middle crank, to which the bottom piston is attached. It should be noticed that for this engine not only is there no resultant primary force, but also there is no resultant primary couple, since the lines of action of the primary forces for the two pistons coincide.

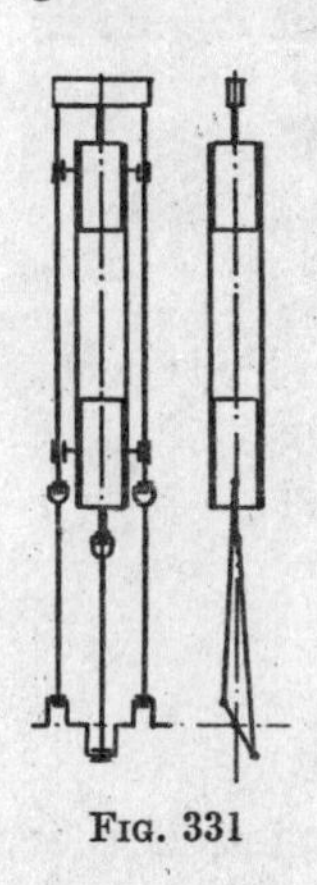

FIG. 331

So far as the secondary forces are concerned, these are unbalanced. Top dead centre is the inner dead centre for the outer cranks as well as for the intermediate crank. The secondary cranks are therefore parallel and the resultant secondary force is the sum of the secondary forces for the two pistons.

In an actual engine of this type the strokes of the top and bottom pistons are respectively 760 mm and 1040 mm and the reciprocating parts weigh respectively 4·84 tons and 3·54 tons. With these values the product Rr is the same for each piston and the engine is in primary balance. The connecting rod lengths are respectively 2080 mm and 3040 mm and the engine speed is 123 r.p.m.

The resultant secondary force is given by:

$$F_s = \frac{R}{g}\omega^2 r\left\{\frac{\cos 2\theta}{n_1}+\frac{\cos 2(\theta+180^\circ)}{n_2}\right\}$$

where n_1, n_2 = connecting rod to crank ratio for the bottom and the top pistons respectively.

$$\therefore\ F_s = \frac{R}{g}\omega^2 r\left(\frac{1}{n_1}+\frac{1}{n_2}\right)\cos 2\theta$$

and, substituting the values given above,

$$F_s = \frac{4\cdot 84}{32\cdot 2}\left(\frac{\pi . 123}{30}\right)^2 \cdot \frac{380}{304\cdot 8}\left(\frac{1}{5\cdot 47}+\frac{1}{5\cdot 85}\right)\cos 2\theta = 11\cdot 0 \cos 2\theta \text{ tons}$$

The maximum value of F_s occurs when the plane of the cranks is either vertical or horizontal.

Example 9. A four-cylinder vertical engine has cranks 1 ft long. The planes of rotation of the first, third and fourth cranks are 2·5, 3·5 and 5·5 ft respectively from that of the second crank and their reciprocating masses weigh 300, 800 and 500 lb respectively. Find the weight of the reciprocating parts for the second cylinder and the relative angular positions of the cranks in order that the engine may be in complete primary balance. If each connecting rod is 4·5 ft long and the speed is 150 r.p.m., find the maximum unbalanced secondary force and couple and the crank positions at which they occur.

Take plane B as reference plane so as to eliminate the couple due to the unknown mass in that plane.

Plane	Weight, R	Radius, r	Force $\div \frac{\omega^2}{g}$, Rr	Distance from plane B, b	Couple $\div \frac{\omega^2}{g}$, Rrb
A	300	1	300	−2·5	−750 (C_a)
B	R_b	1	R_b	0	0
C	800	1	800	+3·5	+2800 (C_c)
D	500	1	500	+5·5	+2750 (C_d)

Since there are three couples of known magnitude (column six), the couple polygon may be drawn, Fig. 332 (a), and the relative directions of the three cranks in planes, A, C and D may be found. Note that since C_a is negative the arrow indicates the radially inward sense of crank OA.

The force polygon (column four) may then be drawn, Fig. 332 (c), and from this may be found the direction of the crank in plane B and the magnitude of R_b. $R_b = 570$ lb and the relative angular positions of the four cranks are shown at (b). The angles α, β and γ are respectively 78·5°, 86° and 30°.

The relative positions of the secondary cranks, Fig. 332 (d), are found from the known condition that each secondary crank must be inclined to the i.d.c. at an angle equal to twice that for the corresponding actual crank. Polygons are then drawn for the forces and couples shown in columns four and six, but with the sides parallel to the corresponding secondary cranks. These two polygons are shown at (e) and (f). The closing line od of the force polygon measures 980 lb ft, and, multiplying this by ω^2/gn,

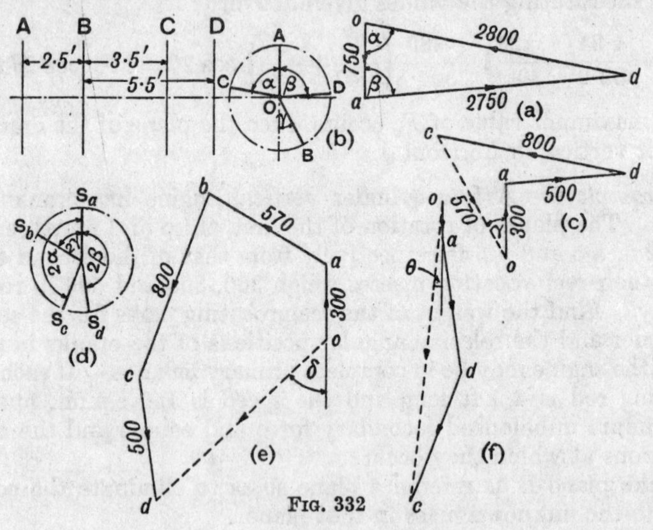

FIG. 332

the actual force is 1670 lb. The unbalanced secondary force is the component of this force parallel to the line of stroke and is given by 1670 cos δ lb. As the crankshaft revolves, the secondary force polygon will revolve and therefore the unbalanced secondary force will be a maximum for the crank positions in which od lies along the line of stroke. Maximum unbalanced force occurs when the crank OA makes an angle $\delta/2 = 24{\cdot}5°$ counter-clockwise with the i.d.c and at successive angular intervals of 90°.

In a similar way the maximum unbalanced secondary couple is given by oc, and scaling it and multiplying by ω^2/gn, this gives 10 370 lb ft. The unbalanced secondary couple is a maximum when the crank OA makes an angle $\theta/2 = 3{\cdot}5°$ counter-clockwise with the i.d.c. and at successive angular intervals of 90°.

It should be noticed that the magnitude of the unbalanced secondary couple depends upon the position of the reference plane. The plane of the crank OB has been taken as reference plane in the above example.

Example 10. The arrangement of the cranks in what is known as the four-cylinder symmetrical engine is shown in Fig. 333. The planes of rotation of the cranks are symmetrically placed with respect to the engine central plane and, as seen in the end view, the line which bisects the angle between cranks 1 and 4 also bisects the angle between cranks 2 and 3. The reciprocating masses attached to cranks 1 and 4 are each equal to R_1 and those attached to cranks 2 and 3 are each equal to R_2. With this arrangement it is possible to obtain a complete balance for primary and secondary forces and also for primary couples, only the secondary couples remaining unbalanced.

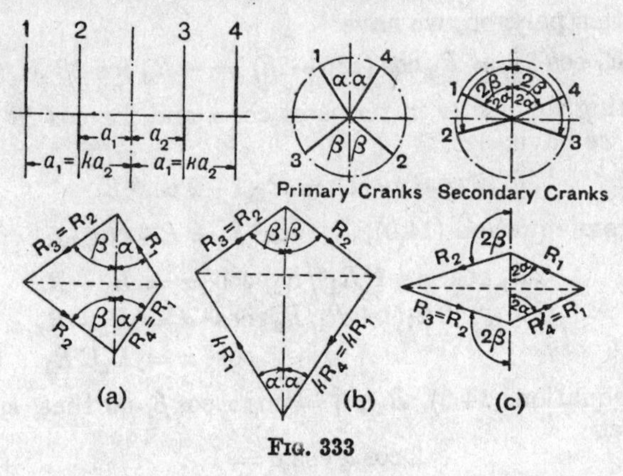

FIG. 333

Let $a_1 = ka_2$, then the angles α and β and the ratio of the reciprocating mass R_2 to the reciprocating mass R_1 may be expressed in terms of k.

For the primary forces and couples to balance, the conditions which have to be satisfied are, as already explained, exactly the same as if the masses were attached to, and revolved with, the crankpins. In other words, the force and couple polygons must be closed figures. The force polygon is obviously as shown at (a) and the couple polygon as shown at (b), Fig. 333. In drawing the couple polygon the central plane of the engine is taken as the reference plane, so that the magnitude of the couple for each extreme crank is proportional to kR_1, while that for each intermediate crank is proportional to R_2; also the vectors which represent the couples for cranks 3 and 4 must be drawn radially inward along the corresponding cranks because the latter lie on the opposite side of the reference plane to cranks 1 and 2.

From the force polygon we have:

$$R_1 \cos \alpha = R_2 \cos \beta \quad . \quad . \quad . \quad (14.9)$$

and from the couple polygon:

$$kR_1 \sin \alpha = R_2 \sin \beta \quad . \quad . \quad . \quad (14.10)$$

Since the secondary forces are also to be balanced, it follows that the system of revolving masses, obtained by attaching to each secondary crank a revolving mass of equal magnitude to the corresponding reciprocating mass, must be balanced. Each imaginary secondary crank is inclined to the i.d.c. at twice the angle of the corresponding actual crank. The relative positions of the secondary cranks are therefore as shown at the extreme right of Fig. 333 and the corresponding force polygon is shown at (c).

From this polygon, we have:

$$R_1 \cos 2\alpha = R_2 \cos (180-2\beta) = -R_2 \cos 2\beta \quad . \quad (14.11)$$

Substituting for $\cos 2\alpha$ in terms of $\cos \alpha$ and for $\cos 2\beta$ in terms of $\cos \beta$, we have:

$$R_1(2 \cos^2 \alpha - 1) = R_2(1 - 2 \cos^2 \beta)$$

But, from equation (14.9), $R_2{}^2 \cos^2 \beta = R_1{}^2 \cos {}^2\alpha$,

$$\therefore \; 2R_1 \cos^2 \alpha + 2(R_1{}^2/R_2) \cos^2 \alpha = R_1 + R_2$$

$$\therefore \; 2R_1(1 + R_1/R_2) \cos^2 \alpha = R_1 + R_2$$

$$\cos^2 \alpha = R_2/2R_1 \quad . \quad (14.12)$$

From equation (14.9), $R_2/R_1 = \cos \alpha/\cos \beta$, so that, substituting, we get

$$2 \cos \alpha \cos \beta = 1 \quad . \quad . \quad . \quad . \quad (14.13)$$

Squaring both sides of equation (14.10),

$$k^2 R_1{}^2 \sin^2 \alpha = R_2{}^2 \sin^2 \beta$$

$$\therefore \; k^2(1 - \cos^2 \alpha) = \left(\frac{R_2}{R_1}\right)^2 (1 - \cos^2 \beta)$$

and, substituting for $\cos \beta$ from equation (14.13) and for R_2/R_1 from equation (14.9),

$$k^2(1 - \cos^2 \alpha) = 4 \cos^4 \alpha \left(1 - \frac{1}{4 \cos^2 \alpha}\right)$$

$$\therefore \; 4 \cos^4 \alpha - (1 - k^2) \cos^2 \alpha - k^2 = 0$$

$$\therefore \; \cos^2 \alpha = \frac{1-k^2}{8} \pm \sqrt{\left\{\left(\frac{1-k^2}{8}\right)^2 + \frac{k^2}{4}\right\}}$$

$$= (1/8)[\sqrt{\{(1-k^2)^2 + 16k^2\}} - (k^2 - 1)] \quad (14.14)$$

The last step follows because $\cos^2 \alpha$ must be positive and by hypothesis k must be greater than unity.

Given the value of k, α may be found from equation (14.14), β from equation (14.13) and R_2/R_1 from equation (14.12).

Thus, let $k = 2{\cdot}0$.

Then: $\cos^2 \alpha = (1/8)\{\sqrt{(9+64)}-3\} = (1/8)5{\cdot}5440 = 0{\cdot}6930$

$\therefore \cos \alpha = 0{\cdot}8325$ and $\alpha = 33° \ 39'$

From equation (14.13), $\cos \beta = 1/2 \cos \alpha = 0{\cdot}6006$ and $\beta = 53° \ 5'$. Also, from equation (14.12), $R_2/R_1 = 2 \cos^2 \alpha = 2.0{\cdot}6930 = 1{\cdot}386$.

187. The Balancing of In-line Engines with Identical Reciprocating Parts for each Cylinder. Multi-cylinder in-line internal combustion engines are very widely used and are capable of high speeds of rotation. In these engines the reciprocating parts are identical for each cylinder and the cranks are arranged as far as possible to provide uniform firing intervals and balance of the reciprocating parts.

Let N_c = no. of cylinders and α = angular spacing of the cranks round the shaft.

Then for a four-stroke cycle engine, in which two revolutions of the crankshaft are required in order to complete the cycle, the firing intervals will be uniform if $\alpha = 4\pi/N_c$.

The inertia force due to the reciprocating parts of one cylinder may be expressed as a Fourier series, equation (3.19):

$$F = (R/g)\omega^2 r(\cos \theta + A_1 \cos 2\theta + B_1 \cos 4\theta + \ldots) \quad (14.15)$$

The inertia forces for the other cylinders may be expressed by similar series in which θ is replaced by:

$$\theta+\alpha, \quad \theta+2\alpha \ldots \theta+\overline{N_c-1}.\alpha$$

The resultant inertia force for the engine is therefore given by the sum of a number of cosine series conforming to the general equation:

$$K_m\{\cos m\theta + \cos m(\theta+\alpha) + \ldots + \cos m(\theta+\overline{N_c-1}.\alpha)\} \quad (14.16)$$

where m has values of 1, 2, 4, etc., and $K_m = (R/g)\omega^2 r$ multiplied by the appropriate constant A_1, B_1, etc., in equation (3.19).

The sum of a cosine series in which the angles are in arithmetical progression is given in text-books on trigonometry. For the above series it may be written:

$$S = K_m \cos m\left(\theta + \frac{N_c - 1}{2}\alpha\right)\frac{\sin mN_c\alpha/2}{\sin m\alpha/2} \quad . \quad (14.17)$$

But for equal firing intervals $N_c\alpha = 4\pi$, so that $mN_c\alpha/2 = 2\pi m$, and, since m is an integer, $\sin 2\pi m = 0$. Hence $S = 0$, except when the denominator $\sin m\alpha/2 = 0$. This will occur when $m\alpha$ is a multiple of 2π, and in these circumstances each term in the cosine series will have exactly the same value, so that:

$$S = K_m N_c \cos m\theta \quad . \quad . \quad . \quad . \quad (14.18)$$

For $m\alpha$ to be a multiple of 2π, m must be a multiple of $N_c/2$.

Hence for a multi-cylinder in-line four-stroke cycle engine with identical cylinders and the cranks spaced so as to give uniform firing intervals, all harmonics will be balanced except those in which m is a multiple of half the number of cylinders.

So far only the harmonic forces have been considered and no reference has been made to the couples. For the couples to be balanced the engine must be symmetrical about a plane normal to the axis of the crankshaft. This means that the cranks must be arranged in pairs, which are parallel to each other and which are situated at equal distances from the central plane. It will, therefore, only be possible for the couples to be balanced in an engine of the type under consideration if there is an even number of cylinders, so that each half of the crankshaft is a mirror-image of the other half. Thus, for a four-cylinder, four-stroke cycle engine, the firing intervals will be uniform if $\alpha = 4\pi/N_c = \pi$, i.e. if the cranks are arranged in two pairs set at 180° to each other. The unbalanced harmonic components of the inertia force due to the reciprocating parts are those for which m is a multiple of $N_c/2$, i.e. a multiple of 2. Hence only the primary forces are balanced, the secondary and all higher harmonic forces being unbalanced. For the primary couples to be balanced the crankshaft must be symmetrical, so that cranks 1 and 4 must be parallel and opposite to cranks 2 and 3; also, the distance between the planes of rotation of cranks 1 and 2 must be the same as that between the planes of rotation of cranks 3 and 4 (Fig. 334 (a)).

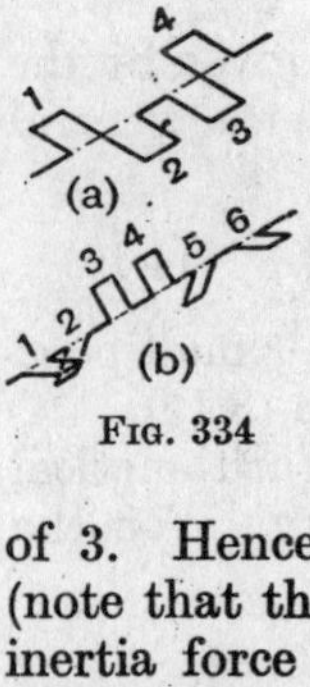

Fig. 334

As a further example, consider a six-cylinder in-line engine. For uniform firing intervals $\alpha = 4\pi/N_c = 2\pi/3$. The cranks must therefore be arranged in three pairs with an angular spacing of 120°. Those harmonic forces are unbalanced for which m is a multiple of $N_c/2$, i.e. a multiple of 3. Hence the sixth, twelfth, etc., harmonics are unbalanced (note that the only odd value of m in the Fourier series for the inertia force is unity) and the primary, secondary and fourth-order forces are balanced in this engine. For the corresponding couples to be balanced, the crankshaft must be symmetrical, so that cranks 1 and 6, 2 and 5, and 3 and 4 will be parallel, and the planes of rotation will be symmetrically spaced with respect to the engine centre line (Fig. 334 (b)).

For a multi-cylinder in-line two-stroke cycle engine with identical cylinders and the cranks so spaced as to give uniform firing intervals, $N_c\alpha = 2\pi$. Hence in equation (14.17)

$$\sin mN_c\alpha/2 = \sin m\pi = 0 \quad \text{and} \quad \sin m\alpha/2 = \sin m\pi/N_c$$

which will also be 0 when m is an integral multiple of N_c.

All harmonic forces will therefore be balanced except those for which m is a multiple of N_c, but even when there is an even number of cylinders it is impossible for one-half of the crankshaft to be a mirror image of the other half, so that there will always be couple unbalance of some harmonics. The choice of firing order will determine the best arrangement.

188. Direct and Reverse Cranks. Each term in the Fourier series for the inertia force of a reciprocating mass conforms to the general expression $K_m \cos m\theta$ (equation (14.16)). Let us suppose that two masses, each of which gives rise to a centrifugal force $K_m/2$, revolve in opposite directions at m times the speed of the crankshaft and that when $\theta = 0°$ the two masses lie on the i.d.c. Then, at every instant, the components of the centrifugal forces of these two masses normal to the line of stroke will be equal and opposite, while the sum of the components parallel to the line of

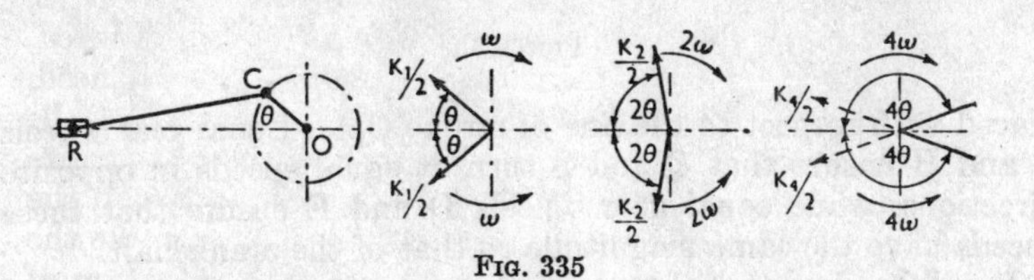

FIG. 335

stroke will be equal to $K_m \cos m\theta$. Hence the resultant disturbing force due to the two revolving masses will be identical with the mth harmonic of the disturbing force due to the reciprocating mass.

For instance, in Fig. 335, the positions of the direct and reverse cranks of the first, second and fourth harmonics are shown, although the forces $K_1/2$, $K_2/2$, and $K_4/2$ are not to scale. Note that K_4 is negative, since the coefficient B_1 in equation (14.15) is negative. The forces $K_4/2$ must therefore act in the opposite directions to the fourth harmonic direct and reverse cranks. The substitution in this way of two revolving masses for a single reciprocating mass simplifies the problem of finding to what extent the different harmonic forces are balanced in certain types of engines. It also helps one to see to what extent any unbalanced harmonics could be balanced, either partially or completely, by means of revolving balance weights.

Example 11. Show how the reciprocating parts of a single cylinder engine may be completely balanced, so far as primary and secondary effects are concerned, by means of revolving balance weights.

The primary effect of the reciprocating mass is equivalent to that of two revolving masses attached to the primary direct and reverse cranks. Each revolving mass is one-half the reciprocating mass, and the direct and reverse cranks turn at equal speeds in opposite senses. It is therefore possible to have revolving balance weights attached to shafts geared to the crankshaft, as shown in Fig. 336. The two shafts Q and S are symmetrically

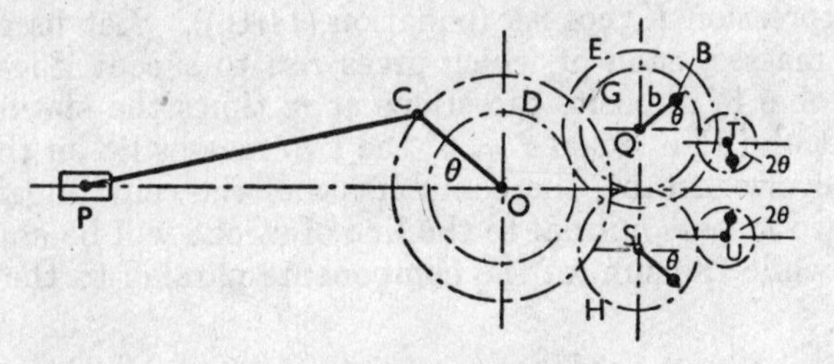

FIG. 336

placed with respect to the line of stroke OP. Equal gear wheels G and H ensure that Q and S turn at equal speeds in opposite directions, while equal gear wheels D and E ensure that these speeds have the same magnitude as that of the crankshaft.

It will be clear that S turns in the same sense as the crank and Q turns in the opposite sense. Further, if the two balance weights are of magnitude B and revolve at radius b, the resultant centrifugal force will act along the line of stroke towards the right, i.e. in the opposite sense to the primary disturbing force of the reciprocating mass. Hence, if

$$2(B/g)b\omega^2 \cos\theta = (R/g)r\omega^2 \cos\theta$$

i.e., if

$$Bb = Rr/2$$

there will be no primary disturbing force on the engine frame. The balance weights must, of course, revolve in the same plane as the crank OC, if there is to be no unbalanced couple.

In a similar way, revolving balance weights could be introduced on shafts T, U, which are geared to run at equal and opposite speeds twice that of the crank. If the balance weights B_s at radius b_s are placed as shown and satisfy the following equation, their resultant centrifugal force will be equal and opposite to the secondary effect of the reciprocating mass for all values of θ.

$$2(B_s b_s/g)(2\omega)^2 \cos 2\theta = A_1(R/g)\omega^2 r \cos 2\theta$$

$$\therefore\ B_s b_s = A_1 Rr/8$$

where A_1 is the coefficient of the second term of equation (14.15) and is approximately equal to the ratio of the crank length to the connecting rod length.

It is not suggested that the above is a practicable method of balancing the disturbing force on the frame caused by the reciprocating mass. The cost and complication would not be justified for a single-cylinder engine, and a partial balance would generally be adopted, see Example 5, p. 500. But the method has been used to balance the secondary effects of the reciprocating masses in a four-cylinder petrol engine, which is inherently balanced for primary effects.

Example 12. A three-cylinder radial engine has the cylinders spaced at angular intervals of 120°. The three connecting rods are coupled directly to a single crank. The stroke is 5 in., the length of each connecting rod is 9 in. and the weight of the

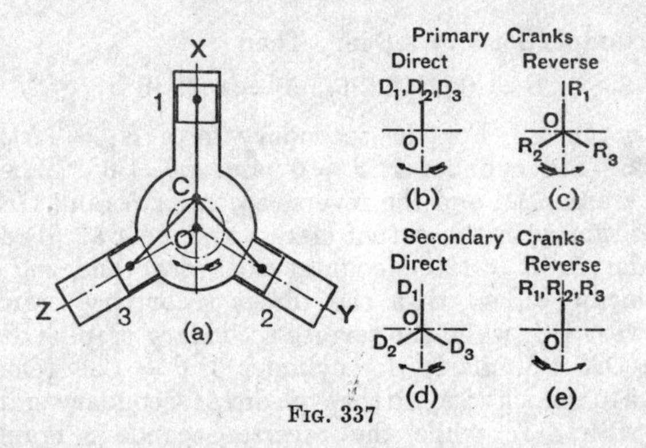

FIG. 337

reciprocating parts per cylinder is 4 lb. Find the resultant primary and secondary forces at a crankshaft speed of 1500 r.p.m.

Primary Force. For the primary force $K = (R/g)\omega^2 r$ and $m = 1$. Let OX, OY and OZ be the three cylinder centre lines, Fig. 337 (a). For convenience let the crank be on the i.d.c. of cylinder I, i.e. along OX. Then, for cylinder 1, $\theta = 0°$ and the direct crank OD_1 and the reverse crank OR_1 coincide in position with the actual crank, as shown at (b) and (c). For cylinder 2, $\theta = 120°$ (counter-clockwise) and, since $m = 1$ for the primary force, the direct crank OD_2 also coincides with the actual crank, while the reverse crank OR_2 makes an angle 120° (clockwise) with the i.d.c. OY, and therefore coincides with OZ. Similarly, for cylinder 3, $\theta = 120°$ (clockwise) and the direct crank OD_3 coincides with the actual crank, while the reverse crank OR_3 makes an angle 120° (counter-clockwise) with the i.d.c. OZ and therefore coincides with OY.

The resultant primary force for the engine is therefore equivalent to three revolving masses coinciding with, and revolving with, the actual crank as shown at (b), together with three revolving masses spaced at intervals of 120° and revolving in the opposite sense to the crank as shown at (c). Obviously the system of reverse cranks forms a balanced system, so that the resultant primary force is equivalent to the combined centrifugal force of the direct cranks, i.e. $(3/2)K_1 = (3/2)(R/g)\omega^2 r$. In other words, the resultant primary force is equal to the centrifugal force of a mass attached to, and revolving with, the crankpin and equal to one-half the total reciprocating mass for the three cylinders. This may be balanced by a mass attached to the crankshaft at 180° to the crank, such that $B.b = (3/2)R.r$, where B and b are respectively the weight of the balancing mass and the radius at which it revolves.

In the example, let $b = 3{\cdot}5$ in. Then

$$B = (3/2)(4.2{\cdot}5/3{\cdot}5) = 4{\cdot}29 \text{ lb}$$

Secondary Force. For the secondary force $K_m = (R/g)\omega^2 rA_1$ and $m = 2$. For cylinder 1, $\theta = 0°$ and $m\theta = 0°$. The direct secondary crank OD_1 and the reverse secondary crank OR_1 coincide in position with the actual crank, as shown at (d) and (e). For cylinder 2 $\theta = 120°$ (counter-clockwise) and $m\theta = 240°$ (counter-clockwise), so that the direct secondary crank OD_2 coincides with OZ, while the reverse secondary crank OR_2 coincides with OX. Similarly, for cylinder 3, $\theta = 120°$ (clockwise) and $m\theta = 240°$ (clockwise), so that the direct secondary crank OD_3 coincides with OY, while the reverse secondary crank OR_3 coincides with OX.

The resultant secondary force for the engine is equivalent to the two systems of revolving masses shown at (d) and (e). Obviously the direct cranks form a balanced system of revolving masses and therefore the resultant secondary force is equivalent to a force $(3/2)(R/g)\omega^2 rA_1$, which revolves in the opposite sense to the crank, and at twice the speed of the crank. It coincides in direction with the crank when the crank is on any of the three inner dead centres.

In the example, $R = 4$ lb, $\omega = \pi . 1500/30 = 50\pi$, $r = 2{\cdot}5$ in. and $A_1 \simeq 1/n = 2{\cdot}5/9 = 1/3{\cdot}6$.

$$\therefore \text{ resultant secondary force} = \frac{3}{2} \cdot \frac{4}{32{\cdot}2} \cdot \frac{(50\pi)^2}{3{\cdot}6} \cdot \frac{2{\cdot}5}{12}$$

$$= 266 \text{ lb}$$

Example 13. As a further illustration of the application of direct and reverse cranks, consider the broad-arrow, or W, engine in which there are three rows of cylinders. An engine of this type has four cylinders in each row and the crankshaft is of the normal "flat" type with one connecting rod from each row coupled directly to each crankpin. The middle row of cylinders is vertical and the other two rows are inclined at 60° to the vertical. The weight of the reciprocating parts is 6 lb per cylinder, the cranks are 3 in. long, the connecting rods 11 in. long and the r.p.m. 2000. Find the maximum and minimum values of the secondary disturbing force on the engine.

The first step in a problem of this kind is to find the resultant secondary force for each row of cylinders and then to combine these by substituting the corresponding direct and reverse cranks. Each row forms a four-cylinder in-line engine and, as we have already seen in Article 187, only the first harmonic or primary force is balanced. The unbalanced secondary force for each row is four times the secondary force for one cylinder and is given by $4K_2 \cos 2\theta$, where $K_2 = (R/g)\omega^2(r/n)$ and θ is the inclination of the crank to the i.d.c. for that row. Hence, so far as the secondary forces are concerned, each row of four cylinders may be replaced by a single cylinder which has a reciprocating mass equal to the total reciprocating mass for the four cylinders. The direct and reverse cranks for any one of the cylinders of the equivalent three-cylinder engine will each give rise to a centrifugal force of magnitude $2K_2$.

For convenience take the crank on the i.d.c. of the middle row 2. Then, for row 2, $\theta = 0°$, and therefore the direct secondary crank OD_2 and the reverse secondary crank OR_2 coincide with OY, as shown in Fig. 338 (b) and (c). For row 1, $\theta = 60°$ (clockwise). Therefore $2\theta = 120°$ (clockwise) and OD_1 coincides with OZ, while OR_1 is inclinded at 120° (counter-clockwise) to OX and therefore lies along the o.d.c. for row 2. Similarly, for row 3, $\theta = 60°$ (counter-clockwise). Therefore $2\theta = 120°$ (counter-clockwise) and OD_3 coincides with OX, while OR_3 is inclined at 120° (clockwise) to OZ and therefore coincides in direction with OR_1. The resultant secondary disturbing force is equivalent to the two systems of revolving masses shown at (b) and (c). The three direct cranks may be combined into the single resultant OD and the three reverse cranks into the single resultant OR, Fig. 338 (d), where OD represents $4K_2$, i.e. twice the centrifugal force for one direct crank of the equivalent three-cylinder engine, and OR represents $2K_2$, i.e. the centrifugal force for one reverse crank of the equivalent three-cylinder engine. The net unbalanced secondary force for the given position of the engine crankshaft is therefore OD−OR or $2K_2$. But the direct and reverse secondary cranks turn at twice the speed of the crankshaft, so that when the actual crank has turned through 45° from the i.d.c. OY, the resultant direct crank OD and the resultant reverse crank OR will both lie along the horizontal OH. The net unbalanced secondary force will then be given by OD+OR or $6K_2$.

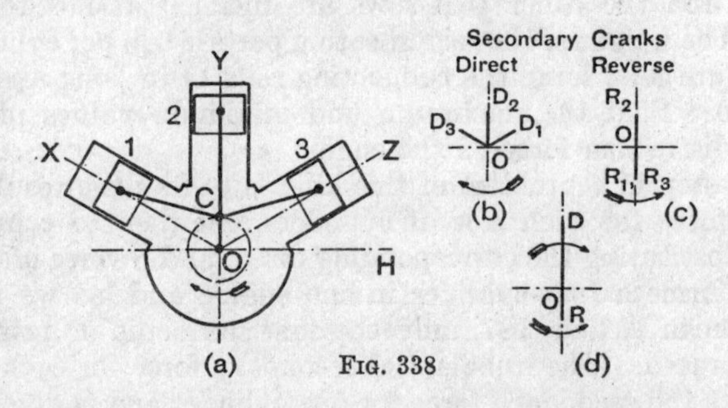

FIG. 338

It is clear that, if the actual crank turns through 90° from OY, the resultant secondary force acts vertically downwards and is again equal to OD−OR or $2K_2$. Hence the resultant unbalanced secondary force is a minimum when the plane of the engine cranks is either vertical or horizontal, and is a maximum when the plane of the engine cranks is inclined at 45° to the vertical or horizontal.

$$\text{Minimum secondary force} = 2\frac{R}{g}\omega^2\frac{r}{n} = \frac{2.6}{32{\cdot}2}\cdot\left(\frac{\pi.200}{3}\right)^2\frac{3}{12}\cdot\frac{3}{11}$$

$$= 1114 \text{ lb}$$

It acts upwards when the plane of the engine cranks is vertical and downwards when the plane of the engine cranks is horizontal.

Maximum secondary force $= 6(R/g)\omega^2(r/n) = 3342$ lb

It acts towards the right when the plane of the engine cranks is inclined 45° clockwise to the vertical and towards the left when the plane of the engine cranks is inclined 45° counter-clockwise to the vertical. It should be noted that the effect of any unbalanced higher harmonic may be examined in exactly the same way.

189. Balancing Machines. Although every care may be taken in the design of a rotating part of a machine to ensure that there is no out-of-balance force or couple, residual errors will always exist in the finished part. These errors may be due to slight variations in the density of the material or to inaccuracies in the casting or machining of the part. Where the rotating part is of large diameter and relatively small axial length, it is often sufficient to ensure that it is statically balanced, since the dynamic couple, if present, will be so small as to be of no practical importance. But in other cases, where the axial length of the part is appreciable, it is not sufficient merely to have static balance, the dynamic couple must also be balanced. As the centrifugal force and couple vary as the square of the speed, even small errors of balance may be serious at high speeds of rotation. It is therefore necessary to measure these residual out-of-balance errors and make suitable corrections to the part so as to reduce the final errors to the smallest possible proportions.

Many different types of machines have been devised in order to measure the extent to which rotating parts are out of balance. Some of these machines measure the static unbalance, some the dynamic unbalance, while others measure both the static and the dynamic unbalance. It is not possible here to do more than indicate the principles on which balancing machines operate and to describe one or two of the simpler machines.

190. Static Balancing Machines. A very simple form of static balancing machine, which relies upon direct weighing, is shown diagrammatically in Fig. 339. The part P to be balanced is

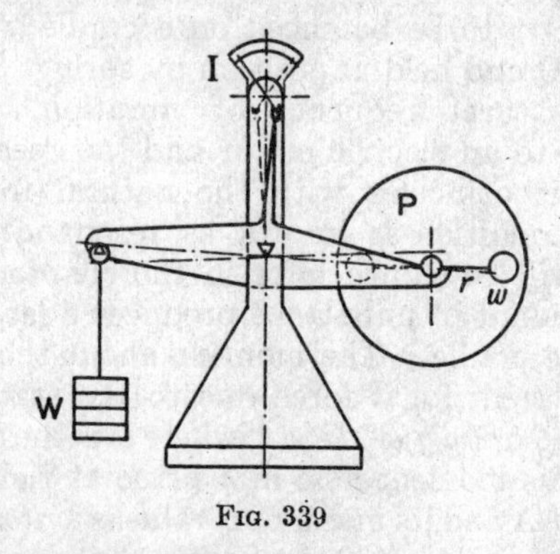

Fig. 339

mounted on a mandrel, which is supported on what is, in effect, one arm of a balance or weighing machine. A deadweight W is suspended from the other arm of the balance so as to bring the beam approximately horizontal. The mandrel is then rotated slowly, either by hand or, in the case of large machines, by an electric motor. If the part is out of balance the beam will move slowly up and down as the mandrel is rotated. Thus, if the amount of unbalance corresponds to a weight w at a radius r, the apparent weight of the body will be greatest when w is in the full-line position and least when it is in the dotted-line position. To facilitate readings of the amount of static unbalance the oscillation of the beam may be made to operate an indicator I, the pointer of which moves over a scale calibrated in ounce-inches or other convenient units. Obviously if the body is in static balance the indicator pointer will remain stationary as the mandrel is rotated.

Another type of static balancing machine is shown diagrammatically in Fig. 340. This is much more sensitive than the machine just described. The greater sensitiveness is obtained by

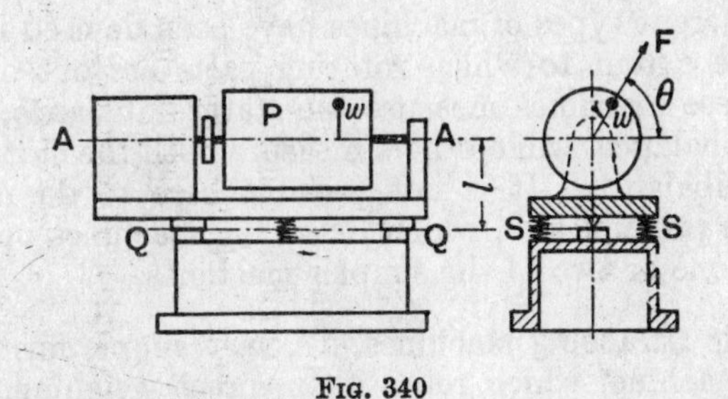

FIG. 340

mounting the part to be balanced on a cradle which is pivoted about an axis QQ and held in position by springs S, S, so that the system has a natural frequency of vibration. The part P is flexibly coupled to an electric motor and the speed of rotation is adjusted until it coincides with the natural frequency of the system. This condition is known as resonance. Under these conditions, as will be shown later in the chapter on vibrations, even a small amount of unbalance produces a large amplitude of oscillation of the cradle. The moment about the axis of oscillation QQ of the centrifugal force due to the static unbalance is given by $F \cdot l \cos\theta$ or $(w/g)\omega^2 rl \cos\theta$, where ω is the angular velocity of rotation. This moment acts in a plane at right angles to the axis of rotation AA and its maximum value is $wrl(\omega^2/g)$. It should be noted that if the part is in static balance but dynamic unbalance, no oscillation of the cradle will take place, since there are then two centrifugal forces F which produce equal and opposite moments about the axis of oscillation.

191. Dynamic Balancing Machines. A dynamic balancing machine is shown diagrammatically in Fig. 341. This machine is similar to the static balancing machine which has just been described, but differs from it in having the axis QQ about which the cradle pivots, at right angles to the axis of rotation AA instead of parallel to it. It is easily seen that the oscillation of the cradle may be excited either by an unbalanced force or an unbalanced

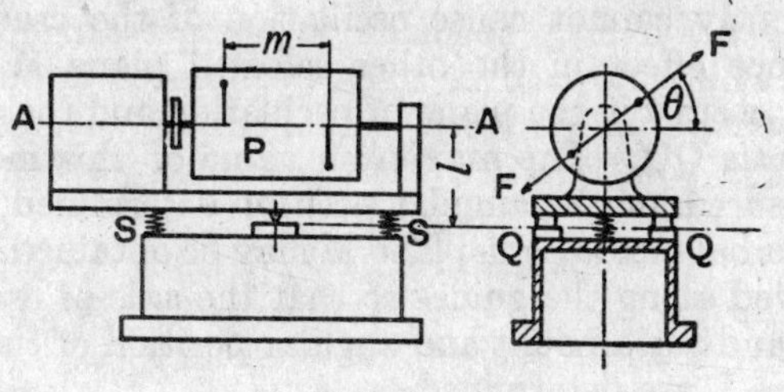

FIG. 341

couple. Hence, if the machine is to be used for measuring the dynamic unbalance of a part, it is essential that the part should first be statically balanced. The complete balance of a part then involves two separate operations in two distinct machines.

Referring to Fig. 341, the dynamic couple $F.m$ may be resolved into two components. One component $F.m \cos \theta$ acts in a horizontal plane and has no effect on the oscillation of the cradle,

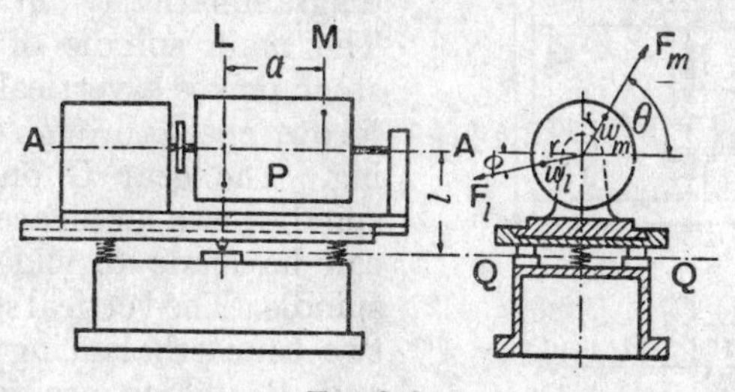

FIG. 342

the other component $F.m \sin \theta$ acts in the vertical plane through AA and causes forced vibrations of the cradle.

A second type of dynamic balancing machine is shown diagrammatically in Fig. 342. This machine, although similar to that just described, has the advantage that no preliminary static balancing of the part is necessary For this reason it is known as a universal balancing machine. The principle on which it works may be understood by referring to Article 180, where it was shown that a given system of revolving masses can always be balanced by introducing two balancing masses in arbitrarily chosen planes of rotation. In order to balance a body on this machine, the planes L and M, in which the out-of-balance effects are to be measured and the necessary corrections applied, are first selected. The body is then mounted on the cradle, as shown in the figure, in such a way that the axis of oscillation QQ lies in one of the two selected planes, say plane L. The out-of-balance effect in this plane obviously cannot cause oscillation of the cradle, but the out-of-balance effect in the other selected plane M produces a moment $F_m.a \sin\theta$ in the plane of oscillation and rocks the cradle about the axis QQ. The maximum value of this moment $F_m.a$ can be measured and its angular position determined, so that the amount of unbalance $w_m.r$ in plane M may be obtained. The cradle is then moved along the guides so that the axis of oscillation lies in plane M and the amount and angular position of the unbalance $w_l.r$ in plane L are determined.

192. Measurement of Unbalanced Force and Couple. So far no indication has been given of the way in which the amount and angular position of the out-of-balance effect may be measured. Various methods are in use but only one will be described. This is the method used on the Olsen-Carwen machine and is shown diagrammatically in Fig. 343. The main spindle of the head-stock drives a vertical shaft V at its own speed through spiral gearing. The gear G on the main spindle has a large face width and can be made to slide along the spindle. The vertical shaft carries two counterbalancing masses B_1 and B_2 which are identical in weight. Both these weights revolve with the vertical shaft and their centres of gravity are on opposite sides of and at equal distances, b, from the axis of rotation. The plane of rotation

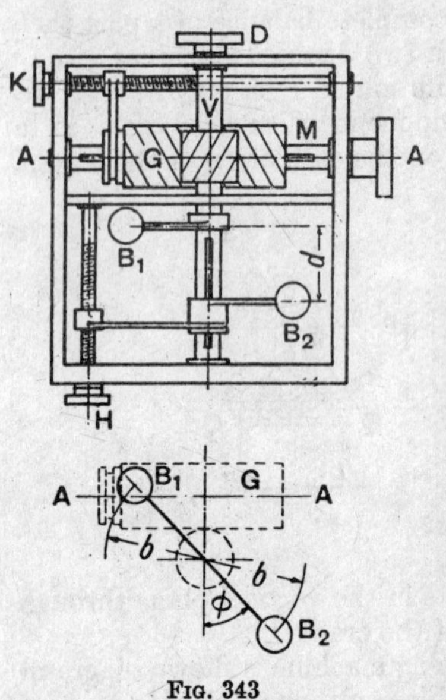

FIG. 343

of one of the weights, B_1, is fixed, but the plane of rotation of the other, B_2, may be altered by turning the handwheel H, which slides the weight along the vertical shaft. In this way the distance, d, between the planes of rotation of the two weights may be altered in order to provide a counterbalancing couple equal in magnitude to the disturbing couple caused by the part to be balanced. The alteration of the angular position of the counterbalancing weights relative to the main spindle is effected by turning the handwheel K which slides the spiral gear G along the main spindle. This rotates the vertical shaft V, and therefore the weights B_1 and B_2, relative to the main spindle. Both the position of the weight B_2 along the vertical shaft and the position of the gear G along the main spindle may be altered while the machine is running. In this way the couple exerted on the cradle by B_1 and B_2 may be adjusted, both as to amount and as to angular position, until it neutralises the couple caused by the unbalance of the part P.

The vertical shaft V and, therefore, the weights B_1 and B_2 revolve at the same speed as the part to be balanced, which is coupled directly to the main spindle. The centrifugal couple of the balance weights is equal to $(B/g)\omega^2 b.d$ and it may be resolved into two component couples, one, of magnitude $(B/g)\omega^2 bd \sin \phi$, in a vertical plane parallel to the axis of rotation AA, and the other, of magnitude $(B/g)\omega^2 bd \cos \phi$, in a vertical plane at right angles to the axis of rotation AA. Which of these two component couples affects the oscillation of the cradle obviously depends upon the position of the axis of oscillation QQ of the cradle. If QQ is parallel to AA, as in the static balancing machine Fig. 340, the effective component is $(B/g)\omega^2 bd \cos \phi$ and this is required to neutralise the rocking moment $(w/g)\omega^2 rl \cos \theta$ caused by the static unbalance of the body mounted in the machine. These two couples must at every instant be equal in magnitude and opposite in sign. This requires that $\phi = \theta$, $Bbd = wrl$ and the sense of the balancing couple must be opposite to that of the rocking moment. By manipulating the handwheels H and K on the balancing head, the oscillation of the cradle is rapidly reduced to a minimum, thus indicating that the above conditions have been satisfied. The machine may then be stopped and the amount and angular position of the static unbalance of the body may be read off.

If, on the other hand, the axis of oscillation QQ is at right angles to the axis of rotation AA (as in the dynamic balancing machines, Figs. 341, 342), the effective component of the centrifugal couple is $(B/g)\omega^2 bd \sin \phi$, in a plane parallel to AA. This couple must neutralise the rocking couple $(w/g)\omega^2 rm \sin \theta$ caused by the dynamic unbalance of the body mounted in the machine. This requires that $\phi = \theta$, $Bbd = wrm$ and that the sense of the balancing couple should be opposite to that of the rocking couple.

EXAMPLES XIV

1. Three masses A, B and C weigh respectively 20 lb, 18 lb and 32 lb and revolve in the same plane at radii of 4, 5 and 2 in. respectively. The angular positions of B and C are 60° and 135° respectively from A. Find the position and magnitude of a mass D at a radius of 6 in. to balance the system.

2. Four masses A, B, C and D are attached to a shaft and revolve in the same plane. The masses are 12, 10 18 and 15 lb respectively and their centres of gravity are respectively 4, 5, 6 and 3 in. from the axis of the shaft. The angular positions of B, C and D are respectively 60°, 135° and 270° from A. Find the position and magnitude of a fifth mass E to revolve at a radius of 10 in. and to balance the system.

3. Five masses A, B, C, D and E revolve in the same plane at equal radii. A, B and C are respectively 10, 5 and 8 lb. The angular positions of B, C, D and E, measured in the same direction from A, are 60°, 135°, 210° and 270°. Find the masses D and E for complete balance.

4. The revolving mass for a single-crank engine is equivalent to a weight of 200 lb at a radius of 9 in. Find the position and magnitude of the balance weights required at a radius of 24 in., in planes L and M distant 12 in. and 36 in. from the plane of the crank, when (a) the planes L and M are on opposite sides of the crank, (b) the planes L and M are both on the same side of the crank.

5. A shaft carries three rotating masses A, B, C of magnitude 20, 18 and 32 lb with their centres of gravity 4 in., 5 in. and 2 in. respectively from the axis of rotation. The distance between the planes of rotation of A and B is 3 ft and between B and C is 5 ft. The angular positions of the masses are B 60° and C 135° from A in the same direction. Find the magnitude and position of the balance weights required in planes L and M situated midway between the planes A and B and the planes B and C respectively. The radius of rotation of the balance weights is 6 in.

6. A shaft carries three pulleys A, B and C at distances apart of 2 ft and 4 ft. The pulleys are out of balance to the extent of 5, 4 and 6 lb respectively, at a radius of 1 in. in each case. The angular positions of the out-of-balance masses in pulleys B and C with respect to that in pulley A are 90° and 210° respectively. Determine, in position and magnitude, the balance weights required in planes L and M midway between planes A and B, and B and C respectively. The radius of rotation of the balance weights is 5 in.

7. Four masses attached to a shaft revolve at equal radii in planes A, B, C and D. The distances between planes A and B, B and C, C and D areres pectively 1·0, 2·5 and 1·5 ft. The revolving masses in planes A, C and D are 200, 300 and 180 lb respectively. Find the magnitude of the fourth revolving mass and the angular positions of the radii to give a complete balance. M.U.

8. Four masses of magnitude 200, 350, 400 and 250 lb are each attached to cranks of 8 in. radius and revolve in planes A, B, C and D. The angular positions of the cranks B, C and D with respect to the crank in plane A are 75°, 135° and 240° taken in order. The distances of planes B, C and D from plane A are 2 ft, 6 ft and 8 ft respectively. Find the magnitude and position of the balance weights required in planes midway between A and B and between C and D. The radius of rotation of the balance weights is 24 in.

9. A shaft carries five masses A, B, C, D and E which revolve at the same radius in equidistant planes. The masses in planes A, C and D weigh respectively 100, 80 and 160 lb. The angle between A and C is 90° and that between C and D is 135°. Find the weights in planes B and E and their angular positions so that the shaft may be completely balanced. M.U.

10. In a three-cylinder engine with cranks at 120°, the distance between the planes of rotation of adjacent cranks is 3 ft. The revolving masses attached to each crank are equivalent to a weight of 400 lb at 1 ft radius. Find the magnitude and position of the balance weights required at 2·5 ft radius in planes 4 ft and 2 ft respectively on opposite sides of the centre crank.

11. Four masses A, B, C and D revolve at equal radii and are equally spaced along a shaft. The mass B weighs 15 lb and the radii of C and D makes angles of 90° and 240° respectively with the radius of B. Find the magnitudes of the masses A, C and D and the angular position of A so that the system may be completely balanced.

12. A shaft rotates in two bearings A and B 6 ft apart and projects 1 ft 6 in. beyond A and B. At the extremities of the shaft are two pulleys of masses 40 lb and 96 lb, their c.g. being respectively $\frac{1}{2}$ in. and $\frac{5}{8}$ in. from the axis of the shaft. Midway between the bearings is a third pulley of mass 112 lb its c.g. being $\frac{5}{8}$ in. from the axis of the shaft.

If the three pulleys have been arranged so as to obtain static balance, find the dynamic forces produced on the bearings when the shaft rotates at 300 r.p.m.

L.U.

13. Particulars of a system of revolving masses rigidly attached to a shaft are given in the table below:

Plane	Mass, lb	Radius, in.	Distance from plane A, in.	Angular position
A	12	4	0	0°
B	10	5	6·5	60°
C	18	6	16·0	135°
D	15	3	24·5	270°

Find the magnitudes and angular positions of two balance weights, revolving at 5 in. radius in planes L and M, the first between A and B and 3 in. from A, and the second between C and D and 7 in. from C.

14. A single-cylinder horizontal oil engine has a crank 7·5 in. long and a connecting rod 33 in. long. The revolving parts are equivalent to 110 lb at crank radius and the piston and gudgeon pin weigh 90 lb. The connecting rod weighs 115 lb and its c.g. is 10·5 in. from the crankpin centre. Revolving balance weights are introduced at a radius of 8·5 in. on extensions of the crank webs in order to balance all the revolving parts and one-half of the reciprocating parts. Find the magnitude of the total balance weight and, neglecting the obliquity of the connecting rod, the nature and magnitude of the residual unbalanced force on the engine. R.p.m. 300 M.U.

15. Explain briefly what advantages are derived from the partial balancing of the reciprocating parts of a locomotive by revolving balance weights. For an inside-cylinder locomotive with the two cranks at right-angles, the reciprocating parts weigh 600 lb per cylinder. The distance between cylinder centre lines is 24 in. and between the planes of rotation of the wheels 60 in. Each crank is 13 in. long and the driving wheels are 6 ft 3 in. dia. Revolving balance weights are introduced in the planes of the wheels partially to balance two-thirds of the reciprocating parts. Find the maximum variation of tractive effort and of wheel load when the locomotive is running at 60 m.p.h. M.U.

16. An inside-cylinder uncoupled locomotive is to be balanced for revolving masses and for two-thirds of the reciprocating masses. The revolving mass per cylinder is 450 lb and the reciprocating mass per cylinder is 540 lb. The cylinder centre lines are 25 in. apart and the wheel centres are 60 in. apart. Find: (a) the magnitude and position of the balance weights required at a radius of 30 in. in the planes of the wheels; (b) the hammer-blow and maximum variation of tractive effort when the cranks make 4 r.p.s. Stroke = 26 in.

17. The following particulars relate to an outside-cylinder uncoupled locomotive:

Revolving weight per cylinder	500 lb
Reciprocating weight per cylinder	600 lb
Length of each crank	13 in.
Distance between wheel centres	58 in.
Distance between cylinder centres	70 in.
Diameter of the driving wheels	72 in.
Radius of balance weights	30 in.

All the revolving parts are balanced and the hammer-blow is limited to 4 tons when the locomotive is running at 50 m.p.h. Find the fraction of the reciprocating parts which must be balanced and the magnitude and position of the balance weights required in the planes of the wheels. M.U.

18. It is required to balance an inside-cylinder uncoupled locomotive in such a way that the driving wheel is on the point of lifting from the rail when the locomotive is running at 75 m.p.h. If all the revolving parts are balanced, what fraction of the reciprocating parts must be balanced?

Load on each driving wheel	6 tons
Wheel centres	5 ft
Cylinder centres	2 ft 4 in.
Diameter of driving wheels	7 ft
Length of each crank	13 in.
Weight of reciprocating parts per cylinder . .	720 lb

M.U.

19. Two locomotives are built with similar sets of reciprocating parts, one with outside cylinders, the other with inside cylinders. The former has driving wheels 7 ft in diameter and the distance between the cylinder centre lines is 6 ft, whilst in the latter the distance between the cylinder centre lines is 2 ft 1 in. In each engine the distance between the wheel centres is 5 ft. If, when the inside-cylinder locomotive is running at 0·75 of the speed of the outside-cylinder locomotive, the hammer-blow is the same for both locomotives, what is the diameter of the driving wheels for the inside-cylinder locomotive? How do the swaying couples compare under these conditions? M.U.

20. Two locomotives are built with identical reciprocating parts; one has outside cylinders, 6 ft centre to centre, and driving wheels 6·5 ft dia.; the other has inside cylinders 2 ft centre to centre and driving wheels 4·75 ft dia. The wheel centres are 5 ft apart in each case and the same fraction of the reciprocating parts is balanced. Compare the hammer-blow and the variation of tractive effort when the two locomotives run at the same speed.

21. In a three-cylinder locomotive the two outside cranks are at 90°, while the inside crank is at 135° to the two outer cranks. The pitch of the cylinders is 3 ft and the stroke 26 in. The planes of the balance weights are 5 ft apart. The reciprocating masses to be balanced are 400 lb for the inside crank and 286 lb for each of the outside cranks. If the driving wheels are 7 ft dia., what is the hammer-blow at 60 m.p.h.? L.U.

22. The following particulars refer to a four-coupled locomotive with two inside cylinders:

Pitch of cylinders	27 in.
Revolving parts per cylinder	500 lb
Reciprocating parts per cylinder	600 lb
Distance between the planes of the driving wheels	60 in.
Diameter of the driving wheels	78 in.
Distance between the planes of the coupling rod cranks	75 in.
Revolving parts for each coupling rod crank .	260 lb

The engine cranks are at right angles and are 13 in. long, while the coupling rod cranks are at 180° to the adjacent engine cranks and are 11 in. long. The whole of the revolving parts and two-thirds of the reciprocating parts are to be balanced by masses in the planes of the wheels at a radius of 32 in. The balance weights required for the reciprocating parts are divided equally between the pairs of coupled wheels. Find: (a) the magnitude and position of the balance weights required and (b) the hammer-blow and maximum variation of tractive effort when the speed is 60 m.p.h.

23. A 4—8—0 locomotive for a 5-ft 6-in. gauge railway has three cylinders with the cranks at 120°. The dimensions are as follows:

Inside cylinder revolving mass	840 lb
Each outside cylinder revolving mass	700 lb
Reciprocating mass for each cylinder	810 lb
Revolving mass for each coupling rod crank . . .	350 lb
Distance between the outside cylinder centre lines .	91 in.
Distance between the coupling rod centre lines . .	79 in.
Distance between the planes of the balance weights .	69 in.
Length of each crank	13 in.
Diameter of coupled wheels	55·5 in.

The whole of the revolving parts and two-thirds of the reciprocating parts are to be balanced. Find the position and magnitude of the balance weights required for the driving axle and for one of the coupled axles and also the magnitude of the hammer-blow at 60 m.p.h. when the balance weights for the reciprocating parts (a) are placed in the driving wheels, (b) are distributed equally between the coupled wheels.

24. Explain what is meant by primary and secondary balancing.

A vertical single-cylinder opposed-piston engine has reciprocating parts weighing 2 tons for the lower piston and 2·75 tons for the upper piston. The lower piston has a stroke of 24 in. and the engine is in primary balance. If the ratio of length of connecting rod to length of crank is 4 for the lower piston and 8 for the upper piston, what is the maximum unbalanced secondary force for a crankshaft speed of 135 r.p.m.? At which crank positions will it occur? M.U.

25. A twin-cylinder V-engine has the cylinder centre lines at 90° and the connecting rods drive on to a single crank. The stroke is 5 in. and the length of each connecting rod is 9·5 in. The crankpin and crank webs are equivalent to 2·5 lb at crank radius and each piston weighs 2 lb. The weight of each connecting rod is 3 lb and the c.g. is 3 in. from the crankpin centre.

Show that the effect of the revolving mass and the primary effect of the reciprocating masses may be balanced by a revolving balance weight. Find its magnitude and position if the distance of the c.g. from the crankshaft centre line is 3 in. What is the nature and magnitude of the resultant secondary force when the crankshaft makes 1600 r.p.m.?

26. A four-cylinder marine oil engine has the cranks arranged at angular intervals of 90°. The inner cranks are 4 ft apart and are placed symmetrically between the outer cranks, which are 10 ft apart. Each crank is 18 in. long, the engine runs at 90 r.p.m., and the weight of the reciprocating parts for each cylinder is 1800 lb. In which order should the cranks be arranged for the best balance of the reciprocating masses, and what will then be the magnitude of the unbalanced primary couple? M.U.

27. A four-cylinder engine with cranks 1 ft long is to be balanced for primary forces and couples. The intermediate cranks are 2 ft apart and are placed symmetrically between the extreme cranks which are 8 ft apart. If the reciprocating masses attached to the intermediate cranks, which are at right angles to each other, weigh 1000 lb, find the angular positions of the extreme cranks and the reciprocating masses attached to them.

With this arrangement what will be the maximum unbalanced secondary force and couple at a speed of 120 r.p.m.? The connecting rods are 4 ft long. M.U.

28. The reciprocating masses for three cylinders of a four-crank engine weigh 3, 5 and 8 tons and the centre lines of these cylinders are 12 ft, $8\frac{1}{2}$ ft and $3\frac{1}{2}$ ft

respectively from that of the fourth cylinder. Find the fourth reciprocating mass and the angles between the cranks so that they may be mutually balanced for primary forces and couples.

If the cranks are each 2 ft long, the connecting rods 9 ft long and the r.p.m. 60, find the maximum value of the secondary disturbing force and the crank positions at which it occurs.

29. A four-cylinder, two-stroke internal-combustion engine with two scavenging pump cylinders is arranged as follows:

Cylinder	Crank angle from No. 1 engine crank	Crank radius in.	Equivalent reciprocating mass, lb	Distance from No. 1 pump cylinder, in.
No. 1 Pump .	?	4·5	50	0
„ 1 Engine .	0	6·0	70	19
„ 2 „ .	90	6·0	70	36
„ 3 „ .	180	6·0	70	53
„ 4 „ .	270	6·0	70	70
„ 2 Pump .	?	4·5	50	89

All the connecting rods are 18 in. long. Determine the angles for the scavenging pump cranks (which must be opposed, i.e. at 180° to each other) relative to No. 1 engine crank, so as to have the least possible unbalanced primary forces and couples.

Then determine the magnitude of any unbalanced primary and secondary forces and couples when the engine is running at 360 r.p.m. Take the reference plane for couples at the mid-point of the engine crankshaft. L.U.A.

30. A six-cylinder, single-acting, two stroke Diesel engine is arranged with cranks at 60° for the firing sequence 1–4–5–2–3–6. The cylinders, numbered 1 to 6 in succession, are pitched 5 ft apart, except Nos. 3 and 4, which are 6 ft apart.

The reciprocating and revolving weights per line are respectively 2·2 and 1·6 tons. The crank length is 15 in., the connecting rod length is 63·75 in. and the speed is 120 r.p.m. The usual rule for primary and secondary forces in one line may be assumed.

Determine, with reference to the central plane between cylinders 3 and 4, the maximum and minimum values of the primary frame couple due to reciprocating and rotational inertia and the maximum value and phase—relative to crank No. 1—of the secondary couple. L.U.A.

31. In a four-crank "symmetrical" engine the angular positions of the consecutive cranks A, B, C, D are in the order A, D, B, C when looking along the shaft from A towards D. The centre lines of cylinders B and C are each a distance a from the middle cross-section of the engine, while those of A and D are each a distance b from this section. The crank angle between A and C is equal to that between B and D and equals α; the crank angle between A and D equals 2β and the remaining crank angle between B and C equals 2γ $(= 2\pi - 2\alpha - 2\beta)$. The reciprocating masses of A and D are each W_1 and those of B and C are each W_2.

Prove that for complete balance of primary forces and couples:

$$(1)\ W_1 \cos \beta = W_2 \cos \gamma, \quad (2)\ b/a = \tan \gamma / \tan \beta$$ L.U.A.

32. In a four-cylinder marine engine the consecutive distances between the cylinder centre lines A, B, C and D are 8 ft, 12 ft and 8 ft. The piston stroke is 4 ft and the reciprocating masses for A and D each weigh 3·07 tons.

Find the masses of the reciprocating parts of B and C and the crank angles so that the reciprocating parts may be in primary balance and may also be balanced for secondary forces.

What is then the out-of-balance couple due to reciprocating parts when running at 120 r.p.m.? L.U.A.

33. In a four-crank symmetrical engine, the reciprocating masses of the two extreme cylinder sets, A and D, are each 0·8 ton and those of the two inner cylinder sets, B and C, are each 1·2 tons. Taking the direction of crank A as 0°, find the angles, measured clockwise, between A and the other three cranks so that the balance of the engine will be complete except for secondary couples. Find also the ratio pitch of outside cylinders to pitch of inside cylinders. L.U.A.

34. Prove that in a radial aero engine with an odd number of cylinders, primary balance may be obtained by the addition of a single weight revolving at a given radius.

In the case of a five-cylinder engine with a single crank find the magnitude and position of the weight required at crank radius in terms of W, the weight of each piston. L.U.A.

35. In a three-cylinder radial engine all three connecting rods act on a single crank. The cylinder centre lines are set at 120°, the weight of the reciprocating parts per line is 5 lb, the crank length is 3 in., the connecting rod length is 11 in. and the r.p.m. are 1800. Determine, with regard to the inertia of the reciprocating parts, (a) the balance weight to be fitted at 4 in. radius to give primary balance; (b) the nature and magnitude of the secondary unbalanced force; (c) whether the fourth and sixth order forces are balanced or unbalanced. L.U.A.

36. The pistons of a 60° V-twin engine have a stroke of 4·5 in. The two connecting rods operate on a common crankpin and each is 8 in. long. If the weight of the reciprocating parts is 2·5 lb per cylinder and the crankshaft speed is 2500 r.p.m., find the maximum and minimum values of (a) the primary force and (b) the secondary force. In each case state the directions in which the forces act and the crank positions at which the maximum and minimum values occur.

37. An eight-cylinder engine is arranged in V-form with the two banks of cylinders at 45° and a flat four-throw crankshaft is used, two rods working on each crankpin. The reciprocating weight per line is 2·75 lb, the crank radius 2 in., the connecting rod length 9 in. and the r.p.m. 2500. Determine the maximum and minimum values of the secondary frame force due to the inertia of the reciprocating parts. Discuss the possibilities of balancing this force. L.U.A.

38. Show how the conception of *direct* and *reverse* cranks may be used to investigate the unbalanced inertia forces caused by the reciprocating parts of a multi-cylinder engine, in which the cylinders are arranged in two, or more, inclined rows.

Determine the maximum and minimum values of the resultant secondary force for a 12-cylinder engine, with three rows of four cylinders. The centre row is vertical and the two outside rows are inclined at 60° to the vertical. A four-throw "flat" crankshaft of the usual type is used and one connecting rod from each row is coupled directly to each crankpin. The weight of the reciprocating parts is 7 lb per cylinder, the cranks are 3 in. long, the connecting rods are 11·5 in. long and the r.p.m. are 2200. M.U

CHAPTER XV

VIBRATIONS

193. When a body which is held in position by elastic constraints is displaced from its equilibrium position by the application of an external force and then released, it commences to vibrate. Work is done by the external force in producing the initial displacement against the internal elastic forces which resist deformation. This work is stored up as elastic or strain energy in the constraints, so that when the external force is removed, the internal elastic forces tend to restore the body to its equilibrium position. Neglecting for the moment any resistances offered to the motion of the vibrating body, the whole of this elastic or strain energy is converted into kinetic energy at the instant the body reaches its original equilibrium position. As a result, the motion of the body continues until the whole of the kinetic energy is absorbed in doing work against the internal elastic forces and the energy in the system is once more strain energy. Again the body begins to return to the equilibrium position and the oscillation or vibration is repeated indefinitely. A vibration of this kind, in which, after the initial displacement, no external forces act and the motion is maintained by the internal elastic forces, is termed a *free* or *natural* vibration.

In practice the energy possessed by the system is gradually dissipated in overcoming internal and external resistances to the motion, and the body finally comes to rest in its original equilibrium position. Such a vibration is said to be *damped*. A third type of vibration, which is of great practical importance, is that in which a *periodic* disturbing force is applied to the body. The vibration then has the same frequency as the applied force and is said to be *forced*.

194. Free Vibrations. Consider the system shown in Fig. 344, in which a shaft, assumed to be weightless and fixed at one end, carries a heavy disc or flywheel at the free end. This system may be made to vibrate in one of three simple ways:

(a) All particles of the flywheel may vibrate along straight paths parallel to the axis of the shaft. This is termed ***longitudinal*** vibration.

(b) All particles of the flywheel may vibrate along straight paths perpendicular to the axis of the shaft. This is termed *transverse* vibration.

Note.—The paths of the particles cannot strictly speaking be either straight or perpendicular to the axis of the shaft, since the latter will bend. In practice the amplitude of the vibrations is generally small in comparison with the length of the shaft, so that the above statement is justified.

(c) All particles of the flywheel may vibrate along circular arcs whose centres lie on the axis of the shaft. This is termed *torsional* vibration.

The effect on the shaft material of the three types of vibration is as follows:

(a) The shaft is alternately extended and compressed.
(b) The shaft is alternately bent and straightened.
(c) The shaft is alternately twisted and untwisted.

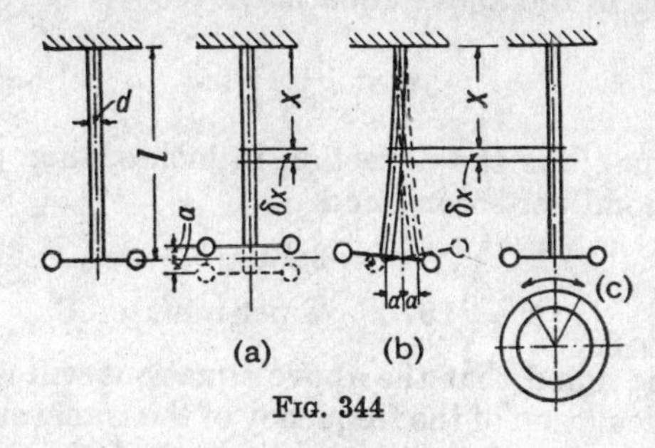

FIG. 344

In each case, if the limit of proportionality—stress proportional to strain—is not exceeded, the restoring force, case (a) or (b), or the restoring couple, case (c), which is exerted on the flywheel by the shaft, is directly proportional to the displacement of the flywheel from its equilibrium position. Hence it follows that the acceleration towards the equilibrium position is directly proportional to the displacement from that position and the vibration is therefore *simple harmonic* (Article 24).

Cases (a) *and* (b).

Let W = weight of flywheel,
a = amplitude of the vibration,
s = stiffness of the shaft, i.e. the force required at the flywheel to produce unit displacement in the direction of the vibration,
and n = frequency of the vibrations per second.

Then the restoring force

$$= s \times \text{displacement from the equilibrium position.}$$

But the restoring force $= W/g \times$ acceleration,
so that

$$\frac{\text{Acceleration}}{\text{Displacement}} = \frac{s.g}{W} = \text{constant}$$

and from equation (2.24):

$$n = \frac{1}{2\pi}\sqrt{\frac{s.g}{W}} \quad . \quad . \quad . \quad . \quad . \quad (15.1)$$

But $W/s = \delta$, the static deflection of the shaft under the gravity load applied by the flywheel.

Note that for case (a) δ is the extension of the shaft produced by the load W, while for case (b) δ is the deflection of the shaft under the load W when the shaft is placed horizontally as a cantilever.

Substituting in the above equation:

$$n = \frac{1}{2\pi}\sqrt{\frac{g}{\delta}} \quad . \quad . \quad . \quad . \quad . \quad (15.2)$$

If, as is usual, δ is the deflection in inches, then g must be in inches per second per second and

$$\therefore \; n = 3{\cdot}13/\sqrt{\delta} \quad . \quad . \quad . \quad . \quad (15.3)$$

or

$$N = 187{\cdot}8/\sqrt{\delta} \text{ per minute} \quad . \quad . \quad . \quad (15.4)$$

It should be noted that the above equations will apply equally well to the calculation of the frequency of the transverse vibrations of any weightless shaft which carries a single heavy disc. The position of the disc along the shaft and the end conditions of the shaft will not affect the above analysis. The deflection must obviously be the deflection under the load and not necessarily the maximum deflection of the shaft.

Case (c). An expression similar in form to equation (15.1) may be deduced for the frequency of the free torsional oscillations of the system.

In addition to the symbols given for cases (a) and (b), let k = the radius of gyration of the flywheel, and q = the torsional stiffness of the shaft, i.e. the torque required on the flywheel to produce an angular displacement of one radian from the equilibrium position. It is proved in text-books on the strength of materials that $q = CJ/l$, where C is the modulus of rigidity for the shaft material, J is the polar second moment of area of the shaft cross-section and l is the length of the shaft.

Then the restoring couple which acts on the flywheel $= q \times$ angular displacement from the equilibrium position.

But the restoring couple is also equal to the product of the mass moment of inertia I of the flywheel and the angular acceleration towards the equilibrium position; so that we have:

$$\frac{\text{Angular acceleration}}{\text{Angular displacement}} = \frac{q}{I} = \text{constant}$$

The motion is therefore simple-harmonic and, from equation (2.24), the frequency per second

$$= n = \frac{1}{2\pi}\sqrt{\frac{q}{I}} \quad . \quad . \quad . \quad . \quad . \quad . \quad (15.5)$$

Since $I = (W/g)k^2$,

$$n = \frac{1}{2\pi}\sqrt{\frac{q.g}{Wk^2}} \quad . \quad . \quad . \quad . \quad . \quad (15.6)$$

Example 1. For the system shown in Fig. 344, the weight of the flywheel is 0·3 ton, the radius of gyration is 15 in., the shaft is 3 in. diameter and 3 ft long to the flywheel boss. For the shaft material Young's modulus is 30×10^6 lb/in^2 and the modulus of rigidity is 12×10^6 lb/in^2. Find the frequencies of the free longitudinal, transverse and torsional vibrations.

(a) *Longitudinal Vibration*:

Static extension of the shaft δ

$$= \frac{Wl}{AE} = \frac{3.224.36.4}{\pi.3^2.30.10^6} = 0{\cdot}000\ 114 \text{ in.}$$

Frequency of the longitudinal vibrations, from equation (15.4),

$$= \frac{187{\cdot}8}{\sqrt{0{\cdot}000\ 114}} = 17\ 580 \text{ per min}$$

(b) *Transverse Vibration*:

Static deflection of a cantilever δ

$$\frac{Wl^3}{3EI} = \frac{3.224.36^3.64}{3.30.10^6.\pi.3^4} = 0{\cdot}0876 \text{ in.}$$

Frequency of the transverse vibrations, from equation (15.4),

$$= \frac{187{\cdot}8}{\sqrt{0{\cdot}0876}} = 634 \text{ per min}$$

(c) *Torsional Vibration*:

$$q = \frac{CJ}{l} = \frac{12.10^6.\pi.3^4}{32.36} = 2{\cdot}652.10^6 \text{ lb in.}$$

Frequency of the torsional vibrations, from equation (15.6),

$$= \frac{1}{2\pi}\sqrt{\frac{qg}{Wk^2}} = \frac{1}{2\pi}\sqrt{\frac{2{\cdot}652.10^6.32{\cdot}2.12}{3.224.15^2}} = 13{\cdot}1 \text{ per sec}$$

$$= 786 \text{ per min}$$

Example 2. A flywheel is mounted on a vertical shaft as shown in Fig. 345, the ends of the shaft being fixed. The shaft is 2 in. dia., the length l_1 is 3 ft and the length l_2 is 2 ft. The flywheel weighs 0·5 ton and its radius of gyration is 20 in. Find the natural frequencies of the longitudinal, the transverse and the torsional vibrations of the system.

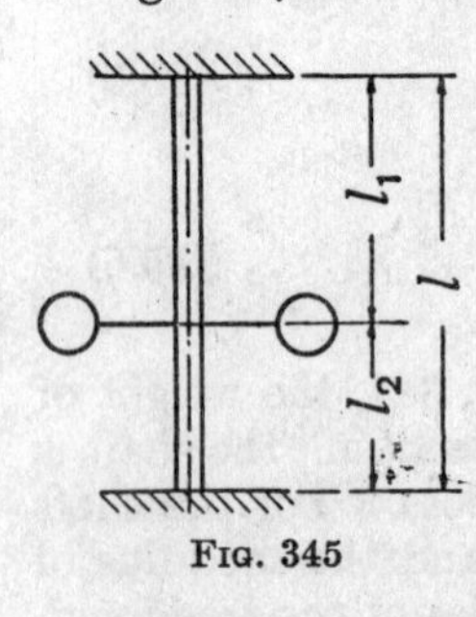

Fig. 345

(a) *Longitudinal Vibration.* Let W_a be that part of the weight W of the flywheel which is carried by the length l_1 of shaft, so that $W-W_a$ is the weight carried by the length l_2.

Then the extension of length $l_1 = W_a l_1/AE$, where A is the cross-sectional area of the shaft and E is Young's modulus.

Similarly, the compression of the length $l_2 = (W-W_a)l_2/AE$.

But the extension of shaft l_1 must clearly be equal to the compression of shaft l_2, so that $W_a l_1 = (W-W_a)l_2$.

$$\therefore\ W_a.3 = (W-W_a)2 \quad \text{and} \quad W_a = \tfrac{2}{5}W$$

$$\therefore \text{ extension of shaft } l_1 = \frac{2}{5}.\frac{0{\cdot}5.2240.3.12}{\pi.30.10^6} = 0{\cdot}000\,171 \text{ in.}$$

$\therefore$ from equation (15.4)

$$N = 187{\cdot}8/\sqrt{0{\cdot}000\,171} = 14\,370 \text{ per min}$$

(b) *Transverse Vibration.* The static deflection under the load, for a horizontal shaft fixed at the ends and loaded at a point which divides the shaft into the two parts l_1 and l_2, is given by:

$$\delta = Wl_1{}^3l_2{}^3/3EIl^3$$

Substituting the given values, we get:

$$\delta = \frac{0{\cdot}5.2240.36^3.24^3.64}{3.30.10^6.\pi.2^4.60^3} = 0{\cdot}0473 \text{ in.}$$

$\therefore$ from equation (15.4)

$$N = 187{\cdot}8/\sqrt{0{\cdot}0473} = 863 \text{ per min}$$

(c) ***Torsional Vibration.*** The torque required to produce a twist of 1 radian in the length $l_1 = q_1 = CJ/l_1$.

Similarly, the torque required to produce a twist of 1 radian in the length $l_2 = q_2 = CJ/l_2$.

$\therefore$ the total torque required at the flywheel to produce an angular displacement of 1 radian

$$= q = q_1 + q_2 = CJ\left(\frac{1}{l_1}+\frac{1}{l_2}\right) = 12 . 10^6 . \frac{\pi . 2^4}{32} . \frac{1}{12}\left(\frac{1}{3}+\frac{1}{2}\right)$$

$$= \frac{5\pi}{12} . 10^6 \text{ lb in.}$$

$\therefore$ from equation (15.6)

$$n = \frac{1}{2\pi}\sqrt{\frac{q . g}{Wk^2}} = \frac{1}{2\pi}\sqrt{\left(\frac{5\pi}{12} . \frac{10^6 . 32{\cdot}2 . 12}{0{\cdot}5 . 2240 . 20^2}\right)}$$

$$= 5{\cdot}35 \text{ per sec} = 321 \text{ per min}$$

195. The Effect of the Inertia of the Shaft. So far we have neglected the effect of the inertia of the shaft on the frequency of the vibrations. In most practical cases the inertia of the shaft is small in comparison with that of the attached mass and the following approximate method of allowing for it will suffice.

(a) *Longitudinal Vibration.* Referring to Fig. 344 (a), let v be the velocity of the free end at a given instant. Then the velocity of the section δx, at the distance x from the fixed end, is approximately equal to $x/l . v$. If w is the weight of the shaft per unit length, then the kinetic energy of the part δx at the given instant

$$= \frac{w\delta x}{2g} . \left(\frac{x}{l}v\right)^2 = \frac{wv^2}{2gl^2}x^2\delta x$$

$$\therefore \text{ Total K.E. of shaft} = \frac{wv^2}{2gl^2}\int_0^l x^2 dx = \frac{wv^2}{2gl^2} . \frac{l^3}{3}$$

$$= \frac{wl}{3} . \frac{v^2}{2g}$$

Hence the whole shaft is dynamically equivalent to one-third of its mass concentrated at the free end.

The inertia of the shaft may therefore be allowed for by adding one-third of its mass to that of the attached disc or flywheel.

(b) *Transverse Vibration.* Referring to Fig. 344 (b), let v be the transverse velocity at a given instant of the free end which carries the flywheel, and assume that the shape of the curve into which the vibrating shaft deflects is identical with the static deflection curve of a cantilever loaded at the end.

Then at the distance x from the fixed end the velocity at the same instant is given by $\{(3lx^2-x^3)/2l^3\}v$, and the kinetic energy of part δx

$$=\frac{1}{2}\frac{w.\delta x}{g}.\left(\frac{3lx^2-x^3}{2l^3}\right)^2 v^2$$

$$\therefore \text{ K.E. of whole shaft } = \frac{w}{8gl^6}v^2\int_0^l (3lx^2-x^3)^2 dx$$

$$=\frac{w}{8gl^6}v^2\int_0^l (9l^2x^4-6lx^5+x^6)dx$$

$$=\frac{w}{8gl^6}v^2.\frac{33}{35}.l^7=\frac{33}{280}.\frac{wl}{g}.v^2$$

Hence the shaft is dynamically equivalent to the fraction 33/140 of its mass concentrated at the free end. The inertia of the shaft may therefore be allowed for by adding 33/140 of its mass to that of the disc or flywheel.

In a similar way it could be shown that, if the flywheel is situated midway along the shaft and the ends are supported, the fraction 17/35 of the mass of the shaft must be added to that of the flywheel. If the ends are fixed, the fraction which must be added is 13/35.

(c) *Torsional Vibration.* Referring to Fig. 344 (c), let ω be the angular velocity at the free end at a given instant and I_s be the mass moment of inertia of the complete shaft.

Then the angular velocity of the section δx, distant x from the fixed end, is given approximately by $x/l.\omega$, the kinetic energy of the section δx

$$=\frac{1}{2}\frac{\delta x}{l}.I_s\left(\frac{x}{l}.\omega\right)^2=\frac{1}{2}\frac{I_s\omega^2}{l^3}x^2\delta x$$

and the total kinetic energy of the shaft

$$=\frac{1}{2}\frac{I_s\omega^2}{l^3}\int_0^l x^2 dx=\frac{1}{3}.\frac{1}{2}I_s\omega^2$$

Hence the kinetic energy of the shaft is equivalent to that of a flywheel of mass moment of inertia $I_s/3$ at the free end.

The inertia of the shaft may therefore be allowed for by adding one-third of its mass moment of inertia to that of the disc or flywheel.

196. Transverse Vibration of a Uniformly Loaded Shaft. (a) *Approximate Solution.* A uniformly loaded shaft may be made to vibrate transversely in just the same way as a shaft which carries a single concentrated load. There is, however, one important difference between the two cases. Whereas the shaft with a single concentrated load has only one natural frequency of transverse vibration, a uniformly loaded shaft has, theoretically, an infinite number. We shall, at present, confine ourselves to the mode of vibration which gives the lowest frequency. For this mode of vibration all particles of the shaft vibrate in phase and the frequency is termed the fundamental frequency.

In his book on the *Theory of Sound*, Lord Rayleigh showed that the calculated frequency is only slightly affected by making different assumptions for the shape of the deflected shaft during vibration, providing that the assumed shape is consistent with the end conditions. It is therefore possible to obtain a close approximation to the true value of the frequency by assuming that the vibrating shaft bends into a curve of simple shape such as a sine curve, a parabola or the static deflection curve of the shaft under the given system of loading.

In what follows it will be assumed that the curve of deflection during vibration is at every instant similar in shape to the static deflection curve. The method of finding an expression for the frequency is first to derive expressions for the maximum additional strain energy of the shaft during the vibration and for the maximum kinetic energy of the shaft. These are then equated, since the total energy of the vibrating system is constant.

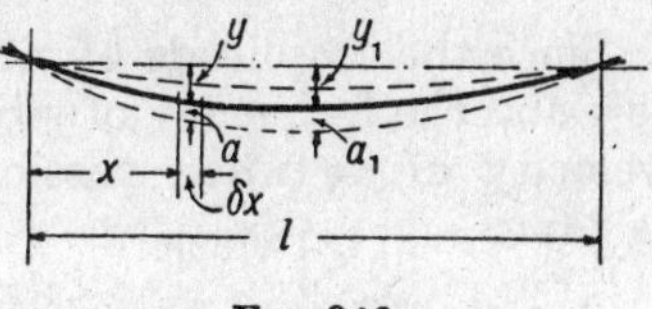

Fig. 346

Referring to Fig. 346, let the full line represent the static deflection curve of the shaft and the dotted lines represent the extreme positions of the shaft when vibrating transversely.

Let y_1 = static deflection at the mid-point,

a_1 = amplitude of vibration at the mid-point,

y, a = corresponding static deflection and amplitude of vibration of the section δx of the shaft, distant x from the support,

w = actual load per unit length

and w_e = the load required per unit length in order to produce unit deflection at the mid-point.

Then, since the deflection curve of the vibrating shaft is assumed at every instant to be similar in shape to the static deflection curve, it follows that the ratio a/y is constant for all points along the shaft and may be denoted by c.

The maximum additional strain energy possessed by the shaft at the end of its swing is equal to the work done in changing the deflection curve from that shown by the full line to that shown by the dotted line. The work done may be found as follows: to produce the additional deflection at every point along the shaft, the load per unit length would have to increase from w to $w+w_e a_1$. Consider the section of length δx.

Then the work done on this section = mean additional load multiplied by the distance through which the load moves $= \frac{1}{2}w_e a_1 \delta x . a$,

$\therefore$ Total work done on the shaft = Maximum additional strain energy

$$= \int_0^l \tfrac{1}{2}w_e . a_1 a \mathrm{d}x$$

But $\qquad a_1/y_1 = a/y = c \quad \text{and} \quad w_e y_1 = w,$

$\therefore$ substituting for a_1, a and $w_e y_1$, we get:

$$\text{Maximum strain energy} = \tfrac{1}{2}c^2 w \int_0^l y \mathrm{d}x$$

Since the amplitude of vibration of the portion δx of the shaft is a and the frequency of vibration is n per second, the maximum velocity of δx when passing through the equilibrium position is $2\pi na$.

$\therefore$ total kinetic energy of the vibrating shaft

$$= \int_0^l \frac{1}{2}\frac{w}{g} . (2\pi na)^2 \mathrm{d}x$$

But $a = c . y$,

$$\therefore \text{ K.E.} = \frac{1}{2}c^2 . w \frac{(2\pi n)^2}{g} \int_0^l y^2 \mathrm{d}x$$

Equating the maximum kinetic energy to the maximum strain energy,

$$\frac{1}{2}c^2 . w \frac{(2\pi n)^2}{g} \int_0^l y^2 \mathrm{d}x = \frac{1}{2}c^2 . w \int_0^l y \mathrm{d}x$$

$$\therefore n = \frac{1}{2\pi}\sqrt{\left(g \int_0^l y \mathrm{d}x \Big/ \int_0^l y^2 \mathrm{d}x\right)} \quad (15.7)$$

Note that: (a) the frequency of vibration is independent of the amplitude of vibration, but, as in the case of a single concentrated load, depends upon the static deflection of the shaft;

(b) equation (15.7) may be applied to any uniformly loaded shaft if the appropriate expression for y in terms of x is substituted;

(c) the equation is based on the assumption that the deflection curve of the vibrating shaft is similar to the static deflection curve and although, as we shall see later, this assumption cannot be true, nevertheless the frequency of vibration calculated from it is very nearly equal to the true frequency;

(d) if the maximum value of y is taken equal to the maximum static deflection of the shaft under the given uniform load and the given end conditions, any expression for y in terms of x which is consistent with the end conditions may be substituted in equation (15.7).

For a uniformly loaded shaft, we have from the ordinary beam theory that $EI.\mathrm{d}^4y/\mathrm{d}x^4 = w$, where E is Young's modulus of elasticity and I is the second moment of area of the shaft cross-section. This equation may be integrated four times in succession in order to derive an expression for y in terms of x. The values of the four constants of integration will be found by making use of the end conditions. Without detailing the steps it will be found that for a shaft with supported ends,

$$y = (w/24EI)(x^4 - 2lx^3 + l^3x)$$

From this equation, it will be found that

$$\int_0^l y\mathrm{d}x = \frac{w}{24EI}\frac{l^5}{5}$$

and

$$\int_0^l y^2\mathrm{d}x = \left(\frac{w}{24EI}\right)^2 \cdot \frac{31}{630} l^9$$

Substituting in equation (15.7):

$$n = \frac{1}{2\pi}\sqrt{\left(g.\frac{24EI}{wl^4}.\frac{630}{155}\right)}$$

But the maximum static deflection of a shaft with supported ends and a uniform load is given by:

$$\delta_s = (5/384)(wl^4/EI)$$

so that

$$EI/wl^4 = 5/384\delta_s \quad . \quad . \quad . \quad . \quad . \quad (15.8)$$

Substituting in the above equation:

$$n = \frac{1}{2\pi}\sqrt{\left(\frac{g}{\delta_s}.\frac{24.5}{384}.\frac{630}{155}\right)} = \frac{3 \cdot 53}{\sqrt{\delta_s}}$$

$$\therefore N = 211 \cdot 8/\sqrt{\delta_s} \quad . \quad . \quad . \quad . \quad (15.9)$$

where δ_s is the static deflection of the shaft in inches.

The constant in the numerator of equation (15.9) varies in magnitude with the end conditions. Its value may be calculated from the equation for the static deflection y in terms of x appropriate to the particular end conditions. The results are given in the following table.

End conditions	Equation for y	Maximum deflection δ	Constant
Both fixed	$y = \frac{w}{24EI}(x^4 - 2lx^3 + x^2l^2)$	$\frac{wl^4}{384EI}$	215·0
One fixed—one free .	$y = \frac{w}{24EI}(x^4 - 4lx^3 + 6x^2l^2)$	$\frac{wl^4}{8EI}$	234·3
One fixed—one supported	$y = \frac{w}{24EI}\left(x^4 - \frac{5}{2}lx^3 + \frac{3}{2}l^2x^2\right)$	$\frac{wl^4}{184{\cdot}6EI}$	213·3

(b) *Exact Solution.* Referring to Fig. 347, the load per unit length of the shaft is w and the frequency of transverse vibration is defined by the angular velocity ω radians per second. Let y be the displacement of the length δx of the shaft from the equilibrium position, then the dynamic, or inertia, load on δx is given by $(w\delta x/g)\omega^2 y$. Note that this is proportional to the product wy, so that the dynamic load, when vibrating, is not a uniform load and therefore the shape of the vibrating shaft cannot be the same as the static deflection curve of the shaft.

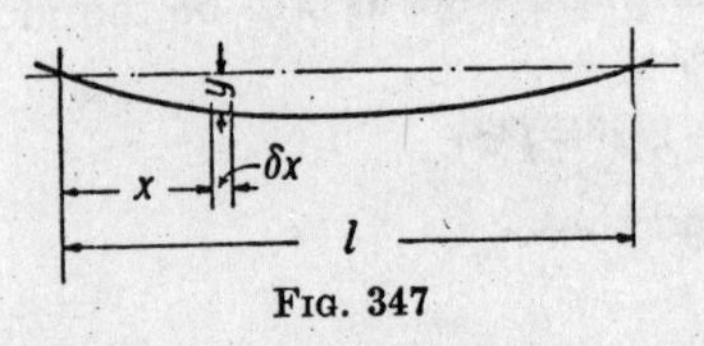

FIG. 347

From the beam theory, we have:

$$EI.\mathrm{d}^4y/\mathrm{d}x^4 = (w/g)\omega^2 y$$

This equation may be written:

$$\mathrm{d}^4y/\mathrm{d}x^4 - m^4y = 0 \quad . \quad . \quad . \quad . \quad (15.10)$$

where

$$m^4 = (w/g)(\omega^2/EI) \quad . \quad . \quad . \quad (15.11)$$

The solution of the differential equation (15.10) is

$$y = A\cos mx + B\sin mx + C\cosh mx + D\sinh mx \quad (15.12)$$

where the constants of integration A, B, C and D are to be determined from the end conditions.

For a shaft with supported ends, $y = 0$ when $x = 0$ or $x = l$.

Substituting in (15.12):

$$0 = A\cos 0 + B\sin 0 + C\cosh 0 + D\sinh 0$$
$$= A + C \quad . \quad . \quad . \quad . \quad . \quad . \quad . \quad . \quad . \quad . \quad (15\text{-}13)$$

and $$0 = A\cos ml + B\sin ml + C\cosh ml + D\sinh ml \quad (15.14)$$

Also $d^2y/dx^2 = 0$ when $x = 0$ or $x = l$, so that, differentiating (15.12) twice and substituting, we get:

$$0 = -A+C \quad . \quad . \quad . \quad . \quad . \quad (15.15)$$

and
$$0 = -Am^2 \cos ml - Bm^2 \sin ml + Cm^2 \cosh ml + Dm^2 \sinh ml \quad (15.16)$$

From (15.13) and (15.15) $\quad A = C = 0$
so that (15.14) becomes $\quad 0 = B \sin ml + D \sinh ml$
and (15.16) becomes $\quad 0 = -B \sin ml + D \sinh ml$
Adding these $\quad 0 = 2D \sinh ml$

But sinh ml cannot be zero. Hence $D = 0$.

$$\therefore \; B \sin ml = 0$$

If the shaft is vibrating, B cannot be zero,

$$\therefore \; \sin ml = 0$$

and
$$ml = \pi, \; 2\pi, \; 3\pi, \text{ etc.}$$

or
$$m = \pi/l, \; 2\pi/l, \; 3\pi/l, \text{ etc.} \quad . \quad . \quad (15.17)$$

Substituting the smallest value of m in equation (15.11):

$$w\omega^2/gEI = (\pi/l)^4$$

$$\therefore \; \omega = \pi^2\sqrt{(gEI/wl^4)}$$

and, substituting for EI/wl^4 from equation (15.8):

$$\omega = \pi^2\sqrt{(g \, . \, 5/384\delta_s)}$$

$$\therefore \; N = 30\omega/\pi = 30/\pi\sqrt{(g \, . \, 5/384\delta_s)} = 211{\cdot}4/\sqrt{\delta_s} \quad . \quad (15.18)$$

where δ_s is the static deflection of the shaft in inches.

If this result is compared with equation (15.9), it will be seen how closely the two expressions for N agree.

Equation (15.18) is obtained by using the smallest value of m from equation (15.17) and therefore gives the lowest or fundamental frequency of vibration. Since ω is proportional to m^2, other frequencies of vibration for the uniformly loaded shaft with supported ends are 4, 9, 16, etc., times the fundamental frequency.

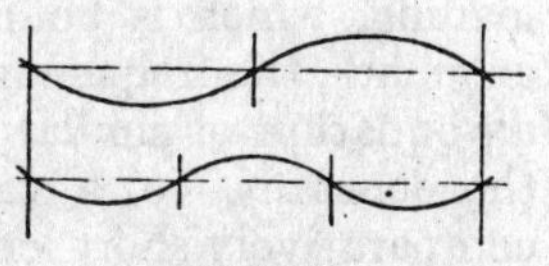

FIG. 348

When vibrating with a frequency equal to four times the fundamental frequency, the shaft will have a node at its mid-point as well as at each support. When vibrating with nine times the fundamental frequency, the shaft will have two intermediate nodes one-third of the length from each support, and so on for the higher frequencies (Fig. 348).

The solution of the differential equation (15.10) for other end conditions may be obtained in a similar way. The results are given below, together with the values of the constant in the numerator of equation (15.18).

End conditions	Solution	Smallest value of ml	Constant
Both supported	$\sin ml = 0$	π	211·4
Both fixed	$\cosh ml = \sec ml$	4·73	214·4
One fixed—one free	$\cosh ml = -\sec ml$	1·875	233·3
One fixed—one supported . .	$\tanh ml = \tan ml$	3·927	213·0

If the values of the constants given in the last column of this table are compared with those in the last column of the table on p. 542, it will be seen how closely they agree.

197. System of Several Loads attached to the Same Shaft. So far the two simple cases in which the shaft carries either a single concentrated load or a uniformly distributed load have been considered. In most practical cases the vibrating system is more complicated and consists of several unequal masses spaced at different intervals along the shaft, together with a more or less uniformly distributed load due to its own weight. There are several methods which may be used to determine the natural frequency of the transverse vibrations of such a system.

First Method. This is an approximate method which requires that the maximum static deflection of the shaft under the given system of loading be first obtained, either by calculation or by the usual graphical method. We have seen that, when the shaft carries a single concentrated load, the frequency of transverse vibration is given by equation (15.4). When the shaft carries a uniformly distributed load, the frequency is given by equation (15.18). We should therefore expect that for any actual system of loading, which is bound to be intermediate between these two extremes, the frequency of transverse vibration would be given by an equation of similar form, but with an intermediate value for the constant, i.e. a value between 187·8 and 211·4. Unless a comparatively short length of the shaft carries very much heavier loading than the rest of the shaft, the appropriate constant is likely to be nearer to the higher of these two figures.

Hence the frequency of the transverse vibrations of a shaft with any arbitrary system of loading is given approximately by

$$N = (187{\cdot}8/\sqrt{\delta_m})K \quad . \quad . \quad . \quad . \quad (15.19)$$

where δ_m is the maximum static deflection of the shaft and K has a value between 1·0 and 1·127. This may be taken for a normal system of loading as, say, 1·10, so that

$$N = 207/\sqrt{\delta_m} \quad . \quad . \quad . \quad . \quad . \qquad (15.20)$$

The above equation must be used with a certain amount of caution. For most systems of loading it will give a close approximation to the true frequency, but, if the loading is mainly concentrated close to the supports and therefore at some distance from the point of maximum deflection, the calculated frequency may be much too low. A moment's reflection will show why this should be so. Although the constants in the numerators of equations (15.4), (15.18) and (15.20) do not differ very widely, the maximum deflection of the shaft is used in the denominator of equations (15.18) and (15.20), whereas the deflection under the load is used in equation (15.4). If the deflection under the main part of the load differs widely from the maximum deflection, it is hardly to be expected that equation (15.20) will give the correct frequency.

Second Method. This is a semi-empirical method suggested by Professor Dunkerley. It can only be applied where the actual shaft is of uniform diameter or where it can be replaced by a shaft of uniform diameter without serious error.

Let W_1, W_2, etc. = concentrated loads at different points along the shaft,

W_s = weight of shaft,

δ_1, δ_2, etc. = static deflections of the shaft under loads W_1, W_2, etc., when each load is acting separately,

δ_s = maximum deflection of the shaft under its own weight,

N_1, N_2, etc. = frequency of the transverse vibrations with each load acting separately,

N_s = frequency of the transverse vibrations of the shaft under its own weight

and N = frequency of the transverse vibrations of the system as a whole.

Then, Dunkerley's empirical equation is:

$$1/N^2 = 1/N_1{}^2 + 1/N_2{}^2 + \ldots + 1/N_s{}^2 \qquad (15.21)$$

But

$N_1 = 187{\cdot}8/\sqrt{\delta_1}$, $N_2 = 187{\cdot}8/\sqrt{\delta_2}$, etc., $N_s = 211{\cdot}4/\sqrt{\delta_s}$

Substituting,

$$1/N^2 = \delta_1/187{\cdot}8^2+\delta_2/187{\cdot}8^2+\ldots+\delta_s/211{\cdot}4^2$$

$$\therefore N = \frac{187{\cdot}8}{\sqrt{(\delta_1+\delta_2+\ldots+\delta_s/1{\cdot}27)}} \quad . \quad . \quad (15.22)$$

If this equation is compared with equation (15.19), it is clear that the frequency as calculated by Dunkerley's method can only have the same value as that calculated by the first method, if

$$K=\sqrt{\frac{\delta_m}{\delta_1+\delta_2+\ldots+\delta_s/1{\cdot}27}}$$

or, taking K as 1·10, if

$$\delta_m = 1{\cdot}21(\delta_1+\delta_2+\ldots+\delta_s/1{\cdot}27)$$

That this relation is approximately satisfied for any normal system of loading may easily be verified. (See Example 3.)

Third Method. This is usually known as the energy method and is essentially the same as that used in deducing the approximate expression for the frequency of the vibrations of a uniformly loaded shaft. If the deflection curve of the vibrating shaft is assumed to be similar in shape to the static deflection curve for the same system of loading, then, proceeding as in Article 196, we shall get, instead of equation (15.7), the following:

$$n = \frac{1}{2\pi}\sqrt{g\frac{\Sigma Wy}{\Sigma Wy^2}} \quad . \quad . \quad . \quad . \quad (15.23)$$

The values of ΣWy and ΣWy^2 may be found by setting down the data in tabular form as shown in Example 3.

Fourth Method. This is essentially a method of arriving at the true shape of the deflection curve of the vibrating shaft and at the true frequency by successive approximations. It was developed by Professor Stodola chiefly for use in determining the whirling speed of a turbine rotor. In applying the method to a particular shaft, a start is made by assuming:

(i) a deflection curve for the shaft, i.e. both as regards the shape and the scale of the deflection,

(ii) a frequency of vibration.

Of these assumptions, the shape of the deflection curve is the only one which affects the accuracy of the first approximation to the true frequency. The assumed shape should be a reasonable one, consistent with the end conditions, the distribution of the masses along the shaft, and the variations of stiffness of the shaft along its length.

From the known masses and the assumed frequency and deflections, the dynamic loads are calculated. These loads are then used to derive a dynamic deflection curve either by calculation or, where the cross-section of the shaft varies from one end to the other, by a graphical construction. The derived deflection curve may, and usually does, give very different values for the deflections along the shaft from those which were assumed. but a first approximation to the true frequency may be obtained as follows:

Let n = assumed frequency of vibration,

y = assumed deflection,

n' = first approximation to the true frequency

and y' = derived deflection due to dynamic loading.

Then the dynamic loads used in obtaining the deflection curve are $(W_1/g)(2\pi n)^2 y_1$, $(W_2/g)(2\pi n)^2 y_2$, etc., and the deflections actually caused by these loads are measured from the derived deflection curve and are given by y_1', y_2', etc. Clearly, then, since the dynamic loads which cause the deflections y_1', y_2', etc., must be equal to $(W_1/g)/(2\pi n)^2 y_1$, $(W_2/g)(2\pi n)^2 y_2$, etc., it follows that $n^2 y_1 = (n')^2 y_1'$, $n^2 y_2 = (n')^2 y_2'$, etc., and

$$n'/n = \sqrt{(y_1/y_1')} = \sqrt{(y_2/y_2')}, \text{ etc.}$$

This can only be true if the ratio y'/y is the same at all points along the shaft, in other words, if the derived deflection curve and the assumed curve are geometrically similar. But where y'/y is not constant the ratio of the maximum deflections for the two curves may be used to find n'.

The process may be repeated, the dynamic loads being calculated for frequency n' and the deflections y' given by the derived curve. A second deflection curve may then be obtained giving deflections y'', and a second approximation to the true frequency will be given by

$$n'' = n'\sqrt{(y'/y'')}$$

The process is convergent and it will be found that the ratio y'/y'' is not only very close to unity, but also much more nearly constant along the length of the shaft. This implies that the second derived deflection curve is very closely similar in shape to the true deflection curve, and the value of n'' is very close to the true frequency.

Example 3. Find the frequency of the natural transverse vibrations of a shaft 10 ft. long and 4 in. dia. which is supported at the ends and loaded as shown in Fig. 349. Young's modulus for the shaft material is 30.10^6 lb/in².

Since the shaft is of uniform diameter, the static deflection under each of the four concentrated loads may be obtained by calculation. Alternatively, the static deflection curve may be obtained by the usual graphical construction. It is not proposed to give in detail either the calculations or the graphical construction. The shape of the deflection curve is shown in the figure and the maximum deflection of the shaft under the given system of loading is 0·0882 in.

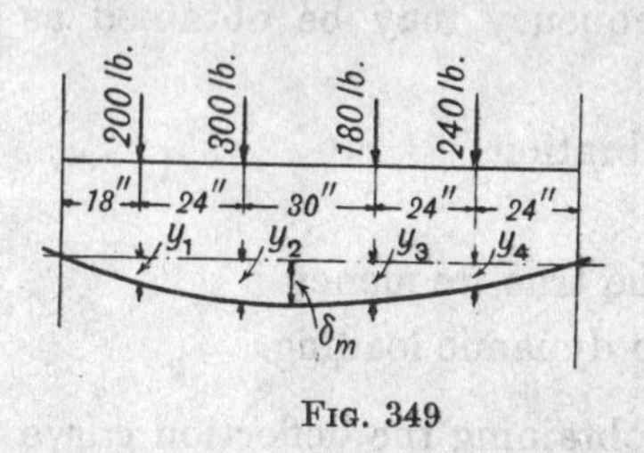

Fig. 349

First Method. The frequency of the transverse vibrations may be calculated from (15.20).

Then $N = 207/\sqrt{0{\cdot}0882} = 694$ vibrations per min.

Second Method. If Dunkerley's method is used, it is necessary first of all to determine the deflection of the shaft under the load when each load acts separately. If a and b are the distances of the load W from the two supports, then the deflection under the load is given by $\delta = Wa^2b^2/3EIl$, where l is the length of the shaft, I is the second moment of area of the shaft cross-section and E is Young's modulus. From this equation the following table may be filled in:

W	*a* in.	*b* in.	δ in.
200	18	102	0·0050
300	42	78	0·0237
180	72	48	0·0159
240	96	24	0·0094

The weight of the shaft $W_s = (\pi/4)d^2l\rho$, where $\rho = 0{\cdot}283$ lb/in³ for steel.

$$\therefore\ W_s = \pi/4.4^2.120.0{\cdot}283 = 427 \text{ lb}$$

The deflection of the shaft under its own weight

$$\delta_s = 5/384.W_sl^3/EI = 0{\cdot}0254 \text{ in.}$$

$$\therefore\ \delta_1+\delta_2+\delta_3+\delta_4+\delta_s/1{\cdot}27 = 0{\cdot}0740 \text{ in.}$$

and, substituting in equation (15.22),

$$N = 187{\cdot}8/\sqrt{0{\cdot}0740} = 691 \text{ vibrations per min}$$

N.B.—In the above example

$$\frac{\delta_m}{\delta_1+\delta_2+\delta_3+\delta_4+\delta_s/1{\cdot}27} = \frac{0{\cdot}0882}{0{\cdot}0740} = 1{\cdot}19$$

Third Method. In order to apply the energy method, the total deflection under each load is either calculated or measured from the deflection curve for the complete system of loading. To each concentrated load is added the weight of that part of the shaft included between the mid-points of the panels into which the concentrated loads divide the shaft. For example, the weight of a 1 ft 9 in. length of shaft is included with the first concentrated load, the weight of a 2 ft 3 in. length with the second load and so on. It should be pointed out that the effect of the shaft would be more accurately taken into account by dividing it into a large number of parts and treating the weight of each part as a concentrated load situated at its mid-point. In the present example, the greater accuracy obtained would not be worth the extra labour involved. The following table is then filled in and the values of ΣWy and ΣWy^2 are determined.

Concentrated load	Shaft weight included	Total load, W	Deflection, y	Wy	Wy^2
200	75	275	0·0410	11·27	0·462
300	96	396	0·0792	31·35	2·487
180	96	276	0·0838	23·12	1·937
240	85	325	0·0521	16·92	0·884
				ΣWy 82·66	ΣWy^2 5·770

Substituting in (15.23),

$$N = 187{\cdot}8\sqrt{(82{\cdot}66/5{\cdot}770)} = 710 \text{ vibrations per min}$$

Fourth Method. The static deflection curve given in Fig. 349 could be used as the assumed deflection curve when applying Stodola's method. Indeed, if this curve is known, it is the best one to use. However, in order to illustrate the general procedure when the static deflection curve is not available, it will be assumed in this example that the dynamic deflection curve is a sine curve with a maximum deflection of one inch, say. The deflections under the loads are then as given in the second line of the following table:

Load lb	200	300	180	240
y in. . . .	0·4540	0·8910	0·9511	0·5878
F lb . . .	90·8	267·3	171·2	141·1
y' in. . .	0·0319	0·0624	0·0659	0·0411
y/y' . . .	14·23	14·27	14·40	14·29

The dynamic load $= F = (W/g)\omega^2 y = (\omega^2/g)Wy$ and, since any value of ω may be chosen, it is convenient to choose that value which makes $F = Wy$ lb, i.e. $\omega^2 = 12g = 386{\cdot}4$. The values of F are given in the third line of the table. The dynamic load due to the mass of the shaft is not uniform and perhaps the simplest plan is to divide the shaft into 2-ft lengths. The mean dynamic load per unit length for each of these sections may then be estimated, and the total dynamic load for each section may be assumed to act at the mid-point.

The deflection curve for the dynamic loading may then be derived by calculation, using Macaulay's method, or graphically.

The values of the deflections under the loads were actually calculated and are given in the fourth line of the table.

The values of the ratio

$$\frac{\text{Assumed deflection}}{\text{Derived deflection}} = \frac{y}{y'}$$

are given in the fifth line, and since they do not differ greatly, the mean value 14·3 may be used to calculate the approximate frequency of the vibrations.

$$n'/n = \omega'/\omega = \sqrt{(y/y')} = \sqrt{14{\cdot}3}$$

$$\text{But } n = \omega/2\pi = (1/2\pi)\sqrt{386{\cdot}4} = 3{\cdot}14 \text{ per sec}$$

$$\therefore\ n' = 3{\cdot}14\sqrt{14{\cdot}3} = 11{\cdot}86 \text{ per sec} = 712 \text{ per min}$$

The ratio y/y' is so nearly constant for all the loads in this example that there would be no point in repeating the process.

198. Damped Vibrations. It is well known that, if a body held in position by elastic constraints is displaced from the equilibrium position and then released, the amplitude of the resulting vibration gradually diminishes as the vibration energy is dissipated in overcoming friction. The vibration is said to be damped.

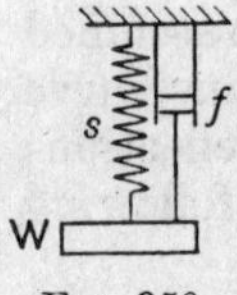

FIG. 350

The resistance to the movement of the body is provided partly by the medium in which the vibration takes place, partly by the internal friction, or hysteresis, of the material of the elastic constraints and partly, in some cases, by a dashpot or other external damping device. It is usual to assume that the damping, whatever its cause, is linear, i.e. that the resistance to the motion of the body is directly proportional to the speed of movement. Under these conditions the resulting motion of the body may be analysed as follows:

Fig. 350 shows a mass suspended from one end of a spiral spring, the other end of which is fixed, and a dashpot is provided between the mass and the rigid support.

Let W = weight of the mass,
s = stiffness of the spring, i.e. the force required to produce unit extension of the spring,
f = damping force per unit velocity,
n = frequency of the *free* vibrations,
n_d = frequency of the *damped* vibrations,
y = displacement of the mass from the equilibrium position at time t
and δ = static deflection of the spring = W/s.

Then, the force required to accelerate the mass $= W/g.\mathrm{d}^2y/\mathrm{d}t^2$,
the force required to move the piston of the dashpot $= f.\mathrm{d}y/\mathrm{d}t$,
and the force required to extend the spring $= s.y$.

The algebraic sum of these three forces is zero.

$$\therefore \frac{W}{g}\frac{\mathrm{d}^2y}{\mathrm{d}t^2}+f\frac{\mathrm{d}y}{\mathrm{d}t}+sy = 0$$

$$\frac{\mathrm{d}^2y}{\mathrm{d}t^2}+a\frac{\mathrm{d}y}{\mathrm{d}t}+by = 0 \quad . \quad . \quad . \quad . \quad (15.24)$$

where $a = fg/W$, $b = sg/W = g/\delta = (2\pi n)^2$, from equation (15.2).

The solution of this equation depends upon whether $b \gtreqless \left(\frac{a}{2}\right)^2$.

If $b > (a/2)^2$, which is true for most practical cases of damping,

$$y = \mathrm{e}^{-(a/2)t}[C_1 \cos \sqrt{\{b-(a/2)^2\}}t+C_2 \sin\sqrt{\{b-(a/2)^2\}t}] \quad (15.25)$$

which may be written in the more convenient form:

$$y = C\mathrm{e}^{-(a/2)t} \cos [\sqrt{\{b-(a/2)^2\}}t-\alpha] \quad . \quad . \quad (15.26)$$

The constants of integration C and α are determined from the initial conditions of the motion. Thus, if t is measured from the instant at which the mass is released after the initial displacement A, then $C = A$ and $\alpha = 0$, so that:

$$y = A\mathrm{e}^{-(a/2)t} \cos \sqrt{\{b-(a/2)^2\}}t \quad . \quad . \quad (15.27)$$

The shape of the curve obtained by plotting this equation is shown in Fig. 351.

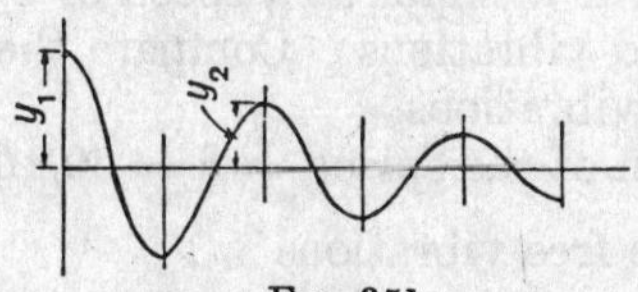

FIG. 351

From equation (15.27) the periodic time is given by:

$$t_p = \frac{2\pi}{\sqrt{\{b-(a/2)^2\}}} \quad . \quad . \quad . \quad (15.28)$$

$$\therefore \text{ frequency } n_d = 1/t_p = (1/2\pi)\sqrt{\{b-(a/2)^2\}} \quad . \quad . \quad (15.29)$$

If there is no damping, $a = 0$ and $n = (1/2\pi)\sqrt{b} = (1/2\pi)\sqrt{(g/\delta)}$, which agrees with equation (15.2).

If y_1, y_2 are successive values of the amplitude on the same side of the equilibrium position, then

$$\frac{y_1}{y_2} = \frac{Ae^{-(a/2)t}}{Ae^{-(a/2)(t+t_p)}} = e^{(a/2)t_p} = \text{constant} \quad . \quad (15.30)$$

since a and t_p are both constant.

The index $(a/2)t_p$ is termed the logarithmic decrement.

If $b = (a/2)^2$ the system is said to be critically damped. There is no true vibration, but the system returns gradually to the equilibrium position.

If $b < (a/2)^2$ conditions are similar except that the return to the equilibrium position takes place more slowly.

In all cases of viscous damping, an infinite time is theoretically required for the system to come to rest, but with critical damping the approach to the rest position takes place most rapidly. With less damping time is wasted in oscillation, while with more damping the rest position is approached more slowly.

Most instruments are required to give as nearly as possible an instantaneous response of the recording pointer to a change in the value of the quantity to be measured. It follows that such instruments should have the pointer critically damped.

For critical damping, $b = a^2/4$, but $b = (2\pi n)^2$, where n is the natural undamped frequency of vibration of the system.

$\therefore$ for critical damping,

$$a/2\pi n = 2$$

Example 4. A mass of 50 lb is suspended from a spring of stiffness 100 lb/in. A dashpot is fitted and it is found that the amplitude of the vibration diminishes from its initial value of 1 in. to a value of 0·25 in. in two complete oscillations. Find the resistance offered by the dashpot at a speed of 1 ft/s and the frequency of the damped vibrations. Compare the latter with the frequency of the free vibrations.

The static deflection of the spring $= \delta = 50/100 = 0{\cdot}5$ in.

$\therefore$ frequency of the free vibrations

$$= n = (1/2\pi)\sqrt{(g/\delta)} = 4{\cdot}424 \text{ per sec}$$

The ratio of the amplitudes of the damped vibrations after two complete oscillations, from equation (15.30),

$$= (e^{(a/2)t_p})^2 = e^{at_p}$$

But this ratio is given as $1/0{\cdot}25 = 4$,

$$\therefore\ at_p = \log_e 4 = 1{\cdot}3863$$

Substituting for t_p from equation (15.28):

$$\frac{2\pi a}{\sqrt{\{b-(a/2)^2\}}} = 1{\cdot}3863$$

$$\therefore\ b-(a/2)^2 = (2\pi a/1{\cdot}3863)^2 = 20{\cdot}54a^2$$

$$\therefore\ 20{\cdot}79a^2 = b \quad \text{and} \quad a/\sqrt{b} = a/2\pi n = 0{\cdot}219$$

But $b = g/\delta = 32{\cdot}2.12/0{\cdot}5 = 772{\cdot}8\ \text{sec}^{-2}$

$$\therefore\ a = \sqrt{(772{\cdot}8/20{\cdot}79)} = 6{\cdot}096\ \text{sec}^{-1}$$

But $a = fg/W$,

$$\therefore\ f = 6{\cdot}096.50/32{\cdot}2 = 9{\cdot}46\ \text{lb per ft/s}$$

The frequency of the damped vibrations, from equation (15.29):

$$n_d = \frac{1}{2\pi}\sqrt{\left\{b-\left(\frac{a}{2}\right)^2\right\}} = \frac{1}{2\pi}\sqrt{(772{\cdot}8-9{\cdot}29)} = 4{\cdot}397 \text{ per sec.}$$

This is $4{\cdot}397/4{\cdot}424 = 0{\cdot}994$ or 99·4% of the frequency of the free vibrations.

Example 5. A flywheel of moment of inertia 500 lb ft² is fixed to one end of a vertical shaft, dia. 1 in. and length 3 ft. The other end of the shaft is fixed. The torsional oscillations of the flywheel are damped by means of a vane V, Fig. 352, which moves in a dash-pot D filled with oil. The amplitude of oscillation is found by experiment to diminish to one-twentieth of its initial value in three complete oscillations. Assuming the damping torque to be directly proportional to the angular velocity, find its magnitude at a speed of 1 rad/s. The modulus of rigidity of the shaft material is 12.10^6 lb/in².

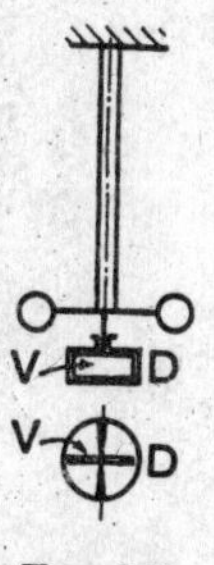

Fig. 352

The frequency of the free vibrations, from equation (15.5),

$$= n = \frac{1}{2\pi}\sqrt{\frac{q}{I}}$$

But $$q = \frac{CJ}{l} = \frac{12.10^6}{36}.\frac{\pi}{32}.1^4 = 3{\cdot}27.10^4 \text{ lb in/rad}$$

$$\therefore\ n = \frac{1}{2\pi}\sqrt{\frac{3{\cdot}27.10^4.32{\cdot}2.12}{500.12^2}} = 2{\cdot}11 \text{ per sec}$$

For damped vibrations, the equation of motion may be found as follows:

Let θ = angular displacement from the equilibrium position at time t,

q = torsional stiffness of the shaft

and T_f = damping torque per unit angular velocity.

Then $$I.\mathrm{d}^2\theta/\mathrm{d}t^2+T_f.\mathrm{d}\theta/\mathrm{d}t+q.\theta = 0$$

or $$\mathrm{d}^2\theta/\mathrm{d}t^2+a_1.\mathrm{d}\theta/\mathrm{d}t+b_1\theta = 0$$

where $a_1 = T_f/I$ and $b_1 = q/I = (2\pi n)^2$.

The solution of this equation will be of the same form as equation (15.27), if A radians is the initial amplitude of the torsional vibrations and a_1, b_1 are substituted for a, b. By analogy, the ratio of the amplitudes at the beginning and at the end of three complete oscillations may be obtained from equation (15.30).

Since this ratio is given as 20 to 1, we have

$$20/1 = \mathrm{e}^{a_1/2.3t_p} \quad \text{or} \quad (3/2)a_1t_p = \log_e 20$$

and, from equation (15.28),

$$t_p = \frac{2\pi}{\sqrt{\{b_1-(a_1/2)^2\}}}$$

$$\therefore \frac{3}{2}.\frac{2\pi a_1}{\sqrt{\{b_1-(a_1/2)^2\}}} = \log_e 20 = 2{\cdot}9957$$

$$\therefore \frac{3\pi a_1}{2{\cdot}9957} = \sqrt{\{b_1-(a_1/2)^2\}}$$

$$\therefore 9{\cdot}895a_1^2 = b_1-a_1^2/4$$

$$\therefore a_1 = \sqrt{(b_1/10{\cdot}15)}$$

Also $$b_1 = \frac{q}{I} = \frac{3{\cdot}27.10^4.32{\cdot}2.12}{500.12^2} = 175{\cdot}5 \text{ sec}^{-2}$$

$$\therefore a_1 = \sqrt{(175{\cdot}5/10{\cdot}15)} = 4{\cdot}16 \text{ sec}^{-1}$$

But $a_1 = T_f/I$, so that $T_f = 4{\cdot}16.500/32{\cdot}2 = 64{\cdot}8$ lb ft per rad/s.

The frequency of the damped vibrations, from equation (15.29):

$$n_d = \frac{1}{2\pi}\sqrt{\left\{b_1-\left(\frac{a_1}{2}\right)^2\right\}} = \frac{1}{2\pi}\sqrt{\left\{175{\cdot}5-\left(\frac{4{\cdot}16}{2}\right)^2\right\}} = 2{\cdot}08 \text{ per sec}$$

$$\text{Ratio } n_d/n = 0{\cdot}988 \quad \text{or} \quad 98{\cdot}8\%$$

199. Forced Vibrations. A type of vibration of great practical importance is that in which the body is subjected to an external periodic force. Referring to Fig. 353, let a harmonic force represented by $F \cos \omega t$ act on the mass. Then the equation of motion may be written:

$$W/g \, . \, \mathrm{d}^2y/\mathrm{d}t^2 + f \, . \, \mathrm{d}y/\mathrm{d}t + s \, . \, y = F \cos \omega t$$

or

$$\mathrm{d}^2y/\mathrm{d}t^2 + a \, . \, \mathrm{d}y/\mathrm{d}t + by = c \cos \omega t \quad . \quad . \quad (15.31)$$

where $a = fg/W$, $b = sg/W$ and $c = Fg/W$.

The complete solution of equation (15.31) is found by adding to the solution of equation (15.24) of the last article a particular solution of equation (15.31) as it stands. The particular solution may be obtained by assuming $y = H \sin \omega t + K \cos \omega t$. Then, on substituting and equating coefficients, we get

$$H = \frac{ca\omega}{(b-\omega^2)^2 + a^2\omega^2}, \quad K = \frac{c(b-\omega^2)}{(b-\omega^2)^2 + a^2\omega^2}$$

$$\therefore \ y = \frac{c}{(b-\omega^2)^2 + a^2\omega^2}\{a\omega \sin \omega t + (b-\omega^2) \cos \omega t\}$$

This may be written in the alternative and more convenient form:

$$y = \frac{c}{\sqrt{\{(b-\omega^2)^2 + a^2\omega^2\}}} \cos (\omega t - \beta) \quad . \quad (15.32)$$

where

$$\tan \beta = \frac{a\omega}{b-\omega^2} \quad . \quad . \quad . \quad . \quad (15.33)$$

The complete solution of equation (15.31) is:

$$y = A\mathrm{e}^{-(a/2)t} \cos \sqrt{\{b-(a/2)^2\}}t + \frac{c}{\sqrt{\{(b-\omega^2)^2 + a^2\omega^2\}}} \cos (\omega t - \beta)$$

In this equation the first term represents the transient vibration which dies out rapidly owing to the damping effect of friction. The second term represents the forced vibration which is maintained by the periodic force. Hence, when a steady state of vibration has been reached, the motion of the mass is represented completely by equation (15.32). From this equation the amplitude of the forced vibration is evidently given by:

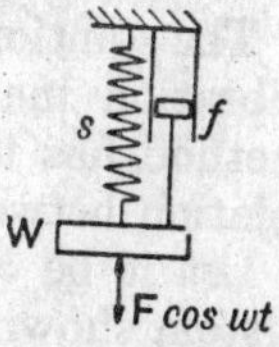

FIG. 353

$$y_{\max} = \frac{c}{\sqrt{\{(b-\omega^2)^2 + a^2\omega^2\}}}$$

Dividing numerator and denominator by b:

$$y_{\max} = \frac{c/b}{\sqrt{\{(1-\omega^2/b)^2 + a^2\omega^2/b^2\}}} \quad . \quad . \quad (15.34)$$

but $c/b = F/s = \Delta$, the deflection produced by a static load F

Also $b = g/\delta = (2\pi n)^2$ and $\omega = 2\pi n_f$, where n_f is the frequency of the applied periodic force.

$$\therefore\ y_{max} = D.\Delta \quad . \quad . \quad . \quad . \quad (15.35)$$

where

$$D = \frac{1}{\sqrt{\{(1-\omega^2/b)^2+a^2\omega^2/b^2\}}}$$

$$= \frac{1}{\sqrt{[\{1-(n_f/n)^2\}^2+(a/2\pi n)^2(n_f/n)^2]}} \quad . \quad (15.36)$$

D is termed the dynamic magnifier, since it gives the factor by which the static deflection produced by a force F must be multiplied in order to obtain the amplitude of the forced vibrations caused by the harmonic force $F \cos \omega t$. If the vibration is undamped, a is zero and the second term under the root sign vanishes. Hence

$$D = \frac{1}{1-(n_f/n)^2} \quad . \quad . \quad . \quad . \quad (15.37)$$

For the undamped vibration, equation (15.37) shows that D is infinitely large when $n_f = n$.

For the damped vibration, equation (15.36) shows that D is always finite even when $n_f = n$.

The phase difference between the force and the displacement is given by equation (15.33). Dividing the numerator and denominator by b and substituting $b = (2\pi n)^2$, this equation may be written:

$$\tan \beta = \frac{a/2\pi n\,.\,n_f/n}{1-(n_f/n)^2} \quad . \quad . \quad . \quad (15.38)$$

The solution of the equation of motion for a forced and damped vibration may also be obtained by the use of vectors. This method has the advantage that it gives a clearer picture of the relation between the physical quantities involved.

Let it be assumed that the displacement of the mass in the system shown in Fig. 353 under the action of the applied simple harmonic force $F \cos \omega t$ is itself simple harmonic, so that it can be represented by the equation

$$y = A \cos (\omega t-\beta)$$

Then

$$dy/dt = -\omega A \sin (\omega t-\beta) = \omega A \cos \{90°+(\omega t-\beta)\}$$

and

$$d^2y/dt^2 = -\omega^2 A \cos (\omega t-\beta) = \omega^2 A \cos\{180°+(\omega t-\beta)\}$$

The force required to extend the spring

$$= s.y = sA \cos(\omega t - \beta) \quad . \quad . \quad . \quad . \quad (15.39)$$

the force required to overcome the resistance of the dashpot

$$= f.\mathrm{d}y/\mathrm{d}t = f\omega A \cos\{90^\circ + (\omega t - \beta)\} \quad . \quad (15.40)$$

and the force required to accelerate the mass

$$= m.\mathrm{d}^2y/\mathrm{d}t^2 = m\omega^2 A \cos\{180^\circ + (\omega t - \beta)\} \quad (15.41)$$

The algebraic sum of these three forces must at the given instant be equal to the applied force $F \cos \omega t$.

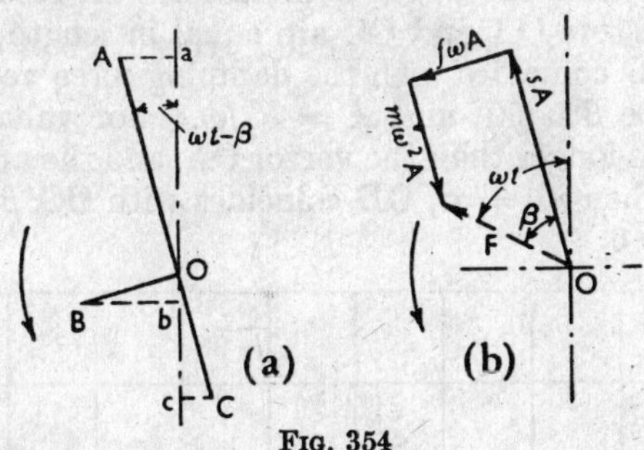

FIG. 354

Set off three vectors OA, OB and OC at successive angular intervals of 90°, as shown in Fig. 354 (a). Let the three vectors represent to the same scale the maximum values sA, $f\omega A$ and $m\omega^2 A$ of the three forces and let the inclination of OA to the vertical represent the angle $\omega t - \beta$. Then the projected lengths Oa, Ob and Oc of the three vectors along the vertical will clearly represent the instantaneous values of the three forces at time t, and Od the algebraic sum of Oa, Ob and Oc must represent the value $F \cos \omega t$ of the applied force at the same instant.

For this to be true at all values of t, the force vector OD must be the vector sum of OA, OB and OC or F must be the vector sum of sA, $f\omega A$ and $m\omega^2 A$, as shown at (b).

From the geometry of the figure, we get

$$F = \mathrm{OD} = \sqrt{\{(\mathrm{OA} - \mathrm{OC})^2 + \mathrm{OB}^2\}}$$

$$= A\sqrt{\{(s - m\omega^2)^2 + f^2\omega^2\}}$$

or

$$A = \frac{F}{\sqrt{\{(s - m\omega^2)^2 + f^2\omega^2\}}}$$

$$= \frac{1}{\sqrt{\{(1 - m\omega^2/s)^2 + f^2\omega^2/s^2\}}} \cdot \frac{F}{s}$$

$$= D.\Delta$$

where as before $\Delta = F/s =$ the deflection produced by the static load F and D is the dynamic magnifier, which may be expressed in the more convenient form of equation (15.36).

The angle DOA $= \beta =$ phase difference between the force vector and the displacement vector and

$$\tan \beta = \frac{\text{OB}}{\text{OA}-\text{OC}} = \frac{f\omega}{s-m\omega^2}$$

which may be expressed in the form of equation (15.38).

The effect of increasing the forcing frequency is easily seen. It increases the length of the vectors OB and OC relative to the vector OA, so that the angle β increases. At resonance, when $n_f = n$, the vectors OA and OC are equal in length, so that the force vector OD coincides with the damping force vector OB, the phase difference β is 90° and $A = F/f\omega$. For values of $n_f > n$ the vector OC is longer than the vector OA, and the angle $\beta > 90°$. In the limit, when $n_f = \infty$, OD coincides with OC, $\beta = 180°$ and $A = F/m\omega^2 = 0$.

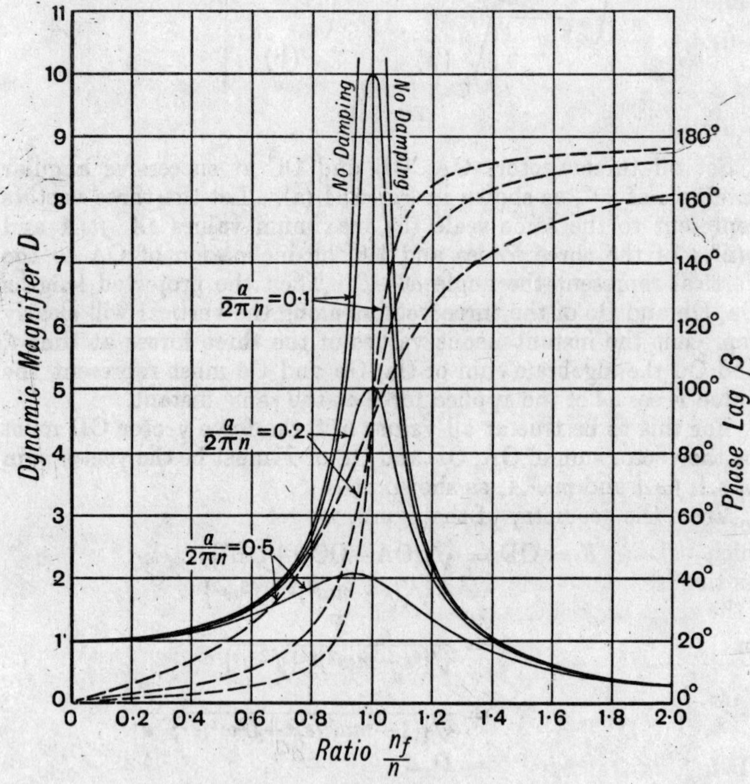

Fig. 355

Values of D and β are plotted in Fig. 355 for different values of n_f/n and for values of $a/2\pi n = 0{\cdot}1$, $0{\cdot}2$ and $0{\cdot}5$.

Values of D for an undamped vibration are also shown. For convenience they are shown positive throughout the range of n_f/n. It will be seen that:

(a) damping has very little effect on D except close to resonance (when $n_f/n = 1$);

(b) the value of β changes from 0° when $n_f/n = 0$, to 90° when $n_f/n = 1$, and to 180° when $n_f/n = \infty$;

(c) the value of D at resonance is slightly less than the maximum value.

From equation (15.36) the dynamic magnifier at resonance is given by:

$$D = 2\pi n/a \quad . \quad . \quad . \quad . \quad . \quad (15.42)$$

There are three important frequencies to be distinguished in connection with a forced and damped vibration. They are:

(i) the free undamped frequency, given by equation (15.2).

$$n = (1/2\pi)\sqrt{b} = (1/2\pi)\sqrt{(g/\delta)}$$

(ii) the damped frequency, given by equation (15.29)

$$\begin{aligned} n_d &= (1/2\pi)\sqrt{\{b-(a/2)^2\}} \\ &= n\sqrt{(1-a^2/4b)} = n\sqrt{\{1-\tfrac{1}{4}(a/2\pi n)^2\}} \quad . \quad (15.43) \end{aligned}$$

(iii) the forcing frequency which corresponds to maximum amplitude. This may be obtained from (15.34) by differentiating with respect to ω and equating to zero, when it will be found that

$$\omega = 2\pi n_f = \sqrt{(b-a^2/2)}$$

or

$$n_f = n\sqrt{(1-a^2/2b)} = n\sqrt{\{1-\tfrac{1}{2}(a/2\pi n)^2\}} \quad . \quad (15.44)$$

It will be seen that the damped frequency n_d is intermediate between the free undamped frequency n and the forcing frequency for maximum amplitude. In practical problems of systems which undergo forced and damped vibrations, $a/2\pi n$ is usually less than $0{\cdot}1$ and seldom exceeds $0{\cdot}4$, so that the relative values of the three frequencies considered are:

	Natural frequency, n	Damped frequency, n_d	Forcing frequency for maximum D, n_f
When $a/2\pi n = 0{\cdot}1$	1	0·9988	0·9975
When $a/2\pi n = 0{\cdot}4$	1	0·9798	0·9592

It follows that the damped frequency and the forcing frequency for maximum amplitude may with little error be assumed to be identical with the free undamped frequency. Note, too, that although the maximum value of the dynamic magnifier D will be slightly greater than its value at resonance, as given by equation (15.42), the difference is small and may often be neglected.

Example 6. A mass of 20 lb is suspended from one end of a spiral spring, the other end of which is fixed. The stiffness of the spring is 50 lb/in. Damping, which may be assumed proportional to the velocity, causes the amplitude to decrease to one-tenth of its initial value in four complete oscillations. If a periodic force of magnitude 30 cos $50t$ lb is applied to the mass, find the amplitude of the forced oscillation. What would be the amplitude if the period of the applied force coincided with the natural period of vibration of the system?

Since the amplitude decreases to one-tenth of its initial value in four oscillations, we have from equation (15.30).

$$e^{2at_p} = 10 \quad \text{or} \quad 2at_p = 2{\cdot}303$$

Substituting for t_p in equation (15.28) and reducing, we get

$$a^2/b = 1/30{\cdot}02$$

The static extension of the spring due to gravity $= \delta = 20/50 = 0{\cdot}4$ in. and the extension of the spring produced by a steady force of 30 lb $= \Delta = 0{\cdot}6$ in.

The dynamic magnifier is given by equation (15.36)

$$D = \frac{1}{\sqrt{\{(1-\omega^2/b)^2+a^2\omega^2/b^2\}}}$$

But $b = g/\delta = 32{\cdot}2.12/0{\cdot}4 = 966$ (sec^{-2}) and $\omega = 50$ (sec^{-1})

$$\therefore\ D = \frac{1}{\sqrt{\{(1-50^2/966)^2+1/30{\cdot}0.50^2/966\}}}$$

$$= \frac{1}{\sqrt{(2{\cdot}521+0{\cdot}086)}} = 0{\cdot}62$$

Hence the amplitude of the forced vibration

$$D.\Delta = 0{\cdot}62.0{\cdot}6 = 0{\cdot}372 \text{ in.}$$

At resonance, from equation (15.42),

$$D = 2\pi n/a = \sqrt{(b/a^2)} = \sqrt{30{\cdot}0} = 5{\cdot}48$$

and the amplitude of the forced vibrations

$$= 5{\cdot}48.0{\cdot}6 = 3{\cdot}29 \text{ in.}$$

200. Elastic Suspension. There are two somewhat similar problems which often arise in practice. (a) A machine may require to be so supported that the periodic forces to which it gives rise when operating are, as far as practicable, prevented from reaching the surrounding structure. (b) An instrument may require to be so supported that it is, as far as practicable, unaffected by the vibrations of the surrounding structure. In both cases the desired effect can be obtained by supporting the machine or instrument on suitable springs.

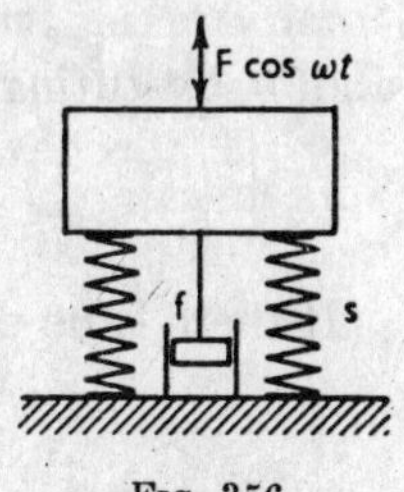

Fig. 356

The arrangement is shown diagrammatically in Fig. 356. The machine is supported on two, or more, springs of total stiffness s and is assumed to have one degree of freedom, i.e. it can move up and down only. The arrangement is clearly analogous to that of Fig. 353 except that the springs are in compression instead of in tension. Hence exactly the same equation of motion will apply and the amplitude of vibration of the machine on the springs will be given by equation (15.35).

Both the springs and the dashpot react on the foundation and the total force transmitted to the foundation is the sum of these reactions.

From equation (15.32)

$$y = A \cos(\omega t - \beta)$$

where y = displacement of machine at time t
and A = amplitude of the forced vibrations.

The springs must in any case transmit the weight W of the machine to the foundation, so that the total reaction of the springs at time t

$$= W + sy = W + sA \cos(\omega t - \beta)$$

Also $$dy/dt = -\omega A \sin(\omega t - \beta)$$

and the reaction of the dashpot on the foundation

$$= -f\omega A \sin(\omega t - \beta)$$

$\therefore$ the total force transmitted to the foundation

$$= W + sA \cos(\omega t - \beta) - f\omega A \sin(\omega t - \beta)$$
$$= W + A\sqrt{(s^2 + f^2\omega^2)} \cos(\omega t - \beta + \gamma)$$

where $\tan \gamma = f\omega/s$.

The maximum value of the periodic force transmitted to the foundation

$$= A\sqrt{(s^2 + f^2\omega^2)}$$

The ratio of the force transmitted to the force applied is termed the transmissibility of the spring support and is given by

$$\epsilon = (A/F)\sqrt{(s^2+f^2\omega^2)}$$

But $A = D.\Delta$, from equation (15.35), and $\Delta = F/s$.

$$\therefore\ \epsilon = (D/s)\sqrt{(s^2+f^2\omega^2)} = D\sqrt{(1+f^2\omega^2/s^2)}$$
$$= D\sqrt{\{1+(a/2\pi n)^2(n_f/n)^2\}}$$

and, substituting for D from equation (15.36),

$$\epsilon = \sqrt{\frac{1+(a/2\pi n)^2(n_f/n)^2}{\{1-(n_f/n)^2\}^2+(a/2\pi n)^2(n_f/n)^2}} \qquad (15.45)$$

If there is no damping this equation reduces to

$$\epsilon = \frac{1}{1-(n_f/n)^2}$$

For values of $n_f/n > 1$, ϵ is negative, which merely means that there is a phase difference of 180° between the transmitted force and the disturbing force. The value of n_f/n must be greater than $\sqrt{2}$, if ϵ is to be less than 1, and it is the numerical value of ϵ, independent of any phase difference between the forces that may exist, which is important. It is therefore more convenient to change the sign on the right-hand side of the equation and write

$$\epsilon = \frac{1}{(n_f/n)^2-1} \quad . \quad . \quad . \quad . \quad (15.46)$$

Values of ϵ for different amounts of damping are plotted against the frequency ratio in Fig. 357. It will be seen that the curves all pass through the point $\epsilon = 1$, when $n_f/n = \sqrt{2}$ and that $\epsilon < 1$ for all values of $n_f/n > \sqrt{2}$. Since the purpose of the spring support is to ensure that the force transmitted to the foundation is less than the periodic force which arises from the operation of the machine, it follows that the stiffness of the spring support should be so chosen that n_f/n is as large as possible. It is also apparent from the curves that for a given value of the frequency ratio, greater than $\sqrt{2}$, the effectiveness of the springs is reduced by an increase of damping. Against this must be set the fact that increased damping reduces the amplitude of vibration of the machine on the springs at resonance and it is, of course, necessary to pass through the resonant condition when the machine is started from rest.

An arrangement of the kind just considered is just as effective in insulating the machine or instrument from vibrations which originate in the structure from which it is supported. The essential requirement is that the ratio n_f/n shall be as large as possible, the natural frequency of vibration of the machine or instrument on its elastic supports being as low as possible relative to the disturbing frequency.

If the curves of ϵ, Fig. 357, and of D, Fig. 355, are compared, it will be seen that they have a similar shape, but for a given value of the damping and of the frequency ratio $\epsilon > D$, and the percentage difference increases with increase of the frequency ratio.

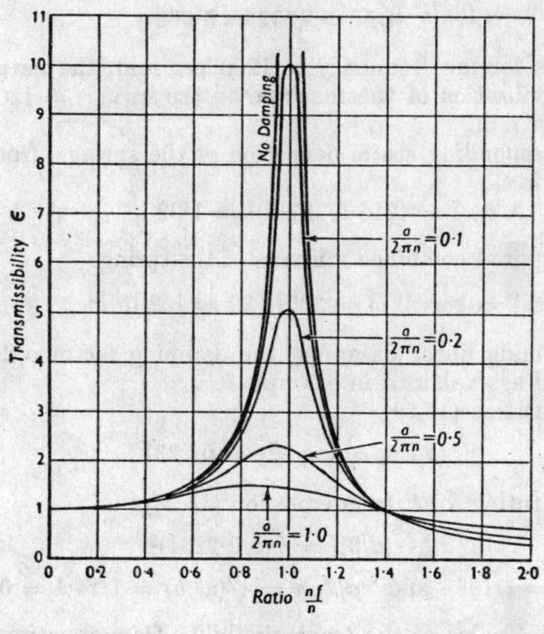

FIG. 357

Example 7. A machine has a total mass of 200 lb and unbalanced reciprocating parts of mass 3 lb which move through a vertical stroke of 3·5 in. with S.H.M. The machine is mounted on four springs, symmetrically arranged with respect to its centre of mass, in such a way that the machine has one degree of freedom and can undergo vertical displacements only.

Neglecting damping, calculate the combined stiffness of the springs in order that the force transmitted to the foundation shall not exceed one-thirtieth of the applied force, when the speed of rotation of the machine crankshaft is 1200 r.p.m.

When the machine is actually supported on the springs, it is found that damping reduces the amplitude of successive free vibrations by 20%. Find: (a) the fraction of the applied force which is actually transmitted at 1200 r.p.m., (b) the amplitude of the forced vibrations of the machine at resonance and the corresponding value of the force transmitted.

The value of ϵ is given as 1/30, so that if damping is neglected equation (15.46) applies and

$$\frac{1}{30} = \frac{1}{(n_f/n)^2-1}$$

$$\therefore \; n_f/n = \sqrt{31} = 5{\cdot}568$$

Since the forcing frequency is 1200 per min, the natural frequency of vibration of the machine on the springs = 1200/5·568 = 215·5 per min.

The corresponding static deflection of the springs, from equation (15.4) is

$$\delta = (215{\cdot}5/187{\cdot}7)^2 = 1{\cdot}32 \text{ in.}$$

and the required combined stiffness of the springs

$$= s = W/\delta = 200/1{\cdot}32 = 152 \text{ lb/in.}$$

(a) Assuming linear damping, the damping factor $a/2\pi n$ may be obtained as explained in example 4.

From equation (15.30)

$$(a/2)t_p = \log_e \tfrac{1}{0\cdot8} = 0{\cdot}2231$$

and, substituting for t_p from equation (15.28),

$$b-(a/2)^2 = (\pi a/0{\cdot}2231)^2$$

$$\therefore \; a^2/b = 1/198 \quad \text{and} \quad a/2\pi n = \sqrt{(a^2/b)} = 1/14{\cdot}1 \doteq 0{\cdot}0709$$

The actual value of the transmissibility from equation (15.45) is given by:

$$\epsilon = \sqrt{\frac{1+0{\cdot}0709^2 . 5{\cdot}568^2}{(1-5{\cdot}568^2)^2+0{\cdot}0709^2 . 5{\cdot}568^2}}$$

$$= \sqrt{\frac{1+0{\cdot}156}{900+0{\cdot}156}} = \frac{1}{27{\cdot}9}$$

The maximum unbalanced force on the machine at 1200 r.p.m. due to the reciprocating parts

$$= \frac{3}{32{\cdot}2}\left(\frac{\pi . 1200}{30}\right)^2 . \frac{7}{48} = 217 \text{ lb}$$

and the maximum force transmitted to the foundation

$$= 217/27{\cdot}9 = 7{\cdot}8 \text{ lb}$$

(b) **At resonance,** $n_f/n = 1$ **and from equation (15.45)**

$$\epsilon = \sqrt{\frac{1+0\cdot0709^2}{0\cdot0709^2}} = \sqrt{\frac{1\cdot0050}{0\cdot00503}} = 14\cdot1$$

The maximum unbalanced force on the machine at the resonant speed 215·5 r.p.m.

$$= 217(215\cdot5/1200)^2 = 7 \text{ lb}$$

and the maximum force transmitted to the foundation $= 14\cdot1 . 7 = 98\cdot7$ lb.

From (15.42) the dynamic magnifier

$$= D = 2\pi n/a = 14\cdot1$$

The static deflection of the springs caused by the force of 7 lb is given by $\Delta = 7/152 = 0\cdot0461$ in. so that the amplitude of the forced vibrations at resonance $= D.\Delta = 14\cdot1 . 0\cdot0461 = 0\cdot650$ in.

201. Critical or Whirling Speeds. In the nature of things the centre of gravity of a loaded shaft will always be displaced from the axis of rotation, although the amount of the displacement may be very small. This displacement may be due to one or more of a number of causes, such as eccentric mounting of the discs or rotors with which the shaft is loaded, lack of straightness of the shaft, bending under the action of gravity in the case of a horizontal shaft, unbalanced magnetic pull in the case of electrical machinery, etc. As a result of this initial displacement the centre of gravity is subjected to a centripetal acceleration as soon as the shaft begins to rotate. The inertia force acts radially outwards and bends the shaft, thus increasing the displacement of the centre of gravity from the axis of rotation. The effect is therefore cumulative.

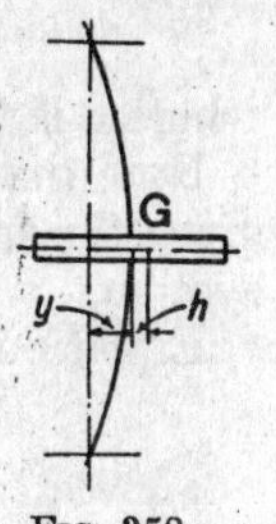

FIG. 358

(a) *Whirling of a Shaft with a Single Rotor.* Fig. 358 shows a rotor of weight W attached to a vertical shaft of negligible mass.

Let s = stiffness of the shaft, i.e. the force required in the plane of rotation of the disc in order to produce unit deflection of the shaft,

h = initial displacement of the c.g. of W from the axis of rotation

and y = additional displacement of the c.g. from the axis of rotation, due to bending of the shaft under the action of the inertia force, when the shaft rotates at a uniform speed ω radians per sec.

Then the radial outward inertia, or centrifugal force

$$= (W/g)\omega^2(y+h)$$

This is balanced by the inward elastic pull exerted by the shaft, the magnitude of which is $s.y$.

$$\therefore \ (W/g)\omega^2(y+h) = s.y$$

$$\therefore \ y = \frac{h}{sg/W\omega^2 - 1} \quad \text{or} \quad \frac{y}{h} = \frac{1}{sg/W\omega^2 - 1}$$

Let $w_c^2 = sg/W = g/\delta$, where δ is the static deflection under the load when the shaft is placed horizontal and is subjected to the pull of gravity. Then

$$\frac{y}{h} = \frac{1}{(\omega_c/\omega)^2 - 1} \quad . \quad . \quad . \quad . \quad (15.47)$$

The ratio y/h will be infinitely large when the denominator of this expression is zero, i.e. when $\omega = \omega_c$.

This value of ω is termed the critical or whirling speed.

Expressed in r.p.m.:

$$N_c = \frac{30}{\pi}\sqrt{\frac{g}{\delta}} = \frac{187{\cdot}8}{\sqrt{\delta}} \quad . \quad . \quad . \quad (15.48)$$

where δ is the static deflection in inches.

Note that this equation is identical with equation (15.4), which gives the frequency of the free transverse vibrations of the same system.

Equation (15.47) may be written:

$$\frac{y}{h} = \frac{1}{(N_c/N)^2 - 1}$$

It will be seen from this equation that when N/N_c is very large, y/h tends towards the limiting value -1. This result is interesting as showing that, if the shaft is run at a speed in excess of the critical, the tendency is for the shaft to deflect so that the axis of rotation passes through the centre of gravity of the rotor. Use is made of this fact in the case of high-speed shafts, such as those of small impulse turbines, where the normal running speed is much in excess of the critical speed, with the result that the shaft runs with exceptional steadiness.

Although not strictly correct, since it assumes much simpler conditions for the motion than those which actually obtain, the following alternative way of looking at the problem may help the student to understand what happens when the running speed coincides with the frequency of the free transverse vibrations.

Let us suppose that the shaft is running at a uniform speed of ω radians per second, where ω is less than the critical value ω_c and that the centre line of the shaft takes up the position shown by the full line of Fig. 358. In this position we have:

$$(W/g)\omega^2(y+h) = s.y$$

If a sharp blow is now given to the shaft, so that the deflection is increased by the amount Δ, the centrifugal force will be increased by the amount $(W/g)\omega^2\Delta$ and the elastic force by the amount $s.\Delta$. Under the assumed conditions, $\omega < \omega_c$, the increase of elastic force is greater than the increase of centrifugal force and an unbalanced force $s\Delta-(W/g)\omega^2\Delta$ or $\{s-(W/g)\omega^2\}\Delta$ acts on the rotor. The shaft therefore begins to move back to the position which it occupied before the blow was struck. It is clear that in these circumstances the shaft will vibrate transversely while at the same time rotating about its axis. The conditions so far as the transverse vibration is concerned are analogous to those of a shaft which is not rotating, but the effective stiffness is reduced from s to $s-(W/g)\omega^2$. The frequency of the vibrations of the shaft is therefore lower when rotating and may be found by substituting the effective stiffness in equation (15.1).

It follows that, as the speed of rotation of the shaft increases, the frequency of the transverse vibrations diminishes until, when $\omega = \omega_c$, $s-(W/g)\omega^2 = 0$, i.e. the effective stiffness of the shaft is zero, and the frequency of the transverse vibrations is zero. At the speed ω_c the shaft is in a state of neutral equilibrium, since for all values of y which are possible without straining the material of the shaft beyond the limit of proportionality, the inertia force is exactly balanced by the elastic force.

In practice, as the speed of rotation of a shaft is gradually increased, the reduction of effective stiffness results in a whirl or region of instability arising in the neighbourhood of the value $\omega = \omega_c$. It is for this reason that this particular speed is termed the *critical* or *whirling* speed.

(b) *Whirling Speeds for Other Systems of Loading.* The whirling speed of a shaft which carries a single disc has been shown to be identical with the frequency of the free transverse vibrations of the same shaft. The whirling speed of a uniformly loaded shaft may be shown to be identical with the frequency of its free transverse vibrations, since the dynamic load per unit length of the rotating shaft is clearly given by the expression $(w/g)\omega^2 y$, so that equation (15.10), Article 196, applies to the uniformly loaded shaft, whether that shaft is vibrating transversely or is rotating. We are therefore justified in assuming that the whirling speed and the frequency of the transverse vibrations are identical for a shaft with any other system of loading. It is therefore possible to use the methods of Article 197 in order to determine the whirling speed of a shaft which carries a number of concentrated loads at different points along its length.

202. Torsional Vibration. Two-rotor System. In practice the problem of torsional vibration seldom presents itself in so simple a form as that dealt with in Article 194. There are often two, and frequently more, rotors attached to the shaft, and the latter is often made up of lengths of different diameters. For instance, consider the two-mass system shown in Fig. 359 (a). The first step is to replace the actual shaft by a shaft of uniform diameter

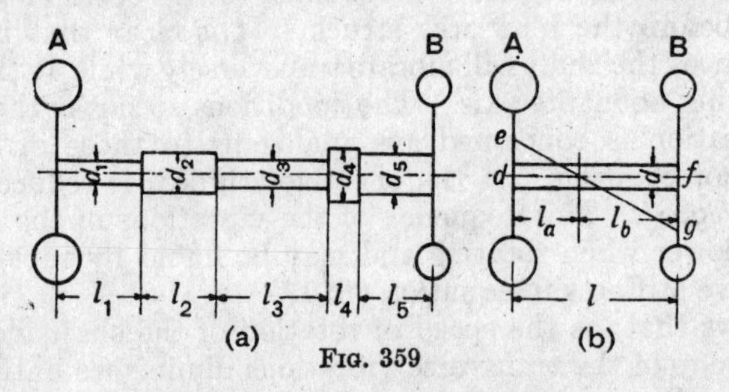

FIG. 359

to which it is torsionally equivalent. A torsionally equivalent shaft is one which twists through exactly the same angle as the actual shaft, when equal and opposite torques of given amount are applied to the two rotors. Let us suppose that l is the length of shaft of uniform diameter d which will satisfy this condition. Let q_1, q_2, etc., be the torques required to produce unit angle of twist on the lengths l_1, l_2, etc., where

$$q_1 = CJ_1/l_1 = (\pi/32)(d_1^4/l_1)C, \quad q_2 = (\pi/32)(d_2^4/l_2)C, \text{ etc.}$$

Then, if equal and opposite unit torques are applied to the rotors A and B, the total angle of twist of B relative to A is given by the sum of the twists produced in each of the lengths l_1, l_2, etc.

The twist of length l_1 under unit torque $= 1/q_1$ and the twist of length l_2 under unit torque $= 1/q_2$, etc.

$$\therefore \text{ total twist of B relative to A} = 1/q_1 + 1/q_2 + \ldots$$
$$= (32/\pi C)(l_1/d_1^4 + l_2/d_2^4 + \ldots)$$

But the twist of B relative to A, when the two rotors are connected by a shaft of length l and uniform diameter d, is given by:

$$1/q = 32/\pi C \,.\, l/d^4$$
$$\therefore\ 32/\pi C \,.\, l/d^4 = (32/\pi C)(l_1/d_1^4 + l_2/d_2^4 + \ldots)$$
$$\therefore\ l = l_1(d/d_1)^4 + l_2(d/d_2)^4 + \ldots$$

To save labour it is convenient to reduce the shaft to an equivalent length of diameter equal to that of one of the sections of the actual shaft. Thus the equivalent length of diameter d_1 is given by:

$$l = l_1 + l_2(d_1/d_2)^4 + l_3(d_1/d_3)^4 + \dots \quad . \quad (15.49)$$

The system (b) is then torsionally equivalent to the system (a). When such a system vibrates torsionally the two flywheels, or rotors, twist in opposite directions and at some point along the shaft there is a node, i.e. a section of the shaft which remains undisturbed by the vibration. The system behaves as though the shaft were clamped at the node, each of the two sections into which the node divides the system vibrating with the same frequency but opposite in phase. In practice the torsional vibrations occur while the system as a whole is turning with a certain mean angular velocity and power is being transmitted along the shaft. But for the purpose of calculating the frequency of the vibrations the shaft may be assumed to be at rest. Referring to Fig. 359 (b), let the node divide the shaft into the two parts l_a and l_b and let I_a, I_b be the mass moments of inertia of the two rotors. Then the frequency of vibration of the system to the left of the node, from equation (15.5), is given by:

$$n_a = \frac{1}{2\pi}\sqrt{\frac{q_a}{I_a}} = \frac{1}{2\pi}\sqrt{\left(\frac{CJ}{l_a}\cdot\frac{1}{I_a}\right)} \quad . \quad . \quad (15.50)$$

and the frequency of the system to the right of the node is given similarly by:

$$n_b = \frac{1}{2\pi}\sqrt{\left(\frac{CJ}{l_b}\cdot\frac{1}{I_b}\right)}$$

But these two frequencies must be equal:

$$\therefore \frac{1}{2\pi}\sqrt{\left(\frac{CJ}{l_a}\cdot\frac{1}{I_a}\right)} = \frac{1}{2\pi}\sqrt{\left(\frac{CJ}{l_b}\cdot\frac{1}{I_b}\right)}$$

$$\therefore l_a I_a = l_b I_b$$

or

$$\frac{l_a}{l_b} = \frac{I_b}{I_a}$$

i.e. the node divides the length of the shaft inversely as the mass moments of inertia of the rotors, or

$$l_a = \frac{I_b}{I_a + I_b} . l \quad . \quad . \quad . \quad . \quad . \quad (15.51)$$

The position of the node having been found from equation (15.51), the frequency of the vibrations may be calculated from equation (15.50).

If de and fg represent to scale the amplitudes of the oscillations of the two rotors A and B, then the straight line eg will pass through the node. The amplitude of oscillation at any section of the shaft will be represented to scale by the distance of eg from the axis df at that section. The line eg is called the elastic line of the shaft.

Example 8. Referring to Fig. 359 (a), let the moments of inertia of the rotors A and B be respectively 1500 lb ft^2 and 1000 lb ft^2, the lengths l_1, l_2, l_3, l_4 and l_5 be respectively 11, 10, 15, 4 and 10 in., and the diameters d_1, d_2, d_3, d_4 and d_5 be respectively 3, 5, 3·5, 7 and 5 in. Find the frequency of the natural torsional oscillations of the system.

The equivalent length of shaft of diameter 3 in. is found from equation (15.49):

$$l = 11+20(3/5)^4+15(3/3{\cdot}5)^4+4(3/7)^4$$
$$= 11+2{\cdot}59+8{\cdot}1+0{\cdot}135 = 21{\cdot}83 \text{ in.}$$

The distance of the node from rotor A is given by equation (15.51):

$$l_a = \frac{I_b}{I_a+I_b}.l = \frac{1000}{1500+1000}.21{\cdot}83 = 8{\cdot}73 \text{ in.}$$

$$\therefore\ q_a = \frac{CJ}{l_a} = \frac{12.10^6}{8{\cdot}73}.\frac{\pi}{32}.3^4 = 10{\cdot}92.10^6 \text{ lb in./rad}$$

$$\therefore\ n = \frac{1}{2\pi}\sqrt{\frac{q_a}{I_a}} = \frac{1}{2\pi}\sqrt{\frac{10{\cdot}92.10^6.32{\cdot}2.12}{1500.12^2}} = \frac{1}{2\pi}\sqrt{(1{\cdot}95.10^4)}$$
$$= 22{\cdot}2 \text{ per sec}$$

203. Torsional Vibration. Three-rotor System. The method of the last article may be applied to systems in which there are three or more rotors. As before, the first step is to reduce the actual shaft to an equivalent shaft of uniform diameter. Referring to Fig. 360 (a), which shows a three-rotor system, there are two possible natural or normal modes of vibration, in which the rotors all reach their extreme positions at the same instant and all pass through their equilibrium positions at the same instant. There will be a different natural frequency for each of these normal modes. In one mode there is a single node between A and B or between B and C and the oscillations of the outside rotors A and C are opposite in phase, while in the other mode there are two nodes, one between A and B and the other between B and C and the oscillations of the outside rotors are in phase. The relative amplitudes of the three rotors for the one-node and the two-node vibrations are shown respectively in Fig. 360 (b) and (c). It is assumed in (b) that the node lies between B and C. For the two-node vibration, let l_a be the distance of one node from A and l_c

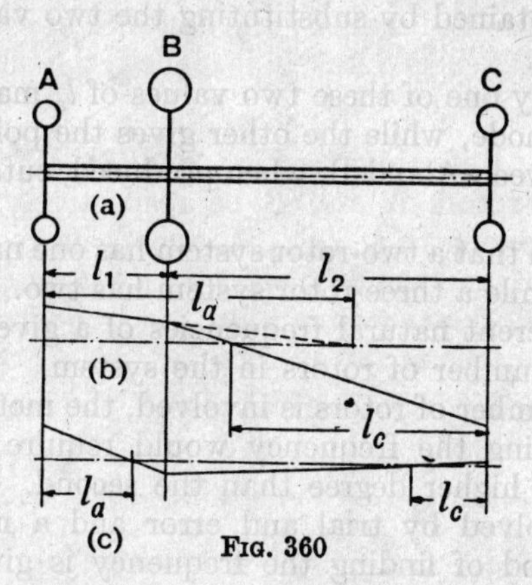

FIG. 360

the distance of the other node from C. Also let I_a, I_b and I_c be the mass moments of inertia of the three rotors. Then the frequency for the left-hand rotor, from equation (15.5):

$$n_a = \frac{1}{2\pi}\sqrt{\frac{q_a}{I_a}} = \frac{1}{2\pi}\sqrt{\left(\frac{CJ}{l_a}\cdot\frac{1}{I_a}\right)}$$

Similarly, for the right-hand rotor:

$$n_c = \frac{1}{2\pi}\sqrt{\left(\frac{CJ}{l_c}\cdot\frac{1}{I_c}\right)}$$

For the middle rotor:

$$n_b = \frac{1}{2\pi}\sqrt{\frac{q_b}{I_b}}$$

where q_b, the torque required to twist B through 1 rad when the shaft is fixed at the nodes, is the sum of the torques required to produce a twist of 1 rad in each of the lengths $l_1 - l_a$ and $l_2 - l_c$.

$$\therefore\ q_b = \frac{CJ}{l_1 - l_a} + \frac{CJ}{l_2 - l_c} = CJ\left(\frac{1}{l_1 - l_a} + \frac{1}{l_2 - l_c}\right)$$

and, substituting for q_b

$$n_b = \frac{1}{2\pi}\sqrt{\left\{\frac{CJ}{I_b}\left(\frac{1}{l_1 - l_a} + \frac{1}{l_2 - l_c}\right)\right\}}$$

Since the frequencies n_a, n_b and n_c must all be equal, the three equations enable a quadratic in l_a (or l_c) to be obtained. The two roots of this quadratic give the positions of the nodes for the one-node and the two-node vibration frequencies. The actual frequencies are obtained by substituting the two values of l_a in the equation for n_a.

Note that only one of these two values of l_a may give the position of a real node, while the other gives the point at which the elastic line between A and B, when produced, cuts the axis of the shaft.

We have seen that a two-rotor system has one natural frequency of vibration, while a three-rotor system has two. In general, the number of different natural frequencies of a given system is one less than the number of rotors in the system. Where a system with a large number of rotors is involved, the method used in this article for finding the frequency would require the solution of an equation of higher degree than the second. Such equations can only be solved by trial and error and a more convenient practical method of finding the frequency is given in the next article.

Example 9. The moments of inertia of 3 rotors A, B and C are respectively 2·5 ton ft², 7·5 ton ft² and 3·0 ton ft². The distance between A and B is 9 ft 6 in. and between B and C is 25 ft. The shaft is 8·5 in. dia. and the modulus of rigidity for the shaft material is 11·8.10⁶ lb/in². Find the frequencies of the free torsional vibrations of the system.

Referring to Fig. 360, if we equate the frequency of vibration of I_a on the length l_a of shaft, to the frequency of vibration of I_c, on the length l_c of shaft, we have:

$$l_a I_a = l_c I_c \qquad (1)$$

Similarly, if we equate the frequencies for I_a and I_b, we get:

$$\frac{1}{l_a I_a} = \frac{1}{I_b}\left(\frac{1}{l_1 - l_a} + \frac{1}{l_2 - l_c}\right)$$

$$\therefore \frac{1}{l_a I_a} = \frac{1}{I_b}\left\{\frac{l_1 + l_2 - (l_a + l_c)}{(l_1 - l_a)(l_2 - l_c)}\right\}$$

$$\therefore (l_1 - l_a)(l_2 - l_c) = (I_a/I_b) . l_a\{l_1 + l_2 - (l_a + l_c)\} \qquad (2)$$

Substituting for I_a and I_c in (1):

$$l_a = (3{\cdot}0/2{\cdot}5)l_c = 1{\cdot}2l_c$$

Also, substituting for l_1, l_2, l_a, I_a and I_b in (2):

$$(9{\cdot}5 - 1{\cdot}2l_c)(25 - l_c) = (2{\cdot}5/7{\cdot}5)1{\cdot}2l_c(34{\cdot}5 - 2{\cdot}2l_c)$$

$$\therefore\ 237{\cdot}5-39{\cdot}5l_c+1{\cdot}2l_c^2 = 13{\cdot}8l_c-0{\cdot}88l_c^2$$

$$\therefore\ 2{\cdot}08l_c^2-53{\cdot}3l_c+237{\cdot}5 = 0$$

$$\therefore\ l_c^2-25{\cdot}6l_c+114{\cdot}1 = 0$$

$$\therefore\ l_c = 19{\cdot}88 \text{ ft} \quad \text{or} \quad 5{\cdot}74 \text{ ft}$$

and $$l_a = 23{\cdot}86 \text{ ft} \quad \text{or} \quad 6{\cdot}89 \text{ ft}$$

The fundamental frequency will be that which corresponds to the larger of these two values of l_c or l_a.

The torsional stiffness of the length l_c of shaft $= q_c = CJ/l_c$.

When $l_c = 19{\cdot}88$ ft,

$$q_c = \frac{11{\cdot}8.10^6}{19{\cdot}88.12}.\frac{\pi}{32}.8{\cdot}5^4 = 2{\cdot}54.10^7 \text{ lb in./rad}$$

Also $$I_c = \frac{3{\cdot}0.2240.12^2}{32{\cdot}2.12} = 2{\cdot}50.10^3 \text{ lb in. sec}^2$$

$\therefore$ fundamental frequency of vibration,

$$= n_c = \frac{1}{2\pi}\sqrt{\frac{q_c}{I_c}} = \frac{1}{2\pi}\sqrt{\frac{2{\cdot}54.10^7}{2{\cdot}50.10^3}} = 16{\cdot}0 \text{ per sec or } 960 \text{ per min}$$

and the two-node frequency

$$= 16{\cdot}0\sqrt{(19{\cdot}88/5{\cdot}74)} = 29{\cdot}8 \text{ per sec} \quad \text{or} \quad 1788 \text{ per min}$$

If the amplitude of rotor A is assumed to be 1 rad, then, for the fundamental vibration, the amplitude of rotor B

$$= \frac{l_a-l_1}{l_a} = \frac{23{\cdot}86-9{\cdot}5}{23{\cdot}86} = 0{\cdot}602 \text{ rad}$$

and the amplitude of rotor C $= \dfrac{l_c}{l_c-l_2}$. amplitude of B

$$= \frac{19{\cdot}88}{19{\cdot}88-25{\cdot}0}.0{\cdot}602 = -2{\cdot}34 \text{ rad}$$

For the two-node vibration, the amplitude of rotor B

$$= \frac{l_a-l_1}{l_a} = \frac{6{\cdot}89-9{\cdot}5}{6{\cdot}89} = -0{\cdot}379 \text{ rad}$$

and the amplitude of rotor C $= \dfrac{l_c}{l_c-l_2}$. amplitude of B

$$= \frac{5{\cdot}74}{5{\cdot}74-25}.-0{\cdot}379 = 0{\cdot}113 \text{ rad.}$$

Fig. 360 is drawn to scale to correspond to the particulars of this example.

A convenient check on the arithmetic is provided by the fact that for each mode of vibration $\Sigma(IA)$ should be zero. (See Article 204.)

204. Torsional Vibration. Multi-rotor Systems. A shaft of uniform diameter which carries several rotors is shown in Fig. 361. Such a system has several normal modes of vibration each with its appropriate frequency. In this case there are five normal modes, equal to the number of sections into which the rotors divide the total length of shaft. For each mode there is a different elastic line which shows the relative amplitudes of vibration of the individual rotors. The elastic line for the fundamental, or lowest, natural frequency intersects the axis of the shaft at one point only, as shown in Fig. 361. The elastic lines for the higher frequencies intersect the axis at two, three, four and five points. Each point of intersection gives a node and the larger the number

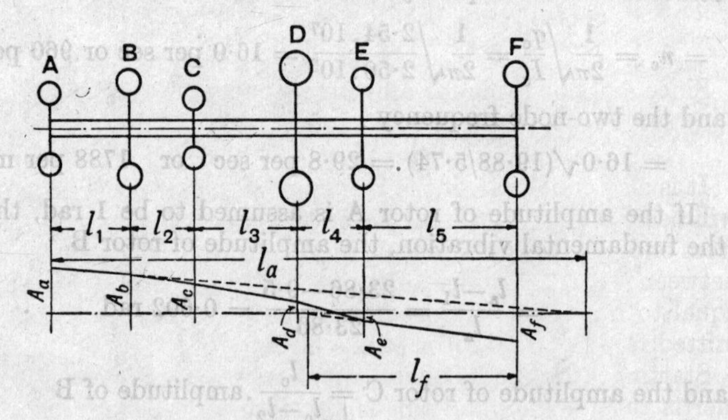

FIG. 361

of nodes, the higher is the frequency. If there is one node, the vibration is said to be of the first degree. If there are two nodes, the vibration is of the second degree, and so on.

If equal and opposite couples are applied to the end rotors, so as to twist the shaft, and the couples are then suddenly removed, the system will not vibrate in a normal mode, since the relative angular displacements of the rotors will not correspond to a normal elastic line. But, after a time, the relative displacements will tend to adjust themselves to correspond to the fundamental frequency. To start the system vibrating in a normal mode it is necessary to give to each rotor the relative angular displacement appropriate to that mode.

In what follows, the method of finding the natural frequency of the fundamental mode is given, but the same procedure is adopted to find the natural frequencies of the other normal modes.

Let A, with the appropriate suffix, be the amplitude in radians of the torsional oscillation of a given rotor. Let $\omega = 2\pi n$, where n is the vibration frequency per second. Then, so long as the stresses in the material of the shaft do not exceed the limit of proportionality, each rotor will oscillate with simple harmonic motion. The maximum angular acceleration of a rotor is $\omega^2 A$, when the rotor is at the extremity of its swing, and the torque which the rotor exerts on the shaft is given by:

$$T = I\omega^2 A \quad . \quad . \quad . \quad . \quad . \quad (15.52)$$

When vibrating freely the only torques which act on the system are the inertia torques and at any instant the algebraic sum of the inertia torques must be zero. Hence, when all the rotors are in their extreme positions,

$$\Sigma T = 0, \quad \therefore \Sigma(I\omega^2 A) = 0$$

Since ω is the same for all rotors, this may be written:

$$\Sigma(IA) = 0 \quad . \quad . \quad . \quad . \quad . \quad (15.53)$$

It is also clear that, if the inertia of the shaft is neglected, there will be no change in the torque as it is transmitted along the shaft from one rotor to the adjacent rotor. Consequently, the difference between the amplitudes of vibration of adjacent rotors will be equal to the angle of twist of the shaft due to the torque transmitted from the one rotor to the other.

Starting with the left-hand rotor of the system, the maximum inertia torque is T_a and this is also the torque transmitted along the length l_1 of shaft. The angle of twist of l_1 is given by T_1/q_1, where $q_1 = CJ/l_1$ = the stiffness of the length l_1 of shaft.

$$\therefore A_a - A_b = T_a/q_1 \quad . \quad . \quad . \quad . \quad . \quad (15.54)$$

$$\therefore A = A_a - T_a/q_1$$

$$= A_a - I_a\omega^2 A_a/q_1 \quad . \quad . \quad (15.55)$$

Similarly, the torque transmitted along l_2 is the algebraic sum of the inertia torques of the rotors A and B and is given by $T_a + T_b$. The angle of twist of l_2 when transmitting this torque is $(T_a + T_b)/q_2$, where $q_2 = CJ/l_2$ = the stiffness of the length l_2 of shaft.

$$\therefore A_b - A_c = (T_a + T_b)/q_2 \quad . \quad . \quad . \quad . \quad (15.56)$$

$$\therefore A_c = A_b - (\omega^2/q_2)(I_a A_a + I_b A_b) \quad . \quad (15.57)$$

A similar expression may be written down by analogy for the amplitude of vibration of any other rotor.

These results are applied to the multi-rotor system as follows. The amplitude of vibration A_a is assumed arbitrarily to have any convenient magnitude, say 1 rad. A value is then arbitrarily assumed for ω. As a rule an approximate value of ω may be obtained by reducing the multi-rotor system to an equivalent two or three-rotor system. The amplitudes of the other rotors may then be calculated from equations (15.55), (15.57), etc. Finally the sum of the products IA for all the rotors is determined.

According to equation (15.53), this sum must be zero for the correct value of ω. If the sum is not zero, then another value of ω must be tried. Two or three trials will generally suffice to fix the correct value. The lowest value of ω which enables (15.53) to be satisfied corresponds to the fundamental or first degree vibration.

If A_a, A_b, etc., Fig. 361, represent to scale the amplitudes of vibration of the rotors A, B, etc., and their extremities are joined by straight lines, the elastic line of the shaft is obtained. It shows how the angular twist of the shaft varies from point to point along its length. The point of intersection of the elastic line and the axis of the shaft determines the position of the node.

It is easily seen that the frequency of vibration of the system is the same as that of the single rotor I_f attached to the shaft of length l_f, where l_f is the distance from the rotor F of the point at which that part of the elastic line between the rotors E and F, produced if necessary, intersects the axis of the shaft.

Similarly, the frequency of vibration of the system is the same as that of the single rotor I_a attached to the shaft of length l_a, where l_a is the distance from the rotor A of the point at which that part of the elastic line between the rotors A and B, produced if necessary, intersects the axis of the shaft.

There will be other and higher values of ω which enable equation (15.53) to be satisfied. These correspond to the second-, third-, etc., degree vibrations. Each value of ω will give a different elastic line. The positions of the nodes for each frequency are determined by the points of intersection of the elastic line with the axis of the shaft.

The calculations may be conveniently set down in a table, as in the following example.

Example 10. A four-cylinder engine drives a motor vessel. The equivalent moment of inertia of the revolving and the reciprocating parts of each cylinder is 0·625 ton ft^2, that of the flywheel is 7·5 ton ft^2 and that of the propeller, with an allowance for entrained water, is 3·0 ton ft^2. The equivalent shaft is 8·5 in. dia.

and the distances between the masses are shown in Fig. 362. Find the frequency of the fundamental torsional vibrations of the system and also that of the two-node vibrations. $C = 11{\cdot}8.10^6$ lb/in^2.

As a first approximation, the four engine masses A, B, C and D may be combined into a single rotor G, shown dotted in the figure. The moment of inertia of G is 2·5 ton ft^2 and its distance from E is 114 in. or 9·5 ft. The equivalent three-rotor system, G, E, F, has already been worked out in Example 9. The frequencies of vibration were found to be 16·0 per sec and 29·8 per sec. The corresponding values of ω are 100·5 and 187 rad/s.

For the fundamental vibration the value of ω may be assumed as a first approximation to be 100 rad/s.

The mass moment of inertia of each engine rotor

$$= \frac{0{\cdot}625 . 2240 . 12^2}{32{\cdot}2 . 12} = 521 \text{ lb in. sec}^2$$

The mass moment of inertia of the flywheel

$$= \frac{7{\cdot}5 . 2240 . 12}{32{\cdot}2} = 6250 \text{ lb in. sec}^2$$

The mass moment of inertia of the propeller

$$= \frac{3{\cdot}0 . 2240 . 12}{32{\cdot}2} = 2500 \text{ lb in. sec}^2$$

The following table may then be filled in.

$\omega = 100$; $\omega^2 = 10\,000$.

$I \div 10^3$, lb in sec^2	l, in.	$q = \frac{CJ}{l} \div 10^7$ lb in. per rad	$\frac{\omega^2}{q} \times 10^3$	A, radians	$IA \div 10^3$	$\Sigma(IA) \div 10^3$	$\frac{\omega^2}{q}\Sigma(IA)$, radians
0·521				1·0000	0·521	0·521	
	46	13·14	0·0761				0·0396
0·521				0·9604	0·500	1·021	
	46	13·14	0·0761				0·0777
0·521				0·8827	0·460	1·481	
	46	13·14	0·0761				0·1127
0·521				0·7700	0·401	1·882	
	45	13·42	0·0745				0·1402
6·25				0·6298	3·936	5·818	
	300	2·02	0·495				2·878
2·50				−2·248	−5·620	0·198	

The first three columns of this table are filled in from the known dimensions of the system, while the fourth column is filled in from the assumed value of ω. The amplitude of the left-hand engine rotor is assumed to be 1 radian in column five and the first entries in columns six and seven follow at once. The first entry in column eight is obtained by multiplying together the first entries in columns four and seven. This gives the twist of the length of shaft between rotors A and B, so that the amplitude of the rotor B is found by subtracting 0·0396 from 1·000. The difference, 0·9604 radian, is entered on the second line of column five. The entry in the second line of column six then follows and is added to the first entry in order to give $\Sigma(IA)$ for the rotors A and B. This sum is entered in the second line of column seven and is multiplied by the second entry in column four to give the

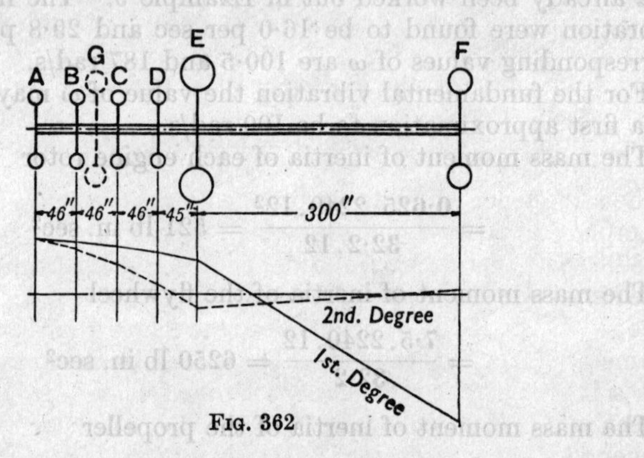

Fig. 362

second entry in column eight. The process is repeated in order to complete the table.

From column seven, $\Sigma(IA)$ for the complete system is $+0·198$. It should of course be zero for the correct value of ω. For odd-degree vibrations, i.e first, third, fifth, etc., a positive value of $\Sigma(IA)$ shows that the assumed value of ω is too low. It will therefore be necessary to repeat the calculations for a higher value of ω. In the table below ω has been assumed to be 102.

$\omega = 102$; $\omega^2 = 10\,400$.

$I \div 10^3$, lb in. sec²	l, in.	$q = \frac{CJ}{l} \div 10^7$ lb in. per rad	$\frac{\omega^2}{q} \times 10^3$	A, radians	$IA \div 10^3$	$\Sigma(IA) \div 10^3$	$\frac{\omega^2}{q}\Sigma(IA)$, radians
0·521				1·0000	0·521	0·521	
0·521	46	13·14	0·0792	0·9587	0·499	1·020	0·0413
0·521	46	13·14	0·0792	0·8779	0·457	1·477	0·0808
0·521	46	13·14	0·0792	0·7610	0·396	1·873	0·1169
6·25	45	13·42	0·0775	0·6159	3·847	5·720	0·1451
2·50	300	2·02	0·515	−2·329	−5·823	−0·103	2·945

For this value of ω, the value of $\Sigma(IA)$ is negative. This indicates that ω is too high.

The calculations may be repeated with $\omega = 101$, or, alternatively, the correct value of ω may be found by simple proportion, since it will be found that, if the values of $\Sigma(IA)$ are plotted against the corresponding values of ω^2 for several different values of ω close to the true value, the resulting curve is practically a straight line.

$$\therefore \text{ correct value of } \omega^2 = 10\,000 + \frac{0 \cdot 198}{0 \cdot 198 + 0 \cdot 103} . 400$$

$$= 10\,260$$

$$\therefore \text{ correct value of } \omega = 101 \cdot 3 \text{ rad/s}$$

$$\therefore \text{ fundamental frequency} = (30/\pi)101 \cdot 3 = 967 \text{ per min}$$

The frequency of the two-node vibrations is found from Example 9 to be 29·8 per sec, which corresponds to a value of ω of 187. As a first approximation, assume $\omega = 200$ and draw up a table similar to those shown above. The value of $\Sigma(IA)$ is found to be $-0 \cdot 238$. For vibrations of even degree this shows that the assumed value of ω is too low. The calculations are then repeated with a slightly higher value of ω and after one or two trials the correct value of ω is found to be 200·7. This corresponds to a frequency of 31·9 per sec or 1916 per min.

The relative amplitudes of the rotors are given by column five of the table, drawn up for the correct value of ω. These amplitudes may be used to plot the elastic lines for the fundamental and the two-node vibrations; as shown in Fig. 362, which is drawn to scale for the particulars of this example.

The usual method of finding the natural frequencies of torsional oscillation of a multi-rotor system by trial and error has just been given, but the following modification will be found to reduce appreciably the amount of arithmetical work involved. It consists in finding the true shape of the elastic line by a process of trial and error. The simplification lies in the fact that the shape of the elastic line is quite independent of the diameter of the shaft and depends only on (a) the relative spacing of the rotors along the shaft, and (b) the relative moments of inertia of the rotors. The system is therefore first reduced to its simplest form by expressing each moment of inertia in terms of one which is taken as unity, and each distance between adjacent rotors in terms of one which is also taken as unity.

Referring to Fig. 361, the frequency of torsional vibration of the multi-rotor system is identical with that of the rotor A on a shaft of length l_a.

$$\therefore \omega^2 = q_a/I_a = CJ/l_a I_a$$

But the maximum angle of twist of the length l_1 of shaft, from equation (15.54),

$$= \delta A_1 = A_a - A_b = T_a/q_1 = I_a\omega^2 A_a/q_1$$

and substituting for ω^2,

$$\delta A_1 = A_a - A_b = (q_a/q_1)A_a = (l_1/l_a)A_a \quad . \quad (15.58)$$

Similarly, the maximum angle of twist of the length l_2 of shaft, from equation (15.56),

$$= \delta A_2 = A_b - A_c = \frac{T_a + T_b}{q_2} = \frac{\omega^2}{q_2}(I_a A_a + I_b A_b)$$

and substituting for ω^2,

$$\delta A_2 = A_b - A_c = \frac{q_a}{q_2}\frac{I_a A_a + I_b A_b}{I_a} = \frac{l_2}{l_a}\cdot\frac{I_a A_a + I_b A_b}{I_a} \quad (15.59)$$

A similar expression may be written down for the angle of twist of any other section of the shaft. Thus for the section between the nth and the $(n+1)$th rotors, we get

$$\delta A_n = A_n - A_{n+1} = \frac{l_n}{l_a}\cdot\frac{\Sigma_0^n(IA)}{I_a} \quad . \quad . \quad (15.60)$$

It will be seen that in evaluating δA_1, δA_2, etc., from these equations, the lengths l_a, l_1, etc., and the moments of inertia I_a, I_b, etc., appear simply as ratios.

To illustrate the application of the method, the particulars of Example 10, p. 576, will be used.

The moment of inertia of each of the engine rotors A, B, C, D, is taken as unity, so that the corresponding moment of inertia of the engine flywheel is $7{\cdot}5/0{\cdot}625 = 12$ and that of the propeller is $3{\cdot}0/0{\cdot}625 = 4{\cdot}8$. Also the cylinder pitch is taken as unity, so that the corresponding distance between the rotors D and E is 45 in./46 in $= 0{\cdot}9781$ and that between the rotors E and F is 300 in./46 in. $= 6{\cdot}521$. The system reduced to its simplest form then appears as shown in Fig. 363.

The approximate distance from the rotor A, at which that part of the elastic line between the rotors A and B when produced intersects the axis of the shaft, may be found from the solution of the equivalent three-rotor system, which is given in Example 9, p. 572. In that example it was found that the fundamental node was 19·88 ft from the propeller rotor F, so that the corresponding distance l_a of the pseudo-node from the engine rotor A in the complete system is evidently given by

$$l_a = 19{\cdot}88 . 4{\cdot}8 = 95{\cdot}42 \text{ ft}$$

or, in terms of the cylinder pitch,

$$l_a = 95{\cdot}42 . 12/46 = 24{\cdot}89$$

This may be used as a first approximation in finding the true shape of the elastic line.

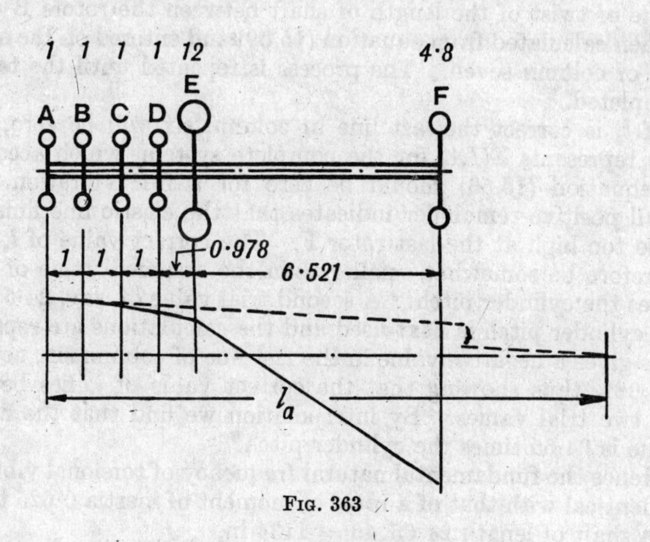

FIG. 363

The following table may then be filled in.

$l_a = 24{\cdot}89$ times the cylinder pitch.

Rotor	I	l	A	IA	$\Sigma(IA)$	δA
A	1·0		1·0000	1·0000	1·0000	
		1·0				0·0402
B	1·0		0·9598	0·9598	1·9598	
		1·0				0·0787
C	1·0		0·8811	0·8811	2·8409	
		1·0				0·1141
D	1·0		0·7670	0·7670	3·6079	
		0·978				0·1418
E	12·0		0·6252	7·5024	11·1103	
		6·521				2·910
F	4·8		−2·285	−10·968	0·142	

The first three columns give the data for the simplified system shown in Fig. 363. The first entry in column four gives the amplitude of the rotor A, which is assumed to be 1 rad. The first entries in columns five and six follow at once. The first entry in column seven is calculated from equation (15.58). This gives the twist of the length of shaft between the rotors A and B, and subtracting it from 1·0000, the first entry in column four, we get the amplitude of the rotor B, 0·9598, which is entered in the second line of this column. The product of I and A for rotor B is entered on the second line of column five, and the sum of the first two items in this column is entered on the second line of column six. The angle of twist of the length of shaft between the rotors B and C is then calculated from equation (15.59) and entered on the second line of column seven. The process is repeated until the table is completed.

If l_a is correct the last line in column six will be zero, since this represents $\Sigma(IA)$ for the complete system, which according to equation (15.53) should be zero for a free vibration. The small positive remainder indicates that the elastic line finishes a little too high at the last rotor F. The correct value of l_a must therefore be somewhat smaller than the assumed value of 24·89 times the cylinder pitch. A second trial value of, say, 24·5 times the cylinder pitch is next used and the calculations are repeated. This gives a negative value in the last line of column six, actually —0·087, thus showing that the correct value of l_a lies between the two trial values. By interpolation we find that the correct value is 24·65 times the cylinder pitch.

Hence the fundamental natural frequency of torsional vibration is identical with that of a rotor of moment of inertia 0·625 ton ft² on a shaft of length 24·65.46 = 1134 in.

Since the diameter of the shaft is 8·5 in., the frequency of the fundamental torsional vibrations of the system is given by

$$N = \frac{30}{\pi}\sqrt{\frac{CJ}{l_a I_a}}$$

$$= \frac{30}{\pi}\sqrt{\left(\frac{11{\cdot}8.10^6}{1134}\cdot\frac{\pi}{32}(8{\cdot}5)^4\cdot\frac{32{\cdot}2.12}{0{\cdot}625.2240.144}\right)} = 965 \text{ per min}$$

This agrees with the value found by the first method.

205. Torsional Vibration. Multi-rotor System. Graphical Method. The following simple graphical method of finding the frequency of vibration of a multi-rotor system is based on an article by K. Waimann.[1] It is essentially a graphical method of obtaining, by a process of trial and error, the true shape of the elastic line of the shaft.

The method will be best explained by considering an actual example. For this purpose the particulars of Example 10 will be

[1] *Z.V.D.I.*, Sept. 15, 1934.

used. For a system of this type the node for the fundamental vibration will lie somewhere between the rotors E and F, Fig. 364. Its actual position may be guessed, or may be found approximately by solving the equivalent three-rotor system. Thus, it follows from Example 9, that the node is approximately 19·88 ft from the rotor F. But, to show more clearly the effect of a wrongly assumed position for the node, we shall take it at the point X, 20·5 ft. (246 in.) from F.

The amplitude of vibration of the rotor F may be represented by any convenient distance ma_1, Fig. 364, preferably so as to give an angle a_1Xm of approximately 45°.

Draw a line py perpendicular to the axis of the shaft, as shown at the right of the figure. Take any convenient polar distance op.

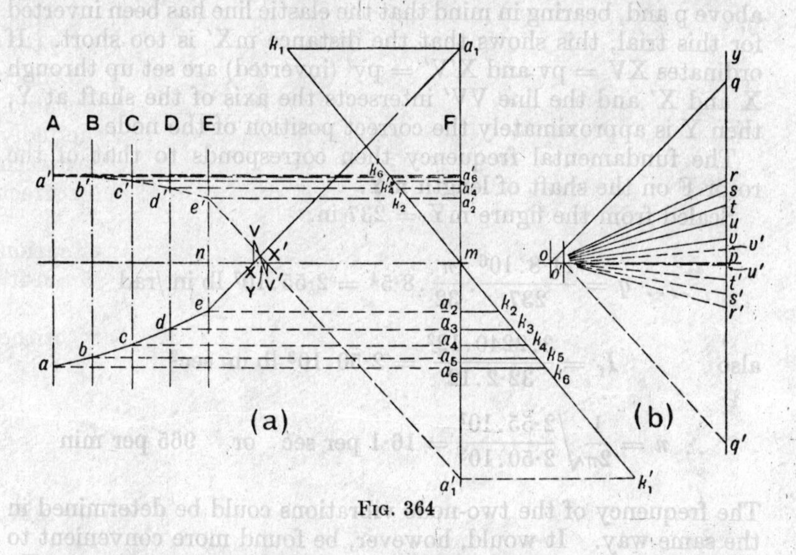

Fig. 364

Through o draw oq parallel to Xa_1 to cut py at q; set off a_1k_1 perpendicular to ma_1 and equal in length to pq; join k_1 to m.

Produce a_1X to cut the plane of rotation of E at e. Draw a line through e parallel to the axis of the shaft and let a_2k_2 be the intercept on this line between a_1m produced and k_1m produced. Set off qr along yp equal in length to $a_2k_2 . I_e/I_f$. Note that, since the amplitude of E is opposite in sign to that of F, qr is set off in the opposite sense to pq.

Join r to o and draw ed parallel to ro; through d draw a line parallel to the axis of the shaft; measure the length of the intercept a_3k_3 and set off rs equal in length to $a_3k_3 . I_d/I_f$.

Repeat the above construction until all the rotors have been dealt with and the final point v on the vertical line py has been obtained.

Then, if the assumed position for the node coincides with its true position, the point v should coincide with p. If the point v lies above p, as in Fig. 364, the distance mX is too large and the construction must be repeated with a shorter distance, say 230 in., in the present example.

To prevent confusion, the elastic line and the construction lines generally have been inverted for the second trial and the corresponding letters are marked with a dash.

Note that $ma_1' = ma_1$, $pq' = pq$ and q'o' is drawn parallel to $a_1'X'$.

Repeating the construction it is found that the point v' lies above p and, bearing in mind that the elastic line has been inverted for this trial, this shows that the distance mX' is too short. If ordinates XV = pv and X'V' = pv' (inverted) are set up through X and X' and the line VV' intersects the axis of the shaft at Y, then Y is approximately the correct position of the node.

The fundamental frequency then corresponds to that of the rotor F on the shaft of length mY.

Scaled from the figure mY = 237 in.

$$\therefore q = \frac{11{\cdot}8 \, . \, 10^6}{237} \cdot \frac{\pi}{32} \, . \, 8{\cdot}5^4 = 2{\cdot}55 \, . \, 10^7 \text{ lb in./rad}$$

also
$$I_f = \frac{3 \, . \, 2240 \, . \, 12^2}{32{\cdot}2 \, . \, 12} = 2{\cdot}50 \, . \, 10^3 \text{ lb in. sec}^2$$

$$\therefore n = \frac{1}{2\pi}\sqrt{\frac{2{\cdot}55 \, . \, 10^7}{2{\cdot}50 \, . \, 10^3}} = 16{\cdot}1 \text{ per sec} \quad \text{or} \quad 965 \text{ per min}$$

The frequency of the two-node vibrations could be determined in the same way. It would, however, be found more convenient to assume the amplitude of the rotor A instead of that of the rotor F., and also to determine the point Z at which that part of the elastic line between the rotors A and B intersects the axis of the shaft. Apart from these changes, the construction is carried out in exactly the same way.

The above graphical solution is easily and quickly obtained. It will be found to be particularly useful as a check on the accuracy of the frequencies determined by the tabular method.

Proof of the Construction. Assume X, Fig. 364, to be the correct position of the node. Then en represents the amplitude of vibration, A_e of the rotor E to the same scale as a_1m represents the amplitude of vibration, A_f, of the rotor F.

But the difference between A_f and A_e arises from the twist of the length l_5 of shaft between E and F under the inertia torque T_f of the rotor F.

$$\therefore\ A_f - A_e = T_f l_5/CJ$$

or

$$(A_f - A_e)/l_5 = T_f/CJ$$

But $(A_f - A_e)/l_5 = a_1a_2/ea_2$ and, from the similar triangles a_1ea_2, qop, we have $a_1a_2/ea_2 = qp/op$.

$$\therefore\ T_f/CJ = qp/op$$

or qp represents T_f to the same scale as op represents CJ.

Since, by construction, $a_1k_1 = pq$, it follows that a_1k_1 represents T_f to the same scale as op represents CJ.

But, from equation (15.52),

$$T_f = I_f\omega^2 A_f \quad \text{and} \quad T_e = I_e\omega^2 A_e$$

so that

$$\frac{T_e}{T_f} = \frac{I_e A_e}{I_f A_f} = \frac{I_e}{I_f}\cdot\frac{ne}{ma_1} = \frac{I_e}{I_f}\cdot\frac{ma_2}{ma_1}$$

Also, from the similar triangles ma_1k_1, ma_2k_2, we have

$$ma_2/ma_1 = a_2k_2/a_1k_1$$

and, since $a_1k_1 = pq$, it follows that

$$T_e/T_f = I_e/I_f\,.\,a_2k_2/pq$$

But, by construction, $qr = I_e/I_f\,.\,a_2k_2$, so that $T_e/T_f = qr/pq$, and, therefore, qr represents T_e to the same scale as pq represents T_f.

In the same way it can be shown that rs represents T_d, st represents T_c, tu represents T_b and uv represents T_a.

But, when the system is vibrating freely, the algebraic sum of the inertia torques must be zero. Hence, v must coincide with p.

206. Torsional Oscillation of a Geared System. Fig. 365 (a) shows diagrammatically a geared system, in which the rotor A on one shaft is connected through the pinion B and the gear wheel C to the rotor D on a second shaft. This system may be reduced to the equivalent single shaft system shown at (b) if it is assumed (a) that there is no backlash in the gearing, (b) that the teeth are rigid and do not distort under the tooth loads, and (c) that the inertia of the shafts and gears is negligible.

If the shafts are not strained beyond the limit of proportionality each rotor in the geared system will oscillate with simple harmonic motion and there will be a node either in the length l_1 or in the length l_2. The two rotors A and D will reach their extreme positions at the same instant and at this instant the whole of the energy in the system will be strain energy due to the twisting of the two shafts. The two rotors will also pass through their equilibrium positions at the same instant, and at this instant the whole of the energy in the system will be kinetic energy of rotation of the rotors. It follows that the single shaft system will be equivalent to the geared system if, for a given amplitude of oscillation of the rotor A, the maximum strain

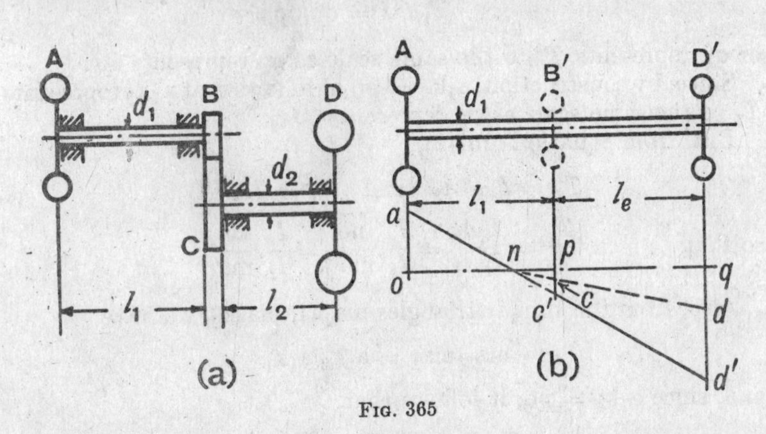

FIG. 365

energies stored in the two systems are identical and also the maximum kinetic energies of the rotors in the two systems are identical.

Let G be the gear ratio, i.e. the ratio of the speed of the pinion B to the speed of the gear wheel C.

Then, if a torque T is applied to the rotor A, a corresponding torque $G.T$ must be applied to the rotor D, and the total angle through which A twists relative to D is evidently given by

$$\theta_a = \theta_1 + G.\theta_2$$

where θ_1 is the angle through which rotor A twists relative to the pinion B and θ_2 is the angle through which the gear wheel C twists relative to the rotor D.

But $\theta_1 = Tl_1/CJ_1$ and $\theta_2 = G.T.l_2/CJ_2$, so that

$$\theta_a = \frac{Tl_1}{CJ_1} + G^2 . \frac{Tl_2}{CJ_2} = \frac{T}{CJ_1}\left\{l_1 + G^2 . l_2 \frac{J_1}{J_2}\right\}$$

$$= \frac{T}{CJ_1}\left\{l_1 + G^2 . l_2 \left(\frac{d_1}{d_2}\right)^4\right\} = \frac{T}{CJ_1}(l_1 + l_e)$$

where l_e is the length of shaft of diameter d_1 which is torsionally equivalent to the shaft of length l_2 and diameter d_2 when the gearing is removed.

$$\therefore\ l_e = G^2 . l_2(d_1/d_2)^4 \quad . \quad . \quad . \quad (15.61)$$

If l_e satisfies this equation, the maximum strain energy stored in the single shaft system will be equal to the maximum strain energy stored in the geared system for equal amplitudes of oscillation of the rotor A in the two systems.

Let n be the position of the node and ao the amplitude of the rotor A, then anc′d′ will be the elastic line for the single shaft system, Fig. 365 (b). The amplitude of the pinion B will be given by c′p and that of the equivalent rotor D′ by d′q. But because of the gearing the actual amplitudes of the wheel C and the rotor D will be given by pc and qd, where pc′/pc = qd′/qd = G.

But the maximum angular velocities of the rotors are directly proportional to their amplitudes of oscillation, so that for the geared system,

$$\text{Maximum kinetic energy} \propto I_a.(\text{oa})^2 + I_d.(\text{qd})^2$$

and, for the single shaft system,

$$\text{Maximum kinetic energy} \propto I_a.(\text{oa})^2 + I_d'.(\text{qd}')^2$$

Since these two expressions must be equal, it follows that

$$I_d'.(\text{qd}')^2 = I_d.(\text{qd})^2$$

$$I_d' = I_d.(\text{qd}/\text{qd}')^2 = I_d/G^2 \quad . \quad (15.62)$$

The single shaft system is therefore equivalent to the geared system if the additional length of shaft l_e satisfies equation (15.61) and if the inertia of the rotor D′ satisfies equation (15.62).

Where the inertia of the gearing is not negligible, there must be an additional rotor B′ on the equivalent single shaft system to allow for the inertia of the pinion and gear wheel. This rotor will be situated at the point p, as shown dotted in Fig. 365 (b), and its inertia will be given by $I_b' = I_b + I_c/G^2$.

When the geared system has been reduced to an equivalent single shaft system with either two or three rotors, the frequency, or frequencies, of natural vibration may be found by the methods already given in Articles 202 and 203.

Example 11. A 12-cylinder aero engine and gear drive to the airscrew is equivalent torsionally to the system shown in Fig. 365 (a). The shafts l_1 and l_2 are respectively $39\frac{1}{2}$ in. and $25\frac{1}{2}$ in. long and $2\frac{3}{4}$ in. and $3\frac{1}{2}$ in. diameter. The airscrew runs at 0·6 of the speed of the engine crankshaft and the moments of inertia of the rotors are: for the combined engine masses A, 1500 lb in², for the pinion B, 54 lb in², for the gear wheel C, 850 lb in² and for the airscrew D, 50 000 lb in². If the modulus of rigidity of the shaft material is 12.10⁶ lb/in², find the natural frequencies of torsional vibration of the system (a) when the inertia of the gearing is neglected, and (b) when the inertia of the gearing is taken into account.

(a) In the equivalent two-rotor system on a shaft $2\frac{3}{4}$ in. dia., we have from equation (15.61),

$$l_e = G^2 . l_2(d_1/d_2)^4 = (1/0{\cdot}6)^2 . 25\tfrac{1}{2} . (2\tfrac{3}{4}/3\tfrac{1}{2})^4 = 27{\cdot}0 \text{ in.}$$

and, from equation (15.62),

$$I_d' = I_d/G^2 = 50\,000 . 0{\cdot}6^2 = 18\,000 \text{ lb in}^2$$

But the node divides the total length of the shaft inversely as the inertias of the two rotors, so that its distance from the rotor A

$$= l_a = \frac{18\,000}{18\,000+1500} . 66\tfrac{1}{2} = \frac{12}{13} . 66\tfrac{1}{2} = 61{\cdot}40 \text{ in.}$$

The torsional stiffness of this length of shaft

$$= q_a = \frac{CJ}{l_a} = \frac{12 . 10^6}{61{\cdot}4} . \frac{\pi}{32}\left(\frac{11}{4}\right)^4 = 1{\cdot}10 . 10^6 \text{ lb in./rad}$$

and the frequency of the torsional vibrations

$$= n = \frac{1}{2\pi}\sqrt{\frac{q_a}{I_a}}$$

$$= \frac{1}{2\pi}\sqrt{\frac{1{\cdot}10 . 10^6 . 32{\cdot}2 . 12}{1500}} = 84.8 \text{ per sec or } 5090 \text{ per min}$$

(b) On a shaft $2\frac{3}{4}$ in. diameter the gearing is equivalent to a rotor B′, as shown dotted at the point p in Fig. 365 (b), where

$$I_b' = I_b + I_c/G^2$$

$$= 54 + 850 . 0{\cdot}6^2 = 54 + 306 = 360 \text{ lb in}^2$$

Thus the equivalent three-rotor system consists of rotors A and D′ at the ends of a shaft $2\frac{3}{4}$ in. diameter and $66\frac{1}{2}$ in. long with the third rotor B′ at a point $39\frac{1}{2}$ in. from A and 27 in. from D′.

If this system is analysed by the method given in Article 203, the distances of the nodes from the rotor A are found to be 63·15 in. and 3·661 in.

The fundamental frequency

$$= 84{\cdot}8\sqrt{(61{\cdot}40/63{\cdot}15)} = 83{\cdot}5 \text{ per sec} \quad \text{or} \quad 5010 \text{ per min}$$

and the two-node frequency

$$= 84{\cdot}8\sqrt{(61{\cdot}40/3{\cdot}661)} = 347 \text{ per sec} \quad \text{or} \quad 20\,800 \text{ per min}$$

It will be seen that when the inertia of the gearing is neglected a close approximation to the fundamental frequency is obtained in this particular example.

EXAMPLES XV

N.B.—Unless otherwise stated, assume $E = 30.10^6$ lb/in² and $C = 12.10^6$ lb/in².

1. Deduce an expression for the frequency of the free vibrations of a mass of weight W hung from a spiral spring which extends 1 in. under a load of s lb. If W is 20 lb and s is 50 lb, find the frequency.

2. A rectangular steel beam, cross-section 1 in. × 2 in., is supported at points 6 ft apart with its 2-in. side vertical. Determine the frequency of the free transverse vibrations.

3. If a concentrated load of 400 lb is placed 2 ft from one of the supports of the beam in Question 2, find the frequency of the free transverse vibrations (a) when the inertia of the beam is neglected, (b) when the inertia of the beam is taken into account.

4. Two parallel floor beams of I-section have a span of 30 ft and support a vertical engine of weight 3 tons which is situated at mid-span. The beams may be taken as simply supported. Each beam weighs 54 lb/ft and the second moment of area of the cross-section is 370 in⁴ units. Calculate the natural frequency of vibration of the system. M.U.

5. A beam of I-section has a span of 20 ft, the ends being simply supported. The weight of the beam is 18 lb/ft run and second moment of area of the cross-section 42 in⁴ units. If a load of 1 ton is carried at a point 8 ft from one support, calculate the natural frequency of vibration of the system. M.U.

6. A shaft 30 in. long has a diameter of 2 in. for the first 10 in. and a diameter of 3 in. for the remaining 20 in. If one end of the shaft is fixed and the other end carries a disc of weight 1500 lb and radius of gyration 20 in., what is the frequency of the free torsional oscillations?

7. Two equal masses, of weight 1000 lb and radius of gyration 15 in., are keyed to opposite ends of a shaft 24 in. long. The shaft is 3 in. dia. for the first 10 in. of its length, 5 in. dia. for the next 4 in. of its length and 3·5 in. dia. for the remainder of its length. Find the frequency of the free torsional vibrations of the system and the position of the node.

8. A steel shaft 4 in. dia. has two flywheels keyed to it at a distance apart of 36 in. The flywheels weigh 3000 and 2500 lb and their radii of gyration are respectively 30 in. and 27 in. Neglecting the inertia of the shaft determine the frequency of the free torsional vibrations.

9. A mass of weight 20 lb is hung from a spiral spring of stiffness 50 lb/in. The vibration is controlled by a dashpot and it is found that the amplitude of the vibration diminishes to one-tenth of its initial value in two complete oscillations. Assuming linear damping, find the damping force at 1 ft/s and the ratio of the frequency of the damped vibration to that of the natural vibration.

10. If a periodic force of 30 cos 25t lb acts on the mass in Question 9, find the amplitude of the forced oscillation. What would be the amplitude at resonance? What damping would be necessary in order to limit the amplitude at resonance to 1·0 in.?

11. A body is held in position by elastic constraints, is subjected to linear damping and is acted upon by a periodic disturbing force. Explain what is meant by the dynamic magnifier and deduce the relation between the dynamic magnifier, the damping force, the frequency of the applied force and the natural frequency of vibration of the system.

12. A flywheel with a moment of inertia of 500 lb ft^2 is fixed to a vertical shaft 1 in. dia. and 3 ft long. A vibration damper is fitted and it is found by experiment that the amplitude of torsional oscillation diminishes to one-twentieth of its initial value in three complete oscillations. Assuming the damping torque to be directly proportional to the angular velocity, find its magnitude at 1 rad/s.

13. If a periodic couple of amount 50 cos 15t lb ft acts on the flywheel in Question 12, find the amplitude of the forced oscillation. What would be the amplitude at resonance and what damping would be required to limit this amplitude to 1·5°? If the amplitude is not to exceed the static twist under the 50 lb ft torque, what is the lowest frequency with which the periodic torque must be applied?

14. A machine weighs 1 ton and is supported on four spiral springs which deflect 0·75 in. under the load. The reciprocating parts of the machine weigh 40 lb and move through a vertical stroke of 8 in. with simple harmonic motion. A dashpot is provided, the resistance of which is proportional to the velocity and amounts to 200 lb at 1 ft/s. Find the amplitude of the forced vibrations and the phase difference between the force and the displacement, when (a) the driving shaft of the machine makes 180 r.p.m., (b) the driving shaft speed coincides with the frequency of the free vibrations of the system. M.U.

15. A small vertical single-cylinder petrol engine is supported at mid-span on a steel joist. The static deflection of the joist is 0·1 in. The weight of the engine, including an allowance for the joist, is 100 lb. The length of the crank is 1·6 in. and the weight of the reciprocating parts, which may be assumed to move with simple harmonic motion, is 1·5 lb. As the speed of rotation of the crankshaft is increased slowly from 300 to 1000 r.p.m. the amplitude of the transverse vibrations at mid-span of the joist is observed to pass through a maximum value of 1·5 in. If it is assumed that damping is directly proportional to velocity, find the magnitude of the damping force in lb per ft/sec. M.U.

16. An engine running at 600 r.p.m. is coupled to a shaft 12 ft long and 4 in. dia. At the end of the shaft remote from the engine there is a flywheel, the moment of inertia of which is 800 lb ft^2, while the rotating parts of the engine, etc., may be considered as equivalent to a flywheel at the engine end with a moment of inertia of 400 lb ft^2. The torque of the engine varies $\pm$500 lb ft. above and below its mean value with a frequency equal to twice the engine speed, and this variation may be assumed to be simple harmonic.

(a) Show that the frequency of the forced torsional oscillation is less than the natural frequency of torsional oscillation of the system.

(b) Deduce the amplitude of oscillation of the flywheel remote from the engine. The mass of the shaft may be neglected. L.U.A.

17. The shaft shown in Fig. 366 carries two heavy masses at A and B. It is driven by a light gear situated at CC. The weight of the mass at A is 800 lb and its radius of gyration is 27 in.; the corresponding values for the mass at B are 1200 lb and 33 in. The shaft diameter between CC and B, marked X, is undecided. Assuming it to be 3·5 in., determine the frequency of the free torsional oscillations of the system. Thereafter determine what X should be if the node of the vibration is to be in the plane CC of the drive. Deduce any formulæ used. L.U.A.

18. A hydraulic dynamometer is fitted with a dashpot to damp out vibrations. It is observed that with a certain oil in the dashpot the amplitudes of successive swings are 15 mm and 13 mm. A second oil is substituted for the first, the viscosity of this being three times as great as that of the first.

Given the same amplitude of the first swing, estimate the amplitude of the second swing with this oil. L.U.A.

19. To reduce vibration an instrument is mounted on a heavy table weighing 100 lb, which is suspended from the roof by three springs each of stiffness 9 lb/in. A dashpot is fitted to assist in damping the vertical vibrations; experiments on the dashpot show that, when a force of 15 lb is applied, it moves at a constant rate of 1 ft/s.

(a) Deduce the time required for the amplitude of any vertical natural vibration of the table to be reduced to 1% of its initial value.

(b) If the roof from which the table is suspended vibrates vertically with an amplitude of 0·05 in. at a frequency of 1200 per min, deduce the amplitude of vibration of the table.

The mass of the springs may be neglected. L.U.A.

20. Deduce an expression for the whirling speed of a shaft which carries a single wheel or disc at a point along its length.

A pulley of weight 150 lb is fixed to a shaft 20 in. long and 1 in. dia. at a point 8 in. from one of the bearings. If the bearings are spherically seated, calculate the whirling speed. Neglect the inertia of the shaft.

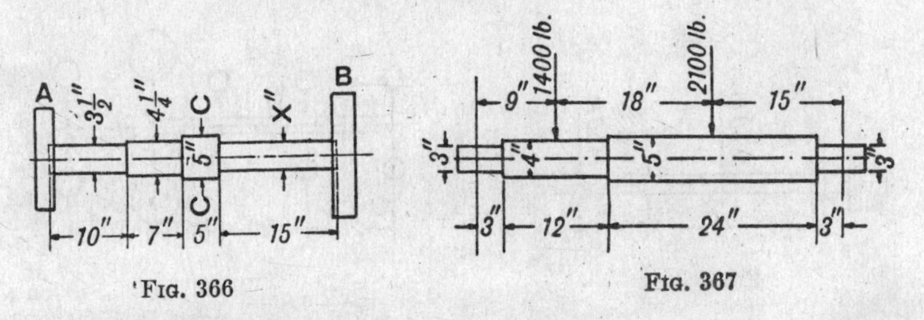

FIG. 366

FIG. 367

21. A uniformly loaded shaft is 3 in. dia. and 10 ft long and is carried in bearings which provide horizontal constraint. Calculate the whirling speed by assuming the whirling shaft to bend to the same shape as the static deflection curve. Compare the result with that obtained by the exact method of Article 196. Weight of material, 0·28 lb/in^3.

22. A shaft 4 in. dia. is supported in bearings 8 ft apart. It carries two pulleys which weigh 500 lb and 300 lb at distances of 3 ft and 6 ft respectively from one bearing. Calculate the whirling speed by (a) Dunkerley's method, (b) the energy method. Find the value of the constant C in the equation $N_c = C\sqrt{\delta_m}$, where N_c is the whirling speed and δ_m is the maximum static deflection of the shaft.

23. A shaft 7·5 in. dia. is supported in bearings 8 ft apart and carries three discs which weigh 500, 1000 and 450 lb and are situated respectively 2 ft, 5 ft and 6·5 ft from one of the bearings. Compare the values of the critical speed as obtained by Dunkerley's method and by the energy method.

24. A shaft 4 in. dia. is supported in bearings 10 ft apart and carries three discs which weigh 200, 300 and 150 lb and are situated respectively 3 ft, 6 ft and 8 ft from one of the bearings. Compare the values of the critical speeds as obtained by Dunkerley's method and by the energy method.

25. Fig. 367 shows a stepped shaft which carries two heavy discs and is supported in spherically seated bearings. Assume that the two concentrated loads shown include an allowance for the weight of the shaft and find the critical or whirling speed of the shaft.

26. Examine the conditions for whirling of a cylindrical shaft of uniform diameter, which is supported horizontally in two bearings, one of which is of sufficient length to ensure that the shaft is horizontal at the bearing and the other is at the end of the shaft and imposes no flexural constraint upon it. Find the whirling speed of a shaft of 1 in. dia. supported in the above manner, when there is a clear length of 5 ft between the bearings. The density of the material is 0·28 lb/in³. L.U.A.

27. Explain briefly why the static deflection curve of a loaded shaft may be used as a whirling form. Establish the rule for critical speed in terms of loads and load point deflections.

The shaft of a high-speed motor is as shown in Fig. 368. The rigid coupling at B, connecting it to a machine, is between the two bearings A and C, which, in effect, give directional fixture to the shaft at C. The masses on the shaft may be taken as equivalent to the two concentrated loads at D and E. The bearing at F is of the swivel type and exercises no bending constraint. Neglecting the influence of the shaft mass, determine approximately the critical speed. The shaft is of uniform $3\frac{1}{2}$-in. dia. $E = 28.10^6$ lb/in². L.U.A.

28. The moments of inertia of three rotors A, B and C are respectively 1·5 ton in², 3 ton in² and 1 ton in². The distance between A and B is 5 ft and between B and C is 3 ft and the shaft is 2 in. dia. If the modulus of rigidity is 5300 ton/in² find the frequencies of the free torsional vibrations of the system.

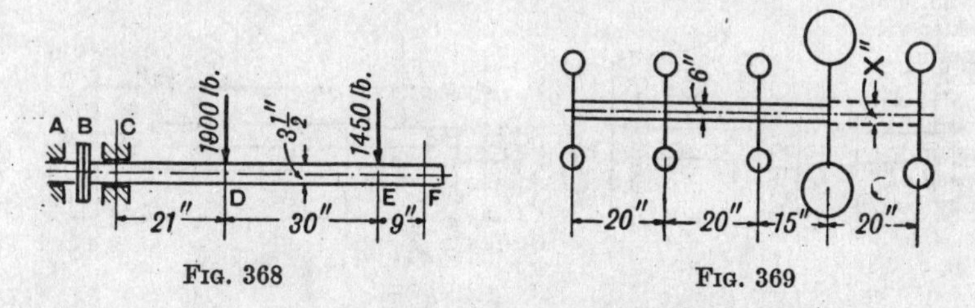

FIG. 368 FIG. 369

29. A marine engine, shaft and propeller are approximately equivalent to the following three-rotor system. The combined moment of inertia of the engine masses is 30·8 ton ft², that of the flywheel is 98 ton ft² and that of the propeller is 56 ton ft². The equivalent shaft between the engine masses and the flywheel is 15 in. dia. and 17·8 ft long and that between the flywheel and the propeller is 14 in. dia. and 37·5 ft long. Find the frequencies of the torsional vibrations of the system and the positions of the nodes.

30. A three-cylinder oil engine is to be rigidly coupled to a generator as shown in Fig. 369. The crankshaft is equivalent torsionally to a solid shaft 6 in. dia. and the moments of inertia are as follows: for each engine crank 280 lb ft², for the flywheel 7000 lb ft² and for the generator rotor 1000 lb ft². In order to prevent a dangerous resonant condition from arising during the running of the plant it is necessary for the frequency of the fundamental torsional vibrations of the system to be adjusted to 3150 per min. Find what diameter X is required for the shaft which connects the engine to the generator. $C = 11{\cdot}8.10^6$ lb/in².
Either a tabular or a graphical method of solution may be used.

31. A four-cylinder Diesel engine is directly coupled to a generator and the torsionally equivalent system is shown in Fig. 370. The moment of inertia of each engine mass is 0·98 ton ft², that of the flywheel is 39·5 ton ft² and that of the generator rotor is 6·52 ton ft². Reduce the system to an equivalent three-rotor system and find the frequencies of the torsional oscillations with one and with two nodes. Then find the frequencies by the tabular method and by the graphical method and sketch the shape of the elastic lines for the two modes of vibration.

32. Fig. 371 shows diagrammatically the engine, transmission shaft and propeller for a motor ship. The crankshaft may be assumed equivalent torsionally to a solid shaft of uniform diameter 14 in. The combined reciprocating and revolving masses for each crank have a moment of inertia of 5·5 ton ft². The moments of inertia of the flywheel and propeller are respectively 70 ton ft² and 40 ton ft².

Reduce the arrangement to an equivalent three-mass system on a shaft of uniform diameter 13 in. and determine the frequencies of the torsional vibrations of the system. $C = 5300$ tons/in². M.U.

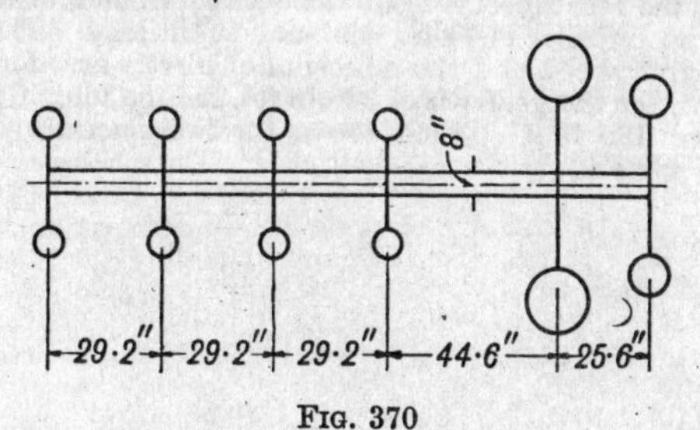

FIG. 370

33. For the system shown in Fig. 371, find the fundamental and the second order vibrations by the tabular method and check the results by the graphical method.

34. A shaft A, of diameter d_a and length l_a, carries at one end a disc of polar moment of inertia I_a; the other end is geared to a second shaft B, carrying at its end a second disc, the corresponding quantities being d_b, l_b and I_b. The speed of B is n times that of A and both the inertia and the flexibility of the gears may be neglected.

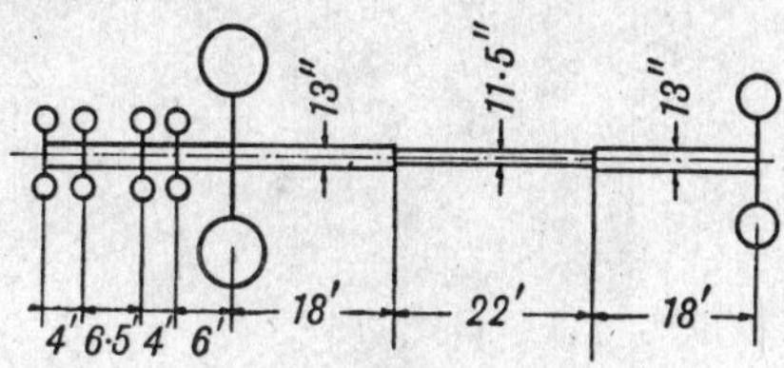

FIG. 371

Show that, for torsional oscillations, the gears and shaft B may be replaced by an extension of shaft A, such that the total equivalent length of shaft A becomes $l_a + (l_b/n^2)(d_a/d_b)^4$ and the disc B being replaced by one having moment of inertia $= n^2 I_b$.

In a particular case in which $n = 3$, the magnitudes of the various quantities are:

	Shaft		Disc	
	Dia., in.	Length, in.	Dia., in.	Length, in.
A	3	18	20	6
B	2	18	15	4

The material has a modulus of rigidity of $12 . 10^6$ lb/in² and weighs 0·28 lb/in³.

Determine the frequency of torsional oscillation of the system, neglecting the inertia of the shafts. L.U.A.

35. If, in the last question, the polar moment of inertia of the gear on shaft A is I_c and that of the gear on shaft B is I_d, show that the arrangement is torsionally equivalent to a system consisting of three rotors, of moments of inertia I_a, $I_c+n^2I_d$ and n^2I_b, the distance between the first and second rotors being l_a, the distance between the second and third rotors being $(l_b/n^2)(d_a/d_b)^4$ and the diameter of the shaft being d_a.

36. An engine drives a centrifugal pump through gearing. The shaft from the engine flywheel to the gear wheel is $2\frac{1}{4}$ in. dia. and 38 in. long, while that from the pinion to the pump impeller is $1\frac{1}{2}$ in. dia. and 12 in. long. The pump runs at four times the engine speed and the moments of inertia are: for the engine flywheel 19 200 lb ft², for the gear wheel 320 lb ft², for the pinion 80 lb ft² and for the pump impeller 400 lb ft². Find the natural frequencies of the torsional oscillations of the system.

MISCELLANEOUS EXAMPLES XVI

1. A winding gear consists of two cages operated by a 6 ft diameter drum and ropes passing over two guide pulleys of 4 ft effective diameter. The drum is driven by a motor through double-reduction gearing, each reduction being $4\frac{1}{2}$ to 1 and the loaded cage ascends as the other descends. The frictional resistance of the whole gear may be taken as equivalent to a torque of 900 lb ft at the drum shaft. Details are as follows:

Part of gear	Weight lb	k in.
Motor armature	500	3·75
Intermediate shaft and gear	850	9·5
Drum shaft and gear	5000	33·0
Guide pulleys, each	500	22·0
Loaded cage	2000	
Empty cage	900	
Rope (vertical)	800	

Find the motor starting torque to give an acceleration of 3·5 ft/s².

L.U. Part II.

2. In order to determine the moment of inertia of a flywheel, of weight 2500 lb, it is suspended with its axis vertical by three parallel equidistant wires, each 12 ft long and attached at a radius of 2 ft 6 in. When the wheel is caused to make small angular oscillations the frequency is observed to be 18 per min.

Derive an expression for the frequency of vibration of the suspended flywheel and hence determine its radius of gyration.

This flywheel is fitted to an engine running at 260 r.p.m. and it is observed that, when the load is disconnected and the fuel shut off, the speed falls from 260 to 230 r.p.m. in 12·5 sec. Assuming that the inertia of the other revolving parts is negligible compared with that of the flywheel, find the horse-power required to overcome the frictional resistances of the engine when running at its normal speed.

L.U.A.

3. The first stage wheel of a marine turbine is, for finding its moment of inertia, suspended on a pin through a pressure balancing hole in the disc at a radius of 14 in. from the wheel centre. The period of small oscillations is found to be 1·87 seconds.

The weight of the wheel is 500 lb and the rotor of which it is part is carried in bearings 6 ft apart. The disc friction effects of the wheel are such as to absorb 5 h.p. at the full speed of 3000 r.p.m. Find (a) the work to be done in speeding up the wheel from 1200 to 3000 r.p.m. with uniform acceleration in 60 sec, if the friction torque varies as the square of the speed, (b) the bearing reactions due to gyroscopic torque of the wheel when the ship takes a turn on a radius of 500 ft at 15 knots, the turbine speed then being 2500 r.p.m.

Establish any formula used. One knot = 6080 ft/h.

L.U. Part II.

4. Two co-axial shafts, A and B, carrying masses of moment of inertia respectively 120 and 20 lb ft², are coupled together by a hydraulic clutch. Initially no torque is being transmitted and the shafts revolve at 2000 r.p.m. A steady torque of 100 lb ft is then applied to A and simultaneously a resisting torque of the same magnitude acts on B. If the torque transmitted by the clutch is given by $2(\omega-\omega_1)^2$ lb ft, in which ω and ω_1 are the instantaneous speeds of A and B in radians per second, find the final steady speeds of the two shafts and the power transmitted.

L.U.A.

5. A shaft A carries a rotor at one end and the internal element of a cone clutch at the other; the total weight is 500 lb and the radius of gyration 6 in. The clutch has a mean radius of 5 in. and a half cone angle of 12 degrees. It is lined with material the limiting coefficient of friction for which is 0·3 and the axial thrust of the operating spring is 120 lb. The external element of the clutch is fixed to a gear wheel of diameter 24 in., weight 100 lb and radius of gyration 8 in., which gears with a pinion 6 in. diameter on a shaft B. The masses on B weigh 60 lb and have a radius of gyration of 3 in.

Initially A is rotating at 300 r.p.m. and the gear wheel and shaft B are at rest. The gear is then clutched in. Determine the speeds of the shafts when slip ceases and, during slipping, the time, the energy lost and the tangential force at the gear teeth. L.U. Part II.

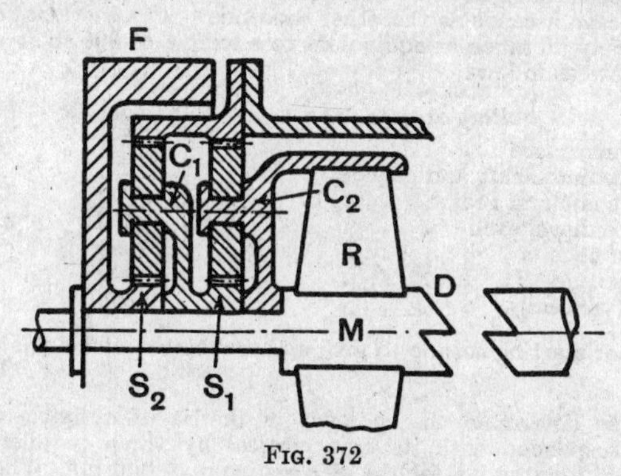

FIG. 372

6. The inertia starter for a Diesel engine is shown diagrammatically in Fig. 372. The small flywheel F has a moment of inertia of 6 lb ft^2 and is connected through epicyclic gearing and a rubber torque member or cushion R to the main shaft. The epicyclic gearing consists of two trains in series, each train having a fixed internal wheel or annulus. The planet carrier C_1 of the first train is rigidly attached to the outer circumference of the torque member R, while its sun wheel S_1 is integral with the planet carrier C_2 of the second train. The sun wheel S_2 of the second train is integral with the flywheel F of the starter. In each train the sun wheel has 17 teeth and the internal wheel or annulus has 83 teeth. The inner circumference of the torque member R is rigidly fixed to the main shaft M. In order to use the starter the main shaft is turned by hand until its speed is 140 r.p.m., when the dog-clutch D on the shaft is engaged with the engine crankshaft by means of a trigger release which is not shown in the sketch.

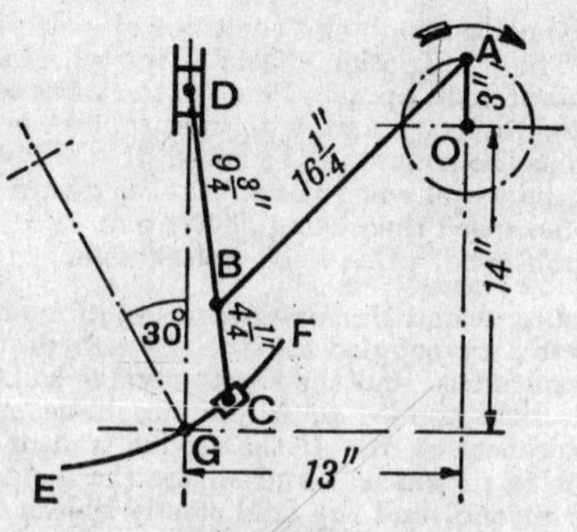

FIG. 373

If the moment of inertia of the engine parts referred to the crankshaft is 5000 lb ft², find (a) the maximum speed reached by the engine crankshaft, (b) the time required to reach this speed, and (c) the maximum twist of the torque member. The stiffness of the torque member may be assumed to be constant and to amount to 100 lb ft per degree and all losses may be neglected. M.U.

7. The mechanism of a variable-stroke feed pump is shown in Fig. 373. The drive is taken from the crank OA to the pin B on the connecting rod CBD. The end C of the connecting rod carries a die-block which moves along the curved slotted link EGF. The radius of curvature of the slot is equal to the length of the connecting rod CD and the stroke of the feed pump may be varied by rotating the slotted link about the fixed fulcrum G. If the crank OA makes 225 r.p.m., find for the given position of the mechanism (a) the velocity and acceleration of the piston D, (b) the angular velocity and angular acceleration of the links AB and CD. M.U. (modified).

8. In the mechanism shown, Fig. 374, D is constrained to move on a horizontal path. Find, for the given configuration, the velocity and acceleration of D and the angular velocity and acceleration of BD when OC is rotating in a counter-clockwise direction at a speed of 180 r.p.m., increasing at the rate of 50 rad/s². L.U. Part II.

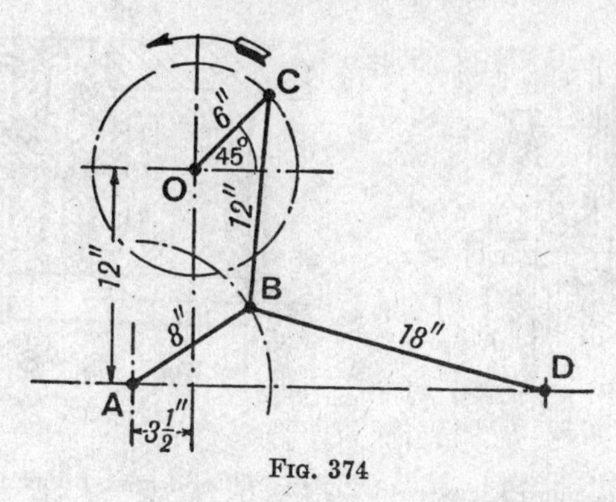

FIG. 374

9. The following are particulars of a single-reduction spur gear. The gear ratio is 10 to 1 and the centre distance is $27\frac{1}{2}$ in. The pinion transmits 500 h.p. at 1800 r.p.m. The teeth are of involute form with addendum 1/(diametral pitch) and pressure angle $22\frac{1}{2}$°. The normal tooth pressure is not to exceed 1000 lb per inch width.

Find (a) the nearest whole number for the diametral pitch if no interference is to occur, (b) the number of teeth in each wheel, (c) the width of the pinion, and (d) the loads on the bearings due to the power transmitted. L.U.

10. Two non-intersecting shafts are inclined at an angle of 72° and are to be connected by spiral gears. The gear reduction ratio is to be $2\frac{1}{2}$ to 1 and the centre distance is to be within the limits $4 \pm 0{\cdot}02$ in. If the minimum number of teeth is not to be less than 20, find, for a spiral angle of 40° on the driving wheel, a suitable *standard* normal diametral pitch. Give the exact centre distance and estimate the efficiency of transmission when the friction angle is 5°. L.U. Part II.

11. Fig. 375 shows diagrammatically the arrangement of the two-speed drive for the supercharger of an aeroplane engine. The wheel A is driven directly from the engine crankshaft; it gears with the pinion B, which in turn gears with the internal wheel C. Wheel D is compound with B and gears with the pinion E on the supercharger shaft. The compound wheel B-D revolves freely on a stud fixed to the carrier F. In top gear the carrier F is prevented from rotating by engaging the clutch G, while the internal wheel C can revolve freely about the axis OO. In low gear C is prevented from rotating by engaging the clutch H,

while F is free to revolve about the axis OO. If the number of teeth on the wheels are A 84, B 30, D 86 E 28, find the two gear ratios provided by the arrangement. M.U.

12. Fig 376 shows a small flywheel F coupled to a shaft A through three trains of epicyclic gears in series. The internal wheels I_1, I_2 and I_3 are all fixed to the casing. The planet carrier B is keyed to the driving shaft A, the sun wheels S_1 and S_2 are respectively integral with the planet carriers C and D, and the sun wheel S_3 is integral with the flywheel F. The numbers of teeth are: S_1 19, I_1 89, S_2 19, I_2 65, S_3 25 and I_3 65. If the shaft A makes 130 r.p.m. find the speed of the flywheel.

13. In a six-cylinder petrol engine the accelerating torque in lb ft. on the crank is given by (a) $A.r(12 \sin 3\theta - 6{\cdot}25 \cos 3\theta)$ due to explosion, and (b) $6{\cdot}3A \sin 3\theta$ due to inertia, where A is the total piston area in square inches and r is the crank radius in feet. If A is 65 in² and r is 2 in., find the fluctuation in speed of the flywheel when its mean speed is 1200 r.p.m. and its moment of inertia is 32 lb ft². Find also the maximum angle which the flywheel is in advance of an imaginary wheel assumed rotating at constant speed. L.U. Part II.

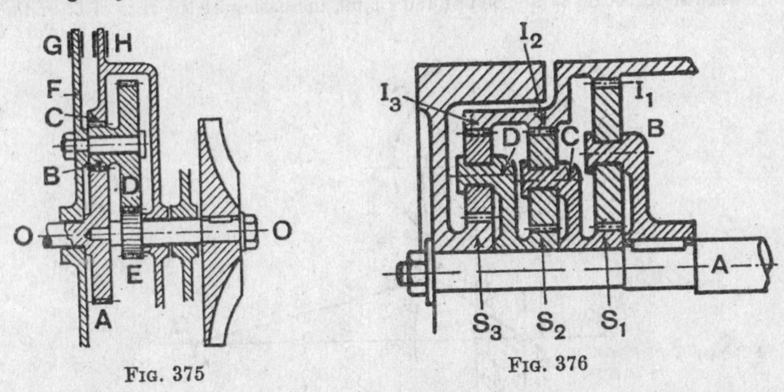

FIG. 375

FIG. 376

14. A gas engine is governed on the hit-and-miss principle and develops 30 i.h.p. at 270 r.p.m. with 125 explosions per minute. The flywheel weighs 2000 lb and has a radius of gyration of 2 ft. Assuming the indicated work done to be identical for each working cycle, the work done by the gases on the explosion stroke to be 2·4 times the work done on the gases during the compression stroke, and the work done on the other two strokes to be negligible, find the maximum percentage fluctuation of speed. M.U.

15. A single-cylinder, four-stroke cycle, oil engine develops 20 i.h.p. at 360 r.p.m. and drives a machine through gearing at 750 r.p.m. The maximum fluctuation of energy is 85% of the indicated energy per cycle. The rotating masses on the engine shaft have a moment of inertia of 2700 lb ft² and the machine shaft masses are equivalent to 400 lb at a radius of gyration of 9 in. Calculate the speed fluctuation of the engine and determine the weight of an additional flywheel, of radius of gyration 18 in., to be fitted to the machine shaft to keep the overall range of speed variation to $\frac{3}{4}$% of the mean speed. L.U.

16. The effective turning moment exerted at the crankshaft of an engine is represented by

$$T = 4{\cdot}0 + 1{\cdot}6 \sin 2\theta - 2{\cdot}2 \cos 2\theta \text{ ton ft}$$

where θ is the crank angle. The weight of the flywheel is 5 tons and its radius of gyration is 3 ft. The engine speed is 120 r.p.m. and the external resistance is constant.

Find (a) the h.p. developed, (b) the maximum fluctuation of energy, (c) the percentage speed fluctuation. L.U.A.

17. A twin-cylinder V-engine has the cylinders set at an angle of 45°, with both pistons connected to the single crank. The crank radius is $2\frac{1}{2}$ in. and the connecting rods are 11 in. long. The reciprocating weight is 3 lb per line and the total rotating weight is equivalent to 4 lb at the crank radius. A balance weight is fitted opposite the crank equivalent to $4\frac{1}{2}$ lb at a radius of $3\frac{1}{2}$ in. Determine, for an engine speed of 1800 r.p.m., the maximum and minimum values of the primary and secondary frame forces, due to inertia of the reciprocating and rotating masses. L.U. Part II.

18. A seven-cylinder vertical engine has reciprocating masses of 400 lb per cylinder, the pitch of the cylinders is 22 in. and the crank order is 1–6–3–4–5–2–7 at equal angular spacing. The stroke is 16 in., the length of each connecting rod is 36 in. and the r.p.m. are 420. Find the magnitudes of any unbalanced primary and secondary effects on the frame.

19. The six cylinders of a single-acting, two-stroke cycle Diesel engine are pitched 40 in. apart and the cranks are spaced at 60° intervals. The crank length is 12 in. and the ratio of connecting rod to crank is 4.5. The reciprocating weight per line is 3000 lb and the rotating weight is 2200 lb. The speed is 200 r.p.m.

Show, with regard to primary and secondary balance, that the firing order 1–5–3–6–2–4 gives unbalance in primary moment only, and the order 1–4–5–2–3–6 gives secondary moment unbalance only. Compare the maximum values of these moments, evaluating with respect to the central plane of the engine. L.U. Part II.

20. A mass weighing 3000 lb is supported symmetrically on four helical springs and its c.g. is subjected to a vertical disturbing force of $1000 \cos 100t$ lb, t being measured in seconds and the angle in radians. Each spring has six free coils, diameter of wire $\frac{3}{4}$ in., mean diameter of coil 5 in. and modulus of the material $12{\cdot}5 \, . \, 10^6$ lb/in^2.

Starting with the appropriate differential equation and assuming that there are no damping actions present, develop the equation for the complete vibration and find: (a) the maximum value of the periodic force transmitted to the foundation, (b) the maximum stress in the spring material.

If the angular frequency of the disturbing force is altered from 100 to 25 rad/s, discuss now the suitability of these springs to reduce the transmission of the periodic forces to the foundation. L.U.A.

21. A four-cylinder Diesel engine drives an electric generator. For the fundamental frequency of torsional vibration, the system may be assumed equivalent to a straight shaft 8 in. diameter and 92 in. long with a rotor at each end. The moments of inertia of the rotors are 4 ton ft^2 and 45 ton ft^2. The smaller rotor, which represents the inertia of the engine masses, is subjected to a sinusoidal torque fluctuation of ± 3000 lb ft which occurs four times per revolution of the crankshaft. Assuming that there is linear damping which amounts to 1000 lb ft at an angular velocity of one radian per second, find the stress due to torsional vibration when the engine speed is 270 r.p.m. What would the stress be under conditions of resonance? M.U.

22. An extension spindle, $1\frac{1}{4}$ in. diameter, is rigidly held in a bracket attached to an engine frame and protrudes horizontally therefrom by 24 in. It carries a wheel of weight 40 lb at the free end. The engine when running at 420 r.p.m. sets up a transverse vibration at the wheel of twice this frequency and of amplitude $0{\cdot}3$ in. Determine the amplitude of vibration of the bracket and estimate the reduction of overhang necessary to reduce the wheel vibration to one-tenth of its observed value. $E = 30 \, . \, 10^6$ lb/in^2. L.U. Part II.

23. A beam section is formed of two channels with flange plates, the moment of inertia of the section being 55 in^4. The beam of total weight 600 lb is freely supported at the ends of a span of 15 ft and carries at the centre of the span a motor of total weight 3000 lb and rotor weight 1000 lb. The rotor is slightly out of balance and at a speed of 370 r.p.m. a forced vibration of $0{\cdot}1$ in. is measured at the motor position. Neglecting damping, estimate the amount by which the rotor is out of balance. To allow for the effect of the mass of the beam one-half of its weight may be added to the motor weight. L.U. Part II.

24. A vibration indicator comprises a light steel cantilever of section 1 in. broad by $\frac{1}{4}$ in. thick firmly clamped to the frame at one end and carrying a weight of 5 lb at a distance of 18 in. from the fixed end. A light recording gear gives a magnification of 10 times the movement relative to the frame. When the cantilever is set freely vibrating the damping is such that successive swings in the same direction are in the ratio of 1 to 0·8. Establish the probable amplitude of the recording pointer when the frame of the instrument is placed on a body vibrating according to the law $x = 0{\cdot}05 \cos 100t$ in., where x is the displacement from the mean position and t is in seconds. Assume that the mass effects of the cantilever and recording gear are equivalent to 0·5 lb at the weight position. $E = 30.10^6$ lb/in².
L.U. Part II.

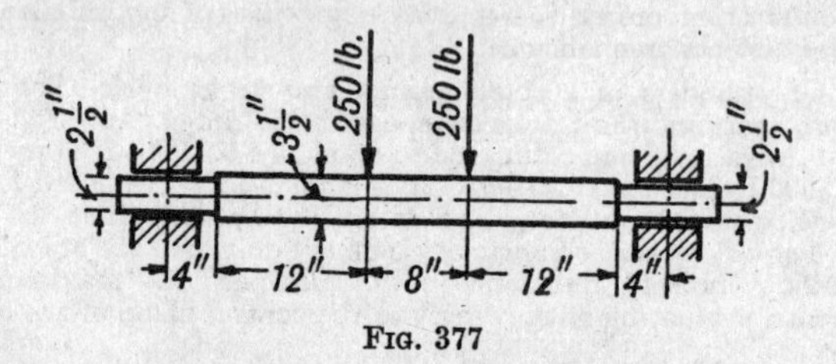

FIG. 377

25. The shaft of a small impulse turbine is shown in Fig. 377. The loading given includes an allowance for the weight of the shaft. If $E = 30.10^6$ lb/in² find the first whirling speed of the shaft.

26. A shaft is simply supported on bearings 10 ft apart and carries five equal concentrated loads equally spaced with the end loads 1 ft from each bearing. If the maximum deflection is 0·1 in., estimate the whirling speed of the shaft when the static deflection curve is assumed to be (a) a sine curve, (b) a parabola.
L.U. Part II.

27. A shaft with similar wheels at the ends is supported in two bearings as shown in Fig. 378. Each wheel has a weight of 600 lb and a radius of gyration

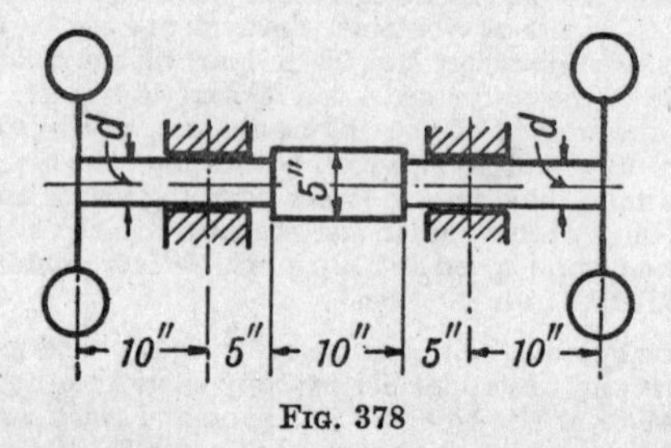

FIG. 378

of $7\frac{1}{2}$ in. The shaft is to run at 1200 r.p.m. and is so driven that there are two torsional impulses per revolution. Determine the diameter d so that the torsional critical speed may be 20% above the running speed. With the diameter so fixed, calculate the whirling speed. Neglect the effect of the shaft mass in both cases. $E = 30.10^6$ and $C = 12.10^6$ lb/in².
L.U. Part II.

28. Establish an expression for the frequency of free torsional vibration of a shaft with concentrated masses of known moments of inertia at the ends, neglecting the effect of the shaft mass.

A motor, connected to one end of a shaft of $2\frac{1}{2}$ in. diameter and 7 ft 6 in. long, has rotating masses of moment of inertia 800 lb ft². At the other end a flywheel and pinion are fitted, the two together having a moment of inertia of 3000 lb ft². The pinion drives a pump at one-quarter of the shaft speed through a gear wheel of moment of inertia 1000 lb ft². The observed frequency of torsional vibration of the system is 7·32 per sec. Determine the effective moment of inertia of the pump impeller and entrained water. $C = 12.10^6$ lb/in². L.U. Part II.

29. A four-cylinder oil engine drives a generator and the system is torsionally equivalent to a straight shaft 11 in. diameter and 165 in. long. The cylinder pitch is 40 in. and the moment of inertia of each engine rotor is 2 ton ft². The combined flywheel-generator rotor is 45 in. from the adjacent cylinder. If C is 5300 tons/in² and the frequency of the fundamental torsional vibrations is to be 1750 per min, find the required moment of inertia of the combined flywheel-generator. Find also the frequency of the two-node vibrations.

30. A three-cylinder oil engine is coupled to the propeller of a motor yacht. The equivalent shaft is $5\frac{1}{2}$ in. diameter and 185 in. long, with the three engine rotors pitched at 20 in. intervals at one end, the flywheel 25 in. from the nearest cylinder rotor and the propeller at the other end. The moment of inertia of each engine rotor is 200 lb ft², that of the flywheel is 1500 lb ft² and that of the propeller is 500 lb ft². Find the frequencies of the fundamental and the two-node vibrations. M.U

31. A six-cylinder oil-engine generator set is torsionally equivalent to a system of seven rotors on a shaft of diameter $8\frac{1}{4}$ in. Each rotor has a moment of inertia of 550 lb ft² and the cylinder pitch is 27 in. The flywheel-generator has a moment of inertia of 28 ton ft² and its distance from the adjacent cylinder is 32 in. If C is 12.10^6 lb/in², find the lowest two frequencies of free torsional vibration.

32. A three-throw compressor is gear-driven by an electric motor, the gear reduction being 7·5 to 1. The motor armature, moment of inertia 540 lb ft², is connected to the pinion, moment of inertia 2·5 lb ft², through a shaft $3\frac{1}{2}$ in. diameter and 32 in. long. The compressor shaft is equivalent to a shaft 6 in. diameter, and 100 in. long, with the gear wheel, moment of inertia 7200 lb ft², at one end and three rotors each of moment of inertia 400 lb ft², pitched at 23 in. from the other end. In addition there is a flywheel on the compressor shaft of moment of inertia 8000 lb ft² and situated 34 in. from the gear wheel. If C is 12.10^6 lb/in², find the frequencies of the fundamental and the two-node torsional vibrations.

ANSWERS TO EXAMPLES

EXAMPLES I (p. 13)

8. 1·87 to 1. **9.** 2·62 ft/s.

EXAMPLES II (pp. 62–67)

3. (a) 2·83 lb; (b) 3·14 lb.
4. 12·39 ft/s; 6·20 ft/s²; 0·578 lb. All 23° 48′ N. of W.
5. Acceleration of c.g. 8·05 ft/s²; angular acceleration 8·95 rad/s².
6. 24.8 sec; 478 lb ft. **7.** 34·1 lb in².
8. 85·6 lb ft². **9.** 146·7 lb ft²; 23·2 in.
10. 0·578 lb and 1·622 lb at 6·6 in. and 2·35 in. resp. from c.g.; 31·0 lb ft.
11. 31·3 lb ft². **12.** 319 lb ft.
13. 164 100 lb ft; 35 900 lb ft. **14.** $6I$; $0{\cdot}4188I$.
15. 2·62 ft/s². **16.** 15·22 m.p.h.; 5·09 ft/s².
17. 4 to 1; α_a 21·48, α_b 5·37 rad/s². **18.** 21·7 to 1; 6·37 ft/s².
19. 374 lb ft. **20.** 3212 lb ft; 29·2 h.p.
21. (a) 2·67 ft/s; (b) −0·667 ft/s and 5·333 ft/s.; 9·45 tons.
24. 548 lb; 18·1 lb. **25.** 9.
26. (a) 5·97 m.p.h.; (b) 7·54 m.p.h. **27.** $\omega_p = 0{\cdot}385$ rad/s.
28. 48·1 lb ft tending to overturn the cycle outwards.
29. 11 580 lb ft tending to turn aeroplane to the left.
30. 4070 lb ft. When the bow is rising it tends to turn the yacht to the right or starboard.
31. 1950 lb ft tending to turn the aeroplane to the left for three-bladed airscrews.
32. 1·8 ton ft; left (port); 0·00503 rad/s².
33. (a) 0·539 rad/s; (b) 24·0 ton ft; (c) 583 ton ft.

EXAMPLES III (pp. 113–120)

1. Maximum speed $\simeq$ 5·0 ft/s; maximum acceleration $\simeq$ 90 ft/s².
2. Maximum speed $\simeq$ 7·8 ft/s; maximum acceleration $\simeq$ 3300 ft/s²; maximum retardation $\simeq$ 1100 ft/s².
3. Cutting stroke, 0·80 and 0·61 ft/s²; return stroke, 6·8 and 7·4 ft/s².
4. Total displacement 6·49 miles. **5.** 29·8 sec; 1890 ft.
6. 148·5 sec; 2034 ft. **7.** a 52·2; b 0·0480.
8. 9·55 ft/s; 2·4 rad/s. **9.** 0·616 ft/s; 0·284 ton ft.
10. 52·5 cm/s. **11.** 4·87 ft/s; 3·69 rad/s; 3·52 rad/s.
12. 0·413 rad/s; 0·26 ft/s. **13.** 2·08; 96·4%.
14. 2·472 in.; 0·618 ft/s; 1·170 ft/s; 0·554 ft/s; 0·957 ft/s.
15. (a) 1·65; (b) 3·11 and 6·84 ft/s; (c) 2·79, 2·86, 5·81, 5·68 ft/s.
16. BC = CD = AB ÷ 4; 0·5 cot θ. **17.** 0·186.
18. 5·64 ft/s; 27·6 ft/s². **19.** CE = 4·36 in.
20. (a) 69·6 ft/s²; 130, 131 rad/s²; (b) 64·2 ft/s²; 81·9, 163 rad/s².
21. 7·8 ft/s along AF, where $\angle$ BAF = 169·5° and $\angle$ CAF = 109·5°, or 16·1 ft/s along AF′, where $\angle$ BAF′ = 118·5° and $\angle$ CAF′ = 178·5°; 1·95 rad/s, 4·88 rad/s²
25. $\theta = 30°$, 7·17 ft/s, 184 ft/s² 3·43 rad/s, 29·7 rad/s²; $\theta = 120°$, 8·92 ft/s, −115·6 ft/s²; 2·01 rad/s, 54·0 rad/s².
26. 35·6 ft/s, 4960 ft/s²; 36·6 ft/s, 5330 ft/s².

27. 4·48 ft/s, 73·7 ft/s²; 7·04 rad/s, 93·4 rad/s²; 6·64 ft/s, 172 ft/s²; 7·79 rad/s, 56·2 rad/s².
28. 140 ft/s²; 300 rad/s².
29. 54·4 ft/s²; 4·49 rad/s².
30. 26·8, 30·5 and 42·3 rad/s².
31. 122 rad/s²; 30·8 ft/s².
32. (a) 56·8 ft/s², 30·5 rad/s²; (b) 58·6 ft/s², 6·9 rad/s².
33. Cutting stroke 0·97 ft/s²; return stroke 5·14 ft/s².
34. 11·9 ft/s²; 20·6 ft/s².
35. 5·37 ft/s, 67·4 ft/s²; 62·4 ft/s².
36. 1560 and 3880 ft/s²; 2120 rad/s².
37. (a) 0; (b) 368 lb.
38. 5·05 ft/s; 14·2 ft/s².

EXAMPLES IV (pp. 139–141)

4. AP÷AB.
5. BP, 7·25 in.
6. Distance of G from CD = 1·29 ft.
7. AP = 1·2 in.
11. OP = 9 in.
13. 11° 38′.
14. 18·9 ft.
15. 22·0 ft.
16. 17° 45′; 571·4 r.p.m.
17. 16° 6′.
18. 136·1 lb ft.
19. 118 lb ft.

EXAMPLES V (pp. 173–176)

4. Cover end: s = 1·30 in., e = 0·23 in., l = 0·13 in. Crank end: s = 0·84 in., e = 0·53 in., l = 0·59 in.
5. Crank end s = 0·98 in., e = 0·49 in.
6. 4·88 in., s = 1·31 in. (cover), 0·87 in. (crank).
7. 2·92 in., 42·5°, l = 0·54 in., s = 0·46 in.
8. 1·316 in., 0·773, 0·24 in.
9. 39°; cover, 1·38 in., 1·62 in.; crank, 1·83 in., 1·17 in.
10. 2·5 in., 37°, 1·25 in., 0·64.
11. 38°, 3·41 in., 1·91 in. (cover), 1·34 in. (crank).
12. 6·5 in., 38°, 1·75 in.
13. Lead 0·25 in., 33°, 1·10 in. (cover), 0·65 in. (crank).
16. 1·39 in. (cover), 1·80 in. (crank).
17. 1·86 in (cover), 2·17 in. (crank).
18. 1·0 in. and 2·57 in. (cover), 1·40 in. and 2·71 in. (crank).
19. 1·44 in. (cover), 1·75 in. (crank).
20. Steam lap of main valve 1·19 in.; exhaust lap, $\frac{1}{16}$ in. (cover), $\frac{5}{16}$ in. (crank) Steam lap of expansion valve, 1·36 in. (cover), 1·77 in. (crank).
21. Difference = 0·28 in.
22. 0·993, 0·974, 0·875; 1·77 in.
23. Cut-off 0·925 (cover), 0·875 (crank); difference = 0·35 in.; 3·19 in.
24. (1) 1·07 in., 1·34 in., 0·39 in. (2) 0·22 in., and 2·36 in. (out); 0·57 in. and 2·36 in. (in). (3) 0·233. (4) 0·26 in. (out); 0·53 in. (in).
25. 0·654, 0·54 in., 0·75 in.
27. 1·97 in., 42·5°.
28. 2·83 in., 50°, 0·62.
29. (a) 1·47 in., 90°; (b) 1·97 in., 44°.
31. 4·26 in., 23° 4′; 356°, 138°, 164°, 330°.
32. (a) 2·97 in., 39°; (b) 2·36 in., 51°.
33. 2·65 in., 28°.
34. 1·27 in., 51·5°.

EXAMPLES VI (pp. 219–222)

3. 21·7 lb.
4. (a) 697·2 lb in.; (b) 209·1 lb in.
5. (a) 228 lb ft., 16·3%, (b) 151·3 lb ft.
6. 56·7 lb ft.
7. 20.
8. $\eta = \dfrac{1-\mu/\sqrt{3}}{1+\mu\sqrt{3}}$.
9. 0·2679, 3·92 lb, 10 lb.
10. 0·48.
11. 31·5.
12. 8·0.
13. 61·9.
14. External dia. 10·5 in., internal dia. 7·5 in.
15. r_1 = 5·67 in., r_2 = 1·89 in.; 2·42.
16. 1·94 in., 633 lb.
17. Mean radius = 7·78 in.
18. 103·3, 453 lb.
20. 710 lb ft.
21. (a) 7·19 lb ft; (b) 4·12 lb ft.; (c) 4·35 lb ft.
22. 1·55 in., 4·0W ft lb/s.
27. Minimum μ = 0·0046.

Examples VII (pp. 245–247)

2. (a) 0·715; (b) 0·862.
3. 5565 lb.
4. Yes.
5. 2255 lb, 14·3, 14·1.
6. 2·72 ft, 0·155 ft/s.
7. 409 lb/in².
8. 8·33 in.
9. 51·3,
10. 4·64 in.
11. 178·7, 63·3 lb.
12. 407·7 lb.
13. 39.
14. 176 lb/in² at zero speed, max h.p. 7·0 at 56·5 ft/s.
17. H.p. 9·9, T_t 125 lb at 120 ft/s.
18. 0·57 r.p.m.
22. 3·62%.

Examples VIII (pp. 277–280)

2. 96·5 lb.
3. 98·2 lb, 90·9 lb; 107 lb, 111·7 lb,
4. 4770 lb, 87 700 lb, 82 400 lb, 42 900 lb.
5. 0·342 ton.
6. 0·346 ton; 5·31, 11·1, 6·98, 6·98, 7·28, 10·66 and 5·75 tons.
7. 136 lb ft.
8. 289 lb ft.
9. 696 lb ft.
10. 9650 lb.
11. 147 lb.
12. 36·1 lb.
13. 43·8 ft, 0·492, 0·508.
14. 151 ft, 72·2 ft.
15. 121 ft, 95·9 ft, 167 ft.
16. 6·21, 10·56 ft/s²; 2·7, 3·81 ft/s².
18. 12·45.
19. 186 lb.
20. 60.

Examples IX (pp. 311–314)

8. (a) 948, (b) 768, (c) 37·7, 48·0.
11. Max. vel. = 16·44 rad/s; max. acc. = 1033 rad/s².
12. 42·5 ft/s², 1·63 ft/s.
13. 894 rad/s².
14. 334, 259·5, −288, −356 ft/s².
15. 3·04 in., 4·51 ft/s, 1766 and 645 ft/s².
16. Inlet: 13·56 ft/s, 14 180 and 4020 ft/s². Exhaust: 12·87 ft/s, 8760 and 4020 ft/s²; 49·9 lb.
17. 59·3 lb.
18. Inlet: 360 cm/s, 3600 and 1336 m/s²; exhaust 341 cm/s, 2400 and 1336 m/s²
19. (a) 0·237 in.; (b) 84° 56′ after b.d.c.; (c) 4·51 ft/s; (d) 2443 ft/s².
20. 12·9 ft/s, 3610 ft/s².
21. Max. vel. 3·35 ft/s. Acc. at the beginning of the lift 74·1 ft/s². Acc. and ret. at the point of maximum velocity, 122 and 78·5 ft/s². Ret. at the end of the lift 89·3 ft/s².
22. 64·2, −90·9 ft/s², 2·56 ft/s, 123·5, −83·6 ft/s².
23. (a) Out: 1895, −3530 rad/s²; (b) 27·0, 18·9 rad/s²; (c) Out: 5190, −5300 rad/s²; In: 1789, −1815 rad/s².
24. Counter-clockwise rotation of the cam: Valve opening: (a) 4·75 ft/s; (b) 500, −1110 ft/s²; (c) 610, −1050 ft/s². Valve closing: (a) 4·80 ft/s; (b) 1110, −500 ft/s²; (c) 1070, −570 ft/s².

Examples X (pp. 362–365)

4. (i) 0·0903ω ft/s; (ii) 0·122ω ft/s.
10. 23, 1·844p.
11. (a) 11, 1·178p; (b) 14, 1·319p; (c) 15, 1·434p.
12. 0·397p.
13. (a) 0·703 in.; (b) 1·872; (c) 21·1°.
14. (a) 32; (b) 18.
15. 1·857m, 0·874m, 2·731÷p_d, 5·824÷p_d, 0·912.
16. (a) 1·026 in.; (b) 1·633; (c) 30·95°.
17. 18, 63, 0·376p.
18. 1·628p.
19. 1·208p.
22. 20·84m, 21° 45′, 1·743m.
23. (a) 21, 42; (b) 50°, 40°; (c) 0·778 in., 0·653 in.; (d) 5·199 in., 8·724 in., 6·962 in., 83·9%.
24. (a) 24, 48; (b) 40°; (c) 0·778 in., 0·653 in.; (d) 7·96 in.
25. 8·48 in.
26. 26° 34′, 63° 26′, 3·913 in., 28, 14, 24·60 in., 6·15 in.
28. (a) 1·009 in., 0·405 in.; (b) 68° 12′, 21° 48′; (c) 18, 45; (d) 5·79 in., 75·86%.

29. (a) 0·56 in., 0·505 in.; (b) 48°, 42°; (c) 20, 50; (d) 5·801 in., 3·568 in., 8·032 in.; (e) 81·08%.

30. Two solutions are possible. The better one is: (a) 0·514 in., 0·548 in.; (b) 43° 10′, 46° 50′; (c) 20, 50; (d) 3·274 in.; 8·726 in.; (e) 80·8%.

31. (a) and (c) 20, 40; 60°, 0°; 84·6%; (b) 23, 46; 33°, 27°; 88·59%.

EXAMPLES XI (pp. 407–411)

1. 14, 56, 8·75 in.; 3·857. **2.** 20, 54, 7·40 in.; 2·684.
3. 10·1 in., 15·0 in., 20·6 in., 33·0 in.
4. Low-speed 19, 59; high-speed 27, 103; centre distance 13·0 in.
5. C = 6·20 in.; 21, 41; 11, 51; 1·95, 4·64.
6. 25, 45; 16, 54; 1·8, 3·375.
7. A 14, B 30, C 18, D 26, E 24, F 20, G 29, H 15.
8. A 18, B 42, C 26, D 34, E 32, F 28, G 41, H 19.
9. 268, M_b 10·3 lb ft, M_e 9·7 lb ft. **10.** −648.
11. 80, M_a 1·796 lb ft, M_e 33·20 lb ft.
12. +38·6, M_f −829 lb ft, M_c +789 lb ft.
13. 5·21 to 1; C 27, E 30; 4·26.
14. 923, M_a −227 lb ft, M_{i_2} 143 lb ft, M_1 61%, M_2 39% of input torque.
15. (a) −294·5, (b) M_b 305·3 lb ft, M_{i_2} −365·3 lb ft; (c) M_c −0·738M_b, M_d 1·738M_b; (d) 1st train 300 lb, 2nd train 707 lb.
16. 3·91 to 1; 2·24 to 1; 1·467 to 1; −5·11 to 1.
17. 2·75 to 1; −4·0 to 1. **18.** 10·53 to 1.
19. 124·2 to 1; −142·3 to 1. **20.** 178·6 to 1.
21. 150 (clockwise); 300 (clockwise); 300 (clockwise); 300 (counter-clockwise).
22. 384 lb ft; 6·6.
23. 4·45 and 5·36 r.p.s. counter-clockwise.
24. 314·7 r.p.m. in opposite sense.

EXAMPLES XII (pp. 446–453)

1. 4180 lb ft. **2.** 10 400 lb ft.
3. 1200 lb ft. **4.** 4·11 rad/s²
5. 10 740 lb ft.
6. Compression −1950 lb ft; expansion 9380 lb ft.
7. 204, 272, 358, 490 lb. **8.** 35·9 lb.
9. (a) 1·68, 0·595; (b) 0·72, 1·40.
10. 0·333V, 0, 0·866V; 0·333Q, 0, 0·866Q; 0·667Q, 0, 1·732Q.
12. 3195 lb, 2170 lb ft; 1496 lb ft.
13. (a) −476 lb ft; (b) 153 lb ft; (c) 378 lb ft.
14. 166·9 lb ft (clockwise).
15. 1·03$\omega^2 r$, 0·237ω^2; (b) 0·69$M\omega^2 r^2$ (counter-clockwise), 1·12$M\omega^2 r$, 0·216 $M\omega^2 r$.
16. (a) 32 700 lb; (b) 4050 lb; (c) 30 800 lb; (d) 20 700 lb ft.
17. (a) 95·1 lb ft; (b) 64·2 lb ft.
18. (a) 3·47 rad/s, 142 ft lb; (b) 212·8 lb ft.
19. (a) 44·5 ft/s², 156 rad/s²; (b) 12·8 lb ft.
20. Neglecting gravity, 18·6 lb in.
21. Neglecting gravity, 18·8 lb in.; allowing for gravity, 10·8 lb in.
22. 2·05 tons. **23.** 6·59 tons.
24. 0·83%. **25.** 15·8 ft, 20·9 tons.
26. 34·4 lb. **27.** 2·03%.
28. 0·0086. **29.** 562 lb.
30. 6·12 ft, 1·10 tons. **31.** 0·176, 5·22 tons.
32. 1·84 tons. **33.** 266·5 r.p.m.
34. 1·97 tons. **35.** 4·07%, 66·6.
36. 0·0199. **37.** 2·66 tons.
38. 1·54 tons. **39.** 48 400 lb ft², 7·5 r.p.m.
40. 1·78%, 1·88%.
41. (a) 114·2; (b) 4·55 tons, (c) Nil and 1·36 rad/s².
42. 251·5 ton ft². **43.** (a) 17·14; (b) 6·92 lb; (c) 180°.

EXAMPLES XIII (pp. 484–488)

2. 160·2, 181·3 r.p.m.
3. 181·2 r.p.m.
4. 216·4 r.p.m.
5. 166·0, 166·6, 168·5, 171·7 r.p.m.
6. 49·0 lb, 23·5 lb/in.
7. 58·2 lb/in., 307·5 r.p.m.
8. 50·1 lb, 99·7 lb/in.
9. 170·3 lb 320 r.p.m.
10. 0·534 in.
11. 79·7 lb, 122·3 lb/in.
12. 34·4 lb, 78·4 lb/in.
13. 365·4, 391 r.p.m.
14. 5·85 in. Stable.
15. 2·24 lb/in., 0·5 in.
16. 66·0 lb/in.
17. (a) 159·3 lb; (b) 70·2 lb/in.
21. (a) 220·4 r.p.m.; (b) 225·3 r.p.m.; (c) 4·8 lb.
22. (a) 227·5 r.p.m.; (b) 1·29%.
23. 151·1, 178·4 r.p.m.
25. 51·7 lb/in., 2·88 lb.
26. (a) 200, 215 r.p.m.; (b) 210·6 r.p.m., 0·0074.
27. 200, 215 r.p.m., 0·022, 0·0109.

EXAMPLES XIV (pp. 526–531)

1. 24·5 lb, 237·5° from A.
2. 7·47 lb, 272° 35′ from A.
3. 7·90, 6·04 lb.
4. (a) B_l 56·25 lb, B_m 18·75 lb; (b) B_l 112·5 lb, B_m 37·5 lb.
5. B_l 27·95 lb, 187° from A; B_m 22·41 lb, 310° 8′ from A.
6. B_l 2·23 lb, 204° 45′ from A; B_m 2·20 lb, 19° 33′ from A.
7. B 301·6 lb, 200° 58′ from A; C 73° 42′ from A; D 269° 11′ from A.
8. B_l 153·8 lb, 232° 16′ from A; B_m 133·9 lb, 354° 21′ from A.
9. B 96·88 lb, 189° 17′ from A; E 119·2 lb, 24° 9′ from A.
10. B_l 138·6 lb, 210° from A; B_m 138·6 lb, 30° from A.
11. A 10·9 lb, 203° 25′ from B; C 12·99 lb, D 10·00 lb.
12. 106·7 lb.
13. B_l 16·42 lb., 233° 16′; B_m 10·64 lb, 350° 32.
14. 222 lb, 1212 lb revolving at the same speed as the crank, but in the opposite sense.
15. 3·36, 3·62 tons.
16. (a) 269 lb, 22° 37′ from bisecting line; (b) 2·62, 2·41 tons.
17. 0·672; 433 lb, 50° 21′ from bisecting line.
18. 0·721.
19. 52·5 in. Outside-cylinder twice that of inside-cylinder.
20. Hammer-blow of O.C. = 0·787 that of I.C. Variation of tractive effort of O.C. = 0·535 that of I.C.
21. 2·3 tons.
22. (a) Driving wheels: 138·4 lb, 4° 46′ to bisecting line; trailing wheels: 53·4 lb, 6° 10′ to bisecting line; (b) 1·70 tons, 3·14 tons.
23. Hammer-blow: (a) 13·43 tons; (b) 3·36 tons.
24. 4·66 tons when plane of cranks horizontal or vertical.
25. 7·96 lb; 199 lb maximum; direction constant and perpendicular to the line bisecting the angle between the cylinder centre lines.
26. 1, 4, 2, 3; 14·1 ton ft.
27. 729·0 lb; C, A and D respectively 90°, 210° 58′ and 239° 2′ from B; 1575 lb, 5820 lb ft about central plane.
28. R_d 5·028 tons; C, B and D respectively 97° 54′, 220° 45′ and 292° 3′ from A; 4·14 tons when A makes 87° 15′ with i.d.c. and at 90° intervals.
29. No. 2 pump 45° and No. 1 pump 225° from No. 1 engine. Unbalanced primary force = 0, couple 54·6 lb ft; secondary force 414 lb, couple 1458 lb ft.
30. 23·3, 9·82 ton ft; 115·5 ton ft when crank No 1 has turned through 15°, 105°, etc., from t.d.c.
32. 4·52 tons, D, B and C respectively 61° 50′, 156° 34′ and 265° 16′ from A; 312 ton ft.
33. D, B and C respectively 60°, 155° 16′ and 264° 44′ from A. 2·45 to 1.
34. 2·5W opposite the crank.
35. (a) 5·63 lb; (b) constant magnitude 564 lb and revolves in opposite sense to the crank at twice crank speed; (c) 4th harmonic unbalanced, 6th balanced.
36. (a) 1945 lb when the crank bisects the angle between the cylinders, 498 lb when the crank has turned through 90° from the above positions; (b) secondary

force constant, 243 lb. It revolves in the same sense as the crank, but at twice crank speed.

37. Maximum 1130 lb, minimum 469 lb.

38. 4510 lb, 1503 lb. Maximum occurs when the plane of the cranks makes 45° to the vertical, and minimum when the plane of the cranks is vertical or horizontal.

EXAMPLES XV (pp. 589–594)

1. 297 v.p.m. **2.** 2125 v.p.m.
3. (a) 535 v.p.m.; (b) 518 v.p.m. **4.** 311 v.p.m.
5. 262 v.p.m. **6.** 282 v.p.m.
7. 1370 v.p.m. **8.** 521 v.p.m.
9. 6·95 lb per ft/s; 0·984.
10. 1·314 in.; 1·668 in.; 11·58 lb per ft/s.
12. 64·7 lb ft.
13. 2·31°, 3·35°; 144 lb ft. per rad/s, 2·98 per sec ($\omega = 18{\cdot}7$).
14. (a) 0·150 in., 18° 47′; (b) 0·563 in., 90°.
15. 3·08.
16. (a) Natural frequency = 1387 per min; (b) 0·195°.
17. 725 v.p.m.; 4·48 in. **18.** 9·75 mm.
19. (a) 1·9 sec; (b) 0·00033 in. **20.** 1500 r.p.m.
21. 2270 r.p.m. **22.** (a) 964 r.p.m.; (b) 980 r.p.m. 203·2.
23. 2156, 2210 r.p.m. **24.** 732, 743, r.p.m.
25. 2680 r.p.m. **26.** 2082 r.p.m.
27. 1600 r.p.m. **28.** 2060 v.p.m.; 3410 v.p.m.
29. 521, 1150 v.p.m. One node: 26·8 ft from propeller. Two node: engine side of flywheel, 4·6 ft from flywheel and propeller side of flywheel, 5·5 ft from propeller. All distances measured on the actual shaft.
30. 5·01 in.
31. Three-rotor system: 1280, 1930 v.p.m. Tabular method: 1384, 1930 v.p.m.
32. 387, 1362 v.p.m. **33.** 387·5, 1466 v.p.m.
34. 2630 v.p.m. **36.** 180, 1277 v.p.m.

MISCELLANEOUS EXAMPLES XVI (p. 595–601)

1. 522 lb ft **2.** 2·17 ft, 4·55 h.p.
3. (a) 1 337 000 ft lb, (b) 67·5 lb.
4. $N_a = 2009{\cdot}65$, $N_b = 1942{\cdot}12$ r.p.m., h.p. = 37.
5. $N_a = 163{\cdot}4$, $N_b = 653{\cdot}6$ r.p.m., 0·769 sec, 872 ft lb, 41·5 lb.
6. (a) 165·3 r.p.m.; (b) 0·40 sec; 106 deg.
7. (a) $v_d = 2{\cdot}78$ ft/s, $f_d = 52$ ft/s²; (b) $\omega_{ab} = 3{\cdot}28$ rad/s (clockwise), $\omega_{cd} = 4{\cdot}0$ rad/s (counter-clockwise), $\alpha_{ab} = 81{\cdot}3$, $\alpha_{cd} = 80{\cdot}5$ rad/s² (both clockwise).
8. $v_d = 5{\cdot}6$ ft/s, $f_d = 38{\cdot}5$ ft/s², $\omega_{bd} = 4{\cdot}43$ rad/s (clockwise), $\alpha_{bd} = 71$ rad/s² (counter-clockwise).
9. (a) $p_d = 3$; (b) 15, 150; (c) 7·58 in.; (d) 2900 lb.
10. $p_d = 16$, 3·988 in., 88·0 %. **11.** 8·60 to 1, 5·80 to 1.
12. 11 770 r.p.m. **13.** 2·31%, 0·22 deg.
14. 9·83%. **15.** 1·93%, 1332 lb ft².
16. (a) 205 h.p.; (b) 2·72 ft tons; (c) 1·23%.
17. Maximum and minimum primary force = 650 lb, 326 lb. Maximum and minimum secondary force = 205 lb, 85 lb.
18. Forces balanced; maximum primary couple = 3·52 ton ft, maximum secondary couple = 2·94 ton ft.
19. 365 ton ft, 93·3 ton ft.
20. (a) 35·1 lb; (b) 22 900 lb/in². Unsuitable, force transmitted = 1190 lb.
21. 1290 lb/in², 10 970 lb/in². **22.** 0·0086 in., 21·3 in.
23. 19·8 lb in. **24.** 0·0822 in.
25. 3570 r.p.m. **26.** (a) 675·6 r.p.m.; (b) 670·2 r.p.m.
27. 3·22 in., 3180 r.p.m. **28.** 6040 lb ft².
29. 48·8 ton ft², 4770 per min. **30.** 2300, 4030 per min.
31. 2260, 6560 per min. **32.** 1145, 3150 per min.

BIBLIOGRAPHY

GENERAL

Text-books on Applied Mechanics and Theory of Machines

Dictionary of Applied Physics. Vol. I. Edited by Sir R. T. Glazebrook. Macmillan.
Applied Mechanics for Engineers. J. Duncan. Macmillan.
Mechanics Applied to Engineering (2 vols.). J. Goodman. Longmans, Green.
Applied Mechanics. D. A. Low. Longmans, Green.
Physical Principles of Mechanics and Acoustics. R. W. Pohl. Blackie.
Mechanics of Particles and Rigid Bodies. J. Prescott. Longmans, Green.
Engineering Mechanics. C. E. Inglis. Oxford University Press.
Mechanism. S. Dunkerley. Longmans, Green.
The Principles of Mechanism. F. Dyson. Oxford University Press.
Kinematics of Machinery. C. D. Albert and F. S. Rogers. Chapman & Hall.
Kinematics and Kinetics of Machinery. J. E. Dent and A. C. Harper. Chapman & Hall.
Kinematics of Machines. G. L. Guillet. Chapman & Hall.
Kinematics of Mechanisms. Rosenhauer and Willis. Associated General Publications.
Mechanics of Machinery. C. W. Ham and E. J. Crane. McGraw-Hill.
Mechanics of Machinery (2 vols.). R. C. H. Heck. McGraw-Hill.
The Mechanics of Machinery. A. B. W. Kennedy. Macmillan.
Kinematics of Machinery. A. W. Klein. McGraw-Hill.
Theory of Machines. R. W. Angus. McGraw-Hill.
The Theory of Machines. R. F. McKay. Arnold.
Theory of Machines. L. Toft and A. T. J. Kersey. Pitman.
Theory of Machines. W. G. Green. Blackie.
Mechanism and the Kinematics of Machines. W. Steeds. Longmans, Green.
Advanced Dynamics. S. Timoshenko and D. H. Young. McGraw-Hill.

CHAPTER I

Kinematic Design in Engineering. A. F. C. Pollard. *Proc. I.M.E.*, 1933.
Power Transmission by Oil. H. S. Hele-Shaw. *Proc. I.M.E.*, 1921.

CHAPTER II

An Elementary Treatment of the Theory of Spinning Tops and Gyroscopic Motion. H. Crabtree. Longmans Green.
The Automatic Stabilisation of Ships. T. W. Chalmers. Chapman & Hall.
Mechanics of the Gyroscope. R. F. Deimel. Macmillan.
Gyrostatics and Rotational Motion. A. Gray. Macmillan.
Spinning Tops. J. Perry. S.P.C.K.
The Gyroscopic Stabilisation of Land Vehicles. J. F. S. Ross. Arnold.

The Gyroscope. P. P. Schilovsky. Spon.
Dynamics of Rotation. A. M. Worthington. Longmans, Green.
The Gyroscope Applied. K. I. T. Richardson. Hutchinson's Technical Press.
The Stability of Gyroscopic Single-Track Vehicles. H. Cousins. *Engineering*, 1913.
The Brennan Mono-rail. *Engineering*, Nov., 1914.
Large Gyroscope for Stabilising a Liner. *Engineer*, Jan., 1932.
Rolling of the S.S. Conte de Savoie. Dr.-Ing. R. de Santis and Dr.-Ing. M. Russo. *Engineer*, Sept., 1936.
Gyroscopic Principles and Applications. C. E. Inglis. *Proc. I.M.E.*, Vol. 151, 1944.

CHAPTER V

Heat Engines. D. A. Low. Longmans, Green.
The Steam Engine—Theory and Practice. W. Ripper and J. Goudie. Longmans, Green.
Elements of Machine Design. Pt. II. W. C. Unwin and A. L. Mellanby. Longmans, Green.
Valves and Valve Gear Mechanisms. W. E. Dalby. Arnold.
Valves and Valve Gearing. C. Hurst. Griffin.

CHAPTER VI

Reports of the Research Committee on Friction. Beauchamp Tower. *Proc. I.M.E.*, 1883, 4, 5, 8 and 1891.
Friction. Sir T. E. Stanton. Longmans, Green.
Mechanical Testing. Vol. II. R. G. Batson and J. H. Hyde. Chapman & Hall.
Rolling Friction. O. Reynolds. *Trans. Royal Society*, Vol. 166.
The Theory of Lubrication and its Application to Beauchamp Tower's Experiments. O. Reynolds. *Trans. Royal Society*, 1886, Series A.
The Theory of Film Lubrication. R. O. Boswall. Longmans, Green.
Theory of Lubrication. M. D. Hersey. Chapman & Hall.
Mechanical Properties of Fluids—Chapter III—Viscosity and Lubrication. A. G. M. Michell. Blackie.
Lubrication and Lubricants. L. Archbutt and R. M. Deeley. Griffin.
Lubrication and Lubricants. J. H. Hyde. Pitman.
The Principles and Practice of Lubrication. A. W. Nash and A. R. Bowen, Chapman & Hall.
Lubrication. Its Principles and Practice. A. G. M. Michell. Blackie.
Friction and Lubrication of Solids. F. P. Bowden and D. Tabor. Oxford University Press.
The Film Lubrication of the Journal Bearing. R. O. Boswall and J. C. Brierley. *Proc. I.M.E.*, 1932.
Hydrodynamic Principles of Journal Bearing Design. H. W. Swift. *Proc. I.M.E.*, 1935.
Relation between Theory, Experiment and Practice in Journal Bearing Design. H. L. Hazlegrave. *Proc. I.M.E.*, 1935.
The Measurement of Attitude and Eccentricity in Complete Clearance Bearings. D. Clayton and C. Jakeman. *Proc. I.M.E.*, 1936.
Investigations in Film Lubrication. A. S. T. Thomson. *Proc. I.M.E.*, 1936.

General Discussion on Lubrication and Lubricants. *Proc. I.M.E.*, 1937.
Ball Bearings. J. Goodman. *Proc. I.C.E.*, 1912.
Ball Bearings. H. Hess. *Trans. A.S.M.E.*, 1907.
Handbook of Ball and Roller Bearings. A. W. Macaulay. Pitman.
Needle Roller Bearings. C. H. Smith. *Proc. I.A.E.*, 1935–6.
Rolling Bearings. R. K. Allan. Pitman.

CHAPTER VII

The Mechanical Transmission of Power. G. F. Charnock. Technical Press, Ltd.
Belting and Its Application. J. Dawson. Chapman & Hall.
Belts for Power Transmission. W. G. Dunkley. Pitman.
The Transmission of Power by Leather Belting. C. G. Barth. *Trans. A.S.M.E.*, Vol. 31, 1909.
Power Transmission by Belts: An Investigation of Fundamentals. H. W. Swift. *Proc. I.M.E.*, 1928.
Cambers for Belt Pulleys. H. W. Swift. *Proc. I.M.E.*, 1932.
Short-centre Belt Drives. H. W. Swift. *Proc. I.M.E.*, 1937.
Transmission of Power by Chains. H. T. Hildage. *Trans. Man. Assoc. of Engrs.*, 1914.

CHAPTER VIII

Mechanical Testing. Vol. II. R. G. Batson and J. H. Hyde. Chapman & Hall.
Engineering Instruments and Meters. E. A. Griffiths. Routledge.
The Testing of Engines, Boilers and Auxiliary Machinery. W. W. F. Pullen. Scientific Publishing Co.
The Testing of Motive Power Engines. R. Royds. Longmans, Green.
Mechanical Braking and Its Influence on Winding Equipment. J. F. Perry and D. M. Smith. *Proc. I.M.E.*, 1932.
Coupled Brakes. *Automobile Engineer*, Jan., 1926.
Dynamometer and Friction Brake. W. Froude. *Proc. I.M.E.*, 1858, 1877.
The Measurement of Torque in Shafts. H. Ford and A. Douglas. *Engineering*, May, 1949.
Torquemeter for Industrial Applications. R. B. Sims and A. D. Morley. *Engineering*, July, 1952.

CHAPTER IX

Cams. Elementary and Advanced. F. D. Furman. Chapman & Hall.
Analytical Approach to Automobile Valve Gear Design. J. L. H. Bishop. *A.D. Proc.*, 1950–51, p. 150.

CHAPTER X

Mechanical Testing. Vol. II. R. G. Batson and J. H. Hyde. Chapman & Hall.
The Mechanical Transmission of Power. G. F. Charnock. Technical Press, Ltd.
Spur Gears. E. Buckingham. McGraw-Hill.
Gears and Gear Cutting. P. Gates. Technical Press, Ltd.

Steam Turbines. J. Goudie. Longmans, Green.
Elements of Machine Design. Pt. I. W. C. Unwin and A. L. Mellanby. Longmans, Green.
The Art of Gear Design. H. E. Merritt. *The Engineer*, July–Dec., 1936.
Worm Gearing. J. J. Guest. *The Automobile Engineer*, 1931.
Worm Gear Contacts. W. Abbott. *Proc. I.M.E.*, 1936.
Worm Gear Performance. H. E. Merritt. *Proc. I.M.E.*, 1935.
A Treatise on Screws and Worm Gears. P. Cormac. Chapman & Hall.
Gears. H. E. Merritt. Pitman.
Analytical Mechanics of Gears. E. Buckingham. McGraw-Hill.

CHAPTER XI

Variable-Speed Gears for Motor Road-Vehicles. R. E. Phillips. *Proc. I.M.E.*, 1917.
Planetary Gearing. F. D. Furman. *Machinery*, Vol. 24, 25, 26 and 27.
Epicyclic Gears. F. W. Lanchester and G. H. Lanchester. *Proc. I.M.E.*, 1924.
Epicyclic Gearing. W. G. Wilson. *Proc. I.A.E.*, 1931–2.
Epicyclic Gearing. P. P. Love. *Proc. I.M.E.*, 1936.
Gear Trains. H. E. Merritt. Pitman.

CHAPTER XII

Inertia Torque in Crankshafts. F. A. S. Acres. *Proc. I.A.E.*, 1919–20.
Graphical Method of Finding Inertia Forces. W. J. Duncan. *Proc. I.M.E.*, 1915.
Dynamics of Mechanisms. E. Eksergian. *Journal of the Franklin Institute*, 1930, 31.

CHAPTER XIV

A Treatise on Engine Balance. P. Cormac. Chapman & Hall.
The Balancing of Engines. W. E. Dalby.. Arnold.
The Balancing of Engines. A. Sharp. Longmans, Green.
The Balancing of Oil Engines. W. Ker Wilson. Griffin.
Mechanical Testing. Vol. II. R. G. Batson and J. H. Hyde. Chapman & Hall.
The Balancing of Machinery. C. N. Fletcher. Emmott.
Engineering Dynamics, Vol. IV. Biezeno and Grammel. Blackie.
Balance and Inertia Torque of Multicylinder Engines. Fearn. *Journal of Royal Aer. Soc.*, Sept., 1928.
The Design of Dynamically-Balanced Crankshafts for Two-Stroke Cycle Engines. P. Cormac. *Engineering*, Oct., 1929.
Static Balancing Machine. *Engineering*, Feb., 1935.
Schenk Direct-Measurement Dynamic Balancing Machine. *Engineering*, July, 1937.

CHAPTER XV

Steam Turbines. J. Goudie. Longmans, Green.
Balancing of Engines. W. E. Dalby. Arnold.
Theory of Vibration for Engineers. E. B. Cole. Crosby Lockwood.
The Prevention of Vibration and Noise. A. B. Eason. Oxford University Press.

Mechanical Vibrations. J. P. D. Hartog. McGraw-Hill.
The Strength of Shafts in Vibration. J. Morris. Technical Press, Ltd.
Dynamics of Engine and Shaft. R. E. Root. Chapman & Hall.
Rayleigh's Principle and its Application to Engineering. G. Temple and W. G. Bickley. Oxford University Press.
Mechanics Applied to Vibrations and Balancing. D. L. Thornton. Chapman & Hall.
Vibration Problems in Engineering. S. Timoshenko and D. H. Young. Macmillan.
Torsional Vibration. W. A. Tuplin. Chapman & Hall.
Practical Solution of Torsional Vibration Problems. W. Ker Wilson. Chapman & Hall.
Fundamentals of Vibration Study. R. G. Manley. Chapman & Hall.
Vibration and Shock Isolation. Creda. Chapman & Hall.
Mechanics of Vibrations. Hansen and Chenea. Chapman & Hall.
Theory of Oscillations. Andronow and Chaikin. Princeton.
Engineering Dynamics, Vol. III and IV. Biezeno and Grammel. Blackie.
Vector Methods of Studying Mechanical Vibrations. D. Robertson. *The Engineer*, 1931.
The Possible Vibration of a Ship's Hull under the Action of an Unbalanced Engine. W. E. Dalby. *Proc. I.M.E.*, 1928.
An Empirical Formula for Crankshaft Stiffness in Torsion. B. C. Carter. *Engineering*, July, 1928.
The Stiffness of Multi-Throw Crankshafts. *Engineering*, Nov., 1929.
On the Stiffness of Crankshafts. H. Constant. H. M. Stationery Office.
Graphical Method of Determining the Frequency of Torsional Vibration. K. Waimann. *Z.V.D.I.*, Sept., 1934.
Elastic Hysteresis in Crankshaft Steels. S. F. Dorey. *Proc. I.M.E.*, 1932.
Damping Influences in Torsional Oscillation. J. F. Shannon. *Proc. I.M.E.*, 1935.
The Cambridge Vibrograph. *Engineering*, Feb., 1925.
Der Torsiograph, Ein neues Instrument zur Untersuchung von Wellen. J. Geiger. *Z.V.D.I.*, 1916.
The R.A.E. Optical Torsiograph. B. C. Carter. *Journal of the Aero. Soc.*, April, 1927.
Eliminating Crankshaft Torsional Vibration in Radial Aircraft Engines. E. S. Taylor. *Journal S.A.E.*, 1936.
The Pendulum Torsional Vibration Damper. J. Dick. *The Engineer*. Dec., 1936.
A Dynamic Damper for Torsional Vibrations. *Sulzer Technical Review*, No. 1, 1938.

[illegible] Vibrations. [illegible] Harvey. *[illegible]*.
The Strength of Plates in Torsion. [illegible] Technical Press, Ltd.
[illegible] Engine and Shafts. [illegible] Chapman & Hall.
[illegible] Fatigue and its Application to Engineering. [illegible] Temple and [illegible] Oxford University Press.
[illegible] of [illegible] Vibrations and Balancing. [illegible] Chapman & Hall.
The [illegible] Engineering. [illegible] and [illegible] Young. [illegible]
[illegible] Vibration. [illegible] Chapman & Hall.
[illegible] Solution of Torsional Vibration Problems. [illegible] Wilson. Chapman & Hall.
[illegible] Chapman & Hall.
[illegible] Chapman & Hall.
[illegible] Theory of Oscillations. [illegible] and Chaikin. [illegible]
[illegible] Vol. [illegible] and [illegible] and Unwin.
[illegible] Methods of [illegible] Vibrations. [illegible] 1951.
[illegible] Engines. [illegible] 1949.
An Empirical Formula for Crankshaft Failures in Torsion. [illegible] 1948.
[illegible] Engineering, Nov. 1939.
On the [illegible] of [illegible] Chamberlain. [illegible]
[illegible] Frequency of Torsional [illegible] Waldon. [illegible] 1934.
[illegible] Crankshaft [illegible] 1952.
[illegible] in Torsional Oscillation. [illegible] 1935.
[illegible] Engineering, [illegible]
The [illegible] for [illegible] Vibration. [illegible] 1946.
The R.A.E. [illegible] Journal of [illegible] April, 1952.
[illegible] Torsional Vibration in [illegible] Aircraft Engines. [illegible] Taylor. *Journal R.Ae.S.*, 1936.
The Pendulum Torsional Vibration Damper. [illegible] Engineer, [illegible]
A Dynamic Damper for Torsional Vibrations. [illegible] No. 7, 1938.

INDEX